Reveal ALGEBRA 1®

mheducation.com/prek-12

Copyright © 2020 McGraw-Hill Education

All rights reserved. No part of this publication may be reproduced or distributed in any form or by any means, or stored in a database or retrieval system, without the prior written consent of McGraw-Hill Education, including, but not limited to, network storage or transmission, or broadcast for distance learning.

Cover: (t to b, l to r) pearleye/E+/Getty Images, Merfin/iStock/Getty Images, fototrav/E+/Getty Images, rickyd/Shutterstock

Send all inquiries to:
McGraw-Hill Education
8787 Orion Place
Columbus, OH 43240

ISBN: 978-0-07-695907-5
MHID: 0-07-695907-4

Printed in the United States of America.

5 6 7 8 9 10 11 LWI 25 24 23 22 21

Contents in Brief

Module 1   Expressions

2   Equations in One Variable

3   Relations and Functions

4   Linear and Nonlinear Functions

5   Creating Linear Equations

6   Linear Inequalities

7   Systems of Linear Equations and Inequalities

8   Exponents and Roots

9   Exponential Functions

10   Polynomials

11   Quadratic Functions

12   Statistics

Reveal AGA® Makes Math Meaningful...

Interactive Student Edition

Learning on the Go!

The flexible approach of *Reveal Algebra 1, Reveal Geometry,* and *Reveal Algebra 2 (Reveal AGA)* can work for you using digital only or digital and your *Student Edition* together.

Student Digital Center

...to Reveal YOUR Full Potential!

Reveal AGA® Brings Math to Life in Every Lesson

Reveal AGA is a blended print and digital program that supports access on the go. Use your student edition as a reference as you work through assignments in the Student Digital Center, access interactive content, animations, videos, eTools, and technology-enhanced practice questions.

Go Online!
my.mheducation.com

Web Sketchpad® Powered by The Geometer's Sketchpad®- Dynamic, exploratory, visual activities embedded at point of use within the lesson.

Animations and Videos — Learn by seeing mathematics in action.

Interactive Tools — Get involved in the content by dragging and dropping, selecting, highlighting, and completing tables.

Personal Tutors — See and hear a teacher explain how to solve problems.

eTools — Math tools are available to help you solve problems and develop concepts.

TABLE OF CONTENTS

Module 1
Expressions

	What Will You Learn?	1
1-1	**Numerical Expressions**	3
	Explore Order of Operations	
1-2	**Algebraic Expressions**	13
	Explore Using Algebraic Expressions in the Real World	
1-3	**Properties of Real Numbers**	23
	Explore Testing the Associative Property	
Expand 1-3	**Operations with Rational Numbers** (Online Only)	
1-4	**Distributive Property**	35
	Explore Using Rectangles with the Distributive Property	
	Explore Modeling the Distributive Property	
1-5	**Expressions Involving Absolute Value**	45
	Explore Distance Between Points on a Number Line	
1-6	**Descriptive Modeling and Accuracy**	49
	Module 1 Review	59

Module 2
Equations in One Variable

	What Will You Learn?	63
2-1	**Writing and Interpreting Equations**	**65**
	Explore Writing Equations by Modeling a Real-World Situation	
2-2	**Solving One-Step Equations**	**75**
	Explore Using Algebra Tiles to Solve One-Step Equations Involving Addition or Subtraction	
	Explore Using Algebra Tiles to Solve One-Step Equations Involving Multiplication	
2-3	**Solving Multi-Step Equations**	**85**
	Explore Using Algebra Tiles to Model Multi-Step Equations	
2-4	**Solving Equations with the Variable on Each Side**	**91**
	Explore Modeling Equations with the Variable on Each Side	
2-5	**Solving Equations Involving Absolute Value**	**101**
	Explore Modeling Absolute Value	
2-6	**Solving Proportions**	**109**
	Explore Comparing Two Quantities	
2-7	**Using Formulas**	**117**
	Explore Centripetal Force	
	Explore Using Dimensional Analysis	
	Module 2 Review	**129**

TABLE OF CONTENTS

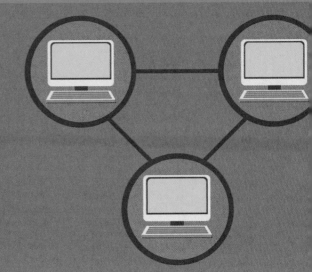

Module 3
Relations and Functions

	What Will You Learn?	133
3-1	**Representing Relations**	135
	Explore Choosing Scales	
3-2	**Functions**	147
	Explore Vertical Line Test	
3-3	**Linearity and Continuity of Graphs**	157
	Explore Representing Discrete and Continuous Functions	
3-4	**Intercepts of Graphs**	167
3-5	**Shapes of Graphs**	179
	Explore Line Symmetry	
	Explore Relative High and Low Points	
3-6	**Sketching Graphs and Comparing Functions**	191
	Explore Modeling Relationships by Using Functions	
	Module 3 Review	203

Module 4
Linear and Nonlinear Functions

	What Will You Learn?	207
4-1	**Graphing Linear Functions**	209
	Explore Points on a Line	
	Explore Lines Through Two Points	
4-2	**Rate of Change and Slope**	219
	Explore Investigating Slope	
4-3	**Slope-Intercept Form**	229
	Explore Graphing Linear Equations by Using the Slope-Intercept Form	
Expand 4-3	**Linear Growth Patterns** (Online Only)	
4-4	**Transformations of Linear Functions**	239
	Explore Transforming Linear Functions	
4-5	**Arithmetic Sequences**	251
	Explore Common Differences	
4-6	**Piecewise and Step Functions**	259
	Explore Age as a Function	
4-7	**Absolute Value Functions**	267
	Explore Parameters of an Absolute Value Function	
	Module 4 Review	281

TABLE OF CONTENTS

Module 5
Creating Linear Equations

	What Will You Learn?	285
5-1	**Writing Equations in Slope-Intercept Form**	287
	Explore Slope-Intercept Form	
5-2	**Writing Equations in Standard and Point-Slope Forms**	295
	Explore Forms of Linear Equations	
5-3	**Scatter Plots and Lines of Fit**	307
	Explore Making Predictions by Using a Scatter Plot	
5-4	**Correlation and Causation**	315
	Explore Collecting Data to Determine Correlation and Causation	
5-5	**Linear Regression**	319
5-6	**Inverses of Linear Functions**	327
	Explore Comparing a Function and its Inverse	
	Module 5 Review	335

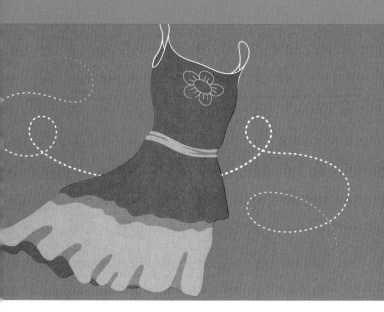

Module 6
Linear Inequalities

	What Will You Learn?	339
6-1	**Solving One-Step Inequalities**	341
	Explore Graphing Inequalities	
	Explore Properties of Inequalities	
6-2	**Solving Multi-Step Inequalities**	351
	Explore Modeling Multi-Step Inequalities	
6-3	**Solving Compound Inequalities**	357
	Explore Guess the Range	
6-4	**Solving Absolute Value Inequalities**	367
	Explore Solving Absolute Value Inequalities	
6-5	**Graphing Inequalities in Two Variables**	375
	Explore Graphing Linear Inequalities on the Coordinate Plane	
	Module 6 Review	381

TABLE OF CONTENTS

Module 7
Systems of Linear Equations and Inequalities

	What Will You Learn?	385
7-1	**Graphing Systems of Equations**	**387**
	Explore Intersections of Graphs	
7-2	**Substitution**	**399**
	Explore Using Substitution	
7-3	**Elimination Using Addition and Subtraction**	**405**
7-4	**Elimination Using Multiplication**	**413**
	Explore Graphing and Elimination Using Multiplication	
7-5	**Systems of Inequalities**	**419**
	Explore Solutions of Systems of Inequalities	
	Module 7 Review	**425**

Module 8
Exponents and Roots

	What Will You Learn?	429
8-1	**Multiplication Properties of Exponents**	431
	Explore Products of Powers	
8-2	**Division Properties of Exponents**	439
	Explore Quotients of Powers	
8-3	**Negative Exponents**	447
	Explore Simplifying Expressions with Negative Exponents	
8-4	**Rational Exponents**	455
	Explore Expressions with Rational Exponents	
8-5	**Simplifying Radical Expressions**	463
	Explore Square Roots and Negative Numbers	
8-6	**Operations with Radical Expressions**	471
Expand 8-6	**Sums and Products of Rational and Irrational Numbers** (Online Only)	
8-7	**Exponential Equations**	479
	Explore Solving Exponential Equations	
	Module 8 Review	483

Table of Contents xiii

TABLE OF CONTENTS

Module 9
Exponential Functions

	What Will You Learn?	**487**
9-1	**Exponential Functions**	**489**
	Explore Exponential Behavior	
	Explore Restrictions on Exponential Functions	
9-2	**Transformations of Exponential Functions**	**497**
	Explore Translating Exponential Functions	
	Explore Dilating Exponential Functions	
	Explore Reflecting Exponential Functions	
9-3	**Writing Exponential Functions**	**509**
	Explore Writing an Exponential Function to Model Population Growth	
9-4	**Transforming Exponential Expressions**	**519**
9-5	**Geometric Sequences**	**523**
	Explore Modeling Geometric Sequences	
9-6	**Recursive Formulas**	**531**
	Explore Writing Recursive Formulas from Sequences	
	Module 9 Review	**539**

Module 10
Polynomials

	What Will You Learn?	543
10-1	**Adding and Subtracting Polynomials**	545
	Explore Using Algebra Tiles to Add and Subtract Polynomials	
10-2	**Multiplying Polynomials by Monomials**	555
	Explore Using Algebra Tiles to Find Products of Polynomials and Monomials	
10-3	**Multiplying Polynomials**	561
	Explore Using Algebra Tiles to Find Products of Two Binomials	
10-4	**Special Products**	569
	Explore Using Algebra Tiles to Find the Squares of Sums	
	Explore Using Algebra Tiles to Find the Squares of Differences	
	Explore Using Algebra Tiles to Find Products of Sums and Differences	
10-5	**Using the Distributive Property**	577
	Explore Using Algebra Tiles to Factor Polynomials	
Expand 10-5	**Proving the Elimination Method** (Online Only)	
10-6	**Factoring Quadratic Trinomials**	583
	Explore Using Algebra Tiles to Factor Trinomials	
10-7	**Factoring Special Products**	591
	Explore Using Algebra Tiles to Factor Differences of Squares	
	Module 10 Review	599

TABLE OF CONTENTS

Module 11
Quadratic Functions

	What Will You Learn?	603
11-1	**Graphing Quadratic Functions**	605
	Explore Graphing Parabolas	
11-2	**Transformations of Quadratic Functions**	615
	Explore Transforming Quadratic Functions	
11-3	**Solving Quadratic Equations by Graphing**	627
	Explore Roots and Zeros of Quadratics	
11-4	**Solving Quadratic Equations by Factoring**	631
	Explore Using Factors to Solve Quadratic Equations	
11-5	**Solving Quadratic Equations by Completing the Square**	641
	Explore Using Algebra Tiles to Complete the Square	
11-6	**Solving Quadratic Equations by Using the Quadratic Formula**	647
	Explore Deriving the Quadratic Formula Algebraically	
	Explore Deriving the Quadratic Formula Visually	
11-7	**Solving Systems of Linear and Quadratic Equations**	655
	Explore Using Algebra Tiles to Solve Systems of Linear and Quadratic Equations	
11-8	**Modeling and Curve Fitting**	663
	Explore Using Differences and Ratios to Model Data	
Expand 11-8	**Exponential Growth Patterns** (Online Only)	
11-9	**Combining Functions**	673
	Explore Using Graphs to Combine Functions	
	Module 11 Review	681

Module 12
Statistics

	What Will You Learn?	685
12-1	**Measures of Center**	687
	Explore Finding Percentiles	
12-2	**Representing Data**	695
12-3	**Using Data**	703
	Explore Phrasing Questions	
12-4	**Measures of Spread**	709
	Explore Using Measures of Spread to Describe Data	
12-5	**Distributions of Data**	715
12-6	**Comparing Sets of Data**	723
	Explore Transforming Sets of Data by Using Addition	
	Explore Transforming Sets of Data by Using Multiplication	
12-7	**Summarizing Categorical Data**	733
	Explore Categorical Data	
	Module 12 Review	741

Selected Answers	SA1
Glossary	G1
Index	IN1
Symbols and Formulas	Inside Back Cover

Module 1
Expressions

e Essential Question
How can mathematical expressions be represented and evaluated?

What Will You Learn?

How much do you already know about each topic **before** starting this module?

KEY
👎 — I don't know.   👉 — I've heard of it.   👍 — I know it!

	Before			After		
	👎	👉	👍	👎	👉	👍
write numerical expressions						
evaluate numerical expressions						
use the order of operations						
write algebraic expressions						
evaluate algebraic expressions						
identify properties of equality						
apply the Identity and Inverse Properties to evaluate expressions						
apply the Commutative, Associative, and Distributive Properties to evaluate expressions						
write and evaluate absolute value expressions						
use descriptive modeling to describe real-world situations						
choose a level of accuracy appropriate to limitations on measurements						

📙 **Foldables** Make this Foldable to help you organize your notes about expressions. Begin with three sheets of notebook paper.

1. **Fold** three sheets of paper in half along the width. Then cut along the crease.
2. **Staple** the six half-sheets together to form a booklet.
3. **Cut** five centimeters from the bottom of the top sheet, four centimeters from the second sheet, and so on.
4. **Label** each tab with a lesson number.

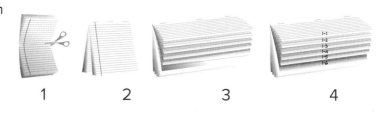

Module 1 · Expressions **1**

What Vocabulary Will You Learn?

- absolute value
- accuracy
- additive identity
- additive inverses
- algebraic expression
- base
- closed
- coefficient
- constant term
- define a variable
- descriptive modeling
- equivalent expressions
- evaluate
- exponent
- like terms
- metric
- multiplicative identity
- multiplicative inverses
- numerical expression
- reciprocals
- simplest form
- term
- variable
- variable term

Are You Ready?

Complete the Quick Review to see if you are ready to start this module.
Then complete the Quick Check.

Quick Review

Example 1

Write $\frac{24}{40}$ in simplest form.

Find the greatest common factor (GCF) of 24 and 40.

factors of 24: 1, 2, 3, 4, 6, 8, 12, 24

factors of 40: 1, 2, 4, 5, 8, 10, 20, 40

The GCF of 24 and 40 is 8.

$\frac{24 \div 8}{40 \div 8} = \frac{3}{5}$ Divide the numerator and denominator by their GCF, 8.

Example 2

Find $2\frac{1}{4} \div 1\frac{1}{2}$.

$2\frac{1}{4} \div 1\frac{1}{2} = \frac{9}{4} \div \frac{3}{2}$ Write mixed numbers as improper fractions.

$= \frac{9}{4} \times \frac{2}{3}$ Multiply by the reciprocal.

$= \frac{18}{12} = \frac{3}{2}$ or $1\frac{1}{2}$ Simplify.

Quick Check

Write each fraction in simplest form.

1. $\frac{24}{36}$
2. $\frac{34}{85}$
3. $\frac{5}{65}$
4. $\frac{64}{88}$

Evaluate.

5. $6 \times \frac{2}{3}$
6. $\frac{7}{8} - \frac{1}{6}$
7. $\frac{3}{8} \div \frac{1}{4}$
8. $\frac{1}{3} + \frac{3}{4}$

How did you do?

Which exercises did you answer correctly in the Quick Check?

Lesson 1-1

Numerical Expressions

Explore Order of Operations

Online Activity Use a real-world situation to complete the Explore.

INQUIRY How can you evaluate a numerical expression?

Today's Goals
- Write numerical expressions for verbal expressions.
- Evaluate numerical expressions.

Today's Vocabulary
numerical expression
exponent
base
evaluate

Learn Writing Numerical Expressions

A **numerical expression** is a mathematical phrase that contains only numbers and mathematical operations. For example, $6 + 2 \div 1$ is a numerical expression.

Some numerical expressions contain multiplication. Multiplication can be represented in several ways, including a raised dot or parentheses. Here are some ways to represent the product of 2 and 3.

$2 \cdot 3 \qquad 2(3) \qquad (2)3 \qquad (2)(3)$

Example 1 Translate a Verbal Expression

Translate *one plus eight divided by three* **into a numerical expression.**

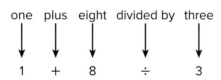

The translation is $1 + 8 \div 3$.

Check

Translate the verbal expression into a numerical expression.
seven times fifteen minus two times nine

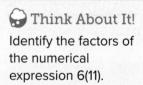 Think About It!
Identify the factors of the numerical expression 6(11).

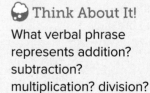 Think About It!
What verbal phrase represents addition? subtraction? multiplication? division?

 Go Online You can complete an Extra Example online.

Lesson 1-1 • Numerical Expressions **3**

Example 2 Translate a Verbal Expression with Grouping Symbols

Translate *the sum of five and nine divided by two* into a numerical expression.

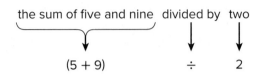

The translation is $\frac{5+9}{2}$.

🌎 **Example 3** Write a Numerical Expression

BOWLING A person's handicap in bowling is usually found by subtracting the person's average from 200, multiplying by 2, and dividing by 3. Marley's bowling average is 170. Write a numerical expression for Marley's handicap in bowling.

The first step to find the handicap is to subtract the average from 200.

Marley's average is 170.	$200 - 170$
Next, multiply the difference by 2.	$2(200 - 170)$
Finally, divide the result by 3.	$\frac{2(200-170)}{3}$

Check

TEMPERATURE To convert a temperature in degrees Celsius to degrees Fahrenheit, multiply the temperature in degrees Celsius by 9, divide by 5, and add 32. Which numerical expression below converts 37° Celsius to Fahrenheit?

A. $\frac{37 \cdot 9}{5} + 32$

B. $\frac{37 \cdot 9}{5 + 32}$

C. $\frac{37 \cdot 9}{5} \cdot 32$

D. $\frac{37 \cdot 5}{9} + 32$

🔎 **Go Online** You can complete an Extra Example online.

💭 **Think About It!**
What would be the first step in finding Marley's handicap? Why?

Watch Out!
Verbal Expressions The first step mentioned in the verbal expression does not always mean it is the first step in writing the numerical expression.

Learn Evaluating Numerical Expressions

An expression of the form x^n is read "x to the nth power." The word *power* is used to refer to the expression, the value, or the exponent of the expression.

When n is a positive integer, the **exponent** indicates the number of times a number is multiplied by itself. In a power, the **base** is the number being multiplied by itself.

To **evaluate** an expression means to find its value. If a numerical expression contains more than one operation, then the rule that lets you know which operation to perform first is called the **order of operations**.

Key Concept • Order of Operations
Step 1 Evaluate expressions inside grouping symbols.
Step 2 Evaluate all powers.
Step 3 Multiply and/or divide from left to right.
Step 4 Add and/or subtract from left to right.

Talk About It!
Explain how the order of operations applies when using the formula $\frac{1}{2}h(b_1 + b_2)$ to find the area of a trapezoid.

Example 4 Evaluate Expressions

Evaluate each expression.

a. 2^4

$2^4 = 2 \cdot 2 \cdot 2 \cdot 2$ Use 2 as a factor 4 times.

$= 16$ Multiply.

b. 4^5

$4^5 = 4 \cdot 4 \cdot 4 \cdot 4 \cdot 4$ Use 4 as a factor 5 times.

$= 1024$ Multiply.

Example 5 Order of Operations

Evaluate $20 - 7 + 8^2 - 7 \cdot 11$.

$20 - 7 + 8^2 - 7 \cdot 11 = 20 - 7 + 64 - 7 \cdot 11$ Evaluate 8^2.

$= 20 - 7 + 64 - 77$ Multiply 7 and 11.

$= 13 + 64 - 77$ Subtract 7 from 20.

$= 77 - 77$ Add 13 and 64.

$= 0$ Subtract 77 from 77.

Think About It!
Write an expression that uses exponents and at least three different operations. Explain the steps you would take to evaluate the expression.

Go Online You can complete an Extra Example online.

Check

Write the steps of the order of operations in the correct order.

Step 1 ? _____

Step 2 ? _____

Step 3 ? _____

Step 4 ? _____

🌎 Example 6 Write and Evaluate a Numerical Expression

ARCADE Mellie is playing a bowling game at an arcade. She rolls two balls into the 30-point hole, four balls into the 20-point hole, and three balls into the 50-point hole. Write and evaluate an expression to find Mellie's total score.

Part A Complete the table to write an expression for Mellie's total score.

To find Mellie's total score, find the number of points scored from each hole and add the products.

Words	two balls rolled into the 30-point hole	plus	four balls rolled into the 20-point hole	plus	three balls rolled into the 50-point hole
Expression	2 · 30	+	4 · 20	+	3 · 50

Part B Evaluate the expression.

2 · 30 + 4 · 20 + 3 · 50 = 60 + 80 + 150   Multiply.

= 290   Add.

Mellie scored 290 points.

Use a Source

Find data about the scoring in a game of interest to you where you can score different numbers of points for different plays. Write and evaluate an expression to represent a possible score.

Check

COMPUTERS A computer technician charges a flat fee of $50 plus $25 per hour. On Monday, he worked on Aika's computer for 2 hours. On Tuesday, he worked on Aika's computer for 3 hours.

Part A Which expression(s) represents Aika's bill? Select all that apply.

A. 50 + 25(2) + 25(3)

B. 50 + 25(2 + 3)

C. 25(2 + 3)

D. 50 + 25(5)

E. 25 + 50(2 + 3)

Part B How much money does Aika owe the technician?

🌎 **Go Online** You can complete an Extra Example online.

6 Module 1 · Expressions

Example 7 Expressions with Grouping Symbols

Evaluate each expression.

a. $\dfrac{(4+5)^2}{3(7-4)}$

$\dfrac{(4+5)^2}{3(7-4)} = \dfrac{(9)^2}{3(3)}$ Evaluate inside parentheses.

$\phantom{\dfrac{(4+5)^2}{3(7-4)}} = \dfrac{81}{3(3)}$ Evaluate the power in the numerator.

$\phantom{\dfrac{(4+5)^2}{3(7-4)}} = \dfrac{81}{9}$ Multiply 3 and 3 in the denominator.

$\phantom{\dfrac{(4+5)^2}{3(7-4)}} = 9$ Divide 81 by 9.

b. $15 - [10 + (3-2)^2] + 6$

$15 - [10 + (3-2)^2] + 6 = 15 - [10 + (1)^2] + 6$ Evaluate innermost parentheses.

$ = 15 - [10 + 1] + 6$ Evaluate power.

$ = 15 - [11] + 6$ Add.

$ = 4 + 6$ Subtract.

$ = 10$ Add.

Check

Evaluate $\dfrac{2^3 - 5}{15 + 9}$.

Learn Plan for Problem Solving

Using a **four-step problem-solving plan** can help you make sense of problems and persevere in solving them.

Key Concept • Four-Step Problem-Solving Plan

Step 1 Identify and understand the task.

Read the task carefully, and make sure you understand what question to answer or problem to solve.

Step 2 Plan your approach.

Choose a strategy. Plan the steps you will use to complete the task.

Step 3 Solve the problem.

Use the strategy you chose in Step 2 to solve the problem.

Step 4 Check the solution.

Make sure that your solution is reasonable and completes the task.

 Go Online You can complete an Extra Example online.

Think About It!
Equivalent expressions have the same value. Are the expressions $(30 + 17) \times 10$ and $10 \times 30 + 10 \times 17$ equivalent? Why or why not?

Study Tip
Grouping Symbols Grouping symbols such as parentheses ( ), brackets [ ], braces { }, and fraction bars are used to clarify or change the order of operations. So, evaluate expressions inside grouping symbols first. For fraction bars, evaluate the numerator and denominator before completing the division.

Think About It!
How can using this four-step problem-solving plan help you effectively solve problems?

Math History Minute

Hungarian mathematician **George Pólya (1887–1985)** is known as the father of mathematical problem solving. One of his books, *How to Solve It*, was a bestseller that was translated into 17 languages. In it, Pólya describes a four-step plan for problem solving.

Example 8 Write and Evaluate Expressions

MONEY Thursdays are Student Days at LSC Theaters. Student tickets are $5 and popcorn refills are free. You will buy a ticket and a 20-ounce bottle of water, and you will split the cost of a large tub of popcorn and large candy with a friend. How much money should you bring?

Popcorn		Drinks		Snacks	
Mini Tub	$3.00	20 oz	$5.00	Fruit Snacks	$3.00
Medium Tub	$5.00	32 oz	$5.50	Hot Dog	$5.00
Large Tub	$7.50	44 oz	$6.00	Candy	$3.50

UNDERSTAND We are given what you will buy, the cost of each item, and the cost of the popcorn and candy that will be split with a friend. We are asked to find the amount of money that will be spent.

PLAN

Step 1 Write an expression to represent the total cost of the ticket, drink, and food.

Step 2 Evaluate the expression to find the cost.

SOLVE

Step 1 Write the expression for the total cost in dollars.

ticket		drink		popcorn		candy
5	+	5	$+\frac{1}{2} \cdot$	7.50	$+\frac{1}{2} \cdot$	3.50

Step 2 Use the order of operations to evaluate the expression.

$5 + 5 + \frac{1}{2} \cdot 7.50 + \frac{1}{2} \cdot 3.50 = 5 + 5 + 3.75 + 1.75$ Multiply.

$\hspace{4.5cm} = 15.50$ Add.

CHECK

The cost of the ticket and water is $10. The total cost of the popcorn and candy is $11. So half of that cost is $5.50. The total is $10 + $5.50 or $15.50.

Check

PETS While she was on a 7-day vacation, Ms. Hernandez boarded her dog at a kennel. If her boarding budget is $250, can Ms. Hernandez afford a daily extra walk and one wash and nail trimming? Explain.

Service	Cost
Board	$24 per day
Wash	$45
Extra walk	$4 per day
Nail trimming	$12
Vitamins	$1 per day

Go Online You can complete an Extra Example online.

Practice

Go Online You can complete your homework online.

Example 1

Write a numerical expression for each verbal expression.

1. two plus twelve divided by four
2. eighteen more than five
3. seven more than three times eleven
4. twenty-five less than one hundred
5. six minus three minus one
6. fourteen decreased by three times four
7. twenty-four divided by six plus seven
8. eight times six divided by two minus nine
9. one hundred sixteen divided by four plus twenty-eight minus thirty-three
10. two hundred fifty-nine minus eighty-five plus sixty-two divided by two

Example 2

Write a numerical expression for each verbal expression.

11. the sum of three and seven divided by two
12. the difference of six and two divided by four
13. the sum of four and nine times three
14. eighteen divided by the sum of two and seven
15. ten divided by the product of four and five
16. the difference of eleven and four times five
17. the sum of one and two divided by twenty
18. the sum of two and four and six times eight
19. the sum of twelve and sixteen divided by the sum of three and four
20. the difference of twenty-two and six divided by the sum of five and three
21. the sum of thirty-six and fourteen divided by the product of two and five
22. the quotient of thirty-two and four divided by the sum of one and three
23. the sum of six and fifteen divided by the difference of thirteen and nine
24. the difference of thirty-one and seventeen divided by the product of ten and four

Example 3

25. SOLAR SYSTEM It takes Earth about 365 days to orbit the Sun. It takes Uranus about 85 times as long. Write a numerical expression to describe the number of days it takes Uranus to orbit the Sun.

26. TEST SCORES To find the average of a student's test scores, add the scores and divide by the number of tests. Suppose Ryan scored 85, 92, 88, and 98 on four tests. Write a numerical expression to describe Ryan's average test score.

27. HOMEWORK It took Carrie five less minutes than twice the amount of time as Hua to complete her homework. It took Hua thirty-five minutes to complete her homework. Write a numerical expression to describe the amount of time it took Carrie to complete her homework.

28. PERIMETER The perimeter of Stephanie's triangle is half the perimeter of Juan's triangle. Juan's triangle is shown. Write a numerical expression to describe the perimeter of Stephanie's triangle.

29. BEDROOM Shenandoah's rectangular bedroom is 12 feet long and 7 feet wide. Write a numerical expression to describe the area of Shenandoah's bedroom.

Examples 4, 5, and 7

Evaluate each expression.

30. 7^2

31. 14^3

32. 2^6

33. $35 - 3 \cdot 8$

34. $18 \div 9 + 2 \cdot 6$

35. $10 + 8^3 \div 16$

36. $[(6^3 - 9) \div 23]4$

37. $\dfrac{8 + 3^3}{12 - 7}$

38. $\dfrac{(1 + 6)9}{5^2 - 4}$

39. $4(16 \div 2 + 6)$

40. $13 - \dfrac{1}{3}(11 - 5)$

41. $(5 \cdot 2 - 9) + 2 \cdot \dfrac{1}{2}$

42. $62 - 3^2 \cdot 8 + 11$

43. $4^3 \div 8$

44. $20 + 3(8 - 5)$

45. $3[4 - 8 + 4^2(2 + 5)]$

46. $\dfrac{2 \cdot 8^2 - 2^2 \cdot 8}{2 \cdot 8}$

47. $25 + \left[(16 - 3 \cdot 5) + \dfrac{12 + 3}{5}\right]$

48. $7^3 - \dfrac{2}{3}(13 \cdot 6 + 9)4$

Example 6

49. BIOLOGY Lavania is studying the growth of a population of fruit flies in her laboratory. After 6 days, she had nine more than five times as many fruit flies as when she began the study. If she observes 20 fruit flies on the first day of the study, write and evaluate an expression to find the population of fruit flies Lavania observed after 6 days.

   a. Write an expression for the population of fruit flies Lavania observed after 6 days.

   b. Find the population of fruit flies Lavania observed after 6 days.

50. PRECISION The table shows how scores are calculated at diving competitions. Each of the five judges scores each dive from 1 to 10 in 0.5-point increments. Tyrell performs a dive with a degree of difficulty of 2.5. His scores from the judges are 8.0, 7.5, 6.5, 7.5, and 7.0.

   a. Write an expression to find Tyrell's score for the dive.

   b. What was Tyrell's score for the dive?

Calculating a Diving Score	
Step 1	Drop the highest and lowest of the five judges' scores.
Step 2	Add the remaining scores to find the raw score.
Step 3	Multiply the raw score by the degree of difficulty.

51. RAMP The side panel of a skateboard ramp is a trapezoid, as shown.

   a. Write an expression to find the amount of wood needed to build the two side panels of a skateboard ramp.

   b. How much wood is needed to build the two side panels of a skateboard ramp?

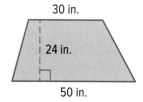

Example 8

52. SKIING The cost of a ski trip is shown. The Sanchez family wants to purchase lift tickets for 2 adults and 3 children. They also need to rent 2 complete pairs of skis. If they also buy a 16-ounce hot chocolate for each person, find the total cost of the ski trip.

Lift Tickets		Rentals		Hot Chocolate	
Children	$34	Skis	$32	12 oz	$3
Seniors	$36	Poles Only	$10	16 oz	$4
Adults	$42	Snowboards	$29	20 oz	$5

Mixed Exercises

Write a numerical expression for each verbal expression.

53. eight to the fourth power increased by six

54. the sum of three and five to the third power times five plus one

55. CONSTRUCT ARGUMENTS Isabel wrote the expression $6 + 3 \times 5 - 6 + 8 \div 2$ and asked Tamara to evaluate it. When Tamara evaluated it, she got a value of 19. Isabel told Tamara that her value was incorrect and said that the value should have been 38. Who is correct? Justify your argument.

56. **REASONING** Write an expression that includes the numbers 2, 4, and 5, has a value of 50, and includes one set of parentheses.

57. **CONSTRUCT ARGUMENTS** Kelly buys 3 video games that cost $18.96 each. She also buys 2 pairs of earbuds that cost $11.50 each. She has a coupon for $2 off the price of each video game. Kelly uses a calculator, as shown, to find that the total cost of the items is $77.85. The cashier tells her that the total cost is $73.85. Who is correct? Justify your argument.

58. **REASONING** The expression $13.25 \times 5 + 6.5$ gives the total cost in dollars of renting a bicycle and helmet for 5 days. The fee for the helmet does not depend upon the number of days.

   a. What does 13.25 represent?

   b. How would the expression be different if the cost of the helmet were doubled?

59. **PERSEVERE** The figure shows a floor plan for a two-room apartment. Write an expression for the area of the apartment, in square feet, by first finding the area of each room and then adding. Then describe how you can write the expression in a different way.

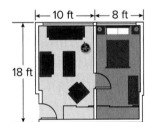

60. **CREATE** Describe a situation that could be represented by each expression.

   a. $9.95 + 0.75 \times 3$

   b. $15 \times 6 - 5 \times 6$

   c. $59 \times 5 - 25 \times 5 - 30$

61. **FIND THE ERROR** A student was asked to evaluate an expression. The student's work is shown. Describe any errors and find the correct value of the expression. Explain your reasoning.

   $6^2 - 5 \times 2 + 2(9-7)$
   $= 6^2 - 5 \times 2 + 2(2)$
   $= 36 - 5 \times 2 + 2(2)$
   $= 36 - 10 + 4$
   $= 36 - 14$
   $= 22$

62. **PERSEVERE** When is $4x < 4$? Use a drawing to justify your reasoning.

12 Module 1 • Expressions

Lesson 1-2

Algebraic Expressions

Learn Writing Algebraic Expressions

A **variable** is a letter used to represent an unspecified number or value.

An **algebraic expression** is an expression that contains at least one variable.

A **term** of an expression is a number, a variable, or a product or quotient of numbers and variables.

A **variable term** is a term that contains a variable.

A **constant term** is a term that does not contain a variable.

Example 1 Write a Verbal Expression

There are common verbal phrases associated with operations. These phrases can be used to interpret an algebraic expression.

Interpreting an Algebraic Expression	
Operation	Verbal Phrases
addition	more than, sum, plus, increased by, added to
subtraction	less than, subtracted from, difference, decreased by, minus
multiplication	product of, multiplied by, times
division	quotient of, divided by

Write a verbal expression for $5x^3 + 2$.

five times x to the third power plus two

Check

Which verbal expression represents $\frac{2}{5}m^2$?

A. two fifths times the product of m and two

B. the quotient of two and five times the product of m and two

C. two fifths of m squared

D. two fifths times m cubed

Go Online You can complete an Extra Example online.

Today's Goals
- Write algebraic expressions for verbal expressions.
- Evaluate algebraic expressions.

Today's Vocabulary
variable

algebraic expression

term

variable term

constant term

define a variable

Think About It!
Write an algebraic expression that uses at least two variables, one constant, one product, and one power. Identify the terms, variables, constant, product, factors, and powers.

Go Online
An alternate method is available for this example.

Watch Out!
Quotient When a verbal expression refers to a quotient, the first term mentioned is the numerator and the second term mentioned is the denominator. For example, *the quotient of 6 and t* is written as $\frac{6}{t}$, while *the quotient of t and 6* is written as $\frac{t}{6}$.

Study Tip
Exponents The exponents 2 and 3 are frequently used in math and have special verbal phrases associated with them: x^2 is read "x squared," and x^3 is read "x cubed."

Example 2 Write a Verbal Expression with Grouping Symbols

Write a verbal expression for $a^4 + \frac{6b}{7}$.

a^4 + $\frac{6b}{7}$

a to the fourth power   plus   the quotient of 6 times *b* and 7

a to the fourth power plus the quotient of 6 times *b* and 7

Try an alternate method.

a^4 + $\frac{6b}{7}$

a to the fourth power   plus   6 times *b* divided by 7

a to the fourth power plus 6 times *b* divided by 7

Compare and contrast the methods.
Quotient implies division, so you can replace quotient with divided by. Thus, the two methods have the same meaning but use a different word to describe division.

Example 3 Write an Algebraic Expression

Write an algebraic expression for each verbal expression.

a. 2 times the quantity y plus 11

The word *times* implies multiplication, *the quantity* implies parentheses, and *plus* implies addition.

So, the expression is written as $2(y + 11)$.

b. n cubed increased by 5

The word *cubed* implies a power of 3, and *increased by* suggests addition.

The expression could be written as $n^3 + 5$ or $5 + n^3$.

Check

Use the verbal phrase *18 increased by* the *product of* 3 and *d*.

Part A The key verbal phrases in the expression are italicized. Identify which operation each key phrase implies.

increased by implies _____?_____

product of implies _____?_____

Part B Write an algebraic expression for the verbal phrase.

Go Online You can complete an Extra Example online.

Practice

🔗 **Go Online** You can complete your homework online.

Example 1

Refer to the figure for Exercises 1-7.

1. Name the lines that are only in plane Q.
2. How many planes are labeled in the figure?
3. Name the plane containing the lines m and t.
4. Name the intersection of lines m and t.
5. Name a point that is *not* coplanar with points A, B, and C.
6. Are points F, M, G, and P coplanar? Explain.
7. Does line n intersect line q? Explain.

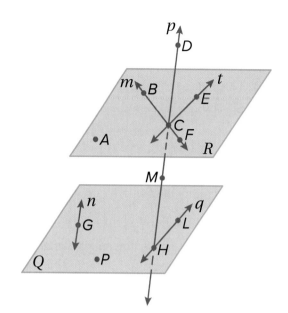

Example 2

Name the geometric terms modeled by each object or phrase.

8. roof of a house
9. a tabletop
10. bridge support beam

11. a chessboard
12.
13.

14. a wall and the floor
15. the edge of a table
16. two connected walls

17. a blanket
18. a telephone pole
19. a tablet computer

Example 3

USE TOOLS Draw and label a figure for each relationship.

20. Points X and Y lie on $\overleftrightarrow{CD}$.

21. Two planes do not intersect.

22. Line m intersects plane R at a single point.

23. Three lines intersect at point J but do not all lie in the same plane.

24. Points $A(2, 3)$, $B(2, -3)$, C, and D are collinear, but $A, B, C, D,$ and F are not.

Example 4

Refer to the figure for Exercises 25–28.

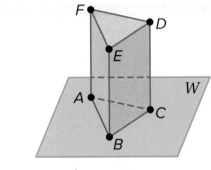

25. How many planes are shown in the figure?

26. How many of the planes contain points F and E?

27. Name four points that are coplanar.

28. Are points A, B, and C coplanar? Explain.

Example 5

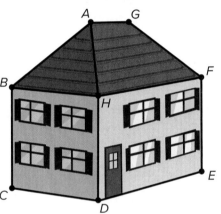

29. **BUILDING** The roof and exterior walls of a house represent intersecting planes. Using the image, name all the lines that are formed by the intersecting planes.

30. If the surface of a lake represents a plane, what geometric term is represented by the intersection of a fishing line and the lake's surface?

31. **ART** Perspective drawing is a method that artists use to create paintings and drawings of three-dimensional objects. The artist first draws the horizon line and two vanishing points along the horizon. Buildings or other objects are created by drawing receding lines and vertical lines.

   a. Where do the receding lines and horizon lines intersect?

   b. Identify examples of planes within this picture.

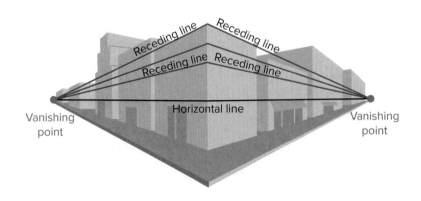

16 Module 1 • Tools of Geometry

Mixed Exercises

USE TOOLS Draw and label a figure for each relationship.

32. $\overleftrightarrow{LM}$ and $\overleftrightarrow{NP}$ are coplanar but do not intersect.

33. $\overleftrightarrow{FG}$ and $\overleftrightarrow{JK}$ intersect at P(4, 3), where point F is at (−2, 5) and point J is at (7, 9).

34. Lines *s* and *t* intersect, and line *v* does not intersect either one.

Refer to the figure for Exercises 35-38.

35. Name a line that contains point E.

36. Name a point contained in line *n*.

37. What is another name for line *p*?

38. Name the plane containing lines *n* and *p*.

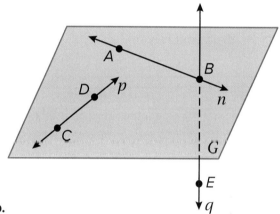

USE TOOLS Draw and label a figure for each relationship.

39. Point K lies on $\overleftrightarrow{RT}$.

40. Plane J contains line *s*.

41. $\overleftrightarrow{YP}$ lies in plane B and contains point C, but does not contain point H.

42. Lines *q* and *f* intersect at point Z in plane U.

Lesson 1-2 • Points, Lines, and Planes 17

43. Name the geometric term modeled by the object.

44. Name the geometric term modeled by a partially-opened folder.

Higher-Order Thinking Skills

45. **CREATE** Sketch three planes that intersect in a point.

46. **ANALYZE** Is it possible for two points on the surface of a prism to be neither collinear nor coplanar? Justify your argument.

47. **FIND THE ERROR** Camille and Hiroshi are trying to determine the greatest number of lines that can be drawn using any two of four random points. Is either correct? Explain your reasoning.

Camille	Hiroshi
Because there are four points, 4 · 3 or 12 lines can be drawn between the points.	You can draw 3 + 2 + 1 or 6 lines between the points.

48. **PERSEVERE** What is the greatest number of planes determined using any three of the points A, B, C, and D if no three points are collinear?

49. **WRITE** A *finite plane* is a plane that has boundaries or does not extend indefinitely. The sides of a cereal box shown are finite planes. Give a real-life example of a finite plane. Is it possible to have a real-life object that is an infinite plane? Explain your reasoning.

50. **CREATE** Sketch three planes that intersect in a line.

Lesson 1-3

Line Segments

Explore Using Tools to Determine Betweenness of Points

Online Activity Use a pencil and straightedge to complete the Explore.

> **INQUIRY** How can a line segment be divided into any number of line segments?

Today's Goals
- Calculate measures of line segments.
- Apply the definition of congruent line segments to find missing values.

Today's Vocabulary
line segment
betweenness of points
congruent
congruent segments

Learn Betweenness of Points

A **line segment** is a measurable part of a line that consists of two points, called endpoints, and all of the points between them. The two endpoints are used to name the segments.

You know that for any two real numbers a and b, there is a real number n between a and b such that $a < n < b$. This relationship also applies to points on a line and is called **betweenness of points**.

Key Concept • Betweenness of Points

Point C is between A and B if and only if A, B, and C are collinear and $AC + CB = AB$.

Talk About It!
What is an example of how the betweenness of points can be applied to the real world?

Example 1 Find Measurements by Adding

Find the measure of $\overline{XZ}$.

XZ is the measure of $\overline{XZ}$. Point Y is between X and Z. Find XZ by adding XY and YZ.

$XY + YZ = XZ$	Betweenness of points
$11.3 + 3.8 = XZ$	Substitution
$15.1 \text{ cm} = XZ$	Add.

(Diagram: X to Y is 11.3 cm, Y to Z is 3.8 cm)

Check

Find the measure of $\overline{DF}$.

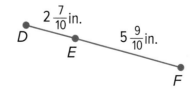

(Diagram: D to E is $2\frac{7}{10}$ in., E to F is $5\frac{9}{10}$ in.)

Go Online You can complete an Extra Example online.

Lesson 1-3 • Line Segments **19**

Problem-Solving Tip

Draw a Diagram Draw a diagram to help you see and correctly interpret a situation that has been described in words.

Example 2 Find Measurements by Subtracting

Find the measure of $\overline{QR}$.

Point Q is between points P and R.

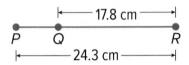

$PQ + QR = PR$	Betweenness of points
$6\frac{5}{8} + QR = 13\frac{3}{4}$	Substitution
$QR = 7\frac{1}{8}$ ft	Subtract $6\frac{5}{8}$ from each side and simplify.

Check

Find the measure of $\overline{PQ}$. Round your answer to the nearest tenth, if necessary.

___?___ cm

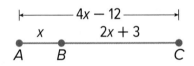

Example 3 Write and Solve Equations to Find Measurements

Find the value of x and BC if B is between A and C, $AC = 4x - 12$, $AB = x$, and $BC = 2x + 3$.

Step 1 Plot two points and label them A and C. Connect the points.

Step 2 Plot point B between points A and C.

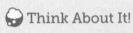

Step 3 Label segments AB, BC, and AC with their given measures.

Step 4 Use betweenness of points to write an equation and solve for x.

$AC = AB + BC$	Betweenness of points
$4x - 12 = x + 2x + 3$	Substitution
$4x - 12 = 3x + 3$	Combine like terms.
$x - 12 = 3$	Subtract $3x$ from each side. Simplify.
$x = 15$	Add 12 to each side. Simplify.

Now find BC.

$BC = 2x + 3$	Given
$= 2(15) + 3$	$x = 15$
$= 33$	

💭 **Think About It!**

How can you check your solution for x?

💭 **Think About It!**

Once you find BC, how could you find AC without evaluating $AC = 4x - 12$?

🅖 **Go Online** You can complete an Extra Example online.

🌐 Apply Example 4 Use Betweenness of Points

SPACE NEEDLE Darrell is visiting the Space Needle in Seattle, Washington. He knows that the total height of the Space Needle is 605 feet. The distance from the ground to the observation deck is 10 feet more than six times the distance from the observation deck to the top of the Space Needle. Help Darrell find the distance from the ground to the observation deck.

1 What is the task?

Describe the task in your own words. Then list any questions that you may have. How can you find answers to your questions?

I need to find the distance from the ground to the observation deck. How does the distance from the ground to the observation deck compare to the total height of the Space Needle? I can express the information that I am given as an equation, solve for any missing information, and then use that information to find the answer.

2 How will you approach the task? What have you learned that you can use to help you complete the task?

I will express the information that I am given into an equation that represents the total height of the Space Needle. I have learned how to convert written information into expressions, and I have learned how to solve equations.

3 What is your solution?

Use your strategy to solve the problem.

What equation represents the distance from the ground to the top of the Space Needle?

$x + 6x + 10 = 605$

What is the distance from the ground to the observation deck?

520 ft

4 How can you know that your solution is reasonable?

✏️ **Write About It!** Write an argument that can be used to defend your solution.

520 feet seems reasonable for the distance from the ground to the observation deck. The distance from the observation deck to the top of the Space Needle is 85 feet. The combined heights are realistic compared to the total height.

Study Tip

Congruent Segments Use a consecutive number of tick marks for each new pair of congruent segments in a figure. The segments with two tick marks are congruent, and the segments with three tick marks are congruent.

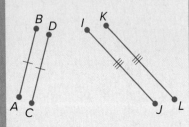

Learn Line Segment Congruence

If two geometric figures have exactly the same shape and size, then they are **congruent**. Two segments that have the same measure are **congruent segments**.

> **Key Concept • Congruent Segments**
>
>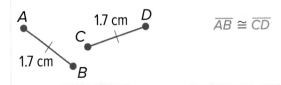
>
> $\overline{AB} \cong \overline{CD}$
>
> $\cong$ is read *is congruent to*. Red slashes on the figure also indicate congruence.
>
> Segment AB is congruent to segment CD.
>
> Congruent segments have the same measure.

Study Tip

Equal vs. Congruent Lengths are *equal*, and segments are *congruent*. It is correct to say that $AB = CD$ and $\overline{AB} \cong \overline{CD}$. However, it is not correct to say that $\overline{AB} = \overline{CD}$ or that $AB \cong CD$.

Example 5 Write and Solve Equations by Using Congruence

Find the value of x if Q is between P and R, $PQ = 6x + 20$, $QR = 2(x + 6)$, and $\overline{PQ} \cong \overline{QR}$.

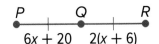

Write the justifications in the correct order. You may use a justification more than once.

Definition of congruence   Distributive Property
Divide each side by 4.   Simplify.   Substitution
Subtract $2x$ from each side.   Subtract 20 from each side.

$PQ = QR$	Definition of congruence
$6x + 20 = 2(x + 6)$	Substitution
$6x + 20 = 2x + 12$	Distributive Property
$6x + 20 - 2x = 2x + 12 - 2x$	Subtract $2x$ from each side.
$4x + 20 = 12$	Simplify.
$4x + 20 - 20 = 12 - 20$	Subtract 20 from each side.
$4x = -8$	Simplify.
$\dfrac{4x}{4} = \dfrac{-8}{4}$	Divide each side by 4.
$x = -2$	Simplify.

Watch Out!

Check Your Answer Sometimes solutions will result in negative segment lengths. If this occurs, review your work carefully. Either an error was made, or there is no solution.

Check

Find the value of x if U is between T and V, $TU = 7x + 35$, $UV = 4(x + 7)$, and $\overline{TU} \cong \overline{UV}$.

$x = \underline{}$

Go Online You may want to complete the construction activities for this lesson.

Practice

Go Online You can complete your homework online.

Examples 1 and 2

Find the measure of each segment.

1. $\overline{PR}$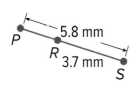

2. $\overline{EF}$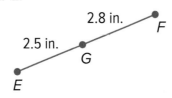

3. $\overline{JL}$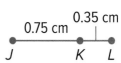

4. $\overline{HJ}$

5. $\overline{AC}$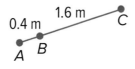

6. $\overline{SV}$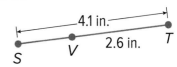

7. $\overline{NQ}$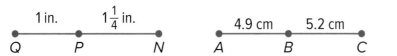

8. $\overline{AC}$

9. $\overline{GH}$

Example 3

Find the value of the variable and YZ if Y is between X and Z.

10. $XY = 11$, $YZ = 4c$, $XZ = 83$

11. $XY = 6b$, $YZ = 8b$, $XZ = 175$

12. $XY = 7a$, $YZ = 5a$, $XZ = 6a + 24$

13. $XY = 5.5$, $YZ = 2c$, $XZ = 8.9$

14. $XY = 5n$, $YZ = 2n$, $XZ = 91$

15. $XY = 4w$, $YZ = 6w$, $XZ = 12w - 8$

16. $XY = 11d$, $YZ = 9d - 2$, $XZ = 5d + 28$

17. $XY = 4n + 3$, $YZ = 2n - 7$, $XZ = 20$

18. $XY = 3a - 4$, $YZ = 6a + 2$, $XZ = 5a + 22$

19. $XY = 3k - 2$, $YZ = 7k + 4$, $XZ = 4k + 38$

20. $XY = 4x$, $YZ = x$, and $XZ = 25$

21. $XY = 4x$, $YZ = 3x$, and $XZ = 42$

22. $XY = 12$, $YZ = 2x$, and $XZ = 28$

23. $XY = 2x + 1$, $YZ = 6x$, and $XZ = 81$

Lesson 1-3 • Line Segments 23

Example 4

24. RAILROADS A straight railroad track is being built to connect two cities. The measured distance of the track between the two cities is 160.5 miles. A mail stop is 28.5 miles from the first city. How far is the mail stop from the second city?

25. CARPENTRY A carpenter has a piece of wood that is 78 inches long. He wants to cut it so that one piece is five times as long as the other piece. What are the lengths of the two pieces?

26. WALKING Marshall lives 2300 yards from school and 1500 yards from the pharmacy. The school, pharmacy, and his home are all collinear, as shown in the figure.

What is the distance from the pharmacy to the school?

27. COFFEE SHOP Chenoa wants to stop for coffee on her way to school. The distance from Chenoa's house to the coffee shop is 3 miles more than twice the distance from the coffee shop to Chenoa's school. The total distance from Chenoa's house to her school is 5 times the distance from the coffee shop to her school.

a. What is the distance from Chenoa's house to the coffee shop? Write your answer as a decimal, if necessary.

b. What assumptions did you make when solving this problem?

Example 5

Find the measure of each segment.

28. $\overline{MO}$

29. $\overline{WY}$

30. $\overline{FG}$

31. $\overline{QT}$

32. $\overline{DE}$

33. $\overline{UX}$

Mixed Exercises

34. Find the length of $\overline{UW}$ if W is between U and V, $UV = 16.8$ centimeters, and $VW = 7.9$ centimeters.

35. Find the value of x if $RS = 24$ centimeters.

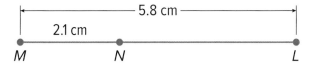

36. Find the length of $\overline{LO}$ if M is between L and O, $LM = 7x - 9$, $MO = 14$ inches, and $LO = 10x - 7$.

37. Find the value of x if $\overline{PQ} \cong \overline{RS}$, $PQ = 9x - 7$, and $RS = 29$.

38. Find the measure of $\overline{NL}$.

39. **PRECISION** If point P is between A and M, write a true statement.

40. **HIKING** A hiking trail is 20 kilometers long. Park organizers want to build 5 rest stops for hikers with one on each end of the trail and the other 3 spaced evenly between. How much distance will separate successive rest stops?

41. **RACE** The map shows the route of a race. You are at Y, 6000 feet from the first checkpoint A. The second checkpoint B is located at the midpoint between A and the end of the race Z. The total race is 3.1 miles. How far apart are the two checkpoints?

42. **FIELD TRIP** The marching band at Jefferson High School is taking a field trip from Lansing, Michigan, to Detroit, Michigan. The bus driver was told to stop 53 miles into the trip. If the rest of the trip is 41 miles and the entire journey can be represented by the expression $3x + 16$, find the value of x.

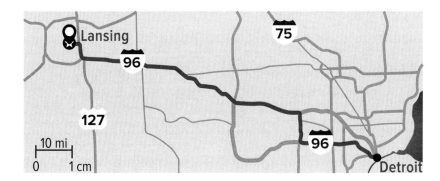

43. **DISTANCE** Madison lives between Anoa and Jamie as depicted on the line segment. The distance between Anoa's house and Madison's house is represented by $3x + 2$ miles, the distance between Madison's house and Jamie's house is represented by $3x + 4$ miles, and the distance between Anoa's house and Jamie's house is represented by $9x - 3$ miles. Find the value of x. Then find the distance between Madison's house and Jamie's house.

44. **FIREFIGHTING** A firefighter training course is taking place in a high-rise building. The high-rise building where they practice is 48 stories high. If the emergency happens on the top floor and the firefighters have already gone 29 stories, how many stories do they still need to go?

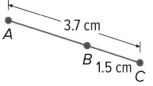

45. **CAFE** You are waiting at the end of a long straight line at Coffee Express. Your friend Denzel is $r + 12$ feet in front of you. Denzel is $2r + 4$ feet away from the front of the line. If Denzel is in the exact middle of the line, how many feet away are you from the front of the line?

46. **REASONING** For $\overline{AC}$, write and solve an equation to find AB.

47. **PERSEVERE** Point K is between points J and L. If $JK = x^2 - 4x$, $KL = 3x - 2$, and $JL = 28$, find JK and KL.

48. **ANALYZE** Determine whether the statement *If point M is between points C and D, then CD is greater than either CM or MD* is *sometimes*, *always*, or *never* true. Justify your argument.

49. **PERSEVERE** Point C is located between points B and D. Also, $BC = 5x + 7$, $CD = 3y + 4$, $BD = 38$, and $BD = 2x + 8y$. Find the values of x and y.

50. **WRITE** If point B is between points A and C, explain how you can find AC if you know AB and BC. Explain how you can find BC if you know AB and AC.

51. **CREATE** Sketch line segment AC. Plot point B between A and C. Use a ruler to find AC and AB. Then write and solve an equation to find BC.

Lesson 1-4

Distance

Learn Distance on a Number Line

The **distance** between two points is the length of the segment between the points. The coordinates of the points can be used to find the length of the segment.

Key Concept • Distance Formula on Number Line

If P has coordinate x_1 and Q has coordinate x_2, then $PQ = |x_2 - x_1|$ or $|x_1 - x_2|$.

Because $\overline{PQ}$ is the same as $\overline{QP}$, the order in which you name the endpoints is not important when calculating distance.

Example 1 Find Distance on a Number Line

Use the number line.

Find CF.

$CF = |x_2 - x_1|$ Distance Formula

      $= |5 - (-1)|$ $x_1 = -1$ and $x_2 = 5$

      $= 6$ Simplify.

Check

Use the number line.

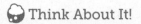

Find AE.

A. -12 B. 2 C. 12 D. 13

Go Online You can complete an Extra Example online.

Today's Goals
- Find the length of a line segment on a number line.
- Find the distance between two points on the coordinate plane.

Today's Vocabulary
distance

Think About It!
Why do you think the Distance Formula uses absolute value?

Think About It!
Compare and contrast the length of $\overline{CF}$ and the length of $\overline{FC}$.

Example 2 Determine Segment Congruence

Determine whether $\overline{CB}$ and $\overline{DF}$ are congruent.

The coordinates of C and B are −1 and −3. The coordinates of D and F are 2 and 5. Find the length of each segment.

$CB = |x_2 - x_1|$ Distance Formula

$ = |-3 - (-1)|$ Substitute.

$ = |-2|$ Subtract.

$ = 2$ Simplify.

The length of $\overline{CB}$ is 2 units.

$DF = |x_2 - x_1|$ Distance Formula

$ = |5 - 2|$ Substitute.

$ = |3|$ Subtract.

$ = 3$ Simplify.

The length of $\overline{DF}$ is 3 units.

Because $CB \neq DF$, the segments are not congruent.

Check

Determine whether $\overline{AC}$ and $\overline{BD}$ are congruent.

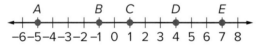

The segments ___?___ congruent.

> **Watch Out!**
> **Subtraction with Negatives** Remember that subtracting a negative number is like adding a positive number.

Explore Use the Pythagorean Theorem to Find Distances

 Online Activity Use dynamic geometry software to complete the Explore.

> **@ INQUIRY** How can you find the distance between two points on the coordinate plane?

 Go Online A derivation of the Distance Formula is available.

Learn Distance on the Coordinate Plane

The endpoints of a segment on the coordinate plane can be used to find the length of that segment by using the Distance Formula.

Key Concept • Distance Formula on the Coordinate Plane

If P has coordinates (x_1, y_1) and Q has coordinates (x_2, y_2), then

$$PQ = \sqrt{(x_2 - x_1)^2 + (y_2 - y_1)^2}.$$

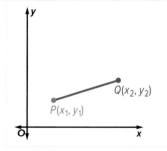

Think About It!

Compare and contrast the Distance Formula on a number line with the Distance Formula on the coordinate plane.

Example 3 Find Distance on the Coordinate Plane

Find the distance between $J(4, 3)$ and $K(-3, -7)$.

Let $J(4, 3)$ be (x_1, y_1) and $K(-3, -7)$ be (x_2, y_2).

$JK = \sqrt{(x_2 - x_1)^2 + (y_2 - y_1)^2}$ Distance Formula

$= \sqrt{(-3 - x_1)^2 + (-7 - y_1)^2}$ Substitute x_2 and y_2.

$= \sqrt{(-3 - 4)^2 + (-7 - 3)^2}$ Substitute x_1 and y_1.

$= \sqrt{(-7)^2 + (-10)^2}$ Subtract.

$= \sqrt{49 + 100}$ Simplify.

$= \sqrt{149}$ Simplify.

The distance between J and K is $\sqrt{149}$ or approximately 12.2 units.

 Go Online An alternate method is available for this example.

Watch Out!

Simplify Radicals Do not forget to leave your answer in simplest radical form when using the Distance Formula or the Pythagorean Theorem.

Check

Find the distance between A and B.

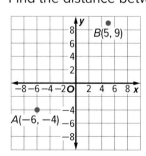

 Go Online You can complete an Extra Example online.

Lesson 1-4 • Distance **29**

Example 4 Calculate Distance

INCLINE Chelsea and Amie are sitting in separate cars on the Monongahela Incline. Chelsea is traveling up Mount Washington and Amie is traveling down. When the two girls notice each other, Chelsea has a horizontal distance of 212.0 feet from the lower station and is at a height of 151.6 feet. Amie has a horizontal distance of 435.3 feet from the lower station and is at a height of 311.3 feet. What is the distance between the two girls?

Step 1 Draw a diagram.

Draw a diagram to represent the situation. Label the x-axis as the "Horizontal Distance from Lower Station (in feet)." Label the y-axis as the "Height (in feet)." Use a scale of 50 on the x-axis and the y-axis.

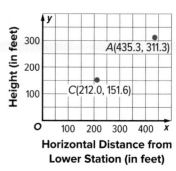

Step 2 Use the Distance Formula.

$(x_1, y_1) = (212.0, 151.6)$ and $(x_2, y_2) = (435.3, 311.3)$

$D = \sqrt{(x_2 - x_1)^2 + (y_2 - y_1)^2}$ Distance Formula

$= \sqrt{(435.3 - 212.0)^2 + (311.3 - 151.6)^2}$ Substitute.

$= \sqrt{223.3^2 + 159.7^2}$ Subtract.

$= \sqrt{49{,}862.89 + 25{,}504.09}$ Square each term.

$= \sqrt{75{,}366.98}$ Add.

≈ 274.5 Take the positive square root.

Chelsea and Amie are approximately 274.5 feet apart.

Check

SNOWBOARDING Manuel wants to go snowboarding with his friend. The closest ski and snowboard resort is approximately 20 miles west and 50 miles north of his house. Manuel picks up his friend who lives 15 miles south and 10 miles east of Manuel's house. How far away are the two boys from the resort?

____?____ mi

Think About It!
Does your answer seem reasonable? Why or why not?

Go Online You can complete an Extra Example online.

Practice

Example 1

Use the number line to find each measure.

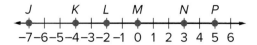

1. JL
2. JK
3. KP

4. NP
5. JP
6. LN

Use the number line to find each measure.

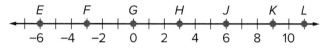

7. JK
8. LK
9. FG

10. JG
11. EH
12. LF

Use the number line to find each measure.

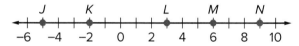

13. LN
14. JL

Example 2

Determine whether the given segments are congruent. Write *yes* or *no*.

15. $\overline{AB}$ and $\overline{EF}$
16. $\overline{BD}$ and $\overline{DF}$
17. $\overline{AC}$ and $\overline{CD}$

18. $\overline{AC}$ and $\overline{DE}$
19. $\overline{BE}$ and $\overline{CF}$
20. $\overline{CD}$ and $\overline{DF}$

Example 3

Find the distance between each pair of points.

21.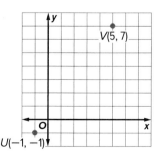

22.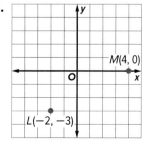

23.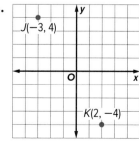

24. A(2, 6), N(5, 10) **25.** R(3, 4), T(7, 2) **26.** X(−3, 8), Z(−5, 1)

Example 4

27. SPIRALS Denise traces the spiral shown in the figure. The spiral begins at the origin. What is the shortest distance between Denise's starting point and her ending point?

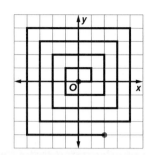

28. ZOOLOGY A tiny songbird called the blackpoll warbler migrates each fall from North America. A tracking study showed one bird flew from Vermont at map coordinates (63, 45) to Venezuela at map coordinates (67, 10) in three days. If each map coordinate represents 75 kilometers, how far did the bird travel?

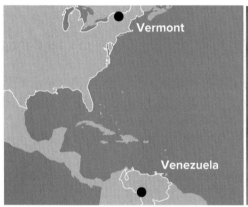 

29. CONSTRUCT ARGUMENTS Mariah is training for a sprint-distance triathlon. She plans on cycling from her house to the library, shown on the grid with a scale in miles. If the cycling portion of the triathlon is 12 miles, will Mariah have cycled at least $\frac{2}{3}$ of that distance during her bike ride? Justify your argument.

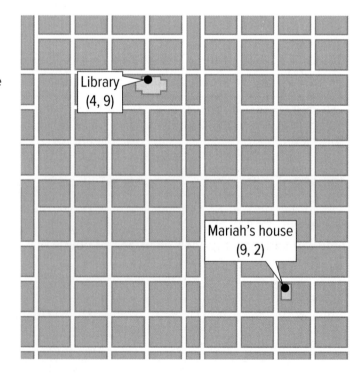

32 Module 1 • Tools of Geometry

30. SPORTS The distance between each base on a baseball infield is 90 feet. The third baseman throws a ball from third base to point P. To the nearest foot, how far did the player throw the ball?

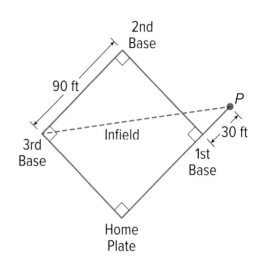

Mixed Exercises

Find the distance between each pair of points. Round to the nearest tenth, if necessary.

31. $M(-4, 9)$, $N(-5, 3)$

32. $C(2, 4)$, $D(5, 7)$

33. $A(5, 1)$, $B(3, 6)$

34. $V(4, 4)$, $X(5, 8)$

35. $S(6, 4)$, $T(3, 2)$

36. $M(-1, 8)$, $N(-3, 3)$

37. $W(-8, 1)$, $Y(0, 6)$

38. $B(3, -4)$, $C(5, -5)$

39. $R(6, 11)$, $T(3, -7)$

40. $A(-3, 8)$ and $B(-1, 4)$

41. $M(4, -3)$ and $N(-2, 1)$

42. $X(3, 5)$ and $Y(4, 2)$

43. Use the number line to determine whether $\overline{SV}$ and $\overline{UX}$ are congruent. Write *yes* or *no*.

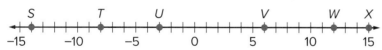

Name the point(s) that satisfy the given condition.

44. two points on the x-axis that are 10 units from (1, 8)

45. two points on the y-axis that are 25 units from (−24, 3)

46. Refer to the figure. Are $\overline{VT}$ and $\overline{SU}$ congruent?

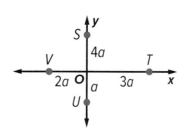

47. KNITTING Mei is knitting a scarf with diagonal stripes. Before she began, she laid out the pattern on a coordinate grid where each unit represented 2 inches. On the grid, the first stripe began at (2, 0) and ended at (5, 4). All the stripes are the same length. How many inches long is each stripe on the scarf?

48. ART A terracotta bowl artifact has a triangular pattern around the top, as shown. All the triangles are about the same size and can be represented on a coordinate plane with vertices at points (0, 6.8), (4.5, 6.8), and (2.25, 0). If each unit represents 1 centimeter, what is the approximate perimeter of each triangle, to the nearest tenth of a centimeter?

49. ANALYZE Consider rectangle *QRST* with $QR = ST = 4$ centimeters and $RS = QT = 2$ centimeters. If point *U* is on $\overline{QR}$ such that $QU = UR$ and point *V* is on $\overline{RS}$ such that $RV = VS$, then is $\overline{QU}$ congruent to $\overline{RV}$? Justify your argument.

50. WRITE Explain how the Pythagorean Theorem and the Distance Formula are related.

51. PERSEVERE Point *P* is located on the segment between point *A*(1, 4) and point *D*(7, 13). The distance from *A* to *P* is twice the distance from *P* to *D*. What are the coordinates of point *P*?

52. CREATE Plot points *Y* and *Z* on a coordinate plane. Then use the Distance Formula to find *YZ*.

53. PERSEVERE Suppose point *A* is located at (1, 3) on a coordinate plane. If *AB* is 10 and the *x*-coordinate of point *B* is 9, explain how to use the Distance Formula to find the *y*-coordinate of point *B*.

54. WRITE Explain how to use the Distance Formula to find the distance between points (*a*, *b*) and (*c*, *d*).

Lesson 1-5

Locating Points on a Number Line

Explore Locating Points on a Number Line with Fractional Distance

Online Activity Use dynamic geometry software to complete the Explore.

INQUIRY What general method can you use to locate a point some fraction of the distance from one point to another point on a number line?

Learn Locating Points on a Number Line with Fractional Distance

While a line segment has two endpoints, a **directed line segment** has an initial endpoint and a terminal endpoint.

Using a directed line segment enables you to calculate the coordinate of an intermediary point some fraction of the length of the segment, or **fractional distance,** from the initial endpoint.

Key Concept • Locating a Point at Fractional Distances on a Number Line

Find the coordinate of a point that is $\frac{a}{b}$ of the distance from point C to point D.

Step 1 Calculate the difference of the coordinates of point C and point D.

$(x_2 - x_1)$

Step 2 Multiply the difference by the given fraction.

The fractional distance is given by $\frac{a}{b}(x_2 - x_1)$.

$\frac{a}{b}(x_2 - x_1)$

Step 3 Add the fractional distance to the coordinate of the initial point x_1.

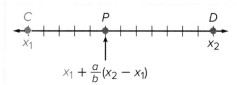

$x_1 + \frac{a}{b}(x_2 - x_1)$

The coordinate of point P is given by $x_1 + \frac{a}{b}(x_2 - x_1)$.

The coordinate of a point on a line segment with endpoints x_1 and x_2 is given by $x_1 + \frac{a}{b}(x_2 - x_1)$, where $\frac{a}{b}$ is the fraction of the distance.

Today's Goals
- Find a point on a directed line segment on a number line that is a given fractional distance from the initial point.
- Find a point that partitions a directed line segment on a number line in a given ratio.

Today's Vocabulary
directed line segment
fractional distance

Watch Out!
Don't Use Absolute Value When finding the distance from an initial endpoint to a terminal endpoint on a directed line segment, don't use absolute value. The difference created by $(x_2 - x_1)$ can be positive or negative. The sign of the difference will indicate the direction of the directed line segment.

 **Talk About It!**
In the Key Concept, what phrase helped you identify the initial endpoint? What phrase helped you identify the terminal endpoint?

Lesson 1-5 • Locating Points on a Number Line **35**

Example 1 Locate a Point at a Fractional Distance

Find B on $\overline{AC}$ that is $\frac{1}{4}$ of the distance from A to C.

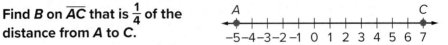

Point A is the initial endpoint, and point C is the terminal endpoint.

Use the equation to calculate the coordinate of point B.

$B = x_1 + \frac{a}{b}(x_2 - x_1)$ Coordinate equation

$= -5 + \frac{1}{4}(7 - (-5))$ $x_1 = -5, x_2 = 7$, and $\frac{a}{b} = \frac{1}{4}$

$= -2$ Simplify.

Point B is located at -2 on the number line.

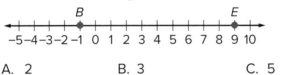

Check

Find X on $\overline{BE}$ that is $\frac{3}{5}$ of the distance from B to E.

A. 2   B. 3   C. 5   D. 6

🌐 Example 2 Locate a Point at a Fractional Distance in the Real World

BIKING Julio is biking from his house to the library. His house is 8 blocks west of the school, and the library is 4 blocks east of the school. If he stops to rest $\frac{1}{3}$ of the distance from his house to the library, at what point does he stop?

Julio's house is the initial endpoint, located at -8, and the library is the terminal endpoint, located at 4. The school is at 0.

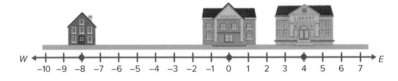

Use the equation to calculate the coordinate of Julio's resting point.

$B = x_1 + \frac{a}{b}(x_2 - x_1)$ Coordinate equation

$= -8 + \frac{1}{3}[4 - (-8)]$ $x_1 = -8, x_2 = 4$, and $\frac{a}{b} = \frac{1}{3}$

$= -4$ Simplify.

Go Online You can complete an Extra Example online.

💭 **Think About It!**
How would you check your solution?

💭 **Think About It!**
What would the coordinate be if Julio wanted to rest $\frac{1}{3}$ of the distance if he is going from the library to his house?

Check

DECORATING Taji is hanging a picture $\frac{5}{8}$ of the distance from the floor to the ceiling. If the distance between the floor and the ceiling is 12 feet, how high should he hang the picture?

___

Learn Locating Points on a Number Line with a Given Ratio

You can calculate the coordinate of an intermediary point that partitions the directed line segment into a given ratio.

> **Go Online**
> You may want to complete the Concept Check to check your understanding.

Key Concept • Section Formula on a Number Line

If C has coordinate x_1 and D has coordinate x_2, then a point P that partitions the line segment in a ratio of $m{:}n$ is located at coordinate $\frac{nx_1 + mx_2}{m + n}$, where $m \neq -n$.

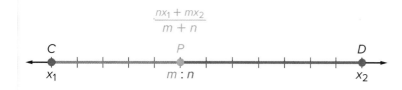

Example 3 Locate a Point on a Number Line When Given a Ratio

Find B on $\overline{AC}$ such that the ratio of AB to BC is 3:4.

Use the Section Formula to determine the coordinate of point B.

$B = \frac{nx_1 + mx_2}{m + n}$ Section Formula

$= \frac{4(-5) + 3(7)}{3 + 4} = \frac{1}{7}$ $m = 3, n = 4, x_1 = -5,$ and $x_2 = 7$

> **Study Tip**
>
> **Checking Solutions** When using the Section Formula you can check your solution by converting the given ratio into a fraction. Use this fraction and the coordinate equation to find the fractional distance from your initial endpoint to your terminal endpoint. If you don't calculate the same coordinate, you have made an error.

So, B is located at $\frac{1}{7}$ on the number line.

(continued on the next page)

Check

Find P on $\overline{AF}$ such that the ratio of AP to PF is 1:3.

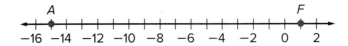

P is located at ___?___ on the number line.

Example 4 Partition a Directed Line Segment

ROAD TRIP Jorge is traveling 2563 miles from New York City to San Francisco by car. He plans on stopping for gas when the ratio of the distance he has already traveled to the distance he still has to travel is 2:5. How far has Jorge traveled when he stops for gas?

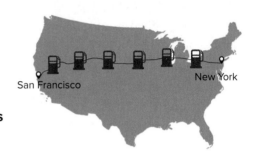

Use the Section Formula to determine how far Jorge will travel before he stops for gas.

$B = \dfrac{nx_1 + mx_2}{m + n}$ Section Formula

$= \dfrac{5(0) + 2(2563)}{2 + 5} = 732.3$ $m = 2, n = 5, x_1 = 0,$ and $x_2 = 2563$

When Jorge has traveled 732.3 miles from New York City, the ratio of the distance he has traveled to the distance that he still has to travel is 2:5.

Check

ERRANDS Eduardo travels 30 miles from his house to the bike shop. When Eduardo goes to the bike shop, he always stops at a local pizza place that is along the way. The ratio of the distance Eduardo travels from his house to the pizza place to the distance he travels from the pizza place to the bike shop is 2:3.

How far is the pizza place from Eduardo's house?

___?___ mi

Think About It!
How can you use estimation to check your answer?

Go Online You can complete an Extra Example online.

Practice

Examples 1 and 3

Refer to the number line.

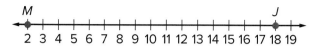

1. Find the coordinate of point B that is $\frac{1}{4}$ of the distance from M to J.

2. Find the coordinate of point C that is $\frac{7}{8}$ of the distance from M to J.

3. Find the coordinate of point D that is $\frac{7}{16}$ of the distance from M to J.

4. Find the coordinate of point X such that the ratio of MX to XJ is 3:1.

5. Find the coordinate of point X such that the ratio of MX to XJ is 2:3.

6. Find the coordinate of point X such that the ratio of MX to XJ is 1:1.

Refer to the number line.

7. Find the coordinate of point G that is $\frac{2}{3}$ of the distance from B to D.

8. Find the coordinate of point H that is $\frac{1}{5}$ of the distance from C to F.

9. Find the coordinate of point J that is $\frac{1}{6}$ of the distance from A to E.

10. Find the coordinate of point K that is $\frac{4}{5}$ of the distance from A to F.

11. Find the coordinate of point X such that the ratio of AX to XF is 1:3.

12. Find the coordinate of point X such that the ratio of BX to XF is 3:2.

13. Find the coordinate of point X such that the ratio of CX to XE is 1:1.

14. Find the coordinate of point X such that the ratio of FX to XD is 5:3.

Refer to the number line.

```
  A  B  C     D  F
 ─┼──┼──┼──┼──┼──┼──
 -5 -4 -3 -2 -1 0 1 2 3 4 5
```

15. Find the coordinate of point X on $\overline{AF}$ that is $\frac{1}{3}$ of the distance from A to F.

16. Find the coordinate of point Y on $\overline{AC}$ that is $\frac{1}{4}$ of the distance from A to C.

Refer to the number line.

```
    A    W     X           Y       Z    E
 ───┼────┼─────┼───────────┼───────┼────┼───
  -10 -9 -8 -7 -6 -5 -4 -3 -2 -1 0 1 2 3 4 5 6 7 8 9 10
```

17. Which point on $\overline{AE}$ is $\frac{2}{3}$ of the distance from A to E?

18. Point X is what fractional distance from E to A?

19. Find the coordinate of point M on $\overline{AE}$ that is $\frac{1}{5}$ of the distance from A to E.

Refer to the number line.

```
  F      G     H        J  K    L
 ─┼──────┼─────┼────────┼──┼────┼──
 -15    -10   -5        0       5
```

20. The ratio of FX to XK is 1:1. Which point is located at X?

21. Find the coordinate of Q on $\overline{FL}$ such that the ratio of FQ to QL is 12:7.

Examples 2 and 4

22. **TRAVEL** Caroline is taking a road trip on I-70 in Kansas. She stops for gas at mile marker 36. Her destination is at mile marker 353 in Topeka, but she decides to stop at an attraction $\frac{3}{4}$ of the way after stopping for gas. At about which mile marker did Caroline stop to visit the attraction?

40 Module 1 • Tools of Geometry

23. **HIKING** A hiking trail is 24 miles from start to finish. There are two rest areas located along the trail.

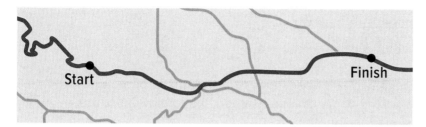

   a. The first rest area is located such that the ratio of the distance from the start of the trail to the rest area and the distance from the rest area to the end of the trail is 2:9. To the nearest hundredth of a mile, how far is the first rest area from the starting point of the trail?

   b. Kadisha claims that the distance she has walked and that the distance she has left to walk has a ratio of 5:7. How many miles has Kadisha walked?

24. Melany wants to hang a canvas, which is 8 feet wide, on his wall. Where on the canvas should Melany mark the location of the hangers if the canvas requires a hanger every $\frac{1}{5}$ of its length, excluding the edges? Justify your answer.

25. **MIGRATION** Many American White Pelicans migrate each year, with hundreds of them stopping to rest in various locations along the way. The ratio of the distance some flocks travel from their summer home to one stopover to the distance from the stopover to the winter home is 3:4. If the total distance that the pelicans migrate is 1680 miles, how long is the distance from the summer home to the stopover?

Mixed Exercises

26. Write an equation that can be used to find the coordinate of point *K* that is $\frac{2}{5}$ of the distance from *Q* to *R*.

27. SOCIAL MEDIA Tito is posting a photo and needs to resize it to fit. The photo's width should fill $\frac{4}{5}$ of the width of the page. On Tito's screen, the total width of the page is 3 inches. How wide should the photo be?

28. NEONATAL At birth, the ratio of a baby's head length to the length of the rest of its body is 1:3. If a baby's total body length is 22 inches, how long is the baby's head?

29. CREATE Draw a segment and label it $\overline{AB}$. Using only a compass and a straightedge, construct a segment $\overline{CD}$ such that $CD = 5\frac{1}{4} AB$. Explain and then justify your construction.

30. WRITE Naoki wants to center a canvas, which is 8 feet wide, on his bedroom wall, which is 17 feet wide. Where on the wall should Naoki mark the location of the nails, if the canvas requires nails every $\frac{1}{5}$ of its length, excluding the edges? Explain your solution process.

31. ANALYZE Determine whether the following statement is *sometimes, always,* or *never* true. Justify your argument.

If $\overline{XY}$ is on a number line and point *W* is $\frac{2}{5}$ of the distance from *X* to *Y*, then the coordinate of point *W* is greater than the coordinate of point *X*.

32. PERSEVERE On a number line, point *A* is at 5, and point *B* is at −10. Point *C* is on $\overline{AB}$ such that the ratio of *AC* to *CB* is 1:3. Find *D* on $\overline{BC}$ that is $\frac{3}{8}$ of the distance from *B* to *C*.

42 Module 1 • Tools of Geometry

Lesson 1-6

Locating Points on a Coordinate Plane

Explore Applying Fractional Distance

🌐 **Online Activity** Use a real-world situation to complete the Explore.

@ **INQUIRY** How do we use fractional distances in the real world?

Today's Goals
- Find a point on a directed line segment on the coordinate plane that is a given fractional distance from the initial point.
- Find a point that partitions a directed line segment on the coordinate plane in a given ratio.

Learn Locating Points on the Coordinate Plane with Fractional Distance

You can find a point on a directed line segment that is a fractional distance from an endpoint on the coordinate plane.

Key Concept • Locating a Point at a Fractional Distance on the Coordinate Plane

The coordinates of a point on a line segment that is $\frac{a}{b}$ of the distance from initial endpoint $A(x_1, y_1)$ to terminal endpoint $C(x_2, y_2)$ are given by $\left(x_1 + \frac{a}{b}(x_2 - x_1), y_1 + \frac{a}{b}(y_2 - y_1)\right)$, where $\frac{a}{b}$ is the fraction of the distance if $b \neq 0$.

Watch Out!

Determine the Initial Endpoint Direction is important when determining a point that is a fractional distance on a directed line segment. Identify the initial endpoint you move from and the terminal endpoint you move toward.

Example 1 Fractional Distances on the Coordinate Plane

Find C on $\overline{AB}$ that is $\frac{3}{4}$ of the distance from A to B.

Step 1 Identify the endpoints.

Identify the initial and terminal endpoints.

$(x_1, y_1) = (-7, -5)$ and $(x_2, y_2) = (6, 8)$

Step 2 Find the x- and y-coordinates.

Find the coordinates of C using the formula for fractional distance.

$\left(x_1 + \frac{a}{b}(x_2 - x_1), y_1 + \frac{a}{b}(y_2 - y_1)\right)$ Fractional Distance Formula

$\left(-7 + \frac{3}{4}[6 - (-7)], -5 + \frac{3}{4}[8 - (-5)]\right)$ Substitution

Point C is located at (2.75, 4.75).

Study Tip

Checking Coordinates You can check that you have computed the coordinates of C correctly by finding the lengths of $\overline{AC}$ and $\overline{AB}$. If $\frac{AC}{AB}$ is not equal to $\frac{3}{4}$, then you have made an error.

💭 **Think About It!**

What are the coordinates of a point that is $\frac{3}{4}$ of the distance from B to A?

 Go Online You can complete an Extra Example online.

Check

Find P on $\overline{QR}$ that is $\frac{1}{6}$ of the distance from Q to R.

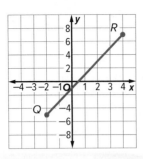

Coordinates of point P ___?___

Learn Locating Points on the Coordinate Plane with a Given Ratio

The Section Formula can be used to locate a point that partitions a directed line segment on the coordinate plane.

Key Concept • Section Formula on the Coordinate Plane

If A has coordinates (x_1, y_1) and C has coordinates (x_2, y_2), then a point B that partitions the line segment in a ratio of $m:n$ has coordinates
$$B\left(\frac{nx_1 + mx_2}{m + n}, \frac{ny_1 + my_2}{m + n}\right),$$
where $m \neq n$.

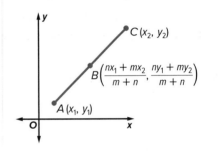

Example 2 Locate a Point on the Coordinate Plane When Given a Ratio

Find C on $\overline{AB}$ such that the ratio of AC to CB is 1:2.

Use the Section Formula to determine the coordinates of point C.

$\left(\dfrac{nx_1 + mx_2}{m + n}, \dfrac{ny_1 + my_2}{m + n}\right)$ Section Formula

$= \left(\dfrac{2(-5) + 1(6)}{1 + 2}, \dfrac{2(-2) + 1(2)}{1 + 2}\right)$ Substitute.

$= \left(-\dfrac{4}{3}, -\dfrac{2}{3}\right)$ Simplify.

Point C is located at $\left(-\dfrac{4}{3}, -\dfrac{2}{3}\right)$.

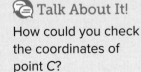

Talk About It!

How could you check the coordinates of point C?

44 Module 1 • Tools of Geometry

Check

Find S on $\overline{QR}$ such that the ratio of QS to SR is 2:1.

A. (4, 8)

B. (2, 3)

C. (1, 1)

D. (0, −1)

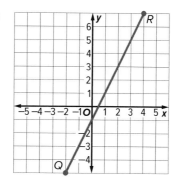

Example 3 Partition a Directed Line Segment on the Coordinate Plane

ZIP LINES Kendrick is riding a zip line. The zip line is 1800 meters long and starts at a platform 600 meters above the ground. After he jumps, someone takes a picture of his descent. When the picture is taken, the ratio of the distance Kendrick has traveled to the distance he has remaining is 1:2. The picture will show the horizontal distance from 400 meters to 1200 meters from the base of the platform and the vertical distance from ground level to a height of 500 meters. Will Kendrick be in the frame of the picture?

To determine whether Kendrick is in the frame of the picture, first, determine the horizontal distance x of the zip line. Then, use this information to determine Kendrick's location using the Section Formula.

Step 1 Determine the horizontal distance x of the zip line.

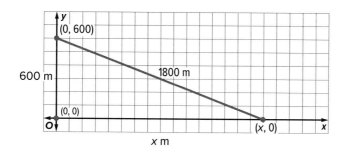

$$a^2 + b^2 = c^2 \quad \text{Pythagorean Theorem}$$
$$600^2 + x^2 = 1800^2 \quad \text{Substitute.}$$
$$x \approx 1697.1 \quad \text{Solve.}$$

The horizontal distance of the zip line is about 1697.1 meters.

(continued on the next page)

Go Online You can complete an Extra Example online.

Step 2 Model the area captured by the photograph.

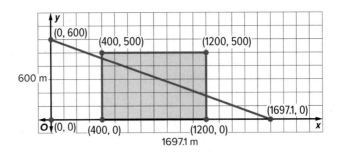

Step 3 Determine Kendrick's location on the zip line.

Use the Section Formula to calculate Kendrick's coordinates.

$\left(\dfrac{nx_1 + mx_2}{m+n}, \dfrac{ny_1 + my_2}{m+n}\right)$ Section Formula

$= \left(\dfrac{2(0) + 1(1697.1)}{1+2}, \dfrac{2(600) + 1(0)}{1+2}\right)$ Substitute.

$= (565.7, 400)$ Simplify.

Kendrick is at (565.7, 400) when the picture is taken.

Step 4 Graph Kendrick's location to determine whether he is in the frame.

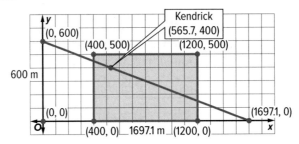

Yes. Kendrick is in the frame when the picture is taken.

Check

TRAVEL Andre is traveling from Jeffersonville to Springfield. He plans to stop for a break when the distance he has traveled and the distance he has left to travel have a ratio of 3:7. Where should Andre stop for his break?

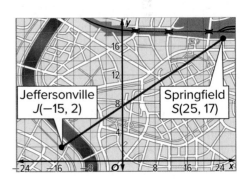

A. (13, 12.5)   B. (22, 12.5)   C. (−3, 6.5)   D. (−12, 6.5)

Go Online You can complete an Extra Example online.

Practice

Go Online You can complete your homework online.

Example 1

1. **POOLS** The accuracy of a pool thermometer is the positive difference between the temperature reading on the thermometer t and the actual temperature of the pool p. Write two absolute value expressions equivalent to the accuracy of a pool thermometer.

2. **ROLLERCOASTER** At a theme park, a person must be a certain height to ride a rollercoaster. A person must be h inches tall, plus or minus 1.5 inches. Anoki says the absolute value expression $|1.5 - h|$ represents an acceptable height. David says the absolute value expression $|h - 1.5|$ represents an acceptable height. Who is correct? Explain.

3. **WATER DEPTH** An *echo sounder* is a device used to determine the depth of water by measuring the time it takes a sound produced just below the water surface to return, or echo, from the bottom of the body of water. The accuracy of an echo sounder is the positive difference between the depth of water reading on the echo sounder r and the actual depth of water w. Write two absolute value expressions equivalent to the accuracy of an echo sounder.

4. **GOLF** A certain company designs and ships boxes of golf balls. Each box must weigh 540 grams. Write two absolute value expressions that represent the number of grams a box weighing g grams is away from the desired weight.

Example 2

Evaluate each expression if $m = -4$, $n = 1$, $p = 2$, $q = -6$, $r = 5$, and $t = -2$.

5. $|-n - 2mp|$

6. $|12 + 2t|$

7. $|q - 2mt|$

8. $|3r + 6m|$

9. $|p + 4q - 3r|$

10. $|16 + 4(3q + p)|$

11. $|2m + 6(q - t)|$

12. $-|10 - 7r + 8m + 2p|$

13. $-|14 - 6n + 7(q + 2t)|$

Example 3

Evaluate each expression if $a = 2$, $b = -3$, $c = -4$, $h = 6$, $y = 4$, and $z = -1$.

14. $|2b - 3y| + 5z$

15. $15 - |2 - 3a|$

16. $|a - 5| - 1$

17. $|b + 1| + 8$

18. $5 - |c + 1|$

19. $|a + b| - c$

20. $5 + |2b|$

21. $|4 - h| - b$

22. $|2 - b - h| - h$

Lesson 1-5 • Expressions Involving Absolute Value 47

Evaluate each expression if $a = -2$, $b = -3$, $c = 2$, $x = 2.1$, $y = 3$, and $z = -4.2$.

23. $|2x + z| + 2y$

24. $4a - |3b + 2c|$

25. $-|5a + c| + |3y + 2z|$

26. $-a + |2x - a|$

27. $|y - 2z| - 3$

28. $3|3b - 8c| - 3$

Evaluate each expression if $a = -\frac{1}{2}$, $b = \frac{3}{4}$, and $c = -\frac{2}{3}$.

29. $-|6c - 16b| + 1$

30. $14 + 2|3c + 10a|$

31. $|-2a - 20b| - 12c$

32. $12a - |-16b|$

33. $|5 - 15c| + a$

34. $|2 - (a - 6b)| + 18c$

Mixed Exercises

35. GPS A golf GPS is a device that can be used to determine the distance a golf ball is from a pin. The accuracy of a golf GPS is the positive difference between the distance a golf ball is from a pin on the golf GPS g and the actual distance a golf ball is from a pin d.

   a. Write two absolute value expressions equivalent to the accuracy of a golf GPS.

   b. Evaluate the expression if the actual distance from the golf ball to the pin is 70 meters and the distance the golf ball is from the pin on the golf GPS is 75 meters.

36. STRUCTURE The students in Mrs. Mangione's class attempt to guess the number of marbles in a jar to earn 2 extra credit points on their next exam. Suppose there are 1206 marbles in a jar and a student makes a guess of m marbles.

   a. Write two absolute value expressions that represent the difference between the guess and the actual number of marbles in the jar.

   b. Evaluate the expression if a student guesses there are 1100 marbles in the jar.

37. CREATE Describe a real-world situation that could be represented by the absolute value expression $|x - 89|$.

38. FIND THE ERROR The accuracy of a rain gauge is the positive difference between the amount of rain in the rain gauge g and the actual amount of rain r. Sam says the absolute value expression $|g| - |r|$ is equivalent to the accuracy of a rain gauge. Is Sam correct? Explain your reasoning.

39. ANALYZE Diaz claims that if a and b are real numbers, then $|a + b|$ is always equal to $|a| + |b|$. Determine whether his claim is *true* or *false*. Justify your argument.

48 Module 1 • Expressions

Lesson 1-6

Descriptive Modeling and Accuracy

Learn Descriptive Modeling

Descriptive modeling is a way to mathematically describe real-world situations and the factors that cause them. A **metric** is a rule for assigning a number to some characteristic or attribute.

Metrics can be used to make comparisons. In sports, earned run average is used to compare baseball pitchers, and the quarterback rating compares the performance of football quarterbacks. In banking, a person's debt-to-income ratio can determine whether the person qualifies for a loan. In a good metric, factors that are important in the situation are considered and included in the metric. For example, the quarterback rating includes measures of passing, running, penalties, and other factors that make a quarterback effective.

Today's Goals
- Define and use appropriate quantities for the purpose of descriptive modeling.
- Choose a level of accuracy appropriate to limitations on measurements when reporting quantities.

Today's Vocabulary
descriptive modeling

metric

accuracy

Example 1 Use Descriptive Modeling

COLLEGE ATHLETICS Some universities use a metric called the Academic Index to qualify high school athletic recruits. A student athlete with a 3.1 G.P.A. received scores of 610 in reading, 640 in writing, and 700 in math on the SAT. Use the expression to determine whether the student athlete qualifies at a university that requires an Academic Index of 186 or greater.

G.P.A.	G.P.A. Value
4.0	80
3.9	79
3.8	78
3.7	77
3.6	75
3.5	73
3.4	71
3.3	70
3.2	69
3.1	68
3.0	67

Think About It!
Some universities require student athletes to take the SAT twice and then use both scores when determining their Academic Index. How do you think this could affect the score?

$$2\left[\frac{\left(\frac{\text{Reading Score} + \text{Writing Score}}{2}\right) + \text{Math Score}}{20}\right] + \text{G.P.A. Value}$$

Step 1 Find all values for the metric.

Use the table to determine the G.P.A. value. For a 3.1 G.P.A., the value is 68.

(continued on the next page)

> **Think About It!**
> What other attributes of high school recruits do you think universities might consider when creating metrics to determine qualification?

Step 2 Substitute values in the metric.

$$2\left[\frac{\left(\frac{\text{Reading Score} + \text{Writing Score}}{2}\right) + \text{Math Score}}{20}\right] + \text{G.P.A. Value}$$

$$= 2\left[\frac{\left(\frac{610 + 640}{2}\right) + 700}{20}\right] + 68 \quad \text{Reading 610, writing 640, math 700, and G.P.A. 68}$$

$$= 2\left[\frac{\left(\frac{1250}{2}\right) + 700}{20}\right] + 68 \quad \text{Add 610 and 640.}$$

$$= 2\left[\frac{625 + 700}{20}\right] + 68 \quad \text{Divide by 2.}$$

$$= 2\left[\frac{1325}{20}\right] + 68 \quad \text{Add 625 and 700.}$$

$$= 2(66.25) + 68 \quad \text{Divide by 20.}$$

$$= 132.5 + 68 \quad \text{Multiply by 2.}$$

$$= 200.5 \quad \text{Simplify.}$$

The Academic Index is 200.5.

Step 3 Evaluate by using the metric.

Because the Academic Index is greater than 186, this student would be qualified to attend the university as an athlete.

Check

PARKS Rachelle wants to determine the best state park for hiking and fishing.

Part A Use the metric to calculate a score for each park. Round to the nearest tenth.

$$\text{Park Score} = 100\left[0.2\left(\frac{\text{online rating}}{5}\right) + 0.4\left(\frac{\text{miles of trails}}{25}\right) + 0.4\left(\frac{\text{fish weight}}{10}\right)\right]$$

State Park	Online Star Rating	Hiking Trails (mi)	Best Fish (lb)	Park Score
Gooseberry Falls	4.8	20	9.8	
Lake Maria	4.3	14	8.2	
Maplewood	4.9	25	7.3	
Camden	4.3	15.8	10.1	

Part B How might someone who enjoys hiking much more than fishing change this metric?

Go Online You can complete an Extra Example online.

Check

LOANS Elan is applying for a home loan. At National Road Bank, Elan's debt-to-income ratio must be 0.36 or less to qualify for a loan, and at New Savings Bank his mortgage-to-income ratio must be 0.28 or less.

Part A Use the two metrics and the information provided to determine whether Elan qualifies for a home loan at each bank. Round to the nearest hundredth.

Monthly Income	Monthly Debt	Monthly Mortgage
$3650	$1165	$1068

National Road Bank: Debt-to-Income Ratio

$$\frac{\text{Monthly Debt}}{\text{Monthly Income}} = \underline{\quad ? \quad}$$

New Savings Bank: Mortgage-to-Income Ratio

$$\frac{\text{Monthly Mortgage}}{\text{Monthly Income}} = \underline{\quad ? \quad}$$

Part B Compare the results of the two metrics. How effective are each of the metrics as measures of whether Elan can afford to buy a house?

> **Study Tip**
>
> **Fractions** Fractions may be more accurate than rounded decimals. For example, the sum of $\frac{3}{7} + \frac{2}{3}$ is more accurately reported as $\frac{23}{21}$ than 1.095.

Learn Accuracy

All measurements are approximations. When you measure something, you are limited by the measurement tool that you are using. **Accuracy** is the nearness of a measurement to the true value of the measure.

The accuracy needed for baking cookies, timing the final seconds of a basketball game, and determining the gold medalist of a 100-meter dash are very different.

Whether measurements should be rounded depends on how the measurement will be used and the limitations of the units in which the measurement is taken.

Go Online You can complete an Extra Example online.

Example 2 Compare Metrics

HEIGHT A child's adult height can be predicted using several metrics. Given the height of the mother, father, and their son at 2 years old, use the metrics to predict the son's height as an adult.

Method 1 The Gray Method

The Gray Method uses the average heights of the parents, adjusted by the gender of the child. For a boy, the mother's height is multiplied by $\frac{13}{12}$, and for a girl, the father's height is multiplied by $\frac{12}{13}$.

$$\frac{\frac{13}{12} \cdot \text{mother's height} + \text{father's height}}{2}$$

$$= \frac{\frac{13}{12} \cdot 65 + 72}{2} \quad \text{mother's height 65, father's height 72}$$

$$= \frac{70.42 + 72}{2} \quad \text{Multiply } \frac{13}{12} \text{ and 65.}$$

$$= \frac{142.42}{2} \quad \text{Add 70.42 and 72.}$$

$$= 71.21 \quad \text{Simplify.}$$

Using the Gray Method, the boy will be about 71 inches tall as an adult.

Method 2 The Doubling Method

The Doubling Method multiplies the height of a child by 2 at a specific age to predict the child's height as an adult. Height at 24 months is used for boys, and height at 18 months is used for girls.

$2 \cdot$ height of child

$= 2 \cdot 35 \quad$ Boy's height of 35 inches

$= 70 \quad$ Simplify.

Using the Doubling Method, the boy will be 70 inches tall as an adult.

> **Use a Source**
> Find information to create a metric to measure something that is important to you. Explain how your metric includes the factors that you think are important to measure.

Go Online You can complete an Extra Example online.

Example 3 Decide Where to Round

ROAD TRIP Damien and two of his friends are taking a road trip. They plan to share the responsibility of driving and will each drive an equal distance. Damien's GPS shows that the total distance is 172 miles. Determine the exact distance that each person should drive. Then determine a more appropriate driving distance for each person given the limitations of the situation.

To determine the exact distance each person should drive, divide the total distance by 3.

$$172 \text{ miles} \div 3 = 57.\overline{3} \text{ miles}$$

Because the distance given by the GPS is accurate to the nearest mile, the distance each driver will drive should be rounded to the nearest mile. Each driver will drive about 57 miles.

Think About It!
The total cost of fuel for the trip was $20. If they split the cost equally, how much should each person pay? What unit of measure limits the accuracy of the solution?

Check

VACATION Inchiro has saved $400.00 to spend on his 7-day vacation. He plans to budget his $400.00 by spending the same amount each day of the vacation. Determine the appropriate amount he should spend each day.

$ __?__

Example 4 Find an Appropriate Level of Accuracy

SPACE SHUTTLE In 2012, NASA's space shuttle *Endeavor* traveled approximately 897 miles from the Kennedy Space Center to Houston, Texas, by a shuttle carrier aircraft. Then it traveled about 1381 miles to the Los Angeles International Airport. Finally, a truck pulled *Endeavor* 12 miles through the streets of Los Angeles to the California Science Center.

If the shuttle carrier aircraft flew at an average speed of 287 miles per hour and the truck pulled *Endeavor* at an average speed of 1.3 miles per hour, determine the total amount of time it took *Endeavor* to travel from the Kennedy Space Center to the California Science Center with a reasonable level of accuracy.

Because the parts of the space shuttle's journey from the Kennedy Space Center to Houston and then from Houston to Los Angeles are at the same speed, add those two distances, 897 + 1381 or 2278.

$$\frac{2278 \text{ miles}}{287 \frac{\text{miles}}{\text{hours}}} + \frac{12 \text{ miles}}{1.3 \frac{\text{miles}}{\text{hours}}} \approx 7.937 \text{ hours} + 9.231 \text{ hours}$$

$$\approx 17.168 \text{ hours}$$

The total travel time for *Endeavor* from the Kennedy Space Center to the California Science Center was about 17 hours.

Talk About It!
Why is it unreasonable to say that it took 17.168 hours for Endeavor to reach the science center?

Go Online You can complete an Extra Example online.

Check

POSTAGE A school is hosting a marching band competition and plans to mail postcards, fliers, and large information packets to other schools. The school will mail 150 postcards, which cost $0.34 in postage, and between 75 to 100 fliers, which cost $0.49 in postage. Forty-three information packets will be mailed. The cost of mailing the information packets varies, but the average is $1.59. Determine the total mailing cost that represents the most reasonable level of accuracy.

Example 5 Determine Accuracy

POPULATION The U.S. Census Bureau Web site shows a counter that displays the population of the United States on a certain day as 329,158,023. How accurate is the reported population? Explain your reasoning.

Because there is no way to count every person in the United States at any given moment, giving an exact population does not make sense. The number of births, deaths, and immigrations varies, so the population does not increase at a steady rate. The Web site uses averages to estimate the population at a specific time.

United States Population
329,158,023

COMPONENTS OF POPULATION CHANGE
One birth every **7 seconds**
One death every **13 seconds**
One international migrant (net) every **29 seconds**
Net gain of one person every **11 seconds**

It would be more appropriate for the Web site to report the population as 329.2 million.

Check

BIOLOGY A science magazine reported that there are, on average, 37 trillion cells that make up the human body. Select the option that best describes the accuracy of the magazine.

A. The magazine is accurate because scientists can count every cell.

B. The magazine is probably accurate because the number is not very specific.

C. The magazine is not accurate because there is no way to count all of the cells of a person.

D. The magazine is not accurate because the number of cells is always changing.

Go Online You can complete an Extra Example online.

Practice

Go Online You can complete your homework online.

Example 1

1. **TEST SCORES** A teacher compares the ratio of the number of questions answered correctly to the total number of questions on a test as a metric. For a student to earn an A or B on a test, the ratio must be greater than or equal to 0.8. The last test given by the teacher had a total of 40 questions. Using this metric, what is the least number of questions a student can answer correctly to earn an A or B on the test?

$$\frac{\text{number of questions answered correctly}}{\text{total number of questions}}$$

2. **DRIVER'S TEST** The Department of Motor Vehicles, DMV, uses a metric to determine whether a person earns a driver's license. In one state, the total number of possible points on the written portion of the driver's exam is 46. A person will pass the written portion of the driver's exam by scoring 84% or greater. The table shows the number of points different people earned on the written portion of the driver's exam. How many people passed the written portion of the driver's exam?

38	40	41	45	39
35	40	46	43	37
41	42	44	41	46
40	38	32	44	45
39	40	30	43	45

3. **TRACK** A college track coach compares the ratio of time it takes a runner to run 100 meters to 12 seconds. For a runner to be on the team, the ratio must be less than or equal to 0.95. What is the slowest time 100 meters can be run to make the team?

Example 2

4. **DEBT-TO-INCOME RATIO** Find Jada's debt-to-income ratio if her monthly expenses are $1850 and her monthly salary is $2500.

5. **DEBT-TO-INCOME RATIO** Find Victoria's debt-to-income ratio if her monthly expenses are $1280 and her monthly salary is $2500.

6. **PLUMBING** Raven is deciding between two plumbing services. Service Provider A multiplies the average number of hours spent at a residence by $50, where the average number of hours spent at a residence is 1.5 hours for a new house and 3.75 hours for an old house. Service Provider B multiplies the exact number of hours spent at a residence by $60. Suppose Raven has a new house and needs a plumber for 1.75 hours. Find the cost of service charge using both methods.

7. **INVESTING** Hector is deciding how much he should invest each year. The Automatic Method multiplies the average income by 10%, where the average income is $50,000 for an employee that has been at the same company for 10 years or less and $60,000 for an employee that has been at the same company for more than 10 years. The Exact Method multiplies the exact income by 7.5%. Suppose Hector has been at the same company for 12 years and his income last year was $75,000. Find the amount Hector should invest using both methods.

Lesson 1-6 • Descriptive Modeling and Accuracy

Example 3

8. **MONEY** Jordan has $20 to share among 3 people. Jordan types 20 ÷ 3 into his calculator and gets 6.666666667. How much should he give to each person?

9. **SNACKS** Ms. Miller has 14 snack bars to share among 6 students. Ms. Miller types 14 ÷ 6 into her calculator and gets 2.333333333. How many snack bars should she give each student?

10. **EVENT PLANNING** Max is planning a banquet for the National Honors Society. Approximately 60 people will be attending the banquet. If 8 people can fit comfortably at a table, how many tables should he have?

11. **GARDEN** Emily wants to plant flowers in a narrow rectangular plot that is 1 foot by 4.5 feet. The flowers she wants to plant need to be spaced at least 8 inches apart. How many plants should she buy for the garden?

12. **LEMONADE** Justin has 64 ounces of lemonade to divide among 9 people. When he types 64 ÷ 9 into his calculator, the number that appears is 7.1111111. How much lemonade should he give to each person?

13. **MONEY** Darnell has $1000 he wants to divide among his 3 children. How much should Darnell give each of his children?

Example 4

14. **FUNDRAISING** At a bake sale, the golf team sold all 50 cupcakes for $1.50 each. They sold almost all of the 100 cookies for $1.00 each. The team also received donations from 7 people averaging $4.25 per donation. Determine the total amount of money the golf team collected from the bake sale with a reasonable level of accuracy.

15. **EXPENSES** Santino spends an average of $125 per month on clothing, not including sales tax. He also spends about $180 per month going out to eat and an average of $130 per week on groceries, including sales tax. If the tax rate is 7.25%, find the total amount he spends on food and clothing in a year, including sales tax, with a reasonable level of accuracy.

16. **TIME MANAGEMENT** Ava spends about 45 minutes studying each day. She also practices the piano for an average of 20 minutes per day, and she practices soccer 1.5 hours three times a week. If Ava decides to reduce the time she does each activity by $\frac{1}{6}$, find the total number of hours she spends studying and practicing in a year with a reasonable level of accuracy.

Example 5

17. **SCHOOLS** The superintendent at Hartgrove High School says there are 3103 students enrolled at the school. How accurate is the reported enrollment? Explain your reasoning.

18. **POPULATION** The U.S. Census Bureau Web site shows that the population of Texas on July 1, 2016 was 27,862,596. How accurate is the reported population? Explain your reasoning.

19. **TRAFFIC LIGHTS** A map maker reported that there were about 12,000 traffic lights in New York City. How accurate is the report? Explain your reasoning.

20. **SAND** A mathematician reported that there are 1,578,932 grains of sand in one cubic foot. How accurate is the report? Explain your reasoning.

Mixed Exercises

21. **USE A SOURCE** A coach compares the ratio of the number of free throws made to the total number of attempted free throws as a metric. For a player to be selected as a free throw shooter when the other team is given a technical foul, the ratio must be greater than or equal to 0.82. Find the number of free throws made and the number of free throws attempted for three former NBA players. Using the metric, which players would and would not be selected as a free throw shooter when the other team is given a technical foul?

22. **REASONING** A carpenter is measuring the length of a living room. Should the carpenter measure the length in feet, inches, meters, or kilometers to be most accurate? Explain.

23. **SPACE** Which unit of measure is the most appropriate for measuring the distance from Earth to the star Polaris: feet, kilometers, or light-years? Explain.

24. **POPULATION** Juanita and Trevor are doing research about the deer population in Ohio. Juanita says there are over 750,000 deer in Ohio. Trevor says there are 734,928 deer in Ohio. Who is more accurate? Explain your reasoning.

The graph shows how the number of visitors at a local zoo is related to the average daily temperature. The line shown is the line that most closely approximates the data in the scatter plot. Use the graph for Exercises 25–27.

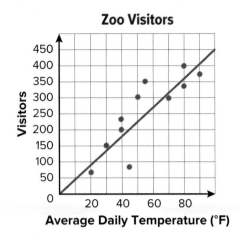

25. STRUCTURE Describe the line in terms of accuracy.

26. USE ESTIMATION Use the line to approximate the number of visitors at the zoo for an average daily temperature of 50°F. Compare this to the actual number of visitors given by the point on the graph for an average daily temperature of 50°F.

27. REASONING Explain why some points are above the line and some points are below the line.

28. METRICS Suppose two mortgage companies compare the ratio of the monthly mortgage payment to the total monthly income as their metric. Suppose 0.3 is the ideal metric for the debt-to-income ratio for Company A, and 0.28 is the ideal metric for the debt-to-income ratio for Company B. Provided the target mortgage payment is the same for either company, then which of these mortgage companies requires a greater monthly income? Explain.

29. WRITE Suppose you start your own company. When hiring employees, you want to set certain metrics, such as typing speed. What other attributes of employees do you think you might consider when creating metrics to determine hiring qualifications?

30. FIND THE ERROR Mr. Moreno's students are weighing materials for a chemistry experiment. Four students weigh the same sample using different scales:

        100 g   104 g   105 g   103.5 g

Mr. Moreno tells the students that they each weighed the amount correctly. Explain how this is possible.

31. WRITE Lamont stops at a gas station that sells gasoline at 3.29\frac{9}{10}$ per gallon. He pumps 8.618 gallons of gasoline into the tank. How much will Lamont pay for gas? How much accuracy is possible? How much accuracy is necessary? Explain.

58 Module 1 • Expressions

Module 1 • Expressions

Review

Essential Question
How can mathematical expressions be represented and evaluated?

Module Summary

Lessons 1-1 and 1-2
Numerical and Algebraic Expressions
- A numerical expression contains only numbers and mathematical operations.
- To evaluate an expression means to find its value. If a numerical expression contains more than one operation, the rule that lets you know which operation to perform first is called the *order of operations*.
- The Substitution Property allows you to evaluate an algebraic expression by replacing the variables with their values.

Lessons 1-3 and 1-4
Properties of Real Numbers and Distributive Property
- The Reflexive Property states that any quantity is equal to itself.
- The Symmetric Property states that if one quantity equals a second quantity, then the second quantity equals the first.
- The Transitive Property states that if one quantity equals a second quantity and the second quantity equals a third quantity, then the first quantity equals the third quantity.
- The Associative Property states that the way you group three or more numbers when adding or multiplying does not change their sum or product.
- The Commutative Property states that the order in which you add or multiply numbers does not change their sum or product.
- The Distributive Property states that multiplying a number by a sum of numbers is the same as doing each multiplication separately and then adding the products.

Lesson 1-5
Expressions Involving Absolute Value
- The absolute value of a number is its distance from 0 on the number line.
- Absolute value is always greater than or equal to zero.

Lesson 1-6
Descriptive Modeling and Accuracy
- Descriptive modeling is a way to mathematically describe real-world situations and the factors that cause them.
- A metric is a rule for assigning a number to a characteristic or attribute.
- Accuracy is the nearness of a measurement to the true value of the measure.

Study Organizer
 Foldables

Use your Foldable to review the module. Working with a partner can be helpful. Ask for clarification of concepts as needed.

Test Practice

1. **MULTIPLE CHOICE** Which is equivalent to 2^5? (Lesson 1-1)

   A. 10

   B. 16

   C. 24

   D. 32

2. **MULTI-SELECT** The table shows the prices of several items at a movie theater. Which expressions represent the total cost of 4 movie tickets, 2 popcorns, and 1 bottled water? Select all that apply. (Lesson 1-1)

Item	Cost
Ticket	$9.75
Popcorn	$6.25
Soda	$5.50
Water	$4.75
Box of Candy	$3.50

   A. $4(9.75) + 2(6.25) + 4.75$

   B. $4(9.75) + 2(5.50) + 4.75$

   C. $39.00 + 12.50 + 4.75$

   D. 54.75

   E. 56.25

3. **OPEN RESPONSE** Write an algebraic expression that represents *five times the quantity x increased by seven, minus four cubed*. (Lesson 1-2)

4. **MULTIPLE CHOICE** What is the value of the expression $9x^2 + 4x - 11$ when $x = 3.2$? Express your answer as a decimal, rounded to the nearest hundredth. (Lesson 1-2)

   A. 20.232

   B. 30.6

   C. 59.4

   D. 93.96

5. **MULTIPLE CHOICE** Which algebraic expression represents the verbal expression *the product of five and a number, decreased by eleven*? (Lesson 1-2)

   A. $5n - 11$

   B. $11 - 5n$

   C. $5(n - 11)$

   D. $11 - (n + 5)$

6. **OPEN RESPONSE** Evaluate the expression $5[13 - (3^2 + 2^2)]$. (Lesson 1-3)

7. **MULTIPLE CHOICE** Which expression is NOT a way to represent $2 \cdot 3\frac{5}{6} + 2 \cdot 12 + 2 \cdot 1\frac{1}{6}$? (Lesson 1-3)

   A. $2\left(3\frac{5}{6} + 12 + 1\frac{1}{6}\right)$

   B. $2 \cdot 5 + 12$

   C. $2(5 + 12)$

   D. 34

60 Module 1 Review • Expressions

8. **OPEN RESPONSE** Ayumi, a chef, wants to determine how many meals she cooked in one evening. The table shows the four meals she made and the number of people that were served each meal. (Lesson 1-3)

Meal	Number of People
Lasagna	27
Spaghetti & Meatballs	21
Steak & Potatoes	19
Shrimp Scampi	13

She uses the following steps to determine how many total meals she cooked.

Step 1: $27 + 21 + 19 + 13$

Step 2: $27 + 13 + 21 + 19$

Step 3: $(27 + 13) + (21 + 19)$

Step 4: $40 + 40$

Step 5: 80

Which property did Ayumi use in Step 3?

9. **OPEN RESPONSE** Indicate whether each of the statements is *true* or *false*. (Lesson 1-3)

A. $4(6 - 2 \times 3) = 0$

B. $11(3^2 - 9) + 2\left(\frac{1}{2}\right) = 0$

C. $4 \cdot 0 + 4^2 - 2^3 - (2 + 2 \cdot 3) = 0$

10. **MULTI-SELECT** Which expressions could be used to evaluate 418(27)? (Lesson 1-4)

A. $418(20 - 7)$

B. $(420 - 2)(27)$

C. $(400 - 18)(27)$

D. $(418)(20 + 7)$

E. $(418)(30 - 3)$

11. **MULTI-SELECT** Indicate whether each expression represents the verbal expression *negative seven times the quantity triple m minus eleven.* (Lesson 1-4)

A. $-7(m^3 - 11)$

B. $-7(3m) - 7(-11)$

C. $-21m - 77$

D. $-21m + 77$

E. $-21m - 11$

F. $-7m^3 + 77$

12. **MULTIPLE CHOICE** Which is the simplified expression of $-8(2m + 9k - 13)$? (Lesson 1-4)

A. $-16m + 9k - 13$

B. $-16m - 72k + 104$

C. $-16m - 72k - 104$

D. $16m - 72k - 104$

13. MULTI-SELECT A group of 8 artists plans to attend a quilting class and purchase lunch. (Lesson 1-4)

Which expression(s) represent(s) the total cost for all 8 artists?

A. 8(10) + 14

B. 10(8 + 14)

C. 8(10 + 14)

D. 80 + 140

E. 80 + 112

14. MULTI-SELECT If x and y are both integers, which expression(s) are equivalent to $|x - y|$? (Lesson 1-5)

A. $|y - x|$

B. $|y + x|$

C. $|y| - |x|$ if $x \leq y$ and $x \geq 0$

D. $|x| - |y|$ if $y > x$

E. $|y| + |x|$ if $y > x$

15. OPEN RESPONSE What is the value of the expression $4^2 + 3|4x - 9|$ when $x = -2$? (Lesson 1-5)

16. OPEN RESPONSE A player's secondary average (SecA) is a way to look at the extra bases gained without regard to batting average. The formula for SecA is $SecA = \frac{T - H + B + S - C}{A}$, where T is total bases, H is hits, B is bases from balls or walks, S is stolen bases, C is number of times caught stealing, and A is times at bat.

Player	T	H	B	S	C	A
Altuve	186	116	39	22	3	330
Murphy	183	110	17	2	3	315
Ortiz	187	94	45	2	0	279
Ramos	139	84	23	0	0	251

Find each player's SecA. Round to the nearest hundredth if necessary. (Lesson 1-6)

17. OPEN RESPONSE A marketing manager bought 4 advertisements for $1345 each. She reported to her supervisor that she spent about $4000 out of her budget. Did the marketing manager report her spending to a reasonable level of accuracy? Explain. (Lesson 1-6)

Module 2
Equations in One Variable

e Essential Question
How can writing and solving equations help you solve problems in the real world?

What Will You Learn?
How much do you already know about each topic **before** starting this module?

KEY

👎 — I don't know.   👍 — I've heard of it.   👍 — I know it!

	Before			After		
	👎	👍	👍	👎	👍	👍
write equations to represent relationships						
interpret equations that represent relationships						
solve one-step equations by using addition and subtraction						
solve one-step equations by using multiplication and division						
solve multi-step equations						
solve equations with variables on each side						
solve equations by applying the Distributive Property						
solve equations that involve absolute value						
solve proportions						
solve an equation with more than one variable for a specific variable						
convert units of measure by using dimensional analysis						

📙 **Foldables** Make this Foldable to help you organize your notes about equations. Begin with four sheets of grid paper.

1. **Fold** four sheets of grid paper in half along the width.
2. **Unfold** each sheet and tape to form one long piece.
3. **Label** each piece with the lesson number as shown. Label the last piece for vocabulary. Refold to form a booklet.

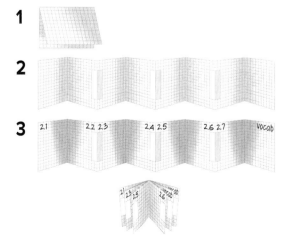

Module 2 • Equations in One Variable 63

What Vocabulary Will You Learn?

- constraint
- dimensional analysis
- equation
- equivalent equations
- formula
- identity
- literal equation
- multi-step equation
- proportion
- solution
- solve an equation

Are You Ready?

Complete the Quick Review to see if you are ready to start this module.
Then complete the Quick Check.

Quick Review

Example 1

Write an algebraic expression for the phrase *the quotient of five and w decreased by eight.*

the quotient of five and *w*   decreased by   eight

$\frac{5}{w}$ — 8

The expression is $\frac{5}{w} - 8$.

Example 2

Evaluate $9 + \frac{4^2}{2} - 2(5 \times 2 - 8)$.

$9 + \frac{4^2}{2} - 2(5 \times 2 - 8)$	Original expression
$= 9 + \frac{4^2}{2} - 2(2)$	Evaluate inside the parentheses.
$= 9 + 8 - 2(2)$	Evaluate the power and divide.
$= 9 + 8 - 4$	Multiply.
$= 13$	Add and subtract.

Quick Check

Write an algebraic expression for each verbal expression.

1. six times a number *n* increased by two
2. a number *d* squared minus three
3. the sum of four times *b* and nine

Evaluate each expression.

4. $(7 + 3)^2 - 4$
5. $4(11 - 5) \div 3$
6. $\frac{1}{3}(21) + \frac{1}{8}(32)$
7. $3 \cdot 2^3 + 64 \div 8$
8. $\frac{11 - 3}{2} + 9$
9. $6[(5 - 3)^2 + 8] \div 2$

How did you do?

Which exercises did you answer correctly in the Quick Check?

Lesson 2-1

Writing and Interpreting Equations

Explore Writing Equations by Modeling a Real-World Situation

Online Activity Use a real-world situation to complete the Explore.

> **INQUIRY** What steps can you use to write equations to represent a real-world situation?

Today's Goals
- Translate sentences into equations.
- Translate equations into sentences.

Today's Vocabulary
equation
constraint

Learn Writing Equations

A mathematical statement that contains two expressions and an equal sign, =, is an **equation**.

Key Concept • Writing Equations
Step 1 Identify each unknown and assign a variable to it.
Step 2 Identify the givens and their relationships.
Step 3 Write the sentence as an equation.

Think About It!
What distinguishes an expression from an equation?

Example 1 Write an Equation for a Sentence

Write an equation for the sentence.

Twenty minus the quotient of 7 and x is the same as twice x.

Recall that a quotient is the result of division.

Twenty minus the quotient of 7 and x is the same as twice x.

$$20 - \quad \frac{7}{x} \quad = \quad 2x$$

The equation is $20 - \frac{7}{x} = 2x$.

Check

Write an equation for the sentence.

Four times a number less 10 is equal to 16.

   A. $4x - 10 = 16$ **B.** $4(x - 10) = 16$

   C. $10 - 4x = 16$ **D.** $4x + 10 = 16$

Study Tip
Verbal Phrases When writing the verbal form of an equation, *is* and *equals* can be used interchangeably.

Go Online You can complete an Extra Example online.

Example 2 Write an Equation

LIFE ONLINE Of 799 teens surveyed about what they do online, some use a social network. Of those on a social network, 430 say people their age are "mostly kind" online and the remaining 193 do not. Write an equation to find the number of teens surveyed who are not on a social network.

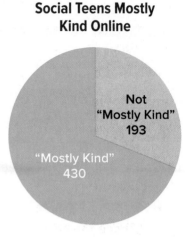

Social Teens Mostly Kind Online

Not "Mostly Kind" 193

"Mostly Kind" 430

Step 1 Identify each unknown and assign a variable to it.

Let $n =$ the number of teens surveyed who are not on a social network.

Step 2 Identify the givens and their relationship.

The givens are:
- 799 teens were surveyed.
- Some number of the teens use a social network.
- 430 of those on a social network say people their age are "mostly kind" online. The other 193 do not.

The 430 and 193 make up the group on a social network. The rest of the 799 surveyed are not on a social network.

Step 3 Write the sentence as an equation.

The sum of the teens on a social network and those not on a social network is 799.

$(430 + 193) + n = 799$

Check

READING Etu has read 12 of the 32 chapters in his assigned book. He plans to finish the book by reading c chapters each for 8 days until the book is due. Which equation best represents the situation?

A. $12 - 8c = 32$

B. $12 + \frac{c}{8} = 32$

C. $12 + 8c = 32$

D. $12 - \frac{c}{8} = 32$

Think About It!

Is there only one equation that represents the situation? Justify your argument.

Go Online You can complete an Extra Example online.

Example 3 Write an Equation with Multiple Variables

GEOMETRY Translate the sentence into a formula.

The perimeter of a rectangle is twice the sum of the length and the width.

Step 1 Identify unknowns.

perimeter, the length, and the width

Step 2 Assign variables.

Let P = perimeter, ℓ = length, and w = width.

Step 3 Identify the givens and their relationships.

Twice means two times.

Twice the sum means you add first, then multiply.

Step 4 Write an equation.

The formula for the perimeter of a rectangle is $P = 2(\ell + w)$.

 Think About It!
Why is it helpful to identify all the unknowns before writing an equation?

Check

Translate the sentence into a formula.

MOTORS The horsepower of a motor is the product of the motor speed and the torque divided by 5252.

A. $H = \dfrac{M}{5252T}$

B. $H = \dfrac{MT}{5252}$

C. $H = \dfrac{5252}{MT}$

D. $H = \dfrac{5252M}{T}$

BAGELS Plain and cinnamon raisin bagels are the most popular flavors. Each year, 24 million more than twice as many packages of plain bagels are sold as cinnamon raisin. There were 136 million packages of plain bagels sold last year. Create an equation that can be used to find the number of millions of packages of cinnamon raisin bagels, c, sold last year.

 Talk About It!
What is an example of a real-life constraint? Explain.

Learn Interpreting Equations

Look for the relationships in an equation by interpreting each part of the expressions in the equation.

As you interpret an equation that represents a real-life situation, consider that the equation may be viewed as a constraint in the situation. In mathematics, a **constraint** is a condition that a solution must satisfy. These conditions limit the number of possible solutions. The solutions of the equation meet the constraints of the problem.

Go Online You can complete an Extra Example online.

Example 4 Write a Sentence for an Equation

Write a sentence for the equation.

$$2z - 1 = 5$$

Two z minus one equals five.

Check

Write a sentence for the equation.

a. $6z - 15 = 45$

   A. Six times a number z, minus fifteen equals forty-five.

   B. Six times the quantity of a number z minus fifteen equals forty-five.

   C. Six times the difference of a number z and fifteen equals forty-five.

   D. Six plus z minus fifteen equals forty-five.

b. $(y + 3)^2 = 25$

   A. y plus 3 squared is 25.

   B. y squared plus 3 is 25.

   C. The quantity y plus 3 squared is 25.

   D. y plus 3 is 25.

Watch Out!

Parentheses When writing a sentence for an equation with parentheses, the phrase *the quantity* should be written immediately before the terms that are contained in the parentheses, not at the beginning of the sentence.

Example 5 Write a Sentence for an Equation with Grouping Symbols

Write a sentence for the equation.

$$3(y + 1) = 12$$

The parentheses tell us that 3 is *three times* the expression in the parentheses.

The parentheses can be written as *the quantity*.

Write $y + 1$ as *y plus 1* or *the sum of y and one*.

Write $= 12$ as *equals twelve* or *is twelve*.

One of the sentences that represents the equation $3(y + 1) = 12$ is:

Three times the quantity y plus one equals twelve.

 Go Online An alternate method is available for this example.

Go Online You can complete an Extra Example online.

Check

Select the sentence(s) that represent(s) the equation.

$7(p + 23) = 102$

- **A.** Seven times the sum of p and twenty-three is the same as one hundred two.
- **B.** Seven times p plus twenty-three equals one hundred two.
- **C.** Seven times the quantity p plus twenty-three equals one hundred two.
- **D.** The quantity seven times p plus twenty-three is the same as one hundred two.
- **E.** Seven times the sum of p and twenty-three is one hundred two.

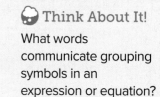
Think About It!
What words communicate grouping symbols in an expression or equation?

Example 6 Interpret an Equation

GEOMETRY Write a sentence for the formula for the surface area of a rectangular prism $S = 2\ell w + 2\ell h + 2wh$. Then interpret the equation in the context of the situation.

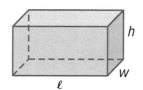

From the equation, we see that the surface area of a rectangular prism depends on the length, width, and height.

S =	$2\ell w +$	$2\ell h +$	$2wh$
Surface area equals	two times length times width plus	Two times length times height plus	two times width times height

- The first term, $2\ell w$, is two times the area of a rectangle. In the prism above, the area of the bottom face is ℓw. The top is the same shape, so it has the same area. This term represents the sum of the areas of the bottom and top faces.
- The second term, $2\ell h$, is the sum of the areas of the front and back faces.
- The third term, $2wh$, is the sum of the areas of the left and right faces.

So, the surface area of the rectangular prism is the sum of the areas of the faces.

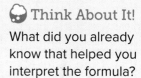
Think About It!
What did you already know that helped you interpret the formula?

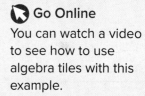
Go Online
You can watch a video to see how to use algebra tiles with this example.

Go Online You can complete an Extra Example online.

Lesson 2-1 · Writing and Interpreting Equations

Check

FINANCE The formula for a loan balance is $b = p\left(1 + \frac{r}{12}\right) - d$ if the previous balance is p, the annual interest rate is r, and a payment of d is made.

Part A Write a sentence for the formula.

A. The balance equals the previous balance multiplied by the quantity one plus the annual interest rate divided by 12 minus the payment.

B. The balance equals the previous balance multiplied by the annual interest rate divided by 12 minus the payment.

C. The balance equals the previous balance multiplied by the sum of one and the annual interest rate minus the payment.

D. The balance equals the previous balance minus the payment.

Part B Select each sentence that is a correct interpretation of the equation in the context of the situation. Select all that apply.

A. The expression $\frac{r}{12}$ represents the monthly interest rate.

B. The expression $p\left(1 + \frac{r}{12}\right)$ represents the previous balance plus interest.

C. If no payments are made, then $d = 0$ and the balance is the same as the previous balance.

D. The expression $\left(1 + \frac{r}{12}\right)$ represents the monthly interest rate.

E. If no payment is made, then $d = 0$ and the balance is the previous balance plus interest.

Go Online You can complete an Extra Example online.

Practice

 Go Online You can complete your homework online.

Example 1

Write an equation for each sentence.

1. Two added to three times a number m is the same as 18.

2. The product of five and the sum of a number x and three is twelve.

3. The quotient of 24 and x equals 14 minus 2 times x.

4. Nine times a number y subtracted from 85 is seven times the sum of four and y.

Example 2

5. **WALKING** Lily has walked 2 miles. Her goal is to walk 6 miles. Lily plans to reach her goal by walking 3 miles each hour h for the rest of her walk. Write an equation to find the number of hours it will take Lily to reach her goal.

6. **MATH** Paulina has completed 24 of the 42 math problems she was assigned for homework. She plans to finish her homework by completing 9 math problems each hour h. Write an equation to find the number of hours it will take Paulina to complete her math homework assignment.

7. **ATHLETICS** Of 107 athletes surveyed about what sport they play, some play basketball. Of those that play basketball, 48 play baseball and the remaining 33 do not play baseball. Write an equation to find the number of athletes surveyed who do not play basketball.

8. **SALES** Cars and trucks are the most popular vehicles. Last year, the number of cars sold was 39,000 more than three times the number of trucks sold. There were 216,000 cars sold last year. Write an equation that can be used to find the number of trucks, t, sold last year.

Example 3

Translate each sentence into an equation or formula.

9. Twice a increased by the cube of a equals b.

10. Seven less than the sum of p and t is as much as 6.

11. The sum of x and its square is equal to y times z.

12. Four times the sum of f and g is identical to six times g.

13. The area A of a square is the length of a side ℓ squared.

14. The perimeter P of a triangle is equal to the sum of the lengths of sides a, b, and c.

Lesson 2-1 • Writing and Interpreting Equations 71

15. The perimeter of a rectangle is equal to 2 times the length plus twice the width.

16. The density of an object is the quotient of its mass and its volume.

17. Simple interest is computed by finding the product of the principal amount p, the interest rate r, and the time t.

18. The surface area of a rectangular prism is 2 times the sum of the width, w, times height, h, and length, l, times width and length times height.

Examples 4 and 5
Write a sentence for each equation.

19. $j + 16 = 35$

20. $4m = 52$

21. $7(p + 23) = 102$

22. $r^2 - 15 = t + 19$

23. $\frac{2}{5}v + \frac{3}{4} = \frac{2}{3}x^2$

24. $\frac{1}{3} - \frac{4}{5}z = \frac{4}{3}y^3$

25. $g + 10 = 3g$

26. $2(t + 4q) = 2q + 4t$

27. $4(a + b) = 9a$

28. $8(2y - 6x) = 4 + 2x$

29. $\frac{1}{2}(f + y) = f - 5$

30. $k^2 - n^2 = 2b$

Example 6
Write a sentence for each formula. Then interpret the equation in the context of the situation.

31. GEOMETRY The formula for the volume of a cylinder is $V = \pi r^2 h$, where V is the volume, r is the length of the radius of the base, and h is the height of the cylinder.

32. GEOMETRY The formula for the volume of a cube is $V = s^3$, where V is the volume and s is the side length.

33. FINANCE The simple interest formula is given by $I = Prt$, where $I =$ interest, $P =$ principal, $r =$ rate, and $t =$ time.

34. FINANCE The compound interest formula is given by $A = P\left(1 + \frac{r}{n}\right)^{nt}$, where A is the amount, P is the principal, r is the rate, n is the number of times interest is compounded per year, and t is the time in years.

35. SCIENCE Newton's second law of motion is $F = ma$, where F is the force acting on an object, m is the mass of the object and a is the acceleration of the object.

36. SCIENCE The formula $d = rt$ relates the distance traveled d, the rate of travel r, and the time spent traveling t.

Mixed Exercises

For Exercises 37–40, match each sentence with an equation.

A. $g^2 = 2(g - 10)$ **B.** $\frac{1}{2}g + 32 = 15 + 6g$ **C.** $g^3 = 24g + 4$ **D.** $3g^2 = 30 + 9g$

37. One half of g plus thirty-two is as much as the sum of fifteen and six times g.
38. A number g to the third power is the same as the product of 24 and g plus 4.
39. The square of g is the same as two times the difference of g and 10.
40. The product of 3 and the square of g equals the sum of thirty and the product of nine and g.

Translate each sentence into an equation.

41. The difference of the square of y and twelve is the same as the product of five and x.
42. The difference of f and five times g is the same as 25 minus f.
43. Three times b less than 100 is equal to the product of 6 and b.
44. Four times the sum of 14 and c is a squared.

Translate each equation into a sentence.

45. $4n = x(5 - n)$ **46.** $2b - 10 = 4$ **47.** $y + 3x^2 = 5x$

Translate each sentence into a formula.

48. The area A of a circle is pi times the radius r squared.
49. The volume V of a rectangular prism equals the product of the length ℓ, the width w, and the height h.
50. REASONING The area of a kitchen is 182 square feet. This is 20% of the area of the first floor of the house. Let F represent the area of the first floor. Write an equation to represent the situation.
51. REASONING Katie is twice as old as her sister Mara. The sum of their ages is 24. Write a one-variable equation to represent the situation.
52. GEOMETRY The formula $F + V = E + 2$ shows the relationship between the number of faces F, edges E, and vertices V of a polyhedron, such as a pyramid. Write the formula in words.

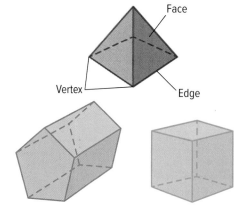

Lesson 2-1 • Writing and Interpreting Equations

53. STRUCTURE A recycling company charges business owners $10 for each cubic yard of waste removed from their facility plus a 10% fuel charge based on the total monthly bill. Let w represent the number of cubic yards of waste removed during the month. Write an equation to describe the total cost c of the recycling service per month.

54. WRITE Determine whether the two sentences describe the same equation. Explain.

The product of x and y plus z equals w.

The product of x and the sum of y and z equals w.

55. ANALYZE Determine whether the equation is an accurate translation of the sentence. Explain.

a. The square of the product of 4 and a number is equal to 8 times the sum of the number and 6. $(4n)^2 = 8(n + 6)$

b. Three more than one-half a number is equal to 2 less than the number. $\frac{n}{\frac{1}{2}} + 3 = n - 2$

56. PERSEVERE Translate the formula $A = \frac{b_1 + b_2}{2} \cdot h$ into words. Let A represent the area. List any constraints on the variables.

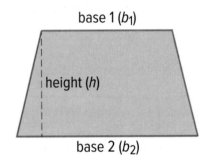

57. CREATE Write a scenario for the equation $12a + 10(a - 1) = 188$.

58. CREATE Write a problem about your favorite television show that uses the equation $x + 8 = 30$.

59. ANALYZE The surface area of a three-dimensional object is the sum of the area of the faces. If ℓ represents the length of the side of a cube, write a formula for the surface area of the cube.

60. ANALYZE Given the perimeter P and width w of a rectangle, write a formula to find the length ℓ.

61. WRITE How can you translate a verbal sentence into an algebraic equation? Explain.

Lesson 2-2

Solving One-Step Equations

Explore Using Algebra Tiles to Solve One-Step Equations Involving Addition or Subtraction

Online Activity Use algebra tiles to complete the Explore.

> **INQUIRY** How can you model and solve addition and subtraction equations?

Explore Using Algebra Tiles to Solve One-Step Equations Involving Multiplication

Online Activity Use algebra tiles to complete the Explore.

> **INQUIRY** How can you use algebra tiles to solve multiplication equations?

Learn Solving One-Step Equations Involving Addition or Subtraction

To **solve an equation** means to find all values of the variable that make the equation true. Each value that makes an equation true is a **solution**. **Equivalent equations** have the same solution.

Key Concept • Addition Property of Equality	
Words	If a number is added to each side of a true equation, the resulting equivalent equation is also true.
Symbols	For any real numbers a, b, and c, if $a = b$, then $a + c = b + c$.

Key Concept • Subtraction Property of Equality	
Words	If a number is subtracted from each side of a true equation, the resulting equivalent equation is also true.
Symbols	For any real numbers a, b, and c, if $a = b$, then $a - c = b - c$.

Today's Goals
- Solve equations by using addition and subtraction.
- Solve equations by using multiplication and division.

Today's Vocabulary
solve an equation

solution

equivalent equations

Think About It!
What happens if you add 5 to each side of $x - 5 = 15$? Which Property of Equality are you using?

Go Online
You may want to complete the Concept Check to check your understanding.

Example 1 Solve by Adding

Use the Addition Property of Equality to solve $g - 25 = 113$.

Horizontal Method		Vertical Method
$g - 25 = 113$	Original equation	$g - 25 = 113$
$g - 25 + 25 = 113 + 25$	Add 25 to each side.	$+ 25 + 25$
$g = 138$	Simplify.	$g = 138$

CHECK

$g - 25 = 113$		Original equation
$138 - 25 \stackrel{?}{=} 113$		Substitute 138 for g.
$113 = 113$		True

Check

Solve $\frac{2}{3} + w = 1\frac{1}{2}$. State which property of equality you used.

A. $\frac{5}{6}$; Subtraction Property of Equality

B. $\frac{5}{6}$; Addition Property of Equality

C. $\frac{13}{6}$; Addition Property of Equality

D. $\frac{13}{6}$; Subtraction Property of Equality

Example 2 Solve by Subtracting

Use the Subtraction Property of Equality to solve $27 + k = 30$.

Horizontal Method		Vertical Method
$27 + k = 30$	Original equation	$27 + k = 30$
$27 - 27 + k = 30 - 27$	Subtract 27 from each side.	$- 27 \quad - 27$
$k = 3$	Simplify.	$k = 3$

CHECK

$27 + k = 30$		Original equation
$27 + 3 \stackrel{?}{=} 30$		Substitute 3 for k.
$30 = 30$		True

Check

Solve $a + 26 = 35$.

$a = $ ___?___

 Go Online You can complete an Extra Example online.

Watch Out!

Equivalent Equations Do not forget to add the same number to each side of the equation so the result is an equivalent equation.

Talk About It!

Ann says that you could also solve $27 + k = 30$ by using the Addition Property of Equality. Explain Ann's reasoning.

Study Tip

Solving Equations When solving equations, you can use either the horizontal method or the vertical method. Both methods will produce the same answer.

76 Module 2 • Equations in One Variable

Example 3 Write a One-Step Equation

TENNIS In tennis, the Grand Slam tournaments are the four most prestigious annual events. At one point in his career, Roger Federer had won three more Grand Slam singles titles than Rafael Nadal. If at that time Roger Federer held the record for the most Grand Slam singles titles won with 17, how many Grand Slam singles titles had Rafael Nadal won?

Complete the table to write an equation that represents the number of Grand Slam singles titles Rafael Nadal won.

Words	Roger Federer	won	three	more than	Rafael Nadal
Variable	Let n = the number of singles Rafael Nadal won.				
Equation	17	=	n	+	3

$17 = n + 3$ Original equation

$17 - 3 = n + 3 - 3$ Subtract 3 from each side.

$14 = n$ Simplify.

Rafael Nadal had won 14 Grand Slam singles titles.

Check

DOGS On average, a male bulldog weighs 15 pounds less than a male golden retriever. If the average male bulldog weighs 50 pounds, write and solve an equation to find the average weight of a male golden retriever.

A. $50 = w - 15$; 65 pounds

B. $50 = w + 15$; 35 pounds

C. $50 = w - 15$; 35 pounds

D. $50 = 15 - w$; 65 pounds

Use a Source

Choose another men's singles tennis player and research the number of Grand Slam singles titles he has won. Write your own equation relating the number of Grand Slam singles titles he has won to the 17 titles of Roger Federer.

Go Online You can complete an Extra Example online.

🍿 **Think About It!**

What happens if you divide each side of $8x = 32$ by 8? Which property of equality does this demonstrate?

Learn Solving One-Step Equations Involving Multiplication or Division

You can also use the Multiplication Property of Equality and the Division Property of Equality to solve equations.

Key Concept • Multiplication Property of Equality	
Words	If an equation is true and each side is multiplied by the same nonzero number, then the resulting equation is equivalent.
Symbols	For any real numbers a, b, and c, if $a = b$, then $ac = bc$.
Example	If $x = 3$, then $8x = 24$.

🍿 **Think About It!**

How could you use the Division Property of Equality to simplify $ax = 32$ to $x = \frac{32}{a}$? How does this relate to using the Division Property of Equality to solve $8x = 32$?

Key Concept • Division Property of Equality	
Words	If an equation is true and each side is divided by the same nonzero number, the resulting equation is equivalent.
Symbols	For any real numbers a, b, and c, $c \neq 0$, if $a = b$, then $\frac{a}{c} = \frac{b}{c}$.
Example	If $x = -35$, then $\frac{x}{7} = \frac{-35}{7}$ or -5.

Example 4 Solve Equations by Multiplying or Dividing

Solve each equation.

a. $\frac{3}{8}x = \frac{9}{4}$

$\frac{3}{8}x = \frac{9}{4}$ Original equation

$\left(\frac{8}{3}\right)\frac{3}{8}x = \left(\frac{8}{3}\right)\frac{9}{4}$ Multiply each side by $\frac{8}{3}$, the reciprocal of $\frac{3}{8}$.

$x = 6$ Simplify.

b. $42 = -14y$

$42 = -14y$ Original equation

$\frac{42}{-14} = \frac{-14y}{-14}$ Divide each side by -14.

$-3 = y$ Simplify.

🍿 **Think About It!**

Describe a method you could use to check your solution for part **a**.

Check

Solve the equation $6y = 54$.

$y = \underline{?}$

📘 **Go Online** You can complete an Extra Example online.

Apply Example 5 Solve by Multiplying

SURVEY Kenji took a survey of the sophomore class. If 96 sophomores, or two-thirds of the class, said they were going to the football game on Saturday, how many sophomores were in the survey?

> **Think About It!**
> How would your equation change if the 96 sophomores planning to attend the game represented three-fourths of the class? What would you multiply each side of the equation by to solve the new equation?

1. What is the task?

Describe the task in your own words. Then list any questions that you may have. How can you find the answers to your questions?

96 sophomores are going to the game on Saturday. They make up $\frac{2}{3}$ of the class. How many sophomores were surveyed? I can find the answer to my question by writing and solving an equation to represent the situation.

2. How will you approach the task? What have you learned that you can use to help you complete the task?

I will write an equation to represent the situation and then solve it. I have learned how to translate a sentence to a mathematical equation. I have learned how to solve equations.

3. What is your solution?

Use your strategy to solve the problem.

What equation represents the number of sophomores surveyed?
$\frac{2}{3}n = 96$ where $n =$ the number of sophomores surveyed

How many sophomores were in the survey?
144

4. How can you know that your solution is reasonable?

✏ Write About It! Write an argument that can be used to defend your solution.

It makes sense that the number of sophomores surveyed is greater than the number of sophomores attending the football game, because the number of sophomores attending the football game is only part of the total number of sophomores surveyed. So, the whole (144) should be greater than the part (96).

Go Online You can complete an Extra Example online.

Check

FASHION Imani is making costumes for a play. She spent $146.58 on 21 yards of fabric. Write and solve an equation to find how much Imani paid for each yard of fabric.

A. $21p = 146.58$; $6.98 per yard

B. $146.58p = 21$; $0.14 per yard

C. $146.58(21) = p$; $3078.18 per yard

D. $21p = 146.58$; $3078.18 per yard

Practice

Go Online You can complete your homework online.

Examples 1, 2, and 4

Solve each equation.

1. $v - 9 = 14$
2. $44 = t - 72$
3. $-61 = d + (-18)$

4. $18 + z = 40$
5. $-4a = 48$
6. $12t = -132$

7. $18 - (-f) = 91$
8. $-16 - (-t) = -45$
9. $\frac{1}{3}v = -5$

10. $\frac{u}{8} = -4$
11. $\frac{a}{6} = -9$
12. $-\frac{k}{5} = \frac{7}{5}$

13. $\frac{3}{4} = w + \frac{2}{5}$
14. $-\frac{1}{2} + a = \frac{5}{8}$
15. $-\frac{t}{7} = \frac{1}{15}$

16. $-\frac{5}{7} = y - 2$
17. $v + 914 = -23$
18. $447 + x = -261$

Solve each equation.

19. $-\frac{1}{7}c = 21$

20. $-\frac{2}{3}v = -22$

21. $\frac{3}{5}q = -15$

22. $\frac{n}{8} = -\frac{1}{4}$

23. $\frac{c}{4} = -\frac{9}{8}$

24. $\frac{2}{3} + r = -\frac{4}{9}$

25. $y - 7 = 8$

26. $w + 14 = -8$

27. $p - 4 = 6$

28. $-13 = 5 + x$

29. $98 = b + 34$

30. $y - 32 = -1$

31. $n + (-28) = 0$

32. $y + (-10) = 6$

33. $-1 = t + (-19)$

34. $j - (-17) = 36$

35. $14 = d + (-10)$

36. $u + (-5) = -15$

37. $11 = -16 + y$

38. $c - (-3) = 100$

39. $47 = w - (-8)$

40. $x - (-74) = -22$

41. $4 - (-h) = 68$

42. $-56 = 20 - (-j)$

43. $12z = 108$

44. $-7t = 49$

45. $18f = -216$

46. $-22 = 11v$

47. $-6d = -42$

48. $96 = -24a$

49. $\frac{c}{4} = 16$

50. $\frac{a}{16} = 9$

51. $-84 = \frac{d}{3}$

52. $-\frac{d}{7} = -13$

53. $\frac{t}{4} = -13$

54. $31 = -\frac{1}{6}n$

Examples 3 and 5

55. SUPREME COURT Chief Justice William Rehnquist served on the Supreme Court for 33 years until his death in 2005. Write and solve an equation to determine the year he was confirmed as a justice on the Supreme Court.

56. SALARY In a recent year, the annual salary of the Governor of New York was $179,000. During the same year, the annual salary of the Governor of Tennessee was $94,000 less than that. Write and solve an equation to find the annual salary of the Governor of Tennessee in that year.

57. WEATHER On a cold January day, Kiara noticed that the temperature dropped 21 degrees over the course of the day to −9°C. Write and solve an equation to determine what the temperature was at the beginning of the day.

58. FARMING The Rolling Hills Farm is 126 acres. This is $\frac{1}{4}$ the size of the Briarwood Farm. Write and solve an equation to determine the number of acres of the Briarwood Farm.

59. SOCCER During the season, 13% of the players who signed up for the soccer league dropped out. A total of 174 players finished the season.

  a. Assign a variable. Write an expression for the number of players who finished the season. Explain your reasoning.

  b. Write an equation to find the number of players who signed up for the soccer league.

  c. Solve the equation to find the number of players who signed up for the soccer league.

Mixed Exercises

Write an equation for each sentence. Then solve the equation.

60. Six times a number is 132.

61. Two thirds equals negative eight times a number.

62. Five elevenths times a number is 55.

63. Four fifths is equal to ten sixteenths of a number.

64. Three and two thirds times a number equals two ninths.

65. Four and four fifths times a number is one and one fifth.

Solve each equation. Check your solution.

66. $\frac{x}{9} = 10$

67. $\frac{b}{7} = -11$

68. $\frac{3}{4} = \frac{c}{24}$

69. $\frac{2}{3} = \frac{1}{8}y$

70. $\frac{2}{3}n = 14$

71. $\frac{3}{5}g = -6$

72. $4\frac{1}{5} = 3p$

73. $-5 = 3\frac{1}{2}x$

74. $6 = -\frac{1}{2}n$

75. $-\frac{2}{5} = -\frac{z}{45}$

76. $-\frac{g}{24} = \frac{5}{12}$

77. $-\frac{v}{5} = -45$

78. $-6 = \frac{2}{3}z$

79. $\frac{2}{7}q = -4$

80. $\frac{5}{9}p = -10$

81. $\frac{a}{10} = \frac{2}{5}$

82. $d - 8 = 6$

83. $-28 = p + 21$

84. $-7x = 63$

85. $-\frac{t}{5} = -8$

86. $y + (-16) = -12$

87. $\frac{3}{5}y = -9$

88. $-8d = -64$

89. $-\frac{3}{4}y = \frac{8}{20}$

90. **VACATION** The Lopez family is on vacation in Tennessee. They drove 210 miles from Memphis to Nashville and continued driving to Knoxville. By the time they reached Knoxville, they had traveled a total of 390 miles.

    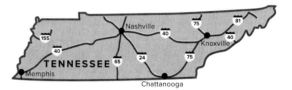

    a. Define the variable and write an equation that represents the distance from Nashville to Knoxville.

    b. Which property of equality could you use to isolate the variable in your equation? Explain your reasoning.

    c. How far is Knoxville from Nashville? How can you verify that your solution is accurate?

    d. If the Lopez family drives from Nashville to Chattanooga instead of Nashville to Knoxville, they will drive 47 fewer miles. Write an equation that represents the distance from Memphis to Chattanooga through Nashville. How far is Chattanooga from Nashville?

91. **TICKETS** Julian and Makayla order season tickets for the local soccer team. The ticket package they choose costs $780 and includes tickets to 12 games.

    a. Write and solve an equation that represents the cost per game.

    b. Single game tickets cost $85. How much do they save per game by using season tickets?

92. **TACOS** Orlando spent $18 at a taco truck. He ordered 4 tacos. Write and solve an equation to find the cost of each taco.

STRUCTURE Solve each equation. State the Property of Equality used.

93. $\dfrac{x}{9} = 24$

94. $m - 183 = -79$

95. $972 + y = 748$

96. $-\dfrac{4}{5}p = 32$

97. $135 = 9b$

98. $45 = \dfrac{3}{2}z$

99. **WHICH ONE DOESN'T BELONG** Identify the equation that does not belong with the other three. Justify your conclusion.

| $n + 14 = 27$ | $12 + n = 25$ | $n - 16 = 29$ | $n - 4 = 9$ |

100. **PERSEVERE** Determine the value for each statement below.

   a. If $x - 9 = 12$, what is the value of $x + 1$?

   b. If $n + 7 = -4$, what is the value of $n + 1$?

101. **CREATE** Write an equation that you would use the Addition Property of Equality to solve.

102. **ANALYZE** Determine whether each sentence is *sometimes*, *always*, or *never* true. Justify your argument.

   a. $x + x = x$

   b. $x + 0 = x$

103. **ANALYZE** How would you solve $5x = 35$? How would you solve $5 + x = 35$? How are the methods similar and how are they different?

104. **WRITE** Consider the Multiplication Property of Equality and the Division Property of Equality. Explain why they can be considered the same property. Which one do you think is easier to use?

84 Module 2 · Equations in One Variable

Lesson 2-3

Solving Multi-Step Equations

Explore Using Algebra Tiles to Model Multi-Step Equations

Online Activity Use algebra tiles to complete the Explore.

> **INQUIRY** How can you model and solve a multi-step equation?

Learn Solving Multi-Step Equations

A **multi-step equation** is an equation that uses more than one property of equality to solve it. To solve this type of equation, you can undo each operation using properties of equality. Working backward in the order of operations makes this process simpler. Each step in this process results in equivalent equations.

Operation	Opposite Operation
Addition	Subtraction
Subtraction	Addition
Multiplication	Division
Division	Multiplication

Example 1 Solve Multi-Step Equations

Use properties of equality to solve each equation. Check your solutions.

a. $2a - 6 = 4$

$2a - 6 = 4$	Original equation.
$2a - 6 + 6 = 4 + 6$	Add 6 to each side.
$2a = 10$	Simplify.
$\frac{2a}{2} = \frac{10}{2}$	Divide each side by 2.
$a = 5$	Simplify.

Today's Goal
- Solve equations involving more than one operation.

Today's Vocabulary
multi-step equation

Think About It!
In $4x - 2 = 5$, which two operations are being used? Which operations would you use to work backward in the order of operations to solve the equation?

(continued on the next page)

Study Tip

Assumptions To solve an equation, you must assume that the original equation has a solution.

CHECK
Check your solution by substituting the result back into the original equation.

$2a - 6 = 4$	Original equation
$2(5) - 6 \stackrel{?}{=} 4$	Substitute 5 for a.
$10 - 6 \stackrel{?}{=} 4$	Simplify.
$4 = 4$	True

b. $\dfrac{n+1}{-2} = 15$

$\dfrac{n+1}{-2} = 15$	Original equation.
$-2\left(\dfrac{n+1}{-2}\right) = -2(15)$	Multiply each side by -2.
$n + 1 = -30$	Simplify.
$n + 1 - 1 = -30 - 1$	Subtract 1 from each side.
$n = -31$	Simplify.

Check your solution by substituting the result back into the original equation.

Check

Solve $3m + 4 = -11$.

$m = \underline{\ ?\ }$

Solve $8 = \dfrac{x-5}{7}$.

$x = \underline{\ ?\ }$

Go Online You can watch a video to see how to use a graphing calculator with this example.

Study Tip

Multiplicative Inverse A number multiplied by its reciprocal is 1.

Talk About It!

Would it work to first multiply each side by $\dfrac{5}{2}$? Explain your reasoning.

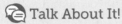 **Example 2** Write and Solve a Multi-Step Equation

FUNDRAISING The student council raised $\dfrac{2}{5}$ of the money they need to cover the cost of the school dance with a bake sale. They raised an additional $150 selling raffle tickets. If the student council has raised $630, what is the cost of the dance? Write an equation for the problem. Then solve the equation.

Go Online You can complete an Extra Example online.

86 Module 2 • Equations in One Variable

Complete the table to write an equation for the cost of the dance.

Words	Two fifths of the cost	plus 150	is 630.
Variable	Let c = the cost of the dance.		
Equation	$\frac{2}{5} \cdot c$	$+ 150$	$= 630$

$$\frac{2}{5}c + 150 = 630 \quad \text{Original equation}$$
$$\frac{2}{5}c + 150 - 150 = 630 - 150 \quad \text{Subtract 150 from each side.}$$
$$\frac{2}{5}c = 480 \quad \text{Simplify.}$$
$$\frac{5}{2}\left(\frac{2}{5}\right)c = \frac{5}{2}(480) \quad \text{Multiply each side by } \frac{5}{2}.$$
$$c = 1200 \quad \text{Simplify.}$$

The dance costs $ 1200.

Check

BASKETBALL A sporting goods store sold $\frac{2}{3}$ of its basketballs, but 8 were returned. Now the store has 38 basketballs. How many were there originally? Write an equation for the problem. Then solve the equation.

A. $\frac{2}{3}b + 8 = 38;\ 45$

B. $\frac{2}{3}b - 8 = 38;\ 69$

C. $\frac{1}{3}b + 8 = 38;\ 90$

D. $\frac{1}{3}b - 8 = 38;\ 138$

Example 3 Solve Multi-Step Equations with Letter Coefficients

Some equations have coefficients that are represented by letters. To solve these equations, apply the process of solving equations to isolate the variable.

Solve $ax + 7 = 5$ for x. Assume that $a \neq 0$.

$$ax + 7 = 5 \quad \text{Original equation}$$
$$ax + 7 - 7 = 5 - 7 \quad \text{Subtract 7 from each side.}$$
$$ax = -2 \quad \text{Simplify.}$$
$$\frac{ax}{a} = \frac{-2}{a} \quad \text{Divide each side by } a.$$
$$x = \frac{-2}{a} \quad \text{Simplify.}$$

💭 **Think About It!**
Why do you have to assume that $a \neq 0$ when solving the equation?

 Go Online You can complete an Extra Example online.

Watch Out!

Isolate the Variable
Make sure that you are isolating x in the equation. Remember that *a* represents a coefficient, not the variable.

Check

Solve $2 - ax = -8$ for x. Assume $a \neq 0$.

A. $x = -\frac{10}{a}$

B. $x = -\frac{6}{a}$

C. $x = \frac{6}{a}$

D. $x = \frac{10}{a}$

Practice

Go Online You can complete your homework online.

Example 1

Use properties of equality to solve each equation. Check your solution.

1. $3t + 7 = -8$
2. $8 = 16 + 8n$
3. $-34 = 6m - 4$

4. $9x + 27 = -72$
5. $\frac{y}{5} - 6 = 8$
6. $\frac{f}{-7} - 8 = 2$

7. $1 + \frac{r}{9} = 4$
8. $\frac{k}{3} + 4 = -16$
9. $\frac{n-2}{7} = 2$

10. $14 = \frac{6+z}{-2}$
11. $-11 = \frac{a-5}{6}$
12. $\frac{22-w}{3} = -7$

Example 2

13. **SHOPPING** Ricardo spent half of his allowance on school supplies. Then he bought a snack for $5.25. When he arrived home, he had $22.50 left. Write and solve an equation to find the amount of Ricardo's allowance *a*.

14. **SHOPPING** Liza earned some money by taking care of her neighbor's pet. She bought a drink for $1.95 and a concert ticket for $30. She bought a ring for $7.20, and then spent two-thirds of the remaining money on a wireless speaker. If Liza has $38.50 left, write and solve an equation to find the amount of money *m* Lisa earned by taking care of her neighbor's pet.

88 Module 2 • Equations in One Variable

15. PET SHELTERS Henry works at a pet shelter after school. He purchases a large package of dog treats. He sets aside 10 treats and distributes the rest equally among the 15 dogs in the shelter. If each dog received 4 treats, write and solve an equation to find the number of treats t that were in the original package.

16. BASKETBALL The average number of points a basketball team scored for three games was 63 points. In the first two games, they scored the same number of points, which was 6 points more than they scored in the third game. Write and solve an equation to find the number of points the team scored in each game.

17. HUMAN HEIGHT Micah's adult height is one less than twice his height at age 2. Micah's adult height is 71 inches. Write and solve an equation to find Micah's height h at age 2.

Example 3

Solve each equation for x. Assume $a \neq 0$.

18. $ax + 3 = 23$

19. $4 = ax - 14$

20. $ax - 5 = 19$

21. $6 + ax = -29$

22. $\frac{8}{ax} - 5 = -3$

23. $18 - ax = 42$

24. $5 = \frac{5}{ax} + 1$

25. $-3 = ax + 11$

26. $-7 = -ax - 16$

Mixed Exercises

Solve each equation. Check your solution.

27. $3x + 8 = 29$

28. $\frac{a}{6} - 5 = 9$

29. $\frac{5r}{2} - 6 = 19$

30. $\frac{n}{3} - 8 = -2$

31. $5 + \frac{x}{4} = 1$

32. $-\frac{h}{3} - 4 = 13$

33. $5(1 + n) = -5$

34. $-27 = -6 - 3p$

35. $-\frac{a}{6} + 5 = 2$

REASONING Write and solve an equation to find each number.

36. A number is divided by 2, and then the quotient is increased by 8. The result is 33.

37. Two is subtracted from a number, and then the difference is divided by 3. The result is 30.

38. PERSEVERE The sum of 4 consecutive odd integers is equal to zero.

   a. Write an equation to model the sentence.

   b. Solve the equation to find the numbers. Check your solution.

39. FIND THE ERROR Kadija and Jorge are solving $\frac{1}{2}n + 5 = \frac{17}{2}$. Jorge uses the Subtraction Property of Equality followed by the Multiplication Property of Equality. Kadija also uses the Subtraction Property of Equality, but because n is multiplied by $\frac{1}{2}$, Kadija claims that the Division Property of Equality can be used to isolate the variable. Which student is correct? Explain your reasoning.

40. CREATE Write a problem that can be represented by the equation $11.9p + 23.1 = 273$. Define the variable and solve the equation.

41. ANALYZE Solve each equation for x. Assume that $a \neq 0$.

   a. $ax + 7 = 5$ **b.** $\frac{1}{a}x - 4 = 9$ **c.** $2 - ax = -8$

42. ANALYZE Determine whether each equation has a solution. Justify your answer.

   a. $\frac{a+4}{5+a} = 1$ **b.** $\frac{1+b}{1-b} = 1$ **c.** $\frac{c-5}{5-c} = 1$

43. ANALYZE Determine whether the following statement is *sometimes*, *always*, or *never* true. Justify your argument.

   The sum of three consecutive odd integers equals an even integer.

44. WRITE Write a paragraph explaining the order of the steps that you would take to solve a multi-step equation.

Lesson 2-4
Solving Equations with the Variable on Each Side

Explore Modeling Equations with the Variable on Each Side

Online Activity Use graphing technology to complete the Explore.

> **INQUIRY** How can you solve an equation with the variable on each side?

Today's Goals
- Solve equations with the variable on each side.
- Solve equations by applying the Distributive Property.
- Prove that equations are identities or have no solution.

Today's Vocabulary
identity

Learn Solving Equations with the Variable on Each Side

Sometimes, the variable will appear on each side of an equation. To solve these equations, use the Addition or Subtraction Property of Equality to write an equivalent equation with the variable terms on one side and the numbers without variables, or constants, on the other side.

Example 1 Solve an Equation with the Variable on Each Side

Solve $5 + 7a = 4a - 13$. Check your solution.

$5 + 7a = 4a - 13$	Original equation
$-4a = -4a$	Subtract $4a$ from each side.
$5 + 3a = -13$	Simplify.
$-5 = -5$	Subtract 5 from each side.
$3a = -18$	Simplify.
$\frac{3a}{3} = \frac{-18}{3}$	Divide each side by 3.
$a = -6$	Simplify.

CHECK

$5 + 7a = 4a - 13$	Original equation
$5 + 7(-6) \stackrel{?}{=} 4(-6) - 13$	Substitution, $a = -6$
$5 + -42 \stackrel{?}{=} -24 - 13$	Multiply.
$-37 = -37$	True

💭 Think About It!
Leon says that when you solve $5 + 7a = 4a - 13$, you can just combine $7a$ and $4a$ because they are like terms. Explain whether Leon is correct.

 Go Online You can complete an Extra Example online.

Study Tip

Solving an Equation You may want to combine the terms with a variable on one side before isolating a constant.

Example 2 Write an Equation with the Variable on Each Side

CONTEST The results of the 2015 Nathan's Hot Dog Eating Contest are shown.

Suppose the men's and women's winners, Matt and Miki, decide to compete against each other. To make the competition more interesting, Matt will not start until Miki has eaten 20 hot dogs. Assume that Matt and Miki eat at a constant rate throughout the competition. Based on the number of hot dogs eaten in 10 minutes by Matt and Miki, how many minutes after Matt starts eating will they have eaten the same number of hot dogs?

Read the Problem

We want to find the number of minutes m for which Matt and Miki have eaten the same number of hot dogs.

Matt eats 62 hot dogs in 10 minutes, which is a rate of 6.2 hot dogs per minute. The number of hot dogs he has eaten m minutes after starting is $6.2m$.

Miki eats 38 hot dogs in 10 minutes, which is a rate of 3.8 hot dogs per minute. She is given a 20-hot dog head start, so the number of hot dogs she has eaten m minutes after Matt starts is $3.8m + 20$.

The equation $6.2m = 3.8m + 20$ represents this situation.

Solve the Problem

$$6.2m = 3.8m + 20 \qquad \text{Original equation}$$

$$6.2m - 3.8m = 3.8m + 20 - 3.8m \qquad \text{Subtract } 3.8m \text{ from each side.}$$

$$2.4m = 20 \qquad \text{Simplify.}$$

$$\frac{2.4m}{2.4} = \frac{20}{2.4} \qquad \text{Divide each side by 2.4.}$$

$$m \approx 8.3 \qquad \text{Simplify.}$$

After approximately $8\frac{1}{3}$ minutes, Matt and Miki will have eaten the same number of hot dogs.

Go Online You can complete an Extra Example online.

Check

BASKETBALL Nolan and Victor were two of the top scoring freshman players in a college basketball conference last season. The table shows how many points Nolan and Victor scored and how many games they played last season. The points Nolan and Victor score this season will be combined with their points from last season to give their total career points. This season, Nolan is hoping to catch up to Victor and have the same number of career points. Assume that Nolan and Victor play every game and score at the same constant rate as last season.

Player	Games Played	Points
Nolan	30	750
Victor	34	782

Part A

Based on each player's average scoring rate, write an equation that represents the number of games it will take Nolan to accumulate the same number of career points as Victor.

$$25p + \underline{\ ?\ } = \underline{\ ?\ }\, p + \underline{\ ?\ }$$

Part B

Based on your equation in Part A, after how many games this season will Nolan and Victor have scored the same number of career points?

$\underline{\ ?\ }$ games

Learn Solving Equations Involving the Distributive Property

Some equations contain grouping symbols. Grouping symbols can include parentheses ( ), brackets [ ], and fraction bars.

The steps for solving an equation can be summarized as follows.

Step 1 Simplify the expressions on each side. Remove any grouping symbols. Use the Distributive Property as needed.

Step 2 Use the Addition and/or Subtraction Properties of Equality to get the variable terms on one side of the equation and the constant terms on the other side. Simplify.

Step 3 Use the Multiplication and Division Properties of Equality to solve.

 Think About It!

Describe the steps you would take to solve $2(1 + t) = 8t$.

Study Tip

Grouping Symbols Some expressions, like $2 - [11 + 5(p - 8)]$, contain grouping symbols inside of grouping symbols. To simplify these expressions, work from the inside out by first simplifying the expression within the innermost grouping symbol.

Go Online You can complete an Extra Example online.

Watch Out!
Keep the Negative
Do not forget to distribute –2 to each term on the right side of the equation.

Example 3 Solve an Equation with Grouping Symbols

Solve $7(n - 1) = -2(3 + n)$.

$7(n - 1) = -2(3 + n)$	Original equation
$7n - 7 = -6 - 2n$	Distributive Property
$7n - 7 + 2n = -6 - 2n + 2n$	Add $2n$ to each side.
$9n - 7 = -6$	Simplify.
$9n - 7 + 7 = -6 + 7$	Add 7 to each side.
$9n = 1$	Simplify.
$\frac{9n}{9} = \frac{1}{9}$	Divide each side by 9.
$n = \frac{1}{9}$	Simplify.

Check

Solve $7(n - 2) + 8 = 3(n - 4) - 2$.

$n = \underline{\ ?\ }$

Math History Minute
During the short life of Indian mathematician **Srinivasa Ramanujan (1887–1920)**, he compiled nearly 3900 results, which included proofs of theorems, equations, and identities, nearly all of which have been proven correct. Ramanujan was known as a genius and an *autodidact*, which is a person who is self-taught.

Example 4 Solve an Equation with a Fraction Bar

Solve $5y = \frac{12y + 16}{4}$.

$5y = \frac{12y + 16}{4}$	Original equation
$4(5y) = 4\left(\frac{12y + 16}{4}\right)$	Multiply each side by 4.
$20y = 12y + 16$	Simplify.
$20y - 12y = 12y + 16 - 12y$	Subtract $12y$ from each side.
$8y = 16$	Simplify.
$\frac{8y}{8} = \frac{16}{8}$	Divide each side by 8.
$y = 2$	Simplify.

Go Online You can complete an Extra Example online.

Example 5 Write an Equation with Grouping Symbols

GEOMETRY Find the value of x so that the figures have the same area.

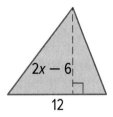

The area of the rectangle is $5(x + 4)$, and the area of the triangle is $\frac{1}{2}(12)(2x - 6)$. The equation $5(x + 4) = \frac{1}{2}(12)(2x - 6)$ represents the situation where the areas of the figures are the same.

$5(x + 4) = \frac{1}{2}(12)(2x - 6)$	Original equation
$5(x + 4) = 6(2x - 6)$	Multiply $\frac{1}{2}$ and 12.
$5x + 20 = 12x - 36$	Distributive Property
$5x + 20 + 36 = 12x - 36 + 36$	Add 36 to each side.
$5x + 56 = 12x$	Simplify.
$5x + 56 - 5x = 12x - 5x$	Subtract $5x$ from each side.
$56 = 7x$	Simplify.
$\frac{56}{7} = \frac{7x}{7}$	Divide each side by 7.
$8 = x$	Simplify.

Check

GEOMETRY Find the value of x so that the figures have the same area.

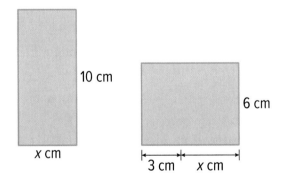

$x = \underline{}$

Go Online You can complete an Extra Example online.

Learn Identities and Equations with No Solutions

One solution	No solution	Identity
Words		
An equation has one solution if exactly one value of the variable makes the equation true.	An equation has no solution if there is no value of the variable that makes the equation true.	An **identity** is an equation that is true for all values its variables.

Example 6 Solve an Equation with No Solution

Solve $6(y - 5) = 2(10 + 3y)$.

$6(y - 5) = 2(10 + 3y)$ Original equation

$6y - 30 = 20 + 6y$ Distributive Property

$6y - 30 - 6y = 20 + 6y - 6y$ Subtract $6y$ from each side.

$-30 \neq 20$ Simplify.

Since $-30 \neq 20$, this equation has no solution.

Example 7 Solve an Identity

Solve $7x + 5(x - 1) = 12x - 5$.

$7x + 5(x - 1) = 12x - 5$ Original equation

$7x + 5x - 5 = 12x - 5$ Distributive Property

$12x - 5 = 12x - 5$ Simplify.

$0 = 0$ Subtract $12x - 5$ from each side.

Since the expressions on each side of the equation are the same, this equation is an identity. It is true for all values of x.

Talk About It!
Could you tell that the equation was an identity before the final step? Explain your reasoning.

Check

Solve each equation and state whether the equation has *one solution*, has *no solution*, or is an *identity*.

A. $8(g + 6) = 5g + 3(g + 16)$

B. $5x + 5 = 3(5x - 4) - 10x$

C. $3w + 2 = 7w$

D. $3(2b - 1) - 7 = 6b - 10$

Go Online to practice what you've learned about solving linear equations in the Put It All Together over Lessons 2-1 through 2-4.

Go Online You can complete an Extra Example online.

Practice

Go Online You can complete your homework online.

Examples 1, 3, and 4

Solve each equation. Check your solution.

1. $7c + 12 = -4c + 78$
2. $2m - 13 = -8m + 27$
3. $9x - 4 = 2x + 3$
4. $6 + 3t = 8t - 14$
5. $\frac{b-4}{6} = \frac{b}{2}$
6. $\frac{3v+12}{6} = \frac{4v}{3}$
7. $2(r + 6) = 4(r + 4)$
8. $6(n + 5) = 3(n + 16)$
9. $5(g + 8) - 7 = 117 - g$
10. $12 - \frac{4}{5}(x + 15) = \frac{2}{5}x + 6$
11. $3(3m - 2) = 2(3m + 3)$
12. $6(3a + 1) - 30 = 3(2a - 4)$
13. $7n + 6 = 4n - 9$
14. $-6(2r + 8) = -10(r - 3)$
15. $5 - 3(w + 4) = w - 7$
16. $2x - 5(x - 3) = 2(x - 10)$

Example 2

17. **OLYMPICS** In the 2010 Winter Olympic Games, the United States won 1 more than 4 times the number of gold medals France won. The United States won 7 more gold metals than France. Write and solve an equation to find the number of gold medals each country won.

18. **REASONING** Diego's sister is twice his age minus 9 years. She is also as old as half the sum of the ages of Diego and both of his 12-year-old twin brothers. Write and solve an equation to find the ages of Diego and his sister.

19. **NATURE** The table shows the current heights and average growth rates of two different species of trees. Write and solve an equation to find how long it will take for the two trees to be the same height.

Tree Species	Current Height	Annual growth
A	38 inches	4 inches
B	45.5 inches	2.5 inches

20. **WEIGHT** A dog weighs two pounds less than three times the weight of a cat. The dog also weighs twenty-two pounds more than the cat. Write and solve an equation to find the weights of the dog and the cat.

Example 5

21. **GEOMETRY** Supplementary angles are two angles with measures that have a sum of 180°. Complementary angles are two angles with measures that have a sum of 90°. The measure of the supplement of an angle is 10° more than twice the measure of the complement of the angle. Let $90 - x$ equal the degree measure of the complement angle and $180 - x$ equal the degree measure of the supplement angle. Write and solve an equation to find the measure of the angle.

22. **GEOMETRY** Write and solve an equation to find the value of x so that the figures have the same area.

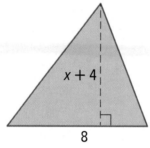
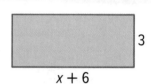

23. **GEOMETRY** Write and solve an equation to find the value of x so that the figures have the same area.

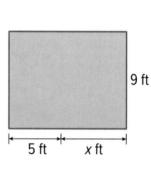
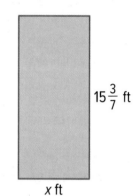

24. **GEOMETRY** Write and solve an equation to find the value of x so that the figures have the same area. The area of a trapezoid is $\frac{1}{2}h(b_1 + b_2)$.

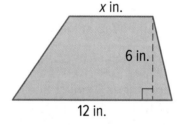
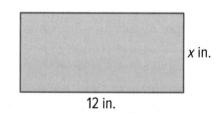

Examples 6 and 7

Solve each equation and state whether the equation has *one solution*, *no solution*, or is an *identity*.

25. $-6y - 3 = 3 - 6y$

26. $\frac{1}{2}(x + 6) = \frac{1}{2}x - 9$

27. $8q + 12 = 4(3 + 2q)$

28. $21(x + 1) - 6x = 15x + 21$

29. $12y + 48 - 4y = 8(y - 6)$

30. $8(z + 6) = 4(2z + 12)$

31. $2a + 2 = 3(a + 2)$

32. $\frac{1}{4}x + 5 = \frac{1}{4}x$

33. $7(c + 9) = 7c + 63$

34. $4k + 3 = \frac{1}{4}(8k + 16)$

35. $3b - 13 + 4b = 7b + 1$

36. $\frac{1}{2}(\frac{1}{2}m - 8) = \frac{1}{4}(m - 16)$

Mixed Exercises

Solve each equation. Check your solution.

37. $2x = 2(x - 3)$

38. $\frac{2}{5}h - 7 = \frac{12}{5}h - 2h + 3$

39. $-5(3 - q) + 4 = 5q - 11$

40. $2(4r + 6) = \frac{2}{3}(12r + 18)$

41. $\frac{3}{5}f + 24 = 4 - \frac{1}{5}f$

42. $\frac{1}{12} + \frac{3}{8}y = \frac{5}{12} + \frac{5}{8}y$

43. $6.78j - 5.2 = 4.33j + 2.15$

44. $14.2t - 25.2 = 3.8t + 26.8$

45. $3.2k - 4.3 = 12.6k + 14.5$

46. $5[2p - 4(p + 5)] = 25$

47. $m - 9 = \frac{2m - 12}{3}$

48. $\frac{3d - 2}{8} = -d + 16\frac{1}{4}$

49. Twice the greater integer of two consecutive odd integers is 13 less than three times the lesser integer.

　　a. Write an equation to find the two consecutive odd integers.

　　b. What are the integers in ascending order?

50. Two times the quantity of eight times a number plus two is equal to three times the quantity of two times the same number minus seven.

　　a. Write an equation to find the number.

　　b. Solve the equation to find the number.

51. USE A MODEL The perimeter of Figure 1 is four times a number minus three. The perimeter of Figure 2 is two times the same number plus five. The perimeters of Figure 1 and Figure 2 are the same.

　　a. Write an equation to find the number.

　　b. Solve the equation to find the number.

　　c. What is the perimeter of Figure 2? Explain.

　　d. Find the perimeter of Figure 1. Compare the perimeter of Figure 1 to the perimeter you found for Figure 2 to justify the value of k is correct. Show your work.

52. STRUCTURE Find two consecutive even integers such that twice the lesser of two integers is 4 less than two times the greater integer.

　　a. Write and solve an equation to find the integers.

　　b. Does the equation have one solution, no solution, or is it an identity? Explain.

53. FIND THE ERROR Anthony and Patty are solving the equation $y - m = m - y + 1$ for y. Is either correct? Explain why or why not.

Anthony
$y - m = m - y + 1$
$2y - m = m + 1$
$2y = 2m + 1$
$y = m + \frac{1}{2}$

Patty
$y - m = m - y + 1$
$2y - m = m + 1$
$2y = 1$
$y = \frac{1}{2}$

54. PERSEVERE Write an equation with variables on each side of the equal sign, at least one fractional coefficient, and a solution of -6. Discuss the steps you used.

55. CREATE Create an equation with at least two grouping symbols for which there is no solution.

56. WRITE Compare and contrast solving equations with variables on both sides of the equation to solving one-step or multi-step equations with a variable on one side of the equation.

57. ANALYZE Determine whether each solution is correct. If it is incorrect, find the correct solution. Justify your argument.

a.
$2(g + 5) = 22$
$2g + 5 = 22$
$2g + 5 - 5 = 22$
$2g = 17$
$g = 8.5$

b.
$5d = 2d - 18$
$5d - 2d = 2d - 18 - 2d$
$3d = -18$
$d = -6$

c.
$-6z + 13 = 7z$
$-6z + 13 - 6z = 7z - 6z$
$13 = z$

58. PERSEVERE Find the value of k for which each equation is an identity.

a. $k(3x - 2) = 4 - 6x$
b. $15y - 10 + k = 2(ky - 1) - y$

59. CREATE Write an equivalent equation to $x = 8$ that has the variable x on both sides.

60. ANALYZE Solve $5x + 2 = ax - 1$ for x. Assume $a \neq 5$. Describe each step.

Lesson 2-5
Solving Equations Involving Absolute Value

Today's Goal
- Solve absolute value expressions.

Explore Modeling Absolute Value

Online Activity Use an infographic to complete the Explore.

 INQUIRY How is margin of error related to absolute value?

Learn Solving Equations Involving Absolute Value

Absolute value equations contain at least one absolute value expression. The simplest form of an absolute value equation is $|x| = n$. Since absolute value represents distance, you must consider the case where the solution is x units from zero in the negative direction and the case where the solution is x units from zero in the positive direction.

Key Concept • Solving Absolute Value Equations	
Words	When solving equations that involve absolute values, there are two cases to consider. **Case 1** The expression inside the absolute value symbol is positive or zero. **Case 2** The expression inside the absolute value symbol is negative.
Symbols	For any real numbers a and b, if $|a| = b$ and $b \geq 0$, then $a = b$ or $a = -b$.
Example	$|d| = 3$, so $d = 3$ or $d = -3$.

Consider the equation $|x| = 3$. This means that the two points on the number line where the distance between 0 and x is 3 are solutions to the equation. The distance between 0 and -3 is 3, so -3 is a solution to the equation. The distance between 0 and 3 is 3, so 3 is also a solution to the equation.

Think About It!
Clark says that the solution set for $|x| = 8$ is 8. Is he correct? Why or why not?

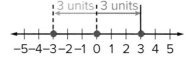

If $|x| = 3$, then $x = -3$ or $x = 3$. Thus the solution set is $\{-3, 3\}$. You can graph the solution set by graphing each solution on the number line.

For each absolute value equation, you must consider both cases. To solve an absolute value equation, first isolate the absolute value on one side of the equal sign if it is not already by itself.

Example 1 Solve an Absolute Value Equation When $n > 0$

Solve $|y + 2| = 4$. Then graph the solution set.

Case 1	Case 2
If y is nonnegative, then $|y| = y$.	If y is negative, then $|y| = -y$.

Case 1		Case 2
$y + 2 = 4$	Original equation	$y + 2 = -4$
$y + 2 - 2 = 4 - 2$	Subtract 2 from each side.	$y + 2 - 2 = -4 - 2$
$y = 2$	Simplify.	$y = -6$

The solution set is $\{-6, 2\}$.

Graph points at -6 and 2 on the number line.

-7-6-5-4-3-2-1 0 1 2 3 4 5 6 7

CHECK

Substitute -6 and 2 into the original equation.

$|y + 2| = 4$	Original equation	$|y + 2| = 4$
$|-6 + 2| \stackrel{?}{=} 4$	Substitute.	$|2 + 2| \stackrel{?}{=} 4$
$|-4| \stackrel{?}{=} 4$	Simplify.	$|4| \stackrel{?}{=} 4$
$4 = 4$ ✓	Take the absolute value.	$4 = 4$ ✓

Check

Graph the solution set of $|2t - 4| = 8$.

-5-4-3-2-1 0 1 2 3 4 5 6 7 8 9 10 11 12 13 14 15

Example 2 Solve an Absolute Value Equation When $n < 0$

Solve $|3x - 4| = -1$.

$|3x - 4| = -1$ means that the distance between $3x$ and 4 is -1. Since distance cannot be negative, the solution is the empty set ∅. The solution set is ∅.

Check

Which statement must be true for the solution of $|ax + b| = c$ to be ∅?

A. If a is negative, the solution will be ∅.

B. If b is negative, the solution will be ∅.

C. If c is negative, the solution will be ∅.

D. If c is positive, the solution will be ∅.

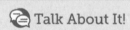 **Go Online** You can complete an Extra Example online.

Study Tip

Scale Keep in mind the values that you will need for graphing. If the values are very large, use a number line and scale that are reasonable for the situation.

Talk About It!

Would the solution change if the equation were changed to $|ax - 4| = -n$, where $n > 0$? Explain your reasoning.

Example 3 Solve an Absolute Value Equation

MUSIC Depending on the size of each song, Luna's phone holds an average of 2000 songs, give or take 250 songs. Write and solve an equation involving absolute value to find the maximum and minimum number of songs that Luna's phone can hold.

The maximum and minimum number of songs will differ from the average by 250 songs. Complete the table to write an equation that represents the maximum and minimum number of songs.

Words	The difference between the number of songs and 2000 is 250.
Variable	Let x = the number of songs on Luna's phone.
Equation	$|x - 2000| = 250$

Case 1

$x - 2000 = 250$ Original equation

$+ 2000 \quad + 2000$ Add 2000.

$x = 2250$ Simplify.

Case 2

$x - 2000 = -250$

$+ 2000 \quad + 2000$

$x = 1750$

The solution set is {1750, 2250}. The maximum and minimum number of songs are 2250 and 1750, respectively.

Think About It!

Does the solution set mean that Luna's phone can hold only 1750 or 2250 songs? If not, how many songs can it hold?

Check

SKYDIVING It takes approximately 6 minutes for a skydiver to land after she jumps out of a plane, give or take 30 seconds. What is the range of time, in seconds, it could take the skydiver to land?

[__?__ , __?__]

Go Online You can complete an Extra Example online.

Study Tip

Find the Midpoint
To find the point midway between two points, add the values together and divide by 2. For the example, $17 + 27 = 44$, and $44 \div 2 = 22$. So 22 is the point halfway between 17 and 27.

Think About It!
Write a general rule that can be used for exercises similar to Example 4.

Example 4 Write an Absolute Value Equation

Write an equation involving absolute value for the graph.

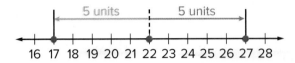

Find the point that is the same distance from 17 and from 27 on the number line. This is the midpoint between 17 and 27, which is 22.

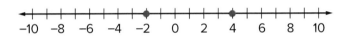

So an equation is $|x - 22| = 5$.

Check

Label each graph with the correct equation.

$|x + 3| = 6 \qquad |x - 1| = 3 \qquad |x - 1| = 6 \qquad |x - 3| = 5$

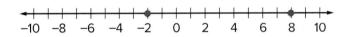

Go Online You can complete an Extra Example online.

Practice

Go Online You can complete your homework online.

Examples 1 and 2
Solve each equation. Then graph the solution set.

1. $|n - 3| = 5$

2. $|f + 10| = 1$

3. $|v - 2| = -5$

4. $|4t - 8| = 20$

5. $|8w + 5| = 21$

6. $|6y - 7| = -1$

7. $|x + 5| = -3$

8. $|-2y + 6| = 6$

9. $\left|\frac{3}{4}a - 3\right| = 9$

10. $|2x - 3| = 7$

Solve each equation.

11. $|7 - 2q| = 3$

12. $|4x - 2| = 26$

13. $|w + 1| = 5$

14. $|n + 2| = -1$

15. $|m - 2| = 2$

16. $|5c - 3| = 1$

17. $|2t + 6| = 4$

18. $|8k - 5| = -4$

Lesson 2-5 • Solving Equations Involving Absolute Value 105

Example 3

19. ENGINEERING *Tolerance* is an allowance made for imperfections in a manufactured object. The manufacturer of an oven specifies a temperature tolerance of ±15°F. This means that the temperature inside the oven will be within 15°F of the temperature to which it is set. Write and solve an absolute value equation to find the maximum and minimum temperatures inside the oven when the thermostat is set to 400°F.

20. POLLS Candidate A and Candidate B are running for mayor. A poll was taken to determine which candidate would likely win the election. The poll is accurate within ±5%. Write and solve an absolute value equation to find the maximum and minimum percent of voters who will vote for Candidate A if 38% of the voters in the poll voted for Candidate A.

21. STATISTICS The most familiar statistical measure is the arithmetic mean, or average. A second important statistical measure is the standard deviation, which is a measure of how far the data are from the mean. For example, the mean score on the Wechsler IQ test is 100 and the standard deviation is 15. This means that people within one standard deviation of the mean have IQ scores that are 15 points higher or lower than the mean.

    a. One year, the mean mathematics score on the ACT test was 20.9 with a standard deviation of 5.3. Write an absolute value equation to find the maximum and minimum scores within one standard deviation of the mean.

    b. What is the range of ACT mathematics scores within one standard deviation of the mean? within two standard deviations of the mean?

22. AVIATION The graph shows the results of a survey that asked 4300 students ages 7 to 18 what they thought would be the most important benefit of air travel in the future. There are about 40 million students in the United States. If the margin of error is ±3%, what is the range of the number of students ages 7 to 18 who would likely say that "finding new resources for Earth" is the most important benefit of future flight?

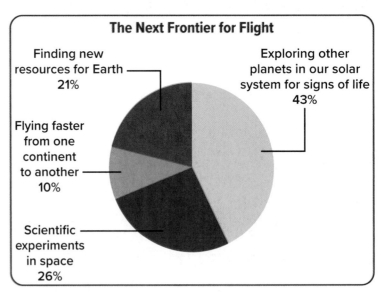

Source: *The World Almanac*

23. MANUFACTURING A hardware store sells bags of rock salt that are labeled as weighing 35 pounds. The equipment used to package the salt produces bags with a weight that is within 8 ounces of the label weight. Write and solve an absolute value equation to determine the maximum and minimum weights for the bag of rock salt. Justify each step in the solution.

Example 4
Write an equation involving absolute value for each graph.

24. (number line with points at -1 and 4, from -5 to 5)

25. (number line with points at -4 and 6, from -10 to 10)

26. (number line with points at -3 and 3, from -5 to 5)

27. (number line with points at -6 and 1, from -7 to 3)

28. (number line with points at -1 and 1, from -5 to 5)

29. (number line with points at -6 and -2, from -7 to 3)

30. (number line with points at 2 and 4, from -5 to 5)

31. (number line with points at -4 and 3, from -5 to 5)

Mixed Exercises

Solve each equation. Then graph the solution set.

32. $\left|-\frac{1}{2}b - 2\right| = 10$

33. $|-4d + 6| = 12$

34. $|5f - 3| = 12$

35. $2|h| - 3 = 8$

36. $4 - 3|q| = 10$

37. $\frac{4}{|p|} + 12 = 14$

Write an equation involving absolute value for each graph.

38. (number line with points at -4 and 2, from -5 to 5)

39. (number line with points at 1 and 3, from -5 to 5)

40. (number line with points at -4 and 2, from -5 to 5)

41. (number line with points at 0.5 and 1.0, from -2.0 to 2.0)

42. (number line with points at 1 and 2, from -2 to 2)

43. (number line with points at -2 and 1, from -3 to 3)

44. **REGULARITY** For tropical fish, aquarium water should be set to 76°F with an allowance of 4°.

   a. Explain how to write an absolute value equation to represent this situation.

   b. Explain the steps to solve the absolute value equation. What do the solutions represent?

45. REASONING The temperature of a refrigerator is 38°F give or take 2°.

    a. Write an equation to find the maximum and minimum temperatures of the refrigerator.

    b. Solve the equation to find the maximum and minimum temperatures of the refrigerator.

46. STRUCTURE A quality control inspector at a bolt factory examines random bolts that come off the assembly line. All bolts being made must be a tolerance of 0.04 mm. The inspector is examining bolts that are to have a diameter of 6.5 mm. Write and solve an absolute value equation to find the maximum and minimum diameters of bolts that will pass his inspection.

47. SWIMMING POOL Chlorine is added to a swimming pool to sanitize the water and make it safe for swimming. The chlorine should be in the range of 2–4 ppm (parts per million). Write an equation that represents the maximum and minimum chlorine concentration.

48. FISH TANK Tom has a 10 gallon fish tank that he wants to fill with neon tetra fish. Tom calculates the number of fish that will fit in the tank using three different methods. Write an equation to represent the maximum and minimum number of fish that will fit in the tank.

Method	Number of Fish
Method 1	5
Method 2	9
Method 3	8

49. PERSEVERE If three points a, b, and c lie on the same line, then b is between a and c if and only if the distance from a to c is equal to the sum of the distances from a to b and from b to c. Write an absolute value equation to represent the definition of betweenness.

50. ANALYZE Translate the sentence $x = 5 \pm 2.3$ into an equation involving absolute value. Explain.

51. FIND THE ERROR Chris and Cami are solving $|x + 3| = -6$. Is either of them correct? Explain your reasoning.

Chris
$|x + 3| = 6$ or $|x + 3| = -6$
$x + 3 = 6$ or $x + 3 = -6$
$x = 3$ or $x = -9$

Cami
$|x + 3| = -6$
The solution is $\varnothing$.

52. WRITE Explain why an absolute value can never be negative.

53. CREATE Describe a real-world situation that could be represented by the absolute value equation $|x - 4| = 10$.

Lesson 2-6

Solving Proportions

Explore Comparing Two Quantities

Online Activity Use graphing technology to complete the Explore.

> **INQUIRY** How can you solve for an unknown value if two quantities have a proportional relationship?

Today's Goal
- Solve proportions.

Today's Vocabulary
proportion

Learn Solving Proportions

A **proportion** is an equation stating that two ratios are equivalent.

Example 1 Solve a Proportion

Solve the proportion. If necessary, round to the nearest hundredth.

$\frac{x}{45} = \frac{15}{25}$

$\frac{x}{45} = \frac{15}{25}$ Original proportion

$45\left(\frac{x}{45}\right) = 45\left(\frac{15}{25}\right)$ Multiply each side by 45.

$x = \frac{45(15)}{25}$ Simplify.

$x = \frac{675}{25}$ Multiply.

$x = 27$ Divide.

CHECK

Check your solution by substituting into the original proportion and check to see if the fractions are equal.

$\frac{x}{45} = \frac{15}{25}$ Original proportion

$\frac{27}{45} \stackrel{?}{=} \frac{15}{25}$ Substitute.

$\frac{3}{5} = \frac{3}{5}$ True.

💭 **Think About It!**
What is another equation that you could write to solve the proportion? Explain your reasoning.

Go Online You can complete an Extra Example online.

Check

Solve $\frac{n-4}{8} = \frac{3}{2}$. If necessary, round to the nearest hundredth.

A. 16

B. 9.33

C. 8

D. 4.75

Example 2 Solve a Proportion with Two Missing Quantities

Solve $\frac{x}{9} = \frac{2x-3}{24}$. If necessary, round to the nearest tenth.

$\frac{x}{9} = \frac{2x-3}{24}$	Original proportion
$9\left(\frac{x}{9}\right) = 9\left(\frac{2x-3}{24}\right)$	Multiply each side by 9.
$x = \frac{9(2x-3)}{24}$	Simplify.
$x = \frac{18x-27}{24}$	Distributive Property.
$24x = 24\left(\frac{18x-27}{24}\right)$	Multiply each side by 24.
$24x = 18x - 27$	Simplify.
$24x - 18x = 18x - 18x - 27$	Subtract 18x from each side.
$6x = -27$	Simplify.
$\frac{6x}{6} = \frac{-27}{6}$	Divide each side by 6.
$x = -4.5$	Simplify.

💭 **Think About It!**
How would the problem differ if the second ratio were 24 over 2x − 3?

Check

Solve $\frac{x}{12} = \frac{2x-5}{18}$. If necessary, round to the nearest hundredth.

A. −10

B. −3.75

C. 0.83

D. 10

Go Online You can complete an Extra Example online.

Example 3 Solve a Proportion by Using a Constant Rate

GEOGRAPHY Parts of Mexico City are sinking at a rate of 140 centimeters every 5 years. If this rate remains constant, how many centimeters will the city sink in the next 12 years?

Step 1 Estimate the solution.

In 10 years, Mexico City will sink 140(2) or 280 centimeters. Because 10 years is slightly less than 12 years, Mexico City will sink more than 280 centimeters in 12 years.

Step 2 Write a proportion.

Let c represent the number of centimeters.

$$\frac{\text{city sinks 140 cm}}{\text{in 5 years}} = \frac{\text{city sinks } c \text{ cm}}{\text{in 12 years}}$$

Step 3 Solve the proportion.

$\frac{140}{5} = \frac{c}{12}$ Original proportion

$12\left(\frac{140}{5}\right) = 12\left(\frac{c}{12}\right)$ Multiply each side by 12.

$\frac{12(140)}{5} = c$ Simplify.

$\frac{1680}{5} = c$ Simplify.

$336 = c$ Divide.

CHECK

How do you know your solution is reasonable?

Sample answer: A rate of 140 centimeters every 5 years is a unit rate of 28 centimeters per year. In 12 years Mexico City would sink 12 years $\cdot \frac{28 \text{ centimeters}}{1 \text{ year}}$ or 336 centimeters. This makes sense with our estimate of more than 280 centimeters.

Talk About It!

Would you really expect the rate of sinking to remain constant over the entire time period? Explain.

Check

MIXTURE Oscar makes fruit punch to sell from his food truck by mixing 8 parts cranberry juice to 3 parts pineapple juice. How many cups of pineapple juice would Oscar need to mix with 48 cups of cranberry juice to make his punch? ___?___ cups

🌐 **Go Online** You can complete an Extra Example online.

Lesson 2-6 • Solving Proportions **111**

Study Tip

Setting Up Ratios
It is a good idea to write the ratio in words to start the problem. Then read the problem to find the numbers or expressions to write each of the two ratios in the proportion.

Think About It!

After multiplying each side by $r + 14$ in Method 1, Raja's resulting equation was $r = \frac{3}{10}r + 14$.

What error did Raja make?

Example 4 Solve a Percent Problem by Using a Proportion

MIXTURES A guide company makes a trail mix of raisins and mixed nuts. How many pounds of raisins does the guide company need to mix with 14 pounds of mixed nuts to make the trail mix 30% raisins?

METHOD 1 : raisins : trail mix

Let r represent the number of pounds of raisins.
Let $r + 14$ represent the number of pounds of the trail mix.
Write and solve a proportion.

$\frac{\text{raisins}}{\text{trail mix}} = \frac{30}{100}$ 30% of the trail mix is raisins.

$\frac{r}{r + 14} = \frac{30}{100}$ Substitute $r + 14$ for the amount of the trail mix.

$(r + 14)\left(\frac{r}{r + 14}\right) = (r + 14)\frac{30}{100}$ Multiply each side by $r + 14$.

$r = (r + 14)\frac{3}{10}$ Simplify.

$r = \frac{3}{10}r + \frac{14 \cdot 3}{10}$ Distributive Property

$r - \frac{3}{10}r = \frac{3}{10}r - \frac{3}{10}r + \frac{42}{10}$ Subtract $\frac{3}{10}r$ from each side.

$\frac{7}{10}r = \frac{42}{10}$ Simplify.

$\frac{10}{7}\left(\frac{7}{10}r\right) = \frac{10}{7}\left(\frac{42}{10}\right)$ Multiply each side by $\frac{10}{7}$.

$r = \frac{42}{7}$ or 6   Simplify.

METHOD 2 : raisins : mixed nuts

Let r represent the number of pounds of raisins, when the number of pounds of mixed nuts is 14.
Write and solve a proportion.

$\frac{\text{raisins}}{\text{mixed nuts}} = \frac{30}{70}$ The ratio of raisins to mixed nuts is 30 : 70.

$\frac{r}{14} = \frac{30}{70}$ Substitute 14 for the amount of mixed nuts.

$14\left(\frac{r}{14}\right) = 14\left(\frac{30}{70}\right)$ Multiply each side by 14.

$r = \frac{14 \cdot 30}{70}$ Simplify.

$r = \frac{420}{70}$ or 6   Simplify.

Check

MIXTURE Ayita is making a plant food mixture to use in her garden. The mixture is to be 20% plant food and 80% water. She needs to make 12 gallons of the mixture to cover her entire garden. Which proportions can be used to find the amount of plant food p she will need? Select all that apply.

A. $\frac{20}{80} = \frac{p}{12}$ B. $\frac{20}{100} = \frac{p}{12}$ C. $\frac{80}{100} = \frac{p}{12}$

D. $\frac{12 - p}{12} = \frac{80}{100}$ E. $\frac{20}{p} = \frac{80}{12}$

Go Online You can complete an Extra Example online.

Practice

Go Online You can complete your homework online.

Example 1

Solve each proportion. If necessary, round to the nearest hundredth.

1. $\dfrac{3}{8} = \dfrac{15}{a}$
2. $\dfrac{t}{2} = \dfrac{6}{12}$
3. $\dfrac{4}{9} = \dfrac{13}{q}$

4. $\dfrac{15}{35} = \dfrac{g}{7}$
5. $\dfrac{7}{10} = \dfrac{m}{14}$
6. $\dfrac{8}{13} = \dfrac{v}{21}$

7. $\dfrac{w}{2} = \dfrac{4.5}{6.8}$
8. $\dfrac{1}{0.19} = \dfrac{12}{n}$
9. $\dfrac{2}{0.21} = \dfrac{8}{n}$

10. $\dfrac{2.4}{3.6} = \dfrac{k}{1.8}$
11. $\dfrac{t}{0.3} = \dfrac{1.7}{0.9}$
12. $\dfrac{7}{1.066} = \dfrac{z}{9.65}$

13. $\dfrac{x-3}{5} = \dfrac{6}{10}$
14. $\dfrac{7}{x+9} = \dfrac{21}{36}$
15. $\dfrac{10}{15} = \dfrac{4}{x-5}$

16. $\dfrac{6}{14} = \dfrac{7}{x-3}$
17. $\dfrac{7}{4} = \dfrac{f-4}{8}$
18. $\dfrac{3-y}{4} = \dfrac{1}{9}$

Example 2

Solve each proportion. If necessary, round to the nearest hundredth.

19. $\dfrac{4v+7}{15} = \dfrac{6v+2}{10}$
20. $\dfrac{9h-3}{9} = \dfrac{5h+5}{3}$

21. $\dfrac{2n-4}{5} = \dfrac{3n+3}{10}$
22. $\dfrac{2}{g+6} = \dfrac{4}{5g+10}$

23. $\dfrac{x}{3} = \dfrac{3x+2}{6}$
24. $\dfrac{w+3}{7} = \dfrac{w-1}{8}$

25. $\dfrac{4q-3}{5} = \dfrac{2q+1}{7}$
26. $\dfrac{5}{7k+4} = \dfrac{2}{2k-3}$

27. $\dfrac{m+1}{9} = \dfrac{m+2}{2}$
28. $\dfrac{j-5}{2} = \dfrac{j+8}{7}$

29. $\dfrac{9f+3}{10} = \dfrac{2f-4}{5}$
30. $\dfrac{2c-1}{3} = \dfrac{c+2}{4}$

31. $\dfrac{5n-2}{8} = \dfrac{n+8}{3}$
32. $\dfrac{h-7}{4} = \dfrac{2h+1}{3}$

33. $\dfrac{14}{3y+5} = \dfrac{3}{y}$
34. $\dfrac{p+10}{8} = \dfrac{2p-7}{4}$

35. $\dfrac{7}{14-d} = \dfrac{3}{18+d}$
36. $\dfrac{2z-4}{5} = \dfrac{3z+3}{10}$

Lesson 2-6 • Solving Proportions

Example 3

37. BOATING Dedra's boat used 5 gallons of gasoline in 4 hours. At this rate, how many gallons of gasoline will the boat use in 10 hours?

38. WATER A dripping faucet wastes 3 cups of water every 24 hours. How much water is wasted in a week?

39. PRECISION In November 2010 the average cost of 5 gallons of regular unleaded gasoline in the United States was $14.46. What was the average cost for 16 gallons of gasoline?

40. SHOPPING Stevenson's Market is selling 3 packs of stylus pens for $5.00. How much will 10 packs of stylus pens cost at this price?

41. STATE YOUR ASSUMPTION During basketball practice, Brent made 36 free throws in 3 minutes.

   a. How many free throws will Brent make in 5 minutes?

   b. What assumption did you make in part **a**? Explain.

42. NAILS Human fingernails grow at an average rate of 3.47 millimeters per month. How much will they grow in 20 months?

43. PICTURE Jasmine enlarged the size of a picture to a height of 15 inches. What is the new width of the picture if it was originally 6 inches wide by 4 inches tall?

44. TRAVEL Roscoe is exchanging $121 for Euros for his upcoming trip to Germany. If $2 can be exchanged for 1.78 Euros, how many Euros will Roscoe have?

Example 4

45. FUNDRAISER Owen is organizing a fundraiser. The proceeds will be split between a charity and the expenses from the fundraiser. Owen would like the cost of the fundraiser to be 15% of the proceeds. If the fundraiser will cost $500, how much money do they need to raise at the fundraiser?

46. COFFEE A barista is mixing a house blend of coffee that is 25% light roast. If there are 8 pounds of the light roast available, how much of the blend can the barista make?

47. CHEMISTRY A chemistry teacher needs to mix an acid solution for an experiment. How much hydrochloric acid needs to be mixed with 1500 milliliters of water to make a solution that is 12% acid?

48. LEMONADE Laronda wants to make fresh lemonade. The recipe she finds online recommends that the fresh lemon juice should be 20% of the total volume. She has 18 ounces of fresh lemon juice. How much water should she mix with the lemon juice?

Mixed Exercises

Solve each proportion. If necessary, round to the nearest hundredth.

49. $\dfrac{9}{g} = \dfrac{15}{10}$

50. $\dfrac{3}{a} = \dfrac{1}{6}$

51. $\dfrac{6}{z} = \dfrac{3}{5}$

52. $\dfrac{5}{f} = \dfrac{35}{21}$

53. $\dfrac{12}{7} = \dfrac{36}{m}$

54. $\dfrac{6}{23} = \dfrac{y}{69}$

55. $\dfrac{42}{56} = \dfrac{6}{f}$

56. $\dfrac{7}{b} = \dfrac{1}{9}$

57. $\dfrac{10}{14} = \dfrac{30}{m}$

58. $\dfrac{3}{4} = \dfrac{n}{20}$

59. $\dfrac{6}{4} = \dfrac{x}{18}$

60. $\dfrac{33}{b} = \dfrac{15}{45}$

61. $\dfrac{m-2}{4} = \dfrac{5}{20}$

62. $\dfrac{9}{5} = \dfrac{3}{x+7}$

63. $\dfrac{5}{b} = \dfrac{3}{b-6}$

64. $\dfrac{2p+3}{3} = \dfrac{4p-7}{2}$

65. $\dfrac{3y+4}{5} = \dfrac{y-1}{4}$

66. $\dfrac{2}{w} = \dfrac{7}{w+5}$

67. $\dfrac{7n-2}{6} = \dfrac{3n-2}{4}$

68. $\dfrac{-a-8}{10} = \dfrac{-a+3}{2}$

69. $\dfrac{c+2}{c-2} = \dfrac{4}{8}$

70. **USE A SOURCE** Find the heights of Willis Tower and the John Hancock Center in Chicago including the tip. Suppose you build a scale model of each building. If you make the model of the Willis Tower 3 meters tall, what would be the approximate height of the John Hancock Center model? Round to the nearest hundredth.

71. **USE TOOLS** A map of Waco, Texas and neighboring towns is shown.

    a. Use a metric ruler to measure the distances between Robinson and Neale on the map.

    b. If the scale on the map is 1 cm = 3 mi, find the actual distance between Robinson and Neale.

    c. How many square miles are shown on this map?

72. **BUDGET** Shawnda spent $259.20 on a scooter. This was 80% of her budget for a new scooter. How much was the total budget?

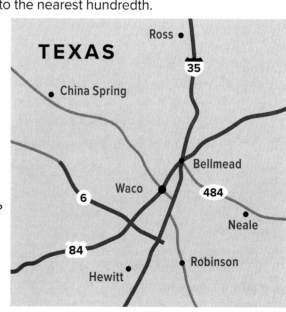

73. **USE TOOLS** On average, 8 potatoes cost $1.50 at a farmer's market. Fernando needs to buy 22 potatoes.

   a. Estimate the cost of 22 potatoes. Explain.

   b. How much will it cost Fernando to buy 22 potatoes? How does this compare to your estimate?

   c. How many potatoes could Fernando buy with $7?

   d. What is the unit cost per potato?

74. **USE A MODEL** Kina and Aiesha started walking from the same location at the same time. Kina walked 6 miles. Aiesha walked 8 miles and walked 1 mile per hour faster than Kina. They each walked for the same amount of time.

   a. Describe how a proportion could be used to find the rate that each person walked.

   b. Solve the proportion. Explain the meaning of the solution.

75. **PERSEVERE** If $\frac{a+1}{b-1} = \frac{5}{1}$ and $\frac{a-1}{b+1} = \frac{1}{1}$, find the value of $\frac{b}{a}$. (Hint: Choose values of a and b for which the proportions are true and evaluate $\frac{b}{a}$.).

76. **CREATE** Describe how a business can use ratios. Include a real-world situation in which a business would use a ratio.

77. **WRITE** Compare and contrast ratios and rates.

78. **PERSEVERE** Find b if $\frac{17}{34} = \frac{a}{32}$ and $\frac{a}{40} = \frac{b}{60}$.

79. **PERSEVERE** A survey showed that $x\%$ of the students at Hoover High school have a job. Write a proportion to find the number of students that have a job, z, if there are y students at Hoover High school.

Lesson 2-7

Using Formulas

Explore Centripetal Force

Online Activity Use a video to complete the Explore.

INQUIRY Why might you want to solve a formula for a specified value?

Learn Solving Equations for Given Variables

A **formula** is an equation that expresses a relationship between certain quantities. A formula or equation that involves more than one variable is called a **literal equation**.

Example 1 Solve for a Specific Variable

Solve $5a - 2b = 15$ for a.

$5a - 2b = 15$	Original equation
$5a - 2b + 2b = 15 + 2b$	Add $2b$ to each side.
$5a = 15 + 2b$	Simplify.
$\dfrac{5a}{5} = \dfrac{15 + 2b}{5}$	Divide each side by 5.
$a = \dfrac{15}{5} + \dfrac{2b}{5}$	Simplify.
$a = 3 + \dfrac{2b}{5}$	Simplify.

Today's Goals
- Solve equations for specific variables.
- Convert units of measure.

Today's Vocabulary
formula
literal equation
dimensional analysis

Think About It!
How do you solve an equation for a specific variable?

Think About It!
Describe how you would solve for b instead of a. How would the answer change?

Go Online You can complete an Extra Example online.

Lesson 2-7 • Using Formulas **117**

Example 2 Solve for a Specific Variable When the Variable Is on Each Side

Solve $4p - 7r = pq + 16$ for p.

$4p - 7r = pq + 16$	Original equation
$4p - 7r + 7r = pq + 16 + 7r$	Add $7r$ to each side.
$4p = pq + 16 + 7r$	Simplify.
$4p - pq = pq + 16 + 7r - pq$	Subtract pq from each side.
$4p - pq = 16 + 7r$	Simplify.
$p(4 - q) = 16 + 7r$	Distributive Property
$\dfrac{p(4-q)}{4-q} = \dfrac{16 + 7r}{4 - q}$	Divide each side by $4 - q$.
$p = \dfrac{16 + 7r}{4 - q}$	Simplify.

> **Talk About It!**
> Would the solution be the same if the variable p were isolated on the right side of the equation instead of the left? Explain your reasoning.

Check

Solve $2v = \dfrac{w + v}{t}$ for v.

$v = \underline{}$

Example 3 Solve Literal Equations for a Given Variable

GEOMETRY The area of a trapezoid is $A = \dfrac{h(b_1 + b_2)}{2}$. A represents the area, h represents the height, b_1 represents the length of one base, and b_2 represents the length of the other base.

Part A

Solve the formula for h.

$A = \dfrac{h(b_1 + b_2)}{2}$	Area Formula
$2A = 2\dfrac{h(b_1 + b_2)}{2}$	Multiply each side by 2.
$2A = h(b_1 + b_2)$	Simplify.
$\dfrac{2A}{b_1 + b_2} = \dfrac{h(b_1 + b_2)}{b_1 + b_2}$	Divide each side by $b_1 + b_2$.
$\dfrac{2A}{b_1 + b_2} = h$	Simplify.

Part B

Find the height of a trapezoid with an area of 70 square feet and bases that are 22 feet and 18 feet.

$h = \dfrac{2A}{(b_1 + b_2)}$ Formula for height

$h = \dfrac{2(70)}{(22 + 18)}$ $A = 70, b_1 = 22, b_2 = 18$

$h = \dfrac{140}{40}$ Simplify.

$h = 3.5$ Divide.

The height of the trapezoid is 3.5 feet.

Check

BUSINESS A Mason jar company wants to increase the volume of its cylindrical jars by 6 cubic inches. The company's designer wants a formula that states the height h of a jar given its volume V and radius r. The volume of a cylindrical jar is modeled by the equation $V = \pi r^2 h$.

Part A
What formula should be used to find the height h?

A. $h = \pi \cdot r^2$

B. $h = \sqrt{\dfrac{V}{\pi r}}$

C. $h = \dfrac{V}{\pi r^2}$

D. $h = \dfrac{V}{r^2}$

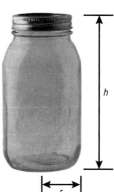

Part B
If the radius of the jar is 1.5 inches and the original volume is 30 cubic inches, then what height should the company make its new jar to increase the volume by 6 cubic inches?

Watch Out!

Dividing by a Quantity
Do not forget to divide each side by the entire quantity $b_1 + b_2$.

Think About It!
How would the height of the trapezoid change if the area were doubled and all other measures remained the same?

Study Tip

Solving for a Specific Variable
When an equation has more than one variable, it can be helpful to highlight the variable for which you are solving on a piece of paper.

Go Online You can complete an Extra Example online.

Example 4 Use Literal Equations

PARTY The total amount of money Kishi spends on pizza for a party is $T = 13.49c + 15.49p$. T represents the total amount of money she spends, c represents the number of cheese pizzas she buys, and p represents the number of pepperoni pizzas she buys.

Part A
If Kishi has $85 to spend on pizza, describe the constraints on $T = 13.49c + 15.49p$.

- The maximum number of cheese pizzas Kishi can buy is 6, because she can buy 6 cheese pizzas and no pepperoni pizzas without exceeding $85.
- The maximum number of pepperoni pizzas Kishi can buy is 5, because she can buy 5 pepperoni and no cheese pizzas without going over her budget.
- The minimum number of each type of pizza she can buy is 0, because you cannot buy a negative number of pizzas.

Part B
Solve $T = 13.49c + 15.49p$ for c.

$T = 13.49c + 15.49p$	Original equation
$T - 15.49p = 13.49c + 15.49p - 15.49p$	Subtract 15.49p.
$T - 15.49p = 13.49c$	Simplify.
$\dfrac{T - 15.49p}{13.49} = \dfrac{13.49c}{13.49}$	Divide by 13.49.
$\dfrac{T - 15.49p}{13.49} = c$	Simplify.

> 🧠 **Think About It!**
> Why did you round your answer in Part C to 2 instead of 3?

Part C
If Kishi has $85 to spend on pizza and she needs to buy 3 pepperoni pizzas, find the maximum number of cheese pizzas she can buy.

$c = \dfrac{T - 15.49p}{13.49}$	Original equation solved for c
$c = \dfrac{85 - 15.49(3)}{13.49}$	$T = 85, p = 3$
$c = \dfrac{38.53}{13.49}$	Simplify.
$c \approx 2.86$	Divide.

Kishi can buy a maximum of 2 cheese pizzas.

Go Online You can complete an Extra Example online.

Check

VIDEO GAMES Ella makes a video game that becomes very popular. She creates a formula, $P = 40c - 300$, to model her profit P given the number of copies sold c, taking into account the $300 fee that she has paid a retailer to sell her game. Which equation would model how many copies c must be sold to yield a specific amount of profit?

A. $c = \frac{P + 40}{300}$

B. $c = \frac{P}{40} + 300$

C. $c = \frac{P}{40} - 300$

D. $c = \frac{P + 300}{40}$

Think About It!
Why might you want to convert units?

Explore Using Dimensional Analysis

Online Activity Use a real-world situation to complete the Explore.

> **INQUIRY** Why might you want to convert the units for a given quantity or measurement?

Learn Dimensional Analysis

When using formulas, you may want to use dimensional analysis. **Dimensional analysis** or **unit analysis** is the process of performing operations with units.

As you plan your solution method, think about

- what units were given,
- what units you need for the solution, and
- the step(s) you need to take to convert your units from what you are given to what you will need for the solution.

Example 5 Multiply by a Conversion Factor

POOLS Mark is purchasing an above-ground swimming pool. The salesperson says that the pool will hold 97,285 liters of water. If **1 gallon = 3.785 liters**, determine approximately how many gallons of water Mark's pool will hold.

Two ratios can be used to compare liters and gallons, $\frac{1 \text{ gallon}}{3.785 \text{ liters}}$ and $\frac{3.785 \text{ liters}}{1 \text{ gallon}}$. To convert from liters to gallons, multiply by $\frac{1 \text{ gallon}}{3.785 \text{ liters}}$.

Number of liters the pool will hold × gallons to liters

$$97{,}285 \text{ liters} \times \frac{1 \text{ gallon}}{3.785 \text{ liters}}$$

$$97{,}285 \text{ liters} \times \frac{1 \text{ gallon}}{3.785 \text{ liters}} \approx 25{,}703 \text{ gallons}$$

Mark's pool will hold approximately 25,703 gallons of water.

Study Tip

Precision Notice that the question asks for an estimate, not an exact answer.

Check

COOKING For the chefs to prepare the dishes for the next hour, Adelina needs to provide them with 96 cloves of garlic. However, she can only purchase bags of whole heads of garlic. If each bag of garlic contains 3 heads and each head has about 8 cloves, then how many bags of garlic should she purchase?

Part A
What assumption must Adelina make when calculating how many bags of garlic to buy?

A. Each head of garlic has exactly 8 cloves.

B. The chefs need 96 heads of garlic.

C. The chefs need 96 cloves of garlic.

D. Each bag has 3 heads of garlic.

Part B
How many bags of garlic should Adelina purchase?

A. 4 B. 12

C. 32 D. 96

Avoid a Common Error
Remember that a unit will only cancel when you divide it by itself. What error does this solution make?

97,285 liters × $\frac{3.785 \text{ liters}}{1 \text{ gallon}}$

≈ 368,224 gallons

Think About It!
When you are converting, how do you know which unit goes in the numerator and which unit goes in the denominator?

🌐 Example 6 Use Dimensional Analysis to Convert Units

COOKING A recipe calls for 20 fluid ounces of milk. If Nita buys a half gallon of milk, how many batches of that recipe can she make?
(Hint: 8 fluid ounces = 1 cup)

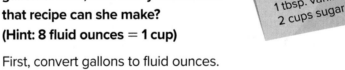

RECIPE
BREAD PUDDING
1.5 tbsp. butter
12 oz. bread
20 fl. oz. milk
3 eggs
1 tbsp. vanilla
2 cups sugar

First, convert gallons to fluid ounces.

total amount × gallons to × quarts to × pint to × cups to
of milk   quarts   pints   cups   fluid ounces

0.5 gallon × $\frac{4 \text{ quarts}}{1 \text{ gallon}}$ × $\frac{2 \text{ pints}}{1 \text{ quart}}$ × $\frac{2 \text{ cups}}{1 \text{ pint}}$ × $\frac{8 \text{ fl. oz.}}{1 \text{ cup}}$ = 64 fl. oz.

Use the following conversion factors to change gallons to ounces.

1 gallon = 4 quarts   1 quart = 2 pints
1 pint = 2 cups   1 cup = 8 fluid ounces

Nita has 64 ounces of milk. Each batch calls for 20 fluid ounces, so to find the number of batches she can make, divide by 20 fluid ounces.

64 fl. oz. × $\frac{1 \text{ batch}}{20 \text{ fl. oz.}}$ = 3.2 batches

Nita has enough milk to make 3 batches with some milk left over.

Go Online You can complete an Extra Example online.

Check

AGRICULTURE On average, a dairy cow produces 832 ounces of milk a day. About how many gallons of milk does a dairy cow produce each year? (Hint: 1 cup = 8 ounces, 1 quart = 4 cups, and 1 gallon = 4 quarts)

A. 6.5 gallons per year

B. 2372.5 gallons per year

C. 4357.2 gallons per year

D. 303,680 gallons per year

Example 7 Use Dimensional Analysis to Convert Rates

SPEED In a novel, the main character, Aiko, can run long distances at 16.5 *kanejaku* per second. Carla knows that the Olympic record for running a marathon distance of 26.2 miles is about 126.5 minutes. She wonders if Aiko could beat that record. If 1 kanejaku = $\frac{10}{33}$ meters, find how far Aiko could run, in miles, in that amount of time. (Hint: 1 mile ≈ 1609.344 meters)

Use the formula $d = rt$ that relates distance d, rate r, and time t to find the distance Aiko could run in 126.5 minutes.

$d = rt$ Distance equation

$d = \left(\frac{16.5 \text{ kanejaku}}{1 \text{ second}}\right) \cdot t$ Substitute Aiko's rate.

In order to compare Aiko to the Olympic runner, convert Aiko's rate in *kanejaku* to miles per minute.

Step 1 Convert distance.

You want distance in miles, but Aiko's distance is in *kanejaku*. Use the given conversion rates that relate to distance to convert Aiko's rate in *kanejaku* per second to miles per second.

$$\frac{16.5 \text{ kanejaku}}{1 \text{ second}} \times \frac{\frac{10}{33} \text{ meters}}{1 \text{ kanejaku}} \times \frac{1 \text{ mile}}{1609.344 \text{ meters}} = \frac{5 \text{ miles}}{1609.344 \text{ seconds}}$$

Step 2 Convert time.

You want time in minutes, but Aiko's time is in seconds. Use the resulting rate from step 2 to convert seconds to minutes.

$$\frac{5 \text{ miles}}{1609.344 \text{ seconds}} \times \frac{60 \text{ seconds}}{1 \text{ minute}} = \frac{300 \text{ miles}}{1609.344 \text{ minutes}}$$

(*continued on the next page*)

> **Problem - Solving Tip**
>
> **Make a Plan**
> Before you solve a problem, think about what the question is asking and what information will apply to the solution.

> **Watch Out!**
>
> **Canceling Units**
> Do not forget to cancel your units as you multiply so that you can see what units are left. The units that are left are the units of your final answer.

Go Online You can complete an Extra Example online.

Step 3 Substitute.

Substitute Aiko's rate in miles per minute and the given time into the formula and simplify.

$$d = \frac{300 \text{ miles}}{1609.344 \text{ minutes}} \times 126.5 \text{ minutes} \approx 23.6 \text{ miles}$$

Aiko would run approximately 23.6 miles in the time it took the Olympic runner to complete 26.2 miles. So, Aiko would not beat the Olympic marathon record time.

> **Go Online** An alternate method is available for this example.

Practice

Examples 1

Solve each equation or formula for the variable indicated.

1. $x - 2y = 1$, for y

2. $d + 3n = 1$, for n

3. $7f + g = 5$, for f

4. $3c - 8d = 12$, for c

5. $7t = x$, for t

6. $r = wp$, for p

7. $q - r = r$, for r

8. $4m - t = m$, for m

9. $7a - b = 15a$, for a

10. $-5c + d = 2c$, for c

Go Online You can complete an Extra Example online.

Example 2

Solve each equation or formula for the variable indicated.

11. $u = vw + z$, for v

12. $x = b - cd$, for c

13. $fg - 9h = 10j$, for g

14. $10m - p = -n$, for m

15. $r = \frac{2}{3}t + v$, for t

16. $\frac{5}{9}v + w = z$, for v

17. $\frac{10ac - x}{11} = -3$, for a

18. $\frac{df + 10}{6} = g$, for f

Example 3

19. RECTANGLES The formula $P = 2\ell + 2w$ represents the perimeter of a rectangle. In this formula, ℓ is the length of the rectangle and w is the width.

  a. Solve the formula for ℓ.

  b. Find the length when the width is 4 meters and the perimeter is 36 meters.

20. BASEBALL The formula $a = \frac{h}{b}$ can be used to find the batting average a of a batter who has h hits in b times at bat.

  a. Solve the formula for b.

  b. If a batter has a batting average of 0.325 and has 39 hits, how many times has the player been at bat?

21. SHOPPING Thomas went to the store to buy videogames for $13.50 each and controllers. The total amount Thomas spent can be represented by $c = 13.50g + p$, where c is the total cost, g is the number of games he bought, and p is the cost of the controllers. The controllers cost $55 and Thomas spent $136 total.

  a. Solve the equation for g.

  b. Find how many games Thomas bought.

22. GEOMETRY The volume of a box V is given by the formula $V = \ell wh$, where ℓ is the length, w is the width, and h is the height.

  a. Solve the formula for h.

  b. What is the height of a box with a volume of 50 cubic meters, length of 10 meters, and width of 2 meters?

Example 4

23. COFFEE SHOP Consuelo is buying flavored coffee and plain coffee. The total amount of money she spends on coffee is $T = 5.50p + 7f$, where p represents the cost of a package of plain coffee and f represents the cost of a package of flavored coffee.

  a. If Consuelo has $40 to spend on coffee, describe the constraints on the formula.

  b. Solve for f.

  c. If Consuelo needs to buy 3 packages of plain coffee, what is the maximum number of packages of flavored coffee she can buy?

24. SHOPPING Kimberly is ordering bath towels and washcloths for the inn where she works. The total cost of the order is $T = 6b + 2w$, where T is the total cost, b is the number of bath towels, and w is the number of washcloths.

  a. Her budget is $85. Describe the constraints.

  b. Solve for b.

  c. She needs to order at least 20 washcloths. How many bath towels can she order and stay under budget?

Examples 5–7

25. ENVIRONMENT The United States released 5.877 billion metric tons of carbon dioxide into the environment through the burning of fossil fuels in a recent year. If 1 trillion pounds = 0.4536 billion metric tons, how many trillion pounds of carbon dioxide did the United States release in that year?

26. EUROS Trent purchases 44 euros worth of souvenirs while on vacation in France. If $1 U.S. = 0.678 euros, find the cost of the souvenirs in United States dollars.

27. LENGTH A pencil is 13.5 centimeters long. If 1 centimeter = 0.39 inch, what is the length of the pencil in feet, to the nearest hundredth?

28. TRACK If a track is 400 meters around, how many laps around the track would it take to run 3.1 miles? Round to the nearest tenth. (*Hint:* 1 foot = 0.3048 meter)

29. BIKING Imelda rode her bicycle 39 kilometers. If 1 meter = 1.094 yards, find the distance Imelda rode her bicycle to the nearest mile. (*Hint:* 1 mi = 1760 yd)

30. MANUFACTURING Aluminum, Inc. produces cans at a rate of 0.04 per hundredth of a second. How many cans can be produced in a 7-hour day?

31. WATER USAGE Each minute, 8.8 quarts of water flow from a shower. If the average person spends 8.2 minutes in the shower, how many gallons of water will the average person have used after taking five showers?

32. TRAVEL The swim team is going to finals. If the meet is in 85 days, determine how many seconds there are until the meet.

33. PRECISION The chemistry teacher set out a 5-pound jar of salt at the beginning of the day. If each student needs 27.6 grams of salt for an experiment, how many students can perform the experiment before the jar is empty? (*Hint:* 1 lb = 454 g)

Mixed Exercises

Solve each equation for the variable indicated.

34. $rt - 2n = y$, for t

35. $bc + 3g = 2k$, for c

36. $kn + 4f = 9v$, for n

37. $8c + 6j = 5p$, for c

38. $\frac{x - c}{2} = d$, for x

39. $\frac{x - c}{2} = d$, for c

40. $-14n + q = rt - 4n$, for n

41. $18t + 11v = w - 13t$, for t

42. $ax + z = aw - y$, for a

43. $10c - f = -13 + cd$, for c

44. STRUCTURE Jethro used dimensional analysis to convert from one rate of speed to another. Two of the conversion factors he used are $\frac{5280 \text{ ft}}{1 \text{ mi}}$ and $\frac{1 \text{ hr}}{60 \text{ min}}$. What could be the units of the initial rate of speed and final rate of speed? Justify your answer.

45. REASONING The formula $A = P(1 + r)$ represents the amount of money A in an account after 1 year, where P is the amount initially deposited and r is the interest rate. Note that the interest rate is written as a decimal. Dennis deposits $2150 into a savings account. After 1 year, he has $2182.25 in his account. Solve the equation for r, and determine the interest rate. Show your work.

46. **REASONING** The regular octagon shown is divided into 8 congruent triangles. Each triangle has an area of 21.7 square centimeters. The perimeter of the octagon is 48 centimeters.

    a. What is the length of each side of the octagon?

    b. Solve the formula for the area of a triangle for *h*.

    c. What is the height of each triangle? Round to the nearest tenth.

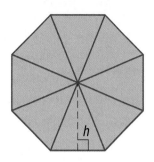

47. Consider the equation $\frac{ry + z}{m} - t = x$.

    a. Solve the equation for *y*.

    b. Would there be any restrictions on the value of each variable? If so, explain the restrictions.

48. **CREATE** Think about the area formula of some geometric figures that involve a fraction.

    a. Write a formula for *A*, the area of a geometric figure that includes a fraction.

    b. Solve the formula for a variable other than *A*.

49. **FIND THE ERROR** The formula represents the relationship between temperatures in degrees Fahrenheit *F* and degrees Celsius *C*. Sasha solves the formula for *C*. Her solution is shown at the right. Is Sasha's solution correct? Explain your reasoning.

    Sasha
    $F = \frac{9}{5}C + 32$
    $\frac{5}{9}F = C + 32$
    $C = \frac{5}{9}F - 32$

50. **PERSEVERE** The formula $A = P(1 + r)^t$ represents the amount of money *A* in an account after *t* years, where *P* is the amount initially deposited and *r* is the interest rate. Patricia currently has $1839.79 in an account that has an interest rate of 2.5%. She opened the account 8 years ago and has made no additional deposits since then.

    a. Solve the formula for *P* and find the amount of Patricia's initial deposit.

    b. Nia says that for the formula in **part a**, *A* is always greater than *P* when *r* is positive and *t* is a positive integer. Do you agree? Why or why not?

128 Module 2 • Equations in One Variable

Module 2 • Equations in One Variable

Review

 Essential Question

How can writing and solving equations help you solve problems in the real world?

Module Summary

Lessons 2-1 and 2-2

One-Step Equations

- To write an equation, first identify each unknown and assign a variable to it. Then identify the givens and their relationships. Finally, write the sentence as an equation.
- Solving an equation is the process of finding all values of the variable that make the equation true.
- If a number is added to or subtracted from each side of a true equation, the resulting equivalent equation is also true.
- If an equation is true and each side is multiplied or divided by the same nonzero number, the resulting equation is equivalent.

Lessons 2-3 and 2-4

Multi-Step Equations

- To solve a multi-step equation, you can undo each operation using properties of equality. Working backward in the order of operations makes this process simpler. Each step in this process results in equivalent equations.
- To solve an equation with the variable on each side, write an equivalent expression with all of the variable terms on one side and the constants on the other side.
- When a grouping symbol appears in an equation, it must first be removed before continuing to solve the equation.

Lessons 2-5 and 2-6

Absolute Value Equations and Proportions

- When solving equations that involve absolute values, there are two cases to consider.

  Case 1: The expression inside the absolute value symbol is positive or zero.

  Case 2: The expression inside the absolute value symbol is negative.
- A proportion is a statement that two ratios are equivalent.

Lesson 2-7

Formulas

- A formula is an equation that expresses a relationship between certain quantities.
- A formula or equation that involves several variables is called a literal equation. To solve a literal equation, solve for a specific variable.
- Dimensional analysis is the process of performing operations with units.

Study Organizer

 Foldables

Use your Foldable to review the module. Working with a partner can be helpful. Ask for clarification of concepts as needed.

Test Practice

1. **MULTIPLE CHOICE** Which equation represents this sentence? (Lesson 2-1)

   The sum of 5 times a number m and 12 is equal to 27.

   A. $5m + 12 = 27$

   B. $5(m + 12) = 27$

   C. $5 + 12m = 27$

   D. $5(12m) = 27$

2. **OPEN RESPONSE** A concert venue surveyed 680 concert attendees about the concession stand. Of those that visited the concession stand, 527 said the concession stand prices are excessive, and the remaining 44 did not. Write an equation to find the number of attendees a who did not visit the concession stand. (Lesson 2-1)

3. **MULTIPLE CHOICE** Write a verbal sentence for the algebraic equation $5x^2 + 2 = 22$. (Lesson 2-1)

   A. 5 times x squared less 2 is 22.

   B. Five plus x squared plus 2 is 22.

   C. The product of 5 times x squared and 2 is 22.

   D. Five times x squared plus 2 is 22.

4. **MULTIPLE CHOICE** Solve $n - 8 = 5$. (Lesson 2-2)

   A. $n = -13$

   B. $n = -3$

   C. $n = 3$

   D. $n = 13$

5. **MULTIPLE CHOICE** Solve $\frac{1}{5}t = 10$ for t. (Lesson 2-2)

   A. 2

   B. 5

   C. 15

   D. 50

6. **OPEN RESPONSE** Solve $z + 12 = -3$ for z. Explain. (Lesson 2-2)

7. **MULTIPLE CHOICE** If $3x - 6 = 42$, what is the value of x? (Lesson 2-3)

   A. 8

   B. 12

   C. 16

   D. 20

8. **OPEN RESPONSE** Solve $8 = 11 - 3v$. (Lesson 2-3)

   $v =$ ___?___

9. **MULTI-SELECT** Jaime bought a notebook and a box of pencils for $5.00. The notebook cost $3.00, and there are 10 pencils in a box. The equation $3 + 10p = 5$ can be used to find the cost of one pencil. Select all equations that are equivalent. (Lesson 2-3)

   A. $10p = 8$

   B. $10p = 2$

   C. $p = 0.80$

   D. $p = 0.20$

   E. $10p = 5$

10. MULTIPLE CHOICE Solve the equation $3x + 1 = 4x - 8$ for x. (Lesson 2-4)

A. $x = -9$

B. $x = -1$

C. $x = 1$

D. $x = 9$

11. MULTIPLE CHOICE Solve the equation $-2(x + 4) + 3x = x - 8$. (Lesson 2-4)

A. $x = 2$

B. $x = 4$

C. all real numbers

D. no solution

12. MULTIPLE CHOICE Solve the equation $5(2y + 1) = 4y + 10$. (Lesson 2-4)

A. $y = \frac{5}{6}$

B. $y = \frac{3}{2}$

C. $y = \frac{15}{14}$

D. $y = \frac{5}{2}$

13. OPEN RESPONSE A park has a ginkgo tree, a dogwood tree, and 2 blue spruce trees. The blue spruce trees are 8 years old. The ginkgo tree is 2 years less than three times the age of the dogwood tree. The ginkgo tree is also half the sum of the ages of the dogwood tree and both of the blue spruce trees. Write and solve an equation to find the ages of the ginkgo and dogwood trees. (Lesson 2-4)

14. MULTI-SELECT Select all the values of x that are solutions of $|3x - 6| = 12$. (Lesson 2-5)

A. -18

B. -6

C. -2

D. 2

E. 6

F. 18

15. MULTIPLE CHOICE A thermometer is accurate to $\pm 2°F$. Which absolute value equation can be used to find the greatest and least possible temperatures if the thermometer reading is 17°F? (Lesson 2-5)

A. $|t - 17| = 2$

B. $|t + 17| = 2$

C. $|t - 2| = 17$

D. $|t + 2| = 17$

16. OPEN RESPONSE Solve the absolute value equation $-5|x + 1| + 2 = 12$. If there is no solution, state no solution. (Lesson 2-5)

17. OPEN RESPONSE Solve $\frac{b}{12} = \frac{10}{15}$. (Lesson 2-6)

18. MULTIPLE CHOICE Solve $\frac{3}{x+4} = \frac{2}{x-4}$. (Lesson 2-6)

A. $x = -4$

B. $x = 4$

C. $x = 8$

D. $x = 20$

19. MULTIPLE CHOICE A biologist estimated that 5% of the seagulls in a flock have been banded. There were 22 seagulls that have been banded. Which equation and solution represent the approximate number of seagulls, g, that were in the flock? (Lesson 2-6)

A. $0.005g = 22; g = 4400$

B. $0.05g = 22; g = 440$

C. $22g = 500; g = 22$

D. $\frac{5}{g} = 22; g = 227$

20. MULTIPLE CHOICE On a map of Texas, the distance between Dallas and Houston is 4.8 inches. If 1 inch = 50 miles, what is the distance, in miles, between the two cities? (Lesson 2-6)

A. 96 miles

B. 104 miles

C. 240 miles

D. 250 miles

21. MULTIPLE CHOICE If 1 foot ≈ 0.305 meter, approximately how many feet are in 8 meters? (Lesson 2-6)

A. 2.44

B. 2.62

C. 24.4

D. 26.2

22. OPEN RESPONSE Solve the formula for the circumference of a circle, $C = 2\pi r$, for r. (Lesson 2-7)

23. OPEN RESPONSE The volume of a right pyramid is given by the formula $V = \frac{1}{3}Bh$, where B is the area of the base and h is the height. (Lesson 2-7)

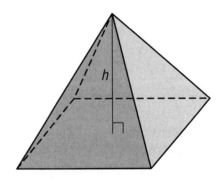

Part A Solve the formula for h.

Part B Find the height, in inches, of a right pyramid with a volume of 900 cubic inches and a base area of 225 square inches.

Module 3
Relations and Functions

Essential Question
Why are representations of relations and functions useful?

What Will You Learn?

How much do you already know about each topic **before** starting this module?

KEY

👎 — I don't know.   👉 — I've heard of it.   👍 — I know it!

	Before			After		
	👎	👉	👍	👎	👉	👍
represent relations using ordered pairs, tables, graphs, and mappings						
analyze graphs of relations						
choose and interpret the scale on a coordinate graph						
determine whether relations are functions						
use function notation						
determine whether a graph is discrete, continuous, or neither						
determine whether a function is linear or nonlinear						
write linear functions in standard form						
find x- and y-intercepts of graphs						
interpret intercepts of graphs of functions						
determine whether a graph has line symmetry						
identify where a graph is increasing and where it is decreasing						
find extrema of a function						
describe the end behavior of a function						
sketch graphs of functions						
solve equations by graphing						

Foldables Make this Foldable to help you organize your notes about expressions. Begin with one sheet of 11" × 17" paper.

1. **Fold** the short sides to meet in the middle.
2. **Fold** the booklet in thirds lengthwise.
3. **Open and cut** the booklet in thirds lengthwise.
4. **Label** the tabs as shown.

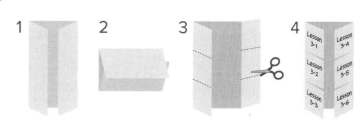

Module 3 • Relations and Functions 133

What Vocabulary Will You Learn?

- continuous function
- decreasing
- dependent variable
- discrete function
- domain
- end behavior
- extrema
- function
- function notation
- increasing
- independent variable
- line symmetry
- linear equation
- linear function
- mapping
- negative
- nonlinear function
- positive
- range
- relation
- relative maximum
- relative minimum
- root
- scale
- x-intercept
- y-intercept
- zero

Are You Ready?

Complete the Quick Review to see if you are ready to start this module. Then complete the Quick Check.

Quick Review

Example 1

Graph and label the point $A(3, 5)$ on the coordinate plane.

Start at the origin. Since the x-coordinate is positive, move 3 units to the right. Then move 5 units up since the y-coordinate is positive. Draw a dot and label it A.

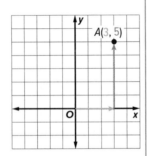

Example 2

Evaluate $5x + 13$ for $x = 4$.

Substitute the known value for x. Then follow the order of operations.

$5x + 13$ Original expression

$= 5(4) + 13$ Substitute 4 for x.

$= 20 + 13$ Multiply.

$= 33$ Add.

Quick Check

Graph and label each point on the coordinate plane.

1. $B(4, 2)$
2. $C(0, 3)$
3. $G(3, 1)$
4. $H(2, 0)$

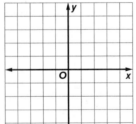

Evaluate each expression for the given value.

5. $3x - 1$ for $x = 7$
6. $\frac{1}{2}x$ for $x = 12$
7. $5x + 2$ for $x = -1$
8. $2x - 9$ for $x = -3$

How Did You Do?

Which exercises did you answer correctly in the Quick Check?

Lesson 3-1

Representing Relations

Learn Relations

You can use math to represent the relationship between two sets of numbers. For example, suppose you recorded the number of minutes you spent driving for your driver's training course each day for a week. You may use the set {1, 2, 3, 4, 5, 6, 7} to represent the days and the set {30, 45, 20, 40, 45, 90, 60} to represent the minutes. The relationship between the sets pairs each day with the time driven that day. This pairing of the numbers is called a **relation**. The set of days is the **domain** of the relation and the set of times is the **range**.

A relation is a set of ordered pairs. The set of the first numbers of the ordered pairs in a relation is called the domain. The set of second numbers of the ordered pairs in a relation is called the range.

A relation can be represented in multiple ways. A **mapping** illustrates the relationship between the domain and range by showing how each element of the domain is paired with an element in the range. An equation shows the relationship between the domain and range of a relation where substituting each value in the domain for x results in the corresponding y-value of the range. The x-values in the domain and the resulting y-values can be written as ordered pairs and are called solutions of the equation because they make the equation true. Below is a mapping, table, graph, and equation of the relation {(3, 5), (−4, −2), (0, 2)}.

Today's Goals
- Represent relations.
- Interpret graphs of relations.
- Choose and interpret appropriate scales for the axes and origins of graphs.

Today's Vocabulary
relation
domain
range
mapping
independent variable
dependent variable
scale

🧠 Think About It!
Compare the table and mapping of the relation. What conclusions can you draw about the x- and y-coordinates and the domain and range?

mapping

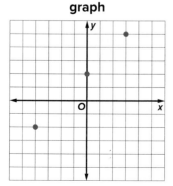

table

x	y
3	5
−4	−2
0	2

graph

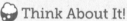

equation

$y = x + 2$

Lesson 3-1 • Representing Relations **135**

Example 1 Representations of a Relation

Use {(2, 0), (0, 4), (3, −5), (−3, −5)}.

Part A Express the relation as a table, a graph, and a mapping.

table

x	y
−3	−5
0	4
2	0
3	−5

graph

mapping

Domain: 2, 0, 3, −3
Range: 0, 4, −5

Part B Determine the domain and the range of the relation.

The domain of the relation is {−3, 0, 2, 3}.

The range of the relation is {−5, 0, 4}.

Check

List the ordered pairs in each relation.

x	y
−1	−5
4	7
−2	3

Domain: 1, 5, −6
Range: 5, −1

Table: _____?_____

Graph: _____?_____

Mapping: _____?_____

Go Online You can complete an Extra Example online.

Learn Analyzing Graphs of Relations

Graphing the total time driven during your driving course can help you visualize the progress.

A relation can be graphed without a scale on either axis to show the relationship between the independent and dependent variables. These graphs can be interpreted by analyzing their shapes.

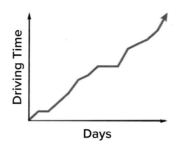

The values in the domain correspond to the independent variable in a relation. The **independent variable**, usually x, has a value that is subject to choice. In the graph above, the independent variable is the days.

The values in the range correspond to the dependent variable of the relation. The **dependent variable** is the variable in a relation, usually y, with values that depend on x. In the graph above, the dependent variable is the driving time.

Example 2 Analyze Graphs

TEXTING The graph represents the number of text messages sent by Nora throughout the day.

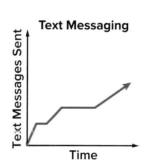

Part A Identify the independent and dependent variables of the relation. independent variable: time dependent variable: number of text messages sent

Part B Describe what happens in the graph.

As you move from left to right along the graph, time increases and the number of text messages sent increases until the graph becomes a horizontal line.

The horizontal line means that time is increasing, but the number of text messages sent remains constant. During this time, Nora stopped sending text messages.

Then she continued to send text messages until she stopped again for a period of time.

Finally, Nora began sending text messages again.

Think About It!
What can you conclude from the graph about the rates at which Nora sent text messages throughout the day?

Check

Identify the independent and dependent variables of each relation.

a. The average price of a ticket to an amusement park has steadily increased over time.

The average price of a ticket is the ___?___ variable.
Time is the ___?___ variable.

b. The air pressure inside a soccer ball decreases with time.

Time is the ___?___ variable.
Air pressure is the ___?___ variable.

Go Online You can complete an Extra Example online.

Lesson 3-1 • Representing Relations **137**

Check

AIRPLANES After an airplane takes off, its altitude increases rapidly. Once it reaches the desired altitude, it continues to fly at that level for a short period of time. Then the plane's altitude fluctuates slightly due to turbulence. Finally, the altitude decreases quickly as the plane lands. Which graph best represents this situation?

A.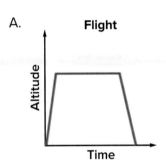

B.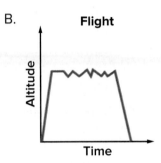

C.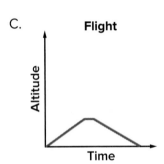

D.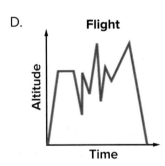

TRAFFIC Kane notices that traffic congestion on his street on weekday mornings depends on the time of day. He draws a graph to represent the relationship between the time of day and the volume of traffic congestion on his street on a weekday morning. Analyze the orange segment. Select the statement that describes what is happening during this time.

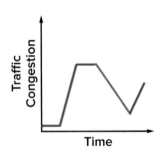

A. Traffic congestion is steadily increasing at a very fast rate.

B. Traffic congestion is constant and very light.

C. Traffic congestion is constant and very heavy.

D. Traffic congestion is decreasing at a steady rate.

Go Online You can complete an Extra Example online.

Explore Choosing Scales

 Online Activity Use a real-world situation to complete the Explore.

> **INQUIRY** How can you tell if an appropriate scale is being used to represent a relationship?

Learn The Coordinate System

When graphing on the coordinate system, the **scale** of a graph refers to the distance, or interval, between tick marks on the x- and y-axes. For example, if one tick mark represents 5 units, then the scale of the graph is 5. Each axis may have a different scale.

A scale of 1 tick mark = 1 unit is frequently used in mathematics. However, using a different scale may make it easier to graph a given set of ordered pairs.

Example 3 Use Appropriate Scales

Graph (5, 80), (−36, 48), (25, −91), (38, 95), (−10, −50), and (1, 22).

Step 1 The x-coordinates are between −36 and 38.
The y-coordinates are between −91 and 95.

Step 2 The x-axis should include values from about −40 to 40.
The y-axis should include values from about −100 to 100.

A scale of 5 on the x-axis is appropriate.
A scale of 10 on the y-axis is appropriate.

Step 3 Graph.

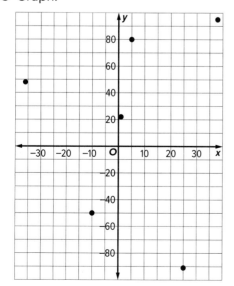

 Go Online You can complete an Extra Example online.

 Talk About It!
Describe a real-world situation in which it might be easier to count by a value other than 1.

Watch Out!
Large Scales Be careful not to choose scales that are too large. If a scale is too large, it can be difficult to accurately graph points.

Think About It!
Gwen used a scale of 4 on the x-axis. Is her scale appropriate? Explain your reasoning.

Lesson 3-1 • Representing Relations 139

Study Tip

Notation Scales of graphs are sometimes written as [−40, 40] scl: 5 or −40 to 40; scale: 5, where −40 and 40 are the minimum and maximum values and 5 is the scale.

Check

When graphing (16, 32), (−10, 11), (4, −27), and (−7, −5), select the most appropriate scale for _____

a. the x-axis
- A. −20 to 20; scale: 1
- B. −12 to 18; scale: 2
- C. −12 to 18; scale: 6
- D. −20 to 20; scale: 10

b. the y-axis
- A. −30 to 35; scale: 1
- B. −30 to 30; scale: 5
- C. −30 to 35; scale: 5
- D. −30 to 40; scale: 10

Example 4 Choose an Appropriate Origin

WEATHER The table shows the total snowfall in January for Boston.

Year	Total Snowfall (inches)
2005	43.30
2006	8.10
2007	1.00
2008	8.30
2009	23.70
2010	13.20
2011	38.30
2012	6.80
2013	5.00
2014	21.80
2015	34.30

Part A Choose an appropriate origin.

Let the x-axis represent the years since 2005 and the y-axis represent the total snowfall. Then the origin (0, 0) represents the year 2005 and 0 inches of snow.

Study Tip

Appropriate Origins When choosing an appropriate origin, you are still using the point (0, 0) as the origin, but are changing what it represents.

Part B Choose an appropriate scale.

The total snowfall is between 1.00 and 43.30 inches, so the y-axis should include values from 0 to 45 and have a scale of 5 inches.

Go Online You can complete an Extra Example online.

Part C Graph the data points on the coordinate plane.

Example 5 Interpret Scales and Origins

SOCIAL MEDIA The average number of posts per day on a social media site each year is given in the table. Interpret the meaning of the axes, scale, and origin of the corresponding graph of the data.

Year	Average Posts Per Day (millions)
2008	15
2009	20
2010	35
2011	50
2012	100
2013	200
2014	340
2015	500
2016	560
2017	625

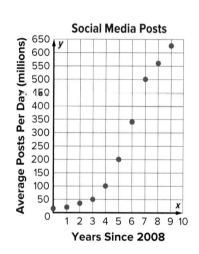

x- and *y*-axes

The *x*-axis represents the number of years since 2008, because it is not appropriate to start at the year 0.

The *y*-axis represents the average number of posts per day (in millions) on the social media site.

Scale

The *x*-axis has a scale of 1 mark = 1 year.
The *y*-axis has a scale of 1 mark = 50 million posts.

Origin

The origin (0, 0) represents the year 2008 and 0 posts per day.

Think About It!
Describe another situation where it might be necessary to choose a different meaning for the axes and origin.

Use a Source
Find data about the number of participants in a sport or other activity of interest to you in recent years. If a graph is provided, interpret the scales of the axes. If no graph is provided, create one with appropriate scales.

Lesson 3-1 • Representing Relations 141

Check

MONEY The United States Mint is responsible for producing and distributing circulating coins that are used by people every day to buy and sell goods. Facilities in Denver and Philadelphia produce billions of coins each year. The table and graph show how many circulating coins were produced each year at these facilities from 2001 to 2011.

Year	Circulating Coins Produced (billions)
2001	19.4
2002	14.4
2003	12
2004	13.2
2005	15.3
2006	15.5
2007	14.4
2008	10.1
2009	3.5
2010	6.4
2011	8.2

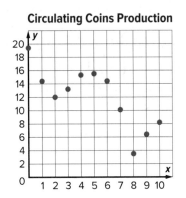

Part A
Interpret the meaning of the x- and y-axes in the context of the situation.

The x-axis represents ____?____ and has a scale of 1 mark = __?__.

The y-axis represents ____?____ and has a scale of 1 mark = __?__.

Part B
Interpret the meaning of the origin in the context of the situation.

The origin (0, 0) represents the year __?__ and __?__ coins produced.

Go Online You can complete an Extra Example online.

Practice

Go Online You can complete your homework online.

Example 1

Express each relation as a table, a graph, and a mapping. Then determine the domain and range.

1. {(−1, −1), (1, 1), (2, 1), (3, 2)}

2. {(0, 4), (− 4, − 4), (−2, 3), (4, 0)}

3. {(3, −2), (1, 0), (−2, 4), (3, 1)}

Example 2

4. **PAYCHECK** The graph represents the amount of Seth's paycheck for different numbers of hours he works.

    a. Identify the independent and dependent variables of the relation.

    b. Describe what happens in the graph.

5. **DEMAND** The graph represents the price of an item and the number of items purchased.

    a. Identify the independent and dependent variables of the relation.

    b. Describe what happens in the graph.

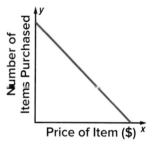

6. **AIRPLANES** The graph represents the number of hours of a flight and the distance an airplane is from the ground.

    a. Identify the independent and dependent variables of the relation.

    b. Describe what happens in the graph.

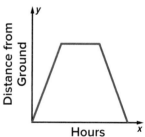

Examples 3 and 4

7. **HEALTH** The American Heart Association recommends that your target heart rate during exercise should be between 50% and 75% of your maximum heart rate. Use the data in the table below to graph the approximate maximum heart rates for people of given ages.
Source: American Heart Association

Age (years)	20	25	30	35	40
Maximum Heart Rate (beats per minute)	200	195	190	185	180

Lesson 3-1 • Representing Relations 143

8. **USE TOOLS** The following ordered pairs give the length in feet and the weight in pounds of five snakes at the reptile house at a zoo. Graph the data. {(5.5, 4.5), (3, 0.5), (3, 2), (8, 4.5), (2, 0.5)}

Example 5

9. **ELEVATOR** The height of an elevator above the ground is given in the table. Interpret the meaning of the axes, scale, and origin of the corresponding graph of the data.

Time (s)	Height (ft)
0	0
1	20
2	40
3	60
4	80

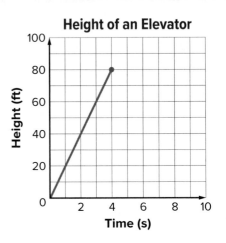

10. **SUPERMARKET** The number of items that eight customers bought at a supermarket and the total cost of the items is given in the table. Interpret the meaning of the axes, scale, and origin of the corresponding graph of the data.

Number of Items	Total Cost ($)
2	2
3	6
5	8
5	10
6	16
8	12
8	18
9	14

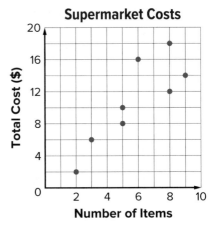

Mixed Exercises

For Exercises 11–14, express the relation in each table, graph, or mapping as a set of ordered pairs.

11.

x	y
1	7
3	45
5	11
13	15

12.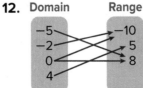

144 Module 3 • Relations and Functions

13.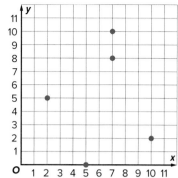

14.
x	y
4	2
7	−7
15	9
12	0

15. **NATURE** Maple syrup is made by collecting sap from sugar maple trees and boiling it down to remove excess water. The graph shows the number of gallons of tree sap required to make different quantities of maple syrup. Express the relation as a set of ordered pairs. Interpret the meaning of the axes, scale, and origin of the corresponding graph of the data.
Source: Vermont Maple Sugar Makers' Association

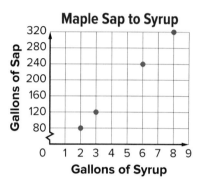

16. **TRAVEL** Omari drives a car that gets 18 miles per gallon of gasoline. The car's gasoline tank holds 15 gallons. The distance Omari drives before refueling is a function of the number of gallons of gasoline in the tank. Identify a reasonable domain for this situation.

17. **DATA COLLECTION** Rafaella collected data to determine the number of books her schoolmates were bringing home each evening. Her data is shown in the mapping. She let x be the number of textbooks brought home after school and y be the number of students with x textbooks.

    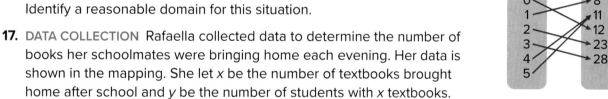

    a. Express the relation as a set of ordered pairs.

    b. What is the domain of the relation?

    c. What is the range of the relation?

18. **COOKIES** Identify the graph that best represents the relationship between the number of cookies and the equivalent number of dozens.

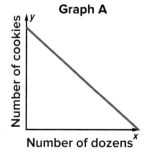
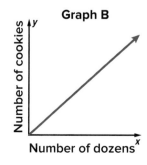
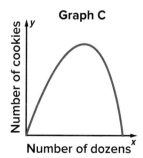

Lesson 3-1 • Representing Relations 145

19. **MOWING** Cordell is mowing his front lawn. His mailbox is on the edge of the lawn. Draw a reasonable graph that shows the distance Cordell is from the mailbox as he mows. Let the horizontal axis show the time and the vertical axis show the distance from the mailbox.

20. **STRUCTURE** Express the relation in the mapping as a set of ordered pairs.

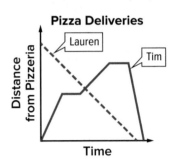

Tim and Lauren use their cars to deliver pizzas. The graph represents their distance from the pizzeria starting at 6 P.M. Use the graph for Exercises 21–24.

21. Describe what happens in Tim's graph.

22. Describe what happens in Lauren's graph.

23. A student said that Tim's and Lauren's graphs intersect, so their cars must have crashed at some time after 6 p.m. Do you agree or disagree? Explain.

24. After 6 p.m., which delivery person was the first to return to the pizzeria? How do you know?

25. **ANALYZE** Cameron said that for any relation, the number of elements in the domain must be greater than or equal to the number of elements in the range. Do you agree? If so, explain why. If not, give a counterexample.

26. **CREATE** Think of a situation that could be modeled by this graph. Then label the axes of the graph and write several sentences describing the situation.

27. **CREATE** Use the set {−1, 0, 1, 2} as a domain and the set {−3, −1, 4, 5} as a range.
    a. Create a relation. Express the relation as a set of ordered pairs.
    b. Express the relation you created in **part a** as a table, a graph, and a mapping.

28. **ANALYZE** Describe a real-life situation where it is reasonable to have a negative number included in the domain or range.

29. **WRITE** Compare and contrast dependent and independent variables.

Lesson 3-2

Functions

Explore Vertical Line Test

 Online Activity Use graphing technology to complete the Explore.

INQUIRY How can you tell whether a relation is a function?

Today's Goals
- Determine whether relations are functions.
- Evaluate functions in function notation for given values.

Today's Vocabulary
function
function notation

Learn Functions

A **function** is a relationship between input and output. In a function, there is exactly one output for each input. The relation shown is a function because each element of the domain is paired with *exactly* one element in the range.

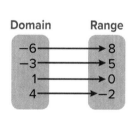
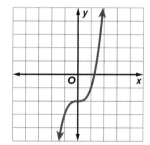

You can use the vertical line test to see if a graph represents a function.

Talk About It!
Describe a way, other than using the vertical line test, that you could use to determine that a relation is not a function.

Key Concept • Vertical Line Test

function	not a function
A relation is a function if it passes the vertical line test, meaning a vertical line intersects the graph no more than once.	A relation is not a function if it fails the vertical line test, meaning that a vertical line intersects the graph more than once.

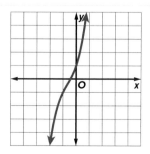
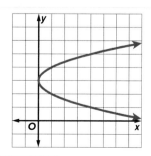

Lesson 3-2 • Functions **147**

Think About It!

Suppose the last element in the domain in part **b** was 4 as shown in the table below. Would this relation be a function? Justify your argument.

Domain	2	6	8	4
Range	3	5	5	−3

Example 1 Identify Functions

Determine whether each relation is a function. Explain.

a.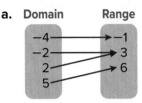

For each element of the domain, there is only one element of the range. So this mapping represents a function.

b.
Domain	2	6	8	2
Range	3	5	5	−3

The element 2 in the domain is paired with both 3 and −3 in the range. This relation is not a function.

c. (2, 5), (4, 7), (8, 11), (4, 13)

The element 4 in the domain is paired with both 7 and 13 in the range. This relation is not a function.

Check

Which of these relations are functions?

I.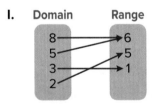

II.
Domain	0	2	4	0	−2
Range	8	6	3	−1	−3

III. {(−4, 5), (−2, 1), (1, −5), (−2, −7), (5, −13)}

A. I only

B. III only

C. I and II

D. I, II, and III

Go Online You can complete an Extra Example online.

Example 2 Analyze Data

LONG JUMP Five schools are competing in the long jump portion of a track meet. The distances of the players with the best jump on each team are as follows: Team 1, 20.6 feet; Team 2, 21.5 feet; Team 3, 20.9 feet; Team 4, 19.4 feet; Team 5, 20.2 feet.

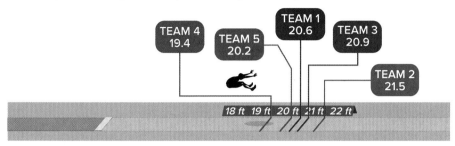

Think About It!
Explain why the domain and range of the function are not all real numbers.

Part A Make a table.

Team Number	1	2	3	4	5
Best Jump (ft)	20.6	21.5	20.9	19.4	20.2

Part B Determine the domain and range of the relation.

Domain: {1, 2, 3, 4, 5}

Range: {20.6, 21.5, 20.9, 19.4, 20.2}

Part C Determine whether the relation is a function.

For each element of the domain, there is only one element of the range. So, this relation is a function.

Check

HEIGHT Bailey recorded the heights of five of her friends. The heights, in inches, are as follows: Hunter, 62; Ling, 66; Ela, 65; Omar, 66; Alma, 67.

Part A Find the domain and range of the relation.

  A. D: {Hunter, 62, Ling, 66, Ela}; R: {65, Omar, 66, Alma, 67}

  B. D: {65, Omar, 66, Alma, 67}; R: {Hunter, 62, Ling, 66, Ela}

  C. D: {Hunter, Ling, Ela, Omar, Alma}; R: {62, 65, 66, 67}

  D. D: {62, 65, 66, 67}; R: {Hunter, Ling, Ela, Omar, Alma}

Part B Is the relation a function? Explain your reasoning.

  A. Yes; for each element of the domain, there is only one element of the range.

  B. No; it fails the vertical line test.

  C. No; an element in the domain is paired with more than one element in the range.

  D. No; the domain and range are all real numbers.

Study Tip

Domain and Range Remember the domain is related to the independent variable, and the range is related to the dependent variable.

Go Online You can complete an Extra Example online.

> **Think About It!**
> Find the domain and range of the function.

> **Study Tip**
> **Vertical Line Test** You can also perform the vertical line test by placing a pencil at the left of the graph and slowly moving it across the graph to represent a vertical line.

Example 3 Equations as Functions

Determine whether $6x + 2y = 14$ is a function. Explain.

For any value *x*, the vertical line passes through no more than one point on the graph. So, the graph and the equation represent a function.

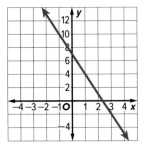

Check

Which relations described represent functions? Copy the chart and select *Yes* or *No*.

A. The relation passes the vertical line test.

B. An element of the domain is paired with only two elements of the range.

C. A vertical line passes through more than one point on the graph.

D. The domain represents each student in class, while the range represents the age of each student.

E. The domain represents the number of a day in May while the range represents the high temperature for the day.

F. The relation is of the form $y = x^2 + b$.

	Yes	No
A		
B		
C		
D		
E		
F		

Learn Function Values

Equations that are functions can be written in a form called **function notation**. Function notation is a way of writing an equation so that $y = f(x)$.

In a function, *x* represents the elements of the domain, and $f(x)$ represents the elements of the range. The graph of $f(x)$ is the graph of the equation $y = f(x)$.

Go Online You can complete an Extra Example online.

Example 4 Find Function Values

For $f(x) = -2x + 9$, find each value.

$f(4) + 3$

$f(4) + 3$	$= [-2(4) + 9] + 3$	$x = 4$
	$= (-8 + 9) + 3$	Multiply.
	$= 1 + 3$	Add.
	$= 4$	Add.

$f(5) - f(1)$

$f(5) - f(1)$	$= [-2(5) + 9] - [-2(1) + 9]$	Substitute for x.
	$= (-10 + 9) - (-2 + 9)$	Multiply.
	$= -1 - 7$	Add.
	$= -8$	Subtract.

Think About It!
Is $f(3) - f(-2)$ the same as $f(3) + f(2)$? Justify your argument.

Check
For $f(x) = -3x + 2$, find each value.

a. $f(-4) = \underline{}$ **b.** $f(5) + 8 = \underline{}$ **c.** $f(3) - f(6) = \underline{}$

Example 5 Evaluate Functions

For $h(x) = -6x^2 + 18x + 36$, find each value.

a. $h(2)$

$h(2)$	$= -6(2)^2 + 18(2) + 36$	$x = 2$
	$= -24 + 36 + 36$	Multiply.
	$= 48$	Add.

b. $h(4) - h(1)$

$h(4) - h(1)$	$= [-6(4)^2 + 18(4) + 36] - [-6(1)^2 + 18(1) + 36]$	
	$= (-96 + 72 + 36) - (-6 + 18 + 36)$	Multiply.
	$= 12 - 48$	Add.
	$= -36$	Subtract.

c. $h(5) - 7$

$h(5) - 7$	$= [-6(5)^2 + 18(5) + 36] - 7$	$x = 5$
	$= (-150 + 90 + 36) - 7$	Multiply.
	$= -24 - 7$	Add.
	$= -31$	Subtract.

Think About It!
Find $h(3)$. Then find $h(-3)$. What was similar and what was different about finding these values?

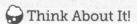 **Go Online** You can complete an Extra Example online.

Check

For $g(x) = 2x^2 - 8x - 10$, find each value.

a. $g(2) = \underline{\ ?\ }$
b. $g(5) + 12 = \underline{\ ?\ }$
c. $g(8) - 18 = \underline{\ ?\ }$
d. $g(-3) - g(4) = \underline{\ ?\ }$

Example 6 Interpret Function Values

EMPLOYMENT Mason works at the movie theater after school. The function $g(x) = 9.25x$ represents the amount of money he makes before taxes for each hour that he works. Evaluate and interpret the function for each value.

a. $g(8)$

$g(x) = 9.25(8) = 74$

This means that if Mason works for 8 hours, he will make $74 before taxes.

b. $g(14)$

$g(x) = 9.25(14) = 129.5$

This means that if Mason works for 14 hours, he will make $129.50 before taxes.

c. $g(27.5)$

$g(x) = 9.25(27.5) = 254.375$

This means that if Mason works for 27.5 hours, he will make $254.38 before taxes.

Think About It!
Find $g(-4)$. What does it mean in the context of the situation?

Watch Out!
Appropriate Domain Remember, it is possible for Mason to work for zero hours and make no money, but it is not possible for him to work a negative number of hours.

Check

POOLS A swimming pool that holds 10,000 gallons of water is being filled at a rate of 600 gallons per hour. The function $h(x) = 600x$ represents the amount of water in the pool after x hours.

Part A Find $h(3.25)$.

$h(3.25) = \underline{\ ?\ }$

Part B Describe the meaning of $h(3.25)$ in the context of the situation.

After $\underline{\ ?\ }$ the total amount of water in the pool will be $\underline{\ ?\ }$.

 Go Online You can complete an Extra Example online.

Practice

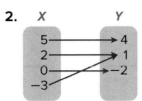 Go Online You can complete your homework online.

Examples 1 and 3

Determine whether each relation is a function. Explain.

1. X → Y: −6→4, −2→1, 1→−3, 3→−5

2. X → Y: 5→4, 2→1, 0→1, −3→−2

3. X → Y: 4→2, 4→−1, 6→−1, 7→3, 7→5

4.
x	y
4	−5
−1	−10
0	−9
1	−7
9	1

5.
x	y
2	7
5	−3
3	5
−4	−2
5	2

6.
x	y
3	7
−1	1
1	0
3	5
7	3

7. {(2, 5), (4, −2), (3, 3), (5, 4), (−2, 5)}

8. {(6, −1), (−4, 2), (5, 2), (4, 6), (6, 5)}

9. $y = 2x - 5$

10. $y = 11$

11.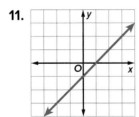

12.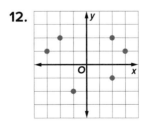

13.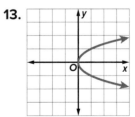

Lesson 3-2 • Functions 153

Example 2

14. **TURNPIKE** The total cost for cars entering President George Bush Turnpike at Beltline Road is related to the number of cars entering the turnpike at Beltline Road. The costs of different numbers of cars entering the turnpike at Beltline Road are as follows: 1 car, $0.75; 2 cars, $1.50; 3 cars, $2.25; 4 cars, $3.00; 5 cars, $3.75.

    a. Make a table.

    b. Determine the domain and range of the relation.

    c. Determine whether the relation is a function.

15. **HOME VALUE** The average value of a house changes over time. The average values of a house in January in Denver, Colorado, for different years are as follows: 2014, $254,000; 2015, $293,000; 2016, $338,000; 2017, $372,000.

    a. Make a table.

    b. Determine the domain and range of the relation.

    c. Determine whether the relation is a function.

Examples 4 and 5

If $f(x) = 3x + 2$ and $g(x) = x^2 - x$, find each value.

16. $f(4)$
17. $f(8)$
18. $f(-2)$
19. $g(2)$
20. $g(-3)$
21. $g(-6)$
22. $f(2) + 1$
23. $f(1) - 1$
24. $g(2) - 2$
25. $g(-1) + 4$
26. $f(x) + 1$
27. $g(3b)$

Example 6

28. **CELL PHONES** Many cell phone plans have an option to include more than one phone. The function for the monthly cost of cell phone service from a wireless company is $f(x) = 25x + 200$, where x is the number of phones on the plan. Find and interpret $f(3)$ and $f(5)$.

29. **GEOMETRY** The area for any square is given by the function $f(x) = x^2$, where x is the length of a side of the square. Find and interpret $f(3.5)$.

30. **TRANSPORTATION** The cost of riding in a cab is $3.00 plus $0.75 per mile. The function that represents this relation is $f(x) = 0.75x + 3$, where x is the number of miles traveled and $f(x)$ is the cost of the trip. Find and interpret $f(17)$.

31. **GYM MEMBERSHIP** The cost of a gym membership is $75 plus $30 per month. The function that represents this relation is $f(x) = 30x + 75$, where x is the number of months and $f(x)$ is the cost. Find and interpret $f(12)$.

Mixed Exercises

32. **USE A MODEL** Mario collected data about some of the players on a women's basketball team. The data are shown in the table, mapping, and graph. Is each relation a function? Why or why not?

Team History	
Years on Team	Games Played
1	24
2	45
3	82
3	88
5	120

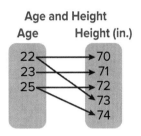

If $f(x) = -2x - 3$ and $g(x) = x^2 + 5x$, find each value.

33. $f(-1)$
34. $f(6)$
35. $g(2)$
36. $g(-3)$
37. $g(-2) + 2$
38. $f(0) - 7$
39. $f(4y)$
40. $g(-6m)$
41. $f(c) - 5$
42. $f(r) + 2$
43. $5[f(d)]$
44. $3[g(n)]$

45. **FINANCIAL LITERACY** Aisha has $40 to spend for her ornithology club. She spends some of it buying birdseed and saves the rest. Her savings is given by $f(x) = 40 - 1.25x$, where x is the number of pounds of birdseed she buys at $1.25 per pound.

   a. Graph the equation.

   b. Is the relation a function? Explain.

   c. Find $f(3)$, $f(18)$, and $f(36)$. What do these values represent?

   d. How many pounds of birdseed can Aisha buy if she wants to save $30?

46. **USE A MODEL** A recipe for homemade pasta dough says that the number of eggs you need is always one more than the number of servings you are making.

   a. Make a graph that shows the relationship between the number of servings and the number of eggs.

   b. Is the relation a function? Explain.

   c. What is the domain of the relation? Describe its meaning in the context of the situation.

Lesson 3-2 • Functions 155

47. **REASONING** The height h of a balloon, in feet, t seconds after it is released is given by the function $h(t) = 2t + 6$.

   a. What is the value of $h(20)$, and what does it mean in the context of the situation?

   b. Explain how to use the function to find the height of the balloon 2 minutes after it is released.

   c. What is the height of the balloon just before it is released? How do you know?

   d. Are there any restrictions on the values of t that can be used as inputs for the function? If so, how would this affect the graph of the function? Explain.

48. **CONSTRUCT ARGUMENTS** The following set of ordered pairs represents a function, but one of the values is missing: {(–4, –1), (–3, –1), (3, 2), (5, 2), (?, 2)}. What conclusions can you make about the missing value? Justify your arguments.

49. **WRITE** How can you determine whether a relation represents a function?

50. **ANALYZE** Feng says that the set of ordered pairs {(0, 1), (3, 2), (3, –5), (5, 4)} represents a function. Determine whether his statement is *true* or *false*. Justify your argument.

51. **PERSEVERE** Consider $f(x) = -4.3x - 2$. Write $f(g + 3.5)$ and simplify by combining like terms.

52. **REASONING** For the function $y = 15x - 4$, assume the domain is only values of x from 0 to 5. What is the range of the function?

53. **PERSEVERE** If $f(3b - 1) = 9b - 1$, find one possible expression for $f(x)$.

Lesson 3-3
Linearity and Continuity of Graphs

Explore Representing Discrete and Continuous Functions

Online Activity Use a real-world situation to complete the Explore.

INQUIRY How can you use the graph of a function to determine whether it is discrete?

Learn Discrete and Continuous Functions

Discrete Function	Continuous Function	Neither Discrete nor Continuous
Points are not connected.	Points are connected to form a line or curve.	Some points are connected by a line or curve, but it is not continuous everywhere.
Domain: set of individual values	Domain: all real numbers	Domain: varies
Range: set of individual values	Range: one interval of real numbers	Range: varies

Today's Goals
- Determine whether functions are continuous, discrete, or neither.
- Determine whether functions are linear or nonlinear.

Today's Vocabulary
discrete function
continuous function
linear function
linear equation
nonlinear function

Think About It!
Can a function be both discrete and continuous? Justify your answer.

Lesson 3-3 • Linearity and Continuity of Graphs **157**

Think About It!

If Bargain Book Barn did not use a sliding scale, that is, 1 book costs $1.50 and 2 books cost $3.00, then would the function still be discrete? Explain your reasoning.

Example 1 Determine Continuity

BOOKS The Bargain Book Barn sells young adult novels on a sliding scale. That is, the more books you buy, the cheaper they are. Let $f(x)$ model the store's prices for given quantities. Is $f(x)$ *discrete* or *continuous*? Explain your reasoning.

Number of Books	Price ($)
1	1.50
2	2.90
3	4.30
4	5.70
5	7.10

Use the table and context of the situation. The quantity and price correspond to ordered pairs like (1, 1.50) which can be graphed. Because books are not sold in fractional quantities, the number of books and their corresponding prices cannot be between the points given. So the points should not be connected and the function is discrete. The domain and range are sets of individual values.

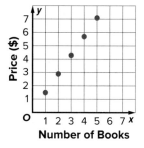

Example 2 Determine Continuity by Using Graphs

Determine whether $f(x)$ and $g(x)$ are *continuous*, *discrete*, or *neither*. Explain your reasoning.

a. $f(x)$

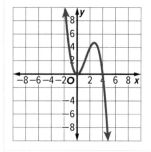

Because $f(x)$ is graphed with a single curve, it is a continuous function. The domain and range are both all real numbers.

b. $g(x)$

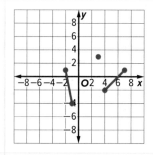

Because $g(x)$ has continuous sections, but is not a single line or curve, it is neither discrete nor continuous. The domain and range are intervals of values.

Go Online You can complete an Extra Example online.

Check

Determine whether f(x) is *discrete*, *continuous*, or *neither*.

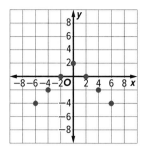

The function is ___?___.

Example 3 Apply Discrete and Continuous Functions

DETECTIVE As a private investigator, Tia charges $25 per hour for any amount of time up to eight hours and then a flat rate of $250 per day. Use the graph to determine if the function that models this situation is *discrete*, *continuous*, or *neither*.

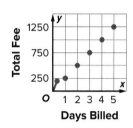

Discrete?

No; the function is not made up entirely of individual points.

Continuous?

No; the function cannot be drawn with a straight line or smooth curve.

Neither?

Yes; the function is neither continuous nor discrete.

Watch Out!

Continuous Intervals Recall that a function can have a continuous interval, but the function itself is not continuous unless it is continuous over its entire domain.

Think About It!

How could Tia alter her pricing model to make it a discrete function? a continuous function?

Check

PLANTS Circadian rhythms are cycles of behavior that occur over twenty-four hours, based on a day-night cycle. One aspect of a plant's circadian rhythm is the percentage of its flowers that are open or closed.

If f(x) is represented by the curve, then is f(x) *discrete*, *continuous*, or *neither*?

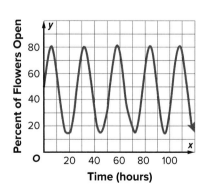

The function is ___?___.

Go Online You can complete an Extra Example online.

Lesson 3-3 • Linearity and Continuity of Graphs 159

Learn Linear and Nonlinear Functions

A **linear function** is a function that has a graph that is a line. If the domain of the function is all real numbers, then the function is continuous.

A **linear equation** can be used to describe a linear function.

Linear equations are often written in standard form.

Key Concept • Standard Form of a Linear Equation	
Words	The standard form of a linear equation is $Ax + By = C$, where $A \geq 0$, A and B are not both zero, and A, B, and C are integers with a greatest common factor of 1.
Examples	In $2x + 5y = 7$, $A = 2$, $B = 5$, and $C = 7$.
	In $x = -3$, $A = 1$, $B = 0$, and $C = -3$.

A **nonlinear function** has a graph with a set of points that cannot all lie on the same line. An equation that represents a nonlinear function cannot be expressed in the form $Ax + By = C$.

The function values of a linear function change at a constant rate for every equivalent change in the x-values. The change in the function values for a nonlinear function will vary for equivalent changes in x.

> **Think About It!**
> Can a function be both linear and nonlinear? Explain your reasoning.

Example 4 Linear and Nonlinear Functions

Determine whether $y = 4x^2 - (2x)^2 + 3x - 5$ is an equation for a *linear* or *nonlinear* function.

Step 1 Simplify the equation.

$y = 4x^2 - (2x)^2 + 3x - 5$ Original equation

$= 4x^2 - 4x^2 + 3x - 5$ Simplify exponents.

$= 0 + 3x - 5$ Subtract.

$= 3x - 5$ Simplify.

Step 2 Rewrite the equation.

$y = 3x - 5$ Simplified equation

$y - 3x = 3x - 3x - 5$ Subtract $3x$ from each side.

$y - 3x = -5$ Simplify.

$3x - y = 5$ Multiply each side by -1.

Because $3x - y = 5$ is in the form $Ax + By = C$, where $A = 3$, $B = -1$, and $C = 5$, $y = 4x^2 - (2x)^2 + 3x - 5$ is linear.

> **Think About It!**
> Can $y = -5x^2$ be written in standard form? If so, write the standard form of the equation.

Check

The function $3x^2 - \sqrt{9}x^2 - y = 17$ is ___?___.

Go Online You can complete an Extra Example online.

Example 5 Identify Linear and Nonlinear Functions

Determine whether $y = 3x^3 - x^3 + 3x + 6$ is an equation for a *linear* or *nonlinear* function.

Step 1 Simplify the equation.

$y = 3x^3 - x^3 + 3x + 6$ Original equation

$ = 2x^3 + 3x + 6$ Subtract.

Step 2 Rewrite the equation.

$y = 2x^3 + 3x + 6$ Simplified equation

$y - 2x^3 - 3x = 2x^3 + 3x + 6 - 2x^3 - 3x$

                     Subtract $2x^3$ and $3x$ from each side.

$y - 2x^3 - 3x = 6$ Simplify.

$2x^3 + 3x - y = -6$ Multiply each side by –1.

Because $2x^3 + 3x - y = -6$ is not in the form $Ax + By = C$, $2x^3 + 3x - y = -6$ is nonlinear.

Check

The function $4x - (2y)^2 = 3$ is _____?_____.

 Think About It!

Is the function represented by $2x + 9 = 5y - 3xy$ a linear or nonlinear function? Justify your argument.

Example 6 Functions in Table Form

SOCCER Salina kicks a soccer ball. The height of the ball after each half second is recorded in the table. Is the function that models the height of the ball a *linear* or *nonlinear* function?

Time (s)	Height (ft)
0	2
0.5	28
1	46
1.5	56
2	58
2.5	52
3	38
3.5	16

First Half-Second Interval

During the first half-second interval, the ball goes from a height of 2 feet to a height of 28 feet. That is an increase of 26 feet.

Second Half-Second Interval

During the second half-second interval, the ball goes from 28 feet to 46 feet. That is an increase of 18 feet.

Since the change in the height varies over the two equivalent intervals, the height of the soccer ball must be modeled by a nonlinear function.

Think About It!

What do you notice about the height of the soccer ball over time that might indicate that the function is not linear?

Go Online You can complete an Extra Example online.

Check

Determine whether the values in each table are best modeled by a *linear* or *nonlinear* function.

x	y
−5	−15
−1	−3
0	0
2	6
3	9
7	21

x	y
0	0
1	−1
2	−8
3	−27
5	−125
7	−343

Example 7 Identify Linear Functions by Graphing

POOL Fernando uses a garden hose to fill his empty pool. The table shows the amount of water in the pool after every five minutes.

Time (min)	Water (gal)
5	60
10	120
15	180
20	240
25	300

Part A Determine linearity.

The amount of water in the pool increases by 60 gallons during the first 5 minutes. It also increases by 60 gallons during each following 5-minute intervals up to 25 minutes. Since the change in the amount of water is constant for every equivalent interval of 5 minutes, the numbers in the table can be modeled by a linear function.

Part B Graph.

The points on the graph can be connected by a straight line.

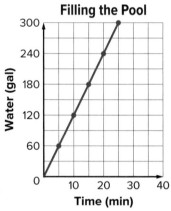

Talk About It!

After 25 minutes, Fernando turns off the hose. He takes 5 minutes to remove any wrinkles from the pool liner. Then he returns to filling the pool at the same rate. Would the function that represents filling the pool from 0 to 40 minutes be a linear function? Explain your reasoning.

Check

MINIMUM WAGE The table shows the federal minimum wage rates during years in which the wage increased. Which statement best describes the function that models the wages over time?

Year	1990	1991	1996	1997	2007	2008	2009
Wage (dollars)	3.80	4.25	4.75	5.15	5.85	6.55	7.25

A. The function is linear because the increase is $0.70 between 1997 and 2007 and $0.70 between 2008 and 2009.

B. The function is nonlinear because the increase is $0.45 between 1990 and 1991 and $0.70 between 2007 and 2008.

C. The function is linear because it is constantly increasing.

D. The function is nonlinear because it is discrete.

Go Online You can complete an Extra Example online.

Practice

Go Online You can complete your homework online.

Examples 1 and 2
Determine whether the function is *discrete*, *continuous*, or *neither*. Explain.

1.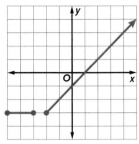

2.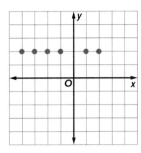

3.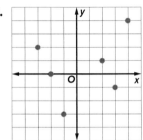

4.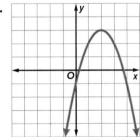

5.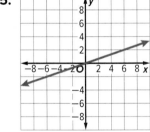

6.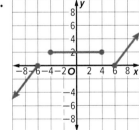

7.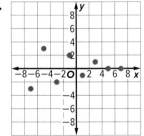

8.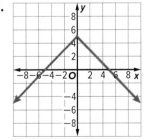

Lesson 3-3 • Linearity and Continuity of Graphs 163

Example 3

9. **EMPLOYMENT** Kylie records the number of hours she works in her after-school job each week and then graphs the function that models the situation. Use the graph to determine if the function that models this situation is *discrete, continuous,* or *neither.*

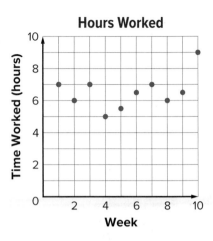

10. **BAKING** Kalynda keeps track of the number of cups of flour she has after baking batches of cookies and then graphs the function that models the situation. Use the graph to determine if the function that models this situation is *discrete, continuous,* or *neither.*

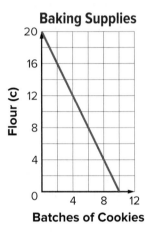

11. **HOMEWORK** Nayati records the number of hours of homework he has each day and then graphs the function that models the situation. Use the graph to determine if the function that models this situation is *discrete, continuous,* or *neither.*

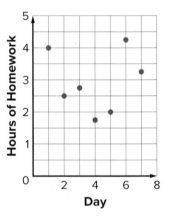

Examples 4, 5, and 6

Determine whether each function is *linear* or *nonlinear.*

12. $y - \frac{1}{x} = 11$

13. $x + (\sqrt{2}x)^2 + y - 2x^2 = 11$

14. $y = 2x + 5$

15. $y = 3x^2 - x + 5$

16. $2y + \frac{x}{3} = -6$

17. $9x - (9x)^2 = 19 - y$

18. $-9x + (\sqrt{4}x)^2 + y = -4 + 7x^2$

19. $y = -x^3 + 2x$

164 Module 3 • Relations and Functions

20.

x	y
1	100
2	125
3	150
4	175
5	200

21.

x	y
1	3
2	5
3	11
4	21
5	35

22.

x	y
−5	−10
−2	−4
0	0
2	4
3	6

23.

x	y
5	35
6	41
7	47
8	53
9	59

Example 7

24. ENDURANCE Tamika wants to improve her running endurance and thus tries to run for longer periods of time each day. The distance she runs, given the day, is provided in the table.

Day	1	2	3	4	5	6	7
Distance	2	2.25	2.75	3.5	4	4.5	5

a. Determine linearity.

b. Graph.

25. SCUBA DIVING Marco wants to know his elevation compared to sea level for periods of time each minute while scuba diving. His elevation is provided in the table.

Minute	1	2	3	4	5	6	7
Elevation	−3	−4.5	−6	−7.5	−9	−10.5	−12

a. Determine linearity.

b. Graph.

Lesson 3-3 • Linearity and Continuity of Graphs

Mixed Exercises

26. **EXERCISE** Rashona keeps track of the total number of minutes she exercises and then graphs the function that models the situation. Use the graph to determine if the function that models this situation is *discrete, continuous,* or *neither.*

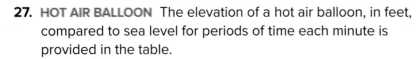

27. **HOT AIR BALLOON** The elevation of a hot air balloon, in feet, compared to sea level for periods of time each minute is provided in the table.

Minute	1	2	3	4	5	6	7	8
Elevation	10	15	12	18	25	45	40	42

  a. Determine linearity.

  b. Graph.

Determine whether the function is *discrete, continuous,* or *neither.* Then determine whether each function is *linear* or *nonlinear.*

28. $y - 14x = 3$

29. $(2x)^2 - 4y = 2$

30.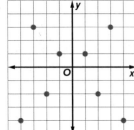

31.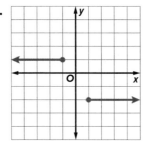

32.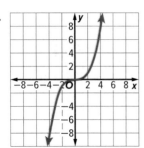

33.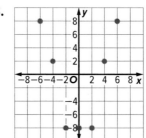

34. **ANALYZE** If $f(x)$ is a linear function, then is $f(x + a)$, where a is a real number, *sometimes, always,* or *never* a continuous function? Justify your argument.

35. **WRITE** Describe a real-world situation where the function that models the situation is neither discrete nor continuous.

36. **ANALYZE** A function that consists of a finite set of ordered pairs is *sometimes, always,* or *never* continuous? Justify your argument.

Lesson 3-4
Intercepts of Graphs

Learn Intercepts of Graphs of Functions

The intercepts of graphs are points where the graph intersects an axis.

The **x-intercept** is the x-coordinate of a point where a graph crosses the x-axis.

The **y-intercept** is the y-coordinate of a point where a graph crosses the y-axis.

A function is **positive** when its graph lies *above* the x-axis.

A function is **negative** when its graph lies *below* the x-axis.

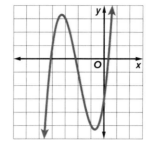

Today's Goals
- Identify the intercepts of functions and intervals where functions are positive and negative.
- Solve equations by graphing.

Today's Vocabulary
x-intercept
y-intercept
positive
negative
root
zero

Example 1 Intercepts of the Graph of a Linear Function

Use the graph to estimate the x- and y-intercepts of the function and describe where the function is positive and negative.

The x-intercept is the point where the graph crosses the x-axis, (3, 0).

The y-intercept is the point where the graph crosses the y-axis, (0, 6).

A function is positive when its graph lies above the x-axis, or when $x < 3$.

A function is negative when its graph lies below the x-axis, or when $x > 3$.

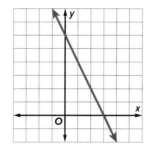

Study Tip
Intercepts Notice that *intercept* can be used to refer to either the point where the graph intersects the axis or the nonzero coordinate of the point where the graph intersects the axis.

Think About It!
Explain why this function is linear.

Check

Use the graph to estimate the x- and y-intercepts of the function and describe where the function is positive and negative.

A. x-intercept: (−2, 0); y-intercept: (0, −6);
positive: $x > -2$; negative: $x < -2$

B. x-intercept: (0, −6); y-intercept: (−2, 0);
positive: $x < -2$; negative: $x > -2$

C. x-intercept: (−2, 0); y-intercept: (0, −6);
positive: $x < -2$; negative: $x > -2$

D. x-intercept: (0, −6); y-intercept: (−2, 0);
positive: $x > -2$; negative: $x < -2$

Go Online You can complete an Extra Example online.

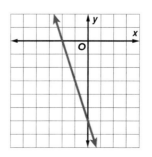

Study Tip
x- and y-intercepts To help remember the difference between the x- and y-intercepts, remember that the x-intercept is where the graph intersects the x-axis, and the y-intercept is where the graph intersects the y-axis.

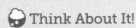
Think About It!
Explain why this function is nonlinear.

Watch Out!
Intercepts of Nonlinear Functions The graphs of nonlinear functions can have more than one x-intercept.

Example 2 Intercepts of the Graph of a Nonlinear Function

Use the graph to estimate the x- and y-intercepts of the function and describe where the function is positive and negative.

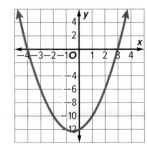

x-intercepts: −4 and 3.

y-intercept: −12.

positive: when $x < -4$ and when $x > 3$.

negative: x is between −4 and 3.

Check

Use the graph of the function to determine key features.

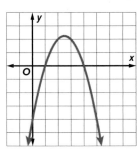

Part A Determine whether each ordered pair represents an x-intercept, a y-intercept, or neither.

(1, 0)

(0, 1)

Part B Describe where the function is positive and negative.

A. positive: $x < 1$ and $x > 4$; negative: x is between 1 and 4

B. positive: $x > 1$ and $x < 4$; negative: x is between 1 and 4

C. positive: x is between 1 and 4; negative: $x > 1$ and $x < 4$

D. positive: x is between 1 and 4; negative: $x < 1$ and $x > 4$

Go Online You can complete an Extra Example online.

Example 3 Find Intercepts from a Graph

SPORTS The graph shows the height of a ball for each second x that it is airborne. Use the graph to estimate the x- and y-intercepts of the function, where the function is positive and negative, and interpret the meanings in the context of the situation.

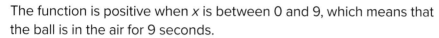

The x-intercept is 9. That means that the ball will hit the ground after 9 seconds.

The y-intercept is 4. This means that at time 4, the ball was at a height of 0 feet.

The function is positive when x is between 0 and 9, which means that the ball is in the air for 9 seconds.

No portion of the graph shows that the function is negative.

Think About It!
The function is only graphed from 0 to 9 seconds. What can you assume about the function when $x > 9$? Interpret this meaning. Does it make sense in the context of the situation?

Check

FITNESS The graph shows the number of people y at a gym x hours after the gym opens.

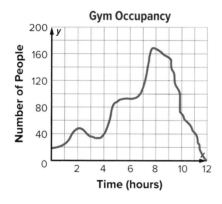

Part A Use the graph to estimate the x- and y-intercepts.

  x-intercept: (? , ?)

  y-intercept: (? , ?)

Part B Which statements describe the meaning of the x- and y-intercepts in the context of the situation? Select all that apply.

A. There were 20 people at the gym when it opened.

B. The gym closed after 20 hours.

C. The gym closed after 12 hours.

D. There were 12 people at the gym when it opened.

E. There was no one at the gym when it opened.

Go Online You can complete an Extra Example online.

Lesson 3-4 • Intercepts of Graphs **169**

Example 4 Find Intercepts from a Table

LUNCH Violet starts the semester with $150 in her student lunch account. Each day she spends $3.75 on lunch. The table shows the function relating the amount of money remaining in her lunch account to the number of days Violet has purchased lunch.

Time (Days) x	Balance ($) y
0	150
2	142.50
5	131.25
10	112.50
15	93.75
30	37.50
40	0

Part A Find the intercepts.

The x-intercept is where y = 0, so the x-intercept is 40.

The y-intercept is where x = 0, so the y-intercept is 150.

Part B Describe what the intercepts mean in the context of the situation.

The x-intercept means that after buying lunch for 40 days, Violet will have $0 left in her lunch account, or it will take Violet 40 days to use all of the money in her lunch account. The y-intercept means that Violet's lunch account has $150 after buying lunch for 0 days, or the beginning balance of her lunch account is $150.

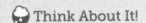 **Think About It!**
Explain why the x-coordinate of the y-intercept is always 0.

Check

MOVIES Ashley received a gift card to the movie theater for her birthday. The table shows the amount of money remaining on her gift card y after x trips to the movie theater.

Number of Trips x	Balance ($) y
0	90
1	81
2	72
3	63
5	45
7	27
10	0

Part A Find the y-intercept. (___?___, ___?___)

Part B Find the x-intercept and describe what it means in the context of the situation.

A. (10, 0); The initial balance on the gift card was $10.

B. (90, 0); The initial balance on the gift card was $90.

C. (10, 0); After 10 trips to the movies, there will be no money left on the gift card.

D. (90, 0); After 90 trips to the movies, there will be no money left on the gift card.

Go Online You can complete an Extra Example online.

Watch Out!

Intercepts The y-coordinate of the x-intercept will always be 0, not the x-coordinate. The x-coordinate of the y-intercept will always be 0, not the y-coordinate.

170 Module 3 • Relations and Functions

Learn Solving Equations by Graphing

The solution, or **root**, of an equation is any value that makes the equation true. A **zero** is an x-intercept of the graph of the function.

For example, the root of $3x = 6$ is 2. A linear equation, like $3x = 6$, has at most one root, while a nonlinear equation, like $x^2 + 4x - 5 = 0$, may have more than one.

Equation	Related Function
$3x = 6$	$f(x) = 3x - 6$ or $y = 3x - 6$
$x^2 + 4x - 5 = 0$	$f(x) = x^2 + 4x - 5$ or $y = x^2 + 4x - 5$

The graph of the related function can be used to find the solutions of an equation. The related function is formed by solving the equation for 0 and then replacing 0 with $f(x)$ or y.

Values of x for which $f(x) = 0$ are located at the x-intercepts of the graph of a function and are called the zeros of the function f. The roots of an equation are the same as the zeros of its related function. The solutions and roots of an equation are the same value as the zeros and x-intercepts of its related function. For the equation $3x = 6$:

- 2 is the solution of $3x = 6$.
- 2 is the root of $3x = 6$.
- 2 is the zero of $f(x) = 3x - 6$.
- 2 is the x-intercept of $f(x) = 3x - 6$.

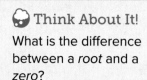

Think About It!
What is the difference between a *root* and a *zero*?

Example 5 Solve a Linear Equation by Graphing

Solve $-2x + 7 = 1$ by graphing. Check your solution.

Find the related function.

$-2x + 7 = 1$ Original equation

$-2x + 7 - 1 = 1 - 1$ Subtract 1 from each side.

$-2x + 6 = 0$ Simplify.

Graph the left side of the equation. The related function is $f(x) = -2x + 6$, which can be graphed.

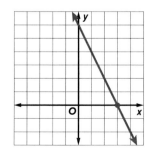

The graph intersects the x-axis at 3. This is the x-intercept, or zero, which is also the root of the equation. So, the solution of the equation is 3.

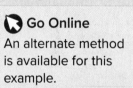

Go Online
An alternate method is available for this example.

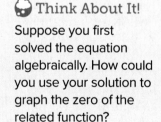

Think About It!
Suppose you first solved the equation algebraically. How could you use your solution to graph the zero of the related function?

Check your solution by solving the equation algebraically.

Go Online You can complete an Extra Example online.

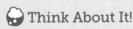**Think About It!**

Name two ways that you can tell that this is a nonlinear function.

Example 6 Solve a Nonlinear Equation by Graphing

Solve $x^2 - 4x = -3$ by graphing. Check your solution.

Find the related function.

$x^2 - 4x = -3$	Original equation
$x^2 - 4x + 3 = -3 + 3$	Add 3 to each side.
$x^2 - 4x + 3 = 0$	Simplify.

Graph the left side of the equation. The related function is $f(x) = x^2 - 4x + 3$, which can be graphed.

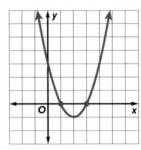

The graph intersects the x-axis at 1 and 3. These are the x-intercepts, or zeros, which are also the roots of the equation. So, the solutions of the equation are 1 and 3.

Go Online
You can watch a video to see how to use a graphing calculator with this example.

Example 7 Solve an Equation of a Horizontal Line by Graphing

Solve $4x + 3 = 4x - 5$ by graphing. Check your solution.

Find the related function.

$4x + 3 = 4x - 5$	Original equation
$4x + 3 + 5 = 4x - 5 + 5$	Add 5 to each side.
$4x + 8 = 4x$	Simplify.
$4x - 4x + 8 = 4x - 4x$	Subtract 4x from each side.
$8 = 0$	Simplify.

Graph the left side of the equation. The related function is $f(x) = 8$, which can be graphed.

Talk About It!
Does solving the equation algebraically give a different solution? Explain your reasoning.

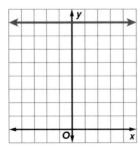

The graph does not intersect the x-axis. This means that there is no x-intercept and, therefore, there is no solution.

Go Online You can complete an Extra Example online.

Check

Equations and the graphs of their related functions are shown. Write the related function and its zero(s) under the appropriate graph.

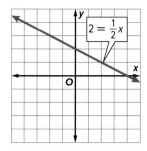

related function:
zeros:

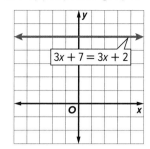

related function:
zeros:

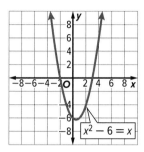

related function:
zeros:

🌎 Apply Example 8 Estimate Solutions by Graphing

PARTY Haley is ordering invitations for her graduation party. She has $40 to spend and each invitation costs $0.96. The function $m = 40 - 0.96p$ represents the amount of money m Haley has left after ordering p party invitations. Find the zero of the function. Describe what this value means in the context of this situation.

1 What is the task?

Describe the task in your own words. Then list any questions that you may have. How can you find answers to your questions?

I need to find the zero of the function and describe what it means. How can I determine the meaning of the zero from a graph of the function? I can review graphing linear functions and labeling axes.

2 How will you approach the task? What have you learned that you can use to help you complete the task?

I will graph the function by making a table of values. I will estimate the x-intercept of the graph to find the zero. I will then check my solution by solving the equation algebraically. I will use the axes labels to help me interpret my solution.

(continued on the next page)

 Go Online You can complete an Extra Example online.

3 What is your solution?

Use your strategy to solve the problem.

Graph the function.

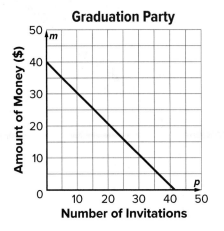

Graduation Party

Estimate the solution.

42 invitations

Check the solution.

≈ 41.67 invitations

What does your solution mean in the context of the situation?

Haley can order 41 invitations with the amount of money she has to spend.

4 How can you know that your solution is reasonable?

✏️ **Write About It!** **Write an argument that can be used to defend your solution.**

This amount is close to the estimated zero of 42 invitations from the graph.

Check

DATA Blair's cell phone plan allows her to use 3 GB of data, and she uses approximately 0.14 GB of data each day. The function $g = 3 - 0.14d$ represents the amount of data g in GB she has left after d days.

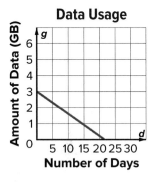

Data Usage

Part A Examine the graph of the function to estimate its zero to the nearest day.

The graph appears to intersect the x-axis at ___?___.

Part B Solve algebraically to check your answer. Round to the nearest tenth.

$x =$ ___?___

Part C Describe what your answer to Part B means in this context.

After ___?___ days, Blair has ___?___ GB left.

🧭 Go Online You can complete an Extra Example online.

174 Module 3 · Relations and Functions

Practice

Go Online You can complete your homework online.

Examples 1 and 2

Use the graph to estimate the *x*- and *y*-intercepts of the function and describe where the function is positive and negative.

1.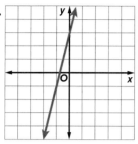

2.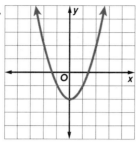

3.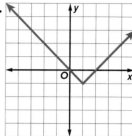

4.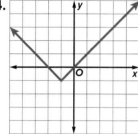

5.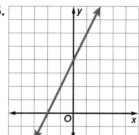

6.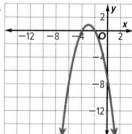

7.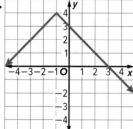

8.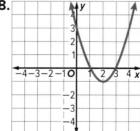

9.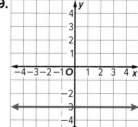

Lesson 3-4 • Intercepts of Graphs **175**

Example 3
10. **FOOTBALL** The graph shows the height of a football after being thrown. Use the graph to estimate the x- and y-intercepts of the function, where the function is positive and negative, and interpret the meanings in the context of the situation.

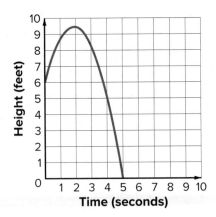

11. **EARNINGS** The graph shows the amount of money Ryan earns. Use the graph to estimate the x- and y-intercepts of the function, where the function is positive and negative, and interpret the meanings in the context of the situation.

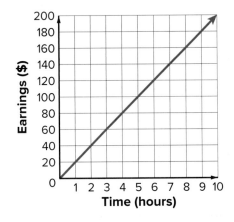

Example 4
12. **CLIMBING** Indira is mountain climbing and starts the day at 182.5 meters above sea level. Each hour she descends 36.5 meters. The table shows the function relating Indira's height to the number of hours she is mountain climbing.

    a. Find the intercepts.

    b. Describe what the intercepts mean in the context of the situation.

Time (hours)	Height (meters)
x	y
0	182.5
2	109.5
4	36.5
5	0

13. **MONEY** Javier borrowed $1950 from his parents. Each month he repaid his parents $325. The table shows the function relating Javier's remaining balance to the number of months.

    a. Find the intercepts.

    b. Describe what the intercepts mean in the context of the situation.

Time (months)	Remaining Balance ($)
x	y
0	1950
2	1300
5	325
6	0

176 Module 3 • Relations and Functions

Examples 5, 6, and 7

Solve each equation by graphing. Check your solution.

14. $2x - 3 = 3$

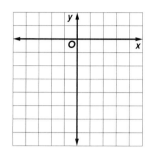

15. $-4x + 2 = -4x + 1$

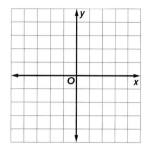

16. $4 = \frac{1}{2}x + 5$

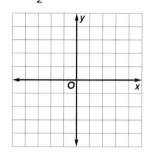

17. $3x + 1 = -5$

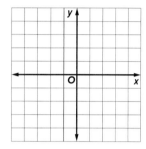

18. $3x - 5 = 3x - 3$

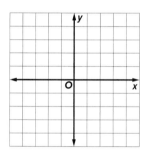

19. $9x - 7 = -6x + 8$

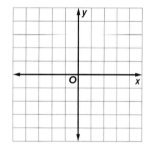

Example 8

20. INVITATIONS Moesha and Keyon are mailing invitations for their wedding. They have $50 to spend and each invitation costs $1.25 to mail. The function $m = 50 - 1.25w$ represents the amount of money m Moesha and Keyon have left after mailing w wedding invitations. Find the zero of the function. Describe what this value means in the context of this situation.

21. GIFT BAGS Juanita is tying ribbon on gift bags. She has 24 feet of ribbon and each gift bag uses 0.75 foot of ribbon. The function $r = 24 - 0.75g$ represents the amount of ribbon r Juanita has left after tying ribbon on g gift bags. Find the zero of the function. Describe what this value means in the context of this situation.

Mixed Exercises

Use the graph to estimate the *x*- and *y*-intercepts of the function and describe where the function is positive and negative.

22.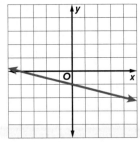

23.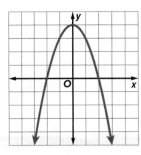

24.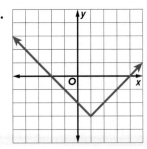

25. **REASONING** The graph shows the height of a bird compared to sea level over time. Use the graph to estimate the *x*- and *y*-intercepts of the function, where the function is positive and negative, and interpret the meanings in the context of the situation.

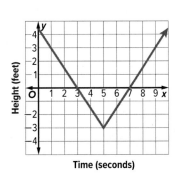

Higher-Order Thinking Skills

26. **ANALYZE** Do linear equations *sometimes, always,* or *never* have *x*- and *y*-intercepts? Justify your argument.

27. **WRITE** Describe how to find *x*- and *y*-intercepts by graphing and by using tables.

28. **CREATE** Write a word problem for the function $y = 60 - 2.5x$. Find the zero of the function. Describe what this value means in the context of your situation.

29. **PERSEVERE** Describe the steps you use to solve the equation $16 = x + 4 + (2^4 - 6)$ by graphing. Then explain how you can check your solution.

178 Module 3 • Relations and Functions

Lesson 3-5

Shapes of Graphs

Explore Line Symmetry

 **Online Activity** Use graphing technology to complete the Explore.

INQUIRY How can you use the graph of a function to determine whether it is symmetric?

Learn Symmetry and Graphs of Functions

The graphs of some functions exhibit a key feature called symmetry. A figure has **line symmetry** if each half of the figure matches the other side exactly.

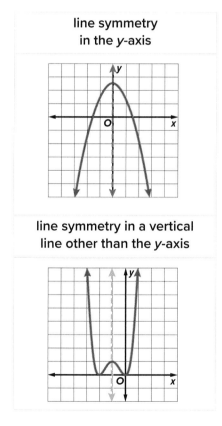

Today's Goals
- Determine whether functions have line symmetry and, if so, find the line of symmetry.
- Identify extrema and where functions are increasing and decreasing.
- Determine the end behaviors of graphs of functions.

Today's Vocabulary
line symmetry
increasing
decreasing
extrema
relative minimum
relative maximum
end behavior

Talk About It!
Can the graph of a function be symmetric about the x-axis? Justify your argument.

Lesson 3-5 • Shapes of Graphs **179**

Example 1 Line Symmetry

Determine whether each function has line symmetry. Explain.

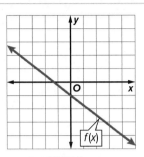

There is no line that can be drawn to make the right half a mirror image of the left half, so the function does not display any line symmetry.

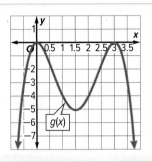

This function is symmetric in the line $x = 1.5$.

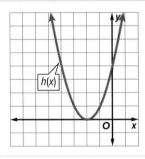

This function is symmetric in the line $x = -2$.

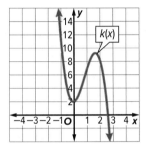

There is no line that can be drawn to make the right half of the function a mirror image of the left half, so the function does not display any line symmetry.

Check

Examine the function.

Part A Does the function possess line symmetry?

Part B Describe the line symmetry, if any, of the function.

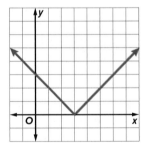

Think About It!
Find the y-intercepts of the functions.

Study Tip
Symmetry Remember a graph can be symmetric in the y-axis or any other vertical line.

Go Online You can complete an Extra Example online.

Example 2 Interpret Symmetry

FOUNTAINS A fountain is spraying a stream of water into the air. The solid portion of the graph represents the path of the water, where x is the distance in feet from the fountain and y is the height in feet of the stream. Find and interpret any symmetry in the graph of the function.

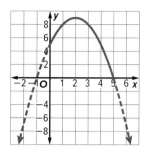

The right half of the graph is the mirror image of the left half in the line $x = 2$.

In the context of the situation, the symmetry of the graph tells you that the height of the stream of water when it is from 0 to 2 feet away from the fountain is the same as the height of the stream of water when it is from 2 to 4 feet away from the fountain.

Think About It!
Find the maximum height of the stream of water.

Problem-Solving Tip
Visualize To help visualize a line of symmetry, imagine folding the graph in half. If the graph lines up perfectly, the graph is symmetric about the line you have created with the fold.

Check

GOLF The solid portion of the graph represents the path of a golf ball after it is hit off of a platform, where x is the distance in feet a golf ball travels and y is the height in feet of the golf ball.

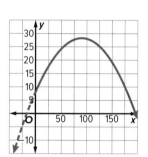

Part A Use the graph to describe any symmetry of the graph of the function.

A. symmetric in the y-axis

B. symmetric in the line $x = 8$

C. symmetric in the line $x = 28.25$

D. symmetric in the line $x = 90$

Part B Interpret the symmetry in the context of the situation.

A. The height of the golf ball when it has traveled a distance of 0 to 8 feet is the same as the height of the golf ball when it has traveled a distance of 8 to 28.25 feet.

B. The height of the golf ball when it has traveled a distance of 0 to 90 feet is the same as the height of the golf ball when it has traveled a distance of 90 to 180 feet.

C. The distance the golf ball has traveled when it is 0 to 8 feet in the air is the same as the distance the golf ball has traveled when it is 8 to 28.25 feet in the air.

D. The distance the golf ball has traveled when it is 0 to 90 feet in the air is the same as the distance the golf ball has traveled when it is 90 to 180 feet in the air.

Go Online You can complete an Extra Example online.

Explore Relative High and Low Points

Study Tip

Reading Math *Extrema* in this context is the plural form of *extreme point*. The plural forms of *maximum* and *minimum* are *maxima* and *minima*, respectively.

 Online Activity Use graphing technology to complete the Explore.

INQUIRY How do the y-values of relatively high and low points on the graph compare to the y-values of nearby points?

Learn Extrema of Graphs of Functions

A function is **increasing** where the graph goes up and **decreasing** where the graph goes down when viewed from left to right.

Points that are the locations of relatively high or low function values are called **extrema**. Point A is a **relative minimum** because no other nearby point has a lesser y-coordinate. Point B is a **relative maximum** because no other nearby point has a greater y-coordinate.

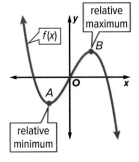

Think About It!

If $f(x)$ has a relative maximum when $x = 2$, then is $f(x)$ increasing or decreasing as x approaches 2? as x moves past 2?

Example 3 Determine Increasing and Decreasing Parts of the Graph of a Function

Determine where $f(x)$ is increasing and/or decreasing.

When $x > 0$, the graph goes up when viewed from left to right. So, the function is increasing for $x > 0$.

When $x < 0$, the graph goes down when viewed from left to right. So, the function is decreasing for $x < 0$.

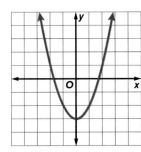

Check

For $x > 1$, $f(x)$ is _____?_____.

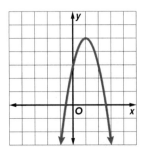

 Go Online You can complete an Extra Example online.

182 Module 3 • Relations and Functions

Example 4 Determine Extrema of the Graph of a Function

Determine the extrema of f(x). Then identify each point as a relative maximum or relative minimum.

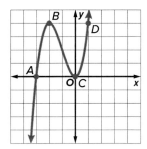

Extrema: Point B and point C are the locations of relatively high or low function values. So, they are the extrema of the function.

Relative Minimum: No other points nearby point C have a lesser y-coordinate. So, point C is a relative minimum.

Relative Maximum: No other points nearby point B have a greater y-coordinate. So, point B is a relative maximum.

Think About It!
Can a point be a relative minimum or relative maximum and not be an extreme point? Explain.

Check

Which point(s) is(are) a relative minimum? Select all that apply.

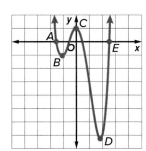

A. A

B. B

C. C

D. D

E. E

Go Online You can complete an Extra Example online.

Lesson 3-5 • Shapes of Graphs   183

Go Online
You can watch a video to see more about the comic book store.

Think About It!
Does this function have a relative minimum? Explain your reasoning.

Example 5 Interpret Extrema of the Graph of a Function

COMIC BOOKS A comic book store uses a function to model its profit in thousands of dollars given the price in dollars that it charges for individual issues. Determine whether point D is a relative minimum, relative maximum, or neither. Then interpret its meaning in the context of the situation.

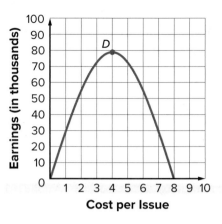

Point D is a relative maximum because all nearby points have a lesser y-coordinate.

Point D represents the greatest profit that the comic book store can earn given the price it charges per issue.

Check

AEROBATICS Aerobatics, or stunt flying, is the practice of intentional maneuvers of an aircraft that are not necessary for normal flight. Lincoln Beachey, an inventor of aerobatics, was known for his stunt called the "Dip of Death" in which his plane would plummet toward the ground from 5000 feet until he leveled the plane. His distance from the ground during the stunt can be approximately modeled by the function $f(x)$. Identify any extrema in the context of the situation.

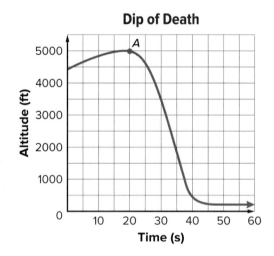

Point A represents that at about 20 seconds, Lincoln Beachey reached a ____?____ relative maximum in height at 5000 feet.

After point A, his height is ____?____ decreasing until he levels out the plane.

Go Online You can complete an Extra Example online.

184 Module 3 · Relations and Functions

Learn End Behavior of Graphs of Functions

End behavior describes the values of a function at the positive and negative extremes in its domain.

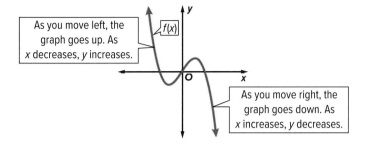

As you move left, the graph goes up. As x decreases, y increases.

As you move right, the graph goes down. As x increases, y decreases.

Example 6 Determine End Behavior of the Graph of a Linear Function

Determine the end behavior of $f(x)$.

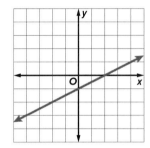

As you move to the left on the graph, the value of y gets increasingly negative. Thus, as x decreases, y decreases.

As you move to the right on the graph, the value of y gets increasingly positive. Thus, as x increases, y increases.

Check

Determine the end behavior of $f(x)$.

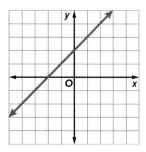

As x increases, y ____?____.

As x decreases, y ____?____.

Go Online You can complete an Extra Example online.

Go Online You may want to complete the Concept Check to check your understanding.

Watch Out!

End Behavior The end behavior of some graphs can be described as approaching a specific value.

Think About It!

Make a conjecture, or educated guess, about the end behavior of a linear function when the slope is positive or negative.

Study Tip

Conjecture A conjecture is an educated guess based on known information.

Lesson 3-5 • Shapes of Graphs 185

Example 7 Determine End Behavior of the Graph of a Nonlinear Function

Determine the end behavior of f(x).

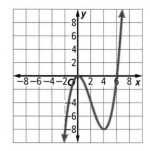

As you move to the left on the graph, the value of y gets increasingly negative. Thus, as x decreases, y decreases.

As you move to the right on the graph, the value of y gets increasingly positive. Thus, as x increases, y increases.

Check

Determine the end behavior of each function.

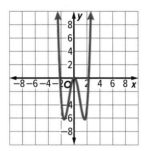

As x increases, y _____?_____.
As x decreases, y _____?_____.

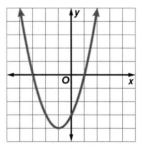

As x increases, y _____?_____.
As x decreases, y _____?_____.

> **Go Online** to practice what you've learned about interpreting graphs in the Put It All Together over Lessons 3-1 through 3-5.

Go Online You can complete an Extra Example online.

Practice

Go Online You can complete your homework online.

Example 1

Determine whether each function has line symmetry. Explain.

1.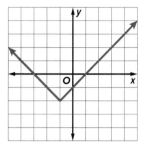

2.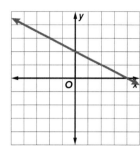

3.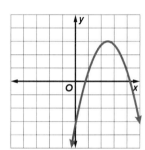

4.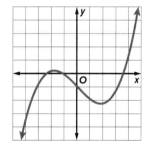

Example 2

5. **GEOMETRY** The solid portion of the graph represents the relationship between the width of a rectangle in centimeters x and the area of the rectangle in centimeters squared y. Find and interpret any symmetry in the graph of the function.

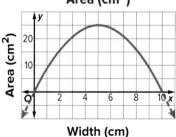

6. **SPRINKLERS** A sprinkler is spraying a stream of water into the air. The solid portion of the graph represents the path of the water, where x represents the distance in feet from the sprinkler and y represents the height in feet of the water. Find and interpret any symmetry in the graph of the function.

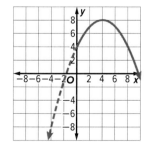

Lesson 3-5 • Shapes of Graphs **187**

Example 3

Determine where f(x) is increasing and/or decreasing.

7.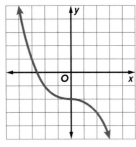

8.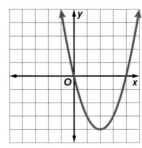

9.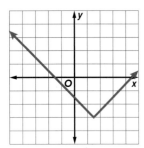

10.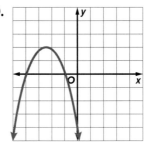

Example 4

Determine the extrema of f(x). Then identify each point as a relative maximum or relative minimum.

11.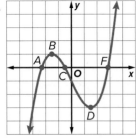

12.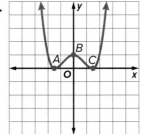

13.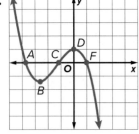

14.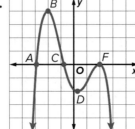

188 Module 3 · Relations and Functions

Example 5

15. GOLF The height of a golf ball compared to the distance the golf ball is from the tee is shown in the graph. Determine whether point A is a *relative minimum, relative maximum,* or *neither*. Describe what this value means in the context of this situation.

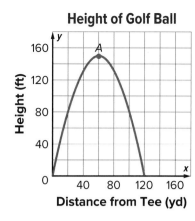

16. ROLLERCOASTER The height of a rollercoaster compared to the distance from start is shown in the graph. Determine whether points B, C, D, and F are *relative minima, relative maxima,* or *neither*. Describe what each value means in the context of this situation.

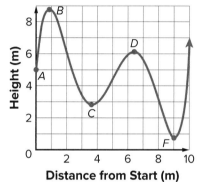

Examples 6 and 7
Determine the end behavior of f(x).

17.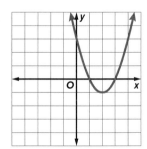

18.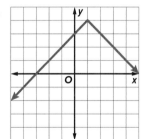

19.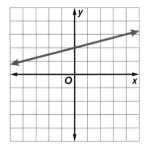

20.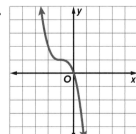

Lesson 3-5 • Shapes of Graphs **189**

Mixed Exercises

Determine whether each function has line symmetry and where the function is *increasing* and/or *decreasing*. Determine the extrema. Then identify each point as a *relative maximum* or *relative minimum*. Determine the end behavior.

21.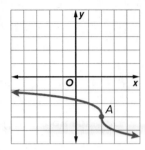

22.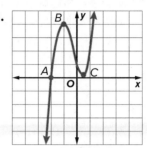

23. **ROCKS** The graph shows the height of a rock after it is thrown into the air over time. Determine the extrema. Then identify each point as a *relative minimum*, *relative maximum*, or *neither*. Describe what each value means in the context of this situation.

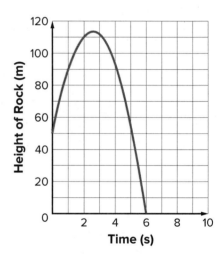

Higher-Order Thinking Skills

24. **WRITE** The graph shows the number of computers that are affected by a virus over time. Determine and interpret the end behavior.

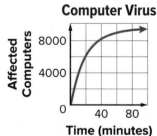

25. **WHICH ONE DOESN'T BELONG?** Which statement about the graph is not true? Justify your conclusion.

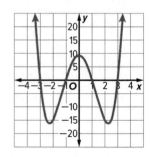

> The graph is symmetric in the line $x = 0$.
> The graph has one relative minimum at about $(-2.25, -16)$.
> The graph has one relative maximum at about $(0, 9)$.
> As x decreases, y increases. As x increases, y increases.

190 Module 3 • Relations and Functions

Lesson 3-6

Sketching Graphs and Comparing Functions

Explore Modeling Relationships by Using Functions

Online Activity Use the infographic to complete the Explore.

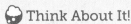

INQUIRY How can you use key features to approximate the graphs of functions?

Learn Sketching Graphs of Functions

You can sketch the graph of a function using its key features. Knowing the domain, range, intercepts, symmetry, end behavior, and extrema of a function as well as intervals where the function is increasing, decreasing, positive, or negative provides a clear idea of what the graph of the function looks like.

Label the graph to identify its key features

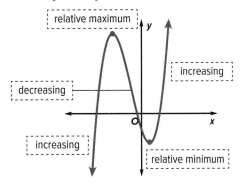

Today's Goal
• Sketch graphs of functions.

Think About It!
What other key features do you see in the graph of the function? Give two examples.

Example 1 Sketch the Graph of a Linear Function

CYCLING In 2015, Christoph Strasser set a new 24-hour cycling record by riding 556 miles in a 24-hour period. The distance he rode over the 24 hours can be represented by a function. Sketch a graph that shows the distance traveled *y* as a function of time *x*.

Before sketching, consider any possible constraints of the situation. It is not possible for Christoph Strasser to ride for a negative amount of time or ride negative miles. Therefore, the domain and range are restricted to nonnegative *x*- and *y*-values, and the graph exists only in the first quadrant.

y-Intercept: No distance traveled when he has ridden for 0 hours.

The point represents 0 hours ridden and 0 miles traveled. Graph this on the coordinate plane.

Linear or Nonlinear: The graph of the function is linear.

Positive: for time greater than 0

Increasing: for time greater than 0

The graph is a straight line that is positive and increasing for all hours ridden.

End Behavior: As the number of hours he has ridden increases, the number of miles he has traveled increases.

As *x* increases, *y* increases.

Think About It!
What assumption is made when graphing Christoph Strasser's record-setting bike ride? Why is it necessary to make this assumption?

Watch Out!
If information about some key features is not provided, do not assume that it is not important. In this example, the missing information about where the function is negative was not necessary because it did not apply in the context of the situation. However, this is not always the case.

Go Online You can complete an Extra Example online.

Example 2 Sketch the Graph of a Symmetric Function

WEATHER A person's happiness can be affected by temperature. Sketch a nonlinear graph that shows the happiness of a person y as a function of temperature x. Interpret the key features.

Positive: between about 25°F and 89°F

Negative: for temperatures less than 25°F and greater than 89°F

Increasing: for temperatures less than about 57°F

Decreasing: for temperatures greater than about 57°F

Relative Maximum: at about 57°F, when a person's happiness is about 85

A relative maximum occurs at 57°F, or $x = 57$, and a happiness of 85, or $y = 85$. This is represented by the point (57, 85), which we can graph on the coordinate plane.

End Behavior: As temperature increases or decreases, a person's happiness decreases.

For temperatures less than 25°F or $x < 25$, the graph is negative. For these temperatures, the graph is also increasing. For temperatures between 25°F and 57°F, the graph is positive and increasing. The graph is positive and decreasing for temperatures between 57°F and 89°F. This interval of the graph is also symmetric to the graph from 25°F to 57°F. This means that the right half of the graph is the mirror image of the left half. For temperatures greater than 89°F, or $x > 89$, the graph is negative, decreasing, and symmetric to the interval of the graph that is less than 25°F. As temperature increases or decreases, a person's happiness decreases. This means that happiness will get increasingly negative to move right and left on the graph.

Symmetry: A person's happiness for temperatures less than 57°F is the same as their happiness for temperatures greater than 57°F.

A person is happiest when it is 57°F. As the temperature gets increasingly cold or hot, a person becomes less happy. When the temperature is below about 25°F or above about 89°F a person is unhappy.

(continued on the next page)

Think About It!
Describe the location of the points where the function changes from positive to negative or negative to positive.

Think About It!
What does a negative y-value represent in the context of this situation?

Go Online You can complete an Extra Example online.

One method for sketching a graph given key features is to first graph any given points. Then analyze the key features of small intervals from left to right to make sketching the graph of the function easier.

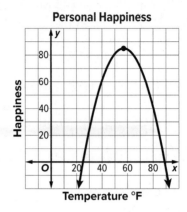

Personal Happiness

Check

Mariana used the key features to sketch the graph of x as a function of y. Examine the key features and graph to identify which key features Mariana graphed correctly and incorrectly.

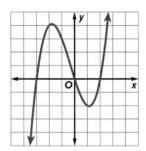

A. **Positive:** between $x = -3$ and $x = 0$ and for $x > 2$

B. **Negative:** for $x < -3$ and between $x = 0$ and $x = 2$

C. **Increasing:** for about $x < -2$ and between about $x = 1$ and $x = 3$

D. **Decreasing:** for between about $x = -2$ and $x = 1$ and for $x > 3$

E. **Intercepts:** The graph intersects the x-axis at $(-3, 0)$, $(0, 0)$, and $(2, 0)$.

F. **Relative Minimum:** $(1, -2)$

G. **Relative Maximum:** $(-2, 4)$ and $(3, 2)$

H. **End Behavior:** As x increases and decreases, the value of y decreases.

	Correct	Incorrect
A		
B		
C		
D		
E		
F		
G		
H		

Example 3 Sketch the Graph of a Nonlinear Function

AMUSEMENT PARK The number of people in line for a rollercoaster throughout the day can be modeled by a function. Use the key features to sketch a graph of the function. Then interpret the key features if x represents the time in hours since the ride opened at 10:00 A.M. and y represents the number of people in line.

Positive: between $x = -0.5$ and $x = 12$

Negative: for $x < -0.5$ and $x > 12$

Increasing: for $x < 1.4$ and between $x = 5.3$ and $x = 9.9$

Decreasing: for between $x = 1.4$ and $x = 5.3$ and for $x > 9.9$

Intercepts: The graph intersects the x-axis at $(-0.5, 0)$ and $(12, 0)$ and intersects the y-axis at $(0, 220)$.

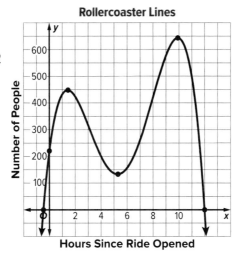

Relative Minimum: at $(5.3, 133)$

Relative Maximum: at $(1.4, 448)$ and $(9.9, 643)$

End Behavior: As x increases or decreases, the value of y decreases

> **Think About It!**
> Why might there be a relative minimum in the number of people in line around 3:00 P.M., 5.3 hours after the ride opened? Why might there be zero people in line 12 hours after the ride opened, at 10:00 P.M.?

The x-intercepts mean that the number of people in line is zero a half hour before the ride opened and 12 hours after it opened. The y-intercept means that 220 people were in line when the ride opened.

The ride experienced a relative low in the number of people in line 5.3 hours after the ride opened and two relative peaks in the number of people in line 1.4 hours and 9.9 hours after it opened.

The number of people in line was negative but increasing until a half hour before the ride opened, positive and increasing from a half hour before the ride opened until 1.4 hours after it opened and again from 5.3 hours after the ride opened until 9.9 hours after it opened, negative and decreasing after the ride had been open for 12 hours, and positive but decreasing from 1.4 hours after the ride opened until 5.3 hours after it opened and again from 9.9 hours after the ride opened until 12 hours after the ride opened.

The graph indicates a period where there is a negative number of people in line. Because it is not possible to have a negative number of people, this graph appears to only model the number of people in line for the ride from a half hour before the ride opened until 12 hours after it opened.

Go Online You can complete an Extra Example online.

Lesson 3-6 • Sketching Graphs and Comparing Functions **195**

Math History Minute

Sometimes the coordinate plane is referred to as the *Cartesian plane*. It is named for French mathematician **Rene Descartes (1596–1650)**, who is known as the father of analytical geometry, the bridge between algebra and geometry.

Check

MARINE LIFE The path of a dolphin jumping out of the ocean can be modeled with a symmetric function.

Part A Which graph could be used to show the height of a dolphin above the water y as a function of time since it emerged from the water x?

Positive: between 0 and 8 seconds

Negative: for time less than 0 seconds and greater than 8 seconds

Decreasing: for time greater than 4 seconds

Relative Maximum: at 4 seconds

End Behavior: As x increases and decreases, the value of y decreases.

Symmetry: The right half of the graph is the mirror image of the left half in approximately the line $x = 4$.

A.
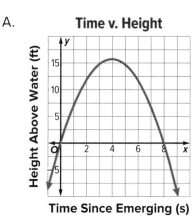

B.
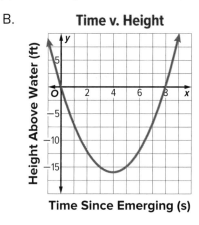

C.
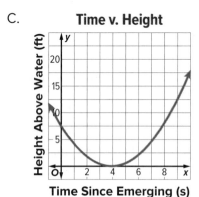

D.
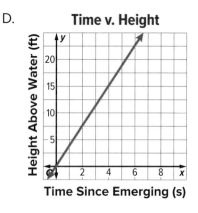

Part B Negative x-values represent _____?_____ and negative y-values represent _____?_____.

196 Module 3 · Relations and Functions

Example 4 Compare Properties of Functions

TENNIS Hawk-Eye is a computer system used in tennis to track the path of the ball. It is used as an officiating aid to locate the landing spot of a tennis ball when players challenge a call. Use the description and graph of a player's forehand and backhand shots to compare the paths of the two shots if y is the vertical height and x is the distance.

Talk About It!

Compare the intervals over which each shot is positive and/or negative. Does this make sense in the context of the situation? Explain your reasoning.

Forehand

During the forehand, the ball leaves the player's racquet at a height of 2.8 feet and travels 29 feet, when it reaches a height of about 10 feet. Then, the height of the ball decreases until it hits the ground 58 feet from where it was hit.

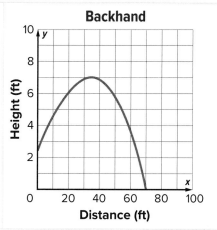

Backhand

	Forehand	Backhand
x-intercept	58	70
y-intercept	2.8	2.5
Extrema	maximum height of 10 feet when $x = 29$.	maximum height of 7 feet when $x = 35$.
Increasing and Decreasing	increases to a height of 10 feet from $x = 0$ to $x = 29$ and then decreases from $x = 29$ to $x = 58$ to a height of 0 feet.	increases to a height of 7 feet from $x = 0$ to $x = 35$ and then decreases from $x = 35$ to $x = 70$ to a height of 0 feet.

x-intercept

The tennis ball travels 12 feet farther during the backhand shot.

y-intercept

The y-intercepts of the two functions mean that the tennis ball is about 0.3 foot higher at the beginning of the forehand shot.

Extrema

The maximum height of the tennis ball is 3 feet higher during the forehand shot.

Increasing and Decreasing

The height of the tennis ball increases over a shorter interval during the forehand shot, but it reaches a higher maximum height. This means that the tennis ball increases at a faster rate during the forehand.

Go Online You can complete an Extra Example online.

Check

CARS Use the description and graph to compare the fuel economy of two cars, where y is the fuel efficiency in miles per gallon and x is the speed in miles per hour.

Car A

The fuel efficiency increases for speeds up to 25 mph when it reaches a relative maximum efficiency of 53 mpg. The fuel efficiency then decreases for speeds between 25 mph and 41 mph, when it gets down to 37 mpg. Above 41 mph, efficiency increases again until it reaches 48 mpg at 60 mph. Finally, the fuel efficiency rapidly decreases for speeds greater than 60 mph until leveling off at 17 mpg.

Car B

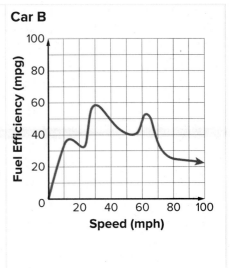

Which statements about the fuel efficiencies of the two cars are true? Select all that apply.

A. Car A has the greatest maximum fuel efficiency.

B. Car B has more relative maximum fuel efficiencies than Car A.

C. As speed increases, Car A levels off at a greater fuel efficiency than Car B.

D. Both cars get 0 mpg when they are traveling at 0 mph.

E. Car A has the least relative minimum fuel efficiency.

F. Both cars increase in fuel efficiency for speeds between 0 mph and about 13 mph.

G. Neither car reaches fuel efficiency below 0 mpg.

Practice

Go Online You can complete your homework online.

Example 1

1. **SAVINGS** David is saving money to buy a new car. The amount he saves can be represented by a function. Sketch a graph that shows the amount in savings y, in dollars, as a function of time x, in weeks.

   x-Intercept: none

   y-Intercept: $1400

   Linear or Nonlinear: The graph of the function is linear.

   Positive: for time greater than 0

   Increasing: for time greater than 0

   End Behavior: As the number of weeks he has saved increases, the amount saved increases.

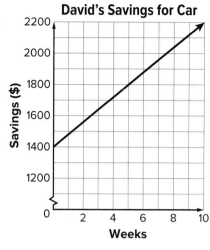

2. **SWIMMING** Yukio is keeping track of the number of calories she burns while swimming freestyle laps. The number of calories she burns can be represented by a function. Sketch a graph that shows the number of calories burned y as a function of time x, in hours.

   y-Intercept: No calories burned when she has swum for 0 hours.

   Linear or Nonlinear: The graph of the function is linear.

   Positive: for time greater than 0

   Increasing: for time greater than 0

   End Behavior: As the number of hours she has swum increases, the number of calories burned increases.

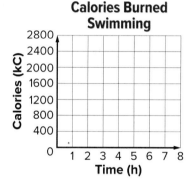

Example 2

3. **FOOTBALL** The flight of a football thrown by a quarterback can be modeled by an interval of a function. Sketch a nonlinear graph that shows the height of a football y, in feet, as a function of time x, in seconds.

   Positive: between 0 seconds and 5 seconds

   Negative: for time greater than 5 seconds (represents time after the ball hits the ground)

   Increasing: for time less than 2 seconds

   Decreasing: for time greater than 2 seconds

   Relative Maximum: at 2 seconds, when the height of the football is 9 feet

   End Behavior: As time increases, the height of the football decreases.

   Symmetry: The height of the football for time between 0 seconds and 2 seconds is the same as the height for time between 2 seconds and 4 seconds.

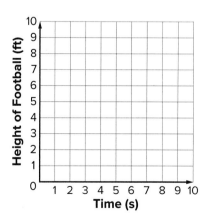

Lesson 3-6 • Sketching Graphs and Comparing Functions **199**

4. **FISH** The height of a fish compared to sea level as it jumps out of the ocean water can be represented by a function. Sketch a nonlinear graph that shows the height of a fish *y*, in inches, as a function of time *x*, in seconds.

Positive: between 2 seconds and 8 seconds

Negative: for time less than 2 seconds and greater than 8 seconds

Increasing: for time less than 5 seconds

Decreasing: for time greater 5 seconds

Relative Maximum: at 5 seconds, when the height of the fish is 9 inches

End Behavior: As time increases or decreases, the height of the fish decreases.

Symmetry: The height of the fish for time less than 5 seconds is the same as the height for time greater than 5 seconds.

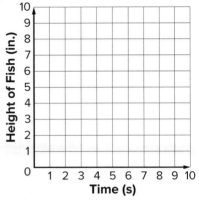

Example 3

5. **TECHNOLOGY** The results of a poll that asks Americans whether they used the Internet yesterday can be modeled as a function. Sketch a graph that shows the number of people polled that responded yes to the survey *y* as a function of time *x*, months since January 2005.

Positive: for time greater than 0 months

Negative: none

Increasing: for all time greater than 0 months

Decreasing: none

Intercepts: The graph intersects the *y*-axis at about (0, 58).

Extrema: none

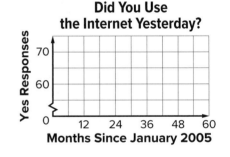

End Behavior: As *x* increases, *y* increases. The data represent people, so the maximum it could ever reach is the maximum number of people surveyed.

6. **MUSIC** The results of a poll that asks Americans whether they have listened to online music can be modeled as a function. Sketch a graph that shows the number of people polled that have listened online *y* as a function of time *x*, months since August 2000.

Positive: for time greater than 0 months

Negative: none

Increasing: between 0 months and 10 months, for time greater than about 65 months

Decreasing: between 10 months and about 65 months

Intercepts: The graph intersects the *y*-axis at about (0, 37).

Relative Minimum: at about (65, 31)

Relative Maximum: at about (10, 39)

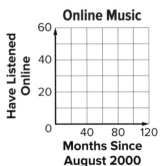

End Behavior: As *x* decreases, *y* decreases. As *x* increases, *y* increases. The data represent people, so the maximum it could ever reach is the maximum number of people surveyed.

Example 4

7. **INTERNET** Use the description and graph to compare Internet use at home and Internet use away from home, where y is the number of people polled, in thousands, that use the Internet several times a day and x is the number of months since March 2004. Use the description and graph to compare Internet use at home and Internet use away from home since March 2004.

Internet Use at Home

About 10,000 of those polled used the Internet at home in March 2004. The number of users decreased to 7000 at 36 months since March 2004. The number of users continued to increase after 36 months since March 2004.

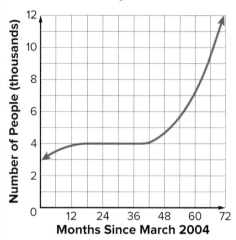

Internet Use Away from Home

8. **SPENDING** Use the description and graph to compare the amount of U.S. spending on electronics and education, where y is the amount spent in billions and x is the number of years since 1949. Write statements to compare U.S. spending on electronics and education since 1949.

U.S. Electronic Spending

In 1949, the U.S. spent $0 on electronics. Twenty years after 1949, the U.S. spent about $4 billion on electronics. Thirty years after 1949, the U.S. spent about $3 billion on electronics. Seventy years after 1949, the U.S. spent about $7.5 billion on electronics and spending continues to increase.

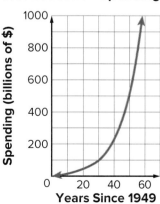

U.S. Education Spending

Lesson 3-6 • Sketching Graphs and Comparing Functions

Mixed Exercises

The costume department of a theatre company is making cone-shaped hats for a play set in medieval times. Each hat will be covered with satin over its entire lateral surface area. The slant height of each hat will remain constant at 20 inches, but the radius of the base will vary to accommodate different head sizes. Using 3.14 for π, the lateral area of the hat can be expressed as a function. Use this information for Exercises 9–11.

9. **USE A MODEL** Sketch a graph that shows the lateral area of the hat y, in square inches, as a function of the radius of its base x, in inches.

   y-Intercept: No lateral area when the radius of its base is 0 inches.

   Linear or Nonlinear: The graph of the function is linear.

   Positive: for radius of a base greater than 0

   Increasing: for radius of a base greater than 0

   End Behavior: As the radius of the base increases, the lateral area increases.

10. **REASONING** Write a function y for the lateral area of the hat as a function of the radius of its base, x. (Hint: The formula for the lateral area of a cone is $y = \pi x s$, where s is the slant height.)

11. **USE TOOLS** Enter the function into your graphing calculator. Press **WINDOW** and enter the following settings: Xmin: −10; Xmax: 10; Ymin: −1000; Ymax: 1000. Then press **GRAPH**. Compare the graph on the calculator to the graph you sketched in Exercise 9.

Aidan buys used bicycles, fixes them up, and sells them. His average cost to buy and fix each bicycle is $47. He also incurred a one-time cost of $840 to purchase tools and a small shed to use as his workshop. He sells bikes for $75 each. Use this information for Exercises 12–15.

12. **WRITE** Write revenue and cost functions $R(x)$ and $C(x)$ for Aidan's situation, where x is the number of bicycles. How do you include the one-time cost in $C(x)$?

13. **WRITE** Write a profit function $P(x)$ such that $P(x) = R(x) - C(x)$. In words, what does $P(x)$ represent?

14. **PERSEVERE** List key features for the profit function $P(x)$. Then use the key features to sketch a graph that shows the profit $P(x)$ as a function of x bicycles.

15. **ANALYZE** Which key feature of the graph represents Aidan's break-even point (profit = 0)? Explain how to use your graph to find the most accurate value for this feature.

16. **CREATE** Research the population in your state over a 10-year period. Sketch a graph to model the data. Then list the key feature of the graph.

Module 3 • Relations and Functions

Review

 Essential Question

Why are representations of relations and functions useful?

Module Summary

Lesson 3-1
Representing Relations
- A relation is a set of ordered pairs.
- Relations can be shown with ordered pairs, with a table, with a graph, or with a mapping.

Lesson 3-2
Functions
- A function is a relationship between input and output. In a function, there is exactly one output for each input.
- If a vertical line intersects the graph of a relation more than once, then the relation is not a function.
- Function notation is a way of writing an equation so that $y = f(x)$.

Lessons 3-3 through 3-5
Interpreting Graphs
- A discrete function is a set of points that are not connected. A continuous function has points that connect to form a line or curve.
- An x-intercept of a graph is a point where the graph intersects the x-axis. The y-intercept of a graph is the point where the graph intersects the y-axis.
- Equations can be solved by using the graphs of related functions.
- A figure has line symmetry if each half of the figure matches the other side exactly.

- A function is increasing where the graph goes up and decreasing where the graph goes down when viewed from left to right.
- Points that are the locations of relatively high or low function values are called extrema.
- A point is a relative minimum when no other nearby point has a lesser y-coordinate.
- A point is a relative maximum when no other nearby point has a greater y-coordinate.

Lesson 3-6
Sketching Graphs and Comparing Functions
- Knowing the intercepts, symmetry, end behavior, and extrema of a function, as well as intervals where the function is increasing, decreasing, positive, or negative, provides a clear idea of what the graph of the function looks like.

Study Organizer

 Foldables
Use your Foldable to review this module. Working with a partner can be helpful. Ask for clarification of concepts as needed.

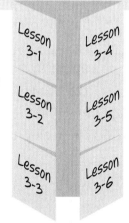

Test Practice

1. MULTI-SELECT The table and graph show the number of pounds of bananas sold at a local grocery store from 2013 to 2018. (Lesson 3-1)

x	y
0	40
1	60
2	55
3	25
4	30
5	50

Number of Pounds of Bananas Sold

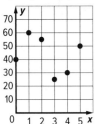

Which of the following statements correctly describes the relation? Select all that apply.

A. The x-axis has a scale mark of 1 mark = 1 pound of bananas.

B. The x-axis has a scale mark of 1 mark = 1 year.

C. The x-axis represents the years since 2013.

D. The x-axis represents the number of pounds of bananas sold.

E. The y-axis represents the years since 2013.

F. The y-axis represents the number of pounds of bananas sold.

G. The y-axis has a scale mark of 1 mark = 10 years.

2. MULTIPLE CHOICE The Hillsborough State Park in Thonotosassa, Florida, charges an admission fee of $4 plus a camping fee of $20 per night. This can be represented by the function $f(x) = 20x + 4$, where $f(x)$ is the total cost and x is the number of nights spent camping. What is the value of $f(5)$, which is the cost of 5 nights of camping? (Lesson 3-2)

A. 80

B. 100

C. 104

D. 120

3. MULTIPLE CHOICE If $f(x) = -9x + 8$, then find $f(-2)$. (Lesson 3-2)

A. −84

B. −8

C. 26

D. 100

4. MULTIPLE CHOICE Which function includes the data set below?
{(2, −2), (6, 10), (13, 31)} (Lesson 3-2)

A. $f(x) = \frac{1}{2}x - 3$

B. $f(x) = -2x + 2$

C. $f(x) = 3x - 8$

D. $f(x) = 4x - 10$

5. OPEN RESPONSE Determine whether the relation shown in the table is a function. Explain. (Lesson 3-2)

Domain	−4	2	−4	5
Range	8	11	13	13

204 Module 3 Review • Relations and Functions

6. OPEN RESPONSE Indicate whether each of the following relations are functions. (Lesson 3-2)

A. 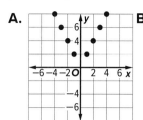   B.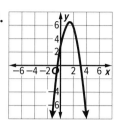

C. 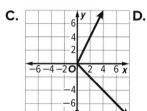   D.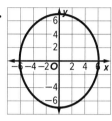

7. MULTIPLE CHOICE What is the domain of this function? (Lesson 3-3)

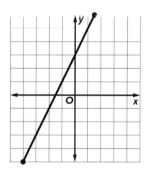

A. $-4 < x < 1.5$

B. $-4 \leq x \leq 1.5$

C. $-6 < y < 6$

D. $-6 \leq y \leq 6$

8. OPEN RESPONSE The graph shows the relationship between the number of Fun Pass tickets sold and the total value of the sales. Use the graph to estimate the x- and y-intercepts of the function, where the function is positive and negative, and interpret the meanings in the context of the situation. (Lesson 3-4)

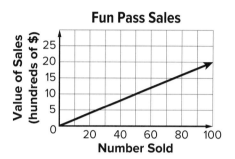

9. OPEN RESPONSE What are the intercepts for the function $y = -x - 1$ graphed below? (Lesson 3-4)

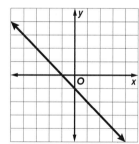

Module 3 Review • Relations and Functions **205**

10. **OPEN RESPONSE** A baker bought a bag of flour to make banana bread. The recipe calls for 1.5 cups of flour per loaf. The number of cups of flour left in the bag y after making x loaves of bread is shown in the table. (Lesson 3-4)

x	y
0	12
2	9
4	6
6	3
8	0

Write the y-intercept as an ordered pair and interpret its meaning in the real-world context.

11. **OPEN RESPONSE** A garden supply store manager found that if she used a certain function she could determine the best price to charge for the shovels she sells to maximize the revenue. The graph represents the revenue ($) y of the store at x price ($) per shovel. Use the graph to find and interpret the symmetry of the function in the context of the situation. (Lesson 3-5)

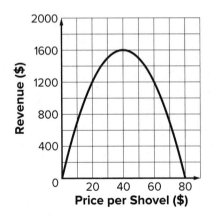

Symmetry: The graph is symmetric in the line $x = $ __?__.

Interpret symmetry: The revenue gained when a shovel is sold for $20 is the same as it is when a shovel is sold for $ __?__.

12. **MULTIPLE CHOICE** Suppose the graph of a function is increasing to the left of $x = 2$ and decreasing to the right of $x = 2$. Which describes the point at $x = 2$? (Lesson 3-5)

A. Unless you know the y-coordinate of the point, you cannot say anything about the point at $x = 2$.

B. It is an x-intercept.

C. It is a relative minimum.

D. It is a relative maximum.

13. **OPEN RESPONSE** Use the description and graph to compare the population data for Ohio and Florida, where y is the population in millions and x is the number of decades since 1900. Write statements about the populations of Ohio and Florida since 1900. (Lesson 3-6)

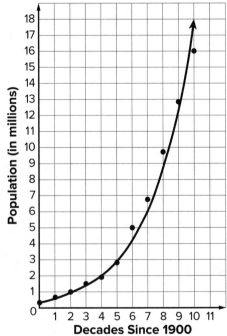

Ohio Population Since 1900
In 1900 the population of Ohio was about 4.2 million. Between 1900 and 1950, the population of Ohio nearly doubled to about 8 million. Then between 1950 and 2000, the population of Ohio grew to approximately 11.4 million. Beyond 2000, the population of Ohio continues to gradually increase.

Module 4
Linear and Nonlinear Functions

Essential Question
What can a function tell you about the relationship that it represents?

What Will You Learn?
How much do you already know about each topic **before** starting this module?

KEY
👎 — I don't know.   👉 — I've heard of it.   👍 — I know it!

	Before 👎	Before 👉	Before 👍	After 👎	After 👉	After 👍
graph linear equations by using a table						
graph linear equations by using intercepts						
find rates of change						
determine slopes of linear equations						
write linear equations in slope-intercept form						
graph linear functions in slope-intercept form						
translate, dilate, and reflect linear functions						
identify and find missing terms in arithmetic sequences						
write arithmetic sequences as linear functions						
model and use piecewise functions, step functions, and absolute value functions						
translate absolute value functions						

Foldables Make this Foldable to help you organize your notes about functions. Begin with five sheets of grid paper.

1. **Fold** five sheets of grid paper in half from top to bottom.
2. **Cut** along fold. Staple the eight half-sheets together to form a booklet.
3. **Cut** tabs into margin. The top tab is 4 lines wide, the next tab is 8 lines wide, and so on. When you reach the bottom of a sheet, start the next tab at the top of the page.
4. **Label** each tab with a lesson number. Use the extra pages for vocabulary.

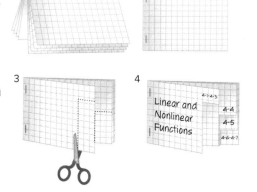

What Vocabulary Will You Learn?

- absolute value function
- arithmetic sequence
- common difference
- constant function
- dilation
- family of graphs
- greatest integer function
- interval
- identity function
- nth term of an arithmetic sequence
- parameter
- parent function
- piecewise-defined function
- piecewise-linear function
- rate of change
- reflection
- sequence
- slope
- step function
- term of a sequence
- transformation
- translation
- vertex

Are You Ready?

Complete the Quick Review to see if you are ready to start this module. Then complete the Quick Check.

Quick Review

Example 1

Graph $A(3, -2)$ on a coordinate grid.

Start at the origin. Because the x-coordinate is positive, move 3 units to the right. Then move 2 units down because the y-coordinate is negative. Draw a dot and label it A.

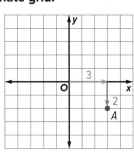

Example 2

Solve $x - 2y = 8$ for y.

$x - 2y = 8$ Original expression

$x - x - 2y = 8 - x$ Subtract x from each side.

$-2y = 8 - x$ Simplify.

$\dfrac{-2y}{-2} = \dfrac{8-x}{-2}$ Divide each side by -2.

$y = \dfrac{1}{2}x - 4$ Simplify.

Quick Check

Graph and label each point on the coordinate plane.

1. $B(-3, 3)$
2. $C(-2, 1)$
3. $D(3, 0)$
4. $E(-5, -4)$
5. $F(0, -3)$
6. $G(2, -1)$

Solve each equation for y.

7. $3x + y = 1$
8. $8 - y = x$
9. $5x - 2y = 12$
10. $3x + 4y = 10$
11. $3 - \dfrac{1}{2}y = 5x$
12. $\dfrac{y+1}{3} = x + 2$

How did you do?

Which exercises did you answer correctly in the Quick Check?

Lesson 4-1
Graphing Linear Functions

Explore Points on a Line

Online Activity Use an interactive tool to complete an Explore.

> **INQUIRY** How is the graph of a linear equation related to its solutions?

Today's Goals
- Graph linear functions by making tables of values.
- Graph linear functions by using the x- and y-intercepts.

Learn Graphing Linear Functions by Using Tables

A table of values can be used to graph a linear function. Every ordered pair that makes the equation true represents a point on its graph. So, a graph represents all the solutions of an equation.

Linear functions can be represented by equations in two variables.

Example 1 Graph by Making a Table

Graph $-2x - 3 = y$ by making a table.

Step 1 Choose any values of x from the domain and make a table.

Step 2 Substitute each x-value into the equation to find the corresponding y-value. Then, write the x- and y-values as an ordered pair.

x	$-2x - 3$	y	(x, y)
-4	$-2(-4) - 3$	5	$(-4, 5)$
-2	$-2(-2) - 3$	1	$(-2, 1)$
0	$-2(0) - 3$	-3	$(0, -3)$
1	$-2(1) - 3$	-5	$(1, -5)$
3	$-2(3) - 3$	-9	$(3, -9)$

Step 3 Graph the ordered pairs in the table and connect them with a line.

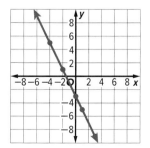

Go Online You can complete an Extra Example online.

Talk About It!
What values of x might be easiest to use when graphing a linear equation when the x-coefficient is a whole number? Justify your argument.

Study Tip

Exactness Although only two points are needed to graph a linear function, choosing three to five x-values that are spaced out can verify that your graph is correct.

Check

Graph $y = 2x + 5$ by using a table. Copy and complete the table. Then graph the function.

x	y
−5	
−3	
−1	
0	
2	

🧠 **Think About It!**
What are some values of x that you might choose in order to graph $y = \frac{1}{7}x - 12$?

Example 2 Choose Appropriate Domain Values

Graph $y = \frac{1}{4}x + 3$ by making a table.

Step 1 Make a table.

Step 2 Find the y-values.

Step 3 Graph the ordered pairs in the table and connect them with a line.

x	$\frac{1}{4}x + 3$	y	(x, y)
−8	$\frac{1}{4}(-8) + 3$	1	(−8, 1)
−4	$\frac{1}{4}(-4) + 3$	2	(−4, 2)
0	$\frac{1}{4}(0) + 3$	3	(0, 3)
4	$\frac{1}{4}(4) + 3$	4	(4, 4)
8	$\frac{1}{4}(8) + 3$	5	(8, 5)

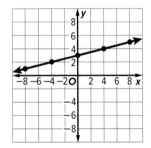

Watch Out!
Equivalent Equations
Sometimes, the variables are on the same side of the equal sign. Rewrite these equations by solving for y to make it easier to find values for y.

Check

Graph $y = \frac{3}{5}x - 2$ by making a table. Copy and complete the table. Then graph the function.

x	y
−10	
−5	
0	
5	
10	

📡 **Go Online** You can complete an Extra Example online.

Example 3 Graph $y = a$

Graph $y = 5$ by making a table.

Step 1 Rewrite the equation.
$y = 0x + 5$

Step 2 Make a table.

x	0x + 5	y	(x, y)
−2	0(−2) + 5	5	(−2, 5)
−1	0(−1) + 5	5	(−1, 5)
0	0(0) + 5	5	(0, 5)
1	0(1) + 5	5	(1, 5)
2	0(2) + 5	5	(2, 5)

Step 3 Graph the line.

The graph of $y = 5$ is a horizontal line through $(x, 5)$ for all values of x in the domain.

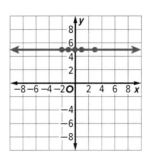

Think About It!
In general, what does the graph of an equation of the form $y = a$, where a is any real number, look like?

Example 4 Graph $x = a$

Graph $x = -2$.

You learned in the previous example that equations of the form $y = a$ have graphs that are horizontal lines. Equations of the form $x = a$ have graphs that are vertical lines.

The graph of $x = -2$ is a vertical line through $(-2, y)$ for all real values of y. Graph ordered pairs that have x-coordinates of -2 and connect them with a vertical line.

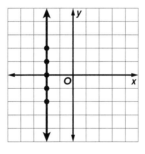

Think About It!
Is $x = a$ a function? Why or why not?

Check

Graph $x = 6$.

Go Online You can complete an Extra Example online.

Explore Lines Through Two Points

Go Online You can watch a video to see how to graph linear functions.

Online Activity Use graphing technology to complete an Explore.

INQUIRY How many lines can be formed with two given points?

Learn Graphing Linear Functions by Using the Intercepts

Think About It! Why are the *x*- and *y*-intercepts easy to find?

You can graph a linear function given only two points on the line. Using the *x*- and *y*-intercepts is common because they are easy to find. The intercepts provide the ordered pairs of two points through which the graph of the linear function passes.

Example 5 Graph by Using Intercepts

Graph $-x + 2y = 8$ by using the *x*- and *y*-intercepts.

To find the *x*-intercept, let $y = 0$.

$$-x + 2y = 8 \quad \text{Original equation}$$
$$-x + 2(0) = 8 \quad \text{Replace } y \text{ with 0.}$$
$$-x = 8 \quad \text{Simplify.}$$
$$x = -8 \quad \text{Divide.}$$

This means that the graph intersects the *x*-axis at $(-8, 0)$.

Think About It! What does a line that only has an *x*-intercept look like? a line that only has a *y*-intercept?

To find the *y*-intercept, let $x = 0$.

$$-x + 2y = 8 \quad \text{Original equation}$$
$$-0 + 2y = 8 \quad \text{Replace } x \text{ with 0.}$$
$$2y = 8 \quad \text{Simplify.}$$
$$y = 4 \quad \text{Divide.}$$

This means that the graph intersects the *y*-axis at $(0, 4)$.

Graph the equation.

Study Tip

Tools When drawing lines by hand, it is helpful to use a straightedge or a ruler.

Step 1 Graph the *x*-intercept.

Step 2 Graph the *y*-intercept.

Step 3 Draw a line through the points.

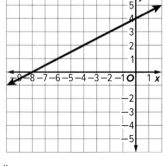

Go Online You can complete an Extra Example online.

212 Module 4 • Linear and Nonlinear Functions

Check

Graph $4y = -12x + 36$ by using the x- and y- intercepts.

x-intercept: __?__

y-intercept: __?__

Example 6 Use Intercepts

PETS Angelina bought a 15-pound bag of food for her dog. The bag contains about 60 cups of food, and she feeds her dog $2\frac{1}{2}$ or $\frac{5}{2}$ cups of food per day. The function $y + \frac{5}{2}x = 60$ represents the amount of food left in the bag y after x days. Graph the amount of dog food left in the bag as a function of time.

Part A
Find the x- and y-intercepts and interpret their meaning in the context of the situation.

To find the x-intercept, let $y = 0$.

$$y + \frac{5}{2}x = 60 \qquad \text{Original equation}$$

$$0 + \frac{5}{2}x = 60 \qquad \text{Replace } y \text{ with 0.}$$

$$\frac{5}{2}x = 60 \qquad \text{Simplify.}$$

$$x = 24 \qquad \text{Multiply each side by } \frac{2}{5}.$$

The x-intercept is 24. This means that the graph intersects the x-axis at (24, 0). So, after 24 days, there is no dog food left in the bag.

To find the y-intercept, let $x = 0$.

$$y + \frac{5}{2}x = 60 \qquad \text{Original equation}$$

$$y + \frac{5}{2}(0) = 60 \qquad \text{Replace } x \text{ with 0.}$$

$$y = 60 \qquad \text{Simplify.}$$

The y-intercept is 60. This means that the graph intersects the y-axis at (0, 60). So, after 0 days, there are 60 cups of food in the bag.

Go Online You can watch a video to see how to use a graphing calculator with this example.

Think About It! Find another point on the graph. What does it mean in the context of the problem?

(continued on the next page)

Lesson 4-1 • Graphing Linear Functions

Think About It!
What assumptions did you make about the amount of food Angelina feeds her dog each day?

Part B
Graph the equation by using the intercepts.

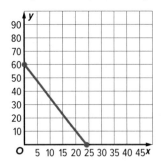

Check

PEANUTS A farm produces about 4362 pounds of peanuts per acre. One cup of peanut butter requires about $\frac{2}{3}$ pound of peanuts. If one acre of peanuts is harvested to make peanut butter, the function $y = -\frac{2}{3}x + 4362$ represents the pounds of peanuts remaining y after x cups of peanut butter are made.

x-intercept: __?__

y-intercept: __?__

Which graph uses the x- and y-intercepts to correctly graph the equation?

A.
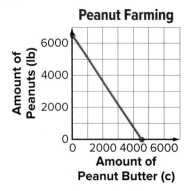

B.
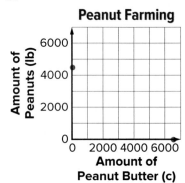

C.
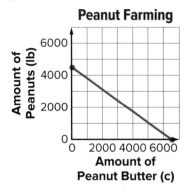

D.
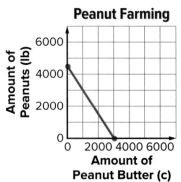

Go Online You can watch a video to see how to graph a linear function using a graphing calculator.

Go Online You can complete an Extra Example online.

Practice

Go Online You can complete your homework online.

Examples 1 through 4

Graph each equation by making a table.

1. $x = -2$

2. $y = -4$

3. $y = -8x$

4. $3x = y$

5. $y - 8 = -x$

6. $x = 10 - y$

7. $y = \frac{1}{2}x + 1$

8. $y + 2 = \frac{1}{4}x$

Lesson 4-1 • Graphing Linear Functions 215

Example 5

Graph each equation by using the x-and y-intercepts.

9. $y = 4 + 2x$

10. $5 - y = -3x$

11. $x = 5y + 5$

12. $x + y = 4$

13. $x - y = -3$

14. $y = 8 - 6x$

Example 6

15. SCHOOL LUNCH Amanda has $210 in her school lunch account. She spends $35 each week on school lunches. The equation $y = 210 - 35x$ represents the total amount in Amanda's school lunch account y for x weeks of purchasing lunches.

    a. Find the x- and y-intercepts and interpret their meaning in the context of the situation.

    b. Graph the equation by using the intercepts.

16. SHIPPING The *OOCL Shenzhen,* one of the world's largest container ships, carries 8063 TEUs (1280-cubic-feet containers). Workers can unload a ship at a rate of 1 TEU every minute. The equation $y = 8063 - 60x$ represents the number of TEUs on the ship y after x hours of the workers unloading the containers from the *Shenzhen.*

    a. Find the x- and y-intercepts and interpret their meaning in the context of the situation.

    b. Graph the equation by using the intercepts.

Mixed Exercises

Graph each equation.

17. $1.25x + 7.5 = y$

18. $2x - 3 = 4y + 6$

19. $3y - 7 = 4x + 1$

Find the x-intercept and y-intercept of the graph of each equation.

20. $5x + 3y = 15$

21. $2x - 7y = 14$

22. $2x - 3y = 5$

23. $6x + 2y = 8$

24. $y = \frac{1}{4}x - 3$

25. $y = \frac{2}{3}x + 1$

26. **HEIGHT** The height of a woman can be predicted by the equation $h = 81.2 + 3.34r$, where h is her height in centimeters and r is the length of her radius bone in centimeters.

   a. What are the r- and h-intercepts of the equation? Do they make sense in the situation? Explain.

   b. Graph the equation by using the intercepts.

   c. Use the graph to find the approximate height of a woman whose radius bone is 25 centimeters long.

27. **TOWING** Pick-M-Up Towing Company charges $40 to hook a car and $1.70 for each mile that it is towed. Write an equation that represents the total cost y for x miles towed. Graph the equation. Find the y-intercept, and interpret its meaning in the context of the situation.

28. **USE A MODEL** Elias has $18 to spend on peanuts and pretzels for a party. Peanuts cost $3 per pound and pretzels cost $2 per pound. Write an equation that relates the number of pounds of pretzels y and the number of pounds of peanuts x. Graph the equation. Find the x- and y-intercepts. What does each intercept represent in terms of context?

29. REASONING One football season, a football team won 4 more games than they lost. The function $y = x + 4$ represents the number of games won y and the number of games lost x. Find the x- and y-intercepts. Are the x- and y-intercepts reasonable in this situation? Explain.

30. WRITE Consider real-world situations that can be modeled by linear functions.
  a. Write a real-world situation that can be modeled by a linear function.

  b. Write an equation to model your real-world situation. Be sure to define variables. Then find the x- and y-intercepts. What does each intercept represent in your context?

  c. Graph your equation by making a table. Include a title for the graph as well as labels and titles for each axis. Explain how you labeled the x- and y-axes. State a reasonable domain for this situation. What does the domain represent?

31. FIND THE ERROR Geroy claims that every line has both an x- and a y-intercept. Is he correct? Explain your reasoning.

32. WHICH ONE DOESN'T BELONG? Which equation does not belong with the other equations? Justify your conclusion.

| $y = 2 - 3x$ | $5x = y - 4$ | $y = 2x + 5$ | $y - 4 = 0$ |

33. ANALYZE Robert sketched a graph of a linear equation $2x + y = 4$. What are the x- and y-intercepts of the graph? Explain how Robert could have graphed this equation using the x- and y-intercepts.

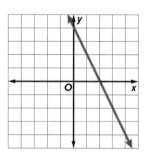

34. ANALYZE Compare and contrast the graph of $y = 2x + 1$ with the domain $\{1, 2, 3, 4\}$ and $y = 2x + 1$ with the domain all real numbers.

CREATE Give an example of a linear equation in the form $Ax + By = C$ for each condition. Then describe the graph of the equation.

35. $A = 0$ **36.** $B = 0$ **37.** $C = 0$

Lesson 4-2

Rate of Change and Slope

Learn Rate of Change of a Linear Function

The **rate of change** is how a quantity is changing with respect to a change in another quantity.

If x is the independent variable and y is the dependent variable, then rate of change $= \dfrac{\text{change in } y}{\text{change in } x}$.

Example 1 Find the Rate of Change

COOKING Find the rate of change of the function by using two points from the table.

Amount of Flour x (cups)	Pancakes y
2	12
4	24
6	36

$$\text{rate of change} = \dfrac{\text{change in } y}{\text{change in } x}$$
$$= \dfrac{\text{change in pancakes}}{\text{change in flour}}$$
$$= \dfrac{24 - 12}{4 - 2}$$
$$= \dfrac{12}{2} \text{ or } \dfrac{6}{1}$$

The rate is $\dfrac{6}{1}$ or 6. This means that you could make 6 pancakes for each cup of flour.

Check

Find the rate of change.

$\dfrac{\;\;?\;\;\text{dollars}}{\text{gallons}}$

Amount of Gasoline Purchased (Gallons)	Cost (Dollars)
4.75	15.77
6	19.92
7.25	24.07
8.5	28.22

Today's Goals
- Calculate and interpret rate of change.
- Calculate and interpret slope.

Today's Vocabulary
rate of change
slope

Think About It!

Suppose you found a new recipe that makes 6 pancakes when using 2 cups of flour, 12 pancakes when using 4 cups of flour, and 18 pancakes when using 6 cups of flour. How does this change the rate you found for the original recipe?

Study Tip

Placement Be sure that the dependent variable is in the numerator and the independent variable is in the denominator. In this example, the number of pancakes you can make *depends* on the amount of flour you can use.

Go Online You can complete an Extra Example online.

Lesson 4-2 • Rate of Change and Slope **219**

Example 2 Compare Rates of Change

STUDENT COUNCIL The Jackson High School Student Council budget varies based on the fundraising of the previous year.

Think About It!
How is a greater increase or decrease of funds represented graphically?

Part A Find the rate of change for 2000–2005 and describe its meaning in the context of the situation.

$$\frac{\text{change in budget}}{\text{change in time}} =$$

$$\frac{1675 - 1350}{2005 - 2000} = \frac{325}{5}, \text{ or } 65$$

This means that the student council's budget increased by $325 over the 5-year period, with a rate of change of $65 per year.

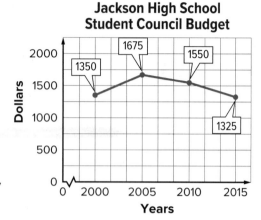
Jackson High School Student Council Budget

Study Tip

Assumptions In this example, we assumed that the rate of change for the budget was constant between each 5-year period. Although the budget might have varied from year to year, analyzing in larger periods of time allows us to see trends within data.

Part B Find the rate of change for 2010–2015 and describe its meaning in the context of the situation.

$$\frac{\text{change in budget}}{\text{change in time}} = \frac{1325 - 1550}{2015 - 2010} = \frac{-225}{5}, \text{ or } = -45$$

This means that the student council's budget was reduced by $225 over the 5-year period, with a rate of change of −$45 per year.

Check

TICKETS The graph shows the average ticket prices for the Miami Dolphins football team.

Part A Find the rate of change in ticket prices between 2009–2010.

$$\frac{\quad ? \quad}{} \frac{\text{dollars}}{\text{year}}$$

Part B The ticket prices have the greatest rate of change between ___?___

Part C Between ___?___ and ___?___, the rate of change is negative.

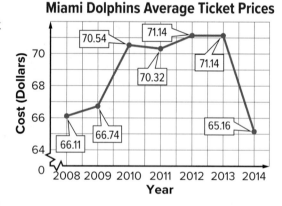
Miami Dolphins Average Ticket Prices

Go Online You can complete an Extra Example online.

Example 3 Constant Rate of Change

Determine whether the function is linear. If it is, state the rate of change.

Find the changes in the x-values and the changes in the y-values.

Notice that the rate of change for each pair of points shown is $-\frac{2}{3}$.

The rates of change are constant, so the function is linear. The rate of change is $-\frac{2}{3}$.

x	y
11	−5
8	−3
5	−1
2	1
−1	3

Example 4 Rate of Change

Determine whether the function is linear. If it is, state the rate of change.

Find the changes in the x-values and the changes in the y-values.

The rates of change are not constant. Between some pairs of points the rate of change is $\frac{3}{7}$, and between the other pairs it is $\frac{2}{7}$. Therefore, this is not a linear function.

x	y
22	−4
29	−1
36	1
43	4
50	6

Study Tip

Linear Versus Not Linear Remember that the word *linear* means that the graph of the function is a straight line. For the graph of a function to be a line, it has to be increasing or decreasing at a constant rate.

Check

Copy and complete the table so that the function is linear.

x	y
	−2.25
	1
11	
10.5	7.5
10	10.75
9.5	

Go Online You can complete an Extra Example online.

Lesson 4-2 • Rate of Change and Slope

Go Online
You can watch a video to see how to find the slope of a nonvertical line.

Think About It!
If the point (1, 3) is on a line, what other point could be on the line to make the slope positive? negative? zero? undefined?

Think About It!
Can a line that passes through two specific points, such as the origin and (2, 4), have more than one slope? Explain your reasoning.

Think About It!
How would lines with slopes of $m = \frac{1}{8}$ and $m = 80$ compare on the same coordinate plane?

Explore Investigating Slope

Online Activity Use graphing technology to complete an Explore.

INQUIRY How does slope help to describe a line?

Learn Slope of a Line

The **slope** of a line is the rate of change in the *y*-coordinates (rise) for the corresponding change in the *x*-coordinates (run) for points on the line.

Key Concept • Slope	
Words	The slope of a nonvertical line is the ratio of the rise to the run.
Symbols	The slope *m* of a nonvertical line through any two points (x_1, y_1) and (x_2, y_2) can be found as follows. $m = \frac{y_2 - y_1}{x_2 - x_1}$
Example	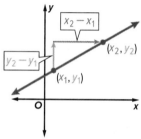

The slope of a line can show how a quantity changes over time. When finding the slope of a line that represents a real-world situation, it is often referred to as the *rate of change*.

Example 5 Positive Slope

Find the slope of a line that passes through (−3, 4) and (1, 7).

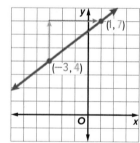

$$m = \frac{y_2 - y_1}{x_2 - x_1}$$

$$= \frac{7 - 4}{1 - (-3)}$$

$$= \frac{3}{4}$$

Check

Determine the slope of a line passing through the given points. If the slope is undefined, write *undefined*. Write your answer as a decimal if necessary.

(−1, 8) and (7, 10)

Go Online You can complete an Extra Example online.

Example 6 Negative Slope

Find the slope of a line that passes through (−1, 3) and (4, 1).

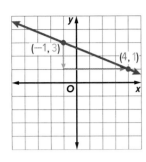

$$m = \frac{y_2 - y_1}{x_2 - x_1}$$

$$= \frac{1 - 3}{4 - (-1)}$$

$$= -\frac{2}{5}$$

Check

Determine the slope of a line passing through the given points. If the slope is undefined, write *undefined*. Write your answer as a decimal if necessary.

(5, −4) and (0, 1)

Example 7 Slopes of Horizontal Lines

Find the slope of a line that passes through (−2, −5) and (4, −5).

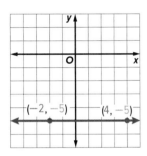

$$m = \frac{y_2 - y_1}{x_2 - x_1}$$

$$= \frac{-5 - (-5)}{4 - (-2)}$$

$$= \frac{0}{6} \text{ or } 0$$

Example 8 Slopes of Vertical Lines

Find the slope of a line that passes through (−3, 4) and (−3, −2).

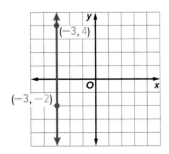

$$m = \frac{y_2 - y_1}{x_2 - x_1}$$

$$= \frac{-2 - 4}{-3 - (-3)}$$

$$= -\frac{6}{0} \text{ or undefined}$$

> **Study Tip**
>
> **Positive and Negative Slope** To know whether a line has a positive or negative slope, read the graph of the line just like you would read a sentence, from left to right. If the line "goes uphill," then the slope is positive. If the line "goes downhill," then the slope is negative.

> **Talk About It!**
>
> Why is the slope for vertical lines always undefined? Justify your argument.

Go Online You can complete an Extra Example online.

Study Tip

Converting Slope
When solving for an unknown coordinate, like the previous example, converting a slope from a decimal or mixed number to an improper fraction might make the problem easier to solve. For example, a slope of $1.\overline{333}$ can be rewritten as $\frac{4}{3}$.

Example 9 Find Coordinates Given the Slope

Find the value of r so that the line passing through $(-4, 5)$ and $(4, r)$ has a slope of $\frac{3}{4}$.

$m = \frac{y_2 - y_1}{x_2 - x_1}$ Use the Slope Formula.

$\frac{3}{4} = \frac{r - 5}{4 - (-4)}$ $(-4, 5) = (x_1, y_1)$ and $(4, r) = (x_2, y_2)$

$\frac{3}{4} = \frac{r - 5}{8}$ Subtract.

$8\left(\frac{3}{4}\right) = \frac{8(r - 5)}{8}$ Multiply each side by 8.

$6 = r - 5$ Simplify.

$6 + 5 = r - 5 + 5$ Add 5 to each side.

$11 = r$ Simplify.

Check

Find the value of r so that the line passing through $(-3, r)$ and $(7, -6)$ has a slope of $2\frac{2}{5}$.

$r = \underline{\quad ? \quad}$

Example 10 Use Slope

OCEANS
What is the slope of the continental slope at Cape Hatteras?

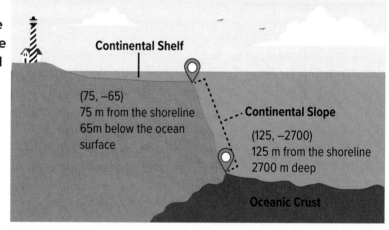

(75, −65)
75 m from the shoreline
65m below the ocean surface

Continental Shelf

Continental Slope

(125, −2700)
125 m from the shoreline
2700 m deep

Oceanic Crust

$m = \frac{y_2 - y_1}{x_2 - x_1}$ Use the Slope Formula

$= \frac{-2700 - (-65)}{125 - 75}$ $(75, -65) = (x_1, y_1)$ and $(125, -2700) = (x_2, y_2)$

$= \frac{-2635}{50}$ or -52.7 Simplify.

The continental slope at Cape Hatteras has a slope of -52.7.

💡 **Think About It!**
If a crab is walking along the ocean floor 112 meters away from the shoreline to 114 meters away from the shoreline, how far does it descend?

Go Online You can complete an Extra Example online.

Practice

Example 1

Find the rate of change of the function by using two points from the table.

1.

x	y
5	2
10	3
15	4
20	5

2.

x	y
1	15
2	9
3	3
4	−3

3. **POPULATION DENSITY** The table shows the population density for the state of Texas in various years. Find the average annual rate of change in the population density from 2000 to 2009.

4. **BAND** In 2012, there were approximately 275 students in the Delaware High School band. In 2018, that number increased to 305. Find the annual rate of change in the number of students in the band.

Population Density	
Year	People Per Square Mile
1930	22.1
1960	36.4
1980	54.3
2000	79.6
2009	96.7

Source: Bureau of the Census, U.S. Dept. of Commerce

Example 2

5. **TEMPERATURE** The graph shows the temperature in a city during different hours of one day.

   a. Find the rate of change in temperature between 6 A.M. and 7 A.M. and describe its meaning in the context of the situation.

   b. Find the rate of change in temperature from 1 P.M. and 2 P.M. and describe its meaning in the context of the situation.

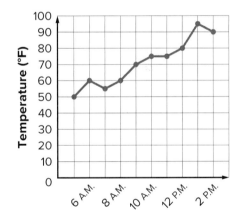

6. **COAL EXPORTS** The graph shows the annual coal exports from U.S. mines in millions of short tons.

   a. Find the rate of change in coal exports between 2000 and 2002 and describe its meaning in the context of the situation.

   b. Find the rate of change in coal exports between 2005 and 2006 and describe its meaning in the context of the situation.

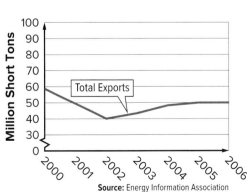

Lesson 4-2 • Rate of Change and Slope **225**

Examples 3 and 4

Determine whether the function is linear. If it is, state the rate of change.

7.
x	4	2	0	−2	−4
y	−1	1	3	5	7

8.
x	−7	−5	−3	−1	0
y	11	14	17	20	23

9.
x	−0.2	0	0.2	0.4	0.6
y	0.7	0.4	0.1	0.3	0.6

10.
x	$\frac{1}{2}$	$\frac{3}{2}$	$\frac{5}{2}$	$\frac{7}{2}$	$\frac{9}{2}$
y	$\frac{1}{2}$	1	$\frac{3}{2}$	2	$\frac{5}{2}$

Examples 5 through 8

Find the slope of the line that passes through each pair of points.

11. (4, 3), (−1, 6)

12. (8, −2), (1, 1)

13. (2, 2), (−2, −2)

14. (6, −10), (6, 14)

15. (5, −4), (9, −4)

16. (11, 7), (−6, 2)

17. (−3, 5), (3, 6)

18. (−3, 2), (7, 2)

19. (8, 10), (−4, −6)

20. (−12, 15), (18, −13)

21. (−8, 6), (−8, 4)

22. (−8, −15), (−2, 5)

23. (2, 5), (3, 6)

24. (6, 1), (−6, 1)

25. (4, 6), (4, 8)

26. (−5, −8), (−8, 1)

27. (2, 5), (−3, −5)

28. (9, 8), (7, −8)

29. (5, 2), (5, −2)

30. (10, 0), (−2, 4)

31. (17, 18), (18, 17)

32. (−6, −4), (4, 1)

33. (−3, 10), (−3, 7)

34. (2, −1), (−8, −2)

35. (5, −9), (3, −2)

36. (12, 6), (3, −5)

37. (−4, 5), (−8, −5)

Example 9

Find the value of r so the line that passes through each pair of points has the given slope.

38. (12, 10), (−2, r), $m = -4$

39. (r, −5), (3, 13), $m = 8$

40. (3, 5), (−3, r), $m = \frac{3}{4}$

41. (−2, 8), (r, 4), $m = -\frac{1}{2}$

42. (r, 3), (5, 9), $m = 2$

43. (5, 9), (r, −3), $m = -4$

44. (r, 2), (6, 3), $m = \frac{1}{2}$

45. (r, 4), (7, 1), $m = \frac{3}{4}$

Example 10

46. ROAD SIGNS Roadway signs such as the one shown are used to warn drivers of an upcoming steep downgrade. What is the grade, or slope, of the hill described on the sign?

47. HOME MAINTENANCE Grading the soil around the foundation of a house can reduce interior home damage from water runoff. For every 6 inches in height, the soil should extend 10 feet from the foundation. What is the slope of the soil grade?

48. USE A SOURCE Research the Americans with Disabilities Act (ADA) regulation for the slope of a wheelchair ramp. What is the maximum slope of an ADA regulation ramp? Use the slope to determine the length and height of an ADA regulation ramp.

49. DIVERS A boat is located at sea level. A scuba diver is 80 feet along the surface of the water from the boat and 30 feet below the water surface. A fish is 20 feet along the horizontal plane from the scuba diver and 10 feet below the scuba diver. What is the slope between the scuba diver and fish?

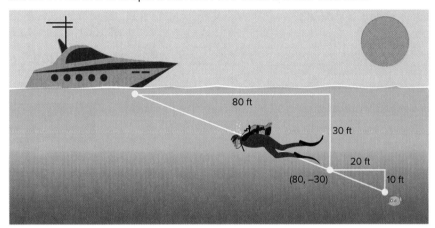

Mixed Exercises

STRUCTURE Find the slope of the line that passes through each pair of points.

50.

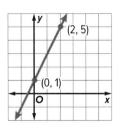

51.

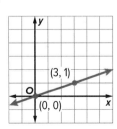

52.

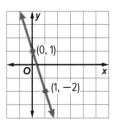

53. (6, −7), (4, −8)

54. (0, 5), (5, 5)

55. (−2, 6), (−5, 9)

56. (5, 8), (−4, 6)

57. (9, 4), (5, −3)

58. (1, 4), (3, −1)

Lesson 4-2 • Rate of Change and Slope

59. REASONING Find the value of r that gives the line passing through (3, 2) and (r, −4) a slope that is undefined.

60. REASONING Find the value of r that gives the line passing through (−5, 2) and (3, r) a slope of 0.

61. CREATE Draw a line on a coordinate plane so that you can determine at least two points on the graph. Describe how you would determine the slope of the graph and justify the slope you found.

62. ARGUMENTS The graph shows median prices for small cottages on a lake since 2005. A real estate agent says that since 2005, the rate of change for house prices is $10,000 each year. Do you agree? Use the graph to justify your answer.

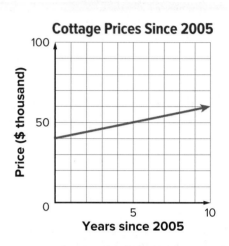

63. CREATE Use what you know about rate of change to describe the function represented by the table.

Time (wk)	Height of Plant (in.)
4	9.0
6	13.5
8	18.0

64. WRITE Explain how the rate of change and slope are related and how to find the slope of a line.

65. FIND THE ERROR Fern is finding the slope of the line that passes through (−2, 8) and (4, 6). Determine in which step she made an error. Explain your reasoning.

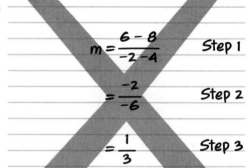

66. PERSEVERE Find the value of d so that the line that passes through (a, b) and (c, d) has a slope of $\frac{1}{2}$.

67. ANALYZE Why is the slope undefined for vertical lines? Explain.

68. WRITE Tarak wants to find the value of a so that the line that passes through (10, a) and (−2, 8) has a slope of $\frac{1}{4}$. Explain how Tarak can find the value of a.

Lesson 4-3

Slope-Intercept Form

Learn Writing Linear Equations in Slope-Intercept Form

An equation of the form $y = mx + b$, where m is the slope and b is the y-intercept, is written in slope-intercept form. When an equation is not in slope-intercept form, it might be easier to rewrite it before graphing. An equation can be rewritten in slope-intercept form by using the properties of equality.

Key Concept • Slope-Intercept Form

Words	The slope-intercept form of a linear equation is $y = mx + b$, where m is the slope and b is the y-intercept.
Example	$y = mx + b$ $y = -2x + 7$

Today's Goals
- Rewrite linear equations in slope-intercept form.
- Graph and interpret linear functions.

Today's Vocabulary
parameter

constant function

Example 1 Write Linear Equations in Slope-Intercept Form

Write an equation in slope-intercept form for the line with a slope of $\frac{4}{7}$ and a y-intercept of 5.

Write the equation in slope-intercept form.

$y = mx + b$ Slope intercept form.

$y = \left(\frac{4}{7}\right)x + 5$ $m = \frac{4}{7}, b = 5$

$y = \frac{4}{7}x + 5$ Simplify.

Check

Write an equation for the line with a slope of −5 and a y-intercept of 12.

💭 Think About It!

Explain why the y-intercept of a linear equation can be written as (0, b), where b is the y-intercept.

 Go Online You can complete an Extra Example online.

> **Think About It!**
> Can $x = 5$ be rewritten in slope-intercept form? Justify your argument.

Example 2 Rewrite Linear Equations in Slope-Intercept Form

Write $-22x + 8y = 4$ in slope-intercept form.

$-22x + 8y = 4$	Original equation
$-22x + 8y + 22x = 4 + 22x$	Add $22x$ to each side.
$8y = 22x + 4$	Simplify.
$\frac{8y}{8} = \frac{22x + 4}{8}$	Divide each side by 8.
$y = 2.75x + 0.5$	Simplify.

Check

What is the slope-intercept form of $-16x - 4y = -56$?

Example 3 Write Linear Equations

JOBS The number of job openings in the United States during a recent year increased by an average of 0.06 million per month since May. In May, there were about 4.61 million job openings in the United States. Write an equation in slope-intercept form to represent the number of job openings in the United States in the months since May.

Use the given information to write an equation in slope-intercept form.

- You are given that there were 4.61 million job openings in May.
- Let x = the number of months since May and y = the number of job openings in millions.
- Because the number of job openings is 4.61 million when $x = 0$, $b = 4.61$, and because the number of job openings has increased by 0.06 million each month, $m = 0.06$.
- So, the equation $y = 0.06x + 4.61$ represents the number of job openings in the United States since May.

> **Think About It!**
> When $x = 2$, describe the meaning of the equation in the context of the situation.

Check

SOCIAL MEDIA In the first quarter of 2012, there were 183 million users of a popular social media site in North America. The number of users increased by an average of 9 million per year since 2012. Write an equation that represents the number of users in millions of the social media site in North America after 2012.

Go Online You can complete an Extra Example online.

Explore Graphing Linear Functions by Using the Slope-Intercept Form

 Online Activity Use graphing technology to complete an Explore.

> **INQUIRY** How do the quantities *m* and *b* affect the graph of a linear function in slope-intercept form?

Learn Graphing Linear Functions in Slope-Intercept Form

The slope-intercept form of a linear equation is $y = mx + b$, where *m* is the slope and *b* is the *y*-intercept. The variables *m* and *b* are called **parameters** of the equation because changing either value changes the graph of the function.

A **constant function** is a linear function of the form $y = b$. Constant functions where $b \neq 0$ do not cross the *x*-axis. The graphs of constant functions have a slope of 0. The domain of a constant function is all real numbers, and the range is *b*.

Example 4 Graph Linear Functions in Slope-Intercept Form

Graph a linear function with a slope of $-\frac{3}{2}$ and a *y*-intercept of 4.

Write the equation in slope-intercept form and graph the function.

$y = mx + b$

$y = \left(-\frac{3}{2}\right)x + 4$

$y = -\frac{3}{2}x + 4$

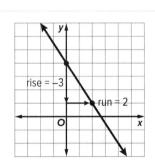

Study Tip

Negative Slope When counting rise and run, a negative sign may be associated with the value in the numerator or denominator. In this case, we associated the negative sign with the numerator. If we had associated it with the denominator, we would have moved up 3 and left 2 to the point (−2, 7). Notice that this point is also on the line. The resulting line will be the same whether the negative sign is associated with the numerator or denominator.

 Think About It!

Use the slope to find another point on the graph. Explain how you found the point.

Talk About It!
Why is it useful to write an equation in slope-intercept form before graphing the function?

Check

Graph a linear function with a slope of −2 and a y-intercept of 7.

Example 5 Graph Linear Functions

Graph 12x − 3y = 18.

Rewrite the equation in slope-intercept form.

$12x - 3y = 18$	Original equation
$12x - 3y - 12x = 18 - 12x$	Subtract 12x from each side.
$-3y = -12x + 18$	Simplify.
$\dfrac{-3y}{-3} = \dfrac{-12x + 18}{-3}$	Divide each side by −3.
$y = 4x - 6$	Simplify.

Graph the function.

Plot the y-intercept (0, −6).

The slope is $\frac{rise}{run} = 4$. From (0, −6), move up 4 units and right 1 unit. Plot the point (1, −2).

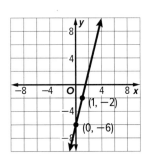

Draw a line through the points (0, −6) and (1, −2).

Go Online You can complete an Extra Example online.

Example 6 Graph Constant Functions

Graph $y = 2$.

Step 1 Plot (0, 2).

Step 2 The slope of $y = 2$ is 0.

Step 3 Draw a line through all the points that have a y-coordinate of 2.

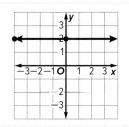

Think About It!
How do you know that the graph of $y = 2$ has a slope of 0?

Check

Graph $y = 1$.

Watch Out!
Slope A line with zero slope is not the same as a line with no slope. A line with zero slope is horizontal, and a line with no slope is vertical.

Match each graph with its function.

___?___ $y = 8$ ___?___ $3x + 7y = -28$ ___?___ $y = \frac{3}{7}x - 4$

___?___ $y = -4$ ___?___ $y = -3x + 8$ ___?___ $3x - y = 8$

A.
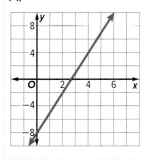

B.
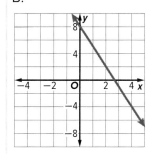

C.
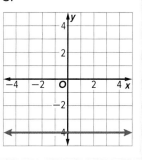

D.
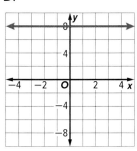

E.
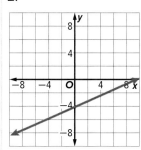

F.
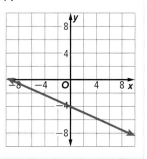

Go Online You can complete an Extra Example online.

Lesson 4-3 • Slope-Intercept Form 233

🌐 Apply Example 7 Use Graphs of Linear Functions

SHOPPING The number of online shoppers in the United States can be modeled by the equation $-5.88x + y = 172.3$, where y represents the number of millions of online shoppers in the United States x years after 2010. Estimate the number of online shoppers in 2020.

1. **What is the task?**
Describe the task in your own words. Then list any questions that you may have. How can you find answers to your questions?

I need to find the number of people who shopped online in 2020.

2. **How will you approach the task? What have you learned that you can use to help you complete the task?**

I will graph the given function. Then I can figure out from the graph how many people shopped online in 2020.

3. **What is your solution?**
Use your strategy to solve the problem. Graph the function.

In 2020, there were approximately 230 million online shoppers in the United States.

Online Shoppers in the United States

4. **How can you know that your solution is reasonable?**
✏️ **Write About It!** **Write an argument that can be used to defend your solution.**

Rewriting the equation in slope-intercept form shows that $b = 172.3$ and $m = 5.88$. This means that there were 172.3 million online shoppers in 2010. The number of online shoppers increased at a rate of 5.88 million per year. The graph of this line shows that in 2020 the number of online shoppers was more than 225 million but less than 250 million. From the graph, there were approximately 230 million online shoppers in 2020.

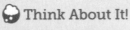 **Think About It!**
Estimate the year when the number of online shoppers in the United States will reach 271 million.

Go Online to learn about intervals in linear growth patterns in Expand 4-3.

Go Online You can complete an Extra Example online.

Practice

> Go Online You can complete your homework online.

Example 1

Write an equation of a line in slope-intercept form with the given slope and y-intercept.

1. slope: 5, y-intercept: −3
2. slope: −2, y-intercept: 7

3. slope: −6, y-intercept: −2
4. slope: 7, y-intercept: 1

5. slope: 3, y-intercept: 2
6. slope: −4, y-intercept: −9

7. slope: 1, y-intercept: −12
8. slope: 0, y-intercept: 8

Example 2

Write each equation in slope-intercept form.

9. $-10x + 2y = 12$
10. $4y + 12x = 16$
11. $-5x + 15y = -30$

12. $6x - 3y = -18$
13. $-2x - 8y = 24$
14. $-4x - 10y = -7$

Example 3

15. **SAVINGS** Wade's grandmother gave him $100 for his birthday. Wade wants to save his money to buy a portable game console. Each month, he adds $25 to his savings. Write an equation in slope-intercept form to represent Wade's savings y after x months.

16. **FITNESS CLASSES** Toshelle wants to take strength training classes at the community center. She has to pay a one-time enrollment fee of $25 to join the community center, and then $45 for each class she wants to take. Write an equation in slope-intercept form for the cost of taking x classes.

17. **EARNINGS** Macario works part time at a clothing store in the mall. He is paid $9 per hour plus 12% commission on the items he sells in the store. Write an equation in slope-intercept form to represent Macario's hourly wage y.

18. **ENERGY** From 2002 to 2005, U.S. consumption of renewable energy increased an average of 0.17 quadrillion BTUs per year. About 6.07 quadrillion BTUs of renewable power were produced in the year 2002. Write an equation in slope-intercept form to find the amount of renewable power P in quadrillion BTUs produced in year y between 2002 and 2005.

Example 4

Graph a linear function with the given slope and y-intercept.

19. slope: 5, y-intercept: 8
20. slope: 3, y-intercept: 10

21. slope: −4, y-intercept: 6
22. slope: −2, y-intercept: 8

Lesson 4-3 • Slope-Intercept Form

Examples 5 and 6

Graph each function.

23. $5x + 2y = 8$

24. $4x + 9y = 27$

25. $y = 7$

26. $y = -\frac{2}{3}$

27. $21 = 7y$

28. $3y - 6 = 2x$

Example 7

29. STREAMING An online company charges $13 per month for the basic plan. They offer premium channels for an additional $8 per month.

  a. Write an equation in slope-intercept form for the total cost c of the basic plan with p premium channels in one month.

  b. Graph the function.

  c. What would the monthly cost be for a basic plan plus 3 premium channels?

30. CAR CARE Suppose regular gasoline costs $2.76 per gallon. You can purchase a car wash at the gas station for $3.

  a. Write an equation in slope-intercept form for the total cost y of purchasing a car wash and x gallons of gasoline.

  b. Graph the function.

  c. Find the cost of purchasing a car wash and 8 gallons of gasoline.

Mixed Exercises

Write an equation of a line in slope-intercept form with the given slope and y-intercept.

31. slope: $\frac{1}{2}$, y-intercept: -3

32. slope: $\frac{2}{3}$, y-intercept: -5

Graph a function of a line with the given slope and y-intercept.

33. slope: 3, y-intercept: -4

34. slope: 4, y-intercept: -6

Graph each function.

35. $-3x + y = 6$

36. $-5x + y = 1$

Write an equation in slope-intercept form for each graph shown.

37.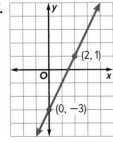

38.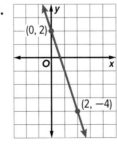

39.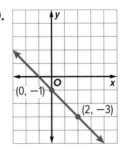

236 Module 4 · Linear and Nonlinear Functions

40. **MOVIES** MovieMania, an online movie rental website charges a one-time fee of $6.85 and $2.99 per movie rental. Let m represent the number of movies you watch and let C represent the total cost to watch the movies.

   a. Write an equation that relates the total cost to the number of movies you watch from MovieMania.

   b. Graph the function.

   c. Explain how to use the graph to estimate the cost of watching 13 movies at MovieMania.

   d. SuperFlix has no sign-up fee, just a flat rate per movie. If renting 13 movies at MovieMania costs the same as renting 9 movies at SuperFlix, what does SuperFlix charge per movie? Explain your reasoning.

   e. Write an equation that relates the total cost to the number of movies you watch from SuperFlix. Round to the nearest whole number.

41. **FACTORY** A factory uses a heater in part of its manufacturing process. The product cannot be heated too quickly, nor can it be cooled too quickly after the heating portion of the process is complete.

   a. The heater is digitally controlled to raise the temperature inside the chamber by 10°F each minute until it reaches the set temperature. Write an equation to represent the temperature, T, inside the chamber after x minutes if the starting temperature is 80°F.

   b. Graph the function.

   c. The heating process takes 22 minutes. Use your graph to find the temperature in the chamber at this point.

   d. After the heater reaches the temperature determined in **part c**, the temperature is kept constant for 20 minutes before cooling begins. Fans within the heater control the cooling so that the temperature inside the chamber decreases by 5°F each minute. Write an equation to represent the temperature, T, inside the chamber x minutes after the cooling begins.

42. **SAVINGS** When Santo was born, his uncle started saving money to help pay for a car when Santo became a teenager. Santo's uncle initially saved $2000. Each year, his uncle saved an additional $200.

   a. Write an equation that represents the amount, in dollars, Santo's uncle saved y after x years.

   b. Graph the function.

   c. Santo starts shopping for a car when he turns 16. The car he wants to buy costs $6000. Does he have enough money in the account to buy the car? Explain.

43. STRUCTURE Jazmin is participating in a 25.5-kilometer charity walk. She walks at a rate of 4.25 km per hour. Jazmin walks at the same pace for the entire event.

    a. Write an equation in slope-intercept form for the remaining distance y in kilometers of walking for x hours.

    b. Graph the function.

    c. What do the x- and y-intercepts represent in this situation?

    d. After Jazmin has walked 17 kilometers, how much longer will it take her to complete the walk? Explain how you can use your graph to answer the question.

For Exercises 44 and 45, refer to the equation $y = -\frac{4}{5}x + \frac{2}{5}$ where $-2 \leq x \leq 5$.

44. ANALYZE Copy and complete the table to help you graph the function $y = -\frac{4}{5}x + \frac{2}{5}$ over the interval. Identify any values of x where maximum or minimum values of y occur.

x	$-\frac{4}{5}x + \frac{2}{5}$	y	(x, y)
-2			
0			
5			

45. WRITE A student says you can find the solution to $-\frac{4}{5}x + \frac{2}{5} = 0$ using the graph. Do you agree? Explain your reasoning. Include the solution to the equation in your response.

46. PERSEVERE Consider three points that lie on the same line, $(3, 7)$, $(-6, 1)$, and $(9, p)$. Find the value of p and explain your reasoning.

47. CREATE Linear equations are useful in predicting future events. Create a linear equation that models a real-world situation. Make a prediction from your equation.

Lesson 4-4

Transformations of Linear Functions

Explore Transforming Linear Functions

 Online Activity Use graphing technology to complete an Explore.

> **INQUIRY** How does performing an operation on a linear function change its graph?

Today's Goals
- Apply translations to linear functions.
- Apply dilations to linear functions.
- Apply reflections to linear functions.

Today's Vocabulary
family of graphs
parent function
identity function
transformation
translation
dilation
reflection

Learn Translations of Linear Functions

A **family of graphs** includes graphs and equations of graphs that have at least one characteristic in common. The **parent function** is the simplest of the functions in a family.

The family of linear functions includes all lines, with the parent function $f(x) = x$, also called the **identity function**. A **transformation** moves the graph on the coordinate plane, which can create new linear functions.

One type of transformation is a translation. A **translation** is a transformation in which a figure is slid from one position to another without being turned. A linear function can be slid up, down, left, right, or in two directions.

Study Tip
Slope When translating a linear function, the graph of the function moves from one location to another, but the slope remains the same.

Vertical Translations

When a constant k is added to a linear function $f(x)$, the result is a vertical translation. The y-intercept of $f(x)$ is translated up or down.

Key Concept • Vertical Translations of Linear Functions
The graph of $g(x) = x + k$ is the graph of $f(x) = x$ translated vertically.

If $k > 0$, the graph of $f(x)$ is translated k units up.	If $k < 0$, the graph of $f(x)$ is translated $|k|$ units down.
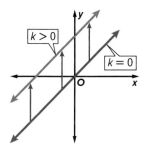	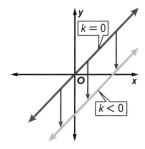
Every point on the graph of $f(x)$ moves k units up.	Every point on the graph of $f(x)$ moves $|k|$ units down.

Watch Out!
Translations of f(x) When a translation is the only transformation performed on the identity function, adding a constant before or after evaluating the function has the same effect on the graph. However, when more than one type of transformation is applied, this will not be the case.

Lesson 4-4 • Transformations of Linear Functions 239

Think About It!

What do you notice about the y-intercepts of vertically translated functions compared to the y-intercept of the parent function?

Example 1 Vertical Translations of Linear Functions

Describe the translation in $g(x) = x - 2$ as it relates to the graph of the parent function.

Graph the parent graph for linear functions.

Because $f(x) = x$, $g(x) = f(x) + k$ where $k = -2$.

$g(x) = x - 2 \rightarrow f(x) + (-2)$

x	f(x)	f(x) − 2	(x, g(x))
−2	−2	−4	(−2, −4)
0	0	−2	(0, −2)
1	1	−1	(1, −1)

The constant k is not grouped with x, so k affects the output, or y-values. The value of k is less than 0, so the graph of $f(x) = x$ is translated $|-2|$ units down, or 2 units down.

$g(x) = x - 2$ is the translation of the graph of the parent function 2 units down.

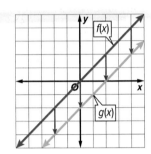

Check

Describe the translation in $g(x) = x - 1$ as it relates to the graph of the parent function.

The graph of $g(x) = x - 1$ is a translation of the graph of the parent function 1 unit ___?___.

Horizontal Translations

When a constant h is subtracted from the x-value before the function $f(x)$ is performed, the result is a horizontal translation. The x-intercept of $f(x)$ is translated right or left.

Go Online
You can watch a video to see how to describe translations of functions.

Go Online
You may want to complete the Concept Check to check your understanding.

Key Concept • Horizontal Translations of Linear Functions

The graph of $g(x) = (x - h)$ is the graph of $f(x) = x$ translated horizontally.

If $h > 0$, the graph of $f(x)$ is translated h units right.

If $h < 0$, the graph of $f(x)$ is translated $|h|$ units left.

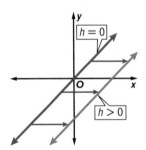 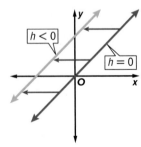

Every point on the graph of $f(x)$ moves h units right.

Every point on the graph of $f(x)$ moves $|h|$ units left.

Go Online You can complete an Extra Example online.

Example 2 Horizontal Translations of Linear Functions

Describe the translation in $g(x) = (x + 5)$ as it relates to the graph of the parent function.

Graph the parent graph for linear functions.

Because $f(x) = x$, $g(x) = f(x-h)$ where $h = -5$.

$g(x) = (x + 5) \rightarrow g(x) = f(x - (-5))$

x	x + 5	f(x + 5)	(x, g(x))
−2	3	3	(−2, 3)
0	5	5	(0, 5)
1	6	6	(1, 6)

The constant h is grouped with x, so h affects the input, or x-values. The value of h is less than 0, so the graph of $f(x) = x$ is translated $|-5|$ units left, or 5 units left.

$g(x) = (x + 5)$ is the translation of the graph of the parent function 5 units left.

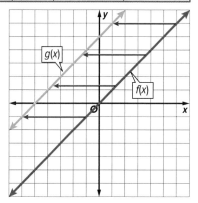

Think About It!
What do you notice about the x-intercepts of horizontally translated functions compared to the x-intercept of the parent function?

Check

Describe the translation in $g(x) = (x + 12)$ as it relates to the graph of the parent function.

The graph of $g(x) = (x + 12)$ is a translation of the graph of the parent function 12 units ___?___.

Example 3 Multiple Translations of Linear Functions

Describe the translation in $g(x) = (x - 6) + 3$ as it relates to the graph of the parent function.

Graph the parent graph for linear functions.

Because $f(x) = x$, $g(x) = f(x - h) + k$ where $h = 6$ and $k = 3$.

x	x − 6	f(x − 6)	f(x − 6) + 3	(x, g(x))
−2	−8	−8	−5	(−2, −5)
0	−6	−6	−3	(0, −3)
1	−5	−5	−2	(1, −2)

$g(x) = (x - 6) + 3 \rightarrow g(x) = f(x - 6) + 3$

The value of h is grouped with x and is greater than 0, so the graph of $f(x) = x$ is translated 6 units right.

The value of k is not grouped with x and is greater than 0, so the graph of $f(x) = x$ is translated 3 units up.

$g(x) = (x - 6) + 3$ is the translation of the graph of the parent function 6 units right and 3 units up.

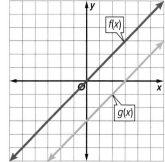

Think About It!
Eleni described the graph of $g(x) = (x - 6) + 3$ as the graph of the parent function translated down 3 units. Is she correct? Explain your reasoning.

Go Online You can complete an Extra Example online.

Example 4 Translations of Linear Functions

TICKETS A Web site sells tickets to concerts and sporting events. The total price of the tickets to a certain game can be modeled by $f(t) = 12t$, where t represents the number of tickets purchased. The Web site then charges a standard service fee of $4 per order. The total price of an order can be modeled by $g(t) = 12t + 4$. Describe the translation of $g(t)$ as it relates to $f(t)$.

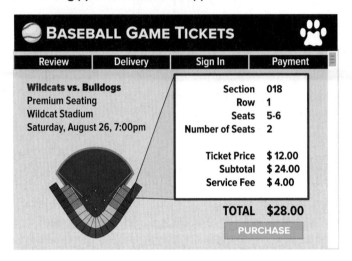

Complete the steps to describe the translation of $g(t)$ as it relates to $f(t)$. Since $f(t) = 12t$, $g(t) = f(t) + k$, where $k = 4$. $g(t) = 12t + 4 \rightarrow f(t) + 4$

The constant k is added to $f(t)$ after the total price of the tickets has been evaluated and is greater than 0, so the graph will be shifted 4 units up. $g(t) = 12t + 4$ is the translation of the graph of $f(t)$ 4 units up.

Graph the parent function and the translated function.

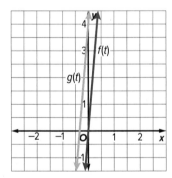

Check

RETAIL Jerome is buying paint for a mural. The total cost of the paint can be modeled by the function $f(p) = 6.99p$. He has a coupon for $5.95 off his purchase at the art supply store, so the final cost of his purchase can be modeled by $g(p) = 6.99p - 5.95$. Describe the translation in $g(p)$ as it relates to $f(p)$.

Go Online You can complete an Extra Example online.

Learn Dilations of Linear Functions

A **dilation** stretches or compresses the graph of a function.

When a linear function $f(x)$ is multiplied by a positive constant a, the result $a \cdot f(x)$ is a vertical dilation.

Key Concept • Vertical Dilations of Linear Functions

The graph of $g(x) = ax$ is the graph of $f(x) = x$ stretched or compressed vertically.

If $|a| > 1$, the graph of $f(x)$ is stretched vertically away from the x-axis.	If $0 < |a| < 1$, the graph of $f(x)$ is compressed vertically toward the x-axis.
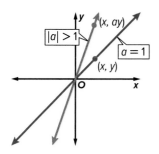	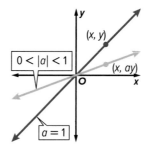
The slope of the graph of $a \cdot f(x)$ is steeper than that of the graph of $f(x)$.	The slope of the graph of $a \cdot f(x)$ is less steep than that of the graph of $f(x)$.

When x is multiplied by a positive constant a before a linear function $f(x)$ is evaluated, the result $f(a \cdot x)$ is a horizontal dilation.

Key Concept • Horizontal Dilations of Linear Functions

The graph of $g(x) = (a \cdot x)$ is the graph of $f(x) = x$ stretched or compressed horizontally.

If $|a| > 1$, the graph of $f(x)$ is compressed horizontally toward the y-axis.	If $0 < |a| < 1$, the graph of $f(x)$ is stretched horizontally away from the y-axis.
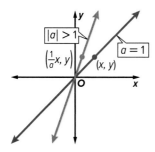	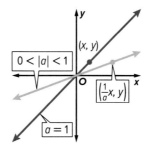
The slope of the graph of $f(a \cdot x)$ is steeper than that of the graph of $f(x)$.	The slope of the graph of $f(a \cdot x)$ is less steep than that of the graph of $f(x)$.

Go Online You can complete an Extra Example online.

Watch Out!

Dilations of $f(x) = x$
When a dilation is the only transformation performed on the identity function, multiplying by a constant before or after evaluating the function has the same effect on the graph. However, when more than one type of transformation is applied, this will not be the case.

Go Online
You can watch a video to see how to describe dilations of functions.

🧠 **Think About It!**
What do you notice about the slope of the vertical dilation g(x) compared to the slope of f(x)?

How does this relate to the constant a in the vertical dilation?

Example 5 Vertical Dilations of Linear Functions

Describe the dilation in $g(x) = 2(x)$ as it relates to the graph of the parent function.

Graph the parent graph for linear functions.

Since $f(x) = x$, $g(x) = a \cdot f(x)$ where $a = 2$.

$g(x) = 2(x) \rightarrow g(x) = 2f(x)$

x	f(x)	2f(x)	(x, g(x))
−2	−2	−4	(−2, −4)
0	0	0	(0, 0)
1	1	2	(1, 2)

The positive constant a is not grouped with x, and $|a|$ is greater than 1, so the graph of $f(x) = x$ is stretched vertically by a factor of a, or 2.

$g(x) = 2(x)$ is a vertical stretch of the graph of the parent function. The slope of the graph of $g(x)$ is steeper than that of $f(x)$.

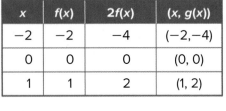

Check

Describe the transformation in $g(x) = 6(x)$ as it relates to the graph of the parent function.

The graph of $g(x) = 6(x)$ is a _____?_____ of the graph of the parent function.

The slope of the graph $g(x)$ is ___?___ than that of the parent function.

🧠 **Think About It!**
What do you notice about the slope of the horizontal dilation g(x) compared to the slope of f(x)?

How does this relate to the constant a in the horizontal dilation?

Example 6 Horizontal Dilations of Linear Functions

Describe the dilation in $g(x) = \left(\frac{1}{4}x\right)$ as it relates to the graph of the parent function.

Graph the parent graph for linear functions.

Since $f(x) = x$, $g(x) = f(a \cdot x)$ where $a = \frac{1}{4}$.

$g(x) = \left(\frac{1}{4}x\right) \rightarrow g(x) = f\left(\frac{1}{4}x\right)$

x	$\frac{1}{4}x$	$f\left(\frac{1}{4}x\right)$	(x, g(x))
−4	−1	−1	(−4, −1)
0	0	0	(0, 0)
4	1	1	(4, 1)

The positive constant a is grouped with x, and $|a|$ is between 0 and 1, so the graph of $f(x) = x$ is stretched horizontally by a factor of $\frac{1}{a}$, or 4.

$g(x) = \left(\frac{1}{4}x\right)$ is a horizontal stretch of the graph of the parent function. The slope of the graph of $g(x)$ is less steep than that of $f(x)$.

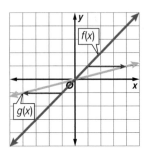

🌐 **Go Online** You can complete an Extra Example online.

Learn Reflections of Linear Functions

A **reflection** is a transformation in which a figure, line, or curve is flipped across a line. When a linear function f(x) is multiplied by −1 before or after the function has been evaluated, the result is a reflection across the x- or y-axis. Every x- or y-coordinate of f(x) is multiplied by −1.

Key Concept • Reflections of Linear Functions

The graph of −f(x) is the reflection of the graph of f(x) = x across the x-axis.

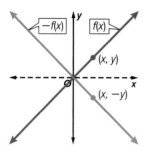

Every y-coordinate of −f(x) is the corresponding y-coordinate of f(x) multiplied by −1.

The graph of f(−x) is the reflection of the graph of f(x) = x across the y-axis.

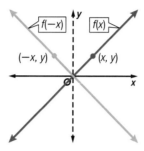

Every x-coordinate of f(−x) is the corresponding x-coordinate of f(x) multiplied by −1.

Go Online You can watch a video to see how to describe reflections of functions.

Watch Out!

Reflections of f(x) = x When a reflection is the only transformation performed on the identity function, multiplying by −1 before or after evaluating the function appears to have the same effect on the graph. However, when more than one type of transformation is applied, this will not be the case.

Example 7 Reflections of Linear Functions Across the x-Axis

Describe how the graph of $g(x) = -\frac{1}{2}(x)$ is related to the graph of the parent function.

Graph the parent graph for linear functions.

Since $f(x) = x$, $g(x) = -1 \cdot a \cdot f(x)$ where $a = \frac{1}{2}$.

$g(x) = -\frac{1}{2}(x) \rightarrow g(x) = -\frac{1}{2}f(x)$

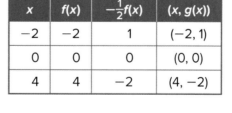

x	f(x)	$-\frac{1}{2}f(x)$	(x, g(x))
−2	−2	1	(−2, 1)
0	0	0	(0, 0)
4	4	−2	(4, −2)

The constant a is not grouped with x, and |a| is less than 1, so the graph of f(x) = x is vertically compressed.

The negative is not grouped with x, so the graph is also reflected across the x-axis.

The graph of $g(x) = -\frac{1}{2}(x)$ is the graph of the parent function vertically compressed and reflected across the x-axis.

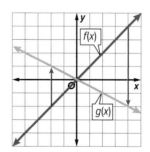

Talk About It!

In the example, the slope of g(x) is negative. Will this always be the case when multiplying a linear function by −1? Justify your argument.

Go Online You can complete an Extra Example online.

Check

How can you tell whether multiplying −1 by the parent function will result in a reflection across the x-axis?

A. If the constant is not grouped with x, the result will be a reflection across the x-axis.

B. If the constant is grouped with x, the result will be a reflection across the x-axis.

C. If the constant is greater than 0, the result will be a reflection across the x-axis.

D. If the constant is less than 0, the result will be a reflection across the x-axis.

Example 8 Reflections of Linear Functions Across the y-Axis

Describe how the graph of $g(x) = (-3x)$ is related to the graph of the parent function.

Graph the parent graph for linear functions.

Since $f(x) = x$, $g(x) = f(-1 \cdot a \cdot x)$ where $a = 3$.

$g(x) = -3x \rightarrow g(x) = f(-3x)$

x	−3x	f(−3x)	(x, g(x))
−1	3	3	(−1, 3)
0	0	0	(0, 0)
1	−3	−3	(1, −3)

The constant a is grouped with x, and $|a|$ is greater than 1, so the graph of $f(x) = x$ is horizontally compressed.

The negative is grouped with x, so the graph is also reflected across the y-axis.

The graph of $g(x) = (-3x)$ is the graph of the parent function horizontally compressed and reflected across the y-axis.

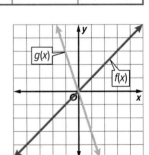

Go Online

You can watch a video to see how to graph transformations of a linear function using a graphing calculator.

Check

Describe how the graph of $g(x) = (-10x)$ is related to the graph of the parent function.

The graph of $g(x) = (-10x)$ is the graph of the parent function compressed horizontally and reflected across the ___?___.

Go Online You can complete an Extra Example online.

Practice

Go Online You can complete your homework online.

Examples 1 through 3
Describe the translation in each function as it relates to the graph of the parent function.

1. $g(x) = x + 11$

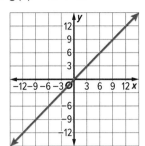

2. $g(x) = x - 8$

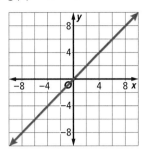

3. $g(x) = (x - 7)$

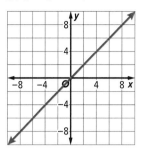

4. $g(x) = (x + 12)$

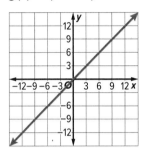

5. $g(x) = (x + 10) - 1$

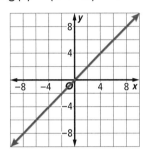

6. $g(x) = (x - 9) + 5$

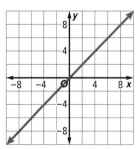

Example 4

7. **BOWLING** The cost for Nobu to go bowling is $4 per game plus an additional flat fee of $3.50 for the rental of bowling shoes. The cost can be modeled by the function $f(x) = 4x + 3.5$, where x represents the number of games bowled. Describe the graph of $g(x)$ as it relates to $f(x)$ if Nobu does not rent bowling shoes.

8. **SAVINGS** Natalie has $250 in her savings account, into which she deposits $10 of her allowance each week. The balance of her savings account can be modeled by the function $f(w) = 250 + 10w$, where w represents the number of weeks. Write a function $g(w)$ to represent the balance of Natalie's savings account if she withdraws $40 to purchase a new pair of shoes. Describe the translation of $f(w)$ that results in $g(w)$.

9. **BOAT RENTAL** The cost to rent a paddle boat at the county park is $8 per hour plus a nonrefundable deposit of $10. The cost can be modeled by the function $f(h) = 8h + 10$, where h represents the number of hours the boat is rented. Describe the graph of $g(h)$ as it relates to $f(h)$ if the nonrefundable deposit increases to $15.

Lesson 4-4 • Transformations of Linear Functions 247

Examples 5 and 6

Describe the dilation in each function as it relates to the graph of the parent function.

10. $g(x) = 5(x)$

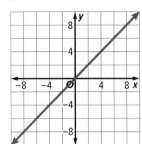

11. $g(x) = \frac{1}{3}(x)$

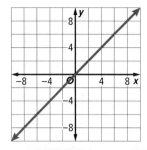

12. $g(x) = 1.5(x)$

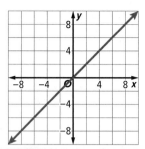

13. $g(x) = (3x)$

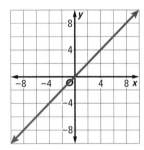

14. $g(x) = \left(\frac{3}{4}x\right)$

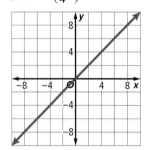

15. $g(x) = (0.4x)$

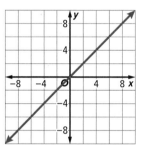

Example 7

Describe how the graph of each function is related to the graph of the parent function.

16. $g(x) = -4(x)$

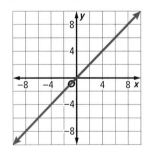

17. $g(x) = -8(x)$

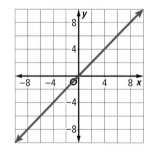

18. $g(x) = -\frac{2}{3}(x)$

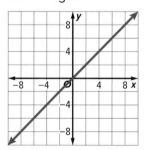

Example 8

Describe how the graph of each function is related to the graph of the parent function.

19. $g(x) = \left(-\frac{4}{5}x\right)$

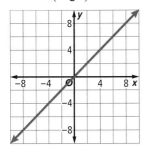

20. $g(x) = (-6x)$

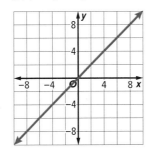

21. $g(x) = (-1.5x)$

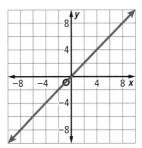

Mixed Exercises

Describe the transformation in each function as it relates to the graph of the parent function.

22. $g(x) = x + 4$

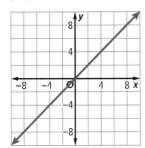

23. $g(x) = (x - 2) - 8$

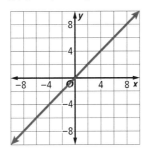

24. $g(x) = \left(-\frac{5}{8}x\right)$

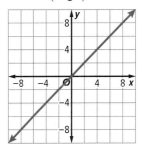

25. $g(x) = \frac{1}{5}(x)$

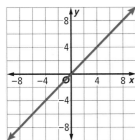

26. $g(x) = -3(x)$

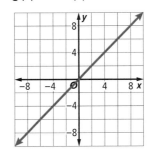

27. $g(x) = (2.5x)$

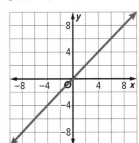

Lesson 4-4 • Transformations of Linear Functions   249

REASONING Write a function g(x) to represent the translated graph.

28. f(x) = 3x + 7 translated 4 units up.   **29.** f(x) = x − 5 translated 2 units down.

30. PERIMETER The function f(s) = 4s represents the perimeter of a square with side length s. Write a function g(s) to represent the perimeter of a square with side lengths that are twice as great. Describe the graph of g(s) compared to f(s).

31. GAMES The function f(x) = 0.50x gives the average cost in dollars for x cell phone game downloads that cost an average of $0.50 each. Write a function g(x) to represent the cost in dollars for x cell phone game downloads that cost $1.50 each. Describe the graph of g(x) compared to f(x).

32. TRAINER The function f(x) = 90x gives the cost of working out with a personal trainer, where $90 is the trainer's hourly rate, and x represents the number of hours spent working out with the trainer. Describe the dilation, g(x) of the function f(x), if the trainer increases her hourly rate to $100.

33. DOWNLOADS Hannah wants to download songs. She researches the price to download songs from Site F. Hannah wrote the function f(x) = x, which represents the cost in dollars for x songs downloaded that cost $1.00 each.

   a. Hannah researches the price to download songs from Site G. Write a function g(x) to represent the cost in dollars for x songs downloaded that cost $1.29 each.

   b. Describe the graph of g(x) compared to the graph of f(x).

34. PERSEVERE For any linear function, replacing f(x) with f(x + k) results in the graph of f(x) being shifted |k| units to the right for k < 0 and shifted k units to the left for k > 0. Does shifting the graph horizontally k units have the same effect as shifting the graph vertically −k units? Justify your answer. Include graphs in your response.

35. CREATE Write an equation that is a vertical compression by a factor of a of the parent function y = x. What can you say about the horizontal dilation of the function?

36. WHICH ONE DOESN'T BELONG Consider the four functions. Which one does not belong in this group? Justify your conclusion.

| f(x) = 2(x + 1) − 3 | f(x) = $\frac{1}{2}$x − 4 | f(x) = −3x + 10 | f(x) = 5(x − 7) + 3 |

250 Module 4 • Linear and Nonlinear Functions

Lesson 4-5
Arithmetic Sequences

Learn Arithmetic Sequences

A **sequence** is a set of numbers that are ordered in a specific way. Each number within a sequence is called a **term of a sequence**.

In an **arithmetic sequence**, each term after the first is found by adding a constant, the **common difference** d, to the previous term.

Words	An arithmetic sequence is a numerical pattern that increases or decreases at a constant rate called the common difference.
Examples	 The common difference is -5. 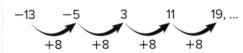 The common difference is 8.

Today's Goals
- Construct arithmetic sequences.
- Apply the arithmetic sequence formula.

Today's Vocabulary
sequence
term of sequence
arithmetic sequence
common difference
nth term of an arithmetic function

Example 1 Identify Arithmetic Sequences

Determine whether the sequence is an arithmetic sequence. Justify your reasoning.

17, 14, 10, 7, 3

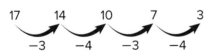

Check the difference between terms.

This sequence does not have a common difference between its terms. This is not an arithmetic sequence.

Check

Determine whether the sequence is an arithmetic sequence. Justify your reasoning.

82, 73, 64, 55, . . .

 Think About It!
How are arithmetic sequences and number patterns alike and different?

Go Online You can complete an Extra Example online.

Talk About It!
Why would it be useful to develop a rule to find terms of a sequence? Explain.

Example 2 Find the Next Term

Determine the next three terms in the sequence.

11, 7, 3, −1

Find the common difference between terms.   −4

Add the common difference to the last term of the sequence to find the next terms.

$-1 + (-4) = -5$ $-5 + (-4) = -9$ $-9 + (-4) = -13$

Check

Determine the next three terms in the sequence.

31, 18, 5, __?__, __?__, __?__

 Go Online You can complete an Extra Example online.

Explore Common Differences

 Online Activity Use a real-world situation to complete the Explore.

> **INQUIRY** How can you tell if a set of numbers models a linear function?

Watch Out!
Subscripts Subscripts are used to indicate a specific term. For example, a_8 means the 8th term of the sequence. It does not mean $a \times 8$.

Learn Arithmetic Sequences as Linear Functions

Each term of an arithmetic sequence can be expressed in terms of the first term a_1 and the common difference d.

> **Key Concept • nth Term of an Arithmetic Sequence**
>
> The **nth term of an arithmetic sequence** with the first term a_1 and common difference d is given by $a_n = a_1 + (n-1)d$, where n is a positive integer.

The graph of an arithmetic sequence includes points that lie along a line. Because there is a constant difference between each pair of points, the function is linear. For the equation of an arithmetic sequence, $a_n = a_1 + (n-1)d$

- n is the independent variable,
- a_n is the dependent variable, and
- d is the slope.

The function of an arithmetic sequence is written as $f(n) = a_1 + (n-1)d$, where n is a counting number.

Think About It!
Why is the domain of a sequence counting numbers instead of all real numbers?

 Go Online You can complete an Extra Example online.

252 Module 4 • Linear and Nonlinear Functions

Example 3 Find the nth Term

Use the arithmetic sequence −4, −1, 2, 5, . . . to complete the following.

Part A Write an equation.

$a_n = 3n - 7$

Part B Find the 16th term of the sequence.

Use the equation from Part A to find the 16th term in the arithmetic sequence.

$a_n = 3n - 7$ Equation from Part A

$a_{16} = 3(16) - 7$ Substitute 16 for n.

$a_{16} = 48 - 7$ Multiply.

$a_{16} = 41$ Simplify.

Check

RUNNING Randi has been training for a marathon, and it is important for her to keep a constant pace. She recorded her time each mile for the first several miles that she ran.

- At 1 mile, her time was 10 minutes and 30 seconds.
- At 2 miles, her time was 21 minutes.
- At 3 miles, her time was 31 minutes and 30 seconds.
- At 4 miles, her time was 42 minutes.

Part A Write a function to represent her sequence of data. Use n as the variable.

Part B How long will it take her to run a whole marathon? Round your answer to the nearest thousandth if necessary. (Hint: a marathon is 26.2 miles.)

Example 4 Apply Arithmetic Sequences as Linear Functions

MONEY Laniqua opened a savings account to save for a trip to Spain. With the cost of plane tickets, food, hotel, and other expenses, she needs to save $1600. She opened the account with $525. Every month, she adds the same amount to her account using the money she earns at her after school job. From her bank statement, Laniqua can write a function that represents the balance of her savings account.

(continued on the next page)

Go Online You can complete an Extra Example online.

Use a Source

Find the cost of a flight from the airport closest to you to Madrid, the capital of Spain. How many months would Laniqua need to save to afford the ticket?

Study Tip

Graphing You might not need to create a table of the sequence first. However, it might serve as a reminder that an arithmetic sequence is a series of points, not a line.

DIXON STATE BANK

Laniqua Jones   Account Number
                     922194075

Current Balance as of 03/01/2019....... $ 690

Balance as of 02/01/2019....... $ 635

Balance as of 01/01/2019....... $ 580

Starting Balance as of 12/01/2018....... $ 525

— End of Statement

Part A Create a function to represent the sequence.

First, find the common difference.

525   580   635   690
  +55   +55   +55

The common difference is 55.

The balance after 1 month is $580, so let $a_1 = 580$. Notice that the starting balance is $525. You can think of this starting point as $a_0 = 525$.

$$f(n) = a_1 + (n - 1)d \quad \text{Formula for the nth term}$$
$$= 580 + (n - 1)(55) \quad a_1 = 580 \text{ and } d = 55$$
$$= 580 + 55n - 55 \quad \text{Simplify.}$$
$$= 55n + 525$$

Part B Graph the function and determine its domain.

n	f(n)
0	525
1	580
2	635
3	690
4	745
5	800
6	855

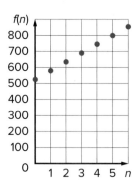

The domain is the number of months since Laniqua opened her savings account. The domain is {0, 1, 2, 3, 4, 5, ...}.

Go Online You can complete an Extra Example online.

Practice

Go Online You can complete your homework online.

Example 1

ARGUMENTS Determine whether each sequence is an arithmetic sequence. Justify your reasoning.

1. −3, 1, 5, 9, ...

2. $\frac{1}{2}, \frac{3}{4}, \frac{5}{8}, \frac{7}{16}, ...$

3. −10, −7, −4, 1, ...

4. −12.3, −9.7, −7.1, −4.5, ...

5. 4, 7, 9, 12, ...

6. 15, 13, 11, 9, ...

7. 7, 10, 13, 16, ...

8. −6, −5, −3, −1, ...

Example 2

Find the common difference of each arithmetic sequence. Then find the next three terms.

9. 0.02, 1.08, 2.14, 3.2, ...

10. 6, 12, 18, 24, ...

11. 21, 19, 17, 15, ...

12. $-\frac{1}{2}, 0, \frac{1}{2}, 1, ...$

13. $2\frac{1}{3}, 2\frac{2}{3}, 3, 3\frac{1}{3}, ...$

14. $\frac{7}{12}, 1\frac{1}{3}, 2\frac{1}{12}, 2\frac{5}{6}, ...$

15. 3, 7, 11, 15, ...

16. 22, 19.5, 17, 14.5, ...

17. −13, −11, −9, −7, ...

18. −2, −5, −8, −11, ...

Example 3

Use the given arithmetic sequence to write an equation and then find the 7th term of the sequence.

19. −3, −8, −13, −18, ...

20. −2, 3, 8, 13, ...

21. −11, −15, −19, −23, ...

22. −0.75, −0.5, −0.25, 0, ...

Lesson 4-5 • Arithmetic Sequences **255**

Example 4

23. **SPORTS** Wanda is the manager for the soccer team. One of her duties is to hand out cups of water at practice. Each cup of water is 4 ounces. She begins practice with a 128-ounce cooler of water.

   a. Create a function to represent the arithmetic sequence.

   b. Graph the function.

   c. How much water is remaining after Wanda hands out the 14th cup?

24. **THEATER** A theater has 20 seats in the first row, 22 in the second row, 24 in the third row, and so on for 25 rows.

   a. Create a function to represent the arithmetic sequence.

   b. Graph the function.

   c. How many seats are in the last row?

25. **POSTAGE** The price to send a large envelope first class mail is 88 cents for the first ounce and 17 cents for each additional ounce. The table shows the cost for weights up to 5 ounces.

Weight (ounces)	1	2	3	4	5
Postage (dollars)	0.88	1.05	1.22	1.39	1.56

Source: United States Postal Service

   a. Create a function to represent the arithmetic sequence.

   b. Graph the function.

   c. How much did a large envelope weigh that cost $2.07 to send?

26. **VIDEO DOWNLOADING** Brian is downloading episodes of his favorite TV show to play on his personal media device. The cost to download 1 episode is $1.99. The cost to download 2 episodes is $3.98. The cost to download 3 episodes is $5.97.

   a. Create a function to represent the arithmetic sequence.

   b. Graph the function.

   c. What is the cost to download 9 episodes?

27. USE A MODEL Chapa is beginning an exercise program that calls for 30 push-ups each day for the first week. Each week thereafter, she has to increase her push-ups by 2.
   a. Write a function to represent the arithmetic sequence.
   b. Graph the function.
   c. Which week of her program will be the first one in which she will do at least 50 push-ups a day?

Mixed Exercises

CONSTRUCT ARGUMENTS Determine whether each sequence is an arithmetic sequence. Justify your argument.

28. −9, −12, −15, −18, ...

29. 10, 15, 25, 40, ...

30. −10, −5, 0, 5, ...

31. −5, −3, −1, 1, ...

Write an equation for the nth term of each arithmetic sequence. Then graph the first five terms of the sequence.

32. 7, 13, 19, 25, ...

33. 30, 26, 22, 18, ...

34. −7, −4, −1, 2, ...

35. SAVINGS Fabiana decides to save the money she's earning from her after-school job for college. She makes an initial contribution of $3000 and each month deposits an additional $500. After one month, she will have contributed $3500.
   a. Write an equation for the nth term of the sequence.
   b. How much money will Fabiana have contributed after 24 months?

36. NUMBER THEORY One of the most famous sequences in mathematics is the Fibonacci sequence. It is named after Leonardo de Pisa (1170–1250) or Filius Bonacci, alias Leonardo Fibonacci. The first several numbers in the Fibonacci sequence are shown.
1, 1, 2, 3, 5, 8, 13, 21, 34, 55, 89, . . .
Does this represent an arithmetic sequence? Why or why not?

37. STRUCTURE Use the arithmetic sequence 2, 5, 8, 11, ...
   a. Write an equation for the nth term of the sequence.
   b. What is the 20th term in the sequence?

Higher-Order Thinking Skills

38. CREATE Write a sequence that is an arithmetic sequence. State the common difference, and find a_6.

39. CREATE Write a sequence that is not an arithmetic sequence. Determine whether the sequence has a pattern, and if so, describe the pattern.

40. REASONING Determine if the sequence 1, 1, 1, 1, . . . is an arithmetic sequence. Explain your reasoning.

41. CREATE Create an arithmetic sequence with a common difference of −10.

42. PERSEVERE Find the value of x that makes $x + 8$, $4x + 6$, and $3x$ the first three terms of an arithmetic sequence.

43. CREATE For each arithmetic sequence described, write a formula for the nth term of a sequence that satisfies the description.

   a. first term is negative, common difference is negative

   b. second term is −5, common difference is 7

   c. $a_2 = 8$, $a_3 = 6$

Andre and Sam are both reading the same novel. Andre reads 30 pages each day. Sam created the table at the right. Refer to this information for Exercises 44–46.

Sam's Reading Progress	
Day	Pages Left to Read
1	430
2	410
3	390
4	370

44. ANALYZE Write arithmetic sequences to represent each boy's daily progress. Then write the function for the nth term of each sequence.

45. PERSEVERE Enter both functions from Exercise 44 into your calculator. Use the table to determine if there is a day when the number of pages Andre has read is equal to the number of pages Sam has left to read. If so, which day is it? Explain how you used the table feature to help you solve the problem.

46. ANALYZE Graph both functions on your calculator, then sketch the graph. How can you use the graph to answer the question from Exercise 45?

Lesson 4-6
Piecewise and Step Functions

Learn Graphing Piecewise-Defined Functions

Some functions cannot be described by a single expression because they are defined differently depending on the interval of *x*. These functions are **piecewise-defined functions**. A **piecewise-linear function** has a graph that is composed of some number of linear pieces.

Example 1 Graph a Piecewise-Defined Function

To graph a piecewise-defined function, graph each "piece" separately.

Graph $f(x) = \begin{cases} 2x + 4 \text{ if } x \leq 1 \\ -x + 3 \text{ if } x > 1 \end{cases}$. **State the domain and range.**

First, graph $f(x) = 2x + 4$ if $x \leq 1$.

- Create a table for $f(x) = 2x + 4$ using values of $x \leq 1$.
- Because *x* is *less than or equal to* 1, place a dot at (1, 6) to indicate that the endpoint is included in the graph.
- Then, plot the points and draw the graph beginning at (1, 6).

x	y
1	6
0	4
−1	2
−2	0
−3	−2

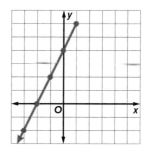

Next, graph $f(x) = -x + 3$ if $x > 1$.

- Create a table for $f(x) = -x + 3$ using values of $x > 1$.
- Because *x* is *greater than but not equal to* 1, place a circle at (1, 2) to indicate that the endpoint is not included in the graph.
- Then, plot the points and draw the graph beginning at (1, 2).

x	y
1	2
2	1
3	0
4	−1
5	−2

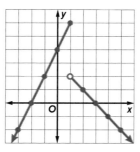

The domain is all real numbers. The range is $y \leq 6$.

Go Online You can complete an Extra Example online.

Today's Goals
- Identify and graph piecewise-defined functions.
- Identify and graph step functions.

Today's Vocabulary
piecewise-defined function
piecewise-linear function
step function
greatest integer function

Think About It!
What would be an advantage of graphing the entire expression and removing the portion that is not in the interval?

Go Online
An alternate method is available for this example.

Watch Out!

Circles and Dots Do not forget to examine the endpoint(s) of each piece to determine whether there should be a circle or a dot. > and < mean that a circle should be used, while ≥ and ≤ mean that a dot should be used.

Study Tip

Piecewise-Defined Functions When graphing piecewise-defined functions, there should be a dot or line that contains each member of the domain.

Go Online

You can watch a video to see how to graph a piecewise-defined function on a graphing calculator

Check

Part A Graph $f(x) = \begin{cases} -x + 1 \text{ if } x \leq -2 \\ -3x - 2 \text{ if } x > -2 \end{cases}$.

Part B Find the domain and range of the function.

Explore Age as a Function

 Online Activity Use a real-world situation to complete an Explore.

> **INQUIRY** When can real-world data be described using a step function?

Learn Graphing Step Functions

A **step function** is a type of piecewise-linear function with a graph that is a series of horizontal line segments. One example of a step function is the **greatest integer function,** written as $f(x) = [\![x]\!]$ in which $f(x)$ is the greatest integer less than or equal to x.

Key Concept • Greatest Integer Function

Type of graph: disjointed line segments

The graph of a step function is a series of disconnected horizontal line segments.

Domain: all real numbers; Because the dots and circles overlap, the domain is all real numbers.

Range: all integers; Because the function represents the greatest integer less than or equal to x, the range is all integers.

Parent function: $f(x) = [\![x]\!]$

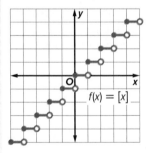

Go Online You can complete an Extra Example online.

260 Module 4 • Linear and Nonlinear Functions

Example 2 Graph a Greatest Integer Function

Graph $f(x) = [\![x + 1]\!]$. State the domain and range.

First, make a table. Select a few values that are between integers.

x	x + 1	$[\![x + 1]\!]$	
−2	−1	−1	−1, −0.75, and −0.25 are greater than or equal to −1 but less than 0. So, −1 is the greatest integer that is not greater than −1, −0.75, or −0.25.
−1.75	−0.75	−1	
−1.25	−0.25	−1	
−1	0	0	0, 0.5, and 0.75 are greater than or equal to 0 but less than 1. So, 0 is the greatest integer that is not greater than 0, 0.5, or 0.75.
−0.5	0.5	0	
−0.25	0.75	0	
0	1	1	1, 1.25, and 1.5 are greater than or equal to 1 but less than 2. So, 1 is the greatest integer that is not greater than 1, 1.25, or 1.5.
0.25	1.25	1	
0.5	1.5	1	
1	2	2	2, 2.25, and 2.75 are greater than or equal to 2 but less than 3. So, 2 is the greatest integer that is not greater than 2, 2.25, or 2.75.
1.25	2.25	2	
1.75	2.75	2	

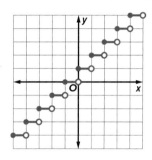

On the graph, dots represent included points. Circles represent points that are not included. The domain is all real numbers. The range is all integers. Note that this is the graph of $f(x) = [\![x]\!]$ shifted 1 unit to the left.

Talk About It!
What do you notice about the symmetry, extrema, and end behavior of the function?

Watch Out!
Greatest Integer Function When finding the value of a greatest integer function, do not round to the nearest integer. Instead, always round nonintegers down to the greatest integer that is not greater than the number.

Check

Graph $f(x) = [\![x - 2]\!]$ by making a table. Copy and complete the table. Then graph the function.

x	x − 2	$[\![x - 2]\!]$
−1	−3	
−0.75	−2.75	
−0.25	−2.25	−3
0	−2	−2
0.25	−1.75	−2
0.5	−1.5	
1	−1	
1.25	−0.75	−1
1.5	−0.5	
2	0	
2.25	0.25	

Go Online You can complete an Extra Example online.

Think About It!

How would the graph change if 1 certified lifeguard could watch up to 59 swimmers? For example, if there are greater than or equal to 60, but fewer than 120 swimmers, there must be 2 lifeguards on duty.

Math History Minute

Oliver Heaviside (1850–1925) was a self-taught electrical engineer, mathematician, and physicist who laid much of the groundwork for telecommunications in the 21st century. Heaviside invented the Heaviside step function,

$$f(x) = \begin{cases} 0 & \text{if } x < 0 \\ \frac{1}{2} & \text{if } x = 0 \\ 1 & \text{if } x > 0 \end{cases}$$

which he used to model the current in an electric circuit.

Example 3 Graph a Step Function

SAFETY A state requires a ratio of 1 lifeguard to 60 swimmers in a swimming pool. This means that 1 lifeguard can watch up to and including 60 swimmers. Make a table and draw a graph that shows the number of lifeguards that must be on duty $f(x)$ based on the number of swimmers in the pool x.

The number of lifeguards that must be on duty can be represented by a step function.

- If the number of swimmers is greater than 0 but fewer than or equal to 60, only 1 lifeguard must be on duty.
- If the number of swimmers is greater than 60 but fewer than or equal to 120, there must be 2 lifeguards on duty.
- If the number of swimmers is greater than 180 but fewer than or equal to 240, there must be 4 lifeguards on duty.

x	$f(x)$
$0 < x \leq 60$	1
$60 < x \leq 120$	2
$120 < x \leq 180$	3
$180 < x \leq 240$	4
$240 < x \leq 300$	5
$300 < x \leq 360$	6
$360 < x \leq 420$	7

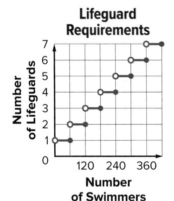

The circles mean that when there are more than a multiple of 60 swimmers, another lifeguard is required.

The dots represent the maximum number of swimmers that can be in the pool for that particular number of lifeguards on duty.

Check

PETS At Luciana's pet boarding facility, it costs $35 per day to board a dog. Every fraction of a day is rounded up to the next day. Copy and complete the table. Then graph the function.

Days	Cost ($)
$0 < x \leq 1$	
$1 < x \leq 2$	
$2 < x \leq 3$	
$3 < x \leq 4$	
$4 < x \leq 5$	
$5 < x \leq 6$	

Go Online You can complete an Extra Example online.

Practice

Go Online You can complete your homework online.

Example 1

Graph each function. State the domain and range.

1. $f(x) = \begin{cases} \frac{1}{2}x - 1 & \text{if } x > 3 \\ -2x + 3 & \text{if } x \leq 3 \end{cases}$

2. $f(x) = \begin{cases} 2x - 5 & \text{if } x > 1 \\ 4x - 3 & \text{if } x \leq 1 \end{cases}$

3. $f(x) = \begin{cases} 2x + 3 & \text{if } x \geq -3 \\ -\frac{1}{3}x + 1 & \text{if } x < -3 \end{cases}$

4. $f(x) = \begin{cases} 3x + 4 & \text{if } x \geq 1 \\ x + 3 & \text{if } x < 1 \end{cases}$

5. $f(x) = \begin{cases} 3x + 2 & \text{if } x > -1 \\ -\frac{1}{2}x - 3 & \text{if } x \leq -1 \end{cases}$

6. $f(x) = \begin{cases} 2x + 1 & \text{if } x < -2 \\ -3x - 1 & \text{if } x \geq -2 \end{cases}$

Example 2

Graph each function. State the domain and range.

7. $f(x) = 3[\![x]\!]$

8. $f(x) = [\![-x]\!]$

9. $g(x) = -2[\![x]\!]$

10. $g(x) = [\![x]\!] + 3$

11. $h(x) = [\![x]\!] - 1$

12. $h(x) = \frac{1}{2}[\![x]\!] + 1$

Example 3

13. BABYSITTING Ariel charges $8 per hour as a babysitter. She rounds every fraction of an hour up to the next half-hour. Draw a graph to represent Ariel's total earnings y after x hours.

14. FUNDRAISING Students are selling boxes of cookies at a fundraiser. The boxes of cookies can only be ordered by the case, with 12 boxes per case. Draw a graph to represent the number of cases needed y when x boxes of cookies are sold.

Mixed Exercises

15. PRECISION A package delivery service determines rates for express shipping by the weight of a package, with every fraction of a pound rounded up to the next pound. The table shows the cost of express shipping packages that weigh no more than 5 pounds. Write a piecewise-linear function representing the cost to ship a package that weighs no more than 5 pounds. State the domain and range.

Weight (pounds)	Rate (dollars)
1	16.20
2	19.30
3	22.40
4	25.50
5	28.60

16. EARNINGS Kelly works in a hospital as a medical assistant. She earns $8 per hour the first 8 hours she works in a day and $11.50 per hour each hour thereafter.

  a. Organize the information into a table. Include a row for hours worked x, and a row for daily earnings $f(x)$.

  b. Write the piecewise equation describing Kelly's daily earnings $f(x)$ for x hours.

  c. Draw a graph to represent Kelly's daily earnings.

17. **REASONING** Write a piecewise function that represents the graph.

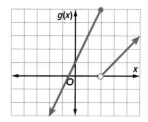

18. **STRUCTURE** Suppose $f(x) = 2[\![x - 1]\!]$.
    a. Find $f(1.5)$.
    b. Find $f(2.2)$.
    c. Find $f(9.7)$.
    d. Find $f(-1.25)$.

19. **RENTAL CARS** Mr. Aronsohn wants to rent a car on vacation. The rate the car rental company charges is $19 per day. If any fraction of a day is counted as a whole day, how much would it cost for Mr. Aronsohn to rent a car for 6.4 days?

20. **USE A MODEL** A roadside fruit and vegetable stand determines rates for selling produce, with every fraction of a pound rounded up to the next pound. The table shows the cost of tomatoes by weight in pounds.

    a. Write a piecewise-linear function representing the cost of purchasing 0 to 5 pounds of tomatoes, where C is the cost in dollars and p is the number of pounds.

Weight (pounds)	Rate (dollars)
1	3.50
2	7.00
3	10.50
4	14.00
5	17.50

    b. Graph the function.

    c. State the domain and range.

    d. What would be the cost of purchasing 4.3 pounds of tomatoes at the roadside stand?

21. **ELECTRONIC REPAIRS** Tech Repairs charges $25 for an electronic device repair that takes up to one hour. For each additional hour of labor, there is a charge of $50. The repair shop charges for the next full hour for any part of an hour.

    a. Copy and complete the table to organize the information. Include a row for hours of repair x, and a row for total cost $f(x)$.

x	0	2	4	6	8
f(x)					

    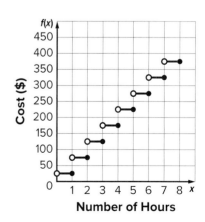

    b. Write a step function to represent the total cost for every hour x of repair.

    c. Graph the function.

    d. Devesh was charged $125 to repair his tablet. How long did the repair take to complete?

22. **INVENTORY** Malik owns a bakery. Every week he orders chocolate chips from a supplier. The supplier's pricing is shown in the table.

Chocolate Chip Pricing	
$4 per pound	Up to 3 pounds
$1.50 per pound	For each pound over 3 pounds

    a. Write a function to represent the cost of chocolate chips.

    b. Malik's budget for chocolate chips for the week is $25. How many whole pounds of chocolate chips can he order?

23. **CREATE** Write a piecewise-defined function with three linear pieces. Then graph the function.

24. **FIND THE ERROR** Amy graphed a function that gives the height of a car on a roller coaster as a function of time. She said her graph is the graph of a step function. Is this possible? Explain your reasoning.

25. **WRITE** What is the difference between a step function and a piecewise-defined function?

26. **ANALYZE** Does the piecewise relation $y = \begin{cases} -2x + 4 \text{ if } x \geq 2 \\ -\frac{1}{2}x - 1 \text{ if } x \leq 4 \end{cases}$ represent a function? Justify your argument.

ANALYZE Refer to the graph for Exercises 27–31.

27. Write a piecewise function to represent the graph.

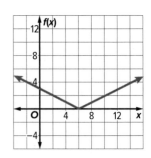

28. What is the domain?

29. What is the range?

30. Find $f(8.5)$.

31. Find $f(1.2)$.

266 Module 4 • Linear and Nonlinear Functions

Lesson 4-7

Absolute Value Functions

Explore Parameters of an Absolute Value Function

🌐 **Online Activity** Use graphing technology to complete the Explore.

> ❓ **INQUIRY** How does performing an operation on an absolute value function change its graph?

Learn Graphing Absolute Value Functions

The **absolute value function** is a type of piecewise-linear function. An absolute value function is written as $f(x) = a|x - h| + k$, where a, h, and k are constants and $f(x) \geq 0$ for all values of x.

The **vertex** is either the lowest point or the highest point of a function. For the parent function, $y = |x|$, the vertex is at the origin.

Key Concept • Absolute Value Function

Parent Function	$f(x) = |x|$, defined as $f(x) = \begin{cases} x \text{ if } x \geq 0 \\ -x \text{ if } x < 0 \end{cases}$
Type of Graph	V-Shaped
Domain:	all real numbers
Range:	all nonnegative real numbers

Learn Translations of Absolute Value Functions

Key Concept • Vertical Translations of Absolute Value Functions

If $k > 0$, the graph of $f(x) = |x|$ is translated k units up.
If $k < 0$, the graph of $f(x) = |x|$ is translated $|k|$ units down.

Key Concept • Horizontal Translations of Linear Functions

If $h > 0$, the graph of $f(x) = |x|$ is translated h units right.
If $h < 0$, the graph of $f(x) = |x|$ is translated $|h|$ units left.

Example 1 Vertical Translations of Absolute Value Functions

Describe the translation in $g(x) = |x| - 3$ as it relates to the graph of the parent function.

Graph the parent function, $f(x) = |x|$, for absolute values.

The constant, k, is outside the absolute value signs, so k affects the y-values. The graph will be a vertical translation.

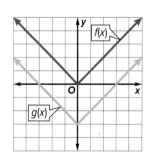

(continued on the next page)

Today's Goals
- Graph absolute value functions.
- Apply translations to absolute value functions.
- Apply dilations to absolute value functions.
- Apply reflections to absolute value functions.
- Interpret constants within equations of absolute value functions.

Today's Vocabulary
absolute value function
vertex

🧠 **Think About It!**
Why does adding a positive value of k shift the graph k units up?

🌐 **Go Online**
You can watch a video to see how to describe translations of functions.

Study Tip

Horizontal Shifts Remember that the general form of an absolute value function is $y = a|x - h| + k$. So, $y = |x + 7|$ is actually $y = |x - (-7)|$ in the function's general form.

Since $f(x) = |x|$, $g(x) = f(x) + k$ where $k = -3$.
$g(x) = |x| - 3 \longrightarrow g(x) = f(x) + (-3)$

The value of k is less than 0, so the graph will be translated $|k|$ units down, or 3 units down.

$g(x) = |x| - 3$ is a translation of the graph of the parent function 3 units down.

Example 2 Horizontal Translations of Absolute Value Functions

Describe the translation in $j(x) = |x - 4|$ as it relates to the parent function.

Graph the parent function, $f(x) = |x|$, for absolute values.

The constant, h, is inside the absolute value signs, so h affects the input, or x-values. The graph will be a horizontal translation.

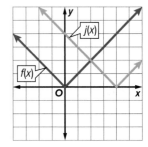

Since $f(x) = |x|$, $j(x) = f(x - h)$, where $h = 4$.
$j(x) = |x - 4| \longrightarrow j(x) = f(x - 4)$

The value of h is greater than 0, so the graph will be translated h units right, or 4 units right.

$j(x) = |x - 4|$ is the translation of the graph of the parent function 4 units right.

Think About It!

Since the vertex of the parent function is at the origin, what is a quick way to determine where the vertex is of $q(x) = |x - h| + k$?

Example 3 Multiple Translations of Absolute Value Functions

Describe the translation in $g(x) = |x - 2| + 3$ as it relates to the graph of the parent function.

The equation has both h and k values. The input and output will be affected by the constants. The graph of $f(x) = |x|$ is vertically and horizontally translated.

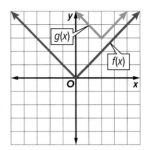

Since $f(x) = |x|$, $g(x) = f(x - h) + k$ where $h = 2$ and $k = 3$.

Because $h = 2$ and $k = 3$, the graph is translated 2 units right and 3 units up.

$g(x) = |x - 2| + 3$ is the translation of the graph of the parent function 2 units right and 3 units up.

Emilio says that the graph of $g(x) = |x + 1| - 1$ is the same graph as $f(x) = |x|$. Is he correct? Why or why not?

Go Online You can complete an Extra Example online.

Example 4 Identify Absolute Value Functions from Graphs

Use the graph of the function to write its equation.

The graph is the translation of the parent graph 1 unit to the right.

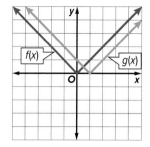

$g(x) = |x - h|$ General equation for a horizontal translation

$g(x) = |x - 1|$ The vertex is 1 unit to the right of the origin.

Example 5 Identify Absolute Value Functions from Graphs (Multiple Translations)

Use the graph of the function to write its equation.

The graph is a translation of the parent graph 2 units to the left and 5 units down.

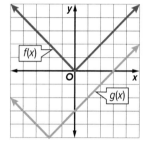

$g(x) = |x - h| + k$ General equation for translations

$g(x) = |x - (-2)| + k$ The vertex is 2 units left of the origin.

$g(x) = |x - (-2)| + (-5)$ The vertex is 5 units down from the origin.

$g(x) = |x + 2| - 5$ Simplify.

Learn Dilations of Absolute Value Functions

Multiplying by a constant a after evaluating an absolute value function creates a vertical change, either a stretch or compression.

Key Concept • Vertical Dilations of Absolute Value Functions

If $|a| > 1$, the graph of $f(x) = |x|$ is stretched vertically.
If $0 < |a| < 1$, the graph of $f(x) = |x|$ is compressed vertically.

When an input is multiplied by a constant a before for the absolute value is evaluated, a horizontal change occurs.

Key Concept • Horizontal Dilations of Absolute Value Functions

If $|a| > 1$, the graph of $f(x) = |x|$ is compressed horizontally.
If $0 < |a| < 1$, the graph of $f(x) = |x|$ is stretched horizontally.

Talk About It!
How is the value of a in an absolute value function related to slope? Explain.

Go Online You can complete an Extra Example online.

Go Online
You can watch a video to see how to describe dilations of functions.

Think About It!
How are $a|x|$ and $|ax|$ evaluated differently?

Example 6 Dilations of the Form $a|x|$ When $a > 1$

Describe the dilation in $g(x) = \frac{5}{2}|x|$ as it relates to the graph of the parent function.

Since $f(x) = |x|$, $g(x) = a \cdot f(x)$, where $a = \frac{5}{2}$.

$g(x) = \frac{5}{2}|x| \longrightarrow g(x) = \frac{5}{2} \cdot f(x)$

$g(x) = \frac{5}{2}|x|$ is a vertical stretch of the graph of the parent graph.

x	$\|x\|$	$\frac{5}{2}\|x\|$	$(x, g(x))$
-4	$\|-4\| = 4$	10	$(-4, 10)$
-2	$\|-2\| = 2$	5	$(-2, 5)$
0	$\|0\| = 0$	0	$(0, 0)$
2	$\|2\| = 2$	5	$(2, 5)$
4	$\|4\| = 4$	10	$(4, 10)$

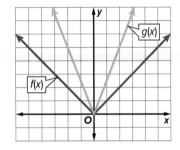

Example 7 Dilations of the Form $|ax|$

Describe the dilation in $p(x) = |2x|$ as it relates to the graph of the parent function.

For $p(x) = |2x|$, $a = 2$. Since a is inside the absolute value symbols, the input is first multiplied by a. Then, the absolute value of ax is evaluated.

x	$\|2x\|$	$p(x)$	$(x, p(x))$
-4	$\|2(-4)\| = \|-8\|$	8	$(-4, 8)$
-2	$\|2(-2)\| = \|-4\|$	4	$(-2, 4)$
0	$\|2(0)\| = \|0\|$	0	$(0, 0)$
2	$\|2(2)\| = \|4\|$	4	$(2, 4)$
4	$\|2(4)\| = \|8\|$	8	$(4, 8)$

Plot the points from the table.

Since $f(x) = |x|$, $p(x) = f(ax)$ where $a = 2$.

$p(x) = |2x| \longrightarrow p(x) = f(2x)$

$p(x) = |2x|$ is a horizontal compression of the graph of the parent graph.

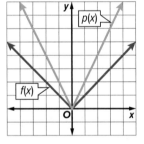

Watch Out!

Differences in Dilations
Although $a|x|$ and $|ax|$ appear to have the same effect on a function, they are evaluated differently and that difference is more apparent when a function is dilated and translated horizontally. For a function with multiple transformations, it is best to first create a table.

Check

Tell whether each equation is an example of a *vertical stretch, vertical compression, horizontal stretch, or horizontal compression.*

A. $j(x) = \left|\frac{4}{3}x\right|$

B. $q(x) = \left|\frac{1}{5}x\right|$

C. $p(x) = 6|x|$

D. $g(x) = \frac{5}{7}|x|$

Go Online You can complete an Extra Example online.

Example 8 Dilations When $0 < a < 1$

Describe the dilation in $j(x) = \frac{1}{3}|x|$ as it relates to the graph of the parent function.

For $j(x) = \frac{1}{3}|x|$, $a = \frac{1}{3}$. Because a is outside the absolute value signs, the absolute value of the input is evaluated first. Then, the function is multiplied by a.

Plot the points from the table.

x	$|x|$	$\frac{1}{3}|x|$	$(x, j(x))$
−6	$|-6| = 6$	2	(−6, 2)
−3	$|-3| = 3$	1	(−3, 1)
0	$|0| = 0$	0	(0, 0)
3	$|3| = 3$	1	(3, 1)
6	$|6| = 6$	2	(6, 2)

Because $f(x) = |x|$, $j(x) = a \cdot f(x)$ where $a = \frac{1}{3}$.

$j(x) = \frac{1}{3}|x| \rightarrow j(x) = \frac{1}{3} \cdot f(x)$

$j(x) = \frac{1}{3}|x|$ is a vertical compression of the graph of the parent function.

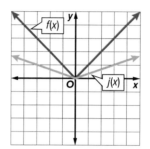

Check

Write an equation for each graph shown.

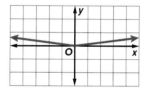
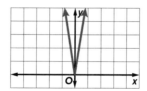

Learn Reflections of Absolute Value Functions

The graph of $-a|x|$ appears to be flipped upside down compared to $a|x|$, and they are symmetric about the x-axis.

Key Concept • Reflections of Absolute Value Functions Across the x-axis

The graph of $-af(x)$ is the reflection of the graph of $af(x) = a|x|$ across the x-axis.

When the only transformation occurring is a reflection or a dilation and reflection, the graphs of $f(ax)$ and $f(-ax)$ appear the same.

Key Concept • Reflections of Absolute Value Functions Across the y-axis

The graph of $f(-ax)$ is the reflection of the graph of $f(ax) = |ax|$ across the y-axis.

Go Online
You can watch a video to see how to describe reflections of functions.

Think About It!
Why would $g(x) = |-2x|$ and $j(x) = |2x|$ appear to be the same graphs?

Go Online You can complete an Extra Example online.

Example 9 Graphs of Reflections with Transformations

Describe how the graph of $j(x) = -|x + 3| + 5$ is related to the graph of the parent function.

x	$|x + 3|$	$-|x + 3|$	$-|x + 3| + 5$	$(x, j(x))$
−5	$|-5 + 3| = |-2| = 2$	−2	$-2 + 5 = 3$	(−5, 3)
−4	$|-4 + 3| = |-1| = 1$	−1	$-1 + 5 = 4$	(−4, 4)
−3	$|-3 + 3| = |0| = 0$	0	$0 + 5 = 5$	(−3, 5)
−2	$|-2 + 3| = |1| = 1$	−1	$-1 + 5 = 4$	(−2, 4)
−1	$|-1 + 3| = |2| = 2$	−2	$-2 + 5 = 3$	(−1, 3)

First, the absolute value of $x + 3$ is evaluated. Then, the function is multiplied by $-1 \cdot a$. Finally, 5 is added to the function.

Plot the points from the table.

Because $f(x) = |x|$, $j(x) = -1 \cdot a \cdot f(x)$ where $a = 1$, $h = -3$, and $k = 5$.

$j(x) = -|x + 3| + 5 \rightarrow j(x) = -1 \cdot f(x + 3) + 5$

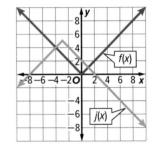

$j(x) = -|x + 3| + 5$ is the graph of the parent function reflected across the x-axis and translated 3 units left and 5 units up.

Example 10 Graphs of $y = -a|x|$

Describe how the graph of $q(x) = -\frac{3}{4}|x|$ is related to the graph of the parent function.

First, the absolute value of x is evaluated. Then, the function is multiplied by $-1 \cdot a$.

Plot the points from the table.

x	$|x|$	$-\frac{3}{4}|x|$	$(x, q(x))$
−8	$|-8| = 8$	−6	(−8, −6)
−4	$|-4| = 4$	−3	(−4, −3)
0	$|0| = 0$	0	(0, 0)
4	$|4| = 4$	3	(4, 3)
8	$|8| = 8$	6	(8, 6)

Because $f(x) = |x|$, $q(x) = -1 \cdot a \cdot f(x)$

where $a = \frac{3}{4}$.

$q(x) = -\frac{3}{4}|x| \rightarrow q(x) = -\frac{3}{4}f(x)$

$q(x) = -\frac{3}{4}|x|$ is the graph of the parent function reflected across the x-axis and vertically compressed.

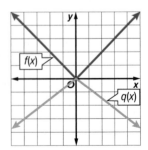

Go Online You can complete an Extra Example online.

Example 11 Graphs of $y = |-ax|$

Describe how the graph of $g(x) = |-4x|$ is related to the graph of the parent function.

First, the input is multiplied by $-1 \cdot a$. Then the absolute value of $-ax$ is evaluated.

Because $f(x) = |x|$, $g(x) = f(-1 \cdot a \cdot x)$ where $a = 4$.

$g(x) = |-4x| \rightarrow g(x) = f(-1 \cdot 4 \cdot x)$

$g(x) = |-4x|$ is the graph of the parent function reflected across the y-axis and horizontally compressed.

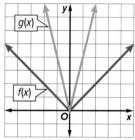

> 💭 **Think About It!**
> Describe how the graph of $y = |-ax|$ is related to the parent function.

Learn Transformations of Absolute Value Functions

You can use the equation of a function to understand the behavior of the function. Because the constants a, h, and k affect the function in different ways, they can help develop an accurate graph of the function.

Concept Summary Transformations of Graphs of Absolute Value Functions

$$g(x) = a|x - h| + k$$

Horizontal Translation, h

If $h > 0$, the graph of $f(x) = |x|$ is translated h units right.

If $h < 0$, the graph of $f(x) = |x|$ is translated $|h|$ units left.

Vertical Translation, k

If $k > 0$, the graph of $f(x) = |x|$ is translated k units up.

If $k < 0$, the graph of $f(x) = |x|$ is translated $|k|$ units down.

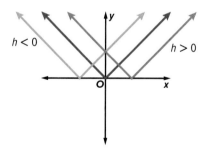
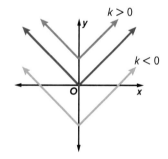

> Why does there appear to be no reflection for the graph of $y = |-ax|$?

Reflection, a

If $a > 0$, the graph opens up.

If $a < 0$, the graph opens down.

Dilation, a

If $|a| > 1$, the graph of $f(x) = |x|$ is stretched vertically.

If $0 < |a| < 1$, the graph is compressed vertically.

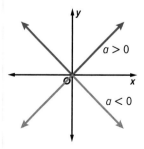
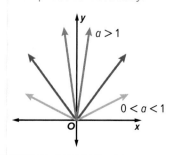

Write the phrase that best describes how each parameter affects the graph of $g(x) = -5|x - 2| + 3$ in relation to the parent function.

−5 Reflects and stretches vertically

2 Translates right

3 Translates up

Example 12 Graph an Absolute Value Function with Multiple Translations

Graph $g(x) = |x + 1| - 4$. State the domain and range.

$a = 1$	The graph is not reflected or dilated in relation to the parent function.	
$h = -1$	The graph is translated 1 unit left from the parent function.	
$k = -4$	The graph is translated 4 units down from the parent function.	

The graph of $g(x) = |x + 1| - 4$ is the graph of the parent function translated 1 unit left and 4 units down without dilation or reflection. The domain is all real numbers. The range is all real numbers greater than or equal to −4.

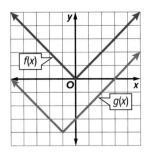

Watch Out!
Dilations and Translations
Don't assume that $j(x) = 2|x - 5| + 1$ and $p(x) = |2x - 5| + 1$ are the same graph. Functions are evaluated differently depending on whether a is inside or outside the absolute value symbols. It might be best to create a table to generate an accurate graph.

Go Online You can watch a video to see how to graph a transformed absolute value function.

Go Online You can complete an Extra Example online.

Example 13 Graph an Absolute Value Function with Translations and Dilation

Graph $j(x) = |3x - 6|$. State the domain and range.

Because a is inside the absolute value symbols, the effect of h on the translation changes.

Evaluate the function for several values of x to find points on the graph.

x	(x, j(x))
0	(0, 6)
1	(1, 3)
2	(2, 0)
3	(3, 3)
4	(4, 6)

The graph of $j(x) = |3x - 6|$ is the graph of the parent function compressed horizontally and translated 2 units right.

The domain is all real numbers. The range is all real numbers greater than or equal to 0.

Example 14 Graph an Absolute Value Function with Translations and Reflection

Graph $p(x) = -|x - 3| + 5$. State the domain and range.

In $p(x) = -|x - 3| + 5$, the parent function is reflected across the x-axis because the absolute value is being multiplied by −1.

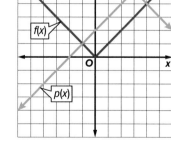

The function is then translated 3 units right.

Finally, the function is translated 5 units up.

$p(x) = -|x - 3| + 5$ is the graph of the parent function translated 3 units right and 5 units up and reflected across the x-axis.

The domain is all real numbers. The range is all real numbers less than or equal to 5.

Think About It!
How is the vertical translation k of an absolute value function related to its range?

Go Online You can complete an Extra Example online.

Example 15 Apply Graphs of Absolute Value Functions

BUILDINGS Determine an absolute value function that models the shape of The Palace of Peace and Reconciliation.

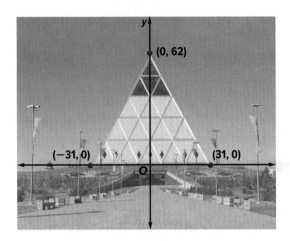

To write the equation for the absolute value function, we must determine the values of a, h, and k in $f(x) = a|x - h| + k$ from the graph.

If we consider the absolute value as a piecewise function, we can find the slope of one side of the graph to determine the value of a.

Because this function opens downward, the graph is a reflection of the parent graph across the x-axis. So we know that the a-value in the equation should be negative.

$$m = \frac{y_2 - y_1}{x_2 - x_1}$$ 	The Slope Formula

$$= \frac{0 - 62}{31 - 0}$$ 	$(0, 62) = (x_1, y_1)$ and $(31, 0) = (x_2, y_2)$

$$= -\frac{62}{31} \text{ or } -2$$

Next, notice that the vertex is not located at the origin. It has been translated. The absolute value function is not shifted left or right, but has been translated 62 units up from the origin.

$y = -2|x - 0| + 62$ 	$a = -2, h = 0, k = 62$

$y = -2|x| + 62$ 	Simplify.

So, $y = -2|x| + 62$ models the shape of The Palace of Peace and Reconciliation.

Check

GLASS PRODUCTION Certain types of glass heat and cool at a nearly constant rate when they are melted to create new glass products. Use the graph to determine the equation that represents this process.

$y = \underline{}|x - \underline{}| + \underline{}$

Go Online You can complete an Extra Example online.

Go Online to practice what you've learned about graphing special functions in the Put It All Together over Lessons 4-6 through 4-7.

Practice

Go Online You can complete your homework online.

Examples 1 through 3

Describe the translation in g(x) as it relates to the graph of the parent function.

1. $g(x) = |x| - 5$
2. $g(x) = |x + 6|$
3. $g(x) = |x - 2| + 7$

4. $g(x) = |x + 1| - 3$
5. $g(x) = |x| + 1$
6. $g(x) = |x - 8|$

Examples 4 and 5

Use the graph of the function to write its equation.

7.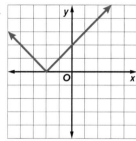
8.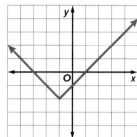
9.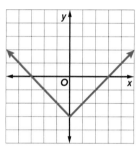

10.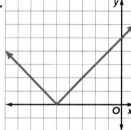
11.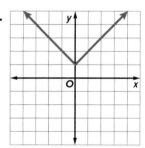
12.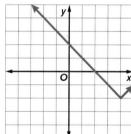

Examples 6 through 8

Describe the dilation in g(x) as it relates to the graph of the parent function.

13. $g(x) = \frac{2}{5}|x|$
14. $g(x) = |0.7x|$
15. $g(x) = 1.3|x|$

16. $g(x) = |3x|$
17. $g(x) = \left|\frac{1}{6}x\right|$
18. $g(x) = \frac{5}{4}|x|$

Lesson 4-7 • Absolute Value Functions 277

Examples 9 through 11

Describe how the graph of g(x) is related to the graph of the parent function.

19. $g(x) = -3|x|$

20. $g(x) = -|x| - 2$

21. $g(x) = \left|-\frac{1}{4}x\right|$

22. $g(x) = -|x - 7| + 3$

23. $g(x) = |-2x|$

24. $g(x) = -\frac{2}{3}|x|$

Examples 12 through 14

USE TOOLS Graph each function. State the domain and range.

25. $g(x) = |x + 2| + 3$

26. $g(x) = |2x - 2| + 1$

27. $f(x) = \left|\frac{1}{2}x - 2\right|$

28. $f(x) = |2x - 1|$

29. $f(x) = \frac{1}{2}|x| + 2$

30. $h(x) = -2|x - 3| + 2$

31. $f(x) = -4|x + 2| - 3$

32. $g(x) = -\frac{2}{3}|x + 6| - 1$

33. $h(x) = -\frac{3}{4}|x - 8| + 1$

Example 15

Determine an absolute value function that models each situation.

34. **ESCALATORS** An escalator travels at a constant speed. The graph models the escalator's distance, in floors, from the second floor x seconds after leaving the ground floor.

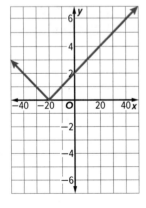

35. **TRAVEL** The graph models the distance, in miles, a car traveling from Chicago, Illinois is from Annapolis, Maryland, where x is the number of hours since the car departed from Chicago, Illinois.

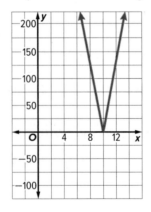

Mixed Exercises

MODELING Graph each function. State the domain and range. Describe how each graph is related to its parent graph.

36. $f(x) = -4|x - 2| + 3$

37. $f(x) = |2x|$

38. $f(x) = |2x + 5|$

Use the graph of the function to write its equation.

39.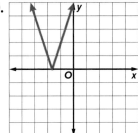

40.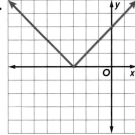

41.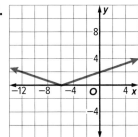

42.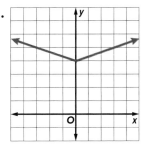

43. **SUNFLOWER SEEDS** A company produces and sells bags of sunflower seeds. A medium-sized bag of sunflower seeds must contain 16 ounces of seeds. If the amount of sunflower seeds s in the medium-sized bag differs from the desired 16 ounces by more than x, the bag cannot be delivered to companies to be sold. Write an equation that can be used to find the highest and lowest amounts of sunflower seeds in a medium-sized bag.

44. **REASONING** The function $y = \frac{5}{4}|x - 5|$ models a car's distance in miles from a parking lot after x minutes. Graph the function. After how many minutes will the car reach the parking lot?

45. **STATE YOUR ASSUMPTION** A track coach set up an agility drill for members of the track team. According to the coach, 21.7 seconds is the target time to complete the agility drill. If the time differs from the desired 21.7 seconds by more than x, the track coach may require members of the track team to change their training. Write an equation that can be used to find the fastest and slowest times members of the track team can complete the agility drill so that their training does not have to change. If $x = 3.2$, what can you assume about the range of times the coach wants the members of the track team to complete the agility drill? Solve your equation for $x = 3.2$ and use the results to justify your assumption.

46. **SCUBA DIVING** The function $y = 3|x - 12| - 36$ models a scuba diver's elevation in feet compared to sea level after x minutes. Graph the function. How far below sea level is the scuba diver at the deepest point in their dive?

Lesson 4-7 • Absolute Value Functions 279

47. MANUFACTURING A manufacturing company produces boxes of cereal. A small box of cereal must have 12 ounces. If the amount of cereal b in a small box differs from the desired 12 ounces by more than x, the box cannot be shipped for selling. Write an equation that can be used to find the highest and lowest amounts of cereal in a small box.

48. STRUCTURE Amelia is competing in a bicycle race. The race is along a circular path. She is 6 miles from the start line. She is approaching the start line at a speed of 0.2 mile per minute. After Amelia reaches the start line, she continues at the same speed, taking another lap around the track.

  a. Organize the information into a table. Include a row for time in minutes x and a row for distance from start line $f(x)$.

  b. Draw a graph to represent Amelia's distance from the start line.

🧠 **Higher-Order Thinking Skills**

49. WRITE Use transformations to describe how the graph of $h(x) = -|x + 2| - 3$ is related to the graph of the parent absolute value function.

50. ANALYZE On a straight highway, the town of Garvey is located at mile marker 200. A car is located at mile marker x and is traveling at an average speed of 50 miles per hour.

  a. Write a function $T(x)$ that gives the time, in hours, it will take the car to reach Garvey. Then graph the function on the coordinate plane.

  b. Does the graph have a maximum or minimum? If so, name it and describe what it represents in the context of the problem.

51. PERSEVERE Write the equation $y = |x - 3| + 2$ as a piecewise-defined function. Then graph the piecewise function.

52. CREATE Write an absolute value function $f(x)$ that has a domain of all real numbers and a range that is greater than or equal to 4. Be sure your function also includes a dilation of the parent function. Describe how your function relates to the parent absolute value graph. Then graph your function.

Module 4 • Linear and Nonlinear Functions

Review

Essential Question
What can a function tell you about the relationship that it represents?

Module Summary

Lessons 4-1 through 4-3

Graphing Linear Functions, Rate of Change, and Slope

- The graph of an equation represents all of its solutions.
- The x-value of the y-intercept is 0. The y-value of the x-intercept is 0.
- The rate of change is how a quantity is changing with respect to a change in another quantity. If x is the independent variable and y is the dependent variable, then rate of change $= \dfrac{\text{change in } y}{\text{change in } x}$.
- The slope m of a nonvertical line through any two points can be found using $m = \dfrac{y_2 - y_1}{x_2 - x_1}$.
- A line with positive slope slopes upward from left to right. A line with negative slope slopes downward from left to right. A horizontal line has a slope of 0. The slope of a vertical line is undefined.

Lesson 4-4

Transformations of Linear Functions

- When a constant k is added to a linear function $f(x)$, the result is a vertical translation.
- When a linear function $f(x)$ is multiplied by a constant a, the result $a \cdot f(x)$ is a vertical dilation.
- When a linear function $f(x)$ is multiplied by -1 before or after the function has been evaluated, the result is a reflection across the x- or y-axis.

Lesson 4-5

Arithmetic Sequences

- An arithmetic sequence is a numerical pattern that increases or decreases at a constant rate called the common difference.
- The nth term of an arithmetic sequence with the first term a_1 and common difference d is given by $a_n = a_1 + (n - 1)d$, where n is a positive integer.

Lessons 4-6, 4-7

Special Functions

- A piecewise-linear function has a graph that is composed of a number of linear pieces.
- A step function is a type of piecewise-linear function with a graph that is a series of horizontal line segments.
- An absolute value function is V-shaped.

Study Organizer

 Foldables

Use your Foldable to review this module. Working with a partner can be helpful. Ask for clarification of concepts as needed.

Test Practice

1. GRAPH Jalyn made a table of how much money she will earn from babysitting. (Lesson 4-1)

Hours Babysitting	Money Earned
1	5
2	10
3	15
4	20

Use the table to graph the function.

2. OPEN RESPONSE Copy and complete the table to find the missing values in the table that show the points on the graph of $f(x) = 2x - 4$. (Lesson 4-1)

x	−2	0	2	4	6
f(x)	−8	−4			

3. OPEN RESPONSE Mr. Hernandez is draining his pool to have it cleaned. At 8:00 A.M., it had 2000 gallons of water and at 11:00 A.M. it had 500 gallons left to drain. What is the rate of change in the amount of water in the pool? (Lesson 4-2)

4. MULTIPLE CHOICE Find the slope of the graphed line. (Lesson 4-2)

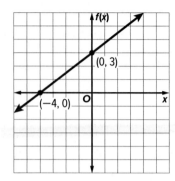

A. $-\frac{4}{3}$

B. $-\frac{3}{4}$

C. $\frac{3}{4}$

D. $\frac{4}{3}$

5. MULTIPLE CHOICE Determine the slope of the line that passes through the points (4, 10) and (2, 10). (Lesson 4-2)

A. −1

B. 0

C. 1

D. undefined

6. GRAPH Graph the equation of a line with a slope of −3 and a y-intercept of 2. (Lesson 4-3)

7. MULTIPLE CHOICE What is the slope of the line that passes through (3, 4) and (−7, 4)? (Lesson 4-3)

A. 0

B. undefined

C. −2

D. −10

8. MULTIPLE CHOICE A teacher buys 100 pencils to keep in her classroom at the beginning of the school year. She allows the students to borrow pencils, but they are not always returned. On average, she loses about 8 pencils a month. Write an equation in slope-intercept form that represents the number of pencils she has left y after a number of x months. (Lesson 4-3)

A. $y = -8x - 100$
B. $y = -8x + 100$
C. $y = 8x + 100$
D. $y = 8x - 100$

9. OPEN RESPONSE Name the transformation that changes the slope, or the steepness, of the graph of a linear function. (Lesson 4-4)

10. OPEN RESPONSE Describe the dilation of $g(x) = \frac{1}{2}(x)$ as it relates to the graph of the parent function, $f(x) = x$. (Lesson 4-4)

11. MULTIPLE CHOICE Arjun begins the calendar year with $40 in his bank account. Each week he receives an allowance of $20, half of which he deposits into his bank account. The situation describes an arithmetic sequence. The situation describes an arithmetic sequence. Which function represents the amount in Arjun's account after n weeks? (Lesson 4-5)

A. $f(n) = 20n + 40$
B. $f(n) = 40n + 20$
C. $f(n) = 40 + 10n$
D. $f(n) = 10 + 40n$

12. OPEN RESPONSE What number can be used to complete the equation below that describes the nth term of the arithmetic sequence $-2, -1.5, -1, -0.5, 0, 0.5, \ldots$? (Lesson 4-5)

$a_n = 0.5n - \underline{\quad ? \quad}$

13. OPEN RESPONSE Write and graph a function to represent the sequence 1, 10, 19, 28, … (Lesson 4-5)

14. OPEN RESPONSE Christa has a box of chocolate candies. The number of chocolates in each row forms an arithmetic sequence, as shown in the table. (Lesson 4-5)

Row	1	2	3	4
Number of Chocolates	3	6	9	12

Write an arithmetic function that can be used to find the number of chocolates in each row.

15. OPEN RESPONSE Daniel earns $9 per hour at his job for the first 40 hours he works each week. However, his pay rate increases to $13.50 per hour thereafter. This situation can be represented with the function

$$f(x) = \begin{cases} 9x, & \text{if } x \leq 40 \\ 360 + 13.5(x - 40), & \text{if } x > 40 \end{cases}$$

Use this function to copy and complete the table with the correct values. (Lesson 4-6)

Hours Worked, x	Money Earned, f(x)
30	
35	315
40	
45	427.5
50	

16. GRAPH Graph the function $f(x) = 2[[x]]$. (Lesson 4-6)

17. MULTIPLE CHOICE Which of the following describes the effect a dilation has upon the graph of the absolute value parent function? (Lesson 4-7)

A. Flipped across axis

B. Stretch or compression

C. Rotated about the origin

D. Shifted horizontally or vertically

18. MULTI-SELECT Describe the transformation(s) of the function graphed below in relation to the absolute value parent function. Select all that apply. (Lesson 4-7)

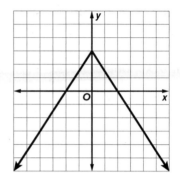

A. Reflected across x-axis

B. Vertical stretch

C. Vertical compression

D. Reflected across y-axis

E. Translated right 3

F. Translated up 3

19. OPEN RESPONSE Describe the graph of $g(x) = |x| + 5$ in relation to the graph of the absolute value parent function. (Lesson 4-7)

20. OPEN RESPONSE Across which axis is the graph of $h(x) = -5|x|$ reflected? (Lesson 4-7)

21. OPEN RESPONSE Use the graph of the function to write its equation. (Lesson 4-7)

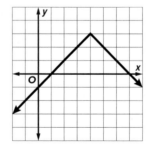

Module 5
Creating Linear Equations

e Essential Question
What can a function tell you about the relationship that it represents?

What Will You Learn?
How much do you already know about each topic **before** starting this module?

KEY
👎 — I don't know.   👉 — I've heard of it.   👍 — I know it!

	Before 👎	Before 👉	Before 👍	After 👎	After 👉	After 👍
write linear equations in slope-intercept form when given the slope and the coordinates of a point						
write linear equations in slope-intercept form when given the coordinates of two points on the line						
write linear equations in standard form						
write linear equations in point-slope form						
write equations of parallel and perpendicular lines						
examine scatter plots to describe relationships between quantities						
make and evaluate predictions by fitting linear functions to sets of data						
distinguish between correlation and causation						
write equations of best-fit lines						
plot and analyze residuals						
find inverses of linear relations and functions						

📙 **Foldables** Make this Foldable to help you organize your notes about linear equations. Begin with one sheet of 11″ × 17″ paper.

1. **Fold** each end of the paper in about 2 inches.
2. **Fold** along the width and the length. Unfold. Cut along the fold line from the top to the center.
3. **Fold** the top flaps down. Then fold in half and turn to form a folder. Staple the flaps down to form pockets.
4. **Label** the front with the chapter title.

1    2 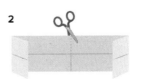   3    4

What Vocabulary Will You Learn?

- best-fit line
- bivariate data
- causation
- correlation coefficient
- inverse functions
- inverse relations
- line of fit
- linear extrapolation
- linear interpolation
- linear regression
- negative correlation
- no correlation
- parallel lines
- perpendicular lines
- positive correlation
- residual
- scatter plot
- trend

Are You Ready?

Complete the Quick Review to see if you are ready to start this module.
Then complete the Quick Check.

Quick Review

Example 1

Solve $5x + 15y = 9$ for x.

$5x + 15y = 9$	Original equation
$5x + 15y - 15y = 9 - 15y$	Subtract 15y from each side.
$5x = 9 - 15y$	Simplify.
$\frac{5x}{5} = \frac{9 - 15y}{5}$	Divide each side by 5.
$x = \frac{9}{5} - 3y$	Simplify.

Example 2

Write the ordered pair for A.

Step 1 Begin at point A.

Step 2 Follow along a vertical line to the x-axis. The x-coordinate is −4.

Step 3 Follow along a horizontal line to the y-axis. The y-coordinate is 2.

The ordered pair for point A is (−4, 2).

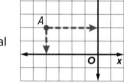

Quick Check

Solve each equation for the given variable.

1. $x + y = 5$ for y
2. $2x - 4y = 6$ for x
3. $y - 2 = x + 3$ for y
4. $4x - 3y = 12$ for x

Write the ordered pair for each point.

5. A
6. B
7. C
8. D
9. E
10. F

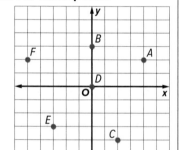

How Did You Do?

Which exercises did you answer correctly in the Quick Check?

286 Module 5 • Creating Linear Equations

Lesson 5-1

Writing Equations in Slope-Intercept Form

Explore Slope-Intercept Form

 Online Activity Use graphing technology to complete the Explore.

> **INQUIRY** How does changing the coordinates of two points on a line affect the slope of the line?

Today's Goals
- Write an equation of a line in slope-intercept form given the slope and one point.
- Write an equation of a line in slope-intercept form given two points.

Learn Creating Linear Equations in Slope-Intercept Form Given the Slope and a Point

If you are given the slope of a line and the coordinates of any point on that line, you can create an equation for that line.

Key Concept • Creating Equations in Slope-Intercept Form Given the Slope and a Point

Step 1	Determine whether the given point is the *y*-intercept. If not, substitute the given information into the slope-intercept form equation to find the *y*-intercept.
Step 2	Use the given slope and *y*-intercept you found in Step 1 to write the equation of the line in slope-intercept form.

Think About It!
How can you determine whether the given point is the *y*-intercept of the line?

Example 1 Write an Equation Given the Slope and a Point

Write an equation of the line that passes through $(-8, 6)$ and has a slope of $-\frac{3}{4}$.

Step 1 Find the *y*-intercept.

$y = mx + b$ Slope-intercept form

$6 = -\frac{3}{4}(-8) + b$ $m = -\frac{3}{4}, x = -8,$ and $y = 6$

$6 = 6 + b$ Simplify.

$0 = b$ Subtract 6 from each side.

Step 2 Write the equation in slope-intercept form.

$y = mx + b$ Slope-intercept form

$y = -\frac{3}{4}x + 0$ $m = -\frac{3}{4}$ and $b = 0$

$y = -\frac{3}{4}x$ Simplify.

Think About It!
What does it mean if $b = 0$ when an equation is written in slope-intercept form?

Study Tip

Slope-Intercept Form Remember, you need two things to write an equation in slope-intercept form: the slope and the *y*-intercept.

Go Online You can complete an Extra Example online.

Example 2 Write an Equation in Slope-Intercept Form

BAKING Marissa is baking a recipe that calls for her to turn down the temperature on her oven for part of the baking time. Write an equation to represent the situation if the temperature in her oven drops 25°F every 30 seconds, and after 2 minutes the temperature is 350°F.

Step 1 Determine a point on the line and the slope.

After 2 minutes the temperature is 350°F.

Let x = the time in minutes and y = the temperature in °F.

So, the point (2, 350) is on the line.

The temperature drops 25°F every 30 seconds.

The change in x is 30 seconds, or 0.5 minute.

"Drops" means a negative change, so the change in y is −25°F.

$$\text{Slope} = \frac{\text{change in } y}{\text{change in } x} = \frac{-25}{0.5} \text{ or } -50°F \text{ per minute.}$$

So, the slope is −50.

Step 2 Find the y-intercept.

$y = mx + b$	Slope-intercept form
$350 = -50(2) + b$	$m = -50$, $x = 2$, and $y = 350$
$350 = -100 + b$	Simplify.
$450 = b$	Add 100 to each side.

This means that the temperature of the oven was 450°F when it was turned off.

Step 3 Write the equation in slope-intercept form.

$y = mx + b$	Slope-intercept form
$y = -50x + 450$	$m = -50$ and $b = 450$

Check

MEMBERSHIP The total monthly cost of Ayzha's gym membership increases by $5 per class she attends. After signing up for 4 classes one month, her total cost is $49.99. Which equation represents Ayzha's total monthly cost y after attending x classes?

A. $y = -5x + 29.99$ B. $y = -5x + 69.99$

C. $y = 5x + 29.99$ D. $y = 5x + 69.99$

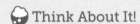 Go Online You can complete an Extra Example online.

Study Tip

Slope When determining the slope, words like "drops" and "decreasing" represent a negative slope, and words like "growing" and "increasing" represent a positive slope.

Think About It!

Find the domain of your equation, and describe the meaning in the context of the situation.

Learn Creating Linear Equations in Slope-Intercept Form Given Two Points

If you are given the coordinates of any two points on a line, you can create an equation for that line.

Key Concept • Creating Equations in Slope-Intercept Form Given Two Points	
Step 1	Use the given points to find the slope of the line containing the points.
Step 2	Use the slope from Step 1 and either of the given points to find the y-intercept of the line.
Step 3	Use the slope you found in Step 1 and the y-intercept you found in Step 2 to write the equation of the line in slope-intercept form.

Talk About It!
Will your equation for the line be different depending on the point you choose in Step 2? Justify your argument.

Example 3 Write Equations Given Two Points

Write an equation of the line that passes through (1.2, −0.7) and (−3.4, 1.6).

Step 1 Find the slope.

$m = \dfrac{y_2 - y_1}{x_2 - x_1}$ Slope Formula

$m = \dfrac{1.6 - (-0.7)}{-3.4 - 1.2}$ $(x_1, y_1) = (1.2, -0.7), (x_2, y_2) = (-3.4, 1.6)$

$m = \dfrac{2.3}{-4.6}$ Simplify.

$m = -0.5$ Simplify.

Watch Out!
Subtraction If the (x_1, y_1) coordinates are negative, be sure to account for both the negative signs and the subtraction symbols in the Slope Formula. Remember, the result of subtracting a negative number is the same as adding its opposite.

Step 2 Use either point to find the y-intercept.

$y = mx + b$ Slope-intercept form

$1.6 = -0.5(-3.4) + b$ $m = -0.5, x = -3.4,$ and $y = 1.6$

$1.6 = 1.7 + b$ Simplify.

$-0.1 = b$ Subtract 1.7 from each side.

Step 3 Write the equation in slope-intercept form.

$y = mx + b$ Slope-intercept form

$y = -0.5x - 0.1$ $m = -0.5$ and $b = -0.1$

Check

Write an equation of the line that passes through (−5, −3) and (−7, −12).

Go Online You can complete an Extra Example online.

Think About It!

Use the table to make an estimate of the number of students enrolled in public high schools in 2030. Then, use the equation to predict the number of students enrolled. How does your estimate compare to the number of students that you calculated?

Problem-Solving Tip

Use a Graph You can also estimate and make predictions using a graph. Plot two points from the table, connect them with a line, and then estimate using the graph.

Study Tip

Units The number of students enrolled is in thousands. While it is impossible to have one-quarter of a student, 14,708.25 thousand students really means 14,708,250 students. So, this solution is within the constraints of the situation.

🌐 Apply Example 4 Write an Equation Given Real-World Data

SCHOOLS The number of students enrolled in public high schools in the United States has risen slightly since 2010. Write an equation that could be used to predict the number of students enrolled in public high schools if enrollment continues to grow at the same rate.

Year	Students (in thousands)
2011	14,749
2012	14,753
2013	14,754
2014	14,826
2015	14,912

1. What is the task?
Describe the task in your own words. Then list any questions that you may have. How can you find answers to your questions?

I need to write an equation to predict enrollment in public high schools. How can I write an equation when given a table? I can review finding rate of change and writing equations in slope-intercept form.

2. How will you approach the task? What have you learned that you can use to help you complete the task?

I will use what I have learned about finding the rate of change from a table to help me find the slope. I will then use the slope and one of the points to find the y-intercept. I will use what I have learned about writing equations in slope-intercept form to write an equation.

3. What is your solution?
Use your strategy to solve the problem.

Find the slope.
$m = 40.75$

Find the y-intercept.
$b = 14,708.25$

Write an equation to predict the number of students enrolled in public high schools if enrollment continues to grow at the same rate.

$y = 40.75x + 14,708.25$

4. How can you know that your solution is reasonable?
✏️ **Write About It!** Write an argument that can be used to defend your solution.

For the year 2015, $x = 5$. I substituted $x = 5$ into my equation to check my solution, and the result matched the number of students enrolled in 2015.

📱 **Go Online** You can complete an Extra Example online.

Practice

Go Online You can complete your homework online.

Example 1
Write an equation of the line that passes through the given point and has the given slope.

1. (4, 2); slope $\frac{1}{2}$
2. (3, −2); slope $\frac{1}{3}$
3. (6, 4); slope $-\frac{3}{4}$

4. (−5, 4); slope −3
5. (4, 3); slope $\frac{1}{2}$
6. (1, −5); slope $-\frac{3}{2}$

Example 2
7. **EXERCISE** Carlos is jogging at a constant speed. He starts a timer when he is 12 feet from his starting position. After 3 seconds, Carlos is 21 feet from his starting position. Write a linear equation to represent the distance d of Carlos from his starting position after t seconds.

8. **JOBS** Mr. Kimball sells computer software. He earns a base salary of $41,250 and 8% commission on his sales. Write an equation to represent Mr. Kimball's total pay p after selling d dollars of software.

9. **USE A MODEL** In 2006, the average ticket price for a National Football League game was $62.38. Since then the cost has increased an average of $2.54 per year. Write a linear equation to represent the cost C of an NFL ticket y years after 2006.

10. **TYPING** Nebi has already typed 250 words. He then starts a timer and finds that he types 150 words in 3 minutes. If Nebi types at a constant rate, write a linear equation to represent the number of words w Nebi types m minutes after starting the timer.

Example 3
Write an equation of the line that passes through each pair of points.

11. (0, −4), (5, −4)
12. (−4, −2), (4, 0)
13. (−2, −3), (4, 5)

14. (0, 1), (5, 3)
15. (−3, 0), (1, −6)
16. (1, 0), (5, −1)

17. (9, 2), (−2, 6)
18. (−6, 5), (−6, −4)
19. (5, −2), (7, −1)

20. (5, −3), (2, 5)
21. $\left(\frac{5}{4}, 1\right), \left(-\frac{1}{4}, \frac{3}{4}\right)$
22. $\left(\frac{5}{12}, -1\right), \left(-\frac{3}{4}, \frac{1}{6}\right)$

Example 4

23. **GUITAR** Lydia wants to purchase guitar lessons. She sees a sign that gives the prices for 7 guitar lessons and 11 guitar lessons. Write a linear equation to find the total cost C for d lessons.

24. **CENSUS** The population of Laredo, Texas, was about 215,500 in 2007. It was about 123,000 in 1990. If we assume that the population growth is constant, write a linear equation with an integer slope to represent p, Laredo's population t years after 1990.

25. **WEATHER** A meteorologist finds that the temperature at the 6000-foot level of a mountain is 76°F and the temperature at the 12,000-foot level of the mountain is 49°F. Write a linear equation to represent the temperature T at an elevation of x, where x is in thousands of feet.

26. **FUNDRAISING** Natalia and her friends held a bake sale to benefit a local charity. The friends sold 15 cakes on the first day and 22 cakes on the second day of the bake sale. They collected $60 on the first day and $88 on the second day. Write an equation to represent the amount R Natalia and her friends raised after selling c cakes.

Mixed Exercises

Write an equation of each line.

27.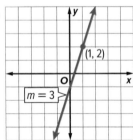

28.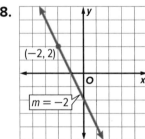

29.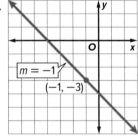

30.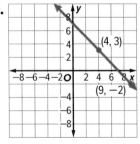

31.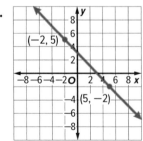

32.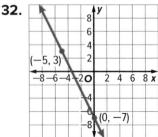

Determine whether the given point is on the line. Explain your reasoning.

33. $(3, -1)$; $y = \frac{1}{3}x + 5$

34. $(6, -2)$; $y = \frac{1}{2}x - 5$

35. $(15, -13)$; $y = -\frac{1}{5}x - 10$

36. $(3, 3)$; $y = -\frac{2}{3}x + 1$

Determine another point on a line given two points on the line.

37. $(2, -4), (4, -2)$

38. $(0, 5), (4, 1)$

39. $(-3, 1), (-1, -3)$

40. $(0, 4), (2, 5)$

41. $(-2, 9), (2, -1)$

42. $(3, 0), (12, 3)$

For Exercises 43–45, determine which equation best represents each situation. Explain the meaning of each variable.

A. $y = \frac{1}{25}x + 300$

B. $y = 25x + 300$

C. $y = 300x + 25$

43. PLANES Plane tickets cost $300 each plus a one-time fee of $25 to select seats.

44. SAVINGS Larry has $300. He saves $25 each week.

45. OIL The current oil level in a tank is 25 feet. The rate that oil is being poured into the tank is $\frac{1}{25}$ inch per hour.

46. USE A MODEL The table of ordered pairs shows the coordinates of the two points on the graph of a function. Write an equation that describes the function.

x	y
-2	2
4	-1

47. USE A SOURCE The table shows how women's shoe sizes in the United Kingdom compare to women's shoe sizes in the United States.

Women's Shoe Sizes

U.K.	3	3.5	4	4.5	5	5.5	6
U.S.	5.5	6	6.5	7	7.5	8	8.5

Source: DanceSport UK

a. Write a linear equation to determine the U.S. size y if you are given the U.K. size x.

b. What would be the U.S. shoe size for a woman who wears a U.K. size 7.5?

c. Research women's shoe sizes in Australia compared to women's shoe sizes in the United States. Write a linear equation to determine the U.S. size y if you are given the Australia size x.

Lesson 5-1 • Writing Equations in Slope-Intercept Form

48. REASONING Shikita borrowed money from her brother and paid back a set amount each week. The table shows how much she owed in a given week.

Week	3	6	8	10	13
Amount Owed	$32.50	$25.00	$20.00	$15.00	$7.50

a. Let x represent the number of weeks and y represent the amount owed. Write an equation in slope-intercept form to model the amount Shikita owed each week.

b. Describe the graph of the equation you found in **part a**. What does its shape tell you about the problem?

49. REASONING Koby tracked the weight of his puppy for 6 months. Her growth is shown in the table where x = age in months and y = weight in pounds.

x	y
2	16
3	23.5
4	31
5	38.5
6	46

a. Write an equation in slope-intercept form to model the growth of Koby's puppy.

b. What is the y-intercept? What does the y-intercept mean in the context of the problem?

c. What is the slope? What does the slope mean in the context of the problem?

Higher-Order Thinking Skills

50. PERSEVERE Write the equation of each line in slope-intercept form.

a. slope: $\frac{4}{5}$, y-intercept: -8

b. $(-3, 0), (3, -16)$

c. What point do the graphs of both equations have in common, and what does this tell you about their graphs?

51. FIND THE ERROR Tess and Jacinta are writing an equation of the line through $(3, -2)$ and $(6, 4)$. Is either of them correct? Explain your reasoning.

Tess
$m = \frac{4 - (-2)}{6 - 3} = \frac{6}{3}$ or 2
$y = mx + b$
$6 = 2(4) + b$
$6 = 8 + b$
$-2 = b$
$y = 2x - 2$

Jacinta
$m = \frac{4 - (-2)}{6 - 3} = \frac{6}{3}$ or 2
$y = mx + b$
$-2 = 2(3) + b$
$-2 = 6 + b$
$-8 = b$
$y = 2x - 8$

52. WRITE Linear equations are useful in predicting future events. Describe some factors in real-world situations that might affect the reliability of the graph in making any predictions.

53. CREATE Create a real-world situation that fits the graph at the right. Define the two quantities and describe the functional relationship between them. Write an equation to represent this relationship, and describe what the slope and y-intercept mean.

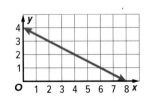

Lesson 5-2
Writing Equations in Standard and Point-Slope Forms

Explore Forms of Linear Equations

Online Activity Use graphing technology to complete the Explore.

> **INQUIRY** How are the point-slope and slope-intercept forms of a linear equation related?

Today's Goals
- Write equations of lines in point-slope form.
- Create and identify equations of parallel or perpendicular lines.

Today's Vocabulary
parallel lines

perpendicular lines

Learn Creating Linear Equations in Point-Slope Form

When the slope and the coordinates of one point of a line are known, an equation for the line can be written in point-slope form.

Key Concept • Point-Slope Form

Words	The linear equation $y - y_1 = m(x - x_1)$ is written in point-slope form, where (x_1, y_1) is a given point on a nonvertical line and m is the slope of the line.
Symbols	$y - y_1 = m(x - x_1)$
Example	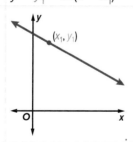

Talk About It!
Why must a line be nonvertical in order to be written in point-slope form? Explain.

If you are given two points on the line or a point on the line and its slope, you can write an equation for the line in point-slope form.

Key Concept • Writing Equations of Lines in Point-Slope Form

Given the Slope and One Point		Given Two Points	
Step 1	Let the x and y coordinates be (x_1, y_1).	Step 1	Find the slope.
Step 2	Substitute the values of m, x_1, and y_1 into the equation of a line in point-slope form.	Step 2	Choose one of the two points to use.
		Step 3	Follow the steps for writing an equation given the slope and one point.

Go Online
You may want to complete the Concept Check to check your understanding.

Example 1 Equation in Point-Slope Form Given Slope and a Point

Write an equation in point-slope form for the line that passes through (−2, 7) with a slope of $-\frac{3}{2}$. Then graph the equation.

$y - y_1 = m(x - x_1)$ Point-slope form

$y - 7 = -\frac{3}{2}[x - (-2)]$ $(x_1, y_1) = (-2, 7)$ and $m = -\frac{3}{2}$

$y - 7 = -\frac{3}{2}(x + 2)$ Simplify.

Step 1 Plot the given point (−2, 7).

Step 2 Use the slope, $-\frac{3}{2}$, to plot another point on the line.

Step 3 Draw a line through the points.

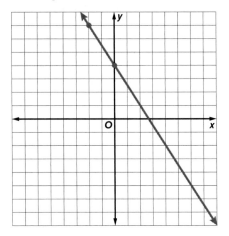

Check

Determine the equation in point-slope form for the line that passes through (7, 5) with a slope of −3. Then graph the equation.

Go Online You can complete an Extra Example online.

Example 2 Equation in Point-Slope Form Given Two Points

Write an equation in point-slope form for the line that passes through the given points.

(2, −7) and (6, −3)

Step 1 Find the slope.

$m = \frac{y_2 - y_1}{x_2 - x_1}$ Slope Formula

$m = \frac{-3 - (-7)}{6 - 2} = \frac{4}{4}$ or 1 $(x_1, y_1) = (2, -7)$ and $(x_2, y_2) = (6, -3)$

Step 2 Write an equation.

You can select either point for (x_1, y_1) in point-slope form.

$y - y_1 = m(x - x_1)$ Point-slope form

$y - (-3) = 1(x - 6)$ $(x_1, y_1) = (6, -3)$ and $m = 1$

$y + 3 = (x - 6)$ Simplify.

Check

Select an equation in point-slope form for the line that passes through (−16, 18) and (−11, −2).

A. $y + 2 = -4(x + 11)$

B. $y + 2 = -\frac{1}{4}(x + 11)$

C. $y - 2 = -4(x + 11)$

D. $y - 2 = -\frac{1}{4}(x - 11)$

E. None of these

> 💭 **Think About It!**
> Write another equation in point-slope form for the line with the points given.
>
> Why are there multiple correct answers with the same given information?

Example 3 Change to Slope-Intercept Form

Write $y + 4 = -2(x - 6)$ in slope-intercept form.

$y + 4 = -2(x - 6)$ Original Equation

$y + 4 = -2x + 12$ Distributive Property

$y = -2x + 8$ Subtract 4 from each side.

Check

Write $y + 3 = -\frac{1}{2}(x - 8)$ in slope-intercept form.

> **Study Tip**
>
> **Checking Your Work**
> To check your work, you can substitute the point from the original point-slope form of the equation, in this case (6, −4), into the slope-intercept form of the equation. If it is a true statement, the equation is correct.
>
> $y = -2x + 8$
> $-4 = -2(6) + 8$
> $-4 = -12 + 8$
> $-4 = -4$ ✓

🌐 **Go Online** You can complete an Extra Example online.

Example 4 Apply Point-Slope Form

READING Nadia's book club is ordering new novels. She knows that the total cost of 5 books is $61.25, and 15 books cost $159.75. Write an equation in point-slope form to represent the total cost *y* of ordering *x* books.

Step 1 Find the slope.

$$m = \frac{y_2 - y_1}{x_2 - x_1} \quad \text{Slope Formula}$$

$$m = \frac{159.75 - 61.25}{15 - 5} = \frac{98.5}{10} \text{ or } 9.85 \quad (x_1, y_1) = (5, 61.25) \text{ and } (x_2, y_2) = (15, 159.75)$$

Step 2 Write an equation.

$$y - y_1 = m(x - x_1) \quad \text{Point-slope form}$$

$$y - 61.25 = 9.85(x - 5) \quad (x_1, y_1) = (5, 61.25) \text{ and } m = 9.85$$

Think About It!
Use the equation to find the cost of purchasing 12 books.

Check

TAXIS The total cost of a taxi fare is given in the table. Determine the equation(s) in point-slope form that model(s) this situation if *x* represents the distance in miles and *y* represents the cost in dollars.

Distance (miles)	1.5	4	7.5	12.25
Cost (dollars)	6.90	13.40	22.50	34.85

A. $y - 13.4 = 2.6(x - 4)$
B. $y - 22.5 = 2.6(x - 7.5)$
C. $y = 2.6x + 3$
D. $y - 6.9 = \frac{5}{13}(x - 1.5)$
E. $y + 34.85 = 2.6(x + 12.25)$

Example 5 Change to Standard Form

Write $y - 1 = -\frac{2}{5}(x + 3)$ in standard form.

$$y - 1 = -\frac{2}{5}(x + 3) \quad \text{Original equation}$$

$$5(y - 1) = -2(x + 3) \quad \text{Multiply each side by 5 to eliminate the fraction.}$$

$$5y - 5 = -2x - 6 \quad \text{Distributive Property}$$

$$5y = -2x - 1 \quad \text{Add 5 to each side.}$$

$$2x + 5y = -1 \quad \text{Add 2x to each side.}$$

Study Tip
Fractional Slopes When working with an equation with a fractional slope, it is often simpler to first multiply each side of the equation by the denominator. This will eliminate distributing a fraction later in the equation.

Check

Write $y = -\frac{7}{2}x + 5$ in standard form.

Go Online You can complete an Extra Example online.

298 Module 5 • Creating Linear Equations

Example 6 Standard Form Given Two Points

Write an equation in standard form for the line that passes through (8, −4) and (−6, −11).

Step 1 Find the slope.

$m = \dfrac{y_2 - y_1}{x_2 - x_1}$ Slope Formula

$m = \dfrac{-11 - (-4)}{-6 - 8} = \dfrac{-7}{-14}$ or $\dfrac{1}{2}$ $(x_1, y_1) = (8, -4)$ and $(x_2, y_2) = (-6, -11)$

Step 2 Write an equation in slope-intercept form.

$y = mx + b$ Slope-intercept form

$-4 = \dfrac{1}{2}(8) + b$ $(x, y) = (8, -4)$ and $m = \dfrac{1}{2}$

$-4 = 4 + b$ Simplify.

$-8 = b$ Subtract 4 from each side.

$y = \dfrac{1}{2}x - 8$ Replace m with $\dfrac{1}{2}$ and b with -8.

Step 3 Write the equation in standard form.

$2y = 2\left(\dfrac{1}{2}x - 8\right)$ Multiply each side by 2.

$2y = x - 16$ Distributive Property

$-x + 2y = -16$ Subtract x from each side.

$x - 2y = 16$ Multiply each side by -1.

Check

Select the equation in standard form for the line that passes through (−9, 8) and (1, −12).

A. $x + 2y = -20$

B. $2x + y = -13$

C. $2x + y = 7$

D. $2x + y = -10$

E. $x + 2y = 20$

 Go Online An alternate method is available for this example.

 Think About It! In Step 2, why is it possible to write an equation in either slope-intercept form or point-slope form and still get the same equation in standard form?

Go Online You can complete an Extra Example online.

Learn Equations of Parallel and Perpendicular Lines

Nonvertical lines in the same plane that have the same slope are called **parallel lines**. Nonvertical lines in the same plane for which the product of the slopes is −1 are called **perpendicular lines**.

Key Concept • Slopes of Parallel and Perpendicular Lines

Parallel Lines	Perpendicular Lines
If two nonvertical lines are parallel, their slopes are the same. 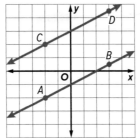 Since both lines have a slope of $\frac{1}{2}$, $\overleftrightarrow{AB} \parallel \overleftrightarrow{CD}$.	If two nonvertical lines are perpendicular, the product of their slopes is −1. 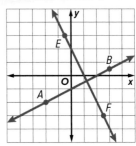 Since $\frac{1}{2}(-2) = -1$, $\overleftrightarrow{AB} \perp \overleftrightarrow{EF}$.

> **Think About It!**
> If the given line is vertical, what is the slope of any line parallel to the given line? perpendicular to the given line?

You can write an equation of a line parallel or perpendicular to a given line if you know a point on the line and an equation of the given line.

Key Concept • Writing Equations of Lines Parallel or Perpendicular to a Given Line

	Parallel Lines		Perpendicular Lines
Step 1	Identify the slope m of the given line.	Step 1	Identify the slope m of the given line. The slope of the line perpendicular to the original line is $-\frac{1}{m}$.
Step 2	Use the point-slope form with slope m and the coordinates of the given point.	Step 2	Use the point-slope form with slope $-\frac{1}{m}$ and the coordinates of the given point.
Step 3	Rewrite the equation in the needed form.	Step 3	Rewrite the equation in the needed form.

Example 7 Parallel Line Through a Given Point

Write an equation in slope-intercept form for the line that passes through (−4, 2) and is parallel to the graph of $y = 3x - 5$.

Step 1 Identify the slope of the given line.

The slope of the line with equation $y = 3x - 5$ is 3. The line parallel to that line has the same slope, 3.

> **Go Online** You may want to complete the Concept Check to check your understanding.

Go Online You can complete an Extra Example online.

Steps 2, 3 Write the equation of the parallel line.

Use the point-slope form to rewrite the equation in slope-intercept form.

$y - y_1 = m(x - x_1)$	Point-slope form
$y - 2 = 3[x - (-4)]$	$(x_1, y_1) = (-4, 2)$ and $m = 3$
$y - 2 = 3(x + 4)$	Simplify.
$y - 2 = 3x + 12$	Distributive Property
$y = 3x + 14$	Add 2 to each side.

> **Study Tip**
>
> **Checking Your Work**
> To check that your equation represents the correct line, graph both lines. Verify that the lines appear to be parallel and that your line passes through the given point.

Check

Write an equation for the line that passes through (8, 2) and is parallel to the graph of $y = \frac{3}{4}x + 2$.

Example 8 Perpendicular Line Through a Given Point

Write an equation in slope-intercept form for the line that passes through (1, −2) and is perpendicular to the graph of $3x + 2y = 12$.

Step 1 Identify the slope of the given line.

Write the equation in slope-intercept form.

$3x + 2y = 12$	Original equation
$3x - 3x + 2y = 12 - 3x$	Subtract $3x$ from each side.
$2y = -3x + 12$	Simplify.
$\frac{2y}{2} = \frac{-3x}{2} + \frac{12}{2}$	Divide each side by 2.
$y = -\frac{3}{2}x + 6$	Simplify.

The slope of the line with equation $3x + 2y = 12$ is $-\frac{3}{2}$. The slope of the line perpendicular to that line is the opposite reciprocal, $\frac{2}{3}$.

Steps 2, 3 Write the equation of the perpendicular line.

Use the point-slope form to rewrite the equation in slope-intercept form.

$y - y_1 = m(x - x_1)$	Point-slope form
$y - (-2) = \frac{2}{3}(x - 1)$	$(x_1, y_1) = (1, -2)$ and $m = \frac{2}{3}$
$y + 2 = \frac{2}{3}(x - 1)$	Simplify.
$y + 2 = \frac{2}{3}x - \frac{2}{3}$	Distributive Property
$y = \frac{2}{3}x - \frac{8}{3}$	Subtract 2 from each side.

Go Online You can complete an Extra Example online.

Check

Select the equation in slope-intercept form for the line that passes through (5, 0) and is perpendicular to the graph of $x - 6y = 1$.

A. $y = -6x + 30$

B. $6x + y = 30$

C. $y = -\frac{1}{6}x + 2$

D. $x - 6y = 5$

Example 9 Determine Line Relationships

Determine whether $\overleftrightarrow{AB}$ and $\overleftrightarrow{EF}$ are *parallel, perpendicular,* or *neither* for A(6, 8), B(2, 5), E(−6, −3), and F(0, 5).

Step 1 Find the slope of each line.

$$\text{slope of } \overleftrightarrow{AB} = \frac{8-5}{6-2} = \frac{3}{4} \qquad \text{slope of } \overleftrightarrow{EF} = \frac{-3-5}{-6-0} = \frac{-8}{-6} \text{ or } \frac{4}{3}$$

Step 2 Determine the relationship.

parallel   To determine whether the lines are parallel, compare their slopes. The two lines do not have the same slope, so they are not parallel.

perpendicular   To determine whether the lines are perpendicular, find the product of their slopes. $\frac{3}{4} \cdot \frac{4}{3} = 1$

Since the product of the slopes is not −1, $\overleftrightarrow{AB}$ and $\overleftrightarrow{EF}$ are not perpendicular.

Check

Determine whether $\overleftrightarrow{CD}$ and $\overleftrightarrow{KL}$ are *parallel, perpendicular,* or *neither* for C(4, 10), D(−1, 12), K(6, −5), and L(1, −3).

$\overleftrightarrow{CD}$ and $\overleftrightarrow{KL}$ are _____?_____.

Complete each sentence given $y = ax - 5$ and $y = bx + 3$.

When $a = 4$ and $b = 4$, the graphs are _____?_____.

When $a = -3$ and $b = 5$, the graphs are _____?_____.

When $a = -2$ and $b = \frac{1}{2}$, the graphs are _____?_____.

Go Online to practice what you've learned about writing linear equations in the Put It All Together over Lessons 5-1 and 5-2.

Go Online You can complete an Extra Example online.

Practice

Go Online You can complete your homework online.

Example 1

Write an equation in point-slope form for the line that passes through each point with the given slope. Then graph the equation.

1. $(-6, -3)$, $m = -1$
2. $(-7, 6)$, $m = 0$
3. $(-2, 11)$, $m = \frac{4}{3}$

Example 2

Write an equation in point-slope form for the line that passes through the given points.

4. $(-4, 6)$, $(-2, 22)$
5. $(1, -3)$, $(4, -15)$

6. $(4, -6)$, $(6, -4)$
7. $(3, 3)$, $(6, 7)$

Example 3

Write each equation in slope-intercept form.

8. $y - 1 = \frac{4}{5}(x + 5)$
9. $y + 5 = -6(x + 7)$

10. $y + 6 = -\frac{3}{4}(x + 8)$
11. $y + 2 = \frac{1}{6}(x - 4)$

Example 4

12. **NATURE** The frequency of a male cricket's chirp is related to the outdoor temperature. The relationship is expressed by the graph, where y is the temperature in degrees Fahrenheit and x is the number of chirps the cricket makes in 14 seconds. Write an equation for the line in point-slope form.

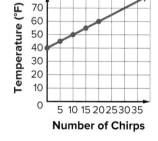
Number of Chirps

13. **CANOEING** Geoff paddles his canoe at an average speed of 3.5 miles per hour. After 5 hours of canoeing, Geoff has traveled 18 miles. Write an equation in point-slope form to find the total distance y Geoff travels after x hours.

14. **GEOMETRY** The perimeter of a square is four times the length of one side. If the side length of a square is 1 centimeter, then the perimeter of the square is 4 centimeters. Write an equation in point-slope form to find the perimeter y of a square with side length x.

Lesson 5-2 • Writing Equations in Standard and Point-Slope Forms   303

Example 5
Write each equation in standard form.

15. $y - 10 = 2(x - 8)$

16. $y + 7 = -\frac{3}{2}(x + 1)$

17. $2y + 3 = -\frac{1}{3}(x - 2)$

18. $4y - 5x = 3(4x - 2y + 1)$

19. $y = x + 1$

20. $y = \frac{1}{3}x - 10$

Example 6
Write an equation in standard form for the line that passes through the given points.

21. $(-2, -3), (4, -7)$

22. $(2, 7)$ and $(-5, 2)$

23. $(-4, 9), (2, -9)$

24. $(-1, 19)$ and $(3, 35)$

Examples 7 and 8
Write an equation in slope-intercept form for the line that passes through the given point and is parallel to the graph of the equation. Then write an equation for the line that passes through the given point and is perpendicular to the graph of the equation.

25. $(3, -2); y = x + 4$

26. $(4, -3); y = 3x - 5$

27. $(0, 2); y = -5x + 8$

28. $(-4, 2); y = -\frac{1}{2}x + 6$

29. $(-2, 3); y = -\frac{3}{4}x + 4$

30. $(9, 12); y = 13x - 4$

Example 9
Determine whether the graphs of each pair of equations are *parallel*, *perpendicular*, or *neither*.

31. $y = 4x + 3$
$4x + y = 3$

32. $y = -2x$
$2x + y = 3$

33. $3x + 5y = 10$
$5x - 3y = -6$

34. $-3x + 4y = 8$
$-4x + 3y = -6$

35. $2x + 5y = 15$
$3x + 5y = 15$

36. $2x + 7y = -35$
$4x + 14y = -42$

Mixed Exercises

37. Write an equation in standard form with an *x*-intercept of 4 and a *y*-intercept of 5.

Write each equation in slope-intercept and standard forms.

38. $y + 3 = -\frac{1}{3}(2x + 6)$

39. $y + 4 = 3(3x + 3)$

40. $y - 6 = -3(x + 2)$

41. $y - 9 = -6(x + 9)$

42. $y + 4 = \frac{2}{3}(x + 7)$

43. $y + 7 = \frac{9}{10}(x + 3)$

44. Consider the graphs of the following equations.

   $y = -2x$ $2y = x$ $4y = 2x + 4$

   a. Which equations are parallel? Explain your reasoning.

   b. Which equations are perpendicular? Explain your reasoning.

45. **INSPECTIONS** Mrs. Sanchez is inspecting a shed to determine if it is safe to use for storing football equipment. Mrs. Sanchez mapped the top view of the ceiling walls of the shed on a coordinate plane. If one of the walls lies from (−6, 11.5) to (2, 9.5) and the second wall lies from (−1, −2.5) to (2, 9.5), are the walls perpendicular? Explain your reasoning.

46. Nya mapped a quadrilateral on a coordinate plane. If she plots one segment from (−3, −3) to (3, 9) and another segment from (−5, 12) to (1, 0), are the segments parallel? Justify your reasoning.

47. **STRUCTURE** Immediately after take-off, a jet plane consistently climbs 20 feet for every 40 feet it moves horizontally. The graph shows the trajectory of the jet.

   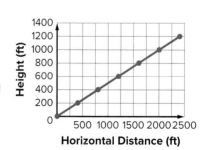

   a. Write an equation in point-slope form for the line representing the jet's trajectory.

   b. Write the equation from **part a** in slope-intercept form.

   c. Write the equation in standard form.

48. CONSTRUCT ARGUMENTS Consider three points, (3, 7), (−6, 1), and (9, p) on the same line. Find the value of p. Justify your argument.

49. WRITE What information is needed to write the equation of a line? Explain.

50. ANALYZE Levy claims that the line through (−6, −2) and (2, 10) is perpendicular to the graph of $3x - 2y = 10$. Do you agree? Justify your argument.

51. ANALYZE Jeremiah says the line through (7, −10) and (3, −2) is parallel to $2x - y = -5$. Do you agree? Justify your argument.

52. FIND THE ERROR Alonae says that the line through (1, −4) and (5, −6) is parallel to the line through (2, −7) and (5, −6). How can you tell she is mistaken without determining the slope? Explain your reasoning.

53. PERSEVERE Write an equation in point-slope form for the line that passes through the points (f, g) and (h, j).

54. WHICH ONE DOESN'T BELONG? Identify the equation that does not belong. Justify your conclusion.

| $y - 5 = 3(x - 1)$ | $y + 1 = 3(x + 1)$ | $y + 4 = 3(x + 1)$ | $y - 8 = 3(x - 2)$ |

55. CREATE Describe a real-life scenario that has a constant rate of change and a value of y for a particular value of x. Represent this situation using an equation in point-slope form and an equation in slope-intercept form.

Lesson 5-3
Scatter Plots and Lines of Fit

Learn Scatter Plots

Bivariate data consists of pairs of values. A **scatter plot** is a graph of bivariate data that consists of ordered pairs on a coordinate plane. Using a scatter plot can help you see the **trend,** or general pattern, in the data. Trends can represent linear or nonlinear associations in the data. Trends can be described as positive or negative correlations.

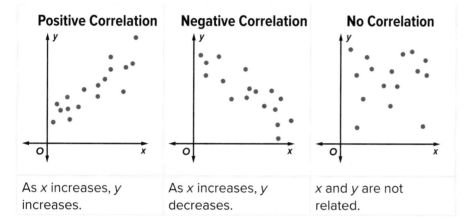

Positive Correlation	Negative Correlation	No Correlation
As *x* increases, *y* increases.	As *x* increases, *y* decreases.	*x* and *y* are not related.

Notice that in the graphs for positive and negative correlations, many of the points form **clusters** of points that slope upward or downward. Points outside of clusters are **outliers.**

Today's Goals
- Categorize the correlation of a set of data in a scatter plot.
- Make and evaluate predictions by fitting linear functions to sets of data.

Today's Vocabulary
bivariate data
scatter plot
trend
positive correlation
negative correlation
no correlation
line of fit
linear extrapolation
linear interpolation

Study Tip
Labeling Axes
Because scatter plots display bivariate data, it is critical to label axes with their corresponding units. Otherwise, the graph may not make sense.

🌐 Example 1 Evaluate Correlation

FOOTBALL The scatter plot displays the height and weight of New Orleans Saints football players. Determine whether the scatter plot shows a *positive, negative,* or *no* correlation. If the correlation is positive or negative, describe its meaning in the situation.

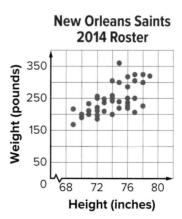

The scatter plot shows a positive correlation. As the height of the football player increases, weight usually increases.

🧠 Think About It!
What type of correlation would you expect between a player's jersey number and his birth month?

 Go Online You can complete an Extra Example online.

🧠 Think About It!

Determining Correlation Similar to slope, when data points are generally increasing from left to right, there is a positive correlation. Negative correlation occurs when the data points generally decrease from left to right. If you are unable to tell if the data are increasing or decreasing, there is probably no correlation.

🧠 Think About It!

How can you ensure that your data predictions that are outside the range of data are as accurate as possible?

Check

TELEPHONES The scatter plot displays the number of landline telephones in the United States, in 100 millions, since 2000.

Determine whether the scatter plot shows a *positive, negative,* or *no correlation*. Describe the correlation's meaning in the situation.

The scatter plot shows ___?___ correlation.

As time increases, the number of landlines generally ___?___.

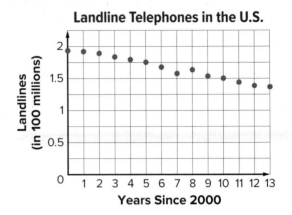

Landline Telephones in the U.S.

Explore Make Predictions by Using a Scatter Plot

🌐 **Online Activity** Use a real-world situation to complete the Explore.

> ❓ **INQUIRY** How can you use a scatter plot to estimate unknown data?

Learn Lines of Fit

A **line of fit** is used to describe the trend of the data in a scatter plot.

Key Concept • Using a Linear Function to Model Data
Step 1 Make a scatter plot. Plot each point of the data and determine whether any relationship exists in the data.
Step 2 Draw a line. Draw a line that closely follows the trend in the data.
Step 3 Write an equation. Use two points on the line of fit to find the slope of the line and create an equation for the line using the slope and a point on the line.
Step 4 Make predictions. Use the equation of the line of fit to make predictions about unknown data.

Linear extrapolation is the use of a linear equation to predict values that are outside of the range of data. **Linear interpolation** is the use of a linear equation to predict values that are inside of the data range.

🌐 **Go Online** You can complete an Extra Example online.

308 Module 5 • Creating Linear Equations

Example 2 Write an Equation for a Line of Fit

BOATS The table shows the average cost of a jet boat in the years after 2000. Write an equation to represent the data. Then, use the equation to predict the cost of a jet boat in 2005 and 2025.

Years Since 2000	Cost ($)
0	17,663
1	19,144
2	21,176
3	20,584
4	23,280
6	24,443
7	27,784

Years Since 2000	Cost ($)
8	28,088
9	29,774
10	32,752
11	34,082
12	35,589
13	37,618

Watch Out!
Variations Equations for scatter plots generally do not have an exact correct solution. Equations will vary depending on how the line of fit was drawn and which points were selected when writing the equation. So, your solutions may not be exactly the same as another student's solutions or the sample answers given.

Step 1 Make a scatter plot.

The independent variable is the number of years since 2000 and the dependent variable is the cost of the jet boats. As the years increase, the cost of the jet boats also increases. This scatter plot shows positive correlation.

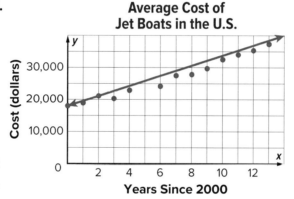

Think About It!
What do the slope and y-intercept mean in the context of this example?

Step 2 Draw a line of fit.

A line is drawn that follows the trend of the data points and passes close to most of the points.

Step 3 Write an equation.

The line of fit passes close to the data points (5, 25,108) and (10, 32,752).

Find the slope.

$m = \dfrac{y_2 - y_1}{x_2 - x_1}$ Slope Formula

$= \dfrac{32{,}752 - 25{,}108}{10 - 5}$ $(x_1, y_1) = (5, 25{,}108)$ and $(x_2, y_2) = (10, 32{,}752)$

$= \dfrac{7644}{5}$ or 1528.8   Simplify.

Use $m = 1528.8$ and a point to write an equation.

$y - y_1 = m(x - x_1)$ Point-Slope Form

$y - 25{,}108 = 1528.8(x - 5)$ $(x_1, y_1) = (5, 25{,}108)$

$y - 25{,}108 = 1528.8x - 7644$ Distribute.

$y = 1528.8x + 17{,}464$ Simplify.

(continued on the next page)

Go Online You can complete an Extra Example online.

Think About It!

What assumptions are made when using a line of fit to make predictions about the cost of a jet boat in a given year?

Step 4 Predict the cost in 2005.

Use evaluation to predict the cost of a jet boat in 2005.

Since the independent variable represents the number of years after 2000, the value of x is $2005 - 2000$ or 5.

$y = 1528.8x + 17{,}464$ Equation of the line of fit
$y = 1528.8(5) + 17{,}464$ $x = 5$
$y = 7644 + 17{,}464$ or $25{,}108$ Simplify.

We can predict that the cost of a jet boat in 2005 was about $25,108.

Step 5 Predict the cost in 2025.

Extrapolate the data to determine the cost of a jet boat in 2025.

$y = 1528.8x + 17{,}464$ Equation of the line of fit
$y = 1528.8(25) + 17{,}464$ $x = 25$
$y = 38{,}220 + 17{,}464$ or $55{,}684$ Simplify.

We can predict that the cost of a jet boat in 2025 will be about $55,684.

Check

ANIMALS The data show the amount of milk that a baby goat needs by its weight.

Weight (pounds)	5	7	10	15	20	25	30	40	50
Milk (ounces)	12	16	20	28	32	40	48	64	80

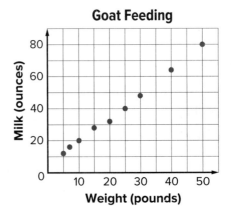

Goat Feeding

Part A Use the data points (10, 20) and (50, 80), which are contained in the line of fit, to write an equation of the line in slope-intercept form.

Part B Use the equation from Part A to predict the amount of milk needed for a 17-pound goat and a 55-pound goat.

17-pound goat: _?_ ounces

55-pound goat: _?_ ounces

Go Online You can complete an Extra Example online.

Practice

Go Online You can complete your homework online.

Example 1

Determine whether each scatter plot shows a *positive, negative,* or *no* correlation. If the correlation is positive or negative, describe its meaning in the situation.

1.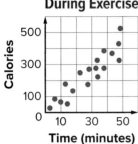
   Calories Burned During Exercise

2.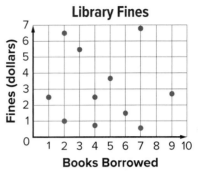
   Library Fines

3.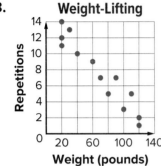
   Weight-Lifting

4.
   Car Dealership Revenue

Example 2

5. **MUSIC** The scatter plot shows the number of CDs in millions that were sold from 2011 to 2016.

   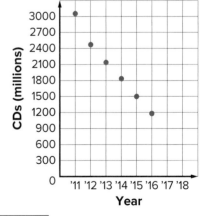

   a. Use the points (2, 2485.6) and (6, 1172.5) to write an equation of the line of fit in slope-intercept form. Let x be the years since 2010.

   b. If the trend continued, about how many CDs were sold in 2019?

6. **HOUSING** The data show the median price of an existing home from 2010 to 2015.

Year	2010	2011	2012	2013	2014	2015
Price	222,900	226,900	238,400	258,400	275,200	296,500

   a. Use the points (1, 226.9) and (4, 275.2) to write an equation for the line of fit in slope-intercept form where x is the number of years since 2010 and y is the median price in thousands of dollars.

   b. If the trend continues, what will be the approximate median price of an existing home in 2025?

Lesson 5-3 • Scatter Plots and Lines of Fit **311**

Mixed Exercises

7. **FAMILY** The table shows the predicted annual cost for a middle-income family to raise a child from birth until adulthood.

Cost of Raising a Child Born in 2013					
Child's Age	2	5	8	11	14
Annual Cost ($)	12,940	12,970	12,800	13,600	14,420

   a. Make a scatter plot and describe what relationship exists within the data.

   b. Use the points (8, 12,800) and (14, 14,420) to write the equation of the line of fit in slope-intercept form.

   c. If the trend continues, what will be the approximate annual cost of raising a child born in 2013 at age 17?

Determine whether each scatter plot shows a *positive*, *negative*, or *no* correlation. If the correlation is positive or negative, describe its meaning in the situation.

8.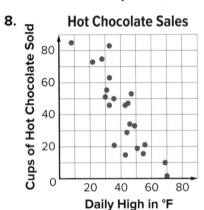

9.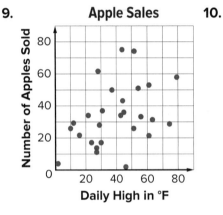

10.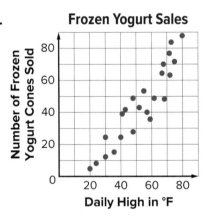

11. **BASEBALL** The table shows the average length in minutes of professional baseball games in selected years.

Average Length of Major League Baseball Games							
Year	2005	2007	2009	2011	2013	2015	2017
Time (min)	169	175	175	176	184	180	189

   a. Make a scatter plot and draw a line of fit.

   b. Write the equation of the line of fit in slope-intercept form where *x* is the number of years since 2005. Explain your process.

   c. If the trend continues, what will be the approximate length of a major league baseball game in 2021?

   d. How reliable is the predicted length of a major league baseball game in 2021? Justify your argument.

12. **INCOME** The table shows the average median income for selected ages.

Age (years)	26	27	28	29	30
Median Income ($1000)	16.8	19.1	23.3	25.8	33.9

   a. Make a scatter plot relating age to median income. Then draw a line of fit for the scatter plot.

   b. Determine whether the graph shows a *positive, negative,* or *no correlation*. If the correlation is positive or negative, describe its meaning in the situation.

   c. Use the table to write an equation of the line of fit.

   d. Use the line of fit to predict the median income for 32-year-olds.

13. **FOOTBALL** The scatter plot shows the average price of a National Football League ticket from 2007 to 2016.

   a. Determine what relationship, if any, exists in the data. Explain.

   b. Use the points (2007, 67.11) and (2016, 92.98) to write the slope-intercept form of an equation for the line of fit shown, where x is the number of years since 2006. Round to the nearest hundredth.

   c. Predict the price of a ticket in 2030.

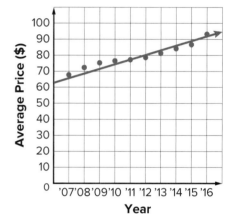

14. **STRUCTURE** Refer to the scatter plot at the right.

   a. Describe the trend in the data shown in the scatter plot and the relationship between x and y.

   b. Describe a real-life situation that could be modeled by the given scatter plot. Explain your reasoning.

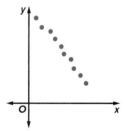

15. **USE A MODEL** The table gives the life expectancy of a child born in the United States in a given year.

   a. Make a scatter plot of the data. Then draw a line of fit.

   b. Use the data to predict the life expectancy of a baby born in 2023. Round to the nearest tenth. Does your answer follow the trend of the data? Explain.

   c. Explain any assumptions you made when using the line of fit to extrapolate and find the life expectancy of a baby born in 2023.

Years of Life Expected at Birth	
Year of Birth	Life Expectancy (years)
1930	59.7
1940	62.9
1950	68.2
1960	69.7
1970	70.8
1980	73.7
1990	75.4
2000	77.0
2010	78.7

16. **USE TOOLS** Several groups volunteered to clean up litter along a mile of the highway near their town. The table shows how many people were in each group and how long it took each group to finish the job.

Workers	9	16	18	8	15	11	9	17	9	15	11	12
Minutes	80	40	35	90	60	60	70	30	70	50	80	70

  a. Graph the data on a scatter plot.

  b. Draw a line of fit to show the trend of the data.

  c. Choose two points on the line of fit. Then find the equation of the line in slope-intercept form.

  d. Another group wants to get done in 45 minutes. About how many workers should they have? Explain your reasoning.

  e. Find the *y*-intercept of the line of fit. Does the *y*-intercept make sense in the context of the situation? Justify your argument.

Higher-Order Thinking Skills

17. **CREATE** Describe a real-life situation that can be modeled using a scatter plot. Describe whether there is *positive, negative,* or *no* correlation.

18. **WHICH ONE DOESN'T BELONG?** Analyze the following situations and determine which one does not belong. Justify your conclusion.

hours worked and amount of money earned	height of an athlete and favorite color
seedlings that grow an average of 2 centimeters each week	number of photos stored on a camera and capacity of camera

19. **ANALYZE** Determine which line of fit shown is a better fit for the data in the scatter plot. Justify your argument.

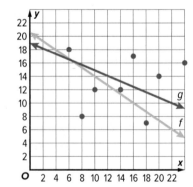

20. **WRITE** Does an accurate line of fit always predict what will happen in the future? Explain your reasoning.

21. **CREATE** Make a scatter plot that shows the height of a person and age. Explain how you could use the scatter plot to predict the age of a person given his or her height. How can the information from a scatter plot be used to identify trends and make decisions?

Lesson 5-4

Correlation and Causation

Explore Collecting Data to Determine Correlation and Causation

Today's Goal
- Determine correlation or causation.

Today's Vocabulary
causation

▶ **Online Activity** Use a real-wolrd situation to complete an Explore.

> **INQUIRY** What is the difference between correlation and causation?

Learn Correlation and Causation

Causation occurs when a change in one variable produces a change in another variable. It is the relationship between cause and effect. Correlation, however, can be observed between many variables.

Think About It!
Why does correlation not prove causation?

Key Concept • Correlation and Causation	
Step 1	Graph ordered pairs to create a scatter plot.
Step 2	Determine whether the scatter plot shows a positive or negative correlation.
Step 3	Determine whether the two sets of data are related. Does one variable *cause* the other? Could other factors be influencing the data results?
Step 4	Decide if the data illustrate correlation or causation.

🌐 Example 1 Correlation and Causation by Graphing

ANALYSIS The data show the per capita consumption of mozzarella cheese and the number of civil engineering doctoral degrees awarded in the United States. Determine whether the data plotted on the graph illustrate a *correlation* or *causation*.

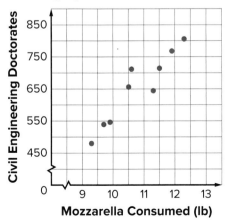

(continued on the next page)

▶ **Go Online** You can complete an Extra Example online.

Lesson 5-4 • Correlation and Causation **315**

> **Talk About It!**
> Describe an experiment that could be conducted to show causation between the number of civil engineers who were awarded a doctoral degree and another factor.

Step 1 Determine the correlation.

As the amount of mozzarella consumed increases, the number of civil engineering doctorates also increases. The scatter plot shows a positive correlation.

Step 2 Determine causation.

Consumption of mozzarella does not cause anyone to obtain a doctoral degree in civil engineering. These two sets of data are not related. Many factors may affect the increase in these two areas. As the demand for more roadways, airports, and water and sewage treatment plants grows, the demand for more civil engineers also increases. An increase in the per capita consumption of mozzarella may be related to increased pizza sales or dairy production. Both variables are affected by a general increase in population.

Step 3 Determine whether the data illustrate a *correlation* or *causation*.

The data exhibit a correlation, but there is no causation.

Check

ANALYSIS Determine whether the data illustrate a *correlation* or *causation*.

Month	March	April	May	June	July	August
Sunscreen Sold	14	37	84	117	135	98
Sunglasses Sold	6	11	28	36	40	39

The data show a ___?___ correlation. As the number of bottles of sunscreen sold increases, the number of sunglasses sold ___?___. These data illustrate a ___?___.

Example 2 Correlation and Causation by Situation

Determine whether the situation illustrates a *correlation* or *causation*. Explain your reasoning, including other factors that might be involved.

A university experiment showed a negative correlation between the average weekly time spent exercising and the probability of developing heart disease.

This situation models causation. Exercise and heart disease are related, and lack of exercise could be a cause of heart disease. Other factors that might have led to heart disease are inherited traits, smoking, or a poor diet.

 Go Online You can complete an Extra Example online.

Practice

Go Online You can complete your homework online.

Example 1

1. **FROZEN DESSERTS** The table shows the number of pounds of frozen yogurt and the number of pounds of sherbet consumed per capita in the United States from 2009 to 2016.

Year	2009	2010	2011	2012	2013	2014	2015	2016
Pounds of Frozen Yogurt	0.9	1	1.2	1.1	1.4	1.3	1.4	1.2
Pounds of Sherbet	1	1	0.9	0.8	0.9	0.9	0.8	0.8

   a. Graph the ordered pairs (pounds of frozen yogurt, pounds of sherbet) to create a scatter plot.

   b. Does the scatter plot show a *positive, negative,* or *no* correlation? Explain.

   c. Determine whether the data illustrate a *correlation* or *causation*. What other factors may influence the data?

2. **LEISURE ACTIVITIES** The table shows the average number of minutes a person reads per weekday and the average number of minutes a person watches television per weekday.

Age	15	25	35	45	55	65
Minutes Reading	7	9	12	17	30	50
Minutes Watching Television	117	115	113	127	155	236

   a. Graph the ordered pairs as a scatter plot (minutes reading, minutes watching television).

   b. Does the scatter plot show a *positive, negative,* or *no* correlation? Explain.

   c. Determine whether the data illustrate a *correlation* or *causation*. What other factors may influence the data?

Example 2

Determine whether each situation illustrates a *correlation* or *causation*. Explain your reasoning.

3. A class experiment shows a negative correlation between the width of a person's palm and the amount of time they spend watching television each day.

4. The larger a person's shoe size, the higher a person's reading level.

5. At a grocery store, there is a negative correlation between the price of cereal and number of boxes of cereal sold.

6. Hae notices that the lower the daily temperature is, the less time she spends outside.

Lesson 5-4 • Correlation and Causation

Mixed Exercises

7. **GARDENING** Jalen weighs each type of fruit his garden produces each week.

Week	1	2	3	4	5
Strawberries (lb)	6.5	8	12	13.5	20
Blueberries (lb)	6	5	4.5	3	2.5

   a. Graph the ordered pairs (pounds of strawberries, pounds of blueberries) to create a scatter plot.

   b. Does the scatter plot show a *positive, negative,* or *no* correlation? Explain.

   c. Determine whether the data illustrate a *correlation* or *causation*. What other factors may influence the data?

8. **SHOES** The table shows the number of pairs of sandals and snow boots sold at a certain store during various months of the year.

Month	January	April	July	December
Sandals	12	153	215	27
Snow Boots	268	34	6	272

   a. Graph the ordered pairs (sandals sold, snow boots sold) to create a scatter plot.

   b. Does the scatter plot show a *positive, negative,* or *no* correlation? Explain.

   c. Determine whether the data illustrate a *correlation* or *causation*. What other factors may influence the data?

Determine whether each situation illustrates a *correlation* or *causation*. If there is a correlation, describe the trend. Explain your reasoning.

9. **PIZZA** The more pizzas a restaurant sells, the more cheese it uses.

10. **BOOKS** Sam notices that as the number of words in a book increases, the number of pages in the book increases.

11. **CONSTRUCT ARGUMENTS** What is meant by this statement: *Correlation does not imply causation*? Justify your argument.

A study compared the average monthly amount spent on swimsuits with the average monthly amount spent on air conditioning for several months in Sunnyside. The data is shown in the scatter plot. Use this information for Exercises 12 and 13.

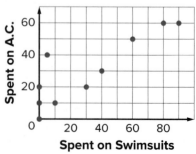

12. **ANALYZE** Explain what the scatter plot shows and describe any correlation.

13. **WRITE** Explain whether this statement is accurate: "There is a strong positive correlation between spending money on swimsuits and spending money on air conditioning. Therefore, to cut down the amount of electricity used in Sunnyside, people should buy fewer swimsuits."

Lesson 5-5

Linear Regression

Learn Linear Regression and Best-Fit Lines

A calculator can find the line that most closely approximates data in a scatter plot, called the **best-fit line**. **Linear regression** is one algorithm used to find a precise line of fit for a set of data.

Calculators may also compute a number r called the **correlation coefficient**. This measure shows how well data are modeled by a linear equation. It will tell you if a correlation is positive or negative and how closely the equation is modeling the data. The closer the correlation coefficient is to 1 or −1, the more closely the equation models the data.

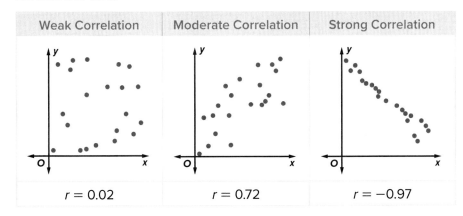

Today's Goals
- Write equations of best-fit lines using linear regressions.
- Determine how well functions fit sets of data.

Today's Vocabulary
best-fit line
linear regression
correlation coefficient
residual

Think About It!
Write the following correlation coefficients in order from weakest to strongest.

0.85   0.3   1   −0.78

0.54   −0.06   −0.9

Example 1 Find a Best-Fit Line

BASEBALL The table shows Jackie Robinson's total hits during each season of his major league career. Use a graphing calculator to write an equation for the best-fit line for the data. Then find and interpret the correlation coefficient.

Year	1947	1948	1949	1950	1951	1952	1953	1954	1955	1956
Total Hits	175	170	203	140	185	157	159	120	81	98

Step 1 Enter the data.

Before you begin, make sure that your Diagnostic setting is on. You can find this under the **CATALOG** menu. Press **D** and then scroll down and click **DiagnosticOn**. Then press enter.

Study Tip
Correlation Coefficient
The table shows a rule of thumb for determining how well the equation models the data based on the correlation coefficient.

Correlation Coefficient	Strength of Correlation
$|r| \geq 0.8$	Strong
$0.5 \leq |r| < 0.8$	Moderate
$|r| < 0.5$	Weak

Go Online to see how to use a graphing calculator with this example.

(continued on the next page)

Math History Minute
One of the areas of interest of British statistician **Florence Nightingale David (1909–1993)**, who was named after family friend Florence Nightingale, was the distribution of correlation coefficients. In 1938, she released a book entitled *Tables of the Correlation Coefficient*, for which all of the calculations were done on a hand-cranked mechanical calculator.

Use a Source
Choose another baseball player and research that player's total number of hits by season. Use a graphing calculator to write an equation for the best-fit line, and decide whether the equation models the data well.

Enter the data by pressing **stat** and selecting the **Edit** option. Let the year 1947 be represented by year 0. Enter the years since 1947 into List 1 (**L1**). These will represent the *x*-values. Enter the total hits into List 2 (**L2**). These will represent the *y*-values.

Step 2 Perform the regression.

Perform the regression by pressing **stat** and selecting the **CALC** option. Scroll down to **LinReg (ax+b)** and press **enter**. Make sure **L1** is the **Xlist** and **L2** is the **Ylist**. Then select **Calculate**.

Step 3 Interpret the results.

Write the equation of the regression line by rounding the *a* and *b* values on the screen. The form that we chose for the regression was $ax + b$, so the equation is $y = -10.32x + 195.22$. The correlation coefficient is about -0.8022, which means that the equation models the data well. Its negative value means that as the years since 1947 increase, the total number of Jackie Robinson's hits decreases.

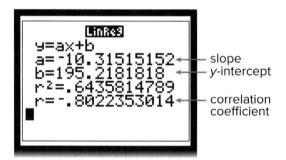

Check

TEMPERATURE The table shows the average annual temperature for the top 10 most populous states in 2014.

Rank	1	2	3	4	5	6	7	8	9	10
Temperature (°F)	59.4	64.8	70.7	45.4	51.8	48.8	50.7	63.5	59	44.4

Part A Use a graphing calculator to write an equation for the best-fit line for the data. Round to the nearest hundredth.

Part B Find the correlation coefficient *r*. Round to the nearest hundredth.

Part C Based on your answer to part **b**, does the equation model the data well? Yes or No?

 Go Online You can complete an Extra Example online.

Best-fit lines can be used to estimate values that are not in the data. Recall that when we estimate values that are between known values, this is called linear interpolation. When we estimate a number outside the range of data, it is called linear extrapolation.

Example 2 Use a Best-Fit Line

SHOPPING The table shows U.S. desktop online sales on Cyber Monday since 2009. Estimate the Cyber Monday sales in 2025.

Year	2009	2010	2011	2012	2013	2014	2015	2016
Sales (millions of dollars)	887	1028	1251	1465	1735	2038	2280	2671

Step 1 Graph the data.

Enter the data from the table into the lists. Let 2009 be represented by 0. Then the years since 2009 are the x-values. Let the sales be the y-values. Graph the scatter plot. Turn on **Plot1** under the **STAT PLOT** menu and choose [scatter icon]. Use **L1** for the **Xlist** and **L2** for the **Ylist**.

[−0.7, 7.7] scl: 1 by [583.72, 2974.28] scl: 1

Change the viewing window so that all data are visible by pressing [zoom] and then selecting **ZoomStat**.

Step 2 Perform the regression.

Perform the regression using the data in the lists. The equation is about $y = 254.51x + 778.58$. The correlation coefficient is 0.9935, which means that the equation models the data well.

[−0.7, 7.7] scl: 1 by [583.72, 2974.28] scl: 1

Step 3 Graph the best-fit line.

Graph the best-fit line. Press [y =] [vars] and choose **Statistics**.

From the **EQ** menu, choose **RegEQ**. Press [graph].

Step 4 Extrapolate.

Use the graph to predict the 2025 Cyber Monday sales. Change the viewing window to include the x-value to be evaluated, 16. Also increase **Ymax** to accommodate the increasing y-values. Press [2nd] **CALC** [enter] 16 [enter] to find that when $x = 16$, $y \approx 4851$.

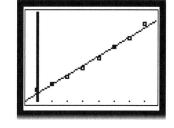

[−0.7, 17] scl: 1 by [583.72, 5500] scl: 1

We can estimate that in 2025, Cyber Monday sales will be about $4,851,000,000.

Go Online You can complete an Extra Example online.

Think About It!

Why is it helpful to define x as *years since 2009* instead of *years*?

Study Tip

Assumptions Using a best-fit line to make predictions requires you to assume that the trend continues at a constant rate and that more people choose to shop on Cyber Monday each year.

Check

SOCIAL MEDIA The table shows the number of daily users on a social media site in various years.

Year	2011	2012	2013	2014	2015
Daily Users (millions)	372	526	665	802	936

Use linear regression to estimate the number of daily users in millions on the site in 2030.

A. 3187.4 users

B. 4591.4 users

C. 285,391.4 users

D. 3047 users

> **Talk About It!**
> Why would a residual plot where the residuals are almost on the line $y = 0$ indicate a very good fit? Explain your reasoning.

Learn Residuals

When finding a best-fit line, not all data will lie on the line. The difference between an observed y-value and its predicted y-value on a regression line is called a **residual**. When residuals are plotted on a scatter plot, they can help assess how well the best-fit line describes the data. If there is no pattern in the residual plot, then the best-fit line is a good fit.

Example 3 Graph and Analyze a Residual Plot

THANKSGIVING The table shows the average price of a 10-person Thanksgiving dinner from 2004 to 2014. Determine whether the best-fit line models the data well by graphing a residual plot.

Year	2004	2005	2006	2007	2008	2009	2010	2011	2012	2013	2014
Price ($)	35.68	36.78	38.10	42.26	44.61	42.91	43.47	49.20	49.48	49.04	49.41

Step 1 Find the best-fit line.

Enter the data from the table into the lists. Let 2004 be represented by 0. Then the years since 2004 are the x-values. Let the prices be the y-values. Perform the linear regression using the data in the lists.

> **Think About It!**
> Use a calculator to find the correlation coefficient of the best-fit line. Does the correlation coefficient also suggest a good fit? Justify your argument.

Step 2 Graph the residual plot.

Turn on **PLOT2** under the **STAT PLOT** menu and choose [icon]. Use **L1** for the **Xlist** and **RESID** for the **Ylist**. You can obtain **RESID** by pressing [2nd] [LIST] and selecting **RESID** from the list of names. Graph the scatter plot of the residuals by pressing [zoom] and choosing **ZoomStat**. The residuals appear to be randomly scattered and centered about the line $y = 0$. Thus, the best-fit line seems to model the data well.

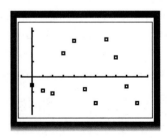

[−1, 1] scl: 1 by [−2.52, 3.21] scl: 1

> **Go Online** to see how to use a graphing calculator with this example.

Go Online You can complete an Extra Example online.

Practice

Go Online You can complete your homework online.

Example 1

1. **SOCCER** The table shows the number of goals a soccer team scored each season since 2010. Let x be the number of years since 2010.

Year	2010	2011	2012	2013	2014	2015
Goals Scored	48	52	50	46	48	42

   a. Write the equation for the best-fit line for the data.

   b. Find and interpret the correlation coefficient.

2. **REVENUE** The table shows the estimated revenue earned for ringtone and ringback purchases, in millions of dollars, each year since 2010.

Year	2010	2011	2012	2013	2014	2015	2016
Revenue ($)	448	276.2	166.9	97.9	66.3	54.5	40.1

   a. Write the equation for the best-fit line for the data.

   b. Find and interpret the correlation coefficient.

3. **SALES** The table shows the sales of a health and beauty supply company, in millions of dollars, for several years. Let x be the number of years since 2010.

Year	2011	2012	2013	2014	2015
Sales	12.2	19.1	29.4	37.3	45.7

   a. Write the equation for the best-fit line for the data.

   b. Find and interpret the correlation coefficient.

Example 2

4. **PURCHASING** A supermarket chain closely monitors how many bottles of sunscreen it sells each year so that it can reasonably predict how many bottles to stock in the following year. Let x be the number of years since 2010.

Year	2013	2014	2015	2016	2017
Bottles of Sunscreen	60,200	65,000	66,300	65,200	70,600

   a. Find the equation for the best-fit line for the data.

   b. How many bottles of sunscreen should the supermarket expect to sell in 2025?

Lesson 5-5 • Linear Regression

5. **GOLD** Ounces of gold are traded by large investment banks in commodity exchanges much the same way that shares of stock are traded. The table below shows the cost of a single ounce of gold on the last day of trading in given years. Let x be the number of years since 2000.

Year	2002	2004	2006	2008	2010	2012	2014
Price	$342.75	$435.60	$635.70	$869.75	$1420.25	$1664.00	$1199.25

a. Find the equation for the best-fit line for the data.

b. According to the equation, what would be the price of an ounce of gold on the last day of trading in 2030?

6. **GOLF SCORES** Emmanuel is practicing golf as part of his school's golf team. Each week he plays a full round of golf and records his total score. His scores for the first five weeks are shown.

Week	1	2	3	4	5
Golf Score	112	107	108	104	98

a. Find the equation for the best-fit line for the data.

b. What score can Emmanuel expect to get after 10 weeks?

Example 3

7. **MODELING** For a science project, Noah measured the effect of light on plant growth. At the end of 3 weeks, he recorded the height of each plant and how many hours of light it received each day.

Hours of Sunlight Per Day (x)	0	3	6	10	4	9	7	8	12	11	5
Height in Inches (y)	1	3	4	8	4	6	7	8	9	6	5

a. Find the equation for the best-fit line for the data.

b. Graph and analyze the residual plot.

8. **STRUCTURE** For his project, Darius measured the effect of fertilizer on plant growth. At the end of 3 weeks, he recorded the height of each plant and how many drops of fertilizer it received each day.

Drops of Fertilizer (x)	5	15	20	25	18	22	21	30	10	13	16
Height in Inches (y)	5	8	9	0	8	0	9	0	7	6	9

a. Find the equation of the best-fit line for the data.

b. Graph and analyze the residual plot.

Mixed Exercises

9. **PHYSICAL FITNESS** The table shows the percentage of students in public school who have met all six of California's physical fitness standards each year since the 2011–2012 school year.

Year	2011–2012	2012–2013	2013–2014	2014–2015
Percentage	20.5%	22.1%	22.6%	21.2%

a. Write the equation for the best-fit line for the data.

b. Find and interpret the correlation coefficient.

c. What constraints are there in the situation? Explain.

10. **FARMING** Some crops, such as barley, are very sensitive to the acidity of the soil. Barley grows best in soil with a pH range of 6 to 7.5. To determine the ideal level of acidity, a farmer measures how many bushels of barley he harvests in different fields with varying acidity levels.

Soil Acidity (pH)	5.7	6.2	6.6	6.8	7.1
Bushels Harvested	3	20	48	61	73

a. Find the equation for the best-fit line for the data and the correlation coefficient.

b. Use the equation of the best-fit line to estimate how many bushels the farmer would harvest if the soil had a pH of 10.

c. Could the equation of the best-fit line be used to extrapolate the data for extremely high levels of soil acidity? Explain.

11. **FOOTBALL** A college running back ran for 1732 total yards in the regular season. The table shows his cumulative total number of yards gained after select games.

Game Number	1	3	6	9	12
Cumulative Yards	184	431	818	1257	1732

a. Find the equation for the best-fit line for the total yards y gained after x games.

b. Find and interpret the correlation coefficient.

c. Use the trend of the data and the table to estimate when the running back will have run for 950 yards. Explain your reasoning.

d. During which game would you expect the running back to reach a total of 1000 yards?

12. **REGULARITY** Consider the linear regression equation that models a set of data very well to be $y = 1.43x - 4.2$. Would there be any restrictions on what the correlation coefficient value could be? Justify your reasoning.

13. **STRUCTURE** The table shows the number of student athletes participating in college athletics since the 2010-2011 school year.

Year	2010-2011	2011-2012	2012-2013	2013-2014	2014-2015
Student Athletes	444,077	453,347	463,202	472,625	482,533

    a. Find the equation for the best-fit line for the data.

    b. Find and interpret the correlation coefficient.

    c. Graph and analyze the residual plot. Does this support your conclusion from **part b**?

    d. Predict the number of college athletes in 2035.

Higher-Order Thinking Skills

14. **WRITE** How are lines of fit and linear regression similar? different?

15. **CREATE** For a class project, the scores that 10 randomly selected students earned on the first 8 tests of the school year are given. Explain how to find a line of best fit. Could it be used to predict the scores of the other students? Explain your reasoning.

16. **ANALYZE** Determine whether the following statement is *sometimes, always,* or *never* true: If the correlation coefficient in a given situation is 0.946, the change in the independent variable **causes** change in the dependent variable. Justify your argument.

17. **PERSEVERE** The table shows the number of participants in high school athletics.

Years Since 1980	0	10	20	25	30
Number of Athletes	5,356,913	5,298,671	6,705,223	7,159,904	7,667,955

    a. Find an equation for the regression line.

    b. According to the equation, how many participated in 2008?

326 Module 5 • Creating Linear Equations

Lesson 5-6

Inverses of Linear Functions

Learn Inverses of Relations

Two relations are **inverse relations** if and only if one relation contains points of the form (a, b) when the other relation contains points of the form (b, a). So, the x-coordinates are exchanged with the y-coordinates for each ordered pair in the relation.

Key Concept • Inverse Relations

Words	If one relation contains the element (a, b), then the inverse relation will contain the element (b, a).
Symbols	(a, b) → (b, a)
Example	A and B are inverse relations. 　　A　　　　B (−8, 12) → (12, −8) (−2, −5) → (−5, −2) 　(0, 4)　 → (4, 0) 　(7, 16)　→ (16, 7)
Graph	The graph of an inverse is the graph of the original relation reflected over the line y = x. For every point (a, b) on the graph of the original relation, the graph of the inverse will include (b, a).

Today's Goals
- Construct the inverses of relations.
- Find inverses of linear functions.

Today's Vocabulary
inverse relations

inverse functions

Example 1 Inverse Relations

Determine the inverse of {(−8, 3), (0, 14), (11, 52), (12, −6)}.

Write the coordinates in the ordered pairs to complete the inverse relation.

(−8, 3) → (3, −8)　　　　(11, 52) → (52, 11)

(0, 14) → (14, 0)　　　　(12, −6) → (−6, 12)

The inverse relation is {(3, −8), (14, 0), (52, 11), (−6, 12)}.

Think About It!
Describe the relationship between the domains and ranges of inverse relations.

Check

Determine the inverse of the relation.

{(−1.4, 5), (1.3, 6.5), (3, −8), (3.05, 9)}

 Go Online You can complete an Extra Example online.

Example 2 Find Inverse Relations from a Table

Find the inverse of the relation shown in the table.

x	−11	0.3	−3	3.5
y	9	−2	−8	2

Write the coordinates in the ordered pairs to complete the inverse relation.

(−11, 9) → (9, −11)   (−3, −8) → (−8, −3)

(0.3, −2) → (−2, 0.3)   (3.5, 2) → (2, 3.5)

The inverse relation is {(9, −11), (−2, 0.3), (−8, −3), (2, 3.5)}.

Example 3 Graph Inverse Relations

Graph the inverse of the relation.

The graph of the relation passes through the points (−1, −5), (0, −2), (1, 1), and (2, 4).

Exchange the x-coordinates and y-coordinates to find points of the inverse relation.

(−1, −5) → (−5, −1)   (1, 1) → (1, 1)

(0, −2) → (−2, 0)   (2, 4) → (4, 2)

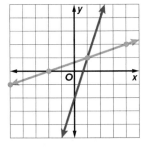

Plot the points of the inverse relation and draw a line passing through them.

> **Study Tip**
>
> **Plotting Points** While it only takes two points to graph a line, use several points when graphing inverse relations to create more accurate graphs.

> **Think About It!**
> Describe the graph of the inverse of the horizontal line y = 3.

Explore Comparing a Function and Its Inverse

Online Activity Use graphing technology to complete the Explore.

> **INQUIRY** How can you graph the inverse of a function?

Learn Inverses of Linear Functions

A linear relation that is described by a function may have an **inverse function** that can generate ordered pairs of the inverse relation. The inverse of the linear function f(x) is written as $f^{-1}(x)$ and is read *f inverse of x* or *the inverse of f of x*.

Go Online You can complete an Extra Example online.

328 Module 5 · Creating Linear Equations

Key Concept • Finding Inverse Functions

To find the inverse function $f^{-1}(x)$ of the linear function $f(x)$, complete the following steps.

Step 1	Replace $f(x)$ with y in the equation for $f(x)$.
Step 2	Interchange y and x in the equation.
Step 3	Solve the equation for y.
Step 4	Replace y with $f^{-1}(x)$ in the new equation.

Watch Out!
In $f^{-1}(x)$, the -1 is not an exponent. It is a way to indicate that $f^{-1}(x)$ is an inverse of another function called $f(x)$.

Example 4 Find an Inverse Linear Function

Find the inverse of $f(x) = 5x + 10$.

Step 1 $f(x) = 5x + 10$ Original equation

$y = 5x + 10$ Replace $f(x)$ with y.

Step 2 $x = 5y + 10$ Interchange y and x.

Step 3 $x - 10 = 5y$ Subtract 10 from each side.

$\dfrac{x - 10}{5} = y$ Divide each side by 5.

Step 4 $\dfrac{x - 10}{5} = f^{-1}(x)$ Replace y with $f^{-1}(x)$.

The inverse of $f(x) = 5x + 10$ is $f^{-1}(x) = \dfrac{x - 10}{5}$ or $f^{-1}(x) = \dfrac{1}{5}x - 2$.

Example 5 Find Inverses of Linear Functions

Find the inverse of $f(x) = -\dfrac{2}{3}x - 8$.

Step 1 $f(x) = -\dfrac{2}{3}x - 8$ Original equation

$y = -\dfrac{2}{3}x - 8$ Replace $f(x)$ with y.

Step 2 $x = -\dfrac{2}{3}y - 8$ Interchange x and y.

Step 3 $x + 8 = -\dfrac{2}{3}y$ Add 8 to each side.

$-\dfrac{3}{2}(x + 8) = y$ Multiply each side by $-\dfrac{3}{2}$.

$-\dfrac{3}{2}x - 12 = y$ Simplify.

Step 4 $-\dfrac{3}{2}x - 12 = f^{-1}(x)$ Replace y with $f^{-1}(x)$.

The inverse of $f(x) = -\dfrac{2}{3}x - 8$ is $f^{-1}(x) = -\dfrac{3}{2}x - 12$.

Talk About It!
What is the inverse of $f(x) = -x$? How could you check your solution?

Go Online to see Example 5.

Example 6 Apply Inverse Linear Functions

BOATING Skyler and Carmen rent a paddle boat at a state park for $15 plus $4 for each hour it is used. The function $C(x) = 4x + 15$ represents the total cost $C(x)$ for x hours.

Part A Determine the inverse function.

Step 1 $C(x) = 4x + 15$ Original equation

           $y = 4x + 15$ Replace $C(x)$ with y.

Step 2 $x = 4y + 15$ Interchange y and x.

Step 3 $x - 15 = 4y$ Subtract 15 from each side.

           $\frac{x - 15}{4} = y$ Divide each side by 4.

Step 4 $\frac{x - 15}{4} = C^{-1}(x)$ Replace y with $C^{-1}(x)$.

Part B Interpret the inverse function.

x is the total cost of renting the paddle boat, and $C^{-1}(x)$ is the number of hours that Skyler and Carmen use the paddle boat.

Part C Evaluate using the inverse function.

Skyler and Carmen have $35 to rent the paddle boat. How long can they rent it?

To find the length of time that they can rent the boat, find $C^{-1}(35)$.

$C^{-1}(x) = \frac{x - 15}{4}$ Original equation

$C^{-1}(35) = \frac{35 - 15}{4}$ Substitute 35 for x.

$= \frac{20}{4}$ or 5   Simplify.

> **Study Tip**
> **Function Notation**
> Function notation is a way to give an equation a name, such as $f(x)$, $g(x)$, or $C(x)$. In Step 1 of finding the inverse function, replace the function notion with y regardless of the name of the function.

Check

CANDLES Javi is making candles to sell at an upcoming festival. He has already made 38 candles, and he makes 24 candles each day. The function $C(x) = 24x + 38$ represents the total number of candles $C(x)$ he has in inventory, where x is the number of days since he began making more candles.

Part A Select the inverse of the function $C(x)$.

A. $C^{-1}(x) = \frac{1}{24}x - \frac{19}{12}$ B. $C^{-1}(x) = \frac{1}{24}x - 38$

C. $C^{-1}(x) = \frac{1}{38}x - \frac{19}{12}$ D. $C^{-1}(x) = \frac{12}{19}x - 24$

Part B Estimate the amount of time it would take Javi to make 350 candles. It would take Javi between __?__ and __?__ days to make 350 candles.

Go Online You can complete an Extra Example online.

Practice

Go Online You can complete your homework online.

Examples 1 and 2

Find the inverse of each relation.

1.

x	y
−9	−1
−7	−4
−5	−7
−3	−10
−1	−13

2.

x	y
1	8
2	6
3	4
4	2
5	0

3.

x	y
−4	−2
−2	−1
0	1
2	0
4	2

4. {(−3, 2), (−1, 8), (1, 14), (3, 20)}

5. {(5, −3), (2, −9), (−1, −15), (−4, −21)}

6. {(4, 6), (3, 1), (2, −4), (1, −9)}

7. {(−1, 16), (−2, 12), (−3, 8), (−4, 4)}

8. {(−5, 13), (6, 10.8), (3, 11.4), (−10, 14)}

9. {(−4, −49), (8, 35), (−1, −28), (4, 7)}

Example 3

Graph the inverse of each function.

10.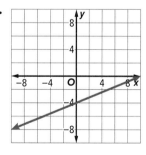

11.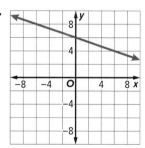

12.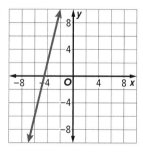

13.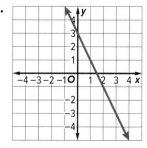

14.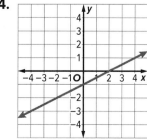

15.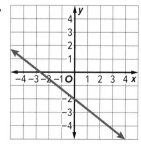

Lesson 5-6 • Inverses of Linear Functions 331

Examples 4 and 5

Find the inverse of each function.

16. $f(x) = 8x - 5$

17. $f(x) = 6(x + 7)$

18. $f(x) = \frac{3}{4}x + 9$

19. $f(x) = -16 + \frac{2}{5}x$

20. $f(x) = \frac{3x + 5}{4}$

21. $f(x) = \frac{-4x + 1}{5}$

Example 6

22. **LEMONADE** Bernardo spent $15 on supplies for his lemonade stand. He charges $1.25 per glass. The function $P(x) = 1.25x - 15$ represents his profit, where x is the number of glasses of lemonade sold.

    a. Find the inverse function, $P^{-1}(x)$.

    b. What do x and $P^{-1}(x)$ represent in the context of the inverse function?

    c. How many glasses must Bernardo sell in order to make $10 in profit?

23. **BUSINESS** Alisha started a baking business. She spent $36 initially on supplies and can make 5 dozen brownies for $12. She charges her customers $10 per dozen brownies. The function $P(x) = 7.6x - 36$ represents her profit, where x is the number of dozens of brownies sold.

    a. Find the inverse function, $P^{-1}(x)$.

    b. What do x and $P^{-1}(x)$ represent in the context of the inverse function?

    c. How many dozens of brownies does Alisha need to sell in order to make a profit?

24. **SEASON PASS** A season pass to an amusement park costs $70 per family member plus an additional $50 fee for parking. The function $C(x) = 70x + 50$ represents the total cost of the season pass for a family, where x is the number of family members on the season pass.

    a. Find the inverse function, $C^{-1}(x)$.

    b. What do x and $C^{-1}(x)$ represent in the context of the inverse function?

    c. How many family members purchased a season pass to the amusement park if the total charge was $470?

25. GARDENING Kara is building raised garden beds for her backyard. The total cost C(x) in dollars is given by C(x) = 125 + 16x, where x is the number of pieces of wood required for the boxes.

  a. Find the inverse function $C^{-1}(x)$.

  b. If the total cost was $269 and each piece of wood was 12 feet long, how many total feet of wood were used?

26. GEOMETRY The area of the base of a cylindrical water tank is 12π square feet. The volume of water in the tank is dependent on the height of the water h and is represented by the function V(h) = 12πh.

  a. Find $V^{-1}(h)$.

  b. What will the height of the water be when the volume reaches 420π cubic feet?

Mixed Exercises

Write the inverse of each function in $f^{-1}(x)$ notation.

27. 3y − 12x = −72

28. x + 3y = 10

29. −42 + 6y = x

30. 3y + 24 = 2x

31. −7y + 2x = −28

32. 12y − x = 7

Write an equation for the inverse function $f^{-1}(x)$ that satisfies the given conditions.

33. slope of f(x) is 7; graph of $f^{-1}(x)$ contains the point (13, 1)

34. graph of f(x) contains the points (−3, 6) and (6, 12)

35. graph of f(x) contains the point (10, 16); graph of $f^{-1}(x)$ contains the point (3, −16)

36. slope of f(x) is 4; $f^{-1}(5) = 2$

Match each function with the graph of its inverse.

37. $f(x) = \frac{1}{2}x + 2$ **38.** $f(x) = \frac{1}{2}x - 2$ **39.** $f(x) = x + 2$

A. 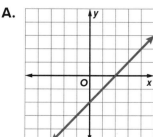   B. 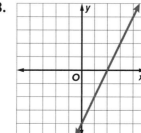   C.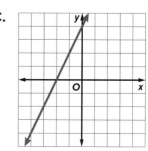

40. **STRUCTURE** Write the inverse of each function.

   a. $f(x) = \dfrac{x - 10}{3}$

   b. $g(x) = \dfrac{3}{4}x + 6$

   c. $h(x) = -5x - 7$

   d. Graph $h(x)$ and $h^{-1}(x)$ on the same coordinate plane to check your answer.

41. **REASONING** Suppose the inverse of a relation is $\{(b, -k), (-g, p), (-w, -m), (r, q)\}$. What is the relation?

42. **ANALYZE** How can you use ordered pairs to check if the inverse of a function is correct?

43. **WRITE** What is the relationship between the slopes of two lines that are inverse functions of one another? Give an example.

44. **WRITE** What is the relationship between the x- and y-intercepts of two lines that are inverse functions of one another? Give an example.

45. **FIND THE ERROR** A student claims that there is a simple method to find the inverse of the function $f(x)$. To find the inverse, $f^{-1}(x)$, we need only remember that raising something to the power of -1 is the same as taking its reciprocal. Is this claim correct? Include an example or counterexample.

46. **PERSEVERE** If $f(x) = 5x + a$ and $f^{-1}(10) = -1$, find a.

47. **PERSEVERE** If $f(x) = \dfrac{1}{a}x + 7$ and $f^{-1}(x) = 2x - b$, find a and b.

ANALYZE Determine whether the following statements are *sometimes*, *always*, or *never* true. Explain your reasoning.

48. If $f(x)$ and $g(x)$ are inverse functions, then $f(a) = b$ and $g(b) = a$.

49. If $f(a) = b$ and $g(b) = a$, then $f(x)$ and $g(x)$ are inverse functions.

50. **CREATE** Give an example of a function and its inverse. Verify that the two functions are inverses by graphing the functions and the line $y = x$ on the same coordinate plane.

51. **WRITE** Explain why it may be helpful to find the inverse of a function.

Module 5 • Creating Linear Equations

Review

Essential Question
What can a function tell you about the relationship that it represents?

Module Summary

Lessons 5-1 and 5-2

Writing Equations
- Slope-intercept form is $y = mx + b$, where m is the slope of the line and b is the y-intercept.
- Point-slope form is $y - y_1 = m(x - x_1)$, where (x_1, y_1) is a given point on a nonvertical line and m is the slope of the line.
- Standard form is $Ax + By = C$, where A, B, and C are integers, $A > 0$, A and B are both not equal to 0, and the GCF of A, B, and C is 1.
- To write a linear equation given two points on a line, first find the slope. Then use either point to write the equation in point-slope form or find the y-intercept to write the equation in slope-intercept form.

Lessons 5-3 through 5-5

Scatter Plots
- A scatter plot shows the relationship between a set of bivariate data, graphed as ordered pairs on a coordinate plane.
- A positive correlation exists when, as x increases, y increases. A negative correlation exists when, as x increases, y decreases. No correlation exists when x and y are not related.
- A line of fit is used to describe the trend of the data in a scatter plot.
- To determine causation, determine whether one variable influences the other variable.
- The correlation coefficient tells you how well the equation for the best-fit line models the data.
- A correlation coefficient close to 1 has a strong positive correlation. A correlation coefficient close to -1 has a strong negative correlation.
- Residuals measure how much the data deviate from the regression line.

Lesson 5-6

Inverses of Linear Functions
- Two relations are inverse relations if and only if one relation contains the element (a, b) when the other relation contains the element (b, a).
- In inverse relations, the x-coordinates are exchanged with the y-coordinates for each ordered pair in the relation.
- To find the inverse of $f(x)$, replace $f(x)$ with y in the equation for $f(x)$. Interchange y and x in the equation. Solve the equation for y. Replace y with $f^{-1}(x)$ in the new equation.

Study Organizer

Foldables
Use your Foldable to review the module. Working with a partner can be helpful. Ask for clarification of concepts as needed.

Creating Linear Equations

Test Practice

1. **MULTIPLE CHOICE** What is the equation of the line that passes through the points $(-2, 1)$ and $(6, 3)$? (Lesson 5-1)
   A. $y = \frac{1}{4}x + \frac{3}{2}$
   B. $y = \frac{3}{2}x + \frac{1}{4}$
   C. $y = \frac{2}{3}x + \frac{1}{4}$
   D. $y = \frac{3}{2}x + \frac{1}{4}$

2. **MULTIPLE CHOICE** Select the equation of a line with a slope of 5 that passes through the point $(2, -3)$. (Lesson 5-1)
   A. $y = 5x + 2$
   B. $y = 5x - 3$
   C. $y = 5x + 7$
   D. $y = 5x - 13$

Use the table for exercises 3 and 4. A movie streaming service charges a set fee for membership each month, plus an additional fee for the number of movies streamed each month. This table shows the total charge for different numbers of movies.

Number of movies streamed (x)	2	4	6
Total cost (y)	$14	$17	$20

3. **OPEN RESPONSE** Write the slope-intercept form of the equation that models the linear relationship in the table. (Lesson 5-1)

4. **OPEN RESPONSE** Interpret the meaning of the slope and y-intercept in the context of the situation. (Lesson 5-1)

5. **MULTIPLE CHOICE** Which equation represents a line that passes through the point $(3, -4)$ with a slope of 7? (Lesson 5-2)
   A. $y + 4 = 7(x - 3)$
   B. $y + 4 = 7(x + 3)$
   C. $y - 4 = 7(x - 3)$
   D. $y - 4 = 7(x + 3)$

6. **MULTI-SELECT** Select all of the equations that represent the line. (Lesson 5-2)

   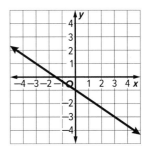

   A. $2x - 3y = -1$
   B. $2x + 3y = -3$
   C. $3x - 2y = 2$
   D. $y + 3 = -\frac{2}{3}(x - 3)$
   E. $y - 1 = -\frac{2}{3}(x + 3)$
   F. $y + 1 = -\frac{3}{2}(x + 3)$

7. **OPEN RESPONSE** A city parking garage charges $4 to park for up to two hours. After that, an additional charge of $2.50 per hour applies. Write an equation in point-slope form that models the total cost y for parking x hours, where $x > 2$. (Lesson 5-2)

8. **MULTIPLE CHOICE** Which scatter plot shows the best line of fit? (Lesson 5-3)

A.
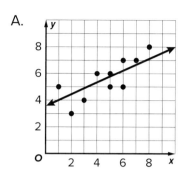

B.
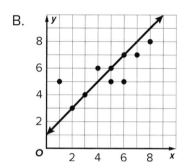

C.
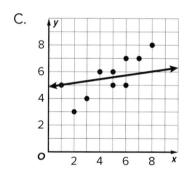

D.
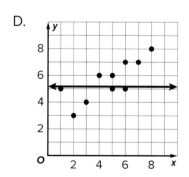

9. **MULTIPLE CHOICE** Which equation represents the best line of fit for the scatter plot? (Lesson 5-3)

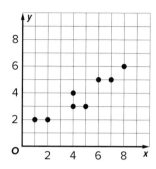

A. $y = 0.6x + 1$
B. $y = 0.5x + 2$
C. $y = x - 2$
D. $y = 0.75x$

10. **OPEN RESPONSE** Adriana keeps the statistics for her favorite basketball team and creates the scatter plot shown. (Lesson 5-3)

Adriana then draws a line of fit for her scatter plot. What does the slope of the line represent?

11. **OPEN RESPONSE** A researcher found that students who spent more time exercising each week also had higher average test scores. Describe the correlation, if any, between time spent exercising and test scores. (Lesson 5-4)

12. **OPEN RESPONSE** Lalita tracked the amount of time she studied each week and her score on a weekly chemistry quiz for eight weeks. She made this scatter plot from the data. Determine whether the data illustrate a *correlation* or *causation*. Explain. (Lesson 5-4)

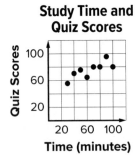

13. **MULTIPLE CHOICE** Use linear regression to estimate the weight, in ounces, of a bluegill that has a length of 9.5 inches. Round your answer to the nearest tenth of an ounce. (Lesson 5-5)

Length (in.)	7	8	9	11	12	13
Weight (oz)	4	7	11	21	26	32

A. 12.9

B. 13.4

C. 14.5

D. 15.1

14. **OPEN RESPONSE** Use a graphing calculator and linear regression to write the equation of a best-fit line for the data in slope-intercept form. Round to the nearest tenth. (Lesson 5-5)

x	2.4	2.8	3.4	4.3	5.1	7.6	8.4	9.1
y	6.2	9.6	8.4	6.5	7.2	2.5	1.8	4.2

15. **OPEN RESPONSE** The graph of a function passes through $(-3, 2)$, $(-1, 1)$, $(1, 0)$, and $(3, -1)$. Find the inverse function. Then graph the inverse function. (Lesson 5-6)

16. **MULTI-SELECT** The table represents the coordinates of a linear function. (Lesson 5-6)

x	y
−6	5
4	1
2	−3

Select the equations that represent the inverse of the function.

A. $f^{-1}(x) = \frac{5}{2}x + \frac{5}{2}$

B. $f^{-1}(x) = -\frac{5}{2}x + \frac{13}{2}$

C. $f(x) = 2.5x - 6.5$

D. $f^{-1}(x) = -1.25x + \frac{13}{2}$

E. $f^{-1}(x) = -2.5x + 6.5$

17. **MULTIPLE CHOICE** Shakir is running in a long-distance race. If he maintains an average speed of 8 miles per hour, then the distance in miles that he has left to run is given by $D(x) = -\frac{2}{15}x + 10$, where x is the number of minutes since Shakir started the race. Which function is the inverse of $D(x)$? (Lesson 5-6)

A. $D^{-1}(x) = -\frac{15}{2}x + 75$

B. $D^{-1}(x) = \frac{2}{15}x - 10$

C. $D^{-1}(x) = -\frac{15}{2}x + \frac{1}{10}$

D. $D^{-1}(x) = -\frac{2}{15}x - \frac{4}{3}$

Module 6
Linear Inequalities

Essential Question
How can writing and solving inequalities help you solve problems in the real world?

What Will You Learn?
How much do you already know about each topic **before** starting this module?

KEY

👎 — I don't know.   👍 — I've heard of it.   👍 — I know it!

	Before			After		
	👎	👍	👍	👎	👍	👍
graph linear inequalities						
solve one-step linear inequalities using addition and subtraction						
solve one-step linear inequalities using multiplication and division						
solve multi-step linear inequalities						
solve compound linear inequalities						
solve absolute value linear inequalities						
graph inequalities in two-variables						

Foldables Make this Foldable to help you organize your notes about linear inequalities. Begin with one sheet of 11" × 17" paper.

1. **Fold** each side so the edges meet in the center.
2. **Fold** in half.
3. **Unfold** and cut from each end until you reach the vertical line.
4. **Label** the front of each flap.

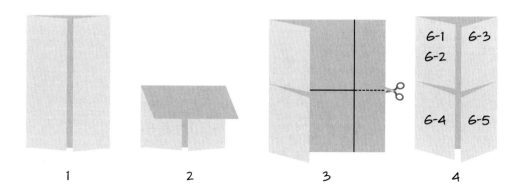

Module 6 • Linear Inequalities

What Vocabulary Will You Learn?

- boundary
- closed half-plane
- compound inequality
- half-plane
- inequality
- intersection
- open half-plane
- set-builder notation
- union

Are you ready?

Complete the Quick Review to see if you are ready to start this module.
Then complete the Quick Check.

Quick Review

Example 1

Solve $-2(x - 4) = 7x - 19$.

$-2(x - 4) = 7x - 19$	Original equation
$-2x + 8 = 7x - 19$	Distributive Property
$-2x + 8 + 2x = 7x - 19 + 2x$	Add $2x$ to each side.
$8 = 9x - 19$	Simplify.
$8 + 19 = 9x - 19 + 19$	Add 19 to each side.
$27 = 9x$	Simplify.
$3 = x$	Divide each side by 3.

Example 2

Solve $|x - 4| = 9$.

if $|x - 4| = 9$, then $x - 4 = 9$ or $x - 4 = -9$.

$x - 4 = 9$ or $x - 4 = -9$
$x - 4 + 4 = 9 + 4$ $x - 4 + 4 = -9 + 4$
$x = 13$ $x = -5$

So, the solution set is $\{-5, 13\}$.

Quick Check

Solve each equation.

1. $2x + 1 = 9$
2. $4x - 5 = 15$
3. $9x + 2 = 3x - 10$
4. $3(x - 2) = -2(x + 13)$

Solve each equation.

5. $|x + 11| = 18$
6. $|3x - 2| = 16$
7. $|x - 7| = 8$
8. $|2x| = -9$

How did you do?

Which exercises did you answer correctly in the Quick Check?

Lesson 6-1
Solving One-Step Inequalities

Explore Graphing Inequalities

Online Activity Use graphing technology to complete the Explore.

> **INQUIRY** How can you graph the solution set of an inequality of the form $x < a$ or $x > a$ for some number a?

Learn Graphing Inequalities

An **inequality** is a mathematical sentence that contains the symbol $<$, $>$, $\leq$, $\geq$, or $\neq$. An inequality compares the value of two numbers or expressions using these symbols.

Example 1 Graph Inequalities

Graph the solution set of $y \leq 4$.

The dot at 4 shows that 4 is a solution. The heavy arrow pointing to the left shows that the solution includes all numbers less than 4.

Check

Graph the solution set of $y > \frac{1}{2}$.

Example 2 Write Inequalities from a Graph

Write an inequality that represents the graph.

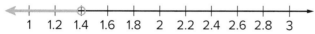

The endpoint is shown with a circle at 1.4, so 1.4 is not included in the solution. The inequality must be $<$ or $>$.

The arrow points to values less than 1.4. The graph represents the solution of $a < 1.4$.

Check

Write an inequality that represents the graph.

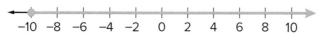

Go Online You can complete an Extra Example online.

Today's Goals
- Graph the solutions of an inequality.
- Solve linear inequalities by using addition.
- Solve linear inequalities by using subtraction.

Today's Vocabulary
inequality
set-builder notation

Think About It!
Compare and contrast the graphs of $x \geq 6$ and $x > 6$.

Think About It!
How do you know which inequality symbol to use by looking at a graph?

Explore Properties of Inequalities

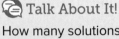 **Online Activity** Use graphing technology to complete the Explore.

> **INQUIRY** Do the properties of equality hold true for inequalities?

Learn Solving Inequalities by Using Addition and Subtraction

Addition and subtraction can be used to solve inequalities.

Key Concept • Addition Property of Inequalities	
Words	If the same number is added to each side of a true inequality, the resulting inequality is also true.
Symbols	For any real numbers a, b, and c, the following are true. If $a > b$, then $a + c > b + c$. If $a < b$ then $a + c < b + c$.

Key Concept • Subtraction Property of Inequalities	
Words	If the same number is subtracted from each side of a true inequality, the resulting inequality is also true.
Symbols	For any real numbers a, b, and c, the following are true. If $a > b$, then $a - c > b - c$. If $a < b$ then $a - c < b - c$.

When solving inequalities, you can write the solution set in a more concise way using **set-builder notation**. For example, $\{x \mid x \geq -4\}$ represents the set of all numbers x such that x is greater than or equal to -4.

Example 3 Solve Inequalities by Adding

Solve $x - 10 < 15$.

$x - 10 < 15$	Original inequality
$x - 10 + 10 < 15 + 10$	Add 10 to each side to isolate x.
$x < 25$	Simplify.

The solution set is $\{x \mid x < 25\}$.

Check

Solve $-9 + b \leq 16$.

Talk About It!
How many solutions of the inequality are there? Justify your argument.

Study Tip
Set-Builder Notation
$\{x \mid x < 25\}$ is read the set of all numbers x such that x is less than 25.

Go Online You can complete an Extra Example online.

Example 4 Solve Inequalities by Subtracting

Solve $x + 24 \geq 61$.

$x + 24 \geq 61$ Original inequality

$x + 24 - 24 \geq 61 - 24$ Subtract 24 from each side.

$x \geq 37$ Simplify.

The solution set is $\{x \mid x \geq 37\}$.

Check

Solve $88 < x + 13$.

> **Think About It!**
> How can you check the solution of the inequality?

Example 5 Add or Subtract to Solve Inequalities with Variables on Each Side

Solve $9y + 3 \geq 10y$.

$9y + 3 \geq 10y$ Original inequality

$9y - 9y + 3 \geq 10y - 9y$ Subtract $9y$ from each side.

$3 \geq y$ Simplify.

Since $3 \geq y$ is the same as $y \leq 3$, $\{y \mid y \leq 3\}$.

Check

Solve $7x + 6 < 8x$.

> **Study Tip**
> **Writing Inequalities** Simplifying the inequality so that the variable is on the left side, as in $y \leq 3$, prepares you to write the solution set in set-builder notation and graph the inequality on a number line.

🌐 Example 6 Use an Inequality to Solve a Problem

DATA USAGE Hassan's wireless contract allows him to use at most 5 gigabytes (GB) of data per month. At this point, Hassan has used 3.7 GB of data. How many gigabytes of data can Hassan use during the rest of the month without exceeding the maximum allowance?

Complete the table to write an inequality to represent how many gigabytes of data Hassan can use. Then solve the inequality.

Words	Hassan can use	at most	5 GB of data.
Variables	Let g = the number of gigabytes that Hassan has left to use.		
Inequality	$3.7 + g$	$\leq$	5

(continued on the next page)

Lesson 6-1 • Solving One-Step Inequalities

> **Use a Source**
> Research data plans for wireless carriers in your area. Write and solve your own inequality to represent the amount of data remaining if you have already used 5.2 GB.

$3.7 + g \leq 5$ Original inequality
$3.7 - 3.7 + g \leq 5 - 3.7$ Subtract 3.7 from each side.
$g \leq 1.3$ Simplify.

The solution set is $\{g \mid g \leq 1.3\}$.

Hassan can use up to 1.3 GB of data without exceeding his maximum allowance. Notice that negative numbers are solutions to the inequality, but they are not viable solutions to the problem because Hassan cannot use a negative amount of data.

Learn Solving Inequalities by Using Multiplication and Division

If you multiply or divide each side of an inequality by a positive number, then the inequality remains true.

If you multiply or divide each side of an inequality by a negative number, the inequality symbol changes direction.

> **Study Tip**
> **Inequalities** Verbal problems containing phrases like *greater than* and *less than* can be solved by using inequalities. Some other phrases that include inequalities are:
>
> < less than; fewer than
> \> greater than; more than
> ≤ less than or equal to; at most; no more than
> ≥ greater than or equal to; at least; no less than

Key Concept • Multiplication Property of Inequalities

Words	If each side of a true inequality is multiplied by a positive number, the resulting inequality is also true.	If each side of a true inequality is multiplied by a negative number, the direction of the inequality sign must be reversed to make the resulting inequality also true.
Symbols	For any real numbers a and b and any positive real number c: If $a > b$, then $ac > bc$. If $a < b$, then $ac < bc$.	For any real numbers a and b and any negative real number c: If $a > b$, then $ac < bc$. If $a < b$, then $ac > bc$.

Key Concept • Division Property of Inequalities

Words	If each side of a true inequality is divided by a positive number, the resulting inequality is also true.	If each side of a true inequality is divided by a negative number, the direction of the inequality sign must be reversed to make the resulting inequality also true.
Symbols	For any real numbers a and b and any positive real number c: If $a > b$, then $\frac{a}{c} > \frac{b}{c}$. If $a < b$, then $\frac{a}{c} < \frac{b}{c}$.	For any real numbers a and b and any negative real number c: If $a > b$, then $\frac{a}{c} < \frac{b}{c}$. If $a < b$, then $\frac{a}{c} > \frac{b}{c}$.

> **Think About It!**
> If a, b, and c are positive real numbers, what must be true if ac is greater than or equal to bc? What must happen to an inequality symbol when you divide each side by a negative number if the inequality is to remain true?

These properties also hold true for inequalities involving $\leq$ and $\geq$.

🌐 Apply Example 7 Write and Solve an Inequality

BOOKS Alisa has read approximately $\frac{1}{4}$ of a novel. If she has read at least 112 pages, how many pages are there in the novel?

1. What is the task?
Describe the task in your own words. Then list any questions that you may have. How can you find answers to your questions?

I know the number of pages read and the fraction of the novel read. I need to find out how many pages are in the novel.

2. How will you approach the task? What have you learned that you can use to help you complete the task?

I will use estimation first. Then I will write an inequality to represent the situation and solve it.

3. What is your solution?
Estimate the number of pages in the novel.
Alisa has read slightly more than 100 pages, which is $\frac{1}{4}$ of the novel, so the novel has more than 4(100), or 400 pages.

Write an inequality to represent this situation. Let $n =$ the number of pages in the novel.

$$\frac{1}{4} \cdot n \geq 112$$
$$4\left(\frac{1}{4}\right)n \geq 4(112)$$
$$n \geq 448$$

There are at least 448 pages in the novel.

4. How can you know that your solution is reasonable?

✏️ **Write About It!** **Write an argument that can be used to defend your solution.**
Use multiplication; $448\left(\frac{1}{4}\right) = 112$, so 448 is reasonable. Also, 448 > 400 which makes sense with our estimate of more than 400 pages.

Math History Minute

German mathematician **Emmy Noether (1882–1935)** has been described as one of the greatest mathematicians of the twentieth century. She devised theorems for several concepts later found in Einstein's theory of relativity and was one of the founders of abstract algebra. One person wrote, "The development of abstract algebra, which is one of the most distinctive innovations of twentieth century mathematics, is largely due to her."

Check

ELECTRIC CAR For every hour x that Eva's electric car charges, she can drive the car 7.5 miles. Eva needs to drive at least 60 miles tomorrow.

Part A What inequality represents the situation in terms of x hours?

Part B What is the least amount of time that Eva will need to charge her car?

🔵 **Go Online** You can complete an Extra Example online.

Study Tip

Multiplicative Inverses
The multiplicative inverse, or reciprocal, of a number can be used to undo multiplication. Multiplying $-\frac{2}{5}x$ by the reciprocal $-\frac{5}{2}$ in the example at the right is the same as dividing by $-\frac{2}{5}$, but is easier to compute mentally.

 Think About It!
Why was the inequality symbol reversed?

 Think About It!
Why is the solution of $-13z \geq 117$ shaded to the left when the original inequality symbol is greater than or equal to?

Study Tip

Negatives A negative sign in an inequality does not necessarily mean that the direction of the inequality symbol should change. For example, when solving $\frac{x}{3} \geq -9$, do not change the direction of the inequality symbol.

Example 8 Solve an Inequality by Multiplying

Solve $-\frac{2}{5}x \leq 11$. Graph the solution set on a number line.

$-\frac{2}{5}x \leq 11$ Original inequality

$\left(-\frac{5}{2}\right)-\frac{2}{5}x \leq 11\left(-\frac{5}{2}\right)$ Multiply each side by $-\frac{5}{2}$. Reverse the inequality symbol.

$x \geq -27.5$ Simplify.

The solution set is $\{x \mid x \geq -27.5\}$.

Example 9 Solve an Inequality by Dividing

Solve $20x < 4$. Graph the solution set on a number line.

$20x < 4$ Original inequality

$\frac{20x}{20} < \frac{4}{20}$ Divide each side by 20.

$x < \frac{1}{5}$ Simplify.

The solution set is $\{x \mid x < \frac{1}{5}\}$.

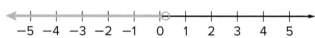

Check

Solve $7x > -161$.

Example 10 Solve an Inequality with a Negative Coefficient

Solve $-13z \geq 117$. Graph the solution set on a number line.

$-13z \geq 117$ Original inequality

$-\frac{13z}{-13} \geq \frac{117}{-13}$ Divide each side by -13.

$z \leq -9$ Simplify.

The solution set is $\{z \mid z \leq -9\}$.

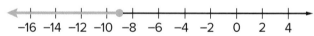

Check

Select the solution set for $-13x > -169$.

A. $\{x \mid x > 13\}$ **B.** $\{x \mid x < 13\}$ **C.** $\{x \mid x > -13\}$ **D.** $\{x \mid x < -13\}$

Go Online You can complete an Extra Example online.

Practice

Go Online You can complete your homework online.

Example 1

Graph the solution set of each inequality.

1. $x \leq -5$
2. $y \geq -2$
3. $g > 5$
4. $h < -6$
5. $a < 7$
6. $b \leq 6$

Example 2

Write an inequality that represents each graph.

7. (open circle at -1, shading left)
8. (open circle at -6, shading left)
9. (open circle at 5, shading left)
10. (closed circle at 1, shading left)
11. (closed circle at -5, shading right)
12. (closed circle at -6, shading right)

Examples 3–5

Solve each inequality.

13. $m - 4 < 3$
14. $p - 6 \geq 3$
15. $r - 8 \leq 7$
16. $t - 3 > -8$
17. $b + 2 \geq 4$
18. $13 > 18 + r$
19. $5 + c \leq 1$
20. $-23 \geq q - 30$
21. $11 + m \geq 15$
22. $h - 26 < 4$
23. $8 \leq r - 14$
24. $-7 > 20 + c$
25. $2a \leq -4 + a$
26. $z + 4 \geq 2z$
27. $w - 5 \leq 2w$
28. $3y \leq 2y - 6$
29. $6x + 5 \geq 7x$
30. $-9 + 2a < 3a$

Lesson 6-1 • Solving One-Step Inequalities 347

Example 6

31. **PIZZA** Tara and friends order a pizza. Tara eats 3 of the 10 slices and pays $4.50 for her share. Assuming that Tara has paid at least her fair share, write and solve an inequality to represent the cost of the pizza.

32. **WEATHER** Theodore Fujita of the University of Chicago developed a classification of tornadoes according to wind speed and damage. The table shows the classification system.

Level	Name	Wind Speed Range (mph)
F0	Gale	40–72
F1	Moderate	73–112
F2	Significant	113–157
F3	Severe	158–206
F4	Devastating	207–260
F5	Incredible	261–318
F6	Inconceivable	319–379

    Source: National Weather Service

    a. Suppose an F3 tornado has winds that are 162 miles per hour. Write and solve an inequality to determine how much the winds would have to increase before the F3 tornado becomes an F4 tornado.

    b. A tornado has wind speeds that are at least 158 miles per hour. Write and solve an inequality that describes how much greater these wind speeds are than the slowest tornado.

Example 7

33. **GARBAGE** The amount of garbage that the average American adds to a landfill each day is 4.6 pounds. If at least 2.5 pounds of a person's daily garbage could be recycled, how much would still go into a landfill?

34. **SUPREME COURT** The first Chief Justice of the U.S. Supreme Court, John Jay, served 2079 days as Chief Justice. He served 10,463 days fewer than John Marshall, who served as Supreme Court Chief Justice for the longest period of time. How many days must the current Supreme Court Chief Justice John Roberts serve to surpass John Marshall's record of service?

35. **AIRLINES** On average, at least 25,000 pieces of luggage are lost or misdirected each day by United States airlines. Of these, 98% are located by the airlines within 5 days. From a given day's lost luggage, at least how many pieces of luggage are still lost after 5 days?

36. **SCHOOL** Gilberto earned these scores on the first three tests in biology this term: 86, 88, and 78. What is the lowest score that Gilberto can earn on the fourth and final test of the term if he wants to have an average of at least 83?

Examples 8–10

Solve each inequality. Graph the solution on a number line.

37. $\frac{1}{4}m \leq -17$

38. $\frac{1}{2}a < 20$

39. $-11 > -\frac{c}{11}$

40. $-2 \geq -\frac{d}{34}$

41. $-10 \leq \frac{x}{-2}$

42. $-72 < \frac{f}{-6}$

43. $\frac{2}{3}h > 14$

44. $-\frac{3}{4}j \geq 12$

45. $-\frac{1}{6}n \leq -18$

46. $6p \leq 96$

47. $4r < 64$

48. $32 > -2y$

49. $-26 < 26t$

50. $-6v > -72$

51. $-33 \geq -3z$

52. $4b \leq -3$

53. $-2d < 5$

54. $-7f > 5$

Mixed Exercises

Match each inequality with its corresponding statement.

55. $3n < 9$ a. Three times a number is at most nine.

56. $\frac{1}{3}n \geq 9$ b. One third of a number is no more than nine.

57. $3n \leq 9$ c. Negative three times a number is more than nine.

58. $-3n > 9$ d. Three times a number is less than nine.

59. $\frac{1}{3}n \leq 9$ e. Negative three times a number is at least nine.

60. $-3n \geq 9$ f. One third of a number is greater than or equal to nine.

Define a variable, write an inequality, and solve each problem. Check your solution.

61. Seven more than a number is less than or equal to –18.

62. Twenty less than a number is at least 15.

63. A number plus 2 is at most 1.

64. One eighth of a number is less than or equal to 3.

65. Negative twelve times a number is no more than 84.

66. Eight times a number is at least 16.

STRUCTURE Solve each inequality. Check your solution, and then graph it on a number line.

67. $14c > 56$
68. $20b \geq -120$
69. $\frac{x}{4} < 9$
70. $\frac{x}{2.5} \leq 8$
71. $m + 3.7 < 9.1$
72. $n - \frac{1}{5} > \frac{4}{5}$
73. $c + (-1.4) \geq 2.3$
74. $k + \frac{3}{4} > \frac{1}{3}$

75. **EVENT PLANNING** The Community Center does not charge a rental fee as long as a rentee orders a minimum of $5000 worth of food. Antonio is planning a banquet. If he is expecting 225 people to attend, what is the minimum he will have to spend on food per person to avoid paying a rental fee?

76. **VITAMINS** The minimum daily requirement of vitamin C for 14-year-olds is at least 50 milligrams per day. An average-sized apple contains 6 milligrams of vitamin C. How many apples would a person have to eat each day to satisfy this requirement? Define a variable and write and solve an inequality to represent this situation.

77. **USE A SOURCE** The loudest insect is the African cicada. It produces sounds as loud as 105 decibels. The blue whale is the loudest mammal. The call of the blue whale can reach levels up to 83 decibels louder than the African cicada. Write and solve an inequality to represent the situation. How loud are the calls of the blue whale? Use a source to verify your answer.

78. **USE A MODEL** In a mathematics exam with a maximum score of 100, Machelle loses less than 27 points. The table shows the grade that matches the exam score. Compare points to grades and identify which grade Machelle can get.
    a. Define a variable. Then write and solve an inequality to represent the number of points Machelle received on her exam.

Grade	Points
A	92–100
B	83–91
C	74–82
D	65–73
F	64 and below

    b. Interpret the solution to your inequality. What do you know about Machelle's grade on the exam?

Higher Order Thinking Skills

79. **WHICH ONE DOESN'T BELONG** Which inequality does *not* have the solution $\{x | x < -2\}$?

    A $-3x > 6$ **B** $-\frac{x}{2} < 1$ **C** $7x < -14$ **D** $\frac{4}{3}x < -\frac{8}{3}$

80. **FIND THE ERROR** Marty and Heath solved the same exercise in different ways. Is either correct? Explain your reasoning.

    Marty:
    $3m \geq -21$
    $\frac{3m}{3} \geq \frac{-21}{3}$
    $m \leq -7$

    Heath:
    $3m \geq -21$
    $\frac{3m}{3} \geq \frac{21}{3}$
    $m \geq -7$

81. Solve each inequality in terms of x. Assume that a does not equal 0.
    a. $ax < 7$
    b. $ax \geq 12$
    c. $ax > 3a$
    d. $ax \geq \frac{a}{4}$

82. **ANALYZE** Determine whether the statement is *sometimes*, *always*, or *never* true. If $a > b$, then $\frac{1}{a} > \frac{1}{b}$. Justify your argument.

Module 6 • Linear Inequalities

Lesson 6-2

Solving Multi-Step Inequalities

Explore Modeling Multi-Step Inequalities

Today's Goals
- Solve multi-step linear inequalities.

Online Activity Use algebra tiles to complete the Explore.

INQUIRY How can you model and solve a multi-step inequality?

Learn Solving Inequalities Involving More Than One Step

Step 1 Isolate the variable terms on one side of the inequality using addition or subtraction.

Step 2 Multiply or divide to isolate the variable.

Study Tip

Negative Numbers When multiplying or dividing by a negative number, the direction of the inequality symbol changes. This holds true for multi-step as well as one-step inequalities.

Example 1 Apply Multi-Step Inequalities

PUBLISHING Suzy wants to self-publish her comic book. One printing company offers to publish the book for a $220 flat rate plus $3 per copy of the book. Her maximum budget is $400.

Part A Write an inequality.

Words	$220 flat rate	Plus	$3 per copy	is at most	$400
Inequality	220	+	$3x$	$\leq$	400

Part B Solve the inequality.

$220 + 3x \leq 400$ Original inequality
$3x \leq 180$ Subtract 220 from each side.
$x \leq 60$ Divide each side by 3.

Suzy can have up to 60 books printed while not exceeding her budget.

Check

TICKETS Jamal has $40 to buy tickets to a performance for himself and his friends. If he buys a $10 membership, he can buy tickets for $5 each. How many tickets can he buy while remaining within his budget?

If x represents the number of tickets Jamal purchases, write an inequality that represents the situation.

Solve the inequality.

 Think About It! Can x be any real number less than or equal to 60? Explain your reasoning.

Go Online You can complete an Extra Example online.

Example 2 Write and Solve a Multi-Step Inequality

Consider the inequality *The opposite of a number divided by two minus seventeen is less than seven.*

Part A Translate the sentence into an inequality.

$$-\frac{x}{2} - 17 < 7$$

Part B Solve the inequality.

$-\frac{x}{2} - 17 < 7$	Original inequality
$-\frac{x}{2} < 24$	Add 17 to each side.
$-x < 48$	Multiply each side by 2.
$x > -48$	Divide each side by -1, reversing the inequality symbol.

Part C Graph the solution on a number line.

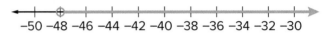

> **Study Tip**
>
> **Empty Set or All Real Numbers** If an inequality simplifies to a false statement, then there is no solution to the inequality. The solution set is the empty set, ∅. However, if all values of a variable make the inequality true, then the solution set is all real numbers.

Example 3 Solve an Inequality with the Distributive Property

Solve the inequality $4(2x - 11) \leq -12 + 2(x - 4)$. Then graph the solution on a number line.

$4(2x - 11) \leq -12 + 2(x - 4)$	Original inequality
$8x - 44 \leq -12 + 2x - 8$	Distributive Property
$8x - 44 \leq -20 + 2x$	Simplify.
$8x \leq 24 + 2x$	Add 44 to each side.
$6x \leq 24$	Subtract $2x$ from each side.
$x \leq 4$	Divide each side by 6.

Graph the solution of $4(2x - 11) \leq -12 + 2(x - 4)$ on a number line.

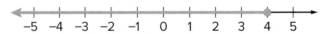

Check

Solve $88 \geq -33 + 11(x + 8)$. Then graph the inequality.

Go Online You can complete an Extra Example online.

Practice

Go Online You can complete your homework online.

Example 1

1. **BEACHCOMBING** Jay wants to rent a metal detector. A rental company charges a one-time rental fee of $15 plus $2 per hour to rent a metal detector. Jay has only $35 to spend.

   a. Write an inequality to represent this situation, where h is the number of hours Jay will rent a metal detector.

   b. Solve the inequality. What is the maximum amount of time he can rent the metal detector?

2. **AGES** Pedro, Sebastian, and Manuel Martinez are each one year apart in age. The sum of their ages is greater than the age of their father, who is 60.

   a. Write an inequality to represent this situation, where x is the age of the youngest brother.

   b. Solve the inequality.

   c. How old can the oldest brother be? Explain your reasoning.

3. **RIDE SHARE** Demetri lives in the city and sometimes uses a ride share service. A ride costs $1.50 for the first $\frac{1}{5}$ mile and $0.25 for each additional $\frac{1}{5}$ mile. Demetri does not want to spend more than $3.75 on a ride.

   a. Write an inequality to represent this situation, where x is the number of miles.

   b. Solve the inequality. What is the maximum distance he can travel if he does not tip the driver?

   c. Generalize a method for writing an inequality for this situation if the service charges $1.50 for the first $\frac{1}{a}$ mile and $0.25 for each additional $\frac{1}{a}$ mile.

4. **POST OFFICE** Keshila goes to the post office to mail a package and a few letters. Stamps cost 49 cents each. It will cost $7.65 to mail the package. Keshila has $10.00.

   a. Write an inequality to represent this situation, where x is the number of stamps.

   b. Solve the inequality. What is the maximum number of stamps Keshila can purchase to mail letters?

5. **BANQUET** A charity is hosting a benefit dinner. They are asking $100 per table plus $40 per person. Nathaniel is purchasing tickets for his friends and does not want to spend more than $250.

   a. Write an inequality to represent this situation, where x is the number of people.

   b. Solve the inequality. What is the maximum number of people Nathaniel can invite to the dinner?

Example 2

Translate each sentence into an inequality. Then solve the inequality and graph the solution on a number line.

6. Five times a number minus one is greater than or equal to negative eleven.

7. Twenty-one is greater than the sum of fifteen and two times a number.

8. Negative nine is greater than or equal to the sum of two-fifths times a number and seven.

9. A number divided by eight minus thirteen is greater than negative six.

10. The sum of the opposite of a number and six is less than or equal to five.

11. Thirty-seven is less than the difference of seven and ten times a number.

12. Eight minus a number divided by three is greater than or equal to eleven.

13. Negative five-fourths times a number plus six is less than twelve.

14. The difference of three times a number and six is greater than or equal to the sum of fifteen and twenty-four times a number.

15. The sum of fifteen times a number and thirty is less than the difference of ten times a number and forty-five.

Example 3

Solve each inequality. Then graph the solution on a number line.

16. $-3(7n + 3) < 6n$

17. $21 \geq 3(a - 7) + 9$

18. $2y + 4 > 2(3 + y)$

19. $3(2 - b) < 10 - 3(b - 6)$

20. $7 + t \leq 2(t + 3) + 2$

21. $8a + 2(1 - 5a) \leq 20$

Mixed Exercises

Solve each inequality. Check your solution.

22. $2(x - 4) \leq 2 + 3(x - 6)$

23. $\frac{2x - 4}{6} \geq -5x + 2$

24. $5.6z + 1.5 < 2.5z - 4.7$

25. $0.7(2m - 5) \geq 21.7$

26. $2(-3m - 5) \geq -28$

27. $-6(w + 1) < 2(w + 5)$

USE TOOLS Use a graphing calculator to solve each inequality.

28. $3x + 7 > 4x + 9$

29. $13x - 11 \leq 7x + 37$

30. $2(x - 3) < 3(2x + 2)$

31. $\frac{1}{2}x - 9 < 2x$

32. $2x - \frac{2}{3} \geq x - 22$

33. $\frac{1}{3}(4x + 3) \geq \frac{2}{3}x + 2$

STRUCTURE Solve each inequality. Then graph it on a number line.

34. $9.1g + 4.5 < 10.1g$

35. $\frac{3}{2}p - \frac{2}{3} \leq \frac{4}{9} + \frac{1}{2}p$

36. $3.3r - 8.3 \geq 5.3r - 12.9$

37. TREEHOUSE DESIGN Devontae is building a treehouse in his backyard. He researches city restrictions on building codes. The height of the treehouse cannot exceed 13 feet. He wants to build a tree house with 2 levels of equal height that is 4 feet off the ground.

  a. Write and solve an inequality.

  b. What is the maximum height of one level?

  c. Devontae decides to build one level higher off the ground. If the level is 8 feet tall, how high can the tree house be off the ground?

38. MEDICINE Clark's Rule is a formula used to determine pediatric dosages of over-the-counter medicines: $\frac{\text{weight of child (lb)}}{150} \times \text{adult dose} = \text{child dose}$.

  a. If an adult dose of acetaminophen is 1000 milligrams and a child weighs no more than 90 pounds, what is the recommended child's dose?

  b. The label below appears on a child's cold medicine. What is the adult minimum dosage in milliliters?

Weight (lb)	Age (yr)	Dose
under 48	under 6	call a doctor
48-95	6-11	2 tsp or 10 mL

  c. What is the maximum adult dosage in milliliters?

39. CONSTRUCT ARGUMENTS Eric says that 15 more than 6 times the number of pencils he has is less than 20. What can you conclude about the number of pencils Eric has? Justify your argument.

40. REASONING The perimeter of a rectangular playground can be no greater than 120 meters. The width of the playground cannot exceed 22 meters. What are the possible lengths of the playground?

41. STRUCTURE Solve $10n - 7(n + 2) > 5n - 12$. Explain each step in your solution.

Higher Order Thinking Skills

42. WRITE What is the solution set of the inequality $2(2x + 4) < 4(x + 1)$? Why? How is the solution set related to the solution set of $2(2x + 4) \geq 4(x + 1)$? Explain.

43. PERSEVERE Mei got scores of 76, 80, and 78 on her last three history exams. Write and solve an inequality to determine the score she needs on the next exam so that her average is at least 82.

44. ANALYZE A triangular carpet has sides of length a feet, b feet, and c feet. The maximum perimeter is 20 feet.

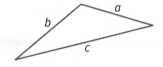

a. Side b is 2 feet longer than a, and c is 2 feet longer than b. Which side is the shortest? Explain.

b. What are the possible lengths of the shortest side of the carpet? Explain.

45. CREATE Write an inequality that has the solution set graphed at the right. Solving the inequality should require the Distributive Property, the Addition Property of Inequalities, and the Division Property of Inequalities.

46. PERSEVERE Let $b > 2$. Describe how you would determine if $ab > 2a$.

47. CREATE Four times the number of baseball cards in Ted's collection is more than five times that number minus 15. Define a variable and write an inequality to represent Ted's baseball cards. Solve the inequality and interpret the results.

48. WRITE Explain how you could solve $-3p + 7 \geq -2$ without multiplying or dividing each side by a negative number.

49. PERSEVERE If $ax + b < ax + c$ is true for all real values x, what will be the solution of $ax + b > ax + c$? Explain your reasoning.

50. WHICH ONE DOESN'T BELONG? Name the inequality that does not belong. Justify your conclusion.

| $4y + 9 > -3$ | $3y - 4 > 5$ | $-2y + 1 < -5$ | $-5y + 2 < -13$ |

51. WRITE Explain when the solution set of an inequality will be the empty set or the set of all real numbers. Show an example of each.

52. PERSEVERE Solve each inequality in terms of a. Assume that a does not equal 0.

a. $ax + 5 < 11$

b. $ax - 4 \geq 12$

c. $ax - 5 > 3a$

d. $ax + 1 \geq \frac{a}{2}$

Lesson 6-3

Solving Compound Inequalities

Explore Guess the Range

⟲ **Online Activity** Use a real-world situation to complete the Explore.

> **INQUIRY** How can you tell if a value will satisfy a compound inequality that includes the word *and*?

Today's Goals
- Solve and graph linear inequalities containing the word *and*.
- Solve and graph linear inequalities containing the word *or*.

Today's Vocabulary
compound inequality
intersection
union

Learn Solving Compound Inequalities Using the Word *and*

A **compound inequality** is two or more inequalities that are connected by the word *and* or *or*. A compound inequality containing the word *and* is only true when both of the inequalities are true. So, its graph is where the graphs of the inequalities overlap. This overlapping section that represents the compound inequality is called the **intersection**. To determine where the graphs intersect, graph each inequality and identify where they overlap.

$x > -4$

$x \leq 3$

$x > -4$ and $x \leq 3$
$-4 < x \leq 3$

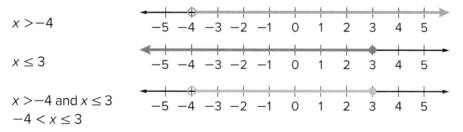

The compound inequality $-4 < x \leq 3$ can be read in two ways. It can be read as *x is greater than -4 and less than or equal to 3* or *x is between -4 and 3 including 3*.

Study Tip

Inequality Solutions For inequalities using the word *and*, a number has to be true for both inequalities in order to be a solution for the compound inequality.

Example 1 Solve and Graph an Intersection

Solve $-8 \leq h - 2 < 1$. Then graph the solution set.

Express the compound inequality as two inequalities joined by the word *and*.

$-8 \leq h - 2$	and	$h - 2 < 1$
	Write the inequality using *and*.	
$-8 + 2 \leq h - 2 + 2$	Add 2 to each side.	$h - 2 + 2 < 1 + 2$
$-6 \leq h$	Simplify.	$h < 3$

⟲ **Go Online**
You can complete an Extra Example online.

Lesson 6-3 • Solving Compound Inequalities 357

The solution set is $\{h|-6 \leq h < 3\}$.

Graph the solution set on a number line.

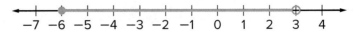

Check

Solve $-7 \leq 3x + 2 \leq 5$. Then graph the solution set.

Think About It!
What are 3 acceptable weights for a cereal box?

Study Tip
Assumptions In this example, we assume that all of the cereal boxes weigh approximately the same amount and that weighing a case of boxes is an accurate test of quality control for the manufacturer. However, it is possible that a case may contain cereal boxes that are heavier or lighter than the desired weight.

Study Tip
Inclusivity and Exclusivity In application problems, *within* implies inclusivity, which means the numbers mentioned will be included and the symbols $\leq$ or $\geq$ will be used. *Between* implies exclusivity and the symbols $<$ or $>$ will be used.

Example 2 Apply Compound Inequalities

MANUFACTURING A cereal manufacturer distributes cases of cereal to grocery stores. Each case contains 12 boxes of cereal. In order to pass the manufacturer's quality assurance test, the case must weigh between 336.4 ounces and 331.6 ounces, which includes 34 ounces for the weight of the case's cardboard box.

Part A
Write an inequality that describes the weight of a box of cereal.

$331.6 < 12b + 34 < 336.4$, where b is the weight of each box.

Part B Solve the inequality.

$331.6 < 12b + 34$	and	$12b + 34 < 336.4$
$297.6 < 12b$	Subtract 34 from each side.	$12b < 302.4$
$24.8 < b$	Divide each side by 12.	$b < 25.2$

Part C Graph the solution set on a number line.

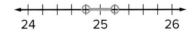

A cereal box must weigh between 24.8 and 25.2 ounces in order to pass the manufacturer's quality assurance test. The compound inequality is $\{b|24.8 < b < 25.2\}$.

Check

CARS Keshawn has been saving to buy his first car. He wants the total cost of the car and fees to be more than $5000 but at most $7000. The fees for buying a used car, such as title, registration, and dealership fees, will be $700.

Graph the list price of the cars Keshawn could buy.

Go Online You can complete an Extra Example online.

Learn Solving Compound Inequalities Using the Word *or*

A compound inequality containing the word *or* is true if at least one of the inequalities is true. A **union** is the graph of a compound inequality containing *or*; the solution is a solution of either inequality, not necessarily both.

$x \geq 1$

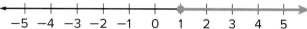

$x \leq -3$

$x \geq 1$ or $x \leq -3$

> **Go Online** You can watch a video to see how to solve inequalities involving *and* and *or*.
>
> **Study Tip**
>
> **Inequality Solutions** For inequalities using the word *or*, a number has to be true for at least one of the inequalities in order for it to be a solution for the compound inequality. The solution must work for the first inequality or the second inequality.

Example 3 Solve and Graph a Union

Solve $4n + 8 \leq 16$ or $-3n + 7 < -11$. Then graph the solution set.

Express the compound inequality as two inequalities joined by the word *or*.

$4n + 8 \leq 16$	or	$-3n + 7 < -11$
$4n + 8 - 8 \leq 16 - 8$	Subtract.	$-3n + 7 - 7 < -11 - 7$
$4n \leq 8$	Simplify.	$-3n < -18$
$\frac{4n}{4} \leq \frac{8}{4}$	Divide.	$\frac{-3n}{-3} > \frac{-18}{-3}$
$n \leq 2$	Simplify.	$n > 6$

Graph the solution set on a number line.

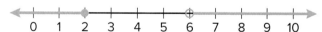

The union contains all points with coordinates less than or equal to 2 and all points with coordinates greater than 6. So, the solution set is $\{n \mid n \leq 2 \text{ or } n > 6\}$.

Go Online You can complete an Extra Example online.

Lesson 6-3 • Solving Compound Inequalities

Check

Solve $5x + 1 < 11$ or $-3x + 10 \leq -11$. Then graph the solution set.

Part A
Write the solution set for $5x + 1 < 11$ or $-3x + 10 \leq -11$.

Part B
Graph the solution set for $5x + 1 < 11$ or $-3x + 10 \leq -11$.

Talk About It!
Why does the union $4k + 12 < 2$ or $4 - 2k > 4$ include $-2\frac{1}{2}$ even though one of the solutions, $k < -2\frac{1}{2}$, does not?

Example 4 Overlapping Intervals

Solve $4k + 12 < 2$ or $4 - 2k > 4$. Then graph the solution set.

Express the compound inequality as two inequalities joined by the word *or*.

$4k + 12 < 2$	or	$4 - 2k > 4$
$4k + 12 - 12 < 2 - 12$	Subtract.	$4 - 2k - 4 > 4 - 4$
$4k < -10$	Simplify.	$-2k > 0$
$\frac{4k}{4} < \frac{-10}{4}$	Divide.	$\frac{-2k}{-2} < \frac{0}{-2}$
$k < \frac{-21}{2}$	Simplify.	$k < 0$

Graph the solution set on a number line.

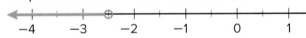

The graph of $k < -2\frac{1}{2}$ contains all points with coordinates less than $-2\frac{1}{2}$.

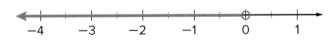

The graph of $k < 0$ contains all points with coordinates less than 0.

The union contains all points with coordinates less than 0.

Because $k < -2\frac{1}{2}$ is contained within $k < 0$, the solution set is $\{k | k < 0\}$.

Go Online You can complete an Extra Example online

Check

Solve $4m + 7 \geq 19$ or $-m + 5 \leq 0$. Then graph the solution set.

Part A
Write the solution set for $4m + 7 \geq 19$ or $-m + 5 \leq 0$.

Part B
Graph the solution set for $4m + 7 \geq 19$ or $-m + 5 \leq 0$.

Example 5 Write a Compound Inequality for an Intersection

Write a compound inequality that describes the graph.

The graph shows an interval between two numbers. Because a compound inequality with the word *and* represents the intersection of two inequalities, its graph shows the overlap as an interval.

Step 1

Analyze the leftmost endpoint of the interval. The endpoint is shown with a circle at -2, so -2 is not included in the solution. Points to the right of the endpoint are shaded, so the graph represents solutions of $x > -2$.

Step 2

Analyze the rightmost endpoint of the interval. The endpoint is shown with a dot at 4, so 4 is included in the solution. Points to the left of the endpoint are shaded, so the graph represents solutions of $x \leq 4$.

Step 3

The shaded interval represents the intersection of the solutions of $x > -2$ and $x \leq 4$, so the compound inequality $-2 < x \leq 4$ describes the graph.

Go Online You can complete an Extra Example online.

Check

Write a compound inequality that describes the graph.

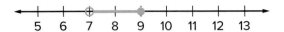

Example 6 Write a Compound Inequality for a Union

Write a compound inequality that describes the graph.

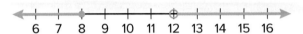

The graph shows the union of two inequalities. Because a compound inequality with the word *or* represents the union of two inequalities, its graph includes the graphs of both inequalities.

The leftmost endpoint is shown with a dot at 8, so 8 is included in the solution. Points to the left of the endpoint are shaded, so the graph represents solutions of $x \leq 8$.

The rightmost endpoint is shown with a circle at 12, so 12 is not included in the solution. Points to the right of the endpoint are shaded, so the graph represents solutions of $x > 12$.

The solutions represented on the graph represent the union of the solutions of $x \leq 8$ and $x > 12$, so the compound inequality $x \leq 8$ or $x > 12$ describes the graph.

Check

Write a compound inequality that describes the graph.

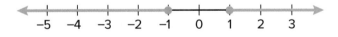

Go Online to practice what you've learned about solving linear inequalities in the Put It All Together over Lessons 6-1 through 6-3.

Go Online You can complete an Extra Example online

Practice

Go Online You can complete your homework online.

Examples 1, 3, and 4

Solve each compound inequality. Then graph the solution set.

1. $f - 6 < 5$ and $f - 4 \geq 2$

2. $n + 2 \leq -5$ and $n + 6 \geq -6$

3. $y - 1 \geq 7$ or $y + 3 < -1$

4. $t + 14 \geq 15$ or $t - 9 < -10$

5. $-5 < 3p + 7 \leq 22$

6. $-3 \leq 7c + 4 < 18$

7. $5h - 4 \geq 6$ and $7h + 11 < 32$

8. $22 \geq 4m - 2$ or $5 - 3m \leq -13$

9. $-y + 5 \geq 9$ or $3y + 4 < -5$

10. $-4a + 13 \geq 29$ and $10 < 6a - 14$

11. $3b + 2 < 5b - 6 \leq 2b + 9$

12. $-2a + 3 \geq 6a - 1 > 3a - 10$

13. $10m - 7 < 17m$ or $-6m > 36$

14. $5n - 1 < -16$ or $-3n - 1 < 8$

15. $m + 3 \geq 5$ and $m + 3 < 7$

16. $y - 5 < -4$ or $y - 5 \geq 1$

Example 2

17. **STORE SIGNS** In Randy's town, all stand-alone signs must be exactly 8 feet high. When mounted atop a pole, the combined height of the sign and pole must be less than 20 feet or greater than 35 feet so that they do not interfere with the power and phone lines.

    a. Write a compound inequality to represent the possible above-ground height of the poles, x.

    b. Solve the inequality. Explain any restrictions.

    c. Graph the inequality.

18. **HEALTH** The human heart circulates from 770,000 to 1,600,000 gallons of blood through a person's body every year.

    a. Write a compound inequality to represent the number of gallons of blood that the heart circulates through the body in one day, x.

    b. Solve the inequality. Round to the nearest whole gallon.

    c. Graph the inequality.

Lesson 6-3 • Solving Compound Inequalities 363

Examples 5 and 6

Write a compound inequality that describes each graph.

19. [number line from -4 to 4, open circle at -3, closed circle at 3]
20. [number line from -2 to 6, closed circle at 1, closed circle at 4]
21. [number line from -4 to 4, open circle at -2, closed circle at 1]
22. [number line from -4 to 4, open circle at -2, open circle at 2]
23. [number line from -4 to 4, closed circle at -1, open circle at 2]
24. [number line from -4 to 4, closed circle at -2, closed circle at 2]
25. [number line from -4 to 4, open circle at -1, closed circle at 1]
26. [number line from -3 to 5, closed circle at -2, open circle at 2]

Mixed Exercises

Solve each compound inequality. Then graph the solution set.

27. $4 < f + 6$ and $f + 6 < 5$

28. $w + 3 \leq 0$ or $w + 7 \geq 9$

29. $-6 < b - 4 < 2$

30. $p - 2 \leq -2$ or $p - 2 > 1$

31. $-5 \leq 2a - 1 < 9$

32. $-1 < 2x - 1 \leq 5$

Define a variable, write an inequality, and solve each problem. Check your solution.

33. A number decreased by two is at most four or at least nine.

34. The sum of a number and three is no more than eight or is more than twelve.

35. **WEATHER** Kenya saw this graph in the local weather forecast. It shows the predicted temperature range for the following day. Write an inequality to represent the number line.

36. **REASONING** The pH of a person's eyes is 7.2. Therefore, the ideal pH for the water in a swimming pool is between 7.0 and 7.6. Write a compound inequality to represent pH levels that could cause physical discomfort to a person's eyes.

37. **FIELD TRIP** It costs $1000 to rent a bus that holds 100 students. A school is planning to rent one of these buses for a field trip to an aquarium. The trip will also have a cost of $15 per student for the tickets to the aquarium. Given that the total expense for the trip must be between $2000 and $3000, find the minimum and maximum number of students who can go on the trip. Explain.

38. **HEALTH** Body mass index (BMI) is a measure of weight status. The BMI of a person over 20 years old is calculated using the following formula.

    $$\text{BMI} = 703 \times \frac{\text{weight in pounds}}{(\text{height in inches})^2}$$

    a. The recommended BMI for a person over 20 years old is 18.5–24.9. Write a compound inequality to represent the recommended BMI range.

    b. Write a compound inequality to represent the weight of an adult who is 6 feet tall that is within the recommended BMI range. Round to the nearest tenth if necessary.

39. **STATE YOUR ASSUMPTION** The Triangle Inequality states that in any triangle, the sum of the lengths of any two sides is greater than the length of the third side. In the figure, this means $a + b > c$, $a + c > b$, and $b + c > a$.

    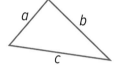

    a. Suppose a triangle has a side that is 5 meters long and a perimeter of 14 meters. Let one of the unknown sides be x. Write a compound inequality that you can use to determine the value of x. Explain.

    b. What assumption about the two unknown side lengths of the triangle can you make? Explain.

40. **CONSTRUCT ARGUMENTS** Bianca said that if k is a real number, then the solution set of the compound inequality $x < k$ or $x > k$ is all real numbers. Do you agree? Justify your argument.

41. **STRUCTURE** Write the solution set of the following compound inequality. Then graph the solution set. $-x + 1 < 8$ and $-x + 1 < 3$ and $x + 1 > -4$?

42. **PHYSICS** The density, in grams per milliliter, of a substance determines whether it will float or sink in a liquid. Any object with a greater density will sink and any object with a lesser density will float. Density is given by the formula $d = \frac{m}{v}$, where m is mass and v is volume. The table shows common chemical solutions and their densities. Plastics vary in density when they are manufactured; therefore, their volumes are variable for a given mass. A tablet of polystyrene (a manufactured plastic) sinks in 70% isopropyl alcohol solution and floats in calcium chloride solution. The tablet has a mass of 0.4 gram. Write an inequality to represent the range of values for v, the volume of the tablet.

Solution	Density (g/mL)
concentrated calcium chloride	1.40
70% isopropyl alcohol	0.92

    Source: American Chemistry Council

43. **COMPUTER SALE** Marietta is shopping during a computer store's sale. She is considering buying computers that range in cost from $500 to $1000.

    a. How much are the computers after the 20% discount?

    b. If sales tax is 7%, how much should Marietta expect to pay?

Higher-Order Thinking Skills

44. CREATE The figure shows the solution set of a compound inequality. Write a compound inequality that has the given solution set. Solving the inequality should require the Distributive Property, Addition Property of Inequalities, and the Division Property of Inequalities.

45. ANALYZE Which value of x is not a solution to $3x - 1 < 5$ or $7 - x \leq 3$?

A 0   B 2   C 4   D 5

46. FIND THE ERROR Sierra solved the compound inequality $-3x + 7x - 1 < -5$ or $-3x + 7x - 1 > 11$ as shown. What error did she make in solving the inequality?

Step	Property	Step	Property
$-3x + 7x - 1 < -5$	Original inequality	$-3x + 7x - 1 > 11$	Original inequality
$4x - 1 < -5$	Combine like terms.	$-4x - 1 > 11$	Combine like terms.
$4x < -4$	Addition Property	$-4x > 12$	Addition Property
$x < -1$	Division Property	$x < -3$	Division Property

Refer to the graphs for Exercises 47–49.

47. WRITE Write a compound inequality whose solution is the union of the two graphs. Then explain how the compound inequality can be expressed as a single inequality.

48. WRITE Write a compound inequality whose solution is the intersection of the two graphs. Then explain how the compound inequality can be expressed as a single inequality.

49. ANALYZE How are the graphs of the solution sets for the inequalities in Exercise 47 and Exercise 48 related to the given graphs?

50. PERSEVERE Jocelyn is planning to place a fence around the triangular flower bed shown. The fence costs $1.50 per foot. Assuming that Jocelyn spends between $60 and $75 for the fence, what is the shortest possible length for a side of the flower bed? Use a compound inequality to explain your answer.

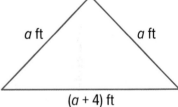

51. PERSEVERE Solve each inequality for x. Assume a is constant and $a > 0$

a. $-3 < ax + 1 \leq 5$ b. $-\frac{1}{a}x + 6 < 1$ or $2 - ax > 8$

52. CREATE Create an example of a compound inequality containing *or* that has infinitely many solutions.

53. ANALYZE Determine whether the following statement is *always*, *sometimes*, or *never* true. Justify your argument. *The graph of a compound inequality that involves an* or *statement is bounded on the left and right by two values of x.*

54. WRITE Give an example of a compound inequality you might encounter at an amusement park. Does the example represent an intersection or a union?

Lesson 6-4

Solving Absolute Value Inequalities

Explore Solving Absolute Value Inequalities

Online Activity Use a graph to complete the Explore.

INQUIRY How is solving an absolute value inequality similar to solving an absolute value equation?

Today's Goals
- Solve absolute value inequalities (<).
- Solve absolute value inequalities (>).

Learn Solving Inequalities Involving < and Absolute Value

For a real number a, the inequality $|x| < a$ means that the distance between x and 0 is less than a.

When solving absolute value inequalities, there are two cases to consider.

Case 1 The expression inside the absolute value symbols is nonnegative. If x is nonnegative, then $|x| = x$.

Case 2 The expression inside the absolute value symbols is negative. If x is negative, then $|x| = -x$.

Study Tip

Absolute Value Inequalities The inequality $|x|$ can be rewritten as $|x - 0| < a$, which is why it is read as the distance between x and 0 is less than a.

Example 1 Solve Absolute Value Inequalities (<)

Solve $|m + 5| < 3$. Then graph the solution set.

Part A Rewrite $|m + 5| < 3$ for Case 1 and Case 2.

Case 1 If $m + 5$ is nonnegative, $|m + 5| = m + 5$.

$$m + 5 < 3 \quad \text{Case 1}$$
$$m < -2 \quad \text{Subtract 5 from each side.}$$

Case 2 If $m + 5$ is negative, $|m + 5| = -(m + 5)$.

$$-(m + 5) < 3 \quad \text{Case 2}$$
$$-m - 5 < 3 \quad \text{Distributive Property}$$
$$-m < 8 \quad \text{Add 5 to each side.}$$
$$m > -8 \quad \text{Divide each side by } -1. \text{ Reverse the inequality symbol.}$$

So, $m < -2$ and $m > -8$. The solution set is $\{m \mid -8 < m < -2\}$.

(*continued on the next page*)

Watch Out

Absolute Value Cases Assigning the correct inequality symbol in Case 2 of an absolute value inequality can be confusing. Think of an inequality like $|x - 72| < 1.8$ as the distance from x to 72 is less than 1.8 units. Visualizing the graph of the inequality as an interval of 1.8 units on each side of the graph of 72 can help you ensure you have the correct symbol.

1.8 units 72 1.8 units

Part B Graph the solution set on a number line.

Check

Solve $|6m + 12| < 12$.
Graph the solution set.

Example 2 Absolute Value Inequalities (<) with No Solutions

Solve $|n - 1| < -5$. Then graph the solution set.

Because $|n - 1|$ is an absolute value expression, it cannot be negative. So it is not possible for $|n - 1|$ to be less than −5. Therefore, there is no solution, and the solution set is the empty set, ∅.

Example 3 Use Absolute Value Inequalities

SURVEY Jonas is a software developer who wants to determine whether the changes he made to his program are popular with users. He releases a survey to get some feedback and finds that 72% of users like the changes. The margin of error is within 1.8%, which means that with a reasonable level of certainty, the actual percentage can be said to fall within 1.8% of 72%.

Part A

Complete the table to write an inequality that represents the percent of users who like the changes.

Words	The difference between the actual percent and 72%	is less than or equal to	1.8%
Variable	Let x be the actual percent of users who like the changes.		
Inequality	$|x - 72|$	≤	1.8

Part B

Solve each case of the inequality.

Case 1 $x - 72$ is nonnegative.

$x - 72 \leq 1.8$ Case 1

$x \leq 73.8$ Add 72 to each side.

Go Online You can complete an Extra Example online.

Study Tip

Absolute Value as Distance The solution set for $|n - 1| < -5$ might be easier to understand if you think of the inequality in terms of distance. The inequality can be read as *the distance between a number and 1 is less than negative five.* However, that would make the distance a negative number. Because distance cannot be negative, the solution set is ∅.

Talk About It!

Is the solution set of $|n - 1| - 6 < -5$ also the empty set? Explain your reasoning.

Go Online An alternate method is available for this example.

368 Module 6 · Linear Inequalities

Case 2 $x - 72$ is negative.

$$-(x - 72) \leq 1.8 \qquad \text{Case 2}$$
$$-x + 72 \leq 1.8 \qquad \text{Distributive Property}$$
$$-x \leq -70.2 \qquad \text{Subtract 72 from each side.}$$
$$x \geq 70.2 \qquad \text{Divide each side by } -1.\text{ Reverse the inequality symbol.}$$

The percent of users who favor the changes Jonas made to his software is between 70.2% and 73.8%, so the solution set is $\{x \mid 70.2 \leq x \leq 73.8\}$. This solution set is a small interval of possible values close to the percent that Jonas found, so the solution set seems reasonable for the situation.

Learn Solving Inequalities Involving > and Absolute Value

For a real number a, the inequality $|x| > a$ means that the distance between x and 0 is greater than a.

When solving absolute value inequalities, there are two cases to consider.

Case 1 The expression inside the absolute value symbols is nonnegative. If x is nonnegative, $|x| = x$.

$$x > a \qquad \text{Case 1}$$

Case 2 The expression inside the absolute value symbols is negative. If x is negative, $|x| = -x$.

$$-x > a \qquad \text{Case 2}$$
$$\frac{-x}{-1} < \frac{a}{-1} \qquad \text{Divide each side by } -1.\text{ Reverse the inequality.}$$
$$x < -a \qquad \text{Simplify.}$$

The solution set is the union of the solutions to these two cases. So, $x > a$ or $x < -a$. The solution set is $\{x \mid x < -a \text{ or } x > a\}$.

> **Study Tip**
> **> and <** If an absolute value inequality involves > or ≥, the solution set uses the word *or*. If an absolute value inequality involves < or ≤, the solution set uses the word *and*.

Example 4 Solve Absolute Value Inequalities (>)

Solve $|2m - 9| > 13$. Then graph the solution set.

Part A Rewrite $|2m - 9| > 13$ for Case 1 and Case 2.

Case 1 If $2m - 9$ is nonnegative, $|2m - 9| = 2m - 9$.

$$2m - 9 > 13 \qquad \text{Case 1}$$
$$2m > 22 \qquad \text{Add 9 to each side.}$$
$$m > 11 \qquad \text{Divide each side by 2.}$$

(continued on the next page)

Go Online You can complete an Extra Example online.

Case 2 If $2m - 9$ is negative, $|2m - 9| = -(2m - 9)$.

$$-(2m - 9) > 13 \quad \text{Case 2}$$
$$-2m + 9 > 13 \quad \text{Distributive Property}$$
$$-2m > 4 \quad \text{Subtract 9 from each side.}$$
$$m < -2 \quad \text{Divide each side by } -2. \text{ Reverse the inequality.}$$

So, $m > 11$ or $m < -2$. The solution set is $\{m|\ m > 11 \text{ or } m < -2\}$.

Part B Graph the solution set on a number line.

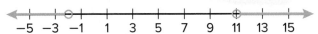

Check
Part A Solve $|4m - 20| \geq 12$.

Part B Graph the solution set.

Study Tip
Overlapping Case Solutions Because the values of n for $n \geq 1$ and $n \leq 11$ overlap and extend infinitely in both directions, the solution set is all real numbers.

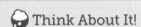 **Think About It!**
Why is the empty set the solution of $|n - 6| \leq -5$, but not the solution of $|n - 6| \geq -5$?

Example 5 Absolute Value Inequalities (>) with Overlapping Case Solutions

Solve $|n - 6| \geq -5$. Then graph the solution set.

Part A Rewrite $|n - 6| \geq -5$ for Case 1 and Case 2.

Case 1 $n - 6$ is nonnegative.

$$n - 6 \geq -5 \quad \text{Case 1}$$
$$n \geq 1 \quad \text{Add 6 to each side.}$$

Case 2 $n - 6$ is negative.

$$-(n - 6) \geq -5 \quad \text{Case 2}$$
$$-n + 6 \geq -5 \quad \text{Distributive Property}$$
$$-n \geq -11 \quad \text{Subtract 6 from each side.}$$
$$n \leq 11 \quad \text{Divide each side by } -1. \text{ Reverse the inequality.}$$

So, $n \geq 1$ or $n \leq 11$. The solution set is $\{n\,|\,n \geq 1 \text{ or } n \leq 11\}$, which is equivalent to $\{n\,|\,n \text{ is a real number}\}$.

Part B Graph the solution set on a number line.

 **Go Online** You can complete an Extra Example online.

Practice

Go Online You can complete your homework online.

Examples 1, 2, 4, 5

Solve each inequality. Then graph the solution set.

1. $|x + 8| < 16$
2. $|r + 1| \leq 2$
3. $|2c - 1| \leq 7$
4. $|3h - 3| < 12$
5. $|m + 4| < -2$
6. $|w + 5| < -8$
7. $|r + 2| > 6$
8. $|k - 4| > 3$
9. $|2h - 3| \geq 9$
10. $|4p + 2| \geq 10$
11. $|5v + 3| > -9$
12. $|-2c - 3| > -4$
13. $|4n + 3| \geq 18$
14. $|5t - 2| \leq 6$
15. $\left|\dfrac{3h + 1}{2}\right| < 8$
16. $\left|\dfrac{2p - 8}{4}\right| \geq 9$
17. $\left|\dfrac{7c + 3}{2}\right| \leq -5$
18. $\left|\dfrac{2g + 3}{2}\right| > -7$
19. $|-6r - 4| < 8$
20. $|-3p - 7| > 5$

Example 3

21. **SPEEDOMETERS** The government requires speedometers on cars sold in the United States to be accurate within ±2.5% of the actual speed of the car. If your speedometer meets this requirement, find the range of possible actual speeds at which your car could be traveling when your speedometer reads 60 miles per hour.

22. **BAKING** Pablo is making muffins for a bake sale. Before he starts baking, he goes online to research different muffin recipes. The recipes that he finds all specify baking temperatures between 350°F and 400°F, inclusive. Write an absolute value inequality to represent the possible temperatures *t* called for in the muffin recipes Pablo is researching.

23. **PAINT** A manufacturer claims that their cans of paint contain exactly 130 fluid ounces of paint. The amount of paint in each can of paint must be accurate within ±3.05 fluid ounces of the actual amount of paint.

    a. Write an absolute value inequality to represent the possible amount of paint, in fluid ounces, *p* for which the manufacturer's claim is correct.

    b. Graph the solution set of the inequality you wrote in part **a**.

24. **CATS** During a recent visit to the veterinarian's office, Mrs. Vasquez was informed that a healthy weight for her cat is approximately 10 pounds, plus or minus one pound. Write an absolute value inequality that represents unhealthy weights *w* for her cat.

25. **STATISTICS** In a recent year, the mean score on the mathematics section of the SAT test was 515 and the standard deviation was 114. This means that people within one deviation of the mean have SAT math scores that are no more than 114 points higher or 114 points lower than the mean.

   a. Write an absolute value inequality to find the range of SAT mathematics test scores within one standard deviation of the mean.

   b. What is the range of SAT mathematics test scores ±2 standard deviation from the mean?

Mixed Exercises

REGULARITY Write an open sentence involving absolute value for each graph.

26.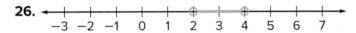

27.

28.

29.

30.

31.

REASONING Match each open sentence with the graph of its solution set.

32. $|x| > 2$

33. $|x - 2| \leq 3$

34. $|x + 1| < 4$

35. $|-x + 1.5| < 3$

a.

b.

c.

d.

USE A MODEL Express each statement using an inequality involving absolute value. Then solve and graph the absolute value inequality.

36. The meteorologist predicted that the temperature would be within 3° of 52°F.

37. Serena will make the B team if she scores within 8 points of the team average of 92.

38. The dance committee expects attendance to number within 25 of last year's 87 students.

Solve each inequality. Then graph the solution set.

39. $\left|\frac{x-1}{2}\right| \leq 1$

40. $|2x - 1| \geq 3$

41. $\left|\frac{x+3}{3}\right| \leq 2$

42. $|x + 7| \geq 4.5$

43. $\left|\frac{2x-1}{7}\right| > 5$

44. $|-4x - 2| < 10$

45. **CONSTRUCT ARGUMENTS** Is the solution to this inequality $|x - 2| > -1$ all real numbers? Justify your argument.

46. **USE TOOLS** Forensic scientists use the equation $h = 2.4f + 46.2$ to estimate the height h of a woman given the length in centimeters f of her femur bone. Suppose the equation has a margin of error of 3 centimeters. Could a female femur bone measuring 47 centimeters be that of a woman who was 170 centimeters tall?

47. **REGULARITY** A box of cereal should weigh 516 grams. The quality control inspector randomly selects boxes to weigh. The inspector sends back any box that is not within 4 grams of the ideal weight.

    a. Explain how to write an absolute value inequality to represent this situation.

    b. Explain the steps to solve this inequality. What do the solutions represent?

48. **REASONING** Write a compound inequality in which the solution set is the given set.

    a. $\{x \mid 4 \leq x\}$

    b. $\{4\}$

49. **ANALYZE** Determine if the open sentence $|x - 2| > 4$ and the compound inequality $-2x < 4$ or $x > 6$ have the same solution set.

50. **ARCHITECTURE** An architect is designing a house for the Frazier family. In the design, she must consider the desires of the family and the local building codes. The rectangular lot on which the house will be built is 158 feet long and 90 feet wide.

    a. The building codes state that one can build no closer than 20 feet to the lot line. Write an inequality to represent the possible widths of the house along the 90-foot dimension. Solve the inequality.

    b. The Fraziers requested that the rectangular house contain no less than 2800 square feet and no more than 3200 square feet of floor space. If the house has only one floor, use the maximum value for the width of the house from part a, and explain how to use an inequality to find the possible lengths.

    c. The Fraziers have asked that the cost of the house be about $175,000 and are willing to deviate from this price no more than $20,000. Write an open sentence involving an absolute value and solve. Explain the meaning of the answer.

Higher-Order Thinking Skills

51. FIND THE ERROR Jordan and Chloe are solving $|x + 3| > 10$.

Jordan
$\|x + 3\| > 10$
$x + 3 > 10$ $-(x + 3) > 10$
$x > 7$ $-x - 3 > 10$
$-x > 13$
$x < -13$

Chloe
$\|x + 3\| > 10$
$x + 3 > 10$ $-(x + 3) > 10$
$x > 7$ $-x + 3 > 10$
$-x > 7$
$x < -7$

Is either correct? Explain your reasoning.

52. CREATE Write an absolute value inequality using the numbers 3, 2, and −7. Then solve the inequality.

53. PERSEVERE Solve $2 < |n + 1| \leq 7$. Explain your reasoning and graph the solution set.

54. ANALYZE Which of the following inequalities could be represented by the graph?

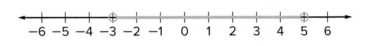

I. $|m - 1| < 4$ **II.** $3x < 15$ or $-x < 1$ **III.** $1 < 2k + 7 < 17$

A. I only   **B.** I and II   **C.** I and III   **D.** II and III

55. FIND THE ERROR Lucita sketched a graph of her solution to $|2a - 3| > 1$. Is she correct? Explain your reasoning.

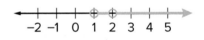

56. ANALYZE The graph of an absolute value inequality is *sometimes*, *always*, or *never* the union of two graphs. Explain.

57. ANALYZE Determine why the solution of $|t| > 0$ is not all real numbers. Explain your reasoning.

58. WRITE How are symbols used to represent mathematical ideas? Use an example to justify your reasoning.

59. WRITE Explain how to determine whether an absolute value inequality uses a compound inequality with *and* or a compound equality with *or*. Then summarize how to solve absolute value inequalities.

60. WHICH ONE DOESN'T BELONG? Which inequality does not belong? Justify your conclusion.

| $\|x + 4\| - 7 \geq 3$ | $\|-6x - 1\| \leq \frac{1}{2}$ | $-2\|10x + 4\| < 6$ | $\|3x + 5\| < -\frac{3}{5}$ |

Lesson 6-5
Graphing Inequalities in Two Variables

Explore Graphing Linear Inequalities on the Coordinate Plane

Online Activity Use graphing technology to complete the Explore.

> **INQUIRY** How is graphing a linear inequality on the coordinate plane similar to and different from graphing on the number line?

Learn Graphing Linear Inequalities in Two Variables

The graph of a linear inequality represents the set of all points that are solutions of the inequality.

The edge of the graph is a **boundary**. Depending on the inequality, the boundary will or will not be included in the solution set.

The boundary divides the coordinate plane into regions called **half-planes**.

When the boundary is included, the solution of the linear inequality is a **closed half-plane**.

When the boundary is not included, it is an **open half-plane**.

Key Concept • Graphing Linear Inequalities	
Step 1	Graph the boundary. Use a solid boundary when the inequality contains $\leq$ or $\geq$. Use a dashed boundary when the inequality contains $<$ or $>$.
Step 2	Use a test point to determine which half-plane should be shaded.
Step 3	Shade the half-plane that contains the solution.

Example 1 Graph an Inequality with an Open Half-Plane

Graph $3x - 2y < 8$.

Step 1 Graph the boundary.

$3x - 2y < 8$ Original inequality

$-2y < -3x + 8$ Subtract $3x$ from each side.

$y > \frac{3}{2}x - 4$ Divide each side by -2.

(continued on the next page)

Today's Goals
- Graph the solutions of linear inequalities in two variables.

Today's Vocabulary
boundary
half-plane
closed half-plane
open half-plane

Go Online
You can watch a video to see how to graph inequalities in two variables.

Watch Out!
Selecting a Test Point When selecting a test point to use, make sure that the point does not lie on the boundary. While using the point (0, 0) will make calculations easier, you cannot use that point if the boundary passes through the origin. Instead, try using (1, 1) or (0, 1).

> **Think About It!**
> What does the shaded area of this graph represent?

Step 2 Use a test point. Select (0, 0) as a test point.

$$3x - 2y < 8 \quad \text{Original inequality}$$
$$3(0) - 2(0) < 8 \quad x = 0 \text{ and } y = 0$$
$$0 < 8 \quad \text{Simplify.}$$

Step 3 Shade the half-plane.
Because the test point is a solution of the inequality, shade the half-plane containing the test point.

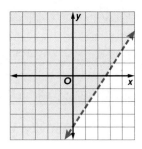

Example 2 Graph an Inequality with a Closed Half-Plane

Graph $3x + 4y \leq 0$.

Step 1 Graph the boundary.

$$3x + 4y \leq 0 \quad \text{Original inequality}$$
$$4y \leq -3x \quad \text{Subtract } 3x \text{ from each side.}$$
$$y \leq \frac{-3}{4}x \quad \text{Divide each side by 4.}$$

Step 2 Use a test point. Use (1, 1) as a test point.

$$3x + 4y \leq 0 \quad \text{Original inequality}$$
$$3(1) + 4(1) \leq 0 \quad x = 1 \text{ and } y = 1$$

7 is not less than or equal to 0   Simplify.

Step 3 Shade the half-plane.
Because the test point *is not* a solution of the inequality, shade the half-plane that does not contain the test point.

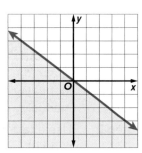

Check

Graph the inequalities.

a. $2x + y < -4$
b. $x - 2y > -4$

> **Go Online** You can watch a video to to see how to use a graphing calculator with this example.

Go Online You can complete an Extra Example online.

Example 3 Apply Graphing Inequalities in Two Variables

REFRESHMENTS Dominique can spend up to $20 to provide the dance squad with drinks after their practice. A bottle of water costs $0.80, and a sports drink costs $1.25. How many bottles of water and sports drinks can Dominique buy for the dance squad?

Step 1 Write an inequality.

Words	$0.80 times the number of bottles of water	plus	$1.25 times the number of sports drinks	is less than or equal to	$20
Variables	Let x = the number of bottles of water and y = the number of sports drinks that Dominique can buy.				
Inequality	$0.8 \cdot x$	+	$1.25 \cdot y$	≤	$20

Step 2 Solve the inequality for y.

$0.8x + 1.25y \leq 20$ Original inequality

$1.25y \leq -0.8x + 20$ Subtract $0.8x$ from each side.

$y \leq -0.64x + 16$ Divide each side by 1.25.

Step 3 Graph the inequality.
Because Dominique cannot buy a negative number of drinks, negative values of x and y are nonviable options. So the domain and range must be nonnegative numbers. Graph the boundary.

x	y
0	16
10	9.6
15	6.4
25	0

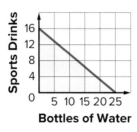

The test point (0, 0) is a solution of the inequality. Shade the closed half-plane that includes (0, 0).

Step 4 Interpret the solution in the context of the situation.

Notice that there are infinitely many solutions of the inequality. Because buying fractional bottles of water or sports drinks is not reasonable, only the solutions in which both x and y are whole numbers are viable. One viable solution is 10 bottles of water and 8 sports drinks.

Problem-Solving Tip

Use a Graph You can use a graph to visualize data, analyze trends, and make predictions.

Study Tip

Specifying Units Because we assigned the variables for the different types of drinks, it is critical to label the axes.

Talk About It!

Are there any viable solutions in which Dominique spends a total of $20? If so, where do those solutions appear on the graph?

Example 4 Solve Linear Inequalities

Graph $-4x + 7 \geq 11$.

Step 1 Graph the boundary.

$-4x + 7 \geq 11$ Original inequality

$-4x \geq 4$ Subtract 7 from each side.

$x \leq -1$ Divide each side by -4. Reverse the inequality.

Step 2 Use a test point. Use (0, 0) as a test point.

$-4x + 7 \geq 11$ Original inequality

$-4(0) + 7 \geq 11$ $x = 0$ and $y = 0$

$7 \ngeq 11$ Simplify.

Step 3 Shade the half-plane.

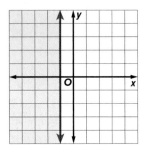

Since the test point *is not* a solution of the inequality, shade the half-plane that does not contain the test point.

Check

The graph of $y = 3x - 4$ is shown.

Consider the solutions of $y > 3x - 4$. Copy and complete the table to write each point in the appropriate column.

(−5, −3)   (−3, 4)
(0, −4)   (0, 0)
(1, −7)   (1, 1)
(2, 2)   (4, 2)

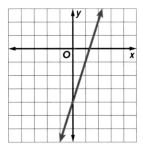

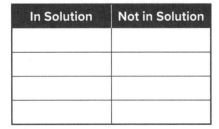

In Solution	Not in Solution

Go Online You can complete an Extra Example online.

Practice

Go Online You can complete your homework online.

Examples 1 and 2

Graph each inequality.

1. $y < x - 3$
2. $y > x + 12$
3. $y \geq 3x - 1$
4. $y \leq -4x + 12$
5. $6x + 3y > 12$
6. $2x + 2y < 18$
7. $5x + y > 10$
8. $2x + y < -3$
9. $-2x + y \geq -4$
10. $8x + y \leq 6$
11. $10x + 2y \leq 14$
12. $-24x + 8y \geq -48$

Example 3

13. **INCOME** In 2006 the median yearly family income was about $48,200 per year. Suppose the average annual rate of change since then is $1240 per year.

    a. Write and graph an inequality for the annual family incomes y that are less than the median for x years after 2006.

    b. Determine whether each of the following points is part of the solution set.

    (2, 51,000)   (8, 69,200)   (5, 50,000)   (10, 61,000)

14. **FUNDRAISING** Troop 200 sold cider and donuts to raise money for charity. They sold small boxes of donut holes for $1.25 and cider for $2.50 a gallon. In order to cover their expenses, they needed to raise at least $100. Write and graph an inequality that represents this situation.

Example 4

Graph each inequality.

15. $2y + 6 \geq 0$
16. $\frac{1}{2}x + 1 < 3$
17. $\frac{2}{3}x - \frac{10}{3} > -4$

Mixed Exercises

Graph each inequality.

18. $y < -1$
19. $y \geq x - 5$
20. $y > 3x$
21. $y \leq 2x + 4$
22. $y + x > 3$
23. $y - x \geq 1$

24. Kumiko has a $50 gift card for a Web site that sells apps and games. Games cost $2.50 each, and apps cost $1.25 each.

    a. Write an inequality that represents the number of apps a and the number of games g that Kumiko can buy and describe any constraints.

    b. Graph the solution of the inequality on a coordinate plane.

    c. Use your graph to find three different combinations of apps and games that Kumiko can buy.

    d. Kumiko decides to buy the same number of apps and games, and she decides to spend as much of the $50 as possible. How many apps and games does she buy? How much money does she have left on her gift card?

Lesson 6-5 • Graphing Inequalities in Two Variables

25. **USE A MODEL** A café sells peach smoothies and berry smoothies. The café makes a profit of $2.25 for each peach smoothie that is sold and a profit of $2 for each berry smoothie that is sold. The owner of the café wants to make a total profit of more than $90 per day from the sales of smoothies.

   a. Write an inequality that represents the number of peach smoothies p and berry smoothies b that the café needs to sell. Describe the constraints on the variables.

   b. Graph the solution of the inequality on a coordinate plane.

   c. On Monday, the café sold 20 peach smoothies and made the daily profit goal. What can you say about the number of berry smoothies that were sold on Monday?

   d. On Tuesday, the café made the daily profit goal by selling the minimum number of smoothies. How many smoothies did they sell? Explain.

26. **REASONING** Oleg is training for a triathlon. One day, he jogged for 2 hours at x miles per hour. Then he bicycled for 2 hours at y miles per hour. Finally, he swam a distance of 2 miles. The total number of miles did not exceed 30 miles.

   a. Write an inequality to represent the distance that he traveled that day. Describe the constraints on the variables.

   b. Graph the solution of the inequality on the coordinate plane shown. Label the axes with a description of the quantity that each axis represents. Include the unit of measure.

   c. What is the greatest possible speed that Oleg could have bicycled that day? How do you know?

27. **CONSTRUCT ARGUMENTS** The solution of the inequality $ax + by < c$ is a half-plane that includes the point (0, 0). What conclusion can you make about the value of c? Justify your argument.

Higher-Order Thinking Skills

28. **FIND THE ERROR** Reiko and Kristin are solving $4y \leq \frac{8}{3}x$ by graphing. Is either of them correct? Explain your reasoning.

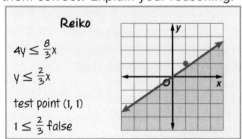

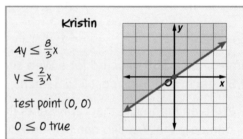

29. **CREATE** Write a linear inequality for which (–1, 2), (0, 1), and (3, –2) are solutions but (1, 1) is not.

30. **ANALYZE** Explain why a point on the boundary should not be used as a test point.

31. **CREATE** Write a two-variable inequality with a restricted domain and range to represent a real-world situation. Give the domain and range, and explain why they are restricted.

32. **WRITE** Summarize the steps to graph an inequality in two variables.

Module 6 • Linear Inequalities

Review

Essential Question
How can writing and solving inequalities help you solve problems in the real world?

Module Summary

Lessons 6-1, 6-2
Solving One-Step and Multi-Step Inequalities
- A solution set can be graphed on a number line. If the endpoint is not included in the solution, use a circle; if the endpoint is included, use a dot.
- If a number is added to or subtracted from each side of a true inequality, the resulting inequality is also true.
- If each side of a true inequality is multiplied or divided by the same positive number, the resulting inequality is also true.
- If each side of a true inequality is multiplied or divided by the same negative number, the direction of the inequality symbol must be changed to make the resulting inequality true.
- Multi-step inequalities can be solved by undoing the operations in the same way you would solve a multi-step equation.

Lesson 6-3
Solving Compound Inequalities
- To determine the solution set of a compound inequality, graph each inequality and identify where they overlap.
- If a compound inequality contains *and*, the overlapping section that represents the compound inequality is an intersection.
- If a compound inequality contains *or*, its graph is a union; the solution is a solution of either inequality, not necessarily both.

Lesson 6-4
Solving Absolute Value Inequalities
- For a real number a, the inequality $|x| < a$ means the distance between x and 0 is less than a. The inequality $|x| > a$ means the distance between x and 0 is greater than a.
- When solving absolute value inequalities, there are two cases to consider. The first case is when the expression inside the absolute value symbols is nonnegative. The second case is when the expression inside the absolute value symbols is negative. The solution set is the intersection of the solutions of their union.

Lesson 6-5
Graphing Linear Inequalities in Two Variables
- To graph a linear Inequality, graph the boundary. Use a solid boundary when the inequality contains $\leq$ or $\geq$. Use a dashed boundary when the inequality contains $<$ or $>$. Then use a test point to determine which half-plane should be shaded. Finally, shade the half-plane that contains the solution.

Study Organizer

Foldables
Use your Foldable to review the module. Working with a partner can be helpful. Ask for clarification of concepts as needed.

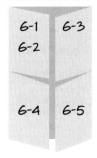

Test Practice

1. **MULTIPLE CHOICE** Select the graph that shows the solution set of $7 \leq n + 5$. (Lesson 6-1)

   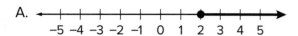

2. **OPEN RESPONSE** Eduardo is writing a historical novel. He wrote 16 pages today, bringing his total number of pages written to more than 50. How many pages p did Eduardo write before today? Complete the inequality that represents this situation. Then solve the inequality. (Lesson 6-1)

   Inequality: ____?____ $+ p >$ ____?____

   Solution: $p >$ ____?____

3. **OPEN RESPONSE** A farmer said that for every row of seeds he plants, he can harvest 6.5 bushels of tomatoes. The farmer needs to harvest at least 52 bushels of tomatoes. What is the least number of rows that the farmer will need to plant? (Lesson 6-1)

4. **OPEN RESPONSE** Find the solution set of $3d - 8 < 4d + 2$. (Lesson 6-1)

5. **MULTIPLE CHOICE** Which inequality has solutions represented by the graph? (Lesson 6-1)

   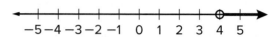

   A. $5 - x > 4$

   B. $2x + 1 \geq 9$

   C. $2x - 5 > 3$

   D. $6x - 7 > 5$

6. **MULTI-SELECT** Consider the inequality *Five plus two times a number n is less than or equal to eleven*. Select all of the representations that are solutions. (Lesson 6-2)

   A. $n \leq 3$

   B. $3 \leq n$

   C. $n \leq 8$

   D. $3 \geq n$

7. **OPEN RESPONSE** Solve $8(t + 2) + 7(t + 2) - 3(t - 2) < 0$. Write the solution using set-builder notation. (Lesson 6-2)

8. **MULTIPLE CHOICE** Solve $-\frac{4}{5}x - 3 \leq 17$. (Lesson 6-2)

   A. $\{x | x \geq -25\}$

   B. $\{x | x \geq -16\}$

   C. $\{x | x \leq -25\}$

   D. $\{x | x \leq -16\}$

9. MULTI-SELECT The science club is planning a car wash fundraiser. Halona writes and graphs an inequality to represent the number of cars c the science club needs to wash in order for the profits to cover their expenses. (Lesson 6-2)

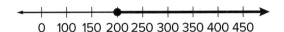

Which inequality could Halona have graphed for this situation? Select all that apply.

A. $2(c + 1) - 50 \geq 352$

B. $310 + c \leq 3(c - 30)$

C. $20(c - 1) \geq 4000$

D. $5(c + 1) - 200 \geq 250$

E. $50(c - 5) \geq 200$

10. OPEN RESPONSE Micaela wants to plant a square garden and enclose it with a fence. She has 120 feet of fencing available, and she wants the sides of her garden to be at least 6 feet long. Complete the inequality to represent the possible side lengths. (Lesson 6-3)

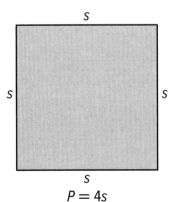

$P = 4s$

$\underline{\ ?\ } \leq s \leq \underline{\ ?\ }$

11. OPEN RESPONSE Solve $4 + g \geq 3$ or $6g \geq -30$. Write the solution using set-builder notation. (Lesson 6-3)

12. MULTIPLE CHOICE Which graph shows the solution of $3n - 1 < 5$ and $-2n + 3 < 11$? (Lesson 6-3)

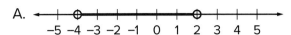

13. OPEN RESPONSE/GRAPH (Lesson 6-4)
Part A Solve $|h + 3| < 5$. Write the solution set using set-builder notation.

Part B Graph the solution set on a number line.

14. MULTIPLE CHOICE Which is NOT a true statement? (Lesson 6-4)

A. The empty set is the solution of $|k - 2| + 1 \leq -2$.

B. The solution of $|k - 2| \geq 2$ is $\{k | k \geq 4 \text{ or } k \leq 0\}$.

C. The solution of $|3k + 3| \leq -9$ is $\{k | -4 \geq k \geq 2\}$.

D. $|k - 7| < -6$ has no solution.

15. MULTIPLE CHOICE The actual weight of a jar of peanuts tends to be within 0.5 ounce of its listed weight. Which graph shows the possible weights of a jar of peanuts that has a listed weight of 6.5 ounces? (Lesson 6-4)

A.

B.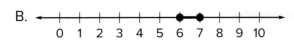

C.

D.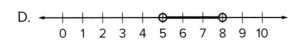

16. OPEN RESPONSE The equation for the boundary of the inequality graphed is $y = 3x - 1$. (Lesson 6-5)

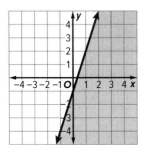

Write the inequality that represents the graph.

17. OPEN RESPONSE Consider the graphs. Determine which graph represents the solution to each of the inequalities. (Lesson 6-5)

Graphs:

A 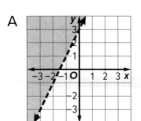 B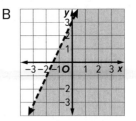

C 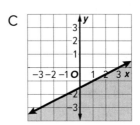 D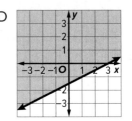

$2y - x \leq -3$

$2y - x \geq -3$

$2x - y > -3$

$2x - y < -3$

18. GRAPH Graph the inequality $\frac{2}{3}x + 5 - y < 6$. (Lesson 6-5)

Module 7
Systems of Linear Equations and Inequalities

Essential Question
How are systems of equations useful in the real world?

What Will You Learn?

How much do you already know about each topic **before** starting this module?

KEY
👎 — I don't know.   👌 — I've heard of it.   👍 — I know it!

	Before 👎	Before 👌	Before 👍	After 👎	After 👌	After 👍
solve systems of equations by graphing						
solve systems of equations by substitution						
solve systems of equations by elimination with addition						
solve systems of equations by elimination with subtraction						
solve systems of equations by elimination with multiplication						
solve systems of inequalities by graphing						

📘 **Foldables** Make this Foldable to help you organize your notes about expressions. Begin with one sheet of paper.

1. **Fold** lengthwise to the holes.
2. **Cut** 6 tabs.
3. **Label** the tabs using the lesson titles.

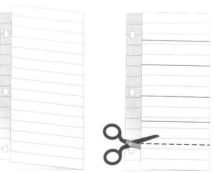

What Vocabulary Will You Learn?

- consistent
- dependent
- elimination
- inconsistent
- independent
- substitution
- system of equations
- system of inequalities

Are You Ready?

Complete the Quick Review to see if you are ready to start this module. Then complete the Quick Check.

Quick Review

Example 1

Name the ordered pair for Q on the coordinate plane.

Follow a vertical line from the point to the x-axis. This gives the x-coordinate, 3.

Follow a horizontal line from the point to the y-axis. This gives the y-coordinate, −2.

The ordered pair is (3, −2).

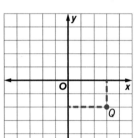

Example 2

Solve $12x + 3y = 36$ for y.

$12x + 3y = 36$ Original equation

$12x + 3y - 12x = 36 - 12x$ Subtract 12x from each side.

$3y = 36 - 12x$ Simplify.

$\frac{3y}{3} = \frac{36 - 12x}{3}$ Divide each side by 3.

$\frac{3y}{3} = \frac{36}{3} - \frac{12x}{3}$ Express as a difference.

$y = 12 - 4x$ Simplify.

Quick Check

Name the ordered pair for each point.

1. A
2. B
3. C
4. D

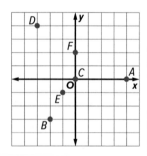

Solve each equation for the variable specified.

5. $2x + 4y = 12$, for x
6. $x = 3y - 9$, for y
7. $m - 2n = 6$, for m
8. $y = mx + b$, for x

How Did You Do?

Which exercises did you answer correctly in the Quick Check?

Lesson 7-1

Graphing Systems of Equations

Explore Intersections of Graphs

 Online Activity Use graphing technology to complete the Explore.

> **@ INQUIRY** How can you solve a linear equation by graphing?

Learn Graphs of Systems of Equations

A set of two or more equations with the same variables is called a **system of equations**. An ordered pair that is a solution of both equations is a solution of the system. A system of two linear equations can have one solution, an infinite number of solutions, or no solution.

- A system of equations is **consistent** if it has at least one ordered pair that satisfies both equations.

- If a consistent system of equations has exactly one solution, it is said to be **independent**. The graphs intersect at one point.

- If a consistent system of equations has an infinite number of solutions, it is **dependent**. The graphs are the same line. This means that there are unlimited solutions that satisfy both equations.

- A system of equations is **inconsistent** if it has no ordered pair that satisfies both equations. The graphs are parallel.

Example 1 Consistent Systems

Use the graph to determine the number of solutions the system has. Then state whether the system of equations is *consistent* or *inconsistent* and if it is *independent* or *dependent*.

$y = -3x + 1$
$y = x - 3$

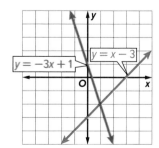

Since the graphs of these two lines intersect at one point, there is exactly one solution. Therefore, the system is consistent and independent.

Go Online You can complete an Extra Example online.

Today's Goals
- Determine the number of solutions of a system of linear equations.
- Solve systems of equations by graphing.
- Solve linear equations by graphing systems of equations.
- Use graphing calculators to solve systems of equations.

Today's Vocabulary
system of equations
consistent
independent
dependent
inconsistent

Go Online You may want to complete the Concept Check to check your understanding.

Lesson 7-1 • Graphing Systems of Equations **387**

Think About It!
How could you change the equations so they form a consistent and dependent system?

Example 2 Inconsistent Systems

Use the graph to determine the number of solutions the system has. Then state whether the system of equations is *consistent* or *inconsistent* and if it is *independent* or *dependent*.

$y = \frac{1}{2}x + 2$
$y = \frac{1}{2}x - 1$

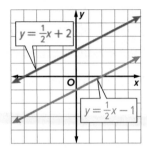

Since the graphs of these two lines are parallel, there is no solution of the system. Therefore, the system is inconsistent.

Check

Determine whether each graph shows a system that is *consistent* or *inconsistent* and if it is *independent* or *dependent*.

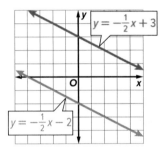 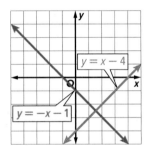

Example 3 Number of Solutions, Equations in Slope-Intercept Form

Determine the number of solutions the system has. Then state whether the system of equations is *consistent* or *inconsistent* and if it is *independent* or *dependent*.

$y = 6x + 10$ Because the slopes are the same and the y-intercepts are different, the lines are parallel.

$y = 6x + 4$ The system has no solution. Therefore, the system is inconsistent.

Check

Determine the number of solutions the system has.

$y = \frac{4}{5}x - 2$

$4y - 5x = 9$

Go Online You can complete an Extra Example online.

Study Tip
Parallel Lines Two lines that have the same slope are parallel and never intersect. A system of parallel lines has no solution.

Example 4 Number of Solutions, Equations in Standard Form

Determine the number of solutions the system has. Then state whether the system of equations is *consistent* or *inconsistent* and if it is *independent* or *dependent*.

$4y - 6x = 16$
$9x - 6y = -24$

Write both equations in slope-intercept form.

$4y - 6x = 16$	Original equation	$9x - 6y = -24$	
$4y - 6x + 6x = +6x + 16$	Isolate the *y*-term.	$9x - 6y - 9x = -9x - 24$	
$4y = 6x + 16$	Simplify.	$-6y = -9x - 24$	
$\frac{4y}{4} = \frac{6x}{4} + \frac{16}{4}$	Divide by coefficient of *y*.	$\frac{-6y}{-6} = \frac{-9x}{-6} + \frac{-24}{-6}$	
$y = \frac{3}{2}x + 4$	Simplify.	$y = \frac{3}{2}x + 4$	

Because the slopes are the same and the *y*-intercepts are the same, this is the same line.

Because the graphs of these two lines are the same, there are infinitely many solutions. Therefore, the system is consistent and dependent.

Check

Determine the number of solutions the system has.

$4x - 8y = 16$

$6x - 12y = 5$

> **Think About It!**
> How many solutions will a system have if the slopes are different?

Learn Solving Systems of Equations by Graphing

You can solve a system of equations by graphing each equation carefully on the same coordinate plane. Every point that lies on the line of one equation represents a solution of that equation. Similarly, every point on the line of the second equation in a system represents a solution of that equation. Therefore, the solution of a system of equations is the point at which the graphs intersect.

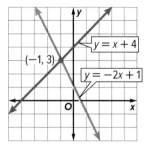

For example, the solution of this system is $(-1, 3)$. That is the point at which the graphs intersect. Since the point of intersection lies on both lines, the ordered pair satisfies each equation in the system.

Lesson 7-1 • Graphing Systems of Equations **389**

Think About It!
Why is it necessary to substitute the values of x and y into both equations to check your solution?

Example 5 Solve a System by Graphing

Graph the system and determine the number of solutions that it has. If it has one solution, determine its coordinates.

$y = -2x + 14$
$y = \frac{3}{5}x + 1$

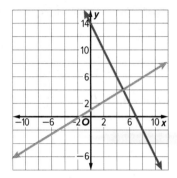

The graphs of the lines appear to intersect at the point (5, 4). If you substitute 5 for x and 4 for y into the equations, both are true. Therefore, (5, 4) is the solution of the system.

Check

Graph the system of equations.

$3x + 5y = 10$
$x - 5y = -10$

What is the solution of the system?

Go Online You can watch a video to see how to use a graphing calculator with this example.

Problem-Solving Tip

Tools When graphing by hand, using graph paper and a straightedge can help you make your graphs more accurate.

Example 6 Graph and Solve a System of Equations

Graph the system and determine the number of solutions that it has. If it has one solution, determine its coordinates.

$-3x + 2y = 12$
$6x - 4y = 8$

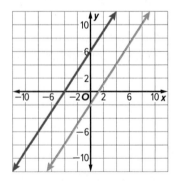

The lines have the same slope but different y-intercepts, so the lines are parallel. Since they do not intersect, this system has no solution.

Go Online You can complete an Extra Example online.

🌐 Apply Example 7 Write a System of Equations

POPULATION China and India are the two most populous countries in the world. The populations of these countries have increased steadily in recent years. In 2010, China and India had populations of about 1.34 billion and 1.19 billion, respectively. By 2016, the populations had grown to about 1.38 billion in China and 1.29 billion in India. Predict the approximate year when the populations of the two countries will be the same.

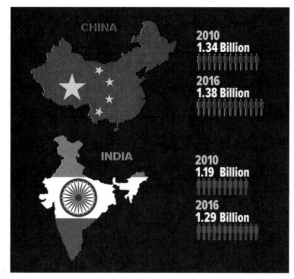

Source: IMF, CEIC

1. **What is the task?**
Describe the task in your own words. Then list any questions that you may have. How can you find answers to your questions?

I need to graph a system of equations that represents this situation and find the intersection. How can I write the system of equations that I need to graph? I can review finding the average rate of change and writing equations in slope-intercept form.

2. **How will you approach the task? What have you learned that you can use to help you complete the task?**

I will find the average rate of change for both countries. Then I will write a system of equations to represent the situation. I will graph the system and find the intersection. I will use what I have learned about graphing equations to help me graph the system.

3. **What is your solution?**
Use your strategy to solve the problem.
Find the average rate of change for the populations of China and India.

China: $\frac{1}{150}$

India: $\frac{1}{60}$

Write a system of equations to represent the situation. Let $x =$ the number of years after 2016. Let $y =$ population.

$y = \frac{1}{150}x + 1.38$

$y = \frac{1}{60}x + 1.29$

(continued on the next page)

Study Tip
Assumptions
Populations do not typically increase at a steady rate. However, assuming that the rate of increase is constant allows you to estimate future data.

Study Tip

Labeling Axes It is important to clearly define and label axes in a real-world situation. The intersection does not mean that the population will be the same in year 9. Since the x-axis represents the number of years after 2016, you must add the x-coordinate of the intersection to 2016 to find the year.

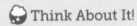 **Think About It!**

Can a value of x that is not a whole number be a viable solution? Justify your argument.

Graph the system of equations.

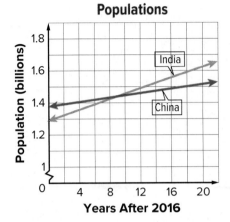

The graphs appear to intersect at (9, 1.44). So, approximately 9 years after 2016, or in 2025, the populations of China and India will be the same, about 1.44 billion.

4. How can you know that your solution is reasonable?

Write About It! Write an argument that can be used to defend your solution.

From the graph, I estimated that the solution was 9 years after 2016, or in 2025. My answer is reasonable because when I substitute 9 in for x in each equation, the value of y is 1.44.

Check

OLYMPICS The number of men and women participating in the Winter Olympic Games has been steadily increasing in recent years. In the 19th Winter Olympics, 1389 men and 787 women participated. 1660 men and 1121 women participated in the 22nd Winter Olympics.

Part A

Write and graph a system to describe the number of men and women participating if x represents the number of Winter Olympics after the 22nd Winter Olympics.

Part B

Use the graph to predict the Winter Olympics when the number of men and women participating will be the same.

Go Online You can complete an Extra Example online.

Learn Using Systems to Solve Linear Equations

Key Concept • Using Systems to Solve Linear Equations

Step 1 Write a system by setting each expression equal to y.
Step 2 Graph the system.
Step 3 Find the intersection.

> **Talk About It!**
> How do you know that the point of intersection satisfies both equations?

Example 8 Use a System to Solve a Linear Equation

Use a system of equations to solve $-6x + 8 = -4$.

Step 1 Write a system.

Write a system of equations. Set each side of $-6x + 8 = -4$ equal to y.

$y = -6x + 8$

$y = -4$

Step 2 Graph the system.

Enter the equations and graph.

Step 3 Find the intersection.

The solution is the x-coordinate of the intersection, 2.

Step 4 Check your solution.

$-6x + 8 = -4$	Original equation
$-6(2) + 8 \stackrel{?}{=} -4$	Substitution
$-12 + 8 \stackrel{?}{=} -4$	Multiply.
$-4 = -4$	Add.

> **Go Online** to see how to use a graphing calculator with this example.

Check

Use a system of equations and your graphing calculator to solve $-3.2x - 5.8 = 2.8x + 7$. Round to the nearest hundredth, if necessary.

 Go Online You can complete an Extra Example online.

Learn Solving Systems of Equations by Using Graphing Technology

You can use a graphing calculator to graph and solve a system of equations by following these steps.

Step 1 Isolate y in each equation.
Step 2 Graph the system.
Step 3 Find the intersection.

> **Go Online** You can watch a video to see how to graph systems of equations on a graphing calculator.

Lesson 7-1 • Graphing Systems of Equations 393

Example 9 Solve a System of Equations

Solve the system of equations.
$-1.38x - y = 5.13$
$0.62x + 2y = 1.60$

Step 1 Isolate y.

Solve each equation for y.

$-1.38x - y = 5.13$	First equation
$-y = 5.13 + 1.38x$	Add 1.38x to each side.
$y = -5.13 - 1.38x$	Multiply each side by -1.
$0.62x + 2y = 1.60$	Second equation
$2y = 1.60 - 0.62x$	Subtract 0.62x from each side.
$y = 0.80 - 0.31x$	Divide each side by 2.

Step 2 Graph the system.
Enter the equations and graph.

Step 3 Find the intersection.
The solution is approximately $(-5.54, 2.52)$.

Check

What is the solution to the system of equations?

$2.29x - 4.41y = 6.52$
$4.16x + 1.11y = 4.72$

🌐 Example 10 Write and Solve a System of Equations

BUSINESS Denzel is starting a food truck business to sell gourmet grilled cheese sandwiches. He has spent $34,000 on the truck, equipment, permits, and other start-up costs. Each sandwich costs about $1.32 to make, and he sells them for $7.

How many sandwiches does Denzel need to sell to start earning a profit?

Let x = the number of grilled cheese sandwiches sold.
Let y = total cost or revenue.

Total Cost: $y = 1.32x + 34{,}000$
Total Revenue: $y = 7x$

Step 1 Graph the system.
Enter the equations and graph.

Step 2 Find the intersection.
The solution is approximately $(5985.92, 41{,}901.41)$.
This means that after Denzel has sold 5986 sandwiches, he will begin to earn a profit.

🌐 **Go Online** You can complete an Extra Example online.

Go Online to see how to use a graphing calculator with this example.

394 Module 7 · Systems of Linear Equations and Inequalities

Practice

Examples 1 and 2

Use the graph to determine the number of solutions the system has. Then state whether the system of equations is *consistent* or *inconsistent* and if it is *independent* or *dependent*.

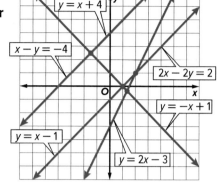

1. $y = x - 1$
   $y = -x + 1$

2. $x - y = -4$
   $y = x + 4$

3. $y = x + 4$
   $2x - 2y = 2$

4. $y = 2x - 3$
   $2x - 2y = 2$

Examples 3 and 4

Determine the number of solutions the system has. Then state whether the system of equations is *consistent* or *inconsistent* and if it is *independent* or *dependent*.

5. $y = \frac{1}{2}x$
   $y = x + 2$

6. $4x - 6y = 12$
   $-2x + 3y = -6$

7. $8x - 4y = 16$
   $-5x - 5y = 5$

8. $2x + 3y = 10$
   $4x + 6y = 12$

9. $y = -\frac{3}{2}x + 5$
   $y = -\frac{2}{3}x + 5$

10. $y = x - 3$
    $y = -4x + 3$

Examples 5 and 6

Graph each system and determine the number of solutions it has. If it has one solution, determine its coordinates.

11. $y = -3$
    $y = x - 3$

12. $y = 4x + 2$
    $y = -2x - 4$

13. $y = x - 6$
    $y = x + 2$

14. $x + y = 4$
    $3x + 3y = 12$

15. $x - y = -2$
    $-x + y = 2$

16. $2x + 3y = 12$
    $2x - y = 4$

Lesson 7-1 • Graphing Systems of Equations **395**

Example 7
17. **AVIATION** An air traffic controller manages the flow of aircraft in and out of airport airspace by guiding pilots during takeoff and landing. An air traffic controller is monitoring two planes that are in flight near a local airport. The first plane is at an altitude of 1000 meters and is ascending at a rate of 400 meters per minute. The second plane is at an altitude of 5900 meters and is descending at a rate of 300 meters per minute.

   a. Write and graph a system of equations that represents the altitude of each plane, where x is the amount of time, in minutes, and y is the altitude, in meters.

   b. Predict when the planes will be at the same altitude.

18. **STRUCTURE** Gustavo sets up tables for a caterer on weekends. Each round table can seat 8 people. Each rectangular table can seat 10 people. One weekend, his boss asked him to set up tables for 124 people. He uses 2 more round tables than rectangular tables. Define variables and write a system of equations to find the number of round tables and rectangular tables Gustavo used. Then solve the system graphically.

Example 8
Write and graph a system of equations to solve each linear equation.

19. $3x + 6 = 6$

20. $2x - 17 = x - 10$

21. $-12x + 90 = 30$

22. $13x - 28 = 24$

23. $2x + 5 = 2x + 5$

24. $x + 1 = x + 3$

Example 9
Solve each system of equations. State the decimal solution to the nearest hundredth.

25. $2.5x + 3.75y = 10.5$
    $1.25x - 8.5y = -5.25$

26. $2.2x + 1.8y = -3.6$
    $-4.8x + 12.4y = 10.6$

27. $1.12x - 2.24y = 4.96$
    $-3.56x - 2.48y = -7.32$

Example 10
28. **USE TOOLS** An office building has two elevators. One elevator starts out on the 4th floor, 35 feet above the ground, and is descending at a rate of 2.2 feet per second. The other elevator starts out at ground level and is rising at a rate of 1.7 feet per second. Write and solve a system of equations to determine when both elevators will be at the same height. Interpret the solution.

29. **USE TOOLS** A bookstore makes a profit of $2.50 on each book they sell, and $0.75 on each magazine they sell. One week, the store sold x books and y magazines, for a weekly profit of $450. The total number of publications sold that week was 260. Write and solve a system of equations to determine the number of books and magazines that the bookstore sold that week. Interpret the solution.

Mixed Exercises

Graph each system and determine the number of solutions it has. If it has one solution, determine its coordinates. Then state whether the system is *consistent* or *inconsistent* and if it is *independent* or *dependent*.

30. $2x - y = -3$
    $2x + y = -1$

31. $2x + y = 4$
    $y = -2x - 2$

32. $2y = 5 + x$
    $3x - 6y = -15$

33. $2x - y = 5$
    $x + y = -2$

34. $2y = 1.2x - 10$
    $4y = 2.4x$

35. $x = 6 - \frac{3}{8}y$
    $4 = \frac{2}{3}x + \frac{1}{4}y$

36. $x - y = 3$
    $x - 2y = 3$

37. $x + 2y = 4$
    $y = -\frac{1}{2}x + 2$

38. $y = 2x + 3$
    $3y = 6x - 6$

39. $y - x = -1$
    $x + y = 3$

40. **BUSINESS** The number of items sold at Store 1 can be represented by $y = 200x + 300$, where x represents the number of days and y represents the number of items sold. The number of items sold at Store 2 can be represented by $y = 200x + 100$, where x represents the number of days and y represents the number of items sold. Look at the graph of the system of equations and determine whether it has *no* solution, *one* solution, or *infinitely many* solutions.

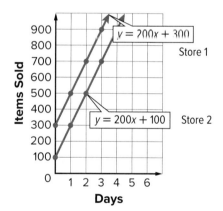

41. **USE A MODEL** Olivia and her brother William had a bicycle race. Olivia rode at a speed of 20 feet per second, while William rode at a speed of 15 feet per second. Olivia gave William a 150-foot head start, and the race ended in a tie.

    a. Define variables and write a system of equations to represent this situation.

    b. Graph the system of equations.

    c. How far away was the finish line from where Olivia started?

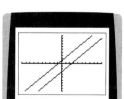

42. **CONSTRUCT ARGUMENTS** Abhijit uses a calculator to graph the system $y = 1.21x - 3$ and $5y - 5 = 6x$. He concludes that the system has no solution. Do you agree or disagree? Explain your reasoning.

Lesson 7-1 • Graphing Systems of Equations

43. **USE A MODEL** Some days, Cayley walks to school. Other days, she rides her bicycle. When she walks to school, she needs to leave home 15 minutes earlier than when she rides her bicycle. Cayley walked 3 days last week and rode her bike on 2 days. Cayley spent a total of 1 hour 10 minutes going to school last week.

    a. Define variables and write a system of equations to represent this situation.

    b. Graph the system of equations.

    c. How long does it take Cayley to walk to school?

44. **REGULARITY** Maureen says that if a system of linear equations has three equations, then there could be exactly two solutions to the system. Is Maureen correct? Use a graph to justify your reasoning.

Higher-Order Thinking Skills

45. **PERSEVERE** Use graphing to find the solution of the system of equations $2x + 3y = 5$, $3x + 4y = 6$, and $4x + 5y = 7$.

46. **ANALYZE** Determine whether a system of two linear equations with (0, 0) and (2, 2) as solutions *sometimes*, *always*, or *never* has other solutions. Justify your argument.

47. **WHICH ONE DOESN'T BELONG?** Which one of the following systems of equations doesn't belong with the other three? Justify your conclusion.

    | $4x - y = 5$ | $-x + 4y = 8$ | $4x + 2y = 14$ | $3x - 2y = 1$ |
    | $-2x + y = -1$ | $3x - 6y = 6$ | $12x + 6y = 18$ | $2x + 3y = 18$ |

48. **CREATE** Write three equations such that they form three systems of equations with $y = 5x - 3$. The three systems should be inconsistent, consistent and independent, and consistent and dependent, respectively.

49. **WRITE** Describe the advantages and disadvantages of solving systems of equations by graphing.

50. **CREATE** Write and graph a system of equations that has the following number of solutions. Then state whether the system of equations is *consistent* or *inconsistent* and if it is *independent* or *dependent*.

    a. one solution   b. infinite solutions   c. no solution

51. **FIND THE ERROR** Store A is offering a 10% discount on the purchase of all electronics in their store. Store B is offering $10 off all the electronics in their store. Francisca and Alan are deciding which offer will save them more money. Is either of them correct? Explain your reasoning.

    Francisca
    You can't determine which store has the better offer unless you know the price of the items you want to buy.

    Alan
    Store A has the better offer because 10% of the sale price is a greater discount than $10.

Lesson 7-2

Substitution

Explore Using Substitution

 Online Activity Use a system of equations to complete the Explore.

> **INQUIRY** How can you rewrite a system of equations as a single equation with only one variable?

Learn Solving Systems of Equations by Substitution

Exact solutions result when algebraic methods are used to solve systems of equations. One algebraic method is called **substitution**.

Key Concept • Substitution Method

Step 1 When necessary, solve at least one equation for one variable.

Step 2 Substitute the resulting expression from Step 1 into the other equation to replace the variable. Then solve the equation.

Step 3 Substitute the value from Step 2 into either equation, and solve for the other variable. Write the solution as an ordered pair.

Example 1 Solve a System by Substitution

Use substitution to solve the system of equations.

$3x - y = -7$; $y = 4x + 11$

Step 1 The second equation is already solved for y.

Step 2 Substitute $4x + 11$ for y in the first equation.

$3x - (4x + 11) = -7$	Substitute $4x + 11$ for y.
$3x - 4x - 11 = -7$	Distributive Property
$-x - 11 = -7$	Combine like terms.
$-x = 4$	Add 11 to each side.
$x = -4$	Multiply each side by -1.

Step 3 Substitute -4 for x in either equation to find y.

$y = 4x + 11$	Second equation
$y = 4(-4) + 11$	Substitution
$y = -5$	Simplify.

The solution is $(-4, -5)$.

 Go Online You can complete an Extra Example online.

Today's Goal
- Solve systems of equations by using the substitution method.

Today's Vocabulary
substitution

Think About It!
In algebra, what does it mean to *substitute*?

Think About It!
How would the substitution process differ if the second equation were $4x - y = -11$?

Go Online
You can watch a video to see how to use algebra tiles with this example.

Check

Refer to the system of equations.

$3x - 2y = -17$
$y = 2x + 2$

Part A Which expression could be substituted for y in the first equation to find the value of x?

A. $2x - 2$ B. $-\frac{3}{2}x + \frac{17}{2}$ C. $2x + 2$ D. $3x - 2y$

Part B What is the solution of the system?

A. $(-13, -24)$ B. $(-13, 24)$ C. $(13, -28)$ D. $(13, 28)$

Example 2 Solve and Then Substitute

Use substitution to solve the system of equations.

$5x - 3y = -25$
$x + 4y = 18$

Step 1 Solve the second equation for x since the coefficient is 1.

$x + 4y = 18$	Second equation
$x = 18 - 4y$	Subtract $4y$ from each side.

Step 2 Substitute $18 - 4y$ for x in the first equation.

$5x - 3y = -25$	First equation
$5(18 - 4y) - 3y = -25$	Substitute $18 - 4y$ for x.
$90 - 20y - 3y = -25$	Distributive Property
$90 - 23y = -25$	Combine like terms.
$-23y = -115$	Subtract 90 from each side.
$y = 5$	Divide each side by -23.

Step 3 Substitute 5 for y in either equation to find x.

$x + 4y = 18$	Second equation
$x + 4(5) = 18$	Substitute 5 for y.
$x + 20 = 18$	Simplify.
$x = -2$	Subtract 20 from each side.

The solution is $(-2, 5)$.

Check

Use substitution to solve the system of equations.

$5x + 3y = 5$
$x + 2y = -13$

 **Go Online** You can complete an Extra Example online.

 **Talk About It!**
Would the solution of the system be the same if the first step used was to solve the second equation for y? Explain.

Study Tip

Slope-Intercept Form If both equations are in the form $y = mx + b$, the expressions can simply be set equal to each other and then solved for x. For example, if $y = 2x - 5$ and $y = -4x + 1$, then $2x - 5 = -4x + 1$. The solution for x can then be used to find the value of y.

400 Module 7 • Systems of Linear Equations and Inequalities

Example 3 Use Substitution When There are No or Many Solutions

Use substitution to solve the system of equations.

$4x + 2y = -8$

$y = -2x - 4$

Substitute $-2x - 4$ for y in the first equation.

$4x + 2y = -8$	First equation
$4x + 2(-2x - 4) = -8$	Substitute $-2x - 4$ for y.
$4x - 4x - 8 = -8$	Distributive Property
$-8 = -8$	Simplify.

The equation $-8 = -8$ is an identity. Thus, there are an infinite number of solutions.

When graphed, the equations are the same line.

Check

Select the correct statement about the system of equations.

$-x + 2y = 2$

$y = \frac{1}{2}x + 1$

A. This system has no solution.

B. This system has one solution at $\left(\frac{2}{3}, \frac{4}{3}\right)$.

C. This system has one solution at $\left(\frac{4}{3}, \frac{2}{3}\right)$.

D. This system has infinitely many solutions.

🌎 Example 4 Write and Solve a System of Equations

TREE PRESERVATION A town ordinance defines an adult tree as having a diameter greater than 10 inches and a sapling as having a diameter less than 10 inches. The ordinance requires that on a new building project, two new trees are planted for each adult tree felled and six new trees are planted for each sapling felled. Last year, there were 167 trees felled, and the community planted 742 replacement trees. How many of each type of tree were felled?

(continued on the next page)

Study Tip

Dependent Systems There are infinitely many solutions of the system because the equations in slope-intercept form are equivalent, and they have the same graph.

💭 Think About It!

What would a solution like $-8 = 0$ mean? What would it look like on a graph?

Study Tip

Assumptions Using systems of equations to solve real-world problems generally requires assuming that the problem actually has a solution. It is a good idea to graph the lines first, to see if a solution exists, before going through the steps to solve for each variable.

Go Online
An alternate method is available for this example.

Use a Source
Research replacing trees cut for construction in an area near you. How could you find the number of trees to plant for a project similar to the one in this community?

Let a = the number of adult trees felled, and let t = the number of sapling trees felled.

$a + t = 167$	This equation represents the total number of adult trees a and sapling trees t felled with a sum of 167.
$2a + 6t = 742$	This equation represents the combinations of adult trees a and sapling trees t replaced with a total of 742.

The solution of the system of equations represents the option that meets both of the constraints.

The first equation can easily be solved for a or t.

Solve for a.

Step 1 Solve the first equation for a.

$a + t = 167$ First equation

$a + t - t = 167 - t$ Subtract t from each side.

$a = 167 - t$ Simplify.

Step 2 Substitute $167 - t$ for a in the second equation.

$2a + 6t = 742$ Second equation

$2(167 - t) + 6t = 742$ Substitute $167 - t$ for a.

$334 - 2t + 6t = 742$ Distributive Property

$4t + 334 = 742$ Combine like terms.

$4t = 408$ Subtract 334 from each side.

$t = 102$ Divide each side by 4.

Step 3 Substitute 102 for t in either equation to find the value of a.

$a + t = 167$ First equation

$a + 102 = 167$ Substitute 102 for t.

$a = 65$ Subtract 102 from each side.

The solution is (65 , 102).

There were 65 adult trees and 102 saplings felled. Because only whole numbers of trees can be felled, this is a viable solution.

Check

GEOMETRY Kymani has two equal-sized large pitchers and two equal-sized small pitchers. All of the pitchers together hold 40 cups of water. The capacity of one large pitcher minus the capacity of one small pitcher is 12 cups. How many cups can each type of pitcher hold?

Small pitcher = __?__ cups

Large pitcher = __?__ cups

Go Online You can complete an Extra Example online.

Practice

Go Online You can complete your homework online.

Examples 1–3

Use substitution to solve each system of equations.

1. $y = 5x + 1$
   $4x + y = 10$

2. $y = 4x + 5$
   $2x + y = 17$

3. $y = 3x - 34$
   $y = 2x - 5$

4. $y = 3x - 2$
   $y = 2x - 5$

5. $2x + y = 3$
   $4x + 4y = 8$

6. $3x + 4y = -3$
   $x + 2y = -1$

7. $y = -3x + 4$
   $-6x - 2y = -8$

8. $-1 = 2x - y$
   $8x - 4y = -4$

9. $x = y - 1$
   $-x + y = -1$

10. $y = -4x + 11$
    $3x + y = 9$

11. $y = -3x + 1$
    $2x + y = 1$

12. $3x + y = -5$
    $6x + 2y = 10$

13. $5x - y = 5$
    $-x + 3y = 13$

14. $2x + y = 4$
    $-2x + y = -4$

15. $-5x + 4y = 20$
    $10x - 8y = -40$

Example 4

16. **MONEY** Harvey has some $1 bills and some $5 bills. In all, he has 6 bills worth $22. Let x be the number of $1 bills, and let y be the number of $5 bills. Write a system of equations to represent the information, and use substitution to determine how many bills of each denomination Harvey has.

17. **REASONING** Shelby and Calvin are conducting an experiment in chemistry class. They need 5 milliliters of a solution that is 65% acid and 35% distilled water. There is no undiluted acid in the chemistry lab, but they do have two beakers of diluted acid. Beaker A contains 70% acid and 30% distilled water. Beaker B contains 20% acid and 80% distilled water.

    a. Write a system of equations that Shelby and Calvin could use to determine how many milliliters they need to pour from each beaker to make their solution.

    b. Solve your system of equations. How many milliliters from each beaker do Shelby and Calvin need?

Mixed Exercises

Use substitution to solve each system of equations.

18. $y = 3.2x + 1.9$
    $2.3x + 2y = 17.72$

19. $y = \frac{1}{4}x - \frac{1}{2}$
    $8x + 12y = -\frac{1}{2}$

20. $y = -10x - 6.8$
    $-50x - 10.5y = 60.4$

21. **USE A SOURCE** Research population trends in South America. Write and solve a system of equations to predict when the population of two countries will be equal.

22. **REGULARITY** Angle A and angle B are complementary, and their measures have a difference of 20°. What are the measures of the angles? Generalize your method.

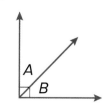

23. **STRUCTURE** A two-digit number is reduced by 45 when the digits are interchanged. The digit in the tens place of the original number is 1 more than 3 times the digit in the units place. Define variables and write a system of equations to find the original number.

24. **USE A MODEL** A zoo keeps track of the number of visitors to each exhibit. The table shows the number of visitors for two exhibits on one day.

	Big Cats	Petting Zoo
Adults	x	21
Children	1024	y

    a. Three times the total number of adults was 17 less than the number of children who visited the petting zoo. Write an equation to model this relationship.

    b. The total number of children who visited the zoo that day was 681 less than 10 times the number of adults who visited the big cats. Write an equation to model this relationship.

    c. Solve the system of equations to find the total number of visitors to those two exhibits on that day.

    d. Tickets to the zoo cost $25 for adults and $10 for children. The total ticket sales one day were $47,750. The number of children who visited the zoo was 270 more than 6 times the number of adults who visited. Write and solve a system of equations to find the total number of visitors to the zoo that day.

25. **FIND THE ERROR** In the system $a + b = 7$ and $1.29a + 0.49b = 6.63$, a represents pounds of apples and b represents pounds of bananas a person bought. Guillermo and Cara are finding and interpreting the solution. Is either correct? Explain your reasoning.

    Guillermo

    $1.29a + 0.49b = 6.63$
    $1.29a + 0.49(a + 7) = 6.63$
    $1.29a + 0.49a + 3.43 = 6.63$
    $1.78a = 3.2$
    $a = 1.8$

    $a + b = 7$, so $b = 5.2$. The solution (1.8, 5.2) means that 1.8 pounds of apples and 5.2 pounds of bananas were bought.

    Cara

    $1.29a + 0.49b = 6.63$
    $1.29(7 - b) + 0.49b = 6.63$
    $9.03 - 1.29b + 0.49b = 6.63$
    $-0.8b = -2.4$
    $b = 3$

    The solution $b = 3$ means that 3 pounds of apples and 3 pounds of bananas were bought.

26. **PERSEVERE** A local charity has 60 volunteers. The ratio of boys to girls is 7:5. Find the number of volunteers who are boys and the number of volunteers who are girls.

27. **ANALYZE** Compare and contrast the solution of a system found by graphing and the solution of the same system found by substitution.

28. **CREATE** Create a system of equations that has one solution. Illustrate how the system could represent a real-world situation, and describe the significance of the solution in the context of the situation.

29. **WRITE** Explain how to determine what to substitute when using the substitution method of solving systems of equations.

Lesson 7-3
Elimination Using Addition and Subtraction

Learn Solving Systems of Equations by Elimination with Addition

The **elimination** method involves eliminating a variable by combining the individual equations within a system of equations. One way to combine equations is by using addition.

Key Concept • Elimination Method Using Addition

Step 1	Write the system so like terms with opposite coefficients are aligned.
Step 2	Add the equations, eliminating one variable. Then solve the equation.
Step 3	Substitute the value from Step 2 into one of the equations and solve for the other variable. Write the solution as an ordered pair.

Example 1 Elimination Using Addition

Use elimination to solve the system of equations.

$3x + 5y = 11$
$5x - 5y = 5$

Step 1 Align terms with opposite coefficients.

Because $5y$ and $-5y$ have opposite coefficients, add the equations to eliminate the variable y.

Step 2 Add the equations.

$$3x + 5y = 11$$
$$(+)\ 5x - 5y = 5$$
$$8x = 16 \qquad \text{The variable } y \text{ is eliminated.}$$
$$\frac{8x}{8} = \frac{16}{8} \qquad \text{Divide each side by 8.}$$
$$x = 2 \qquad \text{Simplify.}$$

Step 3 Solve for the other variable.

$3x + 5y = 11$ First equation
$3(2) + 5y = 11$ Replace x with 2.
$6 + 5y = 11$ Multiply.
$6 - 6 + 5y = 11 - 6$ Subtract 6 from each side.
$5y = 5$ Simplify.
$\frac{5y}{5} = \frac{5}{5}$ Divide each side by 5.
$y = 1$ Simplify.

The solution is (2, 1).

 **Go Online** You can complete an Extra Example online.

Today's Goals
- Solve systems of equations by eliminating a variable using addition.
- Solve systems of equations by eliminating a variable using subtraction.

Today's Vocabulary
elimination

Study Tip

Answer Check You can check your answer by substituting the solution into the equation you did not use in Step 3. If the equality is valid, your solution is correct.

Talk About It!
When graphed, where would these lines intersect? Explain your reasoning.

Check

Use elimination to solve the system of equations.

$5x + 13y = 20$
$-5x - 3y = 30$

Example 2 Write and Solve a System of Equations Using Addition

Seven times a number minus four times another number is thirteen. Negative seven times a number plus seven times another number is fourteen. Find the numbers.

Seven times a number minus four times another number is thirteen.	Negative seven times a number plus seven times another number is fourteen.
$7x - 4y = 13$	$-7x + 7y = 14$

Steps 1 and 2 Write the equations vertically and add.

$$\begin{array}{r} 7x - 4y = 13 \\ (+) \quad -7x + 7y = 14 \\ \hline 3y = 27 \end{array}$$
The variable x is eliminated.

$\dfrac{3y}{3} = \dfrac{27}{3}$ Divide each side by 3.

$y = 9$ Simplify.

Step 3 Substitute 9 for y in either equation to find the value of x.

$-7x + 7y = 14$ Second equation
$-7x + 7(9) = 14$ Replace y with 9.
$-7x + 63 = 14$ Multiply.
$-7x + 63 - 63 = 14 - 63$ Subtract 63 from each side.
$-7x = -49$ Simplify.
$\dfrac{-7x}{-7} = \dfrac{-49}{-7}$ Divide each side by -7.
$x = 7$ Simplify.

The solution is (7, 9). So, $x = 7$ and $y = 9$.

Check

Two times a number minus six times another number is negative six. Negative two times a number plus five times another number is eighteen.

Write the system of equations.

Solve the system of equations.

Go Online You can complete an Extra Example online.

Learn Solving Systems of Equations by Elimination with Subtraction

When the coefficients of a variable are the same in two equations, you can eliminate the variable by subtracting one equation from the other.

Key Concept • Elimination Method Using Subtraction

Step 1	Write the system so like terms with the same coefficients are aligned.
Step 2	Subtract one equation from the other, eliminating one variable. Then solve the equation.
Step 3	Substitute the value from Step 2 into one of the equations and solve for the other variable. Write the solution as an ordered pair.

Example 3 Elimination Using Subtraction

Use elimination to solve the system of equations.

$3x + 6y = 30$
$5x + 6y = 6$

Step 1 Align terms with the same coefficients.

Since $6y$ and $6y$ have the same coefficients, you can subtract the equations to eliminate the variable y.

Step 2 Subtract the equations.

$3x + 6y = 30$
$(-)5x + 6y = 6$
$\overline{-2x = 24}$ The variable y is eliminated.

$\dfrac{-2x}{-2} = \dfrac{24}{-2}$ Divide each side by -2.

$x = -12$ Simplify.

Step 3 Substitute -12 for x in either equation to find the value of y.

$3x + 6y = 30$ First equation
$3(-12) + 6y = 30$ Replace x with -12.
$-36 + 6y = 30$ Multiply.
$-36 + 36 + 6y = 30 + 36$ Add 36 to each side.
$6y = 66$ Simplify.
$\dfrac{6y}{6} = \dfrac{66}{6}$ Divide each side by 6.
$y = 11$ Simplify.

The solution is $(-12, 11)$.

Check

Use elimination to solve the system of equations.

$-2x + 3y = 48$
$7x + 3y = 21$

> **Study Tip**
>
> **Adding and Subtracting Equations** When the variable you want to eliminate has the same coefficient in the two equations, subtract. When the variable you want to eliminate has opposite coefficients, add.

> **Watch Out!**
>
> **Subtracting an Equation** When subtracting one equation from another in order to eliminate a variable, do not forget to distribute the negative sign to each term of the expressions on both sides of the equal sign.

Go Online You can complete an Extra Example online.

🌐 Example 4 Write and Solve a System of Equations Using Subtraction

COMPUTERS Mei and Kara build computers from parts and sell them to make a profit. Mei can build a computer in 0.9 hour, and Kara can build one in 1.2 hours. During a typical week, Mei and Kara spend a total of 15 hours building computers. One week, Mei builds twice as many computers, and the two spend a total of 24 hours on their project. How many computers do Mei and Kara each make during a typical week?

> 💭 **Think About It!**
> What assumption was made about the rates at which Mei and Kara build computers? Why is that assumption made?

Words	Mei's time spent	plus Kara's time spent	is 15 hours.
Variables	Let m = the number of computers that Mei built and k = the number of computers that Kara built.		
Equation	$0.9m$	$+ 1.2k$	$= 15$

Words	Double Mei's time spent	plus Kara's time spent	is 24 hours.
Variables	Let m = the number of computers that Mei built and k = the number of computers that Kara built.		
Equation	$2(0.9m)$	$+ 1.2k$	$= 24$

Steps 1 and 2 Write the equations vertically and subtract.

$$\begin{array}{r} 0.9m + 1.2k = 15 \\ (-)\ 2(0.9m) + 1.2k = 24 \\ \hline -0.9m = -9 \end{array}$$

The variable k is eliminated.

$$\frac{-0.9m}{-0.9} = \frac{-9}{-0.9}$$

Divide each side by -0.9.

$$m = 10$$

Simplify.

Step 3 Substitute 10 for m in either equation to find the value of k.

$0.9m + 1.2k = 15$	First equation
$0.9(10) + 1.2k = 15$	Replace m with 10.
$9 + 1.2k = 15$	Multiply.
$9 - 9 + 1.2k = 15 - 9$	Subtract 9 from each side.
$1.2k = 6$	Simplify.
$\frac{1.2k}{1.2} = \frac{6}{1.2}$	Divide each side by 1.2.
$k = 5$	Simplify.

During a typical week, Mei builds 10 computers and Kara builds 5. Since they cannot sell part of a computer, it makes sense that they would build a whole number of computers in a week. Therefore, 10 computers and 5 computers are viable solutions.

▶ **Go Online** You can complete an Extra Example online.

Practice

Examples 1, 3

Use elimination to solve each system of equations.

1. $-v + w = 7$
   $v + w = 1$

2. $y + z = 4$
   $y - z = 8$

3. $-4x + 5y = 17$
   $4x + 6y = -6$

4. $5m - 2p = 24$
   $3m + 2p = 24$

5. $a + 4b = -4$
   $a + 10b = -16$

6. $6r - 6t = 6$
   $3r - 6t = 15$

7. $6c - 9d = 111$
   $5c - 9d = 103$

8. $11f + 14g = 13$
   $11f + 10g = 25$

9. $9x + 6y = 78$
   $3x - 6y = -30$

10. $3j + 4k = 23.5$
    $8j - 4k = 4$

11. $-3x - 8y = -24$
    $3x - 5y = 4.5$

12. $6x - 2y = 1$
    $10x - 2y = 5$

13. $x - y = 1$
    $x + y = 3$

14. $-x + y = 1$
    $x + y = 11$

15. $x + 4y = 11$
    $x - 6y = 11$

16. $-x + 3y = 6$
    $x + 3y = 18$

17. $3x + 4y = 19$
    $3x + 6y = 33$

18. $x + 4y = -8$
    $x - 4y = -8$

19. $3x + 4y = 2$
    $4x - 4y = 12$

20. $3x - y = -1$
    $-3x - y = 5$

21. $2x - 3y = 9$
    $-5x - 3y = 30$

22. $x - y = 4$
    $2x + y = -4$

23. $3x - y = 26$
    $-2x - y = -24$

24. $5x - y = -6$
    $-x + y = 2$

25. $6x - 2y = 32$
    $4x - 2y = 18$

26. $3x + 2y = -19$
    $-3x - 5y = 25$

27. $7x + 4y = 2$
    $7x + 2y = 8$

Example 2

28. Twice a number added to another number is 15. The sum of the two numbers is 11. Find the numbers.

29. Twice a number added to another number is −8. The difference of the two numbers is 2. Find the numbers.

30. The difference of two numbers is 2. The sum of the same two numbers is 6. Find the numbers.

Example 4

31. **GOVERNMENT** The Texas State Legislature is comprised of state senators and state representatives. There is a greater number of representatives than senators. The sum of the number of representatives and the number of senators is 181. The difference of the number of representatives and number of senators is 119.
    a. Write a system of equations to find the number of state representatives, r, and senators, s.
    b. How many senators and how many representatives make up the Texas State Legislature?

32. **SPORTS** As of 2019, the New York Yankees had won the World Series more than any other team in baseball. The difference of the number of World Series championships won by the Yankees and 2 times the number of World Series championships won by the second-most-winning team, the St. Louis Cardinals, is 5. The sum of the two teams' World Series championships is 38.
    a. Write a system of equations to find the number of World Series championships won by the Yankees, y, and the number of World Series championships won by the Cardinals, x.
    b. How many times has each team won the World Series?

Mixed Exercises

Use elimination to solve each system of equations.

33. $4(x + 2y) = 8$
    $4x + 4y = 12$

34. $3x − 5y = 11$
    $5(x + y) = 5$

35. $4x + 3y = 6$
    $3(x + y) = 7$

36. $0.3x − 2y = −28$
    $0.8x + 2y = 28$

37. $\frac{1}{2}q − 4r = −2$
    $\frac{1}{6}q − 4r = 10$

38. $\frac{1}{2}x + \frac{1}{3}y = −1$
    $-\frac{1}{2}x + \frac{2}{3}y = 10$

39. REASONING At the end of a recent WNBA regular season, the difference of the number of wins and losses by the Phoenix Mercury was 12. The difference of the number of wins and two times the number of losses was 1. How many regular season games did the Phoenix Mercury play during that season?

40. USE A MODEL Marisol works for a florist that sells two types of bouquets, as shown at the right. On Monday, Marisol used 96 tulips to make the bouquets. On Tuesday, she used 192 tulips to make the same number of Spring Mix bouquets as Monday, but 3 times as many Garden Delight bouquets.

Seasonal Bouquets	
Spring Mix	12 tulips
Garden Delight	16 tulips

   a. Write a system of equations that you can use to find how many bouquets of each type Marisol made. Describe what each variable represents.

   b. Find the total number of tulips Marisol used to make Garden Delight bouquets on Monday and Tuesday. Explain your answer.

41. USE A MODEL Jeremy and Kendrick each bought snacks for their friends at a skating rink. The table shows the number of bags of popcorn and the number of plates of nachos each person bought, as well as the total cost of the snacks.

Name	Bags of Popcorn	Plates of Nachos	Total Cost
Jeremy	4	2	$18.50
Kendrick	7	2	$26.75

   a. Write a system of equations that you can use to find the prices of the popcorn and the nachos. Describe what each variable represents.

   b. Solve the system of equations and explain what your solution represents.

42. USE A MODEL The table shows the time Erin spent jogging and walking this weekend and the total distance she covered each day. Erin always jogs at the same rate and always walks at the same rate.

Day	Time Jogging	Time Walking	Total Distance
Saturday	15 min	30 min	3.5 mi
Sunday	1 h	30 min	8 mi

   a. Write a system of equations that you can use to represent this situation. Describe what each variable represents.

   b. Solve the system by elimination. Show your work. Then interpret the solution.

43. **STRUCTURE** Consider the system of equations $0.4x - 2y = 6$ and $0.8x + 2y = 0$.
    a. What is the first step to solving the system of equations by elimination? Explain your reasoning.

    b. What is the solution, as an ordered pair?

    c. Graph the equations on a coordinate plane. Explain how you can use the graph to check your solution.

    d. How would the solution change if the second equation was $x - 5y = 15$? Explain.

    e. How would the solution change if the second equation was $0.4x - 2y = 0$? Explain.

Higher-Order Thinking Skills

44. **ANALYZE** Mikasi says that if you solve a system of equations using elimination by addition and the result is $0 = 0$, then the solution of the system is $(0, 0)$. Provide a counterexample to show that his statement is false. Justify your argument.

45. **ANALYZE** Reece says that if you solve a system of equations using elimination by addition and the result is $0 = 2$, then the solution of the system is $(0, 2)$. Provide a counterexample to show that her statement is false. Justify your argument.

46. **CREATE** Create a system of equations that can be solved by using addition to eliminate one variable. Formulate a general rule for creating such systems.

47. **CREATE** The solution of a system of equations is $(-3, 2)$. One equation in the system is $x + 4y = 5$. Find a second equation for the system such that the system can be solved using elimination by addition. Explain how you derived this equation.

48. **PERSEVERE** The sum of the digits of a two-digit number is 8. The result of subtracting the units digit from the tens digit is -4. Define variables and write the system of equations that you would use to find the number. Then solve the system and find the number.

49. **WRITE** Describe when it would be most beneficial to use elimination to solve a system of equations.

Lesson 7-4

Elimination Using Multiplication

Explore Graphing and Elimination Using Multiplication

Online Activity Use an interactive tool to complete the Explore.

INQUIRY How can you produce a new system of equations with the same solution as the given system?

Learn Solving Systems of Equations by Elimination with Multiplication

Key Concept • Elimination Method Using Multiplication

Step 1 Multiply at least one equation by a constant to get two equations that contain opposite terms.

Step 2 Add the equations, eliminating one variable. Then solve the equation.

Step 3 Substitute the value from Step 2 into one of the equations and solve for the other variable. Write the solution as an ordered pair.

Example 1 Elimination Using Multiplication

Use elimination to solve the system of equations.

$10x + 5y = 30$
$5x - 3y = -7$

Step 1 Multiply an equation by a constant.

The coefficients of x will be opposites if the second equation is multiplied by -2.

$5x - 3y = -7$	Second equation
$(-2)(5x - 3y) = (-2)(-7)$	Multiply each side by -2.
$-10x + 6y = 14$	Simplify.

Step 2 Add the equations.

$10x + 5y = 30$
$(+) \; -10x + 6y = 14$
$ 11y = 44$ — The variable x is eliminated.
$ \dfrac{11y}{11} = \dfrac{44}{11}$ — Divide each side by 11.
$ y = 4$ — Simplify.

(continued on the next page)

Today's Goal
- Solve systems of equations by eliminating a variable using multiplication and addition.

Think About It!
How does the process of solving a system of equations by elimination using multiplication differ from elimination using just addition?

Study Tip
Common Factors If the coefficients of a variable are not the same, or are opposites, and they share a greatest common factor greater than 1, then that variable is the easiest to eliminate using multiplication. For example, the system in this example has two variables, x and y. The coefficients of the y-variable expressions are 5 and -3, which share no common factor greater than 1. However, the coefficients of the x-variable expressions are 10 and 5, which share a common factor of 5. Thus, the x-variable requires fewer steps to eliminate.

Talk About It!
Would you get the solution (1, 4) if you eliminated the y-variable instead of the x-variable? If no, explain your reasoning. If yes, explain why the y-variable was selected for elimination instead of the x-variable.

Step 3 Substitute 4 for y in either equation to find the value of x.

$10x + 5y = 30$	First equation
$10x + 5(4) = 30$	Replace y with 4.
$10x + 20 = 30$	Multiply.
$10x + 20 - 20 = 30 - 20$	Subtract 20 from each side.
$\frac{10x}{10} = \frac{10}{10}$	Divide each side by 10.
$x = 1$	Simplify.

The solution is (1, 4).

Check
Use elimination to solve the system of equations.
$13x + 14y = 59$
$4x + 7y = 37$

Example 2 Multiply Both Equations to Eliminate a Variable

Use elimination to solve the system of equations.

$3x + 4y = -22$
$-2x + 3y = -8$

Step 1 Multiply both equations by a constant.

$3x + 4y = -22$	Original equation	$-2x + 3y = -8$
$2(3x + 4y) = 2(-22)$	Multiply by a constant.	$3(-2x + 3y) = 3(-8)$
$2(3x) + 2(4y) = 2(-22)$	Distributive Property	$3(-2x) + 3(3y) = 3(-8)$
$6x + 8y = -44$	Simplify.	$-6x + 9y = -24$

Step 2 Add the equations.

$6x + 8y = -44$
$(+)\ -6x + 9y = -24$
$\overline{}$
$17y = -68$ The variable x is eliminated.
$\frac{17y}{17} = \frac{-68}{17}$ Divide each side by 17.
$y = -4$ Simplify.

Go Online You can complete an Extra Example online.

Step 3 Use substitution to find the value of x.

$$3x + 4y = -22 \quad \text{First equation}$$
$$3x + 4(-4) = -22 \quad \text{Replace } y \text{ with } -4.$$
$$3x - 16 = -22 \quad \text{Multiply.}$$
$$3x - 16 + 16 = -22 + 16 \quad \text{Add 16 to each side.}$$
$$\frac{3x}{3} = -\frac{6}{3} \quad \text{Divide each side by 3.}$$
$$x = -2 \quad \text{Simplify.}$$

The solution is $(-2, -4)$.

Check

Use elimination to solve the system of equations.
$11x - 6y = 25$
$3x + 9y = 60$

Example 3 Write and Solve a System Using Multiplication

COMICS Jorge's comic book collection consists of single issues that cost $4 each and paperback collections that cost $12 each. He has 100 books in all. His collection cost him $616. Write and solve a system of equations to determine how many single issues and paperbacks Jorge has in his collection.

Write the system of equations. Let $c = $ the number of single issue comics and $p = $ the number of paperback collections.

The number of single issue comics	plus	the number of paperback collections	is	100 books
c	+	p	=	100

$4 per single issue comic	plus	$12 per paperback collection	is	$616
4c	+	12p	=	616

Math History Minute

German mathematician **Carl Friedrich Gauss (1777-1855)** contributed significantly to many fields, including number theory, algebra, and statistics. The elimination method is related to the Gaussian elimination method, an algorithm for solving systems of linear equations that was known to Chinese mathematicians as early as 179 B.C.

(continued on the next page)

> **Think About It!**
> Is (73, 27) a viable solution in the context of the situation? Explain your reasoning.

Step 1 Multiply an equation by a constant.

$$c + p = 100 \qquad \text{First equation}$$
$$-4(c + p) = -4(100) \qquad \text{Multiply each side by } -4.$$
$$-4c - 4p = -4(100) \qquad \text{Distributive Property}$$
$$-4c - 4p = -400 \qquad \text{Simplify.}$$

Step 2 Add the equations.

$$-4c - 4p = -400$$
$$(+)4c + 12p = 616$$
$$\frac{8p}{8} = \frac{216}{8}$$
$$p = 27$$

The variable c is eliminated.
Divide each side by 8.
Simplify.

Step 3 Use substitution to find the value of c.

Substitute 27 for p in either equation to find the value of c.

$$c + p = 100 \qquad \text{First equation.}$$
$$c + 27 = 100 \qquad \text{Replace } p \text{ with 27.}$$
$$c + 27 - 27 = 100 - 27 \qquad \text{Subtract 27 from each side.}$$
$$c = 73 \qquad \text{Simplify.}$$

Jorge has 73 single issue comics and 27 paperback collections.

Check

SOFTWARE A software company releases two products: a home version of their photo editor, which costs $20, and a professional version, which costs $45. The company sells 1000 copies of the photo editing software, earning a total revenue of $38,075. Write and solve a system of equations to determine how many home versions and professional versions of the software the company sold.

Let h = the number of home versions sold and p = the number of professional versions sold.

Part A Write the system of equations.

Part B Solve the system of equations.

> **Go Online** to practice what you've learned about choosing a method to solve a system in the Put It All Together over Lessons 7-1 through 7-4.

The software company sold __?__ home versions and __?__ professional versions of their photo editor.

Go Online You can complete an Extra Example online.

Practice

Go Online You can complete your homework online.

Examples 1 and 2

Use elimination to solve each system of equations.

1. $x + y = 2$
   $-3x + 4y = 15$

2. $x - y = -8$
   $7x + 5y = 16$

3. $x + 5y = 17$
   $-4x + 3y = 24$

4. $6x + y = -39$
   $3x + 2y = -15$

5. $2x + 5y = 11$
   $4x + 3y = 1$

6. $3x - 3y = -6$
   $-5x + 6y = 12$

7. $3x + 4y = 29$
   $6x + 5y = 43$

8. $8x + 3y = 4$
   $-7x + 5y = -34$

9. $8x + 3y = -7$
   $7x + 2y = -3$

10. $4x + 7y = -80$
    $3x + 5y = -58$

11. $12x - 3y = -3$
    $6x + y = 1$

12. $-4x + 2y = 0$
    $10x + 3y = 8$

Example 3

13. **SPORTS** The Fan Cost Index (FCI) tracks the average costs for attending sporting events, including tickets, drinks, food, parking, programs, and souvenirs. According to the FCI, a family of four would spend a total of $592.30 to attend two Major League Baseball (MLB) games and one National Basketball Association (NBA) game. The family would spend $691.31 to attend one MLB and two NBA games.

    a. Write a system of equations to find the family's costs for each kind of game according to the FCI.

    b. Solve the system of equations to find the cost for a family of four to attend each kind of game according to the FCI.

14. **ART** Mr. Santos, the curator of the children's museum, recently made two purchases of firing clay and polymer clay for a visiting artist to sculpt. Use the table to find the cost of each product per kilogram.

Firing Clay (kg)	Polymer Clay (kg)	Total Cost
5	24	$64.05
25	8	$51.45

    a. Write a system of equations to find the cost of each product per kilogram.

    b. Solve the system of equations to find the cost of each product per kilogram.

Mixed Exercises

15. Two times a number plus three times another number equals 13. The sum of the two numbers is 7. What are the numbers?

16. Four times a number minus twice another number is −16. The sum of the two numbers is −1. Find the numbers.

17. **FUNDRAISING** Trisha and Byron are washing and vacuuming cars to raise money for a class trip. Trisha raised $38 by washing 5 cars and vacuuming 4 cars. Byron raised $28 by washing 4 cars and vacuuming 2 cars. Find the amount they charged to wash a car and vacuum a car.

18. **STRUCTURE** Consider the system of equations $-2x + 3y = -5$ and $3x - 4y = 6$.
    a. Describe two different ways to solve the system by elimination.
    b. Explain why multiplying the first equation by 6 and the second equation by 5 and then adding is not useful for solving the system.
    c. Solve the system by elimination.

19. **REASONING** The owner of a juice stand wants to make a new juice drink. He would like to mix Tropical Breeze, t, and Kona Cooler, k, to make 10 quarts of a new drink that is 40% pineapple juice.

Juice Drinks	
Tropical Breeze	20% pineapple juice
Kona Cooler	50% pineapple juice

    a. Make a table to help write a system of equations that the owner of the juice stand can solve to determine the amount of each drink he should use to make the new drink.
    b. Solve the system and explain what your solution represents.
    c. Explain how you know your answer is correct.

20. **USE A MODEL** Marlene works as a cashier at a grocery store. At the end of the day, she has a total of 125 five-dollar bills and ten-dollar bills. The total value of these bills is $990.

    a. Write a system of equations that you can use to find the number of five-dollar bills and the number of ten-dollars bills. Describe what each variable represents.

    b. Solve the system and explain what your solution represents.

21. **FIND THE ERROR** Jason and Daniela are solving a system of equations. Is either of them correct? Explain your reasoning.

22. **ANALYZE** Determine whether the following statement is *true* or *false*: A system of linear equations will only have infinitely many solutions if the equations have the same coefficients. Justify your argument.

    Jason
    $2r + 7t = 11$
    $r - 9t = -7$
    $2r + 7t = 11$
    $(-)\ 2r - 18t = -14$
    $25t = 25$
    $t = 1$
    $2r + 7t = 11$
    $2r + 7(1) = 11$
    $2r + 7 = 11$
    $2r = 4$
    $\frac{2r}{2} = \frac{4}{2}$
    $r = 2$
    The solution is (2, 1).

    Daniela
    $2r + 7t = 11$
    $(-)\ r - 9t = -7$
    $r = 18$
    $2r + 7t = 11$
    $2(18) + 7t = 11$
    $36 + 7t = 11$
    $7t = -25$
    $\frac{7t}{7} = \frac{25}{7}$
    $t = -3.6$
    The solution is (18, -3.6).

23. **CREATE** Write a system of equations that can be solved by multiplying one equation by −3 and then adding the two equations together.

24. **PERSEVERE** The solution of the system $4x + 5y = 2$ and $6x - 2y = b$ is $(3, a)$. Find the values of a and b. Discuss the steps you used.

25. **WRITE** Why is substitution sometimes more helpful than elimination, and vice versa?

Lesson 7-5

Systems of Inequalities

Explore Solutions of Systems of Inequalities

Online Activity Use an interactive tool to complete the Explore.

> **INQUIRY** How are the solutions of a system of inequalities represented on a graph?

Learn Solving Systems of Inequalities by Graphing

A set of two or more inequalities with the same variables is a **system of inequalities**. The solution of a system of inequalities with two variables is the set of ordered pairs that satisfy all of the inequalities in the system. The solution is represented by the overlap, or intersection, of the graphs of the inequalities.

Example 1 Solve by Graphing

Solve the system of inequalities by graphing.

$x - 2y > -6$

$y < 3x$

Step 1 Graph one inequality of the system.
The boundary of $x - 2y > -6$ is dashed and is not included in the solution. The half-plane is shaded below the boundary, in yellow, to indicate solutions of $x - 2y > -6$.

Step 2 Graph the second inequality of the system.
The boundary of $y < 3x$ is dashed and is not included in the solution. The half-plane is shaded to the right of the boundary, in blue, to indicate solutions of $y < 3x$.

Solution
The solution of the system is the set of ordered pairs in the intersection of the graphs of $x - 2y > -6$ and $y < 3x$. The region is shaded green.

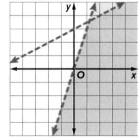

Today's Goal
- Solve systems of linear inequalities by graphing.

Today's Vocabulary
system of inequalities

Go Online
You can watch a video to see how to solve a system of linear inequalities.

Think About It!
The boundaries for the system $y > 3$, $y \leq -2x + 1$ intersect at $(-1, 3)$. Is $(-1, 3)$ included in the solution? Explain.

Go Online
You can watch a video to see how to use a graphing calculator with this example.

Go Online You can complete an Extra Example online.

Check

Graph the system of inequalities.

$\frac{1}{3}x + 2 < y$

$x \geq -3$

Example 2 Solve by Graphing, No Solution

Solve the system of inequalities by graphing.

$-3x + 4y > 0$

$3x - 4y \geq 8$

Step 1 Graph one inequality of the system.

The boundary of $-3x + 4y > 0$ is dashed and is not included in the solution. The half-plane is shaded above the boundary, in yellow, to indicate solutions of $-3x > 4y > 0$.

Step 2 Graph the second inequality of the system.

The boundary of $3x - 4y \geq 8$ is solid and is included in the solution. The half-plane is shaded below the boundary, in blue, to indicate solutions of $3x - 4y \geq 8$.

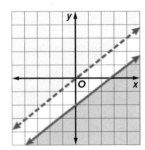

Solution

The graphs of $-3x + 4y = 0$ and $3x - 4y = 8$ are parallel lines. The regions do not intersect at any point, so the system has no solution.

> **Talk About It!**
> Is it possible for a system of inequalities that has boundaries with different slopes to have no solution? Justify your argument.

> **Study Tip**
> **Shaded Regions** When graphing more than one region, it is helpful to use a different color of pencil or a different pattern for each region. This will make it easier to see where the regions intersect and find possible solutions.

Go Online You can complete an Extra Example online.

Check

Graph the system of inequalities.

$x + 3 < y$

$3x - 3y \geq 12$

Example 3 Apply Systems of Inequalities

SEWING A family and consumer sciences class is making pillows and blankets to donate to a local shelter. The class has 40 yards of fabric to use. Pillows require 1.25 yards of fabric and take 1 hour to make. Blankets use 4 yards of fabric and take 2.5 hours to make. The class has 28 hours of class time left for the semester. Determine the number of pillows and blankets the class can make for the shelter.

Think About It!
Can the class make 2 blankets and 24 pillows? Explain.

Part A Define the variables, and write a system of inequalities to represent the situation.

Let p represent the number of pillows the class can make.
Let b represent the number of blankets the class can make.

Fabric and time are two constraints on the numbers of pillows and blankets the class can make.

Because pillows use 1.25 yards of fabric and blankets use 4 yards of fabric, the inequality that represents the fabric constraint is $1.25p + 4b \leq 40$.

Making a pillow takes 1 hour and making a blanket takes 2.5 hours. Because the class has only 28 hours left, the inequality that represents the time constraint is $p + 2.5b \leq 28$.

Part B Graph the system.

Part C Find a viable solution.

Only whole-number solutions make sense in this situation. One possible solution is (4, 15); 4 blankets and 15 pillows can be made.

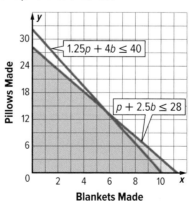

Think About It!
Could the graph be represented with "Pillows Made" as the x-axis and "Blankets Made" as the y-axis? Explain.

Go Online You can complete an Extra Example online.

Check

BAKERY Aisha can work up to 20 hours per week. Working at a bakery, she earns $7 per hour most of the time and $8.50 per hour during the early morning shift. Aisha needs to earn at least $150 this week to pay for a trip with her friends. Determine the number of regular and early morning hours that Aisha could work.

Part A Select the correct system and graph. Let r = regular hours and m = early morning hours.

A. $r < 20$
   $7r + 8.5m \geq 150$

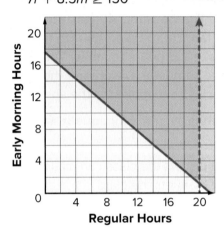

B. $r + m \leq 20$
   $r + m \leq 150$

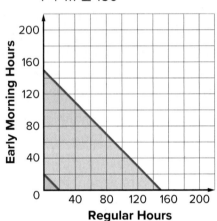

C. $r + m \leq 20$
   $7r + 8.5m \geq 150$

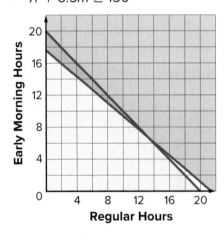

D. $7r + 8.5m > 20$
   $7r + 8.5m \geq 150$

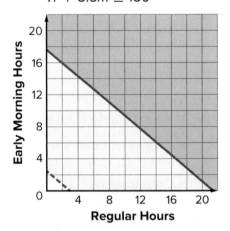

Part B Identify each solution as *viable* or *nonviable*.

Solutions

(2, 17)   (10, 10)

(4, 7)   (18, 6)

(5, 15)   (21, 6)

(8, 13)

Practice

Go Online You can complete your homework online.

Examples 1 and 2

Solve each system of inequalities by graphing.

1. $y < 6$
   $y > x + 3$

2. $y \geq 0$
   $y \leq x - 5$

3. $y \leq x + 10$
   $y > 6x + 2$

4. $y \geq x + 10$
   $y \leq x - 3$

5. $y < 5x - 5$
   $y > 5x + 9$

6. $y \geq 3x - 5$
   $3x - y > -4$

7. $x > -1$
   $y \leq -3$

8. $y > 2$
   $x < -2$

9. $y > x + 3$
   $y \leq -1$

10. $x < 2$
    $y - x \leq 2$

11. $x + y \leq -1$
    $x + y \geq 3$

12. $y - x > 4$
    $x + y > 2$

Example 3

13. **FITNESS** Diego started an exercise program in which each week he walks from 9 to 12 miles and works out at the gym from 4.5 to 6 hours.

    a. Write a system of inequalities to represent this situation. Define your variables.

    b. Graph the system.

    c. List three viable solutions.

14. **SOUVENIRS** Emiliana wants to buy turquoise stones on her trip to New Mexico to give to at least 4 of her friends. The gift shop sells stones for either $4 or $6 per stone. Emiliana has no more than $30 to spend.

    a. Write a system of inequalities to represent this situation. Define your variables.

    b. Graph the system.

    c. List three viable solutions.

Mixed Exercises

Write a system of inequalities for each graph.

15.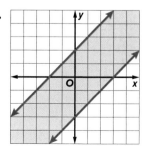

16.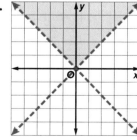

17.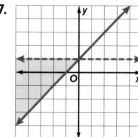

18. **PRECISION** Write a system of inequalities to represent the graph shown at the right.

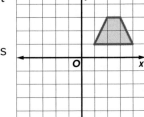

19. **CONSTRUCT ARGUMENTS** Is (2.5, 1) a solution of the system of inequalities $2x + 3y > 8$ and $4x - 5y \geq 2$? Justify your argument. Then explain how you can tell if the point is a solution without graphing the inequality.

20. **PETS** Priya's Pet Store never has more than a combined total of 20 cats and dogs and never more than 8 cats. This is represented by the inequalities $x + y \leq 20$ and $x \leq 8$. Represent the number of cats and dogs that can be at the store on a graph. Solve the system of inequalities by graphing.

21. **FUNDRAISING** The baseball team plans to sell tins of popcorn and peanuts. The players have $900 to spend on products and can order up to 200 tins. They want to order at least as many tins of popcorn as tins of peanuts. A tin of popcorn costs $3, and a tin of peanuts costs $4. Define the variables and write a system of inequalities to represent this situation. Then list any constraints for the variables.

22. **BUSINESS** For maximum efficiency, a factory must have at least 100 workers, but no more than 200 workers on a shift. The factory also must manufacture at least 30 units per worker.

    a. Let x be the number of workers and let y be the number of units. Write a system of inequalities expressing the conditions in the problem.

    b. Graph the systems of inequalities.

    c. Find three possible solutions.

23. **DESIGN** LaShawn designs Web sites for local businesses. He charges $25 an hour to build a Web site and charges $15 an hour to update Web sites once he builds them. He wants to earn at least $100 every week, but he does not want to work more than 6 hours each week. What is a possible number of hours LaShawn can spend each week building Web sites x and updating Web sites y that will allow him to attain his goals? Write your answer as an ordered pair.

Higher-Order Thinking Skills

24. **PERSEVERE** Create a system of inequalities equivalent to $|x| \leq 4$.

25. **ANALYZE** State whether the following statement is *sometimes*, *always*, or *never* true. Justify your argument.

    Systems of inequalities with parallel boundaries have no solutions.

26. **CREATE** One inequality in a system is $3x - y > 4$. Write a second inequality so that the system will have no solution.

27. **PERSEVERE** Graph the system of inequalities $y \geq 1$, $y \leq x + 4$, and $y \leq -x + 4$. Estimate the area that represents the solution.

28. **WRITE** Describe the graph of the solution of the system $6x - 3y \leq -5$ and $6x - 3y \geq -5$ without graphing. Explain your reasoning.

Module 7 • Systems of Linear Equations and Inequalities

Review

 Essential Question

How are systems of equations useful in the real world?

Module Summary

Lesson 7-1
Graphing Systems of Equations

- When you solve a system of equations with $y = f(x)$ and $y = g(x)$, the solution is an ordered pair that satisfies both equations. Thus, the x-coordinate of the intersection of $y = f(x)$ and $y = g(x)$ is the value of x where $f(x) = g(x)$.
- A system of equations is consistent if it has at least one ordered pair that satisfies both equations.
- A system of equations is independent if it has exactly one solution.
- A system of equations is dependent if it has an infinite number of solutions.
- A system of equations is inconsistent if it has no ordered pair that satisfies both equations.

Lessons 7-2 through 7-4
Solving Systems of Equations Algebraically

- To use the substitution method, solve at least one equation for one variable. Substitute the resulting expression into the other equation to replace the variable. Then solve the equation. Substitute this value into either equation, and solve for the other variable. Write the solution as an ordered pair.
- To use the elimination method, write the system so like terms with opposite coefficients are aligned. Add or subtract the equations, eliminating one variable. Then solve the equation. Substitute this value into one of the equations and solve for the other variable. Write the solution as an ordered pair. You may need to multiply at least one equation by a constant to get two equations that contain opposite terms.

Lesson 7-5
Systems of Inequalities

- A set of two or more inequalities with the same variables is called a system of inequalities.
- The solution of a system of inequalities with two variables is the set of ordered pairs that satisfy all of the inequalities in the system. The solution is represented by the overlap, or intersection, of the graphs of the inequalities.

Study Organizer

 Foldables

Use your Foldable to review this module. Working with a partner can be helpful. Ask for clarification of concepts as needed.

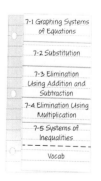

Test Practice

1. **MULTI-SELECT** Use the graph. Which systems of equations are consistent and independent? (Lesson 7-1)

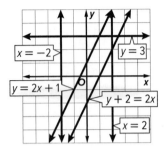

   A. $y = 2x + 1$
      $y + 2 = 2x$

   B. $y = 2x + 1$
      $y = 3$

   C. $y = 3$
      $x = -2$

   D. $x = 2$
      $y + 2 = 2x$

   E. $x = -2$
      $x = 2$

2. **OPEN RESPONSE** Consider the system of equations. (Lesson 7-1)

   $8x + 2y = 8$
   $y = -4x + 4$

   How many solutions are there for the system? Is the system dependent or independent?

3. **MULTIPLE CHOICE** Which system of equations can be entered into a graphing calculator to solve $3.5x + 18 = -5.8x + 30$? (Lesson 7-1)

   A. $y = 3.5x$
      $y = -5.8x$

   B. $y = 3.5x + 18$
      $y = -5.8x + 30$

   C. $0 = 3.5x + 18$
      $0 = -5.8x + 30$

   D. $y = 9.3x - 12$

4. **MULTIPLE CHOICE** Use a system of equations and a graphing calculator to solve $6.9x + 4.3 = 4.7x + 8$. Round your answer to the nearest hundredth, if necessary. (Lesson 7-1)

   A. 1.06
   B. 1.68
   C. 2.14
   D. 5.59

5. **OPEN RESPONSE** Taylan is selling plastic and wooden frames. He sold 7 total frames. The number of plastic frames Taylan sold was 5 less than twice the number of wooden frames. How many of each type of frame did Taylan sell? (Lesson 7-2)

6. **MULTIPLE CHOICE** Consider the system of equations.

   $3x - 2y = 0$
   $x + y = 10$

   What is the solution of the system? (Lesson 7-2)

   A. The solution to the system is (20, −10).
   B. The solution to the system is (3, 7).
   C. The solution to the system is (4, 6).
   D. The solution to the system is (6, 4).

426 Module 7 Review • Systems of Linear Equations and Inequalities

7. **OPEN RESPONSE** Determine whether the system has *no solution, one solution,* or *infinitely many solutions*. If the system has one solution, name it. (Lesson 7-2)

   $x + y = 5$
   $3x + 2y = 8$

8. **OPEN RESPONSE** The sum of the measures of two complementary angles is 90 degrees. Angles *P* and *Q* are complementary, and the measure of angle *P* is 6 degrees more than twice the measure of angle *Q*.

   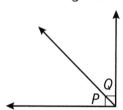

   Write a system of equations and use substitution to find the measure of angles *P* and *Q*. (Lesson 7-2)

9. **OPEN RESPONSE** Solve the system of equations. (Lesson 7-3)

   $3x + y = 34$
   $0.5x - y = 1$

10. **MULTIPLE CHOICE** A rectangle is *x* inches wide and 3*y* inches long. The sum of the length and width is 36 inches and the difference between the length and twice the width is 12 inches. Find the length and width. (Lesson 7-3)

    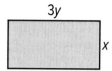

    A. width: 8 inches; length: 28 inches

    B. width: 8 inches; length: 9.3 inches

    C. width: 12 inches; length: 24 inches

    D. width: 15 inches; length: 21 inches

11. **MULTI-SELECT** Select all of the ways the system can be solved. (Lesson 7-4)

    $9x - 2y = 4$
    $3x + 8y = -12$

    A. Multiply the first equation by 4, then add the equations.

    B. Multiply the second equation by 3, then subtract the equations.

    C. Multiply the first equation by 3, then add the equations.

    D. Multiply the second equation by 3, then add the equations.

    E. Multiply the first equation by 3, then subtract the equations.

12. **OPEN RESPONSE** Solve the system of equations. (Lesson 7-4)

    $2x + 5y = 5$
    $3x + 4y = -3$

13. OPEN RESPONSE Solve the system of equations. (Lesson 7-4)

$2r - t = 7$
$r - t = 1$

14. OPEN RESPONSE It takes 3 hours to paddle a kayak 12 miles downstream and 4 hours for the return trip upstream. Find the rate of the kayak in still water.

Let $k =$ the rate of the kayak in still water and $c =$ the rate of the current. (Lesson 7-4)

	r	t	d	rt = d
Downstream	$k + c$	3	12	$3(k + c) = 12$
Upstream	$k - c$	4	12	$4(k - c) = 12$

15. MULTIPLE CHOICE The graph shows the solution to the given system of inequalities. (Lesson 7-5)

$-x + 2y \leq 1$
$-3x + 2y \geq 2$

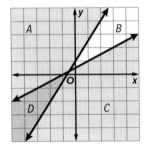

In what region is the solution set?

A. A

B. B

C. C

D. D

16. MULTIPLE CHOICE Which graph represents the solution of the system of inequalities? (Lesson 7-5)

$x - y \geq 2$
$2x + y > -3$

A.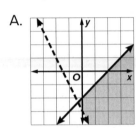

B.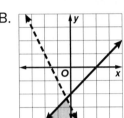

C.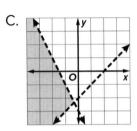

D.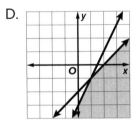

17. MULTIPLE CHOICE Diana wants to build a rectangular pen for her goats. The length of the pen should be at least 50 feet, and the perimeter of the pen should be no more than 190 feet. What is a viable solution for the dimensions of the pen? (Lesson 7-5)

A. 29 feet by 40 feet

B. 29 feet by 76 feet

C. 29 feet by 65 feet

D. 29 feet by 29 feet

Module 8
Exponents and Roots

e Essential Question
How do you perform operations and represent real-world situations with exponents?

What will you learn?
How much do you already know about each topic **before** starting this module?

KEY
👎 — I don't know.   👍 — I've heard of it.   👍 — I know it!

	Before			After		
	👎	👍	👍	👎	👍	👍
use the Product of Powers Property						
use the Power of a Power Property						
use the Power of a Product Property						
use the Quotient of Powers Property						
use the Power of a Quotient Property						
simplify expressions with zero exponents						
simplify expressions with negative exponents						
use rational exponents to solve problems						
simplify radical expressions						
solve exponential equations						

📙 **Foldables** Make this Foldable to help you organize your notes about exponents and roots. Begin with seven sheets of notebook paper.

1. **Arrange** the paper into a stack.
2. **Staple** along the left side. Starting with the second sheet of paper, **cut** along the right side to form tabs.
3. **Label** the cover sheet "Exponents and Roots" and label each tab with a lesson number.

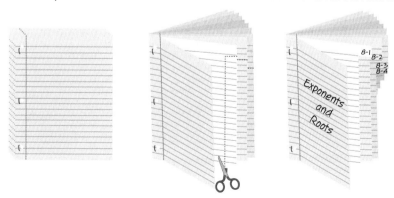

Module 8 • Exponents and Roots

What Vocabulary Will You Learn?

- cube root
- exponential equation
- index
- monomial
- negative exponent
- nth root
- perfect cube
- perfect square
- principal square root
- radical expression
- radicand
- rational exponent
- square root

Are You Ready?

Complete the Quick Review to see if you are ready to start this module.
Then complete the Quick Check.

Quick Review

Example 1

Write $5 \cdot 5 \cdot 5 \cdot 5 + y \cdot y$ using exponents.

There are 4 factors of 5.

There are 2 factors of y.

$5 \cdot 5 \cdot 5 \cdot 5 = 5^4$

$y \cdot y = y^2$

$5 \cdot 5 \cdot 5 \cdot 5 + y \cdot y = 5^4 + y^2$

Example 2

Evaluate $\left(\frac{5}{7}\right)^2$.

$\left(\frac{5}{7}\right)^2 = \frac{5}{7} \cdot \frac{5}{7}$ Expand the expression.

$= \frac{5 \cdot 5}{7 \cdot 7}$ Multiply the numerators and multiply the denominators.

$= \frac{25}{49}$ Simplify.

Quick Check

Write each expression using exponents.

1. $4 \cdot 4 \cdot 4 \cdot 4 \cdot 4$
2. $x \cdot x \cdot x$
3. $m \cdot m \cdot m \cdot p \cdot p \cdot p \cdot p \cdot p$
4. $\left(\frac{1}{5}\right) \cdot \left(\frac{1}{5}\right) \cdot \left(\frac{1}{5}\right) \cdot \left(\frac{1}{5}\right) \cdot \left(\frac{1}{5}\right) \cdot \left(\frac{1}{5}\right)$

Evaluate each expression.

5. 2^5
6. $(-5)^2$
7. $\left(\frac{1}{2}\right)^4$
8. $(-4)^3$

How Did You Do?

Which exercises did you answer correctly in the Quick Check?

Lesson 8-1
Multiplication Properties of Exponents

Today's Goals
- Find products of monomials.
- Find the power of a power.
- Find the power of a product.

Today's Vocabulary
monomial

Explore Products of Powers

 Online Activity Use an interactive tool to complete the Explore.

> **INQUIRY** How can you determine the product of two powers a^m and a^p?

Learn Product of Powers

A **monomial** is a number, a variable, or a product of a number and one or more variables. It has only one term. The term a^4 is a monomial. An expression that involves division by a variable is not a monomial. The term $\frac{ab}{c}$ is not a monomial.

Key Concept • Product of Powers	
Words	To multiply two powers that have the same base, add their exponents.
Symbols	For any real number a and any integers m and p, $a^m \cdot a^p = a^{m+p}$.
Examples	$b^2 \cdot b^4 = b^{2+4}$ or b^6, $d^3 \cdot d^7 = d^{3+7}$ or d^{10}

💭 Think About It!
How does the process for simplifying $b^2 \cdot b^4$ differ from simplifying $b^2 + b^4$?

Example 1 Product of Powers

Simplify each expression.

a. $(3n^4)(4n^7)$

$(3n^4)(4n^7) = (3 \cdot 4)(n^4 \cdot n^7)$ Group coefficients and variables.

$= (3 \cdot 4)(n^{4+7})$ Product of Powers

$= 12n^{11}$ Simplify.

b. $(7xy^2)(2x^4y^3)$

$(7xy^2)(2x^4y^3) = (7 \cdot 2)(x \cdot x^4)(y^2 \cdot y^3)$ Group coefficients and variables.

$= (7 \cdot 2)(x^{1+4})(y^{2+3})$ Product of Powers

$= 14x^5y^5$ Simplify.

Study Tip
Terminology Recall that for the expression b^2, b is called the *base* and the 2 is called the *exponent* or *power*.

💭 Think About It!
What properties allow you to group the coefficients and the variables in the first step of each solution?

Check
Simplify $(7n^7)(-7n)$.   Simplify $(11x^6y^6)(xy^9)$.

Example 2 Product of Powers and Scientific Notation

STARS The fastest recorded star in the Milky Way galaxy is US 708, which travels 2,700,000 miles per hour. How far does US 708 travel in 1 year? Write your answer in scientific notation and round to the nearest tenth. (Hint: 1 year = 8760 hours)

Step 1 Convert the speed of US 708 and the number of hours in a year to scientific notation.

Speed of US 708

$2{,}700{,}000 = 2.7 \times 1{,}000{,}000$ $1 \leq 2.7 < 10$
$\phantom{2{,}700{,}000} = 2.7 \times 10^6$ $1{,}000{,}000 = 10^6$

Hours in a Year

$8760 = 8.76 \times 1000$ $1 \leq 8.76 < 10$
$ = 8.76 \times 10^3$ $1000 = 10^3$

Step 2 Use the formula $d = rt$ to find approximately how far US 708 travels in a year.

$d = rt$	distance = rate · time
$d = (2.7 \times 10^6) \times (8.76 \times 10^3)$	$r = 2.7 \times 10^6$ mph, $t = 8.76 \times 10^3$ hrs
$ = (2.7 \times 8.76) \times (10^6 \times 10^3)$	Comm. and Assoc. Properties
$ = 23.652 \times (10^6 \times 10^3)$	Multiply.
$ = 23.652 \times 10^{6+3}$	Product of Powers
$ = 23.652 \times 10^9$	Simplify.
$ = 2.3652 \times 10^1 \times 10^9$	$23.652 = 2.3652 \times 10^1$
$ = 2.3652 \times 10^{1+9}$	Product of Powers
$ = 2.3652 \times 10^{10}$	Simplify.
$ \approx 2.4 \times 10^{10}$	Round to the nearest tenth.

Check

COMPUTERS Typically, a 4K computer monitor describes a screen with a resolution of 3840 × 2160. This means that the display's width is 3840 pixels and the height is 2160 pixels. How many pixels are there on a typical 4K computer monitor?

A. 2.2×10^3 pixels

B. 3.8×10^3 pixels

C. 8.3×10^6 pixels

D. 8.3×10^7 pixels

Go Online You can complete an Extra Example online.

Problem-Solving Tip
Identify Subgoals Before you can find the distance that US 708 travels, you need to write the rate and time in scientific notation. To convert a number from standard form to scientific notation, rewrite it as a number a such that $1 \leq a < 10$ multiplied by a power of 10. Thus, 100 in scientific notation is 1×10^2. Conversely, 1×10^2 in standard form is 100.

Study Tip
Assumptions Assuming that US 708 travels at exactly 2,700,000 miles per hour for the entire year allows us to approximate the distance it will travel. While 2,700,000 miles per hour is only an approximation of the star's speed, and the speed may not be constant, using this constant rate allows for a reasonable estimate.

Learn Power of a Power

Key Concept • Power of a Power

Words	To find a power of a power, multiply the exponents.
Symbols	For any real number a and any integers m and p, $(a^m)^p = a^{mp}$.
Examples	$(b^2)^4 = b^{2 \cdot 4}$ or b^8; $(d^3)^7 = d^{3 \cdot 7}$ or d^{21}

Example 3 Power of a Power

Simplify each expression.

a. $(n^5)^3$

$(n^5)^3 = n^{5 \cdot 3}$ Power of a Power

$ = n^{15}$ Simplify.

b. $(x^4)^2$

$(x^4)^2 = x^{4 \cdot 2}$ Power of a Power

$ = x^8$ Simplify.

💭 **Think About It!**
Is $(x^4)^2$ equivalent to $(x^2)^4$? Explain your reasoning.

Check

Simplify $(n^4)^{11}$.

Simplify $(x^{-7})^{-3}$.

Learn Power of a Product

Key Concept • Power of a Product

Words	To find a power of a product, find the power of each factor and multiply.
Symbols	For any real numbers a and b and any integer m, $(ab)^m = a^m b^m$.
Example	$(-5x^2y)^3 = (-5)^3 x^6 y^3$ or $-125x^6y^3$

💭 **Think About It!**
Is $2(xy)^3$ equivalent to $(2xy)^3$? Explain your reasoning.

Example 4 Power of a Product

Simplify each expression.

a. $(3x^5y^2)^5$

$(3x^5y^2)^5 = 3^5 x^{5 \cdot 5} y^{2 \cdot 5}$ Power of a Product

$ = 243 x^{25} y^{10}$ Simplify.

(continued on the next page)

🌐 **Go Online** You can complete an Extra Example online.

Think About It!
Is $(-5ab^4)^2$ equivalent to $-(5ab^4)^2$? Explain your reasoning.

b. $(-5ab^4)^2$

$(-5ab^4)^2 = (-5)^2 a^{1 \cdot 2} b^{4 \cdot 2}$ Power of a Product

$\qquad\qquad = 25a^2 b^8$ Simplify.

Check

Simplify $(4x^3 y^{-2})^3$.

Simplify $(-3a^{-3}b)^6$.

🌐 Example 5 Power of a Product and Area

If the side of each smaller square is x inches, and the side of the whole canvas is s inches, then what is the area of the painting in terms of x?

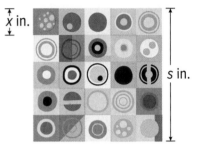

The side length of the canvas, s, can also be described as $5x$.

$\quad A = s^2$ Area of a square

$\quad\; = (5x)^2$ $s = 5x$

$\quad\; = 5^2 x^{1 \cdot 2}$ Power of a Product

$\quad\; = 25x^2$ Simplify.

Talk About It!
Jared argues that increasing the side of each smaller square by a factor of n would increase the area of the canvas by a factor of n. Is he correct? Explain your reasoning.

Watch Out!
Square Units Remember to use square units when measuring area.

Check

GAME DESIGN Jeanine is designing an early version of a game and wants to use a cube to stand in for the character, which she will design later. She bases the dimensions of the cube around the height of the character, which she defines as $\frac{1}{8}x$ pixels, where x is the height of the total game screen. What is the volume of the cube in terms of x?

A. $\frac{1}{64}x^2$ pixels

B. $\frac{1}{512}x^3$ pixels

C. $\frac{1}{8}x^3$ pixels

D. $512x^3$ pixels

🔍 **Go Online** You can complete an Extra Example online.

Practice

Go Online You can complete your homework online.

Examples 1, 3, and 4

Simplify each expression.

1. $(q^2)(2q^4)$
2. $(-2u^2)(6u^6)$
3. $(9w^2x^8)(w^6x^4)$
4. $(y^6z^9)(6y^4z^2)$
5. $(b^8c^6d^5)(7b^6c^2d)$
6. $(14fg^2h^2)(3f^4g^2h^2)$
7. $(j^5k^7)^4$
8. $(n^3p)^4$
9. $[(2^2)^2]^2$
10. $[(3^2)^2]^4$
11. $[(4r^2t)^3]^2$
12. $[(-2xy^2)^3]^2$
13. $(y^2z)(yz^2)$
14. $(\ell^2k^2)(\ell^3k)$
15. $(-5m^3)(3m^8)$
16. $(-2c^4d)(-4cd)$
17. $(3pr^2)^2$
18. $(2b^3c^4)^2$

Example 2

19. **COMMUNITY SERVICE** During the school year, each student planted 1.2×10^2 flowers as part of a community service project. If there are 1.5×10^3 students in the school, how many flowers did they plant in total?

20. **CHERRIES** A farmer has 350 cherry trees on his farm. If each cherry tree yields 6200 cherries per year, how many cherries can the farmer harvest per year? Write your answer in scientific notation.

21. **PANDAS** When born, a panda cub weighs only 2.2×10^{-1} pounds. As they grow older, adult pandas, on average, weigh 1.1×10^3 times more than a newborn panda. How much does an average adult panda weigh?

22. **SALES** An automobile company sold 2.3 million new cars in a year. If the average price per car was $21,000, how much money did the company make that year? Write your answer in scientific notation.

23. **APPLE JUICE** An apple juice company produced 8.3 billion individual-size bottles of apple juice last year. If each bottle contains 355 milliliters of juice, how many milliliters of apple juice did the company produce?

Example 5

24. **BLOCKS** Building blocks are in the shape of a cube. The dimensions of the building blocks are based on the length of the packaging box, which is defined as $\frac{1}{12}x^2$ centimeters, where x is the length of the packaging box. What is the volume of a building block in terms of x?

Lesson 8-1 • Multiplication Properties of Exponents **435**

25. **FIELDS** A field is in the shape of a square. The length of the field is $5x^3y^5$. What is the area of the field in terms of x and y?

26. **PICTURE FRAME** Michelle is purchasing a new picture frame to hang in her bedroom. The picture frame is square shaped. What is the area of the interior of the picture frame in terms of c and d?

27. **SHIPPING** A shipping box is in the shape of a cube. What is the volume of the shipping box in terms of m and n?

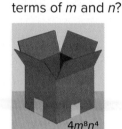
$4m^8n^4$

← $5c^3d^2$ →

Mixed Exercises

Simplify each expression.

28. $(2a^3)^4(a^3)^3$

29. $(c^3)^2(-3c^5)^2$

30. $(2gh^4)^3[(-2g^4h)^3]^2$

31. $(5k^2m)^3[(4km^4)^2]^2$

32. $(p^5r^2)^4(-7p^3r^4)(6pr^3)$

33. $(5x^2y)^2(2xy^3z)^3(4xyz)$

34. $(5a^2b^3c^4)^4(6a^3b^4c^2)$

35. $(10xy^5z^3)(3x^4y^6z^3)$

36. $(0.5x^3)^2$

37. $(0.4h^5)^3$

38. $\left(-\dfrac{3}{4c}\right)^3$

39. $\left(\dfrac{4}{5}a^2\right)^2$

40. $(8y^3)(-3x^2y^2)\left(\dfrac{3}{8}xy^4\right)$

41. $\left(\dfrac{4}{7}m\right)^2(49m)(17p)\left(\dfrac{1}{34}p^5\right)$

42. $(-3r^3w^4)^3(2rw)^2(-3r^2)^3(4rw^2)^3(2r^2w^3)$

43. $(3ab^2c)^2(-2a^2b^4)^2(a^4c^2)^3(a^2b^4c^5)^2(2a^3b^2c^4)^3$

STRUCTURE Determine whether each pair of expressions is equivalent. Write *yes* or *no*.

44. $(a^7)(a^7)(a^7)(a^7)$ and a^{2401}

45. $(-j^9)(-j^9)(-j^9)(-j^9)$ and $-j^{27}$

46. $(5p^3q)(4p^5q^9)$ and $20p^8q^{10}$

47. $(6w^5)^2$ and $36w^{25}$

48. $(x^{10})^3$ and $(x^3)^{10}$

49. $[(2n^2)^3]^2$ and $(4n^4)^3$

436   Module 8 • Exponents and Roots

50. GRAVITY An object that has been falling for x seconds has dropped at an average speed of 16x feet per second. If the object is dropped from a great height, its total distance traveled is the product of the average rate times the time. Write a simplified expression to show the distance the object has traveled after x seconds.

51. ELECTRICITY An electrician uses the formula $W = I^2R$, where W is the power in watts, I is the current in amperes, and R is the resistance in ohms.

  a. Find the power in a household circuit that has $2x^2$ amperes of current and $5x^3$ ohms of resistance.

  b. If the current is reduced by one-half, what happens to the power?

52. MOLECULES A glass of water contains 0.25 liter of water. If 1 milliliter of water contains 3.3×10^{22} water molecules, how many water molecules are there in the glass of water?

53. CIVIL ENGINEERING A developer is planning a sidewalk for a new development. The sidewalk can be installed in rectangular sections that have a fixed width of 3 feet and a length that can vary. Assuming that each section is the same length, express the area of a 4-section sidewalk as a monomial.

54. USE A SOURCE Suppose each student in your school is required to complete 60 hours of community service before they graduate. Find the number of students in the senior class at your school. Based on this number, calculate total minimum number of community service hours completed by the senior class by the time they graduate. Write your answer in scientific notation.

55. REGULARITY Recall that both multiplication and addition are commutative and associative. Multiplication also distributes over addition.

  a. What would it mean for the operation of raising one number to an exponent to be commutative? Decide and explain whether this operation is commutative.

  b. What would it mean for the operation of raising one number to an exponent to be associative? Decide and explain whether this operation is associative.

  c. What would it mean for the process of raising one exponent to another to distribute over addition? Decide and explain whether this is true.

  d. What would it mean for the process of raising one exponent to another to distribute over multiplication? Decide and explain whether this is true.

56. STATE YOUR ASSUMPTION Consider the expressions $(m^p)(m^q)$ and $(m^p)^q$.

a. What must be true about p and q for $(m^p)(m^q) = (m^p)^q$? Give an example.

b. Assuming m, p, and q are all positive integers, can $(m^p)(m^q) > (m^p)^q$ ever be true? Explain.

Higher-Order Thinking Skills

57. PERSEVERE For any nonzero real numbers a and b and any integers m and t, simplify the expression $\left(\dfrac{-a^m}{b^t}\right)^{2t}$. Explain.

58. ANALYZE Consider the equations in the table.

a. For each equation, copy and complete the table to write the related expression and record the power of x.

Equation	Related Expression	Power of x	Linear or Nonlinear
$y = x$			
$y = x^2$			
$y = x^3$			

b. Graph each equation using a graphing calculator.

c. Classify each graph as *linear* or *nonlinear*.

d. Explain how to determine whether an equation, or its related expression, is linear or nonlinear without graphing.

59. CREATE Write three different expressions that can be simplified to x^6.

60. WRITE Write a product of powers that is positive for all nonzero values of the variable. Explain your reasoning.

61. FIND THE ERROR Jade and Sal are each writing an equivalent form of g^5gh^6. Is either correct? Explain your reasoning.

Jade	Sal
$g^5gh^6 =$ $(g \cdot g \cdot g \cdot g \cdot g) \cdot (g) \cdot (h \cdot h \cdot h \cdot h \cdot h \cdot h)$	$g^5gh^6 =$ $(g \cdot g \cdot g \cdot g \cdot g) \cdot (gh \cdot gh \cdot gh \cdot gh \cdot gh \cdot gh)$

Lesson 8-2

Division Properties of Exponents

Explore Quotients of Powers

Online Activity Use an interactive tool to complete the Explore.

> **INQUIRY** How can you determine the quotient of two powers a^m and a^p?

Today's Goals
- Find quotients of monomials.
- Find powers of quotients.

Learn Quotient of Powers

You can use repeated multiplication and the principles for reducing fractions to simplify the quotients of monomials with the same base, like $\frac{2^8}{2^3}$. First, expand the numerator and the denominator. Then, divide the common factors.

$$\frac{2^8}{2^3} = \frac{\overbrace{2 \cdot 2 \cdot 2 \cdot 2 \cdot 2 \cdot 2 \cdot 2 \cdot 2}^{8 \text{ factors}}}{\underbrace{2 \cdot 2 \cdot 2}_{3 \text{ factors}}}$$

$$= 2 \cdot 2 \cdot 2 \cdot 2 \cdot 2$$

$$= 2^5$$

$$\frac{r^5}{r^4} = \frac{\overbrace{r \cdot r \cdot r \cdot r \cdot r}^{5 \text{ factors}}}{\underbrace{r \cdot r \cdot r \cdot r}_{4 \text{ factors}}}$$

$$= r$$

Think About It!
What steps would you take to simplify $\frac{10^{12}}{10^9}$?

These examples demonstrate the Quotient of Powers Property.

Key Concept • Quotient of Powers	
Words	To divide two powers with the same base, subtract the exponents.
Symbols	For any nonzero number a, and any integers m and p, $\frac{a^m}{a^p} = a^{m-p}$.
Examples	$\frac{b^{12}}{b^9} = b^{12-9} = b^3$; $\frac{w^6}{w^2} = w^{6-2} = w^4$

Go Online You can complete an Extra Example online.

🧠 **Think About It!**
In the expression 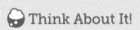$\frac{b^5 c^7}{b^2 c}$, why can you not subtract 5 − 7 in the numerator and 2 − 1 in the denominator and simplify?

🧠 **Think About It!**
Why must you assume that the denominator does not equal 0?

Example 1 Quotient of Powers

Simplify $\frac{b^5 c^7}{b^2 c}$. Assume that the denominator does not equal zero.

Step 1 Group powers with the same base.

$$\frac{b^5 c^7}{b^2 c} = \left(\frac{b^5}{b^2}\right)\left(\frac{c^7}{c}\right)$$

Step 2 Use the Quotient of Powers Property.

$$\left(\frac{b^5}{b^2}\right)\left(\frac{c^7}{c}\right) = (b^{5-2})(c^{7-1}) \quad \text{Subtract the exponents in each group.}$$

$$= b^3 c^6 \quad \text{Simplify.}$$

Check

Simplify $\frac{a^2 b^9 \, cd^4}{b^8 cd}$. Assume that the denominator does not equal zero.

A. abd^3

B. $a^2 bcd^3$

C. $a^2 bcd^4$

D. $a^2 bd^3$

🌐 Example 2 Apply Division of Monomials

CHEMISTRY At sea level, there are about 10^{25} molecules in a cubic liter of air. In the stratosphere, about 30 kilometers above the Earth's surface, the same cubic liter of air has about 10^{23} molecules. Approximately how many times as many molecules are there in a cubic liter of air at sea level as there are in the stratosphere?

$$\frac{a^m}{a^p} = a^{m-p} = 10^{25-23} = 10^2$$

There are about 100 times as many molecules in a cubic liter of air at sea level as there are in the stratosphere.

Check

SHOPPING Canada's West Edmonton Mall claims to have the largest parking lot in the world, with about 3^9 parking spaces. California's Glendale Galleria has about 3^8 parking spaces.

The West Edmonton Mall has ___?___ times as many parking spaces as the Glendale Galleria.

🔎 **Go Online** You can complete an Extra Example online.

440 Module 8 • Exponents and Roots

Learn Power of a Quotient

You can use the Product of Powers Property to find the powers of quotients for monomials.

$$\left(\frac{2}{5}\right)^3 = \overbrace{\left(\frac{2}{5}\right)\left(\frac{2}{5}\right)\left(\frac{2}{5}\right)}^{3 \text{ factors}} = \frac{\overbrace{2 \cdot 2 \cdot 2}^{3 \text{ factors}}}{\underbrace{5 \cdot 5 \cdot 5}_{3 \text{ factors}}} = \frac{2^3}{5^3}$$

$$\left(\frac{p}{q}\right)^2 = \overbrace{\left(\frac{p}{q}\right)\left(\frac{p}{q}\right)}^{2 \text{ factors}} = \frac{\overbrace{p \cdot p}^{2 \text{ factors}}}{\underbrace{q \cdot q}_{2 \text{ factors}}} = \frac{p^2}{q^2}$$

These examples demonstrate the Power of a Quotient Property.

Key Concept • Power of a Quotient	
Words	To find the power of a quotient, find the power of the numerator and the power of the denominator.
Symbols	For any real numbers a and $b \neq 0$, and any integer m, $\left(\frac{a}{b}\right)^m = \frac{a^m}{b^m}$.
Examples	$\left(\frac{1}{4}\right)^5 = \frac{1^5}{4^5}$; $\left(\frac{c}{d}\right)^6 = \frac{c^6}{d^6}$

💭 **Think About It!**

How would you simplify $\left(\frac{1}{2}\right)^2$?

Example 3 Power of a Quotient

Simplify $\left(\frac{5a^2}{6}\right)^3$.

$$\left(\frac{5a^2}{6}\right)^3 = \frac{(5a^2)^3}{6^3} \quad \text{Power of a Quotient}$$

$$= \frac{5^3(a^2)^3}{6^3} \quad \text{Power of a Product}$$

$$= \frac{125a^6}{216} \quad \text{Power of a Power}$$

Study Tip

Power Rules with Variables The power rules apply to variables and numbers. For example, $\left(\frac{4m}{2n}\right)^4 = \frac{(4m)^4}{(2n)^4} = \frac{4^4 m^4}{2^4 n^4} = \frac{256 m^4}{16 n^4}$ or $\frac{16 m^4}{n^4}$.

Check

Simplify $\left(\frac{5j^3 l^2}{7k^5}\right)^4$. Assume that the denominator does not equal zero.

A. $\frac{5j^{12} l^8}{7k^{20}}$

B. $\frac{625 j^{12} l^8}{2401 k^{20}}$

C. $\frac{625 j^7 l^6}{2401 k^9}$

D. $\frac{625 j^{12} l^8}{7k^5}$

Go Online You can complete an Extra Example online.

Think About It!
To how many factors does the exponent 2 need to be applied? Name them.

Example 4 Power of a Quotient with Variables

Simplify $\left(\dfrac{x^4 y}{xyz}\right)^2$. **Assume that the denominator does not equal zero.**

Write the appropriate justification next to each step.

$\left(\dfrac{x^4 y}{xyz}\right)^2 = \dfrac{(x^4 y)^2}{(xyz)^2}$ Power of a Quotient

$= \dfrac{(x^4)^2 y^2}{x^2 y^2 z^2}$ Power of a Product

$= \dfrac{x^8 y^2}{x^2 y^2 z^2}$ Power of a Power

$= \dfrac{x^6}{z^2}$ Quotient of Powers

Check

Simplify $\left(\dfrac{4m^2 n^2 p^2}{3mp}\right)^4$. Assume that the denominator does not equal zero.

A. $\dfrac{256 mn^2 p}{81}$

B. $\dfrac{256 m^4 n^6 p^4}{81}$

C. $\dfrac{256 m^4 n^8 p^4}{81}$

D. $\dfrac{256 m^4 n^8 p^4}{81 mp}$

Go Online to practice what you've learned about simplifying expressions involving exponents in the Put It All Together over Lessons 8-1 and 8-2.

Go Online You can complete an Extra Example online.

Practice

Go Online You can complete your homework online.

Examples 1, 3, and 4

Simplify each expression. Assume that no denominator equals zero.

1. $\dfrac{m^4 p^2}{m^2 p}$

2. $\dfrac{p^{12} t^3 r}{p^2 t r}$

3. $\dfrac{c^4 d^4 f^3}{c^2 d^4 f^3}$

4. $\left(\dfrac{3xy^4}{5z^2}\right)^2$

5. $\left(\dfrac{p^2 t^7}{10}\right)^3$

6. $\dfrac{a^7 b^8 c^8}{a^5 b c^7}$

7. $\left(\dfrac{3np^3}{7q^2}\right)^2$

8. $\left(\dfrac{2r^3 t^6}{5u^9}\right)^4$

9. $\left(\dfrac{3m^5 r^3}{4p^8}\right)^4$

10. $\dfrac{p^{12} t^7 r^2}{p^2 t^7 r}$

11. $\dfrac{k^4 m^3 p^2}{k^2 m^2}$

12. $\dfrac{m^7 p^2}{m^3 p^2}$

13. $\dfrac{32 x^3 y^2 z^5}{-8 xyz^2}$

14. $\left(\dfrac{4p^7}{7r^2}\right)^2$

15. $\dfrac{9d^7}{3d^6}$

16. $\dfrac{12n^5}{36n}$

17. $\dfrac{w^1 x^3}{w^4 x}$

18. $\dfrac{u^3 v^5}{ab^2}$

Example 2

19. **SPACE** The Moon is approximately 25^4 kilometers away from Earth on average. The Olympus Mons volcano on Mars stands 25 kilometers high. How many Olympus Mons volcanoes, stacked on top of one another, would fit between the surface of Earth and the Moon?

20. **GEOMETRY** Write the ratio of the area of a circle with radius r to the circumference of the same circle.

21. **COMBINATIONS** The number of four-letter combinations that can be formed with the English alphabet is 26^4. The number of six-letter combinations that can be formed is 26^6. How many times more six-letter combinations can be formed than four-letter combinations?

22. **BLOOD COUNT** A lab technician draws a sample of blood. A cubic millimeter of the blood contains 22^5 red blood cells and 22^3 white blood cells. How many times more red blood cells are there than white blood cells?

Lesson 8-2 • Division Properties of Exponents   443

Mixed Exercises

Simplify each expression. Assume that no denominator equals zero.

23. $\dfrac{-4w^{12}}{12w^3}$

24. $\dfrac{13r^7}{39r^4}$

25. $\left(\dfrac{2a^4c^3}{5b^2d^2}\right)^2$

26. $\dfrac{m^6 n^3}{m^2 n^6}$

27. $\left(\dfrac{24a^{11}b^{16}c^6}{18a^6 b^6 c^6}\right)^3$

28. $\left(\dfrac{q^2 r^3}{qr^2}\right)^5$

29. $\dfrac{-16x^7 y^2}{-6xy^5}$

30. $\dfrac{21d^{18}e^5}{7d^{11}e^6}$

31. $\left(\dfrac{2x^3 y^2 z^5}{3xyz}\right)^3$

32. $\dfrac{8c^4 d^2 f^9}{4cd^2 f^3}$

33. $\left(\dfrac{7c^4}{14d^2}\right)^6$

34. $\left(\dfrac{6j^5}{7m^6 n^3}\right)^2$

35. **SOUND** Decibels are used to measure sound. The softest sound that can be heard is rated at 0 decibels, or a relative loudness of 1. Ordinary conversation is rated at about 60 decibels, or a relative loudness of 10^6. A stock car race is rated at about 130 decibels, or a relative loudness of 10^{13}. How many times greater is the relative loudness of a stock car race than the relative loudness of ordinary conversation?

36. **COMPUTERS** The byte is the fundamental unit of computer processing. Almost all aspects of a computer's performance and specifications are measured in bytes or multiples of bytes. The byte is based on powers of 2, as shown in the table. How many times greater is a megabyte than a kilobyte?

Memory Term	Number of Bytes
byte	2^0 or 1
kilobyte	2^{10}
megabyte	2^{20}
gigabyte	2^{30}

37. **AREA** The area of the triangle shown is $6x^5 y^3$. Find the base of the triangle.

38. **AREA** The area of the rectangle in the figure is $32xy^3$ square units. Find the width of the rectangle.

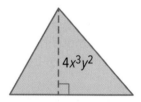

39. **USE A MODEL** An investment is expected to increase in value by 4% every year.

   a. Write an expression that represents the value of the investment after t years if the initial value was n dollars.

   b. By what percent does the value of the investment change between the end of year 2 and the end of year 8? Round your answer to the nearest tenth of a percent, and show your work.

40. **INVESTMENTS** A poor investment is expected to decrease in value by 5% every year.

   a. If the initial value of the investment was $100, what does it mean for it to decrease in value by 5% in the first year?

   b. Erik claims that rather than multiplying by 0.05 and subtracting, we can simply multiply the investment by 0.95. Is he correct? Explain.

   c. Write an expression that represents the value of the investment after t years if the initial value was n dollars.

   d. By what percent does the value of the investment change between the end of year 2 and the end of year 10? Round your answer to the nearest tenth of a percent, and show your work.

41. **PAPER FOLDING** If you fold a sheet of paper in half, you have a thickness of 2 sheets. Folding again, you have a thickness of 4 sheets. Fold the paper in half one more time. How many times thicker is a sheet that has been folded 3 times than a sheet that has not been folded?

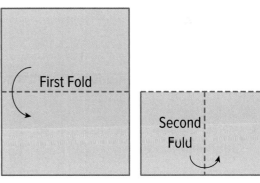

42. **USE TOOLS** Create a table in a computer spreadsheet program to show the powers of 2^0 through 2^{10}. Use the formula functions in the program to show the Quotient of Powers Property is true for five different division problems using the powers you already entered.

43. **CONSTRUCT ARGUMENTS** Is $\left(\frac{x}{y^2}\right)^3$ the same as $\left(\frac{x}{y^3}\right)^2$? Justify your argument.

44. **STRUCTURE** Find the value of x that makes $\frac{8^{5x}}{8^{4x+1}} = 8^9$ true. Explain.

45. **REASONING** Which expression does not have the same answer as the others when simplified using exponent rules?

   A. $\left(\frac{3x^4y}{5}\right)^2$ B. $\frac{-18x^{14}y^{10}}{-50x^6y^8}$ C. $\frac{-3^2x^9y^5}{(-5)^2xy^3}$

46. **PRECISION** What error did the student make in simplifying the expression $\frac{5^4}{5} = 1^4 = 1$. What is the correct value of the expression?

47. REGULARITY Simplify the expression $\dfrac{a^{x+y}}{a^y}$.

48. CREATE Write three quotients of powers expressions that are equivalent to 2^5.

49. REGULARITY Consider the equation $\dfrac{3^h}{3^k} = 3^2$.

   a. Find two numbers h and k that satisfy the equation.

   b. Are there any other pairs of numbers that satisfy the equation? Explain.

50. FIND THE ERROR Kathryn and Salvador used different methods to simplify $\left(\dfrac{p^9}{p^5}\right)^2$. Is either correct? Explain your reasoning.

Kathryn

$\left(\dfrac{p^9}{p^5}\right)^2 = \dfrac{p^{18}}{p^{10}} = p^8$

Salvador

$\left(\dfrac{p^9}{p^5}\right)^2 = (p^4)^2 = p^8$

51. ANALYZE Is $x^y \cdot x^z = x^{yz}$ *sometimes*, *always*, or *never* true? Justify your argument.

52. CREATE Write two monomials with a quotient of $24a^2b^3$.

53. WRITE Explain how to use the Quotient of Powers Property and the Power of a Quotient Property.

54. PERSEVERE Is the expression $\left(\dfrac{p^7}{p^3}\right)^5$ positive or negative for all nonzero values of p? Explain.

55. WHICH ONE DOESN'T BELONG? Which quotient does *not* belong with the other three? Justify your conclusion.

$\dfrac{6^7}{6^2}$ $\dfrac{(-7)^3}{(-7)^2}$ $\dfrac{6^5}{6^3}$ $\dfrac{(-3)^8}{(-2)^4}$

56. FIND THE ERROR Andrew and Mateo are trying to simplify the expression $\dfrac{9^5}{9^3}$. Is either correct? Explain your reasoning.

Andrew

$\dfrac{9^5}{9^3} = 9^2$

Mateo

$\dfrac{9^5}{9^3} = \dfrac{1}{9^2}$

446 Module 8 • Exponents and Roots

Lesson 8-3

Negative Exponents

Learn Zero Exponent

Key Concept • Zero Exponent Property

Words	Any nonzero number raised to the zero power is equal to 1.
Symbols	For any nonzero number a, $a^0 = 1$.
Examples	$30^0 = 1;\ \left(\frac{x}{y}\right)^0 = 1;\ \left(\frac{8}{3}\right)^0 = 1$

Example 1 Zero Exponent

Simplify each expression. Assume that no denominator equals zero.

a. $\left(-\dfrac{8m^2np^8}{9k^2mn^4}\right)^0$

$\left(-\dfrac{8m^2np^8}{9k^2mn^4}\right)^0 = 1 \qquad a^0 = 1$

b. $\dfrac{a^4 b^0}{a^2}$

$\dfrac{a^4 b^0}{a^2} = \dfrac{a^4(1)}{a^2} \qquad b^0 = 1$

$\phantom{\dfrac{a^4 b^0}{a^2}} = a^2 \qquad$ Quotient of Powers

Check

Select the simplified form of $\dfrac{56g^2 h j^{11}}{8g^0 h^0 j^0}$. Assume that the denominator does not equal zero.

A. $7g^2 h j^{11}$

B. $7gj^{10}$

C. $7ghj^{10}$

D. $7g^2 h j^{10}$

Explore Simplifying Expressions with Negative Exponents

 Online Activity Use an interactive tool to complete the Explore.

> @ **INQUIRY** How can you simplify expressions with negative exponents?

 Go Online You can complete an Extra Example online.

Today's Goals
- Simplify expressions containing zero and negative exponents.
- Simplify expressions containing negative exponents.

Today's Vocabulary
negative exponent

 Talk About It!
Why is the answer to part **b** not 1?

 Go Online An alternate method is available for this example.

Lesson 8-3 • Negative Exponents   447

Learn Negative Exponents

Any nonzero real number can be raised to a negative power. That power is called a **negative exponent**.

Key Concept • Negative Exponent Property	
Words	For any nonzero number a and any integer n, a^{-n} is the reciprocal of a^n. Also, the reciprocal of a^{-n} is a^n.
Symbols	For any nonzero number a and any integer n, $a^{-n} = \frac{1}{a^n}$.
Examples	$3^{-5} = \frac{1}{3^5} = \frac{1}{243}$; $\frac{1}{m^{-2}} = m^2$

An expression is considered simplified when:

- it contains only positive exponents.
- each base appears exactly once.
- there are no powers of powers.
- all fractions are in simplest form.

🧠 Think About It!
Describe how to simplify an expression of the form a^{-n}.

Example 2 Negative Exponents

Simplify $\frac{a^3 b^{-4}}{c^{-2}}$. Assume that the denominator does not equal zero.

$$\frac{a^3 b^{-4}}{c^{-2}} = \left(\frac{a^3}{1}\right)\left(\frac{b^{-4}}{1}\right)\left(\frac{1}{c^{-2}}\right) \quad \text{Write as a product of fractions.}$$

$$= \left(\frac{a^3}{1}\right)\left(\frac{1}{b^4}\right)\left(\frac{c^2}{1}\right) \quad a^{-n} \text{ is } \frac{1}{a^n}; \frac{1}{a^{-n}} = a^n$$

$$= \frac{a^3 c^2}{b^4} \quad \text{Multiply.}$$

Check

Select the simplified form of $\frac{42 r^{-2} t^{-6} u^3}{14 r^2 u^{-3}}$. Assume that the denominator does not equal zero.

A. $\frac{3}{r^4 t^6 u^6}$

B. $\frac{3 u^6}{r^4 t^6}$

C. $\frac{3 t^6 u^6}{r^4}$

D. $3 r^4 t^6 u^6$

Math History Minute

In the 15th century, French physician **Nicolas Chuquet (c. 1445–1488)** wrote a book called *Triparty en la science des nombres*, or *A Three-Part Book on the Science of Numbers*, which contained early notation for zero and negative exponents.

▶ **Go Online** You can complete an Extra Example online.

448 Module 8 • Exponents and Roots

Example 3 Simplify an Expression with Negative Exponents

Simplify $\dfrac{3g^{-3}h^2}{-36g^3hj^{-4}}$. Assume that the denominator does not equal zero.

$\dfrac{3g^{-3}h^2}{-36g^3hj^{-4}} = \left(-\dfrac{3}{36}\right)\left(\dfrac{g^{-3}}{g^3}\right)\left(\dfrac{h^2}{h}\right)\left(\dfrac{1}{j^{-4}}\right)$ Group powers with the same base.

$= \left(-\dfrac{1}{12}\right)(g^{-3-3})(h^{2-1})(j^4)$ Quotient of Powers and Negative Exponent Properties

$= \left(-\dfrac{1}{12}\right)g^{-6}hj^4$ Simplify.

$= \left(-\dfrac{1}{12}\right)\left(\dfrac{1}{g^6}\right)hj^4$ Negative Exponent Property

$= -\dfrac{hj^4}{12g^6}$ Multiply.

Check

Select the simplified form of $\dfrac{35h^0j^{-5}k^{-2}}{14h^2m^5}$. Assume that the denominator does not equal zero.

A. $\dfrac{5}{2hj^5m^5k^2}$

B. $\dfrac{5h^2}{2hj^5m^5k^2}$

C. $\dfrac{5}{2h^2j^5m^5k^2}$

D. $\dfrac{5h^2m^5}{2j^5k^2}$

💭 **Think About It!**
Why do you not leave the answer as $\left(-\dfrac{1}{12}\right)g^{-6}hj^4$? Justify your argument.

Order of magnitude is used to compare measures and to estimate and perform rough calculations. The **order of magnitude** of a quantity is the number rounded to the nearest power of 10. For example, the power of 10 closest to 105,000,000 is 10^8, or 100,000,000. So the order of magnitude of 105,000,000 is 10^8.

🌐 Apply Example 4 Apply Properties of Exponents

SPEED The maximum speed of a peregrine falcon is about 90 meters per second. The maximum speed of a garden snail is about 0.01 meter per second. How many orders of magnitude as fast as a peregrine falcon is a garden snail?

1. **What is the task?**
Describe the task in your own words. Then list any questions that you may have. How can you find answers to your questions?

The garden snail is how many orders of magnitude as fast as the falcon? What are the orders of magnitude of each animal? And then what is the ratio of the two orders of magnitude?

(continued on the next page)

🌐 **Go Online** You can complete an Extra Example online.

2. **How will you approach the task? What have you learned that you can use to help you complete the task?**

First, I will compute the orders of magnitude for each animal. Then I will find the ratio of the orders of magnitude of the snail to the falcon. I have learned how to find the order of magnitude.

3. **What is your solution?**
Use your strategy to solve the problem.

The maximum speed of a snail in 0.01 m/s. So, the order of magnitude of the speed of the snail is 10^{-2}.

The maximum speed of a falcon is close to 100 m/s. So, the order of magnitude of the speed of the falcon is 10^2.

What is the ratio of the order of magnitude of the snail to the order of magnitude of the falcon?

$$\frac{\text{order of magnitude of snail}}{\text{order of magnitude of falcon}} = \frac{10^{-2}}{10^2}$$

A snail is approximately 0.0001 times as fast as a falcon, or a snail is −4 orders of magnitude as fast as a falcon.

4. **How can you know that your solution is reasonable?**

✏️ **Write About It!** **Write an argument that can be used to defend your solution.**

The ratio of the snail's speed to the falcon's speed is $\frac{0.01}{90} \approx 0.0001$ or 10^{-4}. Because approaching the problem in a different way yields the same result, the answer is reasonable.

Check

LENGTH The radius of Earth is about 10,000,000 meters. The radius of a virus is about 0.0000001 meter. Approximately how many orders of magnitude as long as the radius of a virus is the radius of Earth?

Part A Using estimation, how many times greater is the radius of Earth than the radius of a virus?

A. 0.0001

B. 10

C. 1,000,000

D. 100,000,000,000,000

Part B The radius of Earth is about ___?___ orders of magnitude greater than the radius of a virus.

▶ **Go Online** You can complete an Extra Example online.

Practice

Go Online You can complete your homework online.

Examples 1–3

Simplify each expression. Assume that no denominator equals zero.

1. $\dfrac{r^6 n^{-7}}{r^4 n^2}$

2. $\dfrac{h^3}{h^{-6}}$

3. $\dfrac{f^{-7}}{f^4}$

4. $\left(\dfrac{16p^5 w^2}{2p^3 w^3}\right)^0$

5. $\dfrac{f^{-5} g^4}{h^{-2}}$

6. $\dfrac{15 x^6 y^{-9}}{5xy^{-11}}$

7. $\dfrac{-15 t^0 u^{-1}}{5 u^3}$

8. $\dfrac{(z^2 w^{-1})^3}{(z^3 w^2)^2}$

9. $\dfrac{-10 m^{-1} y^0 r}{-14 m^{-7} y^{-3} r^{-4}}$

10. $\dfrac{51 x^{-1} y^3}{17 x^2 y}$

11. $\dfrac{3 m^{-3} r^4 p^2}{12 t^4}$

12. $\left(\dfrac{3 t^6 u^2 v^5}{9 t u v^{21}}\right)^0$

13. $\dfrac{x^{-4} y^9}{z^{-2}}$

14. $\left(-\dfrac{5 f^9 g^4 h^2}{f g^2 h^3}\right)^0$

15. $\dfrac{p^4 t^{-3}}{r^{-2}}$

16. $-\dfrac{5 c^2 d^5}{8 c d^5 f^0}$

17. $\dfrac{-2 f^3 g^2 h^0}{8 f^2 g^2}$

18. $\dfrac{g^0 h^7 j^{-2}}{g^{-5} h^0 j^{-2}}$

Example 4

19. **METRIC MEASUREMENT** Consider a dust mite that measures 10^{-3} millimeters in length and a gecko that measures 10 centimeters long. How many orders of magnitude as long as the mite is the gecko?

20. **CHEMISTRY** The nucleus of a certain atom is 10^{-13} centimeters across. If the nucleus of a different atom is 10^{-11} centimeters across, how many orders of magnitude as great is the second nucleus?

21. **WEIGHT** A paper clip weighs about 10^{-3} kilograms. A draft horse weighs about 10^3 kilograms. How many orders of magnitude as heavy is a draft horse than a paper clip?

22. **GDP** Gross Domestic Product (GDP) is a measure of a country's wealth. Indonesia had an estimated GDP in 2017 of 1,020,515,000,000. Madagascar had an estimated GDP of 10,372,000,000 in 2017. About how many orders of magnitude as great is Indonesia's GDP?

23. **GROWTH** An old oak tree is 1012 inches tall. A younger oak tree near it is 98 inches tall. About how many orders of magnitude as tall is the old oak tree?

Lesson 8-3 • Negative Exponents 451

Mixed Exercises

Simplify each expression. Assume that no denominator equals zero.

24. $\dfrac{3wy^{-2}}{(w^{-1}y)^3}$

25. $\dfrac{(4k^3m^2)^3}{(5k^2m^{-3})^{-2}}$

26. $\dfrac{-12c^3d^0f^{-2}}{6c^5d^{-3}f^4}$

27. $\dfrac{20qr^{-2}t^{-5}}{4q^0r^4t^{-2}}$

28. $\dfrac{(5pr^{-2})^{-2}}{(3p^{-1}r)^3}$

29. $\dfrac{(2g^3h^{-2})^2}{(g^2h^0)^{-3}}$

30. $\left(\dfrac{2a^{-2}b^4c^2}{-4a^{-2}b^{-5}c^{-7}}\right)^{-1}$

31. $\left(\dfrac{-3x^{-6}y^{-1}z^{-2}}{6x^{-2}yz^{-5}}\right)^{-2}$

32. $\left(\dfrac{4^0c^2d^3f}{2c^{-4}d^{-5}}\right)^{-3}$

33. $\dfrac{(16x^2y^{-1})^0}{(4x^0y^{-4}z)^{-2}}$

34. **RATIOS** Yvonne is comparing the weights of a semi-truck trailer tire and a mobile home. A semi-truck trailer tire weighs about 10^2 pounds, and a mobile home weighs about 10^4 pounds. What is the ratio of the weight of a semi-truck trailer tire compared to the weight of a mobile home? Write your answer as a monomial.

35. **MARKETING** Jana's marketing plan has a goal that each person who hears of their new product tells 10 people about it, each of those 10 people tells another 10, and so on. The level of spread is the number of times this repeats, as shown in the table. How many orders of magnitude greater is the level 4 spread compared to the level 1 spread?

Level	0	1	2	3	4
Spread	1	10	100	1000	10,000
Powers	10^0	10^1	10^2	10^3	10^4

36. **ASTRONOMY** The diagram shows the distance from Earth to the Sun and the distance from Saturn to the Sun. Approximately what portion of the distance from Saturn to the Sun is the distance from Earth to the Sun?

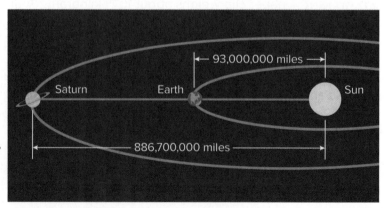

452 Module 8 • Exponents and Roots

37. **COMPUTERS** In 1995, standard capacity for a personal computer hard drive was 40 megabytes (MB). In 2010, a standard hard drive capacity was 500 gigabytes (GB). Refer to the table.

Memory Capacity Approximate Conversions
8 bits = 1 byte
10^3 bytes = 1 kilobyte
10^3 kilobytes = 1 megabyte
10^3 megabytes = 1 gigabyte
10^3 gigabytes = 1 terabyte
10^3 terabytes = 1 petabyte

   a. The newer hard drives have about how many times the capacity of the 1995 drives?

   b. Predict the hard drive capacity in the year 2025 if the capacity continues to grow by the same factor you found in part **a**.

   c. One kilobyte of memory is what fraction of one terabyte?

38. **MICROSCOPES** Cheveyo is looking at a slide of red blood cells under a microscope. One of the red blood cells has an actual radius of about one millionth of a meter, or 10^{-6} meters. Through the microscope, the radius of that red blood cell appears to be 10^{-2} meters. How many times larger does the microscope make the blood cells appear?

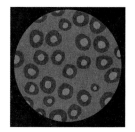

39. **STRUCTURE** Are $4n^5$ and $4n^{-5}$ reciprocals? Explain.

40. **REGULARITY** Explain why $3^0 = 1$.

41. **STRUCTURE** Use the Power of a Power Property to rewrite $\frac{1}{p^{-3}}$ with only positive exponents.

42. **REASONING** Angelo believes that $y^6 \geq y^{-6}$ for all nonzero values of y. Do you agree? Explain.

43. **BACTERIA** The bacterial population in a petri dish was measured at 1500 per cubic centimeter Monday morning. Each day the population doubled from the previous day. This can be modeled by $P = 1500(2^d)$, where d is the day and P is the population. If the population was first measured when $d = 0$, find the population of the bacteria the day before, when $d = -1$.

44. **VEHICLE DEPRECIATION** Jay buys a used car for $12,000. If the car loses about 10% of its value every year, the value of the car can be modeled by the equation $V = 12,000(0.90^y)$, where y is the number of years and V is the value of the car.

   a. Use the equation to determine the value of Jay's car 2 years before he purchased it.

   b. How much money did Jay save by waiting to purchase the car?

45. Consider the expression $\left(\dfrac{4a^3}{2a^{-2}}\right)^4$.

    a. Simplify $\left(\dfrac{4a^3}{2a^{-2}}\right)^4$ by using the Quotient of Powers Property first. Then use the Power of a Power Property.

    b. Simplify $\left(\dfrac{4a^3}{2a^{-2}}\right)^4$ by using the Power of a Quotient Property first. Then use the Quotient of Powers Property.

    c. Write a statement that generalizes the results of part **a** and part **b**.

46. **REASONING** Dalip evaluated -5^0 on his calculator.

    a. What answer did the calculator give Dalip when he evaluated -5^0?

    b. Is this a counterexample that contradicts the Zero Exponent Property? Explain.

47. **FIND THE ERROR** Colleen and Tyler are using different steps to simplify $\left(\dfrac{x^2}{x^9}\right)^3$. Their work is shown in the boxes.

    a. Which student is correct? Explain your reasoning.

    b. Which student's answer is in simplest form?

    Colleen
    $\left(\dfrac{x^2}{x^9}\right)^3 = \dfrac{x^6}{x^{27}}$
    $= \dfrac{1}{x^{21}}$

    Tyler
    $\left(\dfrac{x^2}{x^9}\right)^3 = \left(\dfrac{1}{x^7}\right)^3$
    $= x^{-21}$

48. **PERSEVERE** Use the Quotient of Powers Property to explain why $x^{-n} = \dfrac{1}{x^n}$.

49. **FIND THE ERROR** Lola claims that we never need to use the Division Property of Exponents. She says that exponents in the denominator can be written as negative exponents, and then the Multiplication Property of Exponents can be used instead. Is Lola correct? If so, give an example. If not, explain why.

50. **WRITE** Explain how to evaluate the expression 0^0.

51. **CREATE** Measure something large and something very small in the same units. Write and solve a problem comparing the order of magnitude of the two measurements.

52. **ANALYZE** Consider the expression, $\dfrac{a^b c^{-d}}{w^x y^{-z}}$, where all values are nonnegative. Make a conjecture about why only c and w cannot be zero.

53. **ANALYZE** Dale wrote the value of several powers of 2 on a piece of paper. Describe the pattern shown. Then explain how to use the pattern to find 2^0.

    $2^4 = 2 \times 2 \times 2 \times 2 = 16$

    $2^3 = 2 \times 2 \times 2 = 8$

    $2^2 = 2 \times 2 = 4$

    $2^1 = 2 = 2$

Lesson 8-4

Rational Exponents

Explore Expressions with Rational Exponents

 Online Activity Use an interactive tool to complete the Explore.

> **INQUIRY** How can you simplify expressions with rational exponents?

Learn nth Roots

Not all exponents are integers. Exponents that are expressed as a fraction are called **rational exponents**. One of the most commonly used rational exponents is $\frac{1}{2}$. Expressions with an exponent of $\frac{1}{2}$ can also be represented as a square root.

Key Concept • Powers of One Half

Words For any nonnegative real number b, $b^{\frac{1}{2}} = \sqrt{b}$.

Example $64^{\frac{1}{2}} = \sqrt{64}$ or 8

You are likely familiar with the square root of a number a. In the same way, you can find other roots of numbers. For example, if $a^3 = b$, then a is the cube root of b, and if $a^n = b$ for a positive integer n, then a is the **nth root** of b.

For example, $\sqrt[5]{18}$ is read as *the fifth root of 18*. In this example, 5 is the **index** and 18 is the **radicand**, which is the expression inside the radical symbol.

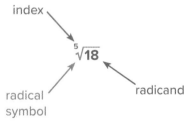

Key Concept • nth Roots

Words For any real numbers a and b and any positive integer n, if $a^n = b$, then a is the nth root of b.

Symbols If $a^n = b$, then $a = \sqrt[n]{b}$.

Example Because $3^4 = 81$, 3 is the fourth root of 81; $\sqrt[4]{81} = 3$.

Today's Goals
- Rewrite expressions involving nth roots and rational exponents.
- Rewrite expressions involving powers of nth roots and rational exponents.

Today's Vocabulary
rational exponent
nth root
index
radicand

Go Online You may want to complete the Concept Check to check your understanding.

Go Online You can complete an Extra Example online.

Lesson 8-4 • Rational Exponents **455**

Since $3^2 = 9$ and $(-3)^2 = 9$, both 3 and -3 are square roots of 9. Similarly, since $2^4 = 16$ and $(-2)^4 = 16$, both 2 and -2 are fourth roots of 16. The positive roots are called *principal roots*. Radical symbols indicate principal roots, so $\sqrt[4]{16} = 2$.

Key Concept • Rational Exponents	
Words	For any nonnegative real number b and any positive integer n, $b^{\frac{1}{n}} = \sqrt[n]{b}$.
Example	$32^{\frac{1}{5}} = \sqrt[5]{32} = \sqrt[5]{2 \cdot 2 \cdot 2 \cdot 2 \cdot 2}$ or 2

Example 1 Radical and Exponential Forms

Write each expression in radical or exponential form.

a. $10^{\frac{1}{2}}$

By the definition of $b^{\frac{1}{2}}$, $10^{\frac{1}{2}} = \sqrt{10}$.

b. $\sqrt{7k^2}$

Because $7k^2$ is all under the radical symbol, both 7 and k^2 must be raised to the $\frac{1}{2}$ power. So, $\sqrt{7k^2} = 7^{\frac{1}{2}}k$.

Check

Write each expression in radical or exponential form.

a. $\sqrt{16p} = $ __?__ b. $34^{\frac{1}{2}} = $ __?__
c. $8\sqrt{p} = $ __?__ d. $(3p)^{\frac{1}{2}} = $ __?__

Example 2 Evaluate *n*th Roots

Evaluate each expression.

a. $\sqrt[5]{243}$

$\sqrt[5]{243} = \sqrt[5]{3 \cdot 3 \cdot 3 \cdot 3 \cdot 3}$ $\qquad 243 = 3 \cdot 3 \cdot 3 \cdot 3 \cdot 3$

$\qquad\quad = 3$ $\qquad\qquad\qquad\qquad\qquad$ Simplify.

b. $\sqrt[4]{625}$

$\sqrt[4]{625} = \sqrt[4]{5 \cdot 5 \cdot 5 \cdot 5}$ $\qquad 625 = 5 \cdot 5 \cdot 5 \cdot 5$

$\qquad\quad = 5$ $\qquad\qquad\qquad\qquad\qquad$ Simplify.

Check

Evaluate each expression.

a. $\sqrt[4]{4096}$ b. $\sqrt[5]{16{,}807}$

Go Online You can complete an Extra Example online.

Think About It!
How do $(6x)^{\frac{1}{2}}$ and $6x^{\frac{1}{2}}$ differ?

Study Tip

Calculators You can use a calculator to find *n*th roots. On many calculators, enter the value of *n*, press MATH, and choose $\sqrt[x]{}$. Then enter the radicand and compute.

Example 3 Evaluate Exponential Expressions with Rational Exponents

Evaluate $4096^{\frac{1}{6}}$.

$4096^{\frac{1}{6}} = \sqrt[6]{4096}$ $b^{\frac{1}{n}} = \sqrt[n]{b}$

$\sqrt[6]{4096} = \sqrt[6]{4 \cdot 4 \cdot 4 \cdot 4 \cdot 4 \cdot 4}$ $2096 = 4 \cdot 4 \cdot 4 \cdot 4 \cdot 4 \cdot 4$

$= 4$ Simplify.

Talk About It!
Malik says that $\sqrt[3]{64} = \sqrt[3]{8 \cdot 8} = 8$. Is he correct? If not, correct his mistake.

Check

Evaluate each expression.

a. $\left(\dfrac{81}{256}\right)^{\frac{1}{4}}$

b. $100{,}000^{\frac{1}{5}}$

Learn Powers of *n*th Roots

Key Concept • Powers of *n*th Roots

Words	For any real number b and any integers m and $n > 1$, $b^{\frac{m}{n}} = \left(\sqrt[n]{b}\right)^m$ or $\sqrt[n]{b^m}$.
Examples	$216^{\frac{4}{3}} = \left(\sqrt[3]{216}\right)^4 = 6^4$ or 1296 $32^{\frac{2}{5}} = \sqrt[5]{32^2} = \sqrt[5]{1024} = \sqrt[5]{4 \cdot 4 \cdot 4 \cdot 4 \cdot 4}$ or 4

Sometimes, the exponent can be simplified before it is evaluated. For example,

$4^{\frac{15}{5}} = 4^3$ $\dfrac{15}{5} = 3$

$= 4 \cdot 4 \cdot 4$ $4^3 = 4 \cdot 4 \cdot 4$

$= 64$ Simplify.

Think About It!
Determine the value of $b^{\frac{m}{n}}$ if $m = n$. Explain your reasoning.

Example 4 Evaluate Expressions of Powers of *n*th Roots

Evaluate each expression.

a. $729^{\frac{5}{6}}$

$729^{\frac{5}{6}} = \left(\sqrt[6]{729}\right)^5$ $b^{\frac{m}{n}} = \left(\sqrt[n]{b}\right)^m$

$= \left(\sqrt[6]{3 \cdot 3 \cdot 3 \cdot 3 \cdot 3 \cdot 3}\right)^5$ $729 = 3 \cdot 3 \cdot 3 \cdot 3 \cdot 3 \cdot 3$

$= \left(\sqrt[6]{3^6}\right)^5$ $3 \cdot 3 \cdot 3 \cdot 3 \cdot 3 \cdot 3 = 3^6$

$= 3^5$ or 243 $\sqrt[6]{3^6} = 3$

(continued on the next page)

Lesson 8-4 • Rational Exponents **457**

Go Online You can watch a video to learn how to evaluate expressions involving rational exponents.

b. $\left(\frac{36}{49}\right)^{\frac{3}{2}}$

$\left(\frac{36}{49}\right)^{\frac{3}{2}} = \frac{36^{\frac{3}{2}}}{49^{\frac{3}{2}}}$ Power of a Quotient

$= \frac{(\sqrt{36})^3}{(\sqrt{49})^3}$ $b^{\frac{m}{n}} = (\sqrt[n]{b})^m$

$= \frac{6^3}{7^3}$ $\sqrt{36} = 6$ and $\sqrt{49} = 7$

$= \frac{216}{343}$ Simplify.

Check

Evaluate each expression.

a. $49^{\frac{3}{2}}$ b. $243^{\frac{3}{5}}$

Example 5 Apply Rational Exponents

BIOLOGY For insects, the resting metabolic rate can be determined by $r = 4.14m^{\frac{2}{3}}$, where r is the resting metabolic rate in cubic milliliters of oxygen per hour and m is the body mass of the insect in milligrams. Determine the resting metabolic rate of a 125-mg ebony jewelwing damselfly.

$r = 4.14m^{\frac{2}{3}}$ Original equation

$= 4.14(125)^{\frac{2}{3}}$ $m = 125$

$= 4.14\,(\sqrt[3]{125})^2$ $125^{\frac{2}{3}} = (\sqrt[3]{125})^2$

$= 4.14(5)^2$ $\sqrt[3]{125} = 5$

$= 4.14 \cdot 25$ Simplify.

$= 103.5$ Simplify.

The resting metabolic rate of a ebony jewelwing damselfly is 103.5 cubic milliliters of oxygen per hour.

Use a Source

Find the weight of another insect, in milligrams, and determine its resting metabolism. Provide the name, weight, and resting metabolism of the insect, and compare it to the ebony jewelwing damselfly.

Check

PROFITS The profit p, in thousands of dollars, of a company is modeled by the equation $p = 3.7c^{\frac{6}{5}}$, where c is the number of customers in thousands. Determine the profits of the company if they have 32,000 customers.

A. $142,080

B. $236,800

C. $307,625.29

D. $942,717,779.90

Go Online You can complete an Extra Example online.

458 Module 8 · Exponents and Roots

Practice

Go Online You can complete your homework online.

Example 1

Write each expression in radical or exponential form.

1. $15^{\frac{1}{2}}$
2. $24^{\frac{1}{2}}$
3. $4k^{\frac{1}{2}}$
4. $(12y)^{\frac{1}{2}}$
5. $\sqrt{26}$
6. $\sqrt{44}$
7. $2\sqrt{ab}$
8. $\sqrt{3xyz}$

Examples 2–4

Evaluate each expression.

9. $\left(\frac{1}{16}\right)^{\frac{1}{4}}$
10. $\sqrt[5]{3125}$
11. $729^{\frac{1}{3}}$
12. $\left(\frac{1}{32}\right)^{\frac{1}{5}}$
13. $\sqrt[6]{4096}$
14. $1024^{\frac{1}{5}}$
15. $\left(\frac{16}{625}\right)^{\frac{1}{4}}$
16. $\sqrt[6]{15{,}625}$
17. $117{,}649^{\frac{1}{6}}$
18. $\sqrt[4]{\frac{16}{81}}$
19. $\left(\frac{1}{81}\right)^{\frac{1}{4}}$
20. $\left(\frac{3125}{32}\right)^{\frac{1}{5}}$
21. $729^{\frac{5}{6}}$
22. $256^{\frac{3}{8}}$
23. $125^{\frac{4}{3}}$
24. $49^{\frac{5}{2}}$
25. $\left(\frac{9}{100}\right)^{\frac{3}{2}}$
26. $\left(\frac{8}{125}\right)^{\frac{4}{3}}$

Example 5

27. **VELOCITY** The velocity v in feet per second of a freely falling object that has fallen h feet can be represented by $v = 8h^{\frac{1}{2}}$. Find the velocity of an object if it has fallen a distance of 144 feet.

28. **GEOMETRY** The surface area S of a cube in square inches can be determined by $S = 6V^{\frac{2}{3}}$, where V is the volume of the cube in cubic inches. Find the surface area of a cube that has a volume of 4096 cubic inches.

29. **PLANETS** The average distance d in astronomical units that a planet is from the Sun can be modeled by $d = t^{\frac{2}{3}}$, where t is the number of Earth years that it takes for the planet to orbit the Sun. Find the average distance a planet is from the Sun if the planet has an orbit of 27 Earth years.

30. **BIOLOGY** The relationship between the mass m in kilograms of an organism and its metabolism P in Calories per day can be represented by $P = 73.3\sqrt[4]{m^3}$. Find the metabolism of an organism that has a mass of 16 kilograms.

31. **PROFIT** The profit P of a company, in thousands of dollars, can be modeled by $P = 12.75\sqrt[5]{c^2}$, where c is the number of customers in hundreds. What is the profit of the company if the company has 3200 customers?

32. **TIRE MARKS** When a driver applies the brakes, the tires lock but the car will continue to slide, leaving skid marks on the road. You can approximate the speed at which a car was traveling on a dry road based on the length of a skid mark left by the car using the formula Speed = $(30 \cdot \text{length} \cdot 0.75)^{\frac{1}{2}}$, where speed is measured in miles per hour and length is measured in feet. At approximately what speed was a car traveling if it left a 50-foot long skid mark? Round to the nearest tenth.

Mixed Exercises

Write each expression in radical form, or write each radical in exponential form.

33. $17^{\frac{1}{3}}$

34. $q^{\frac{1}{4}}$

35. $7b^{\frac{1}{3}}$

36. $m^{\frac{2}{3}}$

37. $\sqrt[3]{29}$

38. $\sqrt[5]{h}$

39. $2\sqrt[3]{a}$

40. $\sqrt[3]{xy^2}$

Simplify.

41. $\sqrt[3]{0.027}$

42. $\sqrt[4]{\dfrac{n^4}{16}}$

43. $a^{\frac{1}{3}} \cdot a^{\frac{2}{3}}$

44. $c^{\frac{1}{2}} \cdot c^{\frac{3}{2}}$

45. $(8^2)^{\frac{2}{3}}$

46. $\left(y^{\frac{3}{4}}\right)^{\frac{1}{2}}$

47. $9^{-\frac{1}{2}}$

48. $16^{-\frac{3}{2}}$

49. $(3^2)^{-\frac{3}{2}}$

50. $\left(81^{\frac{1}{4}}\right)^{-2}$

51. $k^{-\frac{1}{2}}$

52. $\left(d^{\frac{4}{3}}\right)^0$

53. **USE A MODEL** In economics, the Cobb-Douglas production function is commonly used to relate input to output. The general form of the function is $P = bL^{ax}K^y$. The values in the table are given for the U.S. economy for the years 1900 and 1920. For both parts, assume a is 1.

		1900	1920
L	Labor input	105	194
K	Capital output	107	407
b	Total factor productivity	1.01	1.01
x	Output elasticity of labor	$\frac{3}{4}$	$\frac{3}{4}$
y	Output elasticity of capital	$\frac{1}{4}$	$\frac{1}{4}$

a. Find the total production P for 1900, to the nearest whole number.
b. Find the total production P for 1920, to the nearest whole number.

54. **BASKETBALL** The formula $S = 4\pi\left(\frac{3V}{4\pi}\right)^{\frac{2}{3}}$ can be used to find the surface area of a sphere, where V represents its volume. A regulation basketball has a volume of about 456 cubic inches. How much leather is needed (surface area) to make a regulation basketball? Round your answer to the nearest tenth.

55. **BIOLOGY** The function $h(x) = 0.4x^{\frac{2}{3}}$ can be used to find the height h in meters of a female giraffe with mass x kilograms. What is the height of a female giraffe with a mass of 32.8 kilograms? Round your answer to the nearest tenth of a meter.

56. **GRAPH** The graph of $y = x^{\frac{2}{5}}$ is shown over the interval $0 < x < 10$. Copy and complete the table to find the values. Round to the nearest tenth if necessary.

x	0		5	
y		1		2.3

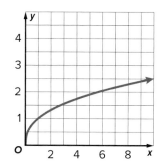

57. **CONSTRUCT ARGUMENTS** Use the properties of exponents to show which of the expressions are equivalent. Explain your reasoning.

a. $p^{\frac{f}{k}}$ b. $p^{\frac{f}{5}}$ c. $\sqrt[k]{p^f}$ d. $\left(\sqrt[k]{p}\right)^f$

58. **USE TOOLS** The amount of material needed to make a party hat in the shape of a cone can be found by calculating the hat's lateral area. The formula $LA = \pi r(h^2 + r^2)^{\frac{1}{2}}$ represents the lateral area of a cone for a given height h and radius r. Use your calculator to find the lateral area of a party hat that has a radius of 2 inches and a height of 7 inches. Round the answer to the nearest tenth of a square inch.

59. GEOMETRY The surface area S of a cylinder in square centimeters can be determined by $S = 2\left(\frac{V}{h}\right) + 2\pi\left(\frac{V}{\pi h}\right)^{\frac{1}{2}} h$, where V is the volume of the cylinder in cubic centimeters and h is the height of the cylinder is centimeters. Find the surface area of a cylinder that has a volume of 160π cubic centimeters and a height of 10 centimeters. Write the surface area in terms of π.

60. PENDULUMS A pendulum is a weight hanging from a point of suspension so that it can swing freely. The formula $T = 2\pi\left(\frac{L}{32}\right)^{\frac{1}{2}}$ gives the time T in seconds it takes for a pendulum of length L in feet to complete one full cycle. How long will it take the pendulum shown to complete one full cycle? Write the length of time in terms of π.

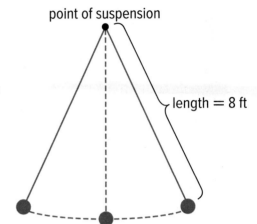

Higher-Order Thinking Skills

61. CREATE Write two different expressions with rational exponents equal to $\sqrt{2}$.

62. ANALYZE Determine whether each statement is *sometimes*, *always*, or *never* true. Assume that $x > 0$. Justify your argument.

a. $x^2 = x^{\frac{1}{2}}$

b. $x^{-2} = x^{\frac{1}{2}}$

c. $x^{\frac{1}{3}} = x^{\frac{1}{2}}$

d. $\sqrt{x} = x^{\frac{1}{2}}$

e. $\left(x^{\frac{1}{2}}\right)^2 = x$

f. $x^{\frac{1}{2}} \cdot x^2 = x$

63. PERSEVERE For what values of x is $x = x^{\frac{1}{3}}$?

64. WRITE Explain why $16^{\frac{1}{4}}$ will be less than 3.

65. FIND THE ERROR Sachi and Makayla are evaluating $27^{\frac{2}{3}}$. Is either correct? Explain your reasoning.

Sachi
$27^{\frac{2}{3}} = \sqrt[3]{27^2}$
$= \sqrt[3]{729}$
$= 9$

Makayla
$27^{\frac{2}{3}} = \sqrt[3]{27^2}$
$= 3^2$
$= 9$

66. WHICH ONE DOESN'T BELONG? Consider the four expressions shown. Which one does not belong? Explain your reasoning.

$\left(\sqrt[4]{w}\right)^5$ $\sqrt[4]{w^5}$ $w^{\frac{4}{5}}$ $w^{\frac{5}{4}}$

462 Module 8 · Exponents and Roots

Lesson 8-5

Simplifying Radical Expressions

Explore Square Roots and Negative Numbers

 Online Activity Use an interactive tool to complete the Explore.

> **INQUIRY** How can you simplify algebraic expressions with square roots?

Today's Goals
- Simplify square roots.
- Simplify cube roots.

Today's Vocabulary
radical expression
square root
perfect square
principal square root
cube root
perfect cube

Learn Simplifying Square Root Expressions

A **radical expression** contains a radical such as a square root. A **square root** of a number is a value that, when multiplied by itself, gives the number. So, because $7 \cdot 7 = 49$, a square root of 49 is 7. A number like 49 is sometimes called a perfect square. A **perfect square** is a rational number with a square root that is a rational number.

Recall that the square root of a positive number has two possible solutions. For $\sqrt{a^2}$ where $a > 0$, $\sqrt{a^2} = \pm a$, because $a^2 = (-a)^2$. The positive solution a is called the **principal square root**.

Key Concept • Product Property of Square Roots

Words	For any nonnegative real numbers a and b, the square root of ab is the square root of a times the square root of b.
Symbols	$\sqrt{ab} = \sqrt{a} \cdot \sqrt{b}$ if $a \geq 0$ and $b \geq 0$
Example	$\sqrt{4 \cdot 16} = \sqrt{4} \cdot \sqrt{16} = 2 \cdot 4$ or 8

Key Concept • Quotient Property of Square Roots

Words	For any real numbers a and b, where $a \geq 0$ and $b > 0$, the square root of $\frac{a}{b}$ is equal to the square root of a divided by the square root of b.
Symbols	$\sqrt{\frac{a}{b}} = \frac{\sqrt{a}}{\sqrt{b}}$ if $a \geq 0$ and $b > 0$
Example	$\sqrt{\frac{100}{4}} = \sqrt{25} = 5$ or $\sqrt{\frac{100}{4}} = \frac{\sqrt{100}}{\sqrt{4}} = \frac{10}{2} = 5$

A square root is in simplest form if these three conditions are true.

- The square root contains no perfect square factors other than 1.
- The square root contains no fractions.
- There are no square roots in the denominator of a fraction.

Study Tip

Prime Factorization If you are trying to determine the prime factorization of a number and the number is even, then you know that the prime factorization contains at least one 2. Divide the number by 2 and repeat until you have an odd number, and then see if the odd number can be factored any further as a product of primes.

Go Online You can complete an Extra Example online.

Lesson 8-5 • Simplifying Radical Expressions **463**

Example 1 Simplify Square Roots

Simplify $\sqrt{72}$.

$$\begin{aligned}\sqrt{72} &= \sqrt{2 \cdot 2 \cdot 2 \cdot 3 \cdot 3} && \text{Prime factorization of 72}\\ &= \sqrt{2} \cdot \sqrt{2} \cdot \sqrt{2} \cdot \sqrt{3} \cdot \sqrt{3} && \text{Product Property of Square Roots}\\ &= \sqrt{2^2} \cdot \sqrt{2} \cdot \sqrt{3^2} && \text{Product Property of Square Roots}\\ &= 2 \cdot \sqrt{2} \cdot 3 = 6\sqrt{2} && \text{Simplify.}\end{aligned}$$

> **Think About It!**
> Why can $\sqrt{2^4}$ be simplified to 4?

Example 2 Multiply Square Roots

Simplify $\sqrt{5} \cdot \sqrt{48}$.

$$\begin{aligned}\sqrt{5} \cdot \sqrt{48} &= \sqrt{5} \cdot (\sqrt{2} \cdot \sqrt{2} \cdot \sqrt{2} \cdot \sqrt{2} \cdot \sqrt{3}) && \text{Prime factorizations of 5 and 48}\\ &= \sqrt{5} \cdot (\sqrt{2^2} \cdot \sqrt{2^2} \cdot \sqrt{3}) && \text{Product Property of Square Roots}\\ &= 4\sqrt{5} \cdot \sqrt{3} = 4\sqrt{15} && \text{Simplify.}\end{aligned}$$

Check

Simplify $\sqrt{8} \cdot \sqrt{14}$.

Example 3 Divide Square Roots

Simplify $\sqrt{\frac{24}{27}}$.

$$\begin{aligned}\sqrt{\frac{24}{27}} &= \sqrt{\frac{8}{9}} && \text{Reduce the radicand.}\\ &= \frac{\sqrt{8}}{\sqrt{9}} && \text{Quotient Property of Square Roots}\\ &= \frac{2\sqrt{2}}{3} && \text{Product Property of Square Roots}\end{aligned}$$

Example 4 Simplify Square Roots with Variables

Simplify $\sqrt{525x^4y^5z^5}$.

$$\begin{aligned}\sqrt{525x^4y^5z^5} &= \sqrt{3 \cdot 5^2 \cdot 7 \cdot x^4 \cdot y^5 \cdot z^5}\\ &= \sqrt{3} \cdot \sqrt{5^2} \cdot \sqrt{7} \cdot \sqrt{x^4} \cdot \sqrt{y^5} \cdot \sqrt{z^5}\\ &= \sqrt{3} \cdot 5 \cdot \sqrt{7} \cdot \sqrt{x^4} \cdot \sqrt{y^4} \cdot \sqrt{y} \cdot \sqrt{z^4} \cdot \sqrt{z}\\ &= \sqrt{3} \cdot 5 \cdot \sqrt{7} \cdot x^2 \cdot y^2 \cdot \sqrt{y} \cdot z^2 \cdot \sqrt{z}\\ &= 5x^2y^2z^2\sqrt{21yz}\end{aligned}$$

> **Think About It!**
> Why does the final answer include $\sqrt{21}$ when $\sqrt{21}$ does not appear in the previous step?

Check

Simplify $\sqrt{147x^3y^4z^5}$.

Go Online You can complete an Extra Example online.

Example 5 Write and Solve a Radical Equation

FINANCE Sarah uses an online calculator to determine how much interest she would accrue in a savings account after two years. She plans to invest $750, and the calculator determines that she could have $780 after two years. She can find the interest rate r of the savings account by using the formula $r = \sqrt{\frac{a}{p}} - 1$, where a is the amount after two years and p is the initial investment. What is the interest rate?

$$r = \sqrt{\frac{a}{p}} - 1 \qquad \text{Original formula}$$

$$= \sqrt{\frac{780}{750}} - 1 \qquad a = 780;\ p = 750$$

$$= \sqrt{\frac{26}{25}} - 1 \qquad \text{Simplify the radicand.}$$

$$= \frac{\sqrt{26}}{\sqrt{25}} - 1 \qquad \text{Quotient Property of Square Roots}$$

$$= \frac{\sqrt{26}}{5} - 1 \qquad \text{Simplify.}$$

$$\approx 0.02 \qquad \text{Simplify.}$$

The interest rate is approximately 2%.

Talk About It!
Could this equation be solved if $\sqrt{\frac{a}{p}}$ could not be simplified?

Learn Simplifying Cube Root Expressions

The **cube root** of a number is the value that, when multiplied by itself twice, gives the number. A **perfect cube** is a rational number with a cube root that is an integer. So, because $6 \cdot 6 \cdot 6 = 216$, the cube root of 216 is 6 and 216 is a perfect cube.

Talk About It!
What must be true for a cube root expression to be in simplest form?

Key Concept • Product Property of Radicals

Words	For any nonnegative real numbers a and b, the nth root of ab is equal to the nth root of a times the nth root of b.
Symbols	$\sqrt[n]{ab} = \sqrt[n]{a} \cdot \sqrt[n]{b}$ if $a \geq 0$ and $b \geq 0$
Examples	$\sqrt[3]{8 \cdot 64} = \sqrt[3]{512}$ or 8; $\sqrt[3]{8 \cdot 64} = \sqrt[3]{8} \cdot \sqrt[3]{64} = 2 \cdot 4$ or 8

Key Concept • Quotient Property of Radicals

Words	For any real numbers a and b, where $a \geq 0$ and $b > 0$, the nth root of $\frac{a}{b}$ is equal to the nth root of a divided by the nth root of b.
Symbols	$\sqrt[n]{\frac{a}{b}} = \frac{\sqrt[n]{a}}{\sqrt[n]{b}}$, if $a \geq 0$ and $b > 0$
Examples	$\sqrt[3]{\frac{216}{8}} = \sqrt[3]{27}$ or 3; $\sqrt[3]{\frac{216}{8}} = \frac{\sqrt[3]{216}}{\sqrt[3]{8}} = \frac{6}{2}$ or 3

Watch Out!
Cube Roots Remember, $\sqrt[3]{64}$ is not the same as $3\sqrt{64}$. The former means *the cube root of 64* or 4, and the latter means *3 times the square root of 64* or 24.

Example 6 Find Cube Roots

Simplify $\sqrt[3]{343}$.

$$\sqrt[3]{343} = \sqrt[3]{7 \cdot 7 \cdot 7} \qquad \text{Prime factorization of 343}$$

$$= 7 \qquad \text{Simplify.}$$

Go Online You can complete an Extra Example online.

Think About It!
Explain why $\sqrt[3]{3} \cdot \sqrt[3]{3} \cdot \sqrt[3]{3}$ simplifies to 3.

Example 7 Simplify Cube Roots

Simplify $\sqrt[3]{135}$.

$$\sqrt[3]{135} = \sqrt[3]{3 \cdot 3 \cdot 3 \cdot 5} \quad \text{Prime factorization of 135}$$
$$= \sqrt[3]{3} \cdot \sqrt[3]{3} \cdot \sqrt[3]{3} \cdot \sqrt[3]{5} \quad \text{Product Property of Radicals}$$
$$= 3\sqrt[3]{5} \quad \text{Simplify.}$$

Think About It!
Why are you able to simplify the radicand before finding the cube root?

Example 8 Multiply Cube Roots

Simplify $\sqrt[3]{2} \cdot \sqrt[3]{36}$.

$$\sqrt[3]{2} \cdot \sqrt[3]{36} = \sqrt[3]{2} \cdot \left(\sqrt[3]{2 \cdot 2 \cdot 3 \cdot 3}\right) \quad \text{Prime factorizations of 2 and 36}$$
$$= \sqrt[3]{2} \cdot \sqrt[3]{2} \cdot \sqrt[3]{2} \cdot \sqrt[3]{3} \cdot \sqrt[3]{3} \quad \text{Product Property of Radicals}$$
$$= \sqrt[3]{8} \cdot \sqrt[3]{9} \quad \text{Product Property of Radicals}$$
$$= 2\sqrt[3]{9} \quad \text{Simplify.}$$

Example 9 Divide Cube Roots

Simplify $\sqrt[3]{\dfrac{168}{375}}$.

$$\sqrt[3]{\dfrac{168}{375}} = \sqrt[3]{\dfrac{2 \cdot 2 \cdot 2 \cdot 3 \cdot 7}{3 \cdot 5 \cdot 5 \cdot 5}} \quad \text{Prime factorizations of 168 and 375}$$
$$= \sqrt[3]{\dfrac{2 \cdot 2 \cdot 2 \cdot 7}{5 \cdot 5 \cdot 5}} \quad \text{Simplify the radicand.}$$
$$= \dfrac{\sqrt[3]{2 \cdot 2 \cdot 2 \cdot 7}}{\sqrt[3]{5 \cdot 5 \cdot 5}} \quad \text{Quotient Property of Radicals}$$
$$= \dfrac{\sqrt[3]{2} \cdot \sqrt[3]{2} \cdot \sqrt[3]{2} \cdot \sqrt[3]{7}}{\sqrt[3]{5} \cdot \sqrt[3]{5} \cdot \sqrt[3]{5}} \quad \text{Product Property of Radicals}$$
$$= \dfrac{2\sqrt[3]{7}}{5} \quad \text{Simplify.}$$

Think About It!
How would your answer change if you simplified $\sqrt[3]{\dfrac{15x^2y^3z^4}{81x^4}}$ instead of $\sqrt[3]{\dfrac{15x^2y^3z^4}{81x^5}}$? Is this answer in simplest form? Why or why not?

Example 10 Simplify Cube Roots with Variables

Simplify $\sqrt[3]{\dfrac{15x^2y^3z^4}{81x^5}}$.

$$\sqrt[3]{\dfrac{15x^2y^3z^4}{81x^5}} = \sqrt[3]{\dfrac{3 \cdot 5 \cdot x \cdot x \cdot y \cdot y \cdot y \cdot z \cdot z \cdot z \cdot z}{3 \cdot 3 \cdot 3 \cdot 3 \cdot x \cdot x \cdot x \cdot x \cdot x}}$$
$$= \sqrt[3]{\dfrac{5 \cdot y \cdot y \cdot y \cdot z \cdot z \cdot z \cdot z}{3 \cdot 3 \cdot 3 \cdot x \cdot x \cdot x}}$$
$$= \sqrt[3]{\dfrac{5 \cdot y \cdot y \cdot y \cdot z \cdot z \cdot z \cdot z}{3 \cdot 3 \cdot 3 \cdot x \cdot x \cdot x}}$$
$$= \dfrac{\sqrt[3]{5 \cdot y \cdot y \cdot y \cdot z \cdot z \cdot z \cdot z}}{\sqrt[3]{3 \cdot 3 \cdot 3 \cdot x \cdot x \cdot x}}$$
$$= \dfrac{\sqrt[3]{5} \cdot \sqrt[3]{y} \cdot \sqrt[3]{y} \cdot \sqrt[3]{y} \cdot \sqrt[3]{z} \cdot \sqrt[3]{z} \cdot \sqrt[3]{z} \cdot \sqrt[3]{z}}{\sqrt[3]{3} \cdot \sqrt[3]{3} \cdot \sqrt[3]{3} \cdot \sqrt[3]{x} \cdot \sqrt[3]{x} \cdot \sqrt[3]{x}}$$
$$= \dfrac{yz\sqrt[3]{5z}}{3x}$$

Watch Out!
Simplifying Cube Roots Be sure to simplify the fraction before finding the cube root.

Go Online You can complete an Extra Example online.

Practice

Go Online You can complete your homework online.

Examples 1–4

Simplify each expression.

1. $\sqrt{52}$
2. $\sqrt{56}$
3. $\sqrt{72}$
4. $\sqrt{162}$
5. $\sqrt{243}$
6. $\sqrt{245}$
7. $\sqrt{5} \cdot \sqrt{10}$
8. $\sqrt{10} \cdot \sqrt{20}$
9. $\sqrt{2} \cdot \sqrt{10}$
10. $\sqrt{5} \cdot \sqrt{60}$
11. $\sqrt{8} \cdot \sqrt{12}$
12. $\sqrt{11} \cdot \sqrt{22}$
13. $\sqrt{\frac{75}{49}}$
14. $\sqrt{\frac{8}{81}}$
15. $\sqrt{\frac{192}{216}}$
16. $\sqrt{\frac{88}{18}}$
17. $\sqrt{\frac{84}{121}}$
18. $\sqrt{\frac{75}{20}}$
19. $\sqrt{16b^4}$
20. $\sqrt{40x^4y^8}$
21. $\sqrt{81a^{12}d^4}$
22. $\sqrt{75qr^3}$
23. $\sqrt{28a^4b^3}$
24. $\sqrt{w^{13}x^5z^8}$

Example 5

25. **NATURE** In 2010, an earthquake below the ocean floor initiated a devastating tsunami in Sumatra. Scientists can approximate the velocity V in feet per second of a tsunami in water of depth d feet with the formula $V = \sqrt{16} \cdot \sqrt{d}$. Determine the velocity of a tsunami in 300 feet of water. Write your answer in simplified radical form.

26. **PHYSICAL SCIENCE** The average velocity V of gas molecules is represented by the formula $V = \sqrt{\frac{3kt}{m}}$. The velocity is measured in meters per second, t is the temperature in Kelvin, and m is the molar mass of the gas in kilograms per mole. The variable k represents a value called the molar gas constant, which is about 8.3. Write a simplified radical expression that represents the average velocity of gas molecules that have a temperature of 300 Kelvin and a mass of 0.045 kilogram per mole. Round to the nearest meter per second.

Lesson 8-5 • Simplifying Radical Expressions **467**

27. **CLOCKS** A grandfather clock has a pendulum that swings back and forth. The time t in seconds it takes the pendulum to complete one full cycle is given by the formula $t = 0.2\sqrt{p}$, where p is the length of the pendulum in centimeters. How long is a full cycle if the pendulum is 22 cm long? Round your answer to the nearest tenth of a second.

28. **HORIZON** The distance, d, in miles that a person can see to the horizon can be modeled by the formula $d = \sqrt{\frac{3h}{2}}$ where h is the person's height above sea level in feet. How far, in miles, to the horizon can a person see if they are 60 feet above sea level? Write your answer in simplified radical form.

Examples 6–10
Simplify each expression.

29. $\sqrt[3]{8}$
30. $\sqrt[3]{125}$

31. $\sqrt[3]{216}$
32. $\sqrt[3]{64}$

33. $\sqrt[3]{27}$
34. $\sqrt[3]{1000}$

35. $\sqrt[3]{243}$
36. $\sqrt[3]{4000}$

37. $\sqrt[3]{162}$
38. $\sqrt[3]{875}$

39. $\sqrt[3]{80}$
40. $\sqrt[3]{384}$

41. $\sqrt[3]{3} \cdot \sqrt[3]{32}$
42. $\sqrt[3]{24} \cdot \sqrt[3]{18}$

43. $\sqrt[3]{12} \cdot \sqrt[3]{6}$
44. $\sqrt[3]{16} \cdot \sqrt[3]{20}$

45. $\sqrt[3]{9} \cdot \sqrt[3]{3}$
46. $\sqrt[3]{2} \cdot \sqrt[3]{6}$

47. $\sqrt[3]{\frac{162}{375}}$
48. $\sqrt[3]{\frac{648}{750}}$

49. $\sqrt[3]{\frac{18}{128}}$
50. $\sqrt[3]{\frac{162}{3}}$

51. $\sqrt[3]{\frac{192}{3}}$
52. $\sqrt[3]{\frac{500}{2}}$

53. $\sqrt[3]{54b^8}$
54. $\sqrt[3]{10yz^5} \cdot \sqrt[3]{4y^3}$

55. $\sqrt[3]{\frac{192x^7}{1029x^{10}}}$
56. $\sqrt[3]{\frac{640a^3b^8}{5ab^4}}$

57. $\sqrt[3]{54x^2y^8} \cdot \sqrt[3]{5x^5y^4}$
58. $\sqrt[3]{25p^2r^5} \cdot \sqrt[3]{30q^3r}$

Mixed Exercises

Simplify each expression.

59. $\sqrt{245m^9n^5}$

60. $2\sqrt{5} \cdot 7\sqrt{10}$

61. $\sqrt{\dfrac{96d^4e^2f^8}{75de^6}}$

62. $-11\sqrt[3]{250}$

63. $\sqrt[3]{320x^{14}y^{17}z^{20}}$

64. $\sqrt[3]{45k^4m^{10}} \cdot \sqrt[3]{32k^7m^3}$

65. $\sqrt[3]{\dfrac{264w^6xy^7}{3w^6x^4y^3}}$

66. $\sqrt[3]{\dfrac{48a^7}{125b^9}}$

67. AREA The base of a triangle measures $6\sqrt{2}$ meters and the height measures $3\sqrt{6}$ meters. What is the area?

68. AREA The length of a rectangle measures $8\sqrt{12}$ centimeters and the width measures $4\sqrt{8}$ centimeters. What is the area of the rectangle?

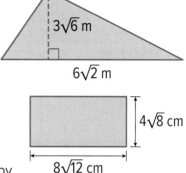

69. MEAN The geometric mean of two numbers h and k can be found by evaluating $\sqrt{h \cdot k}$. Find the geometric mean of 32 and 14 in simplified radical form.

70. THEATRE CREW Fernanda oversees painting props for a theatre production. She needs to create a cube that will be used as a prop by the actors. She has enough paint to cover 5 square yards. The formula $s = \sqrt{\dfrac{A}{6}}$ gives the longest side length s in yards of a cubic prop Fernanda can make, where A is the surface area to be covered with paint. What is the longest side length of a cube Fernanda could make and not have to purchase any more paint? Round to the nearest tenth of a yard.

71. SCALE MODEL While on vacation, Deon decided to purchase a scale model of the Empire State Building. Before making the purchase, Deon wants to determine the maximum height of a model that will fit inside his suitcase. Deon's suitcase measures 22 inches long by 14 inches wide by 9 inches tall. The diagonal length D is given by $D = \sqrt{L^2 + W^2 + H^2}$, where L is the length in inches, W is the width in inches, and H is the height in inches. Find the length of the tallest scale model that will fit inside Deon's suitcase. Round your answer to the nearest tenth of an inch.

72. BALLOONS The radius of a sphere r is given in terms of its volume V by the formula $r = \sqrt[3]{\dfrac{0.75V}{\pi}}$. By how many inches has the radius of a spherical balloon increased when the amount of air in the balloon is increased from 4.5 cubic feet to 4.7 cubic feet? Round your answer to the nearest hundredth.

Lesson 8-5 • Simplifying Radical Expressions

73. **REGULARITY** The Product Property of Square Roots and the Quotient Property of Square Roots can be written in symbols as $\sqrt{ab} = \sqrt{a} \cdot \sqrt{b}$ and $\sqrt{\frac{a}{b}} = \frac{\sqrt{a}}{\sqrt{b}}$, respectively.

   a. Explain the Product Property of Square Roots and discuss any limitations of a and b for this property.

   b. Explain the Quotient Property of Square Roots and discuss any limitations of a and b for this property.

   c. Discuss any similarities of the two properties.

74. **PERSEVERE** Use rational exponents to find an equivalent radical expression in simplest form.

   a. $\sqrt[4]{16m^{32}}$
   b. $(\sqrt{x})(\sqrt[3]{x})$
   c. $\sqrt[3]{\sqrt{b}}$

75. **CREATE** Create a problem where two square roots are being either multiplied or divided. Be sure to include at least one variable in your problem. Solve your problem.

76. **PERSEVERE** Margarita takes a number, subtracts 4, multiplies by 4, takes the square root, and takes the reciprocal to get $\frac{1}{2}$. What number did she start with? Write a formula to describe the process.

77. **ANALYZE** Find a counterexample to show that the following statement is false. *If you take the square root of a number, the result will always be less than the original number.*

78. **ANALYZE** Order the expressions from least to greatest. $\sqrt{47}$, 9, $\sqrt[3]{421}$, $\sqrt{85}$

79. **PERSEVERE** If the area of a rectangle is $144\sqrt{5}$ square inches, what are possible dimensions of the rectangle? Explain your reasoning.

80. **WRITE** Describe the required conditions for a radical expression to be in simplest form.

Lesson 8-6
Operations With Radical Expressions

Today's Goals
- Add and subtract radical expressions.
- Multiply radical expressions.

Learn Adding and Subtracting Radical Expressions

To add or subtract radical expressions, the radicands must be alike in the same way that monomial terms must be alike to add or subtract.

Monomials

$5a + 3a = (5 + 3)a$
$= 8a$

$7b - 2b = (7 - 2)b$
$= 5b$

Radical Expressions

$5\sqrt{3} + 3\sqrt{3} = (5 + 3)\sqrt{3}$
$= 8\sqrt{3}$

$7\sqrt{5} - 2\sqrt{5} = (7 - 2)\sqrt{5}$
$= 5\sqrt{5}$

Notice that when adding and subtracting radical expressions, the radicand does not change. This is the same as when adding or subtracting monomials.

Think About It!
Can you simplify $6\sqrt{3} + 7\sqrt{5}$? If so, simplify it, and if not, explain why.

Watch Out!
Radicands Before applying the Distributive Property, be certain that you group expressions with like radicands.

Example 1 Add and Subtract Expressions with Like Radicands

Simplify $4\sqrt{5} + 3\sqrt{7} - 2\sqrt{5} + 7\sqrt{7}$.

$4\sqrt{5} + 3\sqrt{7} - 2\sqrt{5} + 7\sqrt{7} = (4 - 2)\sqrt{5} + (3 + 7)\sqrt{7}$
$= 2\sqrt{5} + 10\sqrt{7}$

Check
Simplify $17\sqrt{19} - 14\sqrt{6} + 11\sqrt{6} - 3\sqrt{19}$.

A. 0

B. $14\sqrt{19} - 3\sqrt{6}$

C. $11\sqrt{25}$

D. $28\sqrt{19} - 17\sqrt{6}$

Go Online You can complete an Extra Example online.

Think About It!
Jaylen tries to simplify $3\sqrt{5} - 3\sqrt{5}$ and determines that the simplest form is $0\sqrt{5}$. Is he correct? Why or why not?

Talk About It!

If a radical expression has addition or subtraction with unlike radicands, what must you check before determining whether the terms can be added or subtracted? Explain your reasoning.

Example 2 Add and Subtract Expressions with Unlike Radicands

Simplify $3\sqrt{75} - 6\sqrt{48} - \sqrt{27}$.

$3\sqrt{75} - 6\sqrt{48} - \sqrt{27}$

$= 3(\sqrt{5^2}) \cdot \sqrt{3} - 6(\sqrt{4^2} \cdot \sqrt{3}) - (\sqrt{3^2}) \cdot \sqrt{3}$ Product Property of Square Roots

$= 3(5\sqrt{3}) - 6(4\sqrt{3}) - 3\sqrt{3}$ Simplify.

$= 15\sqrt{3} - 24\sqrt{3} - 3\sqrt{3}$ Multiply.

$= (15 - 24 - 3)\sqrt{3}$ Distributive Property

$= -12\sqrt{3}$ Simplify.

Check

Simplify $-11\sqrt{50} + 2\sqrt{32} - 18\sqrt{8}$.

A. $99\sqrt{2}$

B. $-83\sqrt{2}$

C. $-27\sqrt{2}$

D. $-11\sqrt{50} - 14\sqrt{8}$

Example 3 Use Radical Expressions

DECORATIONS Kate is decorating a rectangular pavilion for a party and wants to string lights along the edge of the roof. Two of the sides have a length of $13\sqrt{5}$ feet, and the other two sides have a length of $6\sqrt{45}$ feet. How many feet of lights will Kate need to decorate the entire pavilion?

Because there are two sides $13\sqrt{5}$ feet long and two sides $6\sqrt{45}$ feet long, the expression $13\sqrt{5} + 13\sqrt{5} + 6\sqrt{45} + 6\sqrt{45}$ represents the total perimeter of the roof.

$13\sqrt{5} + 13\sqrt{5} + 6\sqrt{45} + 6\sqrt{45} = (13 + 13)\sqrt{5} + (6 + 6)\sqrt{45}$

$= 26\sqrt{5} + 12\sqrt{45}$

$= 26\sqrt{5} + 12(\sqrt{5} \cdot \sqrt{3^2})$

$= 26\sqrt{5} + 36\sqrt{5}$

$= 62\sqrt{5}$

Kate will need $62\sqrt{5}$ feet of lights to decorate the pavilion.

Go Online You can complete an Extra Example online.

472 Module 8 • Exponents and Roots

Check

CONSTRUCTION A business is adding wheelchair ramps to their building. The three ramps will require $10\sqrt{28}$ ft³, $13\sqrt{7}$ ft³, and $4\sqrt{112}$ ft³ of concrete. How much concrete will be needed to build the three ramps?

A. 27 ft³

B. $27\sqrt{147}$ ft³

C. $49\sqrt{7}$ ft³

D. $117\sqrt{7}$ ft³

Learn Multiplying Radical Expressions

Multiplying radical expressions is similar to multiplying monomials. Let $x \geq 0$.

$$\text{Monomials}$$
$$5x(3x) = 5 \cdot 3 \cdot x \cdot x$$
$$= 15x^2$$

$$\text{Radical Expressions}$$
$$5\sqrt{x}(3\sqrt{x}) = 5 \cdot 3 \cdot \sqrt{x} \cdot \sqrt{x}$$
$$= 15x$$

You can apply the Distributive Property to radical expressions. You can also multiply radical expressions with more than one term in each factor. This is similar to multiplying two binomials.

$$\text{Binomials}$$
$$(2x + 3x)(4x + 5x) = 2x(4x) + 2x(5x) + 3x(4x) + 3x(5x)$$
$$= 8x^2 + 10x^2 + 12x^2 + 15x^2$$
$$= 45x^2$$

$$\text{Radical Expressions}$$
$$(2\sqrt{2} + 3\sqrt{2})(4\sqrt{2} + 5\sqrt{2})$$
$$= 2\sqrt{2}(4\sqrt{2}) + 2\sqrt{2}(5\sqrt{2}) + 3\sqrt{2}(4\sqrt{2}) + 3\sqrt{2}(5\sqrt{2})$$
$$= 8\sqrt{2^2} + 10\sqrt{2^2} + 12\sqrt{2^2} + 15\sqrt{2^2}$$
$$= 45\sqrt{2^2}$$
$$= 90$$

Notice that when multiplying radical expressions, you multiply the radicands.

Go Online You can complete an Extra Example online.

Example 4 Multiply Radical Expressions

Simplify $5\sqrt{3} \cdot 4\sqrt{6}$.

$$5\sqrt{3} \cdot 4\sqrt{6} = (5 \cdot 4)(\sqrt{3} \cdot \sqrt{6})$$
$$= 20\sqrt{18}$$
$$= 20(3\sqrt{2})$$
$$= 60\sqrt{2}$$

Check

Simplify $3\sqrt{10} \cdot (-9\sqrt{6})$.

Example 5 Multiply Radical Expressions by Using the Distributive Property

Simplify $4\sqrt{7}(2\sqrt{8} + 3\sqrt{7})$.

$$4\sqrt{7}(2\sqrt{8} + 3\sqrt{7}) = (4\sqrt{7} \cdot 2\sqrt{8}) + (4\sqrt{7} \cdot 3\sqrt{7})$$
$$= [(4 \cdot 2)(\sqrt{7} \cdot \sqrt{8})] + [(4 \cdot 3)(\sqrt{7} \cdot \sqrt{7})]$$
$$= 8\sqrt{56} + 12\sqrt{49}$$
$$= 16\sqrt{14} + 12(7)$$
$$= 16\sqrt{14} + 84$$

Check

Simplify $-4\sqrt{6}(7\sqrt{12} - 4\sqrt{8})$.

Go Online to learn how to find the sums and products of rational and irrational numbers in Expand 8-6.

Go Online You can complete an Extra Example online.

Practice

Go Online You can complete your homework online.

Examples 1 and 2

Simplify.

1. $7\sqrt{7} - 2\sqrt{7}$

2. $3\sqrt{13} + 7\sqrt{13}$

3. $7\sqrt{5} + 4\sqrt{5}$

4. $2\sqrt{6} + 9\sqrt{6}$

5. $12\sqrt{r} - 9\sqrt{r}$

6. $9\sqrt{6a} - 11\sqrt{6a} + 4\sqrt{6a}$

7. $3\sqrt{5} - 2\sqrt{20}$

8. $3\sqrt{50} - 3\sqrt{32}$

9. $2\sqrt{13} + 4\sqrt{2} - 5\sqrt{13} + \sqrt{2}$

10. $5\sqrt{8} + 2\sqrt{20} - \sqrt{8}$

11. $7\sqrt{3} - 2\sqrt{2} + 3\sqrt{2} + 5\sqrt{3}$

12. $8\sqrt{12} - \sqrt{5} - 4\sqrt{3}$

Example 3

13. **ARCHITECTURE** The Pentagon is the building that houses the U.S. Department of Defense. If the building is a regular pentagon with each side measuring $23\sqrt{149}$ meters, find the perimeter. Leave your answer as a radical expression.

$23\sqrt{149}$ m

14. **BIKING** Iker rode his bike on three trails this week. The first was $2\sqrt{3}$ kilometers, the second was $4\sqrt{3}$ kilometers, and the third was $3\sqrt{3}$ kilometers. How long did Iker ride this week? Give your answer as a radical expression.

15. **RAMP** A ramp at a dog park is made of three sections. The incline and decline pieces are the same length, $13\sqrt{37}$ inches. The center is $10\sqrt{41}$ inches long. What is the total length of the ramp? Give your answer as a radical expression.

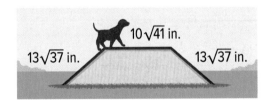

16. **DISTANCE** The diagonal length of a football field, including both end zones, is $40\sqrt{97}$ feet. The diagonal length of a practice field is $29\sqrt{97}$ feet. How much longer is the diagonal length of a football field than the diagonal length of the practice field? Give your answer as a radical expression.

17. **CIRCLES** A large circle has an area of 400 cm² and a small circle has an area of 200 cm². The radius of the large circle is $\sqrt{\frac{400}{\pi}}$ cm. The radius of the small circle is $\sqrt{\frac{200}{\pi}}$ cm. Find the difference of the radius of the large circle and the radius of the small circle.

Lesson 8-6 • Operations With Radical Expressions 475

18. **GARDEN** Camila put a frame around a triangular garden in her yard. The two legs of the triangle are 8 feet long each, and the third side is equal to the length of a leg times $\sqrt{2}$. What is the perimeter of the triangle? Give your answer as a simplified radical expression.

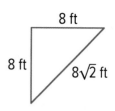

Examples 4 and 5
Simplify.

19. $2\sqrt{3} \cdot 3\sqrt{15}$

20. $5\sqrt{3} \cdot 2\sqrt{21}$

21. $6\sqrt{7} \cdot 2\sqrt{8}$

22. $7\sqrt{10} \cdot 4\sqrt{10}$

23. $11\sqrt{6} \cdot 3\sqrt{12}$

24. $10\sqrt{5} \cdot 5\sqrt{11}$

25. $\sqrt{2}(\sqrt{8} + \sqrt{6})$

26. $\sqrt{5}(\sqrt{10} - \sqrt{3})$

27. $\sqrt{5}(\sqrt{2} + 4\sqrt{2})$

28. $\sqrt{6}(2\sqrt{10} + 3\sqrt{2})$

29. $4\sqrt{5}(3\sqrt{5} + 8\sqrt{2})$

30. $5\sqrt{3}(6\sqrt{10} - 6\sqrt{3})$

Mixed Exercises

Simplify.

31. $\sqrt{\frac{1}{25}} - \sqrt{5}$

32. $\sqrt{\frac{2}{9}} + \sqrt{6}$

33. $2 + 2\sqrt{2} - \sqrt{8}$

34. $8\sqrt{\frac{5}{4}} + 3\sqrt{20} - \sqrt{45}$

35. $\sqrt{24} - 5\sqrt{12} + 4\sqrt{2} - 3\sqrt{3}$

36. $\frac{1}{2}\sqrt{8} - \sqrt{\frac{2}{9}} + 9\sqrt{2}$

37. $5\sqrt{3} \cdot 3\sqrt{5}$

38. $8\sqrt{7} \cdot \frac{2}{3}\sqrt{7}$

39. $\sqrt{\frac{3}{4}}(\sqrt{6} + 2\sqrt{20})$

40. $6\sqrt{6}(9\sqrt{2} - 5\sqrt{12})$

41. $(a\sqrt{3} + b\sqrt{5})(c\sqrt{8} + d\sqrt{5})$

42. $(6\sqrt{2} + 2\sqrt{3})(\sqrt{10} - 4\sqrt{7})$

43. **GEOMETRY** The area of a trapezoid is found by multiplying its height by the average length of its bases. Find the area of the deck attached to Mr. Wilson's house. Give your answer as a simplified radical expression.

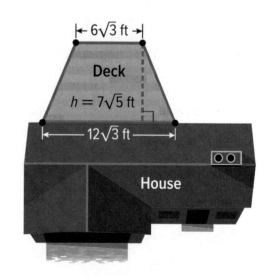

44. **TRAVEL** Lucia used the straight-line distance between Lincoln, Nebraska, and Houston, Texas, and the straight-line distance between Houston and Tallahassee, Florida, to estimate the straight-line distance between Lincoln and Tallahassee, as shown on the map. Lucia traveled from Lincoln to Tallahassee for vacation. A week after returning home from vacation, Lucia traveled from Lincoln to Houston to visit a relative. About how far did Lucia travel after returning home from Houston? Give your answer as a radical expression.

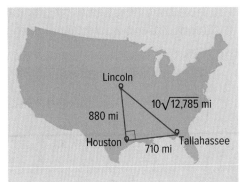

45. **FREE FALL** A ball is dropped from a building window 800 feet above the ground. Another ball is dropped from a lower window 288 feet high. Both balls are released at the same time. Assume air resistance is not a factor, and use the formula $t = \frac{1}{4}\sqrt{h}$ to find how many seconds t it will take a ball to fall h feet.

   a. How much longer after the first ball hits the ground does the second ball land?

   b. Find a decimal approximation of the answer for part **a**. Round your answer to the nearest tenth.

46. **REASONING** Deja is putting a fence around her yard. The long side of the yard is $12\sqrt{2}$ meters, the two shorter sides are each $5\sqrt{2}$ meters, and there is a $\sqrt{2}$-meter section that runs up to the house. When Deja buys the fencing, she approximates the length of fencing she needs. Write and solve an equation to find the total length of fencing f needed for the yard. Give your answer as a radical expression. Then approximate the amount of fencing Deja should buy to the nearest meter.

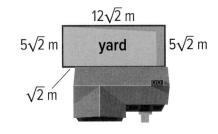

47. **SHADOWS** The distance from the top of Gino's head to the end of his head's shadow is the hypotenuse of a 45°-45°-90° triangle, or Gino's height times $\sqrt{2}$. Gino is 4.5 feet tall. The distance from a nearby tree to the end of its shadow is 2.5 times as long as the distance from the top of Gino's head to the end of his shadow. How far is it from the top of the tree to the end of its shadow? Give your answer as a radical expression.

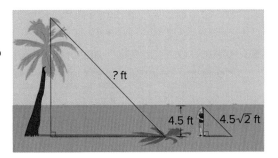

48. **STRUCTURE** The table shows several radical expressions. Copy and complete the table by writing *yes* or *no* in each row to indicate whether the expression in column A and the expression in column B have the same radicand when simplified.

Expression A	Expression B	Simplify to have same radicand? (Write Yes or No)
$4\sqrt{24}$	$6\sqrt{28}$	
$3\sqrt{18}$	$2\sqrt{32}$	
$5\sqrt{15}$	$12\sqrt{30}$	
$10\sqrt{20}$	$7\sqrt{45}$	

49. **SCALING** Austin has a model car that uses a 1:24 scale. The real car's hood is $20\sqrt{5}$ inches long. How long is the model car's hood? Give your answer as a radical expression.

50. **FIND THE ERROR** Jhumpa said $3\sqrt{32}$ cannot be subtracted from $8\sqrt{8}$ because they have different radicands. Kenyon said they can be subtracted. Is either correct? Explain your reasoning.

51. **PERSEVERE** Determine whether the following statement is *true* or *false*. Provide a proof or a counterexample to support your answer.
$(x + y)^2 > \left(\sqrt{x^2 + y^2}\right)^2$ when $x > 0$ and $y > 0$

52. **CONSTRUCT ARGUMENTS** Make a conjecture about the sum of a rational number and an irrational number. Is the sum *rational* or *irrational*? Is the product of a nonzero rational number and an irrational number *rational* or *irrational*? Explain your reasoning.

53. **CREATE** Write an equation that shows a sum of two radicals with different radicands. Explain how you could combine these terms.

54. **WRITE** A radical expression is an exact value. Rounding a radical gives an approximate value. Give examples of when an exact answer is preferred and when an approximate answer is acceptable.

55. **WHICH ONE DOESN'T BELONG?** Consider the expressions. Identify which one does not belong and explain your reasoning.

| $10\sqrt{5}$ | $2\sqrt{20}$ | $6\sqrt{12}$ | $8\sqrt{45}$ |

Module 8 • Exponents and Roots

Lesson 8-7

Exponential Equations

Explore Solving Exponential Equations

Online Activity Use an interactive tool to complete the Explore.

INQUIRY How can you solve an equation where the variable is in the exponent?

Today's Goals
- Solve exponential equations.

Today's Vocabulary
exponential equation

Learn Exponential Equations

In an **exponential equation**, the independent variable is an exponent. The Power Property of Equality and the other properties of exponents can be used to solve exponential equations.

Key Concept • Power Property of Equality	
Words	For any real number $b > 0$ and $b \neq 1$, $b^x = b^y$ if and only if $x = y$.
Examples	If $4^x = 4^5$, then $x = 5$. If $n = \frac{2}{5}$, $5^n = 5^{\frac{2}{5}}$. If $x^{\frac{1}{2}} = 3^{\frac{1}{2}}$, then $x = 3$.

Example 1 Solve a One-Step Exponential Equation

Solve $5^x = 625$.

$5^x = 625$ Original equation

$5^x = 5^4$ Rewrite 625 as 5^4.

$x = 4$ Simplify.

Go Online You may want to complete the Concept Check to check your understanding.

Check

Solve $2^x = 512$.

$x =$ __?__

Go Online You can complete an Extra Example online.

Think About It!
Can you use the Power Property of Equality if the two bases are not equal? Justify your argument.

Need Help?
In order to apply the Power Property of Equality in this example, the two bases must be raised to the same exponent. Thus, you rewrite 8^2 as $(8^4)^{\frac{1}{2}}$ so that you can set the bases equal to each other.

Example 2 Solve a Multi-Step Exponential Equation

Solve $27^{x-1} = 3$.

$27^{x-1} = 3$	Original equation
$(3^3)^{x-1} = 3$	Rewrite 27 as 3^3.
$3^{3x-3} = 3$	Power of a Power Property, Distributive Property
$3^{3x-3} = 3^1$	Rewrite 3 as 3^1.
$3x - 3 = 1$	Power Property of Equality
$3x = 4$	Add 3 to each side.
$x = \frac{4}{3}$	Divide each side by 3.

Example 3 Solve a Real-World Exponential Equation

AEROSPACE In the 1950s, scientists proposed a space station that could house a crew of approximately 80 people. The station could produce artificial gravity by rotating at a speed of $w = \sqrt{gr}$, where g is 32 feet per second squared, and r is the radius of the station. If the station design required a rotating speed of approximately 64 feet per second to simulate gravity on Earth, what would the radius need to be?

$w = \sqrt{gr}$	Original equation
$64 = \sqrt{32r}$	$w = 64, g = 32$
$64 = (32r)^{\frac{1}{2}}$	$\sqrt[n]{b} = b^{\frac{1}{n}}$
$8^2 = (32r)^{\frac{1}{2}}$	$64 = 8^2$
$(8^4)^{\frac{1}{2}} = (32r)^{\frac{1}{2}}$	$8^2 = (8^4)^{\frac{1}{2}}$
$8^4 = 32r$	Power Property of Equality
$4096 = 32r$	$8^4 = 4096$
$128 = r$	Divide each side by 32.

The radius of the space station would need to be approximately 128 feet.

Check

SUNSCREEN The degree to which sunscreen protects your skin is measured by its sun protection factor (SPF). For a sunscreen with SPF f, the percentage of UV-B rays absorbed p is $p = 50f^{\frac{1}{5}}$. What SPF absorbs 100% of UV-B rays?

Go Online You can complete an Extra Example online.

Practice

Go Online You can complete your homework online.

Examples 1 and 2

Solve each equation.

1. $2^x = 512$
2. $3^x = 243$
3. $6^x = 46{,}656$
4. $5^x = 125$
5. $3^x = 6561$
6. $16^x = 4$
7. $3^{x-3} = 243$
8. $4^{x-1} = 1024$
9. $6^{x-1} = 1296$
10. $4^{2x+1} = 1024$
11. $2^{4x+3} = 2048$
12. $3^{3x+3} = 6561$

Example 3

13. **ELECTRICITY** The relationship of the current, power, and resistance of an appliance can be modeled by $I\sqrt{R} = \sqrt{P}$, where I is the current in amperes, P is the power in watts, and R is the resistance in ohms. Find the resistance that an appliance is using if the current is 2.5 amps and the power is 100 watts.

14. **VIDEO** Felipe uploaded a funny video of his dog. The relationship between the elapsed time in days, d, since the video was first uploaded and the total number of views, v, that the video received is modeled by $v = 4^{1.25d}$. Find the number of days it took Felipe's video to get 1024 views.

15. **CONSTRUCTION** A large plot of land has been purchased by developers. They roll out a schedule of construction. The relationship between the area of the undeveloped land in hectares, A, and the elapsed time in months, t, since the construction began is modeled by the function $A = 6250 \cdot 10^{-0.1t}$. How many months of construction will there be before the area of the undeveloped land decreases to 62.5 hectares?

Mixed Exercises

Solve each equation.

16. $2^{5x} = 8^{2x-4}$
17. $81^{2x-3} = 9^{x+3}$
18. $2^{4x} = 32^{x+1}$
19. $16^x = \frac{1}{2}$
20. $25^x = \frac{1}{125}$
21. $6^{8-x} = \frac{1}{216}$
22. $5^x = 125$
23. $2^{5x-4} = 64$
24. $4^{x+1} = 256$
25. $3^{4x-2} = 729$

Lesson 8-7 • Exponential Equations 481

26. **USE A MODEL** Without advertising, a Web site had 96 total visits. Today, the owners of the site are starting a new promotion, which is expected to double the total number of visits to their Web site every 5 days.
    a. Write an equation that relates the total number of visits, v, to the number of days the promotion has been running, d.
    b. Use your equation from part **a** to find how many days the promotion should be run in order to increase the traffic to the Web site to 12,288 total visits.

27. **PHYSICS** The velocity v of an object dropped from a tall building is given by the formula $v = \sqrt{64d}$, where d is the distance the object has dropped. What distance was the object dropped from if it has a velocity of 49 feet per second? Round your answer to the nearest hundredth.

28. **FENCING** Representatives from the neighborhood have requested that the city install a fence around a newly-built playground. The equation $f = 4\sqrt{A}$ represents the amount of fence f needed based on the area A of the playground. If the playground has 324 feet of fencing, find the area of the playground.

29. **CREATE** Write an equation equivalent to $8^{(2 + x)} = 16^{(2.5 - 0.5x)}$ with a base of 4. Then solve for x. Justify your answer.

30. **FIND THE ERROR** Zari and Jenell are solving $128^x = 4$. Is either correct? Explain your reasoning.

Zari	Jenell
$128^x = 4$	$128^x = 4$
$(2^7)^x = 2^2$	$(2^7)^x = 4$
$2^{7x} = 2^2$	$2^{7x} = 4^1$
$7^x = 2$	$7^x = 1$
$x = \frac{2}{7}$	$x = \frac{1}{7}$

31. **CONSTRUCT ARGUMENTS** Make a conjecture about why the equation $2^b = 5^{b-1}$ cannot be solved by the methods used in this section.

32. **CREATE** Write and solve an exponential equation where both bases need to be changed.

33. **PERSEVERE** Find the solutions to the equation $32^{(x^2 + 4x)} = 16^{(x^2 + 4x)}$.

34. **CREATE** Write an exponential equation that has a solution of $x = -1$.

35. **WHICH ONE DOESN'T BELONG?** Which exponential equation does not belong? Justify your conclusion.

| $3^{2x+1} = 243$ | $4^{2x-1} = 1024$ | $2^{3x+2} = 2048$ | $5^{x+2} = 3125$ |

36. **WRITE** Explain how to solve the exponential equation $9^{4n-3} = 3^6$. Include the solution in your explanation.

Module 8 • Exponents and Roots
Review

> **Essential Question**
> How do you perform operations and represent real-world situations with exponents?

Module Summary

Lessons 8-1 and 8-2

Properties of Exponents

- To multiply two powers with the same base, add their exponents.
- To find the power of a power, multiply the exponents.
- To find the power of a product, find the power of each factor and multiply.
- To divide two powers with the same base, subtract their exponents.
- To find the power of a quotient, find the power of the numerator and the power of the denominator.

Lessons 8-3 and 8-4

Negative and Rational Exponents

- Any nonzero number raised to the zero power is 1.
- A negative exponent is the reciprocal of a number raised to the positive exponent.
- A rational exponent is an exponent that is a rational number.
- If $a^2 = b$, then a is the square root of b.
- If $a^3 = b$, then a is the cube root of b.
- If $a^n = b$ for any positive integer n, then a is an nth root of b.
- The nth root of b can be written as $b^{\frac{1}{n}}$ or $\sqrt[n]{b}$.
- For any positive real number b and any integers m and $n > 1$, $b^{\frac{m}{n}} = \left(\sqrt[n]{b}\right)^m$ or $\sqrt[n]{b^m}$.

Lessons 8-5 and 8-6

Radical Expressions

- For any nonnegative real numbers a and b, $\sqrt{ab} = \sqrt{a} \cdot \sqrt{b}$.
- For any nonnegative real numbers a and b, $\sqrt{\frac{a}{b}} = \frac{\sqrt{a}}{\sqrt{b}}$, if $b \neq 0$.
- A square root is in simplest form if it contains no perfect square factors other than 1, it contains no fractions, and there are no square roots in the denominator of a fraction.

Lesson 8-7

Exponential Equations

- In an exponential equation, the independent variable is an exponent.
- The Power Property of Equality and the other properties of exponents can be used to solve exponential equations.
- If you square both sides of an equation, the resulting equation is still true.

Study Organizer

Foldables

Use your Foldable to review this module. Working with a partner can be helpful. Ask for clarification of concepts as needed.

Test Practice

1. **MULTI-SELECT** Select all expressions equivalent to $(5xy^3)(3xy^3)$. (Lesson 8-1)

   A. $(5 \cdot 3)(xy^3 \cdot xy^3)$
   B. $15x^2y^6$
   C. $15xy^6$
   D. $(5 + 3)(xy^3 + xy^3)$
   E. $15xy^9$

2. **OPEN RESPONSE** Earth's mass is about 5,973,600,000,000,000,000,000,000 kg. Write this mass in scientific notation. (Lesson 8-1)

3. **MULTIPLE CHOICE** The volume of a cube is $V = s^3$, where s is the length of one side. Which expression represents the volume of the cube? (Lesson 8-1)

   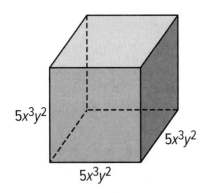

   A. $5x^9y^6$
   B. $125x^9y^6$
   C. $25x^9y^6$
   D. $15x^3y^2$

4. **MULTIPLE CHOICE** Simplify $(w^3)^5$. (Lesson 8-1)

   A. w^8
   B. w^{15}
   C. w^{27}
   D. w^{243}

5. **OPEN RESPONSE** A manufacturer sells square trivets in various sizes. In the corner of each trivet sold, they engrave the company's square logo. The diagram shows the ratio of the size of each trivet to the size of the company's logo.

   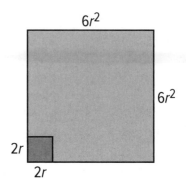

   What is the ratio of the area of the trivet to the area of the logo in simplified form? (Lesson 8-2)

   Ratio of area of trivet to area of logo:
   __?__ r^2 to 1.

6. **OPEN RESPONSE** Simplify $\dfrac{k^6 r^3 t^5}{k r^2 t^4}$. (Lesson 8-2)

7. **MULTIPLE CHOICE** Simplify $\left(\dfrac{2x^2y^3}{3x^5}\right)^0$. (Lesson 8-3)

   A. 0
   B. $\dfrac{2}{3}$
   C. 1
   D. $3x^5$

484 Module 8 Review • Exponents and Roots

8. MULTIPLE CHOICE Simplify $\left(\dfrac{2m^2j^3}{3p^5}\right)^4$. (Lesson 8-2)

A. $\dfrac{8m^8j^{12}}{12p^{20}}$

B. $\dfrac{16m^8j^{12}}{81p^{20}}$

C. $\dfrac{16m^8j^{12}}{3p^5}$

D. $\dfrac{16m^6j^7}{81p^9}$

9. OPEN RESPONSE The maximum speed of an object made of material A when pushed by wind is about 0.001 foot per minute, so the order of magnitude is about 10^{-3}.

The maximum speed of an object made of material B when pushed by the same wind is about 1000 feet per minute, so the order of magnitude is about 10^3 feet per minute.

What is the ratio of the maximum speed of the object made of material A to the maximum speed of the object made of material B? Write your answer in decimal form. (Lesson 8-3)

10. OPEN RESPONSE Explain whether the following equation is true or false. If false, explain how to make it true. (Lesson 8-3)

$$\dfrac{49x^{-3}y^4z^{-1}}{4x^2y^{-2}} = \dfrac{7y^6}{2x^5yz}$$

11. OPEN RESPONSE Explain how to simplify the following expression in your own words. $729^{\frac{1}{6}}$ (Lesson 8-4)

12. MULTIPLE CHOICE Evaluate $\sqrt[4]{81}$. (Lesson 8-4)

A. 4

B. $\dfrac{1}{4}$

C. 81

D. 3

13. MULTIPLE CHOICE Evaluate $\sqrt[6]{4096}$. (Lesson 8-4)

A. 2

B. 4

C. 16

D. 64

14. MULTIPLE CHOICE Simplify $25^{-\frac{3}{2}}$. (Lesson 8-4)

A. -125

B. -37.5

C. $\dfrac{1}{125}$

D. $\dfrac{1}{15}$

15. OPEN RESPONSE Evaluate $81^{\frac{3}{4}}$. (Lesson 8-4)

16. OPEN RESPONSE Simplify $\sqrt{98}$. (Lesson 8-5)

17. MULTIPLE CHOICE Simplify $\sqrt{121a^5b^3c}$. (Lesson 8-5)

A. $11a^2b\sqrt{abc}$

B. $11a^2b^4$

C. $11ab\sqrt{ab^3c}$

D. $11b\sqrt{a^3b^3c}$

18. MULTIPLE CHOICE Simplify $\sqrt[3]{\dfrac{125a^6}{27b^3}}$. (Lesson 8-5)

A. $\dfrac{a^2}{b}$

B. $\dfrac{5a^6}{3b^3}$

C. $\dfrac{\sqrt[3]{a^2}}{\sqrt[3]{b}}$

D. $\dfrac{5a^2}{3b}$

19. OPEN RESPONSE Simplify $8\sqrt{5} + 2\sqrt{7} - 6\sqrt{5} + 4\sqrt{7}$. (Lesson 8-6)

20. OPEN RESPONSE Simplify $4\sqrt{3}(5\sqrt{8} + 2\sqrt{3})$. (Lesson 8-6)

21. MULTIPLE CHOICE Solve $7 = 343^{x-2}$. (Lesson 8-7)

A. 5

B. $\dfrac{7}{3}$

C. 2

D. 1

22. MULTIPLE CHOICE Solve $4 = 16^{x+2}$. (Lesson 8-7)

A. -4

B. $-\dfrac{3}{2}$

C. $\dfrac{2}{3}$

D. 4

23. OPEN RESPONSE The volume of a pyramid is $V = \dfrac{1}{3}Bh$, where B is the area of the base and h is the height. What is the value of x if the volume of the pyramid, in simplest form, is equivalent to $6^{\frac{x}{3}}$ cm³? (Lesson 8-7)

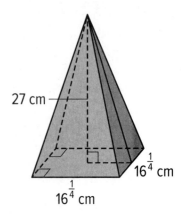

27 cm

$16^{\frac{1}{4}}$ cm

$16^{\frac{1}{4}}$ cm

Module 9
Exponential Functions

Essential Question
When and how can exponential functions represent real-world situations?

What Will You Learn?
How much do you already know about each topic **before** starting this module?

KEY

👎 — I don't know. 👊 — I've heard of it. 👍 — I know it!

	Before			After		
	👎	👊	👍	👎	👊	👍
graph exponential growth functions						
graph exponential decay functions						
translate exponential functions						
dilate exponential functions						
reflect exponential functions						
solve problems involving exponential growth and decay						
transform exponential expressions						
generate geometric sequences						
write recursive formulas						
translate between recursive and explicit formulas						

Foldables Make this Foldable to help you organize your notes about exponential functions. Begin with a sheet of 11" × 17" paper and six index cards.

1. **Fold** lengthwise about 3" from the bottom.
2. **Fold** the paper in thirds.
3. **Open** and staple the edges on either side to form three pockets.
4. **Label** the pockets as shown. Place two index cards in each pocket.

 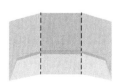 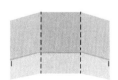 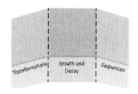

What Vocabulary Will You Learn?
- asymptote
- common ratio
- compound interest
- explicit formula
- exponential decay function
- exponential function
- exponential growth function
- geometric sequence
- recursive formula

Are You Ready?
Complete the Quick Review to see if you are ready to start this module.
Then complete the Quick Check.

Quick Review

Example 1

Evaluate $2x^3$ for $x = 5$.

$2x^3$	Original expression
$= 2(5)^3$	Substitute 5 for x.
$= 2(125)$	Evaluate the exponent.
$= 250$	Multiply.

Example 2

Divide $\frac{5}{6} \div \frac{1}{3}$.

$\frac{5}{6} \div \frac{1}{3} = \frac{5}{6} \cdot \frac{3}{1}$	Multiply by the reciprocal.
$= \frac{15}{6}$	Multiply the numerators and multiply the denominators.
$= \frac{5}{2}$	Find the simplest form.

Quick Check

Evaluate each expression for the given value.

1. $-4x^2$ for $x = 7$
2. $3x^2$ for $x = 3$
3. $0.25x^4$ for $x = 1$
4. x^5 for $x = 3$

Divide.

5. $128 \div 4$
6. $\frac{1}{3} \div 2$
7. $-9 \div 3$
8. $\frac{1}{8} \div \frac{1}{2}$

How did you do?
Which exercises did you answer correctly in the Quick Check?

Lesson 9-1

Exponential Functions

Explore Exponential Behavior

 Online Activity Use a table to complete the Explore.

@ **INQUIRY** How does exponential behavior differ from linear behavior?

Learn Identifying Exponential Behavior

An **exponential function** is a function of the form $y = ab^x$, where $a \neq 0$, $b > 0$, and $b \neq 1$. Some examples of exponential functions are $y = 2(3)^x$, $y = 4^x$, and $y = \left(\frac{1}{2}\right)^x$.

Linear

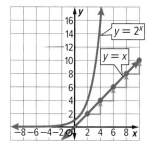

the rate of change remains constant

Exponential

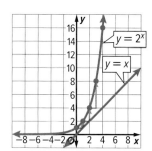

the rate of change increases by the same factor

Note that in the linear function, the rate of change remains constant. In the exponential function, the rate of change increases by the same factor.

Today's Goals
- Recognize situations modeled by linear or exponential functions.
- Graph exponential functions, showing intercepts and end behavior.

Today's Vocabulary
exponential function

exponential growth function

exponential decay function

asymptote

Lesson 9-1 • Exponential Functions **489**

Think About It!
Is a function that relates an independent variable to a change in order of magnitude always exponential? Explain your reasoning.

Example 1 Identify Exponential Behavior

EARTHQUAKES The Richter Scale measures the energy that an earthquake releases and assigns a magnitude to it. These orders of magnitude can be approximated by comparing them to the explosive power of TNT. Determine whether the set of data displays exponential behavior.

Magnitude	TNT (tons)
1	0.6
2	6
3	60
4	600
5	6000
6	60,000
7	600,000
8	6,000,000

Magnitudes 1 and 2

As the order of magnitude increases from 1 to 2 the amount of TNT that is approximately equal in magnitude increases from 0.6 ton to 6 tons. That is an increase by a factor of 10.

Magnitudes 2 and 3

As the order of magnitude increases from 2 to 3, the amount of TNT that is approximately equal in magnitude increases from 6 tons to 60 tons. That is an increase by a factor of 10.

Since the change in the amount of TNT increases by the same factor given an equal change in magnitude, the data set displays exponential behavior.

Think About It!
Can you confirm that the entire data set displays exponential behavior by looking at the first two intervals?

Check

MEMORY In the 19th century, psychologist Hermann Ebbinghaus created a formula to approximate how quickly people forget information over time. The approximate percentage of the newly learned information a person retains over time is shown in the table. Determine whether the data displays exponential behavior.

Time (days)	% Retained
0	100
1	80
2	64
3	51.2
4	40.96
5	32.768

Source: Indiana University

The data set ___?___ display exponential behavior.

Go Online
An alternate method is available for this example.

Explore Restrictions on Exponential Functions

 Online Activity Use graphing technology to complete the Explore.

> **INQUIRY** Why are exponential functions defined such that $a \neq 0$, $b > 0$, and $b \neq 1$?

Learn Graphing Exponential Functions

Functions of the form $f(x) = ab^x$, where $a > 0$ and $b > 1$, are called **exponential growth functions**. Functions of the form $f(x) = ab^x$, where $a > 0$ and $0 < b < 1$, are called **exponential decay functions**.

The graphs of exponential functions have an **asymptote**. An asymptote is a line that a graph approaches.

Go Online You can watch a video to see how to graph exponential functions.

Key Concept • Types of Exponential Functions

Exponential Growth Functions	Exponential Decay Functions
Equation	
$f(x) = ab^x$, $a > 0$, $b > 1$	$f(x) = ab^x$, $a > 0$, $0 < b < 1$
Domain, Range	
D = all real numbers; R = $\{y \mid y > 0\}$	D = all real numbers; R = $\{y > 0\}$
Intercepts	
one y-intercept, no x-intercepts	one y-intercept, no x-intercepts
End Behavior	
as x increases, $f(x)$ increases; as x decreases, $f(x)$ approaches 0	as x increases, $f(x)$ approaches 0; as x decreases, $f(x)$ increases
Graph	

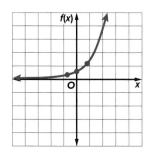 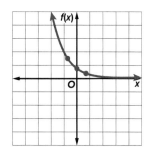

Lesson 9-1 • Exponential Functions 491

Talk About It!
How can you determine the *y*-intercept without substituting into the original equation? (Hint: Consider what $x = 0$ and the *y*-intercept mean in the context of the situation.)

Problem-Solving Tip
Make an Organized List Making an organized list of *x*-values and corresponding *y*-values is helpful in graphing the function. It can also help you identify patterns in the data.

Go Online
You can watch a video to see how to use a graphing calculator with this example.

Example 2 Exponential Growth Function

FOLDING Each time you fold a piece of paper in half, it doubles in thickness. If a piece of paper is 0.05 millimeter thick, then you can determine the thickness *y* of a piece of paper given the number of folds *x* with the function $y = 0.05(2)^x$. Identify the key features of the function, graph it, and then identify the relevant domain and range in the context of the situation.

Part A Identify key features.

Because $a > 0$ and $b > 1$, $y = 0.05(2)^x$ is an exponential growth function. The domain is all real numbers and the range is $y > 0$.

The *y*-intercept is the value of *y* when $x = 0$.

$y = 0.05(2)^0$

$y = 0.05(1)$

$= 0.05$

The *y*-intercept is 0.05.

Because $y = 0.05(2)^x$ is on the exponential growth function, as *x* increases, *y* increases, and as *x* decreases, *y* approaches 0.

Part B Graph the function.

Make a table of values. Then, plot the points and draw a curve to approximate it.

x	$y = 0.05(2)^x$	y
−2	$y = 0.05(2)^{-2}$	0.0125
−1	$y = 0.05(2)^{-1}$	0.025
0	$y = 0.05(2)^0$	0.05
1	$y = 0.05(2)^1$	0.1
2	$y = 0.05(2)^2$	0.2
3	$y = 0.05(2)^3$	0.4
4	$y = 0.05(2)^4$	0.8

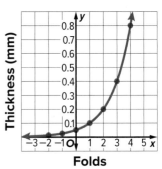

Part C Identify relevant domain and range.

Because the number of folds cannot be negative and folds must be counted in integers, the potential domain is the set of whole numbers and the potential range is the set of real numbers greater than or equal to 0.05. However, because the paper cannot be folded indefinitely, the thickness of the paper cannot continue to grow to infinity. So the domain will be restricted to the greatest possible number of folds, and the range will be restricted to the greatest thickness of the paper.

Go Online You can complete an Extra Example online.

Check

Consider $y = 3^x$.

Part A

List the key features that apply to $y = 3^x$. Include the domain, range, y-intercept, and end behavior of the function.

FILM The function $y = 3^x$ can be used to model a real-world situation. Sarah wants to crowdfund a film project. To spread the word, she shares the page with 3 friends, and requests that each friend share it with 3 more friends. The function that models the number of people with a link to the crowdfunding page y given the number of cycles x is defined by the function $y = 3^x$.

Part B

Select the correct graph of $y = 3^x$.

A.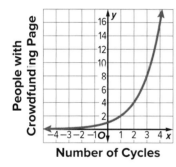

B.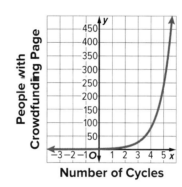

C.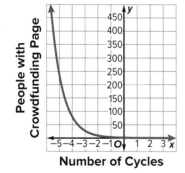

D.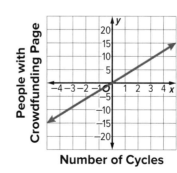

Part C

Describe the relevant range in the context of the situation.

Math History Minute

While a junior in high school, **Britney Gallivan (1985-)** showed that a single length of toilet paper 4000 feet long can be folded in half a maximum of twelve times. Britney's proof involved the exponential equation $L = \frac{\pi t}{6}(2^n + 4)(2^n - 1)$, where L is the length of the paper, t is the thickness of the material to be folded, and n represents the number of folds desired.

Use a Source

Research the amount of caffeine in another caffeinated drink. Write a function that models the amount of caffeine left in your system after x hours, and identify the key features of the function.

🌐 Example 3 Exponential Decay Function

CAFFEINE The half-life of a substance describes how long it takes for the substance to deplete by half. The half-life of caffeine in the body of a healthy adult is approximately 5 hours, meaning that it takes 5 hours for the body to break down half of the caffeine. Suppose an energy drink contains 160 milligrams of caffeine. The amount of caffeine y left in your system after x hours is modeled by the function $y = 160\left(\frac{1}{2}\right)^{\frac{x}{5}}$. Identify the key features of the function, graph it, and then identify the relevant domain and range in the context of the situation.

Part A Identify key features.

Because $a > 0$ and $0 < b < 1$, $y = 160\left(\frac{1}{2}\right)^{\frac{x}{5}}$ is an exponential decay function. The domain is all real numbers and the range is $y > 0$.

The y-intercept is the value of y when $x = 0$.

$$y = 160\left(\frac{1}{2}\right)^{\frac{0}{5}}$$
$$= 160(1) = 160$$

The y-intercept is 160.

Part B Graph the function.

x	$160\left(\frac{1}{2}\right)^{\frac{x}{5}}$	y
−1	$160\left(\frac{1}{2}\right)^{\frac{-1}{5}}$	184
0	$160\left(\frac{1}{2}\right)^{\frac{0}{5}}$	160
1	$160\left(\frac{1}{2}\right)^{\frac{1}{5}}$	139
2	$160\left(\frac{1}{2}\right)^{\frac{2}{5}}$	121
3	$160\left(\frac{1}{2}\right)^{\frac{3}{5}}$	106
4	$160\left(\frac{1}{2}\right)^{\frac{4}{5}}$	92
5	$160\left(\frac{1}{2}\right)^{\frac{5}{5}}$	80

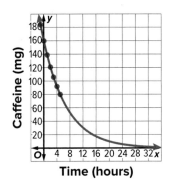

Part C Identify relevant domain and range.

Because time cannot be negative, the relevant domain is $\{x \geq 0\}$. Because the amount of caffeine cannot be negative and the amount of caffeine when $x = 0$ is 160 mg, the relevant range is $\{0 < y \leq 160\}$.

▶ **Go Online** You can complete an Extra Example online.

Practice

Go Online You can complete your homework online.

Example 1

Determine whether the set of data displays exponential behavior. Write *yes* or *no*. Explain why or why not.

1.
x	−3	−2	−1	0
y	9	12	15	18

2.
x	0	5	10	15
y	20	10	5	2.5

3.
x	4	8	12	16
y	20	40	80	160

4.
x	50	30	10	−10
y	90	70	50	30

5. **PICTURE FRAMES** Since a picture frame includes a border, the picture must be smaller in area than the entire frame. The table shows the relationship between picture area and frame length for a particular line of frames. Is this an exponential relationship? Explain.

Frame Length (in.)	Picture Area (in²)
5	6
6	12
7	20
8	30
9	42

Examples 2 and 3

6. **WASTE** Suppose the waste generated by nonrecycled paper and cardboard products in tons y after x days can be approximated by the function $y = 1000(2)^{0.3x}$.

   a. Identify key features.

   b. Graph the function.

   c. Identify relevant domain and range.

7. **IODINE** Iodine 131 is a radioisotope that is related to nuclear energy, medical diagnostic and treatment procedures, and natural gas production. A scientist is testing 50 milligrams of Iodine 131. The scientist knows that the half-life of Iodine 131 is about 8.02 days. The function $y = 50\left(\frac{1}{2}\right)^{\frac{x}{8.02}}$ represents the amount of Iodine 131 remaining in milligrams y after x days.

   a. Identify key features.

   b. Graph the function.

   c. Identify relevant domain and range.

8. **DEPRECIATION** Suppose a company's computer equipment is decreasing in value according to the function $y = 4000(0.87)^x$. In the equation, x represents the number of years that have elapsed since the equipment was purchased and y represents the value in dollars. What was the value 5 years after the computer equipment was purchased? Round your answer to the nearest dollar.

Lesson 9-1 • Exponential Functions 495

Mixed Exercises

Graph each function. Find the y-intercept and state the domain, range, and the equation of the asymptote.

9. $y = 2\left(\frac{1}{6}\right)^x$

10. $y = \left(\frac{1}{12}\right)^x$

11. $y = -3(9^x)$

12. $y = -4(10^x)$

13. $y = 3(11^x)$

14. $y = 4^x + 3$

15. **METEOROLOGY** The atmospheric pressure in millibars at altitude x meters above sea level can be approximated by the function $f(x) = 1038(1.000134)^{-x}$ when x is between 0 and 10,000.
    a. What is the atmospheric pressure at sea level?
    b. The McDonald Observatory in Texas is at an altitude of 2000 meters. What is the approximate atmospheric pressure there?
    c. As altitude increases, what happens to atmospheric pressure?

16. **PERSEVERE** Use tables and graphs to compare and contrast an exponential function $f(x) = ab^x + c$, where $a \neq 0$, $b > 0$, and $b \neq 1$, and a linear function $g(x) = ax + c$. Include intercepts, symmetry, end behavior, extrema, and intervals where the functions are increasing, decreasing, positive, or negative.

17. **CREATE** Write an exponential function that passes through (0, 3) and (1, 6).

18. **ANALYZE** Determine whether the graph of $y = ab^x$, where $a \neq 0$, $b > 0$, and $b \neq 1$, *sometimes*, *always*, or *never* has an *x*-intercept. Justify your argument.

19. **WRITE** Find an exponential function that represents a real-world situation, and graph the function. Analyze the graph, and explain why the situation is modeled by an exponential function rather than a linear function.

Lesson 9-2

Transformations of Exponential Functions

Explore Translating Exponential Functions

 Online Activity Use graphing technology to complete the Explore.

> **INQUIRY** What effect does adding to or subtracting from a function before or after it has been evaluated have on the function?

Today's Goals
- Apply translations to exponential functions.
- Apply dilations to exponential functions.
- Apply reflections to exponential functions.
- Use transformations to identify exponential functions from graphs and write equations of exponential functions.

Learn Translations of Exponential Functions

Key Concept • Vertical Translations of Exponential Functions

- The graph of $g(x) = b^x + k$ is the graph of $f(x) = b^x$ translated vertically.
- If $k > 0$, the graph of $f(x)$ is translated k units up.
- If $k < 0$, the graph of $f(x)$ is translated $|k|$ units down.

Key Concept • Horizontal Translations of Exponential Functions

- The graph of $g(x) = b^{x-h}$ is the graph of $f(x) = b^x$ translated horizontally.
- If $h > 0$, the graph of $f(x)$ is translated h units right.
- If $h < 0$, the graph of $f(x)$ is translated $|h|$ units left.

Go Online You can watch a video to see how to describe translations of functions.

Think About It! For $h < 0$, why must you move $|h|$ units left instead of h units?

Example 1 Vertical Translations of Exponential Functions

Describe the translation in $g(x) = 2^x + 3$ as it relates to the graph of the parent function $f(x) = 2^x$.

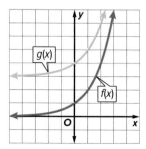

The constant k is added to the function after it has been evaluated, so k affects the output values.

The value of k is greater than 0, so the graph of $f(x) = 2^x$ is translated 3 units up.

Think About It! What do you notice about the asymptote of a vertically translated exponential function compared to the asymptote of the parent function?

Lesson 9-2 • Transformations of Exponential Functions **497**

🧠 **Think About It!**
What do you notice about the asymptote of a horizontally translated exponential function compared to the asymptote of the parent function?

Example 2 Horizontal Translations of Exponential Functions

Describe the translation in $g(x) = 3^{x+1}$ as it relates to the graph of the parent function $f(x) = 3^x$.

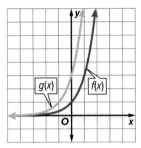

The constant h is subtracted from x before the function is performed, so h affects the input values.

The value of h is less than 0, so the graph of $f(x) = 3^x$ is translated 1 unit left.

🧠 **Think About It!**
How can the placement of the constant tell you if the resulting transformation will be a vertical or horizontal translation?

Example 3 Multiple Translations of Exponential Functions

Describe the translation in $g(x) = \left(\frac{1}{2}\right)^{x-2} - 4$ as it relates to the graph of the parent function $f(x) = \left(\frac{1}{2}\right)^x$.

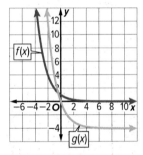

The value of h is subtracted from x before the function is performed and is greater than 0, so the graph of $f(x) = \left(\frac{1}{2}\right)^x$ is translated 2 units right.

The value of k is added to the function after it has been evaluated and is less than 0, so the graph of $f(x) = \left(\frac{1}{2}\right)^x$ is also translated 4 units down.

Check

Describe the translation in $g(x) = 2^{x+1} - 8$ as it relates to the graph of the parent function $f(x) = 2^x$.

The graph of $g(x) = 2^{x+1} - 8$ is the translation of the graph of the parent function 1 unit __?__ and 8 units __?__.

▶ **Go Online** You can complete an Extra Example online.

498 Module 9 • Exponential Functions

Example 4 Identify Exponential Functions from Graphs (Vertical Translations)

The given graph is a translation of the parent function $f(x) = \left(\frac{1}{4}\right)^x$. Use the graph of the function to write its equation.

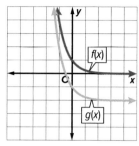

$$g(x) = \left(\frac{1}{4}\right)^x - 2$$

The horizontal asymptote of $g(x)$ is different from the horizontal asymptote of $f(x)$, implying a vertical translation of the form $g(x) = \left(\frac{1}{4}\right)^x + k$. The parent graph has a y-intercept at (0, 1). The translated graph has a y-intercept at (0, −1). The y-intercept is shifted 2 units down, so $k = -2$.

Study Tip

Vertical Translations Any exponential parent function has a y-intercept at (0, 1) and an asymptote at $y = 0$. By examining how far these features are shifted up or down, you can easily determine the value of k when identifying exponential functions.

Example 5 Identify Exponential Functions from Graphs (Horizontal Translations)

The given graph is a translation of the parent function $f(x) = 2^x$. Use the graph of the function to write its equation.

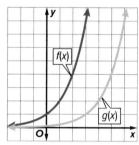

$$g(x) = 2^{x-3}$$

The horizontal asymptote of $g(x)$ is the same as the horizontal asymptote of $f(x)$, implying a horizontal translation of the form $g(x) = 2^{x-h}$. The parent graph passes through (0, 1). The translated graph has a y-value of 1 at (3, 1). The graph is shifted 3 units right, so $h = 3$.

Study Tip

Horizontal Translations Any exponential parent function has a y-intercept at (0, 1). By examining how far this point is shifted right or left, you can easily determine the value of h when identifying exponential functions.

Check

The given graph is a translation of $f(x) = 5^x$. Which is the equation for the function shown in the graph?

A. $g(x) = 5^x + 4$

B. $g(x) = 5^{x-4}$

C. $g(x) = 5^x - 4$

D. $g(x) = 5^{x+4}$

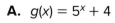 Go Online You can complete an Extra Example online.

Explore Dilating Exponential Functions

 Online Activity Use graphing technology to complete the Explore.

> **INQUIRY** What effect does multiplying a function by a value before or after it has been evaluated have on the function?

Learn Dilations of Exponential Functions

Key Concept • Vertical Dilations of Exponential Functions

- The graph of $g(x) = ab^x$ is the graph of $f(x) = b^x$ stretched or compressed vertically by a factor of $|a|$.
- If $|a| > 1$, the graph of $f(x)$ is stretched vertically away from the x-axis.
- If $0 < |a| < 1$, the graph of $f(x)$ is compressed vertically toward the x-axis.

Key Concept • Horizontal Dilations of Exponential Functions

- The graph of $g(x) = b^{ax}$ is the graph of $f(x) = b^x$ stretched or compressed horizontally by a factor of $\frac{1}{|a|}$.
- If $|a| > 1$, the graph of $f(x)$ is compressed horizontally toward the y-axis.
- If $0 < |a| < 1$, the graph of $f(x)$ is stretched horizontally away from the y-axis.

Go Online
You can watch a video to see how to describe dilations of functions.

Study Tip
Vertical Dilations For a vertical dilation, if you multiply each y-coordinate of the function $f(x)$ by a, you'll get the corresponding y-coordinate of the function $g(x)$. For the function in the example, the point $(1, 3)$ on $f(x)$ corresponds to the point $\left(1, \frac{3}{4}\right)$ on $g(x)$. The y-coordinate of $f(x)$, 3, is multiplied by a.

Example 6 Vertical Dilations of Exponential Functions

Describe the dilation in $g(x) = \frac{1}{4}(3)^x$ as it relates to the graph of the parent function $f(x) = 3^x$.

Since $f(x) = 3x$, $g(x) = a \cdot f(x)$, where $a = \frac{1}{4}$.

$g(x) = \frac{1}{4}(3)^x \rightarrow g(x) = \frac{1}{4} f(x)$

The function is multiplied by the positive constant a after it has been evaluated, and $|a|$ is between 0 and 1, so the graph of $f(x) = 3^x$ is compressed vertically by a factor of $|a|$, or $\frac{1}{4}$.

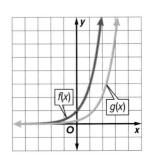

 Go Online You can complete an Extra Example online.

Example 7 Horizontal Dilations of Exponential Functions

Describe the dilation in $g(x) = \left(\frac{5}{3}\right)^{2x}$ as it relates to the graph of the parent function $f(x) = \left(\frac{5}{3}\right)^{x}$.

x is multiplied by the positive constant a before it has been evaluated, and |a| is greater than 1, so the graph of $f(x) = \left(\frac{5}{3}\right)^{x}$ is compressed horizontally by a factor of $\frac{1}{|a|}$, or $\frac{1}{2}$.

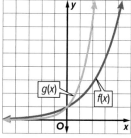

Check

Identify the dilation in each function as it relates to the parent function $f(x) = 4^x$ by copying and completing the table and writing the type of dilation and dilation factor next to each equation.

$g(x) = 4^{\frac{2}{3}x}$	↔	? ?
$h(x) = 3(4)^x$	↔	? ?
$k(x) = \frac{6}{7}(4)^x$	↔	? ?
$g(x) = 4^{\frac{5}{2}x}$	↔	? ?

🌎 Example 8 Describe Dilations of Exponential Functions

ENERGY Since 2000, solar PV capacity in the world has been growing exponentially. It can be approximated by the function $c(x) = 0.897(1.46)^x$, where $c(x)$ is the solar PV capacity in gigawatts, x is the number of years since 2000, and 0.897 is the initial capacity. Describe the dilation in $c(x) = 0.897(1.46)^x$ as it is related to the parent function $f(x) = (1.46)^x$.

The parent function is $f(x) = (1.46)^x$.

Then $c(x) = af(x)$, where $a = 0.897$.

$c(x) = 0.897(1.46)^x \rightarrow c(x) = 0.897f(x)$

The function is multiplied by the positive constant a after it has been evaluated and |a| is between 0 and 1, so the graph of $f(x) = (1.46)^x$ is compressed vertically by a factor of |a|, or 0.897.

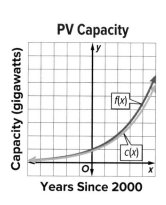

💭 Think About It!
How can you easily tell if an exponential function is going to be horizontally dilated?

💭 Think About It!
Why does a horizontal dilation not change the y-intercept of an exponential function? Justify your argument.

Study Tip
Assumptions Assuming that the rate at which PV capacity increases remains the same allows us to represent the situation with an exponential function.

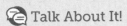**Talk About It!**

Make a conjecture about the y-intercept of g(x) and the value of a. Will this hold true for any vertical dilation of an exponential function? Explain your reasoning.

Example 9 Identify Exponential Functions from Graphs (Dilations)

The given graph is a dilation of the parent function $f(x) = 2^x$. Use the graph of the function to write its equation.

Notice that every point on the graph of g(x) is closer to the x-axis, implying a vertical compression of the form $g(x) = a(2)^x$.

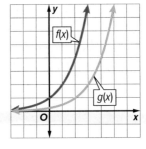

The point (2, 1) lies on the graph. Solve for a.

$$1 = a(2)^2$$
$$1 = 4a$$
$$\frac{1}{4} = a$$

$g(x) = \frac{1}{4}(2)^x$

Check

The given graph is a dilation of $f(x) = 3^x$. Which is the equation for the function shown in the graph?

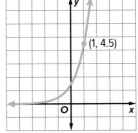

(1, 4.5)

A. $g(x) = 0.007(3)^x$

B. $g(x) = \frac{2}{3}(3)^x$

C. $g(x) = 3^x + \frac{3}{2}$

D. $g(x) = \frac{3}{2}(3)^x$

Explore Reflecting Exponential Functions

Online Activity Use graphing technology to complete the Explore.

INQUIRY What effect does multiplying a function by −1 before or after it has been evaluated have on the function?

Think About It!

What does the general form of an exponential function look like once it has been reflected across the x-axis?

Learn Reflections of Exponential Functions

Key Concept • Reflections of Exponential Functions Across the x-axis

- The graph of −f(x) is the reflection of the graph of $f(x) = b^x$ across the x-axis.
- Every y-coordinate of −f(x) is the corresponding y-coordinate of f(x) multiplied by −1.

Go Online
You can watch a video to see how to describe reflections of functions.

Key Concept • Reflections of Exponential Functions Across the y-axis

- The graph of f(−x) is the reflection of the graph of $f(x) = b^x$ across the y-axis.
- Every x-coordinate of f(−x) is the corresponding x-coordinate of f(x) multiplied by −1.

Example 10 Vertical Reflections of Exponential Functions

Describe how the graph of $g(x) = -3(2)^x$ is related to the graph of the parent function $f(x) = 2^x$.

The function is multiplied by −1 and the positive constant a after it has been evaluated and $|a|$ is greater than 1, so the graph of $f(x) = 2^x$ is stretched vertically and reflected across the x-axis.

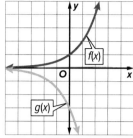

Example 11 Horizontal Reflections of Exponential Functions

Describe how the graph of $g(x) = (3)^{-2x}$ is related to the graph of the parent function $f(x) = 3^x$.

The function is multiplied by −1 and the constant a before it is evaluated and $|a|$ is greater than 1, so the graph of $f(x) = 3^x$ is compressed horizontally and reflected across the y-axis.

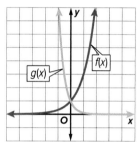

Think About It!

The example shows the reflection of an exponential function of the form $f(x) = ab^x$ over the y-axis for the case where $b > 1$. Examine the following cases and describe the effect a reflection across the y-axis would have on the end behavior of the parent function $f(x) = ab^x$.

Case 1: $g(x) = ab^{-x}$ where $b > 1$

Case 2: $g(x) = ab^{-x}$ where $0 < b < 1$

Check

Match each function with its graph.

1. $g(x) = -3^x$
2. $h(x) = \left(\frac{1}{3}\right)^{-x}$
3. $k(x) = -\left(\frac{1}{3}\right)^x$
4. $g(x) = 3^{-x}$

A.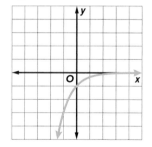

B.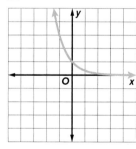

C.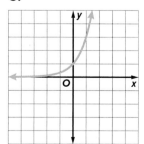

D.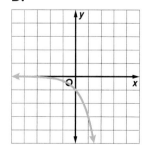

Go Online You can complete an Extra Example online.

Go Online
You can watch a video to see how to graph transformations of exponential functions using a graphing calculator.

Think About It!
Write an exponential function that is the parent function $f(x) = 2^x$ stretched vertically and translated 3 units left and 6 units up.

Learn Transformations of Exponential Functions

The general form of an exponential function is $f(x) = ab^{x-h} + k$, where a, h, and k are parameters that dilate, reflect, or translate a parent function with base b.

- The value of $|a|$ stretches or compresses (dilates) the parent graph.
- When a is negative, the graph is reflected across the x-axis.
- The value of h shifts (translates) the parent graph right or left.
- The value of k shifts (translates) the parent graph up or down.

Example 12 Multiple Transformations of Exponential Functions

Describe how the graph of $g(x) = -\frac{1}{2}(3)^{x-2} - 1$ is related to the graph of the parent function $f(x) = 3^x$.

$a < 0$ and $0 < |a| < 1$, so the graph of $f(x) = 3^x$ is reflected across the x-axis and compressed vertically by a factor of $|a|$, or $\frac{1}{2}$.

$h > 0$, so the graph is then translated h units right, or 2 units right.

$k < 0$, so the graph is then translated $|k|$ units down, or 1 unit down.

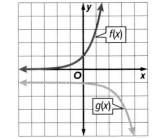

$g(x) = -\frac{1}{2}(3)^{x-2} - 1$ is the graph of the parent function compressed vertically, reflected across the x-axis, and translated 2 units right and 1 unit down.

Check

Part A
Describe how the graph of $g(x) = -4\left(\frac{1}{2}\right)^{x+5} - 2$ is related to the graph of the parent function $f(x) = \left(\frac{1}{2}\right)^x$.

The graph of $f(x) = \left(\frac{1}{2}\right)^x$ is reflected across the __?__ and __?__ vertically. The graph is translated 5 units __?__ and 2 units __?__.

Part B
Sketch the graph of $g(x) = -4\left(\frac{1}{2}\right)^{x+5} - 2$.

Go Online You can complete an Extra Example online.

Practice

Go Online You can complete your homework online.

Examples 1–3, 6–7, 10–12

Describe the transformation of g(x) as it relates to the parent function f(x).

1. $f(x) = 6^x$; $g(x) = 6^x + 8$

2. $f(x) = 5^x$; $g(x) = -5^x$

3. $f(x) = 3^x + 1$; $g(x) = 3^{2x} + 1$

4. $f(x) = 4^x - 3$; $g(x) = 4^{0.5x} - 3$

5. $f(x) = 2.3^x$; $g(x) = -2.3^{x-1}$

6. $f(x) = 2^x$; $g(x) = 2^{-x} + 1$

7. $f(x) = 5^x + 2$; $g(x) = 5^{-x} + 6$

8. $f(x) = 1.4^x - 1$; $g(x) = -1.4^x + 6$

9. $f(x) = 3^x + 1$; $g(x) = 2(3^x + 1)$

10. $f(x) = -4^x$; $g(x) = \frac{1}{3}(-4^x)$

11. $f(x) = 4^x$; $g(x) = 4^{x-3}$

12. $f(x) = \left(\frac{1}{2}\right)^x + 5$; $g(x) = \left(\frac{1}{2}\right)^x$

Examples 4–5, 9

Each graph is a transformation of the parent function $y = 2^x$. Use the graph of the function to write its equation.

13.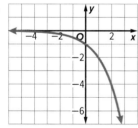

14.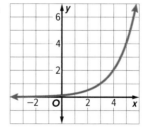

15.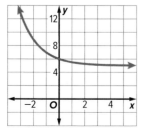

16.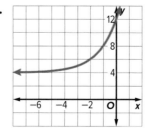

Example 8

17. SAVING Celia invests $2000 in a savings account that earns 1.25% interest per year compounded annually. The amount of money in her bank account after x years can be modeled by $g(x) = 2000(1.0125)^x$. Describe the dilation in $g(x)$ as it relates to the parent function $f(x) = 1.0125^x$.

18. CAFFEINE Suppose an 8-ounce cup of coffee contains 100 milligrams of caffeine. The rate at which caffeine is eliminated from an adult's body is 11% per hour. The function $f(x) = 100(0.89)^x$ can be used to model the amount of caffeine left in a person's bloodstream after x hours of consuming the cup of coffee. Suppose the function $g(x) = 25(0.89)^x$ represents the amount of caffeine left in a person's bloodstream after x hours of consuming an 8-ounce cup of green tea. Describe $g(x)$ as a transformation of $f(x)$.

19. VISITORS The number of visitors to a new skateboarding park can be modeled by the exponential function $g(x) = 20(2^x)$, where x represents the number of months since the park's grand opening. Explain how the number of visitors during that first month is a dilation of the parent function $f(x) = 2^x$.

20. DEPRECIATION Depreciation is the decrease in the value of an item resulting from its age or wear. When an item loses about the same percent of its value each year, an exponential function can be used to model its decreasing value over time. The function $g(x) = 12{,}000(0.85)^x$ can be used to model the value of a $12,000 car as it depreciates at an annual rate of 15% over x years. $g(x)$ is a dilation of the parent function $f(x) = 0.85^x$. The graph shows the function $g(x)$.

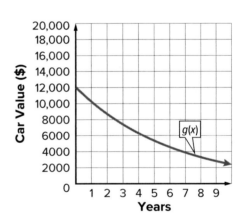

a. Write the equation of a function $h(x)$ that represents the depreciation of a $20,000 car depreciating at the same rate over x years.

b. Describe $h(x)$ as it relates to the parent function.

c. What is the difference between the values of the $12,000 car and the $20,000 car after 5 years?

Mixed Exercises

Describe the transformation of $g(x)$ as it relates to the parent function $f(x) = 2^x$.

21. $g(x) = 2^x + 6$

22. $g(x) = 3(2)^x$

23. $g(x) = -\frac{1}{4}(2)^x$

24. $g(x) = -3 + 2^x$

25. $g(x) = 2^{-x}$

26. $g(x) = -5(2)^x$

Write a function g(x) to represent the transformation of the parent function f(x).

27. $f(x) = 2^x$ translated 3 units up

28. $f(x) = 8^x$ translated 1 unit down

29. $f(x) = 5^x$ translated 2 units right

30. $f(x) = 3^x$ translated 4 units left

31. $f(x) = 6^x + 7$ translated 2 units down

32. $f(x) = -2^x + 3$ translated 5 units right

33. $f(x) = 4^x$ is compressed vertically by a factor of $\frac{1}{2}$

34. $f(x) = 3^x$ is stretched vertically by a factor of 5

35. $f(x) = 2^x$ is compressed horizontally by a factor of $\frac{1}{3}$

36. $f(x) = 5^x$ is stretched horizontally by a factor of 4

Each graph is a transformation of the parent function $f(x) = 5^x$. Use the graph of the function to write its equation.

37.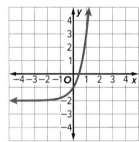

38.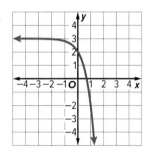

39.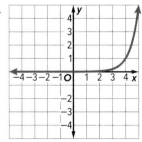

40.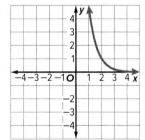

41. ASSETS Thomas and Rebecca each put $1000 into a bank account that earns 1.5% interest per year compounded annually. Thomas also has an antique toy automobile. The graph shows the amount of their assets over time.

  a. Describe the graph of Thomas' assets as a transformation of Rebecca's assets.

  b. Use the graph to extrapolate the value of Thomas' antique toy automobile.

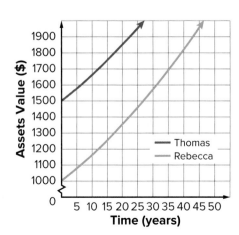

Lesson 9-2 • Transformations of Exponential Functions

The graph of g(x) is a transformation of the parent function f(x). Graph g(x) and describe the transformation in each function as it relates to the parent function.

42. $f(x) = 3^x$
    $g(x) = 5 \cdot 3^{x+2} - 4$

43. $f(x) = 2^x$
    $g(x) = -3 \cdot 2^{-x} + 1$

44. **CONSTRUCT ARGUMENTS** Name the coordinates of the point at which the graphs of $g(x) = 2^x + 3$ and $h(x) = 5^x + 3$ intersect. Explain your reasoning.

45. **STRUCTURE** Describe the similarities between the graph of $f(x) = 4^{x+2}$ and the graph of $g(x) = 16 \cdot 4^x$. Use the properties of exponents to justify your answer.

Higher-Order Thinking Skills

46. **ANALYZE** What would happen to the shape of the graph of an exponential function if the function is multiplied by a number between 0 and −1? What would happen to its shape if the exponent is multiplied by a number between 0 and −1? Justify your argument.

47. **FIND THE ERROR** Jennifer claims that the graph of $g(x) = 2(2^x)$ is a graph that rises more rapidly than its parent function $f(x) = 2^x$. James claims that it is actually the parent graph shifted to the left 2 units. Who is correct? Explain your reasoning.

48. **WRITE** A deficit is a negative amount of some quantity, such as money. A deficit that is growing exponentially can be modeled by $y = ab^{c(x-h)} + k$. Describe the constraints on a, b, and c.

49. **WHICH ONE DOESN'T BELONG?** Consider each pair of transformations of the function f(x) to g(x). Which one does not belong? Justify your conclusion.

$f(x) = 3^{x+2}$	$f(x) = 2^x$	$f(x) = 4^{x+1}$
$g(x) = 3^{x-1} + 2$	$g(x) = 2^{x+3} + 2$	$g(x) = 4^{x+4} + 2$

50. **CREATE** The graph shows a parent function f(x).

    a. Write a function to represent the parent function f(x).

    b. Write a function to represent a transformation of the parent function g(x).

    c. Describe the transformation.

    d. Graph the transformed function g(x).

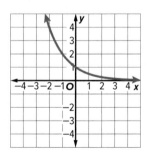

Lesson 9-3

Writing Exponential Functions

Explore Writing an Exponential Function to Model Population Growth

Online Activity Use a real-world situation to complete the Explore.

INQUIRY How can you find an equation that models the population growth of a colony of organisms that grows exponentially?

Learn Constructing Exponential Functions

If you are given two points, a graph, or a description of an exponential function, you can write an exponential function to model the data.

Given Two Points

Substitute the values of x and y into the equation $y = ab^x$. The result will be a system of equations in two variables, each with a and b as unknowns. You can then solve the system using substitution.

Given a Graph

Use any two points to write an equation in the form $y = ab^x$. Substitute the values of x and y into the equation, and solve the resulting system of equations for a and b using substitution.

Given a Description

Use the information to create a table or a graph. Then look for a pattern between the input and output values. Keep an eye out for words such as *exponential*, *linear*, *multiple*, *constant*, and *factor*, which can help you determine whether the function is exponential.

Example 1 Write an Exponential Function Given Two Points

Write an exponential function for the graph that passes through (1, 6) and (3, 24).

Substitute x and y into $y = ab^x$ to get a system of two equations.

$y = ab^x$ General form $y = ab^x$ General form
$6 = ab^1$ $y = 6$ and $x = 1$ $24 = ab^3$ $y = 24$ and $x = 3$

(continued on the next page)

Today's Goals
- Construct exponential functions by using a graph, a description, or two points.
- Create equations and solve problems involving exponential growth.
- Create equations and solve problems involving exponential decay.

Today's Vocabulary
compound interest

Study Tip
Graphs If you write an exponential equation based on a graph that includes the y-intercept, you can substitute the y-value of the intercept for a in $y = ab^x$. Since $b^0 = 1$, $y = a$. You can then use another point to find the common ratio b.

Solve the system of equations using substitution.

Solve the first equation for a.

$6 = ab^1$ First equation

$\frac{6}{b} = a$ Division Property of Equality

Substitute $\frac{6}{b}$ for a in the second equation to find b.

$24 = ab^3$ Second equation

$24 = \frac{6}{b}b^3$ Substitute $\frac{6}{b}$ for a.

$24 = \frac{6b^3}{b}$ Multiply.

$24 = 6b^2$ Quotient of Powers

$4 = b^2$ Division Property of Equality; then simplify

$2 = b$ Definition of exponent

Substitute 2 for b in either equation to find a.

$6 = ab^1$ Second equation

$6 = a(2)^1$ Substitute 2 for b.

$3 = a$ Simplify.

Write the equation.

$y = 3 \times 2^x$

Check

Write an exponential function that passes through $(-1, 20)$ and $(1, 5)$.

$y = \underline{\quad ? \quad} \times \underline{\quad ? \quad}$

> **Study Tip**
>
> **Assumptions** The graphs of multiple functions pass through the points (1, 6) and (3, 24). For example, the linear equation $y = 9x - 3$ also passes through these points. However, because you are told to find the exponential function that passes through those points, you can assume that the exponential relationship is the correct one.

> 💭 **Think About It!**
>
> Use the graph to estimate the value of y when $x = 2$. Then use the function to find y when $x = 2$.

Example 2 Write an Exponential Function Given a Graph

Write an exponential function for the graph.

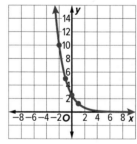

x	y
−2	10
−1	5
0	2.5
1	1.25

You can use any two points to write a system of two equations using $y = ab^x$.

$2.5 = ab^0$

$1.25 = ab^1$

Since there is a zero exponent in the first equation, solve it for a.

$a = 2.5$

Then substitute this value into the second equation to find b.

$b = 0.5$

The graph can be modeled by the function $y = 2.5(0.5)^x$.

🔵 **Go Online** You can complete an Extra Example online.

Check

Part A Use the graph to estimate the value of y when x = 4.

Part B Write an exponential function that models the graph.

Part C Use the function in Part B to find y when x = 4.

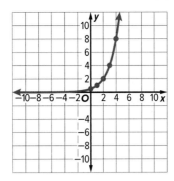

Example 3 Write an Exponential Function Given a Description

CONTEST A radio station is giving away $1000 to the first listener who answers a question correctly. If the question goes unanswered for one hour, the prize increases by 10% until it is answered correctly. Write a function to describe this situation.

Hour (h)	Prize Total (P)
0	1000
1	1100
2	1210
3	1331

Talk About It!
Why is the common ratio 1.1 and not 0.1? Explain.

This is an example of exponential growth, so $b > 1$. Divide each value of P by the preceding term to find a common ratio of 1.1.

The value of a can be found by identifying the y-intercept.

$a = 1000$

Write the function that best models the situation.

$P = 1000(1.1)^h$

Learn Solving Problems Involving Exponential Growth

Key Concept • Equation for Exponential Growth

$$y = a(1 + r)^t$$

y is the final amount.
t is time.
a is the initial amount.
r is the rate of growth expressed as a decimal, $r > 0$.

Think About It!
Why is the constant 1 in the exponential growth formula? What does it represent?

Key Concept • Equation for Compound Interest

$$A = P\left(1 + \frac{r}{n}\right)^{nt}$$

A is the current amount.
r is the annual interest rate expressed as a decimal, $r > 0$.
n is the number of times the interest is compounded each year.
t is time in years.
P is the principal, or the initial amount.

Think About It!
Why do you not substitute 2016 for t to determine the GDP per capita in 2016?

Study Tip
Assumptions The actual rates of change for the GDP are calculated annually and have varied from −11% to 18%, depending on the current economy. For the purposes of this lesson, we will assume a constant rate of change is present.

Example 4 Exponential Growth

GOODS The gross domestic product (GDP) is the monetary value of all the goods and services produced within a country in a specific time period. The GDP per capita is this value divided by the population. One model says that the GDP per capita in the United States was $13,513 in 1946, and it has increased by 2% every year.

Part A Write an equation to represent the GDP per capita after t years, where a represents the initial GDP in 1946, and r represents the rate of growth each year.

If the initial year was 1946, the initial GDP was $13,513. So $a = 13{,}513$. The GDP grew 2% each year, so $r = 2\%$ or 0.02.

$y = a(1 + r)^t$ Equation for exponential growth

$y = 13{,}513(1 + 0.02)^t$ $a = 13{,}513$ and $r = 2\%$ or 0.02

$y = 13{,}513(1.02)^t$ Simplify.

In this equation, y is the GDP per capita, and t is the time in years since 1946.

Part B If this trend continued, calculate the GDP per capita in 2016.

$t = 70$

Using $y = 13{,}513(1.02)^t$, the GDP per capita in 2016 was 54,046.

Check

POPULATION From 2013 to 2014, the city of Austin, Texas, saw one of the highest population growth rates in the country at 2.9%. The population of Austin in 2014 was estimated to be about 912,000.

Part A If the trend were to continue, which equation represents the estimated population t years after 2014?

A. $y = 912{,}000(0.029)^t$

B. $y = 912{,}000(3.9)^t$

C. $y = 1.029(912{,}000)^t$

D. $y = 912{,}000(1.029)^t$

Part B To the nearest person, predict the population of Austin in 5 years.

_____?_____ people

Go Online You can complete an Extra Example online.

Apply Example 5 Compound Interest

COLLEGE PLANNING Maria invests $5500 into a college savings account that pays 3.25% compounded quarterly. How much money will there be in the account after 5 years?

1 What is the task?

Describe the task in your own words. Then list any questions that you may have. How can you find answers to your questions?

I need to determine how much money will be in Maria's account after 5 years. How can I write an equation to represent this situation? I can apply what I have learned about different types of functions.

2 How will you approach the task? What have you learned that you can use to help you complete the task?

I will substitute the information I know into the compound interest equation and simplify. Then I will find the amount of money in Maria's account after 5 years. I will use the properties of exponents to simplify my equation.

3 What is your solution?

Use your strategy to solve the problem.

Write an equation to represent the amount of money in Maria's account after t years.

$A = 5500(1.008125)^{4t}$

How much money will be in Maria's account after 5 years?

$6466.22

4 How can you know that your solution is reasonable?

✏️ **Write About It!** **Write an argument that can be used to defend your solution.**

My solution is reasonable because if I find the amount of money in the account after each quarter, I get the same answer as I do when I use the compound interest equation.

Check

BANKING Twin brothers Amare and Jermaine each received $1000 for graduation. Amare invests his money in an account that pays 2.25% compounded daily. Jermaine invests his money in an account that pays 2.25% compounded annually.

Part A Which brother will have more money at the end of 10 years?

    A. Amare

    B. Jermaine

    C. The accounts will be equal.

Part B To the nearest cent, how much more money? $ ___?___

🌐 **Go Online** You can complete an Extra Example online.

🧠 Think About It!
What major difference do you notice between the equation for exponential growth and the equation for exponential decay?

Learn Solving Problems Involving Exponential Decay

Key Concept • Equation for Exponential Decay

$$y = a(1 - r)^t$$

y is the final amount.
t is time.
a is the initial amount.
r is the rate of decay expressed as a decimal, $0 < r < 1$.

🌐 Example 6 Exponential Decay

BANKS In banking, a dormant account is one that has not been used in over a year. A bank charges a monthly fee on dormant accounts of 0.8% of the account balance. One dormant account initially had a balance of $1609.

Part A Write an equation to represent the balance in the account after t months.

$y = a(1-r)^t$	Equation for exponential decay
$y = 1609(1 - 0.008)^t$	$a = 1609$ and $r = 0.8\%$ or 0.008
$y = 1609(0.992)^t$	Simplify.

The equation is $y = 1609(0.992)^t$, where y is the balance in the account after t months.

Part B Estimate the balance in the account after a year.

$t = 12$

$y = 1609(0.992)^t = 1461$

🧠 Think About It!
Why is the rate subtracted from 1 in the equation for exponential decay, but added to 1 in the equation for exponential growth?

Check

CITY PLANNING A city has been experiencing a slight population loss over the last few years. In 2014, the population was 1.8503 million, representing a 0.18% decrease from the previous year.

Part A If the trend were to continue, which equation represents the estimated population in millions after t years?

A. $y = 1.8503(0.18)^t$
B. $y = 1.8503(0.9982)^t$
C. $y = 1.8503(1.0018)^t$
D. $y = 1.8503(1.9982)^t$

Part B To the nearest ten thousandth, predict the population in 15 years. ___?___ million

🔎 **Go Online** You can complete an Extra Example online.

🚀 Go Online
to practice what you've learned about exponential growth and decay in the Put It All Together over Lessons 9-1 through 9-3.

Practice

Example 1
Write an exponential function for a graph that passes through the points.

1. (2, 16) and (3, 32)
2. (1, 1) and (3, 0.25)
3. (2, 90) and (4, 810)

4. (−2, 4) and (1, 0.5)
5. (1, 12) and (3, 192)
6. (1, 18) and (3, 72)

Example 2
Write an exponential function for the graph.

7.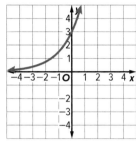

8.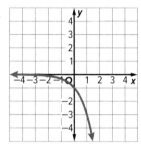

9.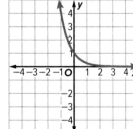

10.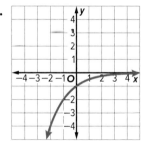

Example 3

11. **BIOLOGY** A certain species of bacteria in a laboratory culture begins with 50 cells and doubles in number every 30 minutes. Write a function to model the situation.

12. **DEPRECIATION** Amrita bought a new delivery van for $32,500. The value of this van depreciates at a rate of 12% each year. Write a function to model the value of the van after x years of ownership.

13. **COMMUNICATION** Cell phone usage grew about 23% each year from 2010 to 2016. If cell phone usage in 2010 was 43 million, write a function to model U.S. cell phone usage over that time period.

Examples 4–6

14. INVESTING Robyn invests $1500 at 4.85% compounded quarterly. Write an equation to represent the amount of money she will have in *t* years.

15. POPULATION The population of New York City increased from 8,192,426 in 2010 to 8,550,405 in 2015. The annual rate of population increase for the period was about 0.9%.

   a. Write an equation for the population, *P*, *t* years after 2010.

   b. Use the equation to predict the population of New York City in 2025.

16. SAVINGS A company has a bonus incentive for its employees. The company pays employees an initial signing bonus of $1000 and invests that amount for the employees. Suppose the investment earns 8% interest compounded quarterly.

   a. If an employee receiving this incentive withdraws the balance of the account after 5 years, how much will be in the account?

   b. If an employee receiving this incentive withdraws the balance of the account after 35 years, how much will be in the account?

17. MANUFACTURING A textile company bought a piece of weaving equipment for $60,000. It is expected to depreciate at an average rate of 10% per year.

   a. Write an equation for the value of the piece of equipment *Z* after *t* years.

   b. Find the value of the piece of equipment after 6 years.

18. HIGHER EDUCATION The table lists the average annual costs of attending a four-year college in the United States during a recent year.

College Sector	Tuition and Fees	Room and Board
Four-year Public	$9,410	$10,138
Four-year Private	$32,410	$11,516

Source: College Board

Rayelle's parents plan to invest $15,000 in a mutual fund earning an average of 4.5 percent interest, compounded monthly. After 15 years, for how many years will this investment be able to cover the tuition, fees, room, and board for Rayelle at a public college if costs stay the same? Round your answer to the nearest month.

19. **DEPRECIATION** The value of a home theater system depreciates by about 7% each year. Aeryn purchases a home theater system for $3000. What is its value 4 years after purchase? Round your answer to the nearest hundred.

20. **MONEY** Hans opens a savings account by depositing $1200. The account earns 0.2 percent interest compounded weekly. How much will be in the account in 10 years if he makes no more deposits? Assume that there are exactly 52 weeks in a year, and round your answer to the nearest cent.

21. **POPULATION** In 2016 the U.S. Census Bureau estimated the population of the United States at 322 million. If the annual rate of growth was about 0.81%, find the expected population at the time of the 2030 census. Round your answer to the nearest ten million.

Mixed Exercises

Write an exponential function for a graph that passes through the points.

22. (2, 1.4) and (4, 5.6)

23. (1, 10.4) and (4, 665.6)

24. (1, 42) and (3, 2688)

25. **POPULATION** The population of Camden, New Jersey, has been decreasing by 0.12% a year on average. If this trend continues, and the population was 79,318 in 2006, estimate Camden's population in 2025.

26. **MEDICINE** When doctors prescribe medication, they have to consider the rate at which the body filters a drug from the bloodstream. Suppose it takes the human body 6 days to filter out half of a certain vaccine. The amount of the vaccine remaining in the bloodstream x days after an injection is given by the equation $y = y_0(0.5)^{\frac{x}{6}}$, where y_0 is the initial amount. Suppose a doctor injects a patient with 20 μg (micrograms) of the vaccine.

   a. How much of the vaccine will remain after 1 day? Round your answer to the nearest tenth, if necessary.

   b. How much of the vaccine will remain after 12 days? Round your answer to the nearest tenth, if necessary.

   c. After how many days will the amount of vaccine be less than 1 μg?

27. **USE TOOLS** Graham invested money to save for a car. After x years, the value of Graham's investment can be modeled by the equation $y = 2400(0.95)^x$. How much did Graham originally invest? Is the value of his investment increasing or decreasing? Explain your reasoning. Use technology to find when the investment will be worth half of its starting value.

28. USE A MODEL There is a leak in a container that holds a certain nontoxic gas. Each hour, it loses 10% of its volume.

   a. Write an equation that models the amount of gas left in the container after *x* hours, assuming there were 300 cubic centimeters in the container before the leak. Then use your equation to determine the amount of gas left in the container after 11 hours. Round your answer to the nearest tenth.

   b. Dewanda believes a graph of this function should be a scatter plot instead of a continuous curve. Do you agree? Explain how this relates to the domain of the function.

29. STRUCTURE A wildlife researcher is studying the population of deer in a forest.

Years of Study	0	1	2	3
Population	128	160	200	250

   a. The table shows the estimated number of deer in the forest over a 3-year period. Write an exponential function that fits this data and can be used to predict the deer population in future years.

   b. The average rate of change is the change in the value of the dependent variable divided by the change in the value of the independent variable. What was the average rate of change in population during those three years?

   c. If the population growth follows the model from **part a**, do you expect the deer population to continue to increase by the value you came up with in **part b**? Explain.

   d. Use the values in the table to show how you know the function is exponential, not linear.

Higher-Order Thinking Skills

30. ANALYZE Determine the growth rate (as a percent) of a population that quadruples every year. Justify your argument.

31. PERSEVERE Santos invested $1200 into an account with an interest rate of 8% compounded monthly. Use a calculator to approximate how long it will take for Santos's investment to reach $2500.

32. ANALYZE The amount of water in a container doubles every minute. After 8 minutes, the container is full. After how many minutes was the container half-full? Justify your argument.

33. WRITE What should you consider when using exponential models to make decisions?

34. WRITE Compare and contrast the exponential growth formula and the exponential decay formula.

35. CREATE Honovi purchased a new car for $25,000 and has $5000 left to invest.

   a. Choose an interest rate between 4% and 7% for Honovi's investment, and find the length of time it would take for the investment to double.

   b. Choose an annual depreciation rate from 8% to 10% for the new car that Honovi purchased, and find the length of time it would take for the car's value to be equal to one-half of the purchase price.

   c. Using the rates from **part a** and **part b**, find the length of time it would take for the investment to be equal to the value of the car. What is the value at that time?

Lesson 9-4

Transforming Exponential Expressions

Example 1 Write Equivalent Exponential Expressions

BANKING Savewell Bank offers a savings account with 0.15% interest compounded monthly, and Second Local Bank offers a savings account with 2% interest compounded annually.

To compare the two accounts, we need to compare rates with the same compounding frequency. One way to do this is to compare the approximate monthly interest rates offered by each bank, which is also called the *effective* monthly interest rate.

Part A Compare monthly rates.

Write a function to represent the amount A that would be earned after t years with Second Local Bank. Then write an equivalent function that represents monthly compounding.

For convenience, let the initial amount of the investment be $1.

$y = a(1 + r)^t$ Equation for exponential growth

$A(t) = 1(1 + 0.02)^t$ $y = A(t)$, $a = 1$, $r = 2\%$ or 0.02

$A(t) = 1.02^t$ Simplify.

Then write a function that represents 12 compoundings per year, a power of $12t$, instead of 1 compounding per year, a power of $1t$.

$A(t) = 1.02^{1t}$ Original function

$A(t) = 1.02^{\left(\frac{1 \cdot 12}{12}\right)t}$ 1 year $= \frac{1 \text{ year}}{12 \text{ months}} \cdot 12$ months

$A(t) = \left(1.02^{\frac{1}{12}}\right)^{12t}$ Power of a Power

$A(t) \approx 1.00165^{12t}$ $(1.02)^{\frac{1}{12}} = \sqrt[12]{1.02}$ or about 1.00165

The effective monthly interest rate offered by Second Local Bank is about 0.00165 or about 0.165% per month. It is slightly more than the 0.15% offered by Savewell Bank. So, Second Local Bank is a better choice.

Part B Compare annual rates.

Write a function to represent the amount A earned after t months by Savewell Bank. Then write an equivalent function that represents annual compounding.

$y = a(1 + r)^t$ Equation for exponential growth

$A(t) = 1(1 + 0.0015)^t$ $y = A(t)$, $a = 1$, $r = 0.15\%$ or 0.0015

$A(t) = 1.0015^t$ Simplify.

(continued on the next page)

Today's Goal
- Use the properties of exponents to transform expressions for exponential functions.

Go Online You can watch a video to see how to compare savings accounts.

$A(t) = 1.0015^t$ represents the amount earned with a savings account at Savewell Bank after t months.

Write an equivalent function that represents 1 compounding per year. Because there are 12 months in a year, the exponent should be $\frac{1}{12}t$.

$A(t) = 1.0015^{1t}$	Original function
$A(t) = 1.0015^{(12 \cdot \frac{1}{12})t}$	1 year = 12 months $\cdot \frac{1 \text{ year}}{12 \text{ months}}$
$A(t) = (1.0015^{12})^{\frac{1}{12}t}$	Power of a Power
$A(t) = (1.0181)^{\frac{1}{12}t}$	$1.0015^{12} \approx 1.0181$

From this expression, we can determine that the effective annual interest rate of Savewell Bank is about 0.0181, or about 1.81%, which is less than the 2% interest rate offered by Second Local Bank.

Talk About It
Does the result in **Part B** make sense compared to the result of **Part A**? Explain.

Check

SAVINGS Tareq is planning to invest money into a savings account. Oak Hills Financial offers 3.1% interest compounded annually. First City Bank has savings accounts with a quarterly compounded interest rate of 0.7%.

Part A Write the function $A(t)$ that represents the amount that Tareq would earn after t years through Oak Hills Financial if interest were compounded quarterly.

$A(t) \approx$ __?__

What is the effective quarterly interest rate of Oak Hills Financial, rounded to the nearest hundredth?

Part B Write the function $A(t)$ that represents the amount that Tareq would earn after t quarters through First City Bank if interest were compounded annually.

$A(t) \approx$ __?__

What is the effective annual rate of First City Bank, rounded to the nearest hundredth?

__?__ is the better bank for Tareq's savings account.

Go Online You can complete an Extra Example online.

Practice

> Go Online You can complete your homework online.

Example 1

1. **INVESTING** Kimiyo is planning to invest money in a savings account. She is comparing the interest rates of savings accounts at two banks. Bank A offers a savings account with 2.1% interest compounded annually. Bank B offers a savings account with a quarterly compounded interest rate of 0.8%.

   a. Write a function to represent the amount A that Kimiyo would earn after t years through Bank A, assuming an initial investment of $1. Then write an equivalent function that represents quarterly compounding.

   b. Which is the better plan? Explain.

   c. What is the approximate effective annual interest rate at Bank B? How does your result relate to your answer to part **b**?

2. **COLLECTIONS** Keandra is comparing the growth rates in the value of two items in a collection. The value of a necklace increases by 3.2% per year. The value of a ring increases by 0.33% per month.

   a. Write a function to represent the value A of the necklace after t years, assuming an initial value of $1. Then write an equivalent function that represents monthly compounding.

   b. Which item is increasing in value at a faster rate? Explain.

   c. What is the approximate annual rate of growth of the ring? How does your result relate to your answer to part **b**?

3. **SAVINGS** Amir is trying to decide between two savings account plans at two different banks. He finds that Bank A offers a quarterly compounded interest rate of 0.95%, while Bank B offers 3.75% interest compounded annually. Which is the better plan? Explain.

4. **BACTERIA** The scientist found that Bacteria A has a growth rate of 0.99% per minute, while Bacteria B has a growth rate of 0.018% per second. Determine which bacterium has a faster growth rate. Explain.

5. **POPULATION** The population of Species A is decreasing at a rate of about 0.25% per quarter. The population of Species B is decreasing at a rate of about 1.34% per year. Determine which species has a population that is decreasing at a faster rate. Explain.

Mixed Exercises

6. **POPULATION** The table shows the population of two small towns that experience increases in population.

Year	Population Town A	Population Town B
2012	8,000	9,500
2013	8,480	9,975
2014	8,989	10,474
2015	9,528	10,997
2016	10,100	11,547

   a. Write a function that can be used to estimate the population $P(t)$ of Town A t years after 2012.

   b. Write a function that can be used to estimate the population $P(t)$ of Town B t years after 2012.

   c. Use your equations and properties of exponents to find the approximate effective monthly increase in the populations of Town A and Town B.

Lesson 9-4 • Transforming Exponential Expressions

7. **ACCOUNTS** Dominic is trying to decide between two checking account plans. Plan A offers a monthly compounded interest rate of 0.05%, while Plan B offers 0.5% interest compounded annually. Which is the better plan?

8. **CAR DEPRECIATION** Juana is deciding between two cars to purchase. Car A depreciates annually at a rate of 3.5%, while Car B depreciates monthly at a rate of 0.32%. Which car has a better effective rate of depreciation?

9. **INVESTMENT** As a wedding gift, Dotty and Brad received $10,000 cash from Dotty's grandparents. The couple is trying to decide where to invest the money. Account A offers 2.3% interest compounded semi-annually. Account B offers 4.2% interest compounded annually. Which account has the better rate? Explain.

10. **SAVINGS** Hernando is deciding between two certificate of deposit accounts. Account Y offers 4.5% interest compounded annually. Account Z offers 1.13% interest compounded quarterly. Which is the better deal? Explain.

11. **FINANCE** Gita is deciding between two retirement accounts. Account A offers 0.5% interest compounded monthly. Account B offers 2.5% interest compounded annually. Which is the better deal? Explain.

12. **WILDLIFE** The table shows that the population of hawks in two different nature preserves has been decreasing.

Year	Hawk Population (Nature Preserve A)	Hawk Population (Nature Preserve B)
2013	114	120
2014	111	115
2015	108	110
2016	105	106

    a. Write a function that can be used to estimate the population $P(t)$ of the hawks in Nature Preserve A t years after 2013.

    b. Write a function that can be used to estimate the population $P(t)$ of the hawks in Nature Preserve B t years after 2013.

    c. Use your equations and properties of exponents to find the approximate effective quarterly decrease in population of hawks in Nature Preserve A and Nature Preserve B.

Higher-Order Thinking Skills

13. **PERSEVERE** The rate at which an object cools is related to the temperature of the surrounding environment. At the time of an experiment, Mrs. Haubner's lab temperature was 72°F. The approximate temperature of the water at time t in minutes in Mrs. Haubner's lab is predicted by the function $T(t) = 72 + (212 - 72)2.72^{-0.4t}$, where −0.4° per minute is defined as the rate of cooling. Rewrite this function so that the coefficient of t in the exponent is 1.

14. **WRITE** Explain why it is important for a consumer to compare rates in the same unit before making a purchase.

15. **CREATE** Write a scenario that compares two accounts with interest rates compounded at different rate units. Then determine which account has the better rate.

16. **FIND THE ERROR** Marsha is opening a savings account. Eagle Savings Bank is offering her an account with a 0.13% monthly interest rate, while Admiral Savings Bank is offering an account with a 1% annual interest rate. Marsha believes the account at Admiral Savings bank is better because 1% is a greater interest rate than 0.13%. Why is Marsha incorrect? Explain your reasoning.

Lesson 9-5

Geometric Sequences

Explore Modeling Geometric Sequences

Online Activity Use a real-world situation to complete the Explore.

> **INQUIRY** How can you create a formula to predict how a ball bounces?

Today's Goals
- Identify and generate geometric sequences.
- Construct and use exponential functions for geometric sequences.

Today's Vocabulary
geometric sequence
common ratio

Learn Geometric Sequences

A **geometric sequence** is a pattern of numbers that begins with a nonzero term and each term after is found by multiplying the previous term by a nonzero constant r. This constant is called the **common ratio**. Dividing a term by the previous term results in the common ratio.

To find the common ratio, divide each term by the previous term. Then, write the ratio in simplest form.

Talk About It!
Why must neither the first term nor the common ratio of a geometric sequence be zero?

Example 1 Geometric Sequences

Determine whether the sequence −432, 144, −48, 16, … is geometric. Explain.

$$\frac{144}{-432} = -\frac{1}{3} \qquad \frac{-48}{144} = -\frac{1}{3} \qquad \frac{16}{-48} = -\frac{1}{3}$$

Since the ratio is the same for all of the terms, $-\frac{1}{3}$, the sequence is geometric.

Watch Out!

Find the Common Ratio Be sure to write the term as the numerator and the previous term as the denominator when finding the common ratio. Otherwise, you will be calculating the reciprocal of the common ratio.

Example 2 Identify Geometric Sequences

Determine whether the sequence 16, 12, 8, 4, … is geometric. Explain.

$$\frac{12}{16} = \frac{3}{4} \qquad \frac{8}{12} = \frac{2}{3} \qquad \frac{4}{8} = \frac{1}{2}$$

The ratios are not the same, so the sequence is not geometric.

Go Online You can complete an Extra Example online.

Lesson 9-5 • Geometric Sequences 523

Check

Determine whether each sequence is geometric. If so, determine its common ratio.

a.

n	1	2	3	4	...
a_n	8	20	50	125	...

This sequence __?__ a geometric sequence. It has a common ratio of __?__.

b. −0.7, 0.07, −0.007, 0.0007, ...

This sequence __?__ a geometric sequence. It has a common ratio of __?__.

> **Think About It!**
> How could you determine a term prior to any given term of a geometric sequence?

Example 3 Find Terms of Geometric Sequences

Find the next three terms in each geometric sequence.

a. 64, 16, 4, 1, ...

Step 1 Find the common ratio.

$$\frac{16}{64} = \frac{1}{4} \qquad \frac{4}{16} = \frac{1}{4} \qquad \frac{1}{4} = \frac{1}{4}$$

The common ratio is $\frac{1}{4}$.

Step 2 Multiply by the common ratio.

$$1 \times \frac{1}{4} = \frac{1}{4} \qquad \frac{1}{4} \times \frac{1}{4} = \frac{1}{16} \qquad \frac{1}{16} \times \frac{1}{4} = \frac{1}{64}$$

The next three terms are $\frac{1}{4}$, $\frac{1}{16}$, and $\frac{1}{64}$.

b.

n	1	2	3	4	...
a_n	8	12	18	27	...

Step 1 Find the common ratio.

$$\frac{12}{8} = 1.5 \qquad \frac{18}{12} = 1.5 \qquad \frac{27}{18} = 1.5$$

The common ratio is 1.5.

Step 2 Multiply by the common ratio.

$27 \times 1.5 = 40.5 \qquad 40.5 \times 1.5 = 60.75 \qquad 60.75 \times 1.5 = 91.125$

The next three terms are 40.5, 60.75, and 91.125.

Check

Find the next three terms in each geometric sequence.

a. 729, 243, 81, __?__, __?__, __?__, ...

b. −4, −44, −484, __?__, __?__, __?__, ...

Go Online You can complete an Extra Example online.

Learn Geometric Sequences as Exponential Functions

Key Concept • nth Term of a Geometric Sequence

The nth term a_n of a geometric sequence with first term a_1 and common ratio r is given by the following formula, where n is any positive integer and $a_1, r \neq 0$.

$$a_n = a_1 r^{n-1}$$

> **Think About It!**
> When finding the nth term of a geometric sequence, why is r raised to the $n-1$ power instead of to the nth power?

Example 4 Find the nth Term of a Geometric Sequence

Use an explicit formula to find the 11th term of the geometric sequence.

512, 256, 128, 64, ...

The first term of the sequence is 512. So, $a_1 = 512$.

Find the common ratio.

$\frac{256}{512} = \frac{1}{2}$ $\qquad$ $\frac{128}{256} = \frac{1}{2}$ $\qquad$ $\frac{64}{128} = \frac{1}{2}$

The common ratio is $\frac{1}{2}$.

Use the common ratio to find the 11th term of the sequence.

$a_n = a_1 r^{n-1}$ $\qquad$ Formula for the nth term

$a_n = 512\left(\frac{1}{2}\right)^{n-1}$ $\qquad$ $a_1 = 512$ and $r = \frac{1}{2}$

$a_{11} = 512\left(\frac{1}{2}\right)^{11-1}$ $\qquad$ To find the eleventh term, $n = 11$.

$= 512\left(\frac{1}{2}\right)^{10}$ $\qquad$ Simplify.

$= 512\left(\frac{1}{1024}\right)$ $\qquad$ $\left(\frac{1}{2}\right)^{10} = \left(\frac{1}{1024}\right)$

$= \frac{1}{2}$ $\qquad$ Simplify.

> **Watch Out!**
> **Exponents** Remember that the base, which is the common ratio, is raised to $n-1$ instead of n.

Check

Write the equation for the nth term of the geometric sequence.

n	1	2	3	4	...
a_n	729	243	81	27	...

$a_n = \underline{}$

Find the 8th term of the sequence.

Go Online You can complete an Extra Example online.

Think About It!
What assumption is made when calculating the population of North Dakota in 2030?

Example 5 Use a Geometric Sequence

POPULATION North Dakota's population is increasing more quickly than any other state's population. In 2011, the population was 685,242, and it has been increasing by an average of 2.5% each year. If this trend continues, determine the estimated population in 2030.

Since the population is growing exponentially, we can apply the equation for exponential growth, $y = a(1 + r)^t$ to determine the common ratio. An increase of 2.5% means that the population is being multiplied by $1 + 0.025$, or 1.025, each year. So, $r = 1.025$. Since $a_1 = 685,242$ in 2011, the population in 2030 is represented by the twentieth term, a_{20}.

$a_n = a_1 r^{n-1}$ Formula for the nth term

$a_{20} = 685,242(1.025)^{20-1}$ $a_1 = 685,242, r = 1.025, n = 20$

$a_{20} = 685,242(1.025)^{19}$ Simplify.

$a_{20} = 1,095,462$ Use a calculator.

In 2030, the estimated population of North Dakota will be 1,095,462.

Check

BOUNCES A rubber bouncy ball is dropped from a height of 5 feet. Each time the ball bounces back to 85% of the height from which it fell. Determine the height of the ball after 6 bounces and after 10 bounces. Round to the nearest hundredth.

$a_6 = \underline{\quad ? \quad}$ ft

$a_{10} = \underline{\quad ? \quad}$ ft

Go Online You can complete an Extra Example online.

Practice

Go Online You can complete your homework online.

Examples 1 and 2

Determine whether each sequence is geometric. Explain.

1. 4, 1, 2, ...

2. 10, 20, 30, 40, ...

3. 4, 20, 100, ...

4. 212, 106, 53, ...

5. −10, −8, −6, −4, ...

6. 5, −10, 20, 40, ...

7. −96, −48, −24, −12, ...

8. 7, 13, 19, 25, ...

9. 3, 9, 81, 6561, ...

10. 108, 66, 141, 99, ...

11. $\frac{3}{8}, -\frac{1}{8}, -\frac{5}{8}, -\frac{9}{8}, ...$

12. $\frac{7}{3}$, 14, 84, 504, ...

Example 3

Find the next three terms in each geometric sequence.

13. 2, −10, 50, ...

14. 36, 12, 4, ...

15. 4, 12, 36, ...

16. 400, 100, 25, ...

17. −6, −42, −294, ...

18. 1024, −128, 16, ...

19. 2, 6, 18, ...

20. 2500, 500, 100, ...

21. $\frac{4}{5}, \frac{2}{5}, \frac{1}{5}, ...$

22. −4, 24, −144, ...

23. 72, 12, 2, ...

24. −3, −12, −48, ...

Example 4

Use an explicit formula to find the 10th term of each geometric sequence.

25. 1, 9, 81, 729, ...

26. 2, 8, 32, 128, ...

27. −9, 27, −81, 243, ...

28. 6, −24, 96, −384, ...

Example 5

29. MUSEUMS The table shows the annual visitors to a museum in millions. Write an equation for the projected number of visitors after n years.

Year	Visitors (millions)
1	4
2	6
3	9
4	$13\frac{1}{2}$
n	?

30. WORLD POPULATION The CIA estimates that the world population is growing at a rate of 1.167% each year. The world population in 2015 was about 7.3 billion.

  a. Write an equation for the world population after n years.

  b. Find the estimated world population in 2025.

31. DEPRECIATION Te'Andra has a computer system that she bought for $5000. Each year, the computer system loses one-fifth of its then-current value. How much money will the computer system be worth after 6 years?

Mixed Exercises

32. POPULATION The table shows the projected population of the United States through 2060. Does this table show an *arithmetic sequence*, a *geometric sequence*, or *neither*? Explain.

Year	Population
2020	334,503,000
2030	359,402,000
2040	380,219,000
2050	398,328,000
2060	416,795,000

Source: U.S. Census Bureau

33. SAVINGS ACCOUNTS A bank offers a savings account that earns 0.5% interest each month.

  a. Write an equation for the balance of the savings account after n months.

  b. Given an initial deposit of $500, what will the account balance be after 15 months?

34. Write an equation for the nth term of the geometric sequence 3, −24, 192, ... Then find the 9th term of this sequence.

35. Write an equation for the nth term of the geometric sequence $\frac{9}{16}, \frac{3}{8}, \frac{1}{4}, ...$. Then find the 7th term of this sequence.

36. Write an equation for the nth term of the geometric sequence 1000, 200, 40, ... Then find the 5th term of this sequence.

37. Write an equation for the nth term of the geometric sequence $-8, -2, -\frac{1}{2}, \ldots$. Find the 8th term of this sequence.

38. Write an equation for the nth term of the geometric sequence $32, 48, 72, \ldots$. Find the 6th term of this sequence.

39. USE A SOURCE Research the average annual salary for a 25-year-old and the average rate of increase in salary per year. Then write an equation for the nth year of employment. Find the 20th term of this sequence, and explain what it means.

40. STRUCTURE For each of the geometric sequences below, fill in the missing terms, write the corresponding exponential equation, and use the exponential equation to determine the 10th term in the sequence.

    a. $0.5, 6,$ _____; $f(x) =$ _____; 10th term: _____

    b. __, $10,$ __, $40,$ __; $g(x) =$ _____; 10th term: _____

41. REASONING Find the previous three terms of the geometric sequence $-192, -768, -3072, \ldots$.

42. STATE YOUR ASSUMPTION Consider two different geometric sequences. Each starts with the same constant. The common ratio producing subsequent terms in the first is positive and is the reciprocal of the common ratio producing subsequent terms in the second. How would the graphs of the two sequences compare? Think about intercepts, asymptotes, and end behavior. Then graph an example of the situation.

43. REASONING You have just been offered a part-time job. The employer offers two different methods of payment. They are shown in the table.

Month	Method 1 Payment	Method 2 Payment
1	$100.00	$0.01
2	$108.00	$0.02
3	$116.00	$0.04
4	$124.00	$0.08

    a. Describe the two different methods of payment being offered.

    b. What kind of mathematical equations can you use to model each situation? How do you know? Write each equation.

    c. You are planning to work at this job for two years. Your manager promises to raise your salary the way it is described in the table, as long as you meet the minimum performance rating each month. Which payment plan would you choose? Explain your reasoning.

44. CONSTRUCT ARGUMENTS The terms of a geometric sequence are defined by the equation $a_n = 512(0.5)^x$. A second sequence contains the terms $b_3 = 7168$ and $b_7 = 28$.

    a. Determine which sequence has the greater common ratio.

    b. What is the initial term of each sequence? Explain your reasoning.

45. **REGULARITY** The sum of the interior angles of a triangle is 180°. The interior angles of a pentagon add to 540°. Is the relationship between the number of sides in a polygon and the sum of interior angles a geometric sequence? Use the sum of the measures of the interior angles of a square to justify your answer.

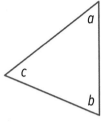
$a + b + c = 180°$

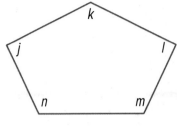
$j + k + l + m + n = 540°$

Higher-Order Thinking Skills

46. **PERSEVERE** Write a sequence that is both geometric and arithmetic. Explain your answer.

47. **FIND THE ERROR** Haro and Matthew are finding the ninth term of the geometric sequence −5, 10, −20, … . Is either of them correct? Explain your reasoning.

Haro	Matthew
$r = \frac{10}{-5}$ or -2	$r = \frac{10}{-5}$ or -2
$a_9 = -5(-2)^{9-1}$	$a_9 = -5 \cdot (-2)^{9-1}$
$= -5(512)$	$= -5 \cdot -256$
$= -2560$	$= 1280$

48. **ANALYZE** Write a sequence of numbers that form a pattern but are neither arithmetic nor geometric. Justify your argument.

49. **WRITE** How are graphs of geometric sequences and exponential functions similar? How are they different?

50. **WRITE** Summarize how to find a specific term of a geometric sequence.

51. **CREATE** Give a counterexample for the following statement:
As n increases in a geometric sequence, the value of a_n will move farther away from zero.

52. **CREATE** Write a geometric sequence. Then explain why your sequence is geometric.

Lesson 9-6

Recursive Formulas

Today's Goals
- Calculate terms in sequences by using recursive formulas.
- Write arithmetic and geometric sequences recursively and use them to model situations.

Today's Vocabulary
explicit formula
recursive formula

Learn Using Recursive Formulas

An **explicit formula** allows you to find any term of a sequence by using a formula written in terms of n. A **recursive formula** allows you to find the nth term of a sequence by performing operations to one or more of the preceding terms.

Example 1 Recursive Formula for an Arithmetic Sequence

Find the first five terms of the sequence $a_1 = 7$ and $a_n = a_{n-1} - 9$ if $n \geq 2$.

Use $a_1 = 7$ and the recursive formula to find the next four terms.

$a_2 = a_{2-1} - 9$	$n = 2$	$a_3 = a_{3-1} - 9$	$n = 3$
$= a_1 - 9$	Simplify.	$= a_2 - 9$	Simplify.
$= 7 - 9$	$a_1 = 7$	$= -2 - 9$	$a_2 = -2$
$= -2$	Simplify.	$= -11$	Simplify.
$a_4 = a_{4-1} - 9$	$n = 4$	$a_5 = a_{5-1} - 9$	$n = 5$
$= a_3 - 9$	Simplify.	$= a_4 - 9$	Simplify.
$= -11 - 9$	$a_3 = -11$	$= -20 - 9$	$a_4 = -20$
$= -20$	Simplify.	$= -29$	Simplify.

The first five terms of the sequence are 7, −2, −11, −20, and −29.

Example 2 Recursive Formula for a Geometric Sequence

Find the first five terms of the sequence $a_1 = 5$ and $a_n = 3a_{n-1}$ if $n \geq 2$.

n	$a_n = 3a_{n-1}$	a_n
1	—	5
2	$a_n = 3(5)$	15
3	$a_n = 3(15)$	45
4	$a_n = 3a_{4-1}$	135
5	$a_n = 3a_{5-1}$	405

The first five terms of the sequence are 5, 15, 45, 135, and 405.

Go Online You can complete an Extra Example online.

Study Tip
Recursive and Explicit Formulas Recursive formulas are used for generating sequences of numbers. They are not as useful for finding, for example, the fiftieth term of a sequence since you would first have to find terms one through forty-nine. For this type of calculation, it is better to use an explicit formula.

Lesson 9-6 • Recursive Formulas 531

Explore Writing Recursive Formulas from Sequences

Online Activity Use an interactive tool to complete the Explore.

> **INQUIRY** How can you write a formula that relates the numbers in a geometric sequence?

Talk About It!
When writing a recursive formula for an arithmetic or geometric sequence, how do you know which formula to use?

Learn Writing Recursive Formulas

Key Concept • Writing Recursive Formulas

Step 1 Determine whether the sequence is arithmetic or geometric by finding a common difference or a common ratio.

Step 2 Write a recursive formula.

Arithmetic Sequence

$a_n = a_{n-1} + d$, where d is the common difference

Geometric Sequence

$a_n = r \cdot a_{n-1}$, where r is the common ratio

Step 3 State the first term and domain for n.

Example 3 Write a Recursive Formula Using a List

Write a recursive formula for 16, 48, 144, 432, …

Step 1 Determine whether a common difference or ratio exists.

Subtract each term from the term that follows it to check for a common difference.

$48 - 16 = 32 \quad 144 - 48 = 96 \quad 432 - 144 = 288$

There is no common difference.

Study Tip
*n*th Term For the *n*th term of a sequence, the value of *n* must be a positive integer. Although we must still state the domain of *n* from this point forward, we will assume that *n* is an integer.

Check for a common ratio by dividing each term by the term that precedes it.

$\frac{48}{16} = 3 \qquad \frac{144}{48} = 3 \qquad \frac{432}{144} = 3$

The common ratio is 3. The sequence is geometric.

Step 2 Write a recursive formula.

$a_n = r \cdot a_{n-1}$ Recursive formula for geometric sequence

$a_n = 3 \cdot a_{n-1}$ $r = 3$

Step 3 State the first term and domain for n.

The first term a_1 is 16, and the domain of the function is $n \geq 1$.

A recursive formula for the sequence is $a_1 = 16$, $a_n = 3\, a_{n-1}$, $n \geq 2$.

Notice that n must be greater than or equal to 2 in the recursive formula.

Go Online
You may want to complete the Concept Check to check your understanding.

Check

SOCIAL MEDIA The table shows the total number of views at the end of each day for a video.

Day	Views
1	100
2	9000
3	17,900
4	26,800

Write a recursive formula for the sequence.

$a_1 = $ __?__

$a_n = a_{n-1}$ __?__

Example 4 Write a Recursive Formula Using a Graph

Write a recursive formula for the graph.

Step 1 Find a common difference or common ratio, or determine that neither exists.

$84 - 109 = -25$

$59 - 84 = -25$

$34 - 59 = -25$

The common difference is -25. The sequence is arithmetic.

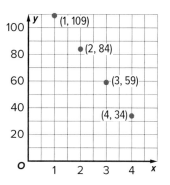

Step 2 Write a recursive formula.

$a_n = a_{n-1} + d$ Recursive formula for arithmetic sequence

$a_n = a_{n-1} + (-25)$ $d = -25$

Step 3 State the first term and domain for n.

The first term a_1 is 109, and $n \geq 1$.

A recursive formula for the sequence is $a_1 = 109$, $a_n = a_{n-1} - 25, n \geq 2$.

Think About It!
How can you make sure that your recursive formula is correct?

Example 5 Write Recursive and Explicit Formulas

MOVIES The premise of a movie is that a new virus is spreading, turning infected persons into zombie-like creatures. The table outlines the total number of infected persons at the end of each day.

Day	Infected Persons
1	3
2	12
3	48
4	192
5	768

a. Write a recursive formula for the sequence.

Step 1 Find a common difference or common ratio.

$12 - 3 = 9$ $48 - 12 = 36$ $192 - 48 = 144$

There is no common difference. Check for a common ratio by dividing each term by the term that precedes it.

$\frac{12}{3} = 4$ $\frac{48}{12} = 4$ $\frac{192}{48} = 4$ $\frac{768}{192} = 4$

Math History Minute

Hungarian mathematician **Rózsa Péter** (1905–1977) was the first Hungarian female mathematician to become an Academic Doctor of Mathematics. She helped to establish the modern field of recursive function theory, and she was the author of *Playing with Infinity: Mathematical Explorations and Excursions*.

(continued on the next page)

There is a common ratio of 4. The sequence is geometric.

Step 2 Write a recursive formula.

$a_n = r \cdot a_{n-1}$ Recursive formula for geometric sequence
$a_n = 4a_{n-1}$ $r = 4$

Step 3 State the first term and domain for n.

The first term a_1 is 3, and $n \geq 1$. A recursive formula for the sequence is $a_1 = 3$, $a_n = 4a_{n-1}$, $n \geq 2$.

b. Write an explicit formula for the sequence.

Steps 1 and 2 The common ratio is 4.

Step 3 Use the formula for the nth term of a geometric sequence.

$a_n = a_1 r^{n-1}$ Formula for the nth term
$= 3(4)^{n-1}$ $a_1 = 3$ and $r = 4$

An explicit formula for the sequence is $a_n = 3(4)^{n-1}$.

Think About It!
For the sequence in part **b**, find a_2.

Example 6 Translate Between Recursive and Explicit Formulas

A recursive formula is useful when finding a number of successive terms in a sequence. An explicit formula is useful when finding the nth term of a sequence. Therefore, it may be necessary to translate between the two forms.

a. Write a recursive formula for $a_n = 0.5n + 2$.

$a_n = 0.5n + 2$ is an explicit formula for an arithmetic sequence with $d = 0.5$ and $a_1 = 0.5(1) + 2$ or 2.5.

Therefore, a recursive formula for a_n is $a_1 = 2.5$, $a_n = a_{n-1} + 0.5$, $n \geq 2$.

b. Write an explicit formula for $a_1 = 1011$, $a_n = 1.25a_{n-1}$, $n \geq 2$.

$a_n = 1.25a_{n-1}$ is a recursive formula for a geometric sequence with $a_1 = 1011$ and $r = 1.25$.

Therefore, an explicit formula for a_n is $a_n = 1011 \cdot (1.25)^{n-1}$.

Study Tip
Geometric Sequences Recall that the formula for the nth term of a geometric sequence is $a_n = a_1 r^{n-1}$.

Check
Part A Write a recursive formula for $a_n = \frac{1}{2} + (n-1)10$.

$a_1 = \underline{\ ?\ }$
$a_n = 1.25a_{n-1} \underline{\ ?\ }$

Part B Write an explicit formula for $a_1 = -60$, $a_n = 1.5a_{n-1}$, $n \geq 2$.

$a_n = \underline{\ ?\ }(\underline{\ ?\ })^{n-1}$

Go Online You can complete an Extra Example online.

Practice

Go Online You can complete your homework online.

Examples 1 and 2

Find the first five terms of each sequence.

1. $a_1 = 23, a_n = a_{n-1} + 7, n \geq 2$

2. $a_1 = 48, a_n = -0.5a_{n-1} + 8, n \geq 2$

3. $a_1 = 8, a_n = 2.5a_{n-1}, n \geq 2$

4. $a_1 = 12, a_n = 3a_{n-1} - 21, n \geq 2$

5. $a_1 = 13, a_n = -2a_{n-1} - 3, n \geq 2$

6. $a_1 = \frac{1}{2}, a_n = a_{n-1} + \frac{3}{2}, n \geq 2$

Example 3

Write a recursive formula for each sequence.

7. 12, −1, −14, −27, ...

8. 27, 41, 55, 69, ...

9. 2, 11, 20, 29, ...

10. 100, 80, 64, 51.2, ...

11. 40, −60, 90, −135, ...

12. 81, 27, 9, 3, ...

Example 4

Write a recursive formula for each graph.

13.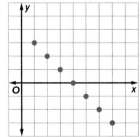

14.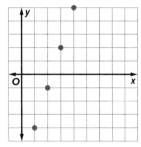

15.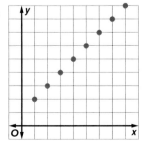

16.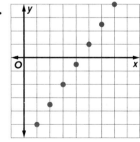

Lesson 9-6 • Recursive Formulas 535

Write a recursive formula for each graph.

17.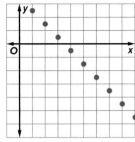

18.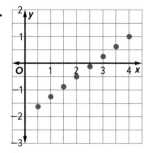

Example 5

19. **VIRAL VIDEOS** A viral video got 175 views in one hour, 350 views in two hours, 525 views in three hours, 700 views in four hours, and so on.

   a. Find the next 5 terms in the sequence.

   b. Write a recursive formula for the sequence.

   c. Write an explicit formula for the sequence.

20. **PAPER** A piece of paper is folded several times. The number of sections into which the piece of paper is divided after each fold is shown.

   a. Write a recursive formula for the sequence.

   b. Write an explicit formula for the sequence.

Number of Folds	Sections
1	2
2	4
3	8
4	16
5	32

21. **SNOW** A snowman begins to melt as the temperature rises. The height of the snowman in feet after each hour is shown.

   a. Write a recursive formula for the sequence.

   b. Write an explicit formula for the sequence.

Hour	Height (ft)
1	6.0
2	5.4
3	4.86
4	4.374

Example 6

For each recursive formula, write an explicit formula. For each explicit formula, write a recursive formula.

22. $a_n = 3(4)^{n-1}$

23. $a_1 = -2, a_n = a_{n-1} - 12, n \geq 2$

24. $a_1 = 38, a_n = \frac{1}{2}a_{n-1}, n \geq 2$

25. $a_n = -7n + 52$

26. $a_1 = 38, a_n = a_{n-1} - 17, n \geq 2$

27. $a_n = 5n - 16$

28. $a_n = 50(0.75)^{n-1}$

29. $a_1 = 16, a_n = 4a_{n-1}, n \geq 2$

536 Module 9 • Exponential Functions

Mixed Exercises

30. CLEANING An equation for the cost a_n in dollars that a carpet cleaning company charges for cleaning n rooms is $a_n = 50 + 25(n - 1)$. Write a recursive formula to represent the cost a_n.

31. SAVINGS A recursive formula for the balance of a savings account a_n in dollars at the beginning of year n is $a_1 = 500$, $a_n = 1.05a_{n-1}$, $n \geq 2$. Write an explicit formula to represent the balance of the savings account a_n.

32. USE TOOLS In 2010, County A had a population of 1.3 million people. The largest factory in the area produced 1700 million widgets per year. The population of County A is projected to grow at 1.2% per year, and the number of widgets produced is expected to grow by 10 million per year.

  a. Develop explicit formulas for the population and annual widget production, in millions, as functions of the number of years n after 2010.

  b. The graph of $y = \frac{1700+10x}{1.3(1.012)^x}$ represents the annual widget production per person for County A from 2010 to 2020, where x is the number of years after 2010. The next-highest widget-producing county produces widgets at a constant rate of 1200 widgets per person. Use a graphing calculator to extend the graph and find the year when County A will no longer be the leader in widget production. Explain your results.

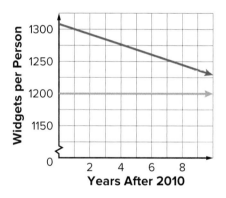

33. USE A MODEL Ramon has been tracing his family tree with his parents. He claims that he has over 250 great- great- great- great- great- great-grandparents. Is this possible? Write both an explicit and recursive formula for this situation.

34. REASONING Carl Friedrich Gauss, a German mathematician of the 1700s, was asked as a young boy for the sum of the integers from 1 to 100, and he unhesitatingly replied with the correct answer.

  a. Identify the type of the sequence 1, 2, 3, ... 100, and explore a way to find its sum based on grouping pairs of numbers from each end of the sequence. Then explain how Gauss was able to find the sum so quickly.

  b. Find an explicit formula for the sum S of n terms of an arithmetic sequence in which n is an even number, the first term is a_1, and the nth, or last term is a_n.

35. **REGULARITY** The first ten numbers in the Fibonacci sequence can be defined by $a_{n+1} = a_n + a_{n-1}$, and each ratio $\frac{a_n}{a_{n-1}}$ can be computed using a spreadsheet (see column C).

   a. Which spreadsheet formulas could have been used to calculate the entries in cells B3 and C2?

   b. Compute the ratio $\frac{a_n}{a_{n-1}}$ up to $n = 50$. What do you observe?

	A	B	C
1	1	1	
2	2	1	1
3	3	2	2
4	4	3	1.5
5	5	5	1.666667
6	6	8	1.6
7	7	13	1.625
8	8	21	1.615385
9	9	34	1.619048
10	10	55	1.617647

36. **STRUCTURE** There is a famous puzzle called the "Tower of Hanoi." There are three pegs, and a certain number of disks of varying sizes can be set on each peg. The puzzle starts with the disks in a stack on the left-most peg, with the largest disk on the bottom and the disks getting smaller as they are stacked. The goal is to move the disks from the left-most peg to the right-most peg while obeying three rules. First, only one disk can be moved at a time. Second, only the top disk on any peg can be moved. Third, at no time can a larger disk be placed on a smaller disk.

   a. If a_n is the number of moves it takes to solve a puzzle consisting of n disks, discuss why the recursive formula $a_n = a_{n-1} + 1 + a_{n-1}$ makes sense.

   b. Simplify the recursive formula. What is a_1? Why?

37. **FIND THE ERROR** Pati and Linda are working on a math problem that involves the sequence 2, −2, 2, −2, 2, … . Pati thinks that the sequence can be written as a recursive formula. Linda believes that the sequence can be written as an explicit formula. Is either of them correct? Explain your reasoning.

38. **PERSEVERE** Find a_1 for the sequence in which $a_4 = 1104$ and $a_n = 4a_{n-1} + 16$.

39. **ANALYZE** Determine whether the following statement is *true* or *false*. Justify your argument. *There is only one recursive formula for every sequence.*

40. **PERSEVERE** Find a recursive formula for 4, 9, 19, 39, 79, … .

41. **WRITE** Explain the difference between an explicit formula and a recursive formula.

42. **CREATE** Give a counterexample for the following statement: In a recursive sequence, if $a_1 = a_2$, then $a_2 = a_3$, and so on.

Module 9 • Exponential Functions
Review

Essential Question
When and how can exponential functions represent real-world situations?

Module Summary

Lesson 9-1
Exponential Functions
- Functions of the form $y = ab^x$, where $a \neq 0$ and $b > 1$, are exponential growth functions.
- Functions of the form $y = ab^x$, where $a \neq 0$ and $0 < b < 1$, are exponential decay functions.
- The graphs of exponential functions have an asymptote.

Lessons 9-2 through 9-4
Transforming and Writing Exponential Functions
- The graph $f(x) = b^x$ is a parent graph of an exponential function.
- The graph of $g(x) = b^x + k$ is the graph of $f(x) = b^x$ translated vertically.
- The graph of $g(x) = b^{x-h}$ is the graph of $f(x) = b^x$ translated horizontally.
- The graph $g(x) = ab^x$ is the graph of $f(x) = b^x$ stretched or compressed vertically by a factor of $|a|$.
- The graph $g(x) = b^{ax}$ is the graph of $f(x) = b^x$ stretched or compressed horizontally by a factor of $\frac{1}{|a|}$.
- When an exponential function $f(x)$ is multiplied by -1, the result is a reflection across the x- or y-axis.
- In the equation $y = a(1 + r)^t$, y is the final amount, a is the initial amount, r is the rate of change expressed as a decimal, and t is time.

Lesson 9-5
Geometric Sequences
- A geometric sequence is a pattern of numbers that begins with a nonzero term and each term after is found by multiplying the previous term by a nonzero constant r.
- The nth term a_n of a geometric sequence with first term a_1 and common ratio r is given by the formula $a_n = a_1 r^{n-1}$, where n is any positive integer, $a_1 \neq 0$, and $r \neq 0$.

Lesson 9-6
Recursive Functions
- An explicit formula allows you to find any term a_n of a sequence by using a formula written in terms of n.
- To write a recursive formula for an arithmetic or geometric sequence, determine whether the sequence is arithmetic or geometric by finding a common difference or a common ratio.

Study Organizer
Foldables
Use your Foldable to review this module. Working with a partner can be helpful. Ask for clarification of concepts as needed.

Test Practice

1. GRAPH The table shows the function $y = 2^x - 1$. (Lesson 9-1)

x	y
0	0
1	1
2	3
3	7

Graph the function.

2. MULTIPLE CHOICE The table shows the number of text messages Ernesto sent each month. (Lesson 9-1)

Month	Text Messages
April	2
May	6
June	18
July	54

What type of behavior is shown in the table?

A. linear

B. piece-wise

C. exponential

D. none of the above

3. OPEN RESPONSE Describe the end behavior of the graph of the exponential function shown on the graph. (Lesson 9-1)

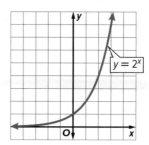

4. MULTIPLE CHOICE Consider the graph. Which function represents the reflection of the parent function $f(x) = 3^x$ across the y-axis? (Lesson 9-2)

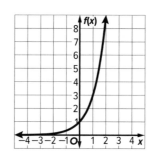

A. $f(x) = -3^x$

B. $f(x) = 3^{-x}$

C. $f(x) = 3^{-2x}$

D. $f(x) = -2(3)^x$

5. **MULTIPLE CHOICE** Describe the translation in $h(x) = 2^x + 5$ as it relates to the parent function $h(x) = 2^x$. (Lesson 9-2)

   A. Up 5 units

   B. Down 5 units

   C. Right 5 units

   D. Left 5 units

6. **OPEN RESPONSE** Horticulturists can estimate the number of hybrid plants of a certain type they will sell based on the parent function $b(x) = 2.5^x$. Suppose a new facility starts with 4 of these plants to hybridize, which can be modeled with the function $b(x) = 4(2.5)^x$. Describe the effect on the graph as it relates to the parent function. (Lesson 9-2)

7. **MULTIPLE CHOICE** Which exponential function models the graph? (Lesson 9-3)

   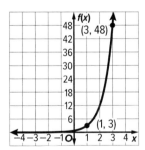

   A. $y = \frac{3}{4}(4)^x$

   B. $y = 4\left(\frac{3}{4}\right)^x$

   C. $y = \frac{3}{4}(x)^4$

   D. $y = 4(x)^x$

8. **OPEN RESPONSE** A population, $f(x)$, after x years may be modeled with $f(x) = 2(3)^x$. What is the initial amount, growth rate, domain and range? (Lesson 9-3)

Use the table below for Exercises 9–11.

Joey wants to invest money in a savings account. The table compares two banks he is considering. Joey needs to decide which is the better deal for investing his money.

	Interest Rate	Compound Frequency
First & Loan	0.6%	monthly
Local Credit Union	9%	annually

9. **MULTIPLE CHOICE** What is the effective monthly interest rate offered by Local Credit Union? (Lesson 9-4)

   A. 5.5%

   B. 2.2%

   C. 0.75%

   D. 0.72%

10. **MULTIPLE CHOICE** What is the effective annual interest rate offered by First & Loan? (Lesson 9-4)

    A. 7.4%

    B. 7.2%

    C. 1.006%

    D. 0.6%

11. **OPEN RESPONSE** Which bank gives Joey the better savings plan? Justify your answer. (Lesson 9-4)

12. MULTIPLE CHOICE Whitney invests $3000 in an account earning 4.5% interest that is compounded annually. How much money will be in Whitney's account after 10 years? (Lesson 9-3)

A. $1893.02

B. $4658.91

C. $4700.98

D. $123,254.07

13. OPEN RESPONSE Attendance for local baseball games has been increasing by an average of 10% per year for the last few years. In 2018, the average attendance was 100 people.

Predict the average number of people attending local baseball games in 2022 if this trend continues. Round to the nearest whole number. (Lesson 9-5)

14. MULTIPLE CHOICE What equation can be written for the nth term of this geometric sequence? (Lesson 9-5)

n	1	2	3	4
a_n	100	−50	25	−12.5

A. $a_n = 100(2)^{n-1}$

B. $a_n = 100(-2)^{n-1}$

C. $a_n = 100\left(-\frac{1}{2}\right)^{n-1}$

D. $a_n = 100\left(\frac{1}{2}\right)^{n-1}$

15. OPEN RESPONSE The table shows the number of pages Aaron read in his book each day. Write a recursive formula for the sequence. (Lesson 9-6)

Day	1	2	3	4
Pages Read	20	35	50	65

16. OPEN RESPONSE What are the first five terms of the sequence for $a_1 = -2$ and $a_n = 2a_{n-1} + 5$ if $n \geq 2$. (Lesson 9-6)

17. OPEN RESPONSE Copy and complete the table for the geometric sequence. (Lesson 9-6)
$a_1 = 3$ and $a_n = 4a_{n-1}$, if $n \geq 2$

n	formula	a_n
1	—	3
2	$a_n = 4(3)$	?
3	$a_n = 4(\_?\_)$	?
4	$a_n = 4(\_?\_)$	?

Module 10
Polynomials

e Essential Question
How can you perform operations on polynomials and use them to represent real-world situations?

What Will You Learn?
How much do you already know about each topic **before** starting this module?

KEY

👎 — I don't know.   👍 — I've heard of it.   👍 — I know it!

	Before			After		
write polynomials in standard form						
add polynomials						
subtract polynomials						
multiply polynomials by a monomial						
solve equations with polynomial expressions						
multiply binomials						
multiply polynomials						
factor polynomials using the Distributive Property						
factor quadratic trinomials by grouping						
factor polynomials that are the result of special products						

📓 **Foldables** Make this Foldable to help you organize your notes about polynomials. Begin with four sheets of grid paper.

1. **Fold** in half along the width. On the first two sheets, cut 5 centimeters along the fold at the ends. On the second two sheets cut in the center, stopping 5 centimeters from the ends.

2. **Insert** the first sheets through the second sheets and align the folds. Label the front Module 10, Polynomials. Label the pages with lesson numbers and the last page vocabulary.

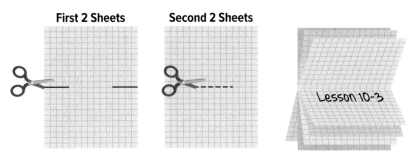

What Vocabulary Will You Learn?

- binomial
- degree of a monomial
- degree of a polynomial
- difference of two squares
- factoring
- factoring by grouping
- leading coefficient
- perfect square trinomials
- polynomial
- prime polynomial
- quadratic expression
- standard form of a polynomial
- trinomial

Are You Ready?

Complete the Quick Review to see if you are ready to start this module.
Then complete the Quick Check.

Quick Review

Example 1

Rewrite $6x(-3x - 5x - 5x^2 + x^3)$ using the Distributive Property. Then simplify.

$6x(-3x - 5x - 5x^2 + x^3)$
$= 6x(-3x) + 6x(-5x) + 6x(-5x^2) + 6x(x^3)$
$= -18x^2 + (-30x^2) - 30x^3 + 6x^4$
$= -48x^2 - 30x^3 + 6x^4$

Example 2

Simplify $8c + 6 - 4c + 2c^2$.

$8c + 6 - 4c + 2c^2$	Original expression
$= 2c^2 + 8c - 4c + 6$	Rewrite in descending order.
$= 2c^2 + (8 - 4)c + 6$	Use the Distributive Property.
$= 2c^2 + 4c + 6$	Combine like terms.

Quick Check

Rewrite each expression using the Distributive Property. Then simplify.

1. $a(a + 5)$
2. $2(3 + x)$
3. $n(n - 3n^2 + 2)$
4. $-6(x^2 - 5x + 4)$

Simplify each expression. If not possible, write *simplified*.

5. $3x + 10x$
6. $4w^2 + w + 15w^2$
7. $6m^2 - 8m$
8. $2x^2 + 5 + 11x^2 + 7$

How did you do?

Which exercises did you answer correctly in the Quick Check?

Adding and Subtracting Polynomials

Lesson 10-1

Learn Types of Polynomials

A **polynomial** is a monomial or the sum of two or more monomials. Some polynomials have special names.

- A **monomial** is a number, a variable, or a product of a number and one or more variables.
- A **binomial** is the sum of *two* monomials.
- A **trinomial** is the sum of *three* monomials.

The **degree of a monomial** is the sum of the exponents of all its variables. A nonzero constant term has degree 0, and zero has no degree.

The **degree of a polynomial** is the greatest degree of any term in the polynomial. You can find the degree of a polynomial by finding the degree of each term. Polynomials are named by their degree.

Degree	Name
0	constant
1	linear
2	quadratic
3	cubic
4	quartic
5	quintic
6 or more	6th degree, 7th degree, . . .

Addition is commutative, and therefore the terms of a polynomial can be written in any order. However, the **standard form of a polynomial** has the terms written in order from greatest degree to least degree. When a polynomial is in standard form, the coefficient of the first term is called the **leading coefficient**.

Example 1 Identify Polynomials

Determine whether each expression is a polynomial. If it is a polynomial, find the degree and determine whether it is a monomial, binomial, or trinomial.

a. $8ab - 2c$

$8ab - 2c$ is the sum of two monomials, $8ab$ and $-2c$, so this is a polynomial. Degree: 2; binomial.

Today's Goals
- Identify and write polynomials by using the standard form.
- Add polynomials.
- Subtract polynomials.

Today's Vocabulary
polynomial
binomial
trinomial
degree of a monomial
degree of a polynomial
standard form of a polynomial
leading coefficient

Think About It!
Is $4x - 2x^2 + 9$ written in standard form? Justify your argument.

Talk About It!
Explain why $8ab - 2c$ is a 2nd degree polynomial and not a 1st degree polynomial.

b. −11.25

−11.25 is a real number, so this is a polynomial. Degree: 0; monomial.

c. $2x^{-2} + 3xy$

$2x^{-2} = \frac{2}{x^2}$, which is not a monomial, so this is not a polynomial.

d. $9x^3 − 8x + 5x − 27$

The simplified form is $9x^3 − 3x − 27$, which is the sum of three monomials, so this is a polynomial. Degree: 3; trinomial.

e. $2m^2 + 2mn − n^2$

$2m^2 + 2mn − n^2$ is the sum of three monomials, so this is a polynomial. Degree: 2; trinomial.

Watch Out!
Degree of a Polynomial Remember that the degree of a monomial is the sum of the exponents of all its variables, and the degree of the polynomial in which it is a term is the greatest degree of any term in it. Be careful not to identify the degree of a polynomial by looking only at the greatest exponent.

Check

Copy and complete the table. Determine whether each expression is a polynomial. If it is a polynomial, find the degree and determine whether it is a *monomial*, *binomial*, or *trinomial*.

Expression	Is it a polynomial?	Degree	Classification
a. $3z^{-2}$			
b. $2x^3 + x − 12$			
c. $9b$			
d. $9x − 2$			

Think About It!
In standard form, why is the constant term at the end of the polynomial rather than the beginning?

Example 2 Standard Form of a Polynomial

Write $4x + 12 + 2x^3 − 3x^2$ in standard form. Identify the leading coefficient.

To write the polynomial in standard form, rewrite the terms in order from greatest degree, 3, to least degree, 0. The polynomial can be rewritten as $2x^3 − 3x^2 + 4x + 12$ with a leading coefficient of 2.

Check

Part A Write $5b − 10b^2 + 35 − b^3$ in standard form.

Part B Identify the leading coefficient in $5b − 10b^2 + 35 − b^3$.

The leading coefficient of the polynomial is ____?____.

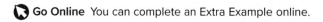 **Go Online** You can complete an Extra Example online.

Explore Using Algebra Tiles to Add and Subtract Polynomials

Online Activity Use algebra tiles to complete the Explore.

> **INQUIRY** How are the processes for adding and subtracting polynomials similar?

Learn Adding Polynomials

Adding polynomials involves adding like terms. When adding polynomials, you can group like terms by using a horizontal or vertical format.

Method 1 Horizontal Method

Group and combine like terms.

$(3x^2 + 9x + 27) + (2x^2 + 4x - 12)$
$= [3x^2 + 2x^2] + [9x + 4x] + [27 + (-12)]$ Group like terms.
$= 5x^2 + 13x + 15$ Combine like terms.

Method 2 Vertical Method

Align like terms in columns and combine.

$\quad\quad 3x^2 + 9x + 27$
$(+)\ \ 2x^2 + 4x - 12$
$\quad\quad 5x^2 + 13x + 15$ Align like terms. Combine like terms.

Think About It!
How can writing polynomials in standard form be helpful when adding?

Example 3 Add Polynomials

Find each sum.

a. $(3x^2 - 4) + (x^2 - 9)$

$(3x^2 - 4) + (x^2 - 9) = [3x^2 + x^2] + [-4 + (-9)]$ Group like terms.
$\quad\quad\quad\quad\quad\quad\quad\quad = 4x^2 - 13$ Combine like terms.

b. $(8 - x^2) + (4x + 2x^2 - 9)$

$-x^2 + 8 \quad\to\quad\quad -x^2 + 0x + 8$ Insert a placeholder to align the terms.
$2x^2 + 4x - 9 \quad\to\quad (+)\ 2x^2 + 4x - 9$ Align and combine like terms.
$\quad\quad\quad\quad\quad\quad\quad\quad x^2 + 4x - 1$

Study Tip

Placeholders When adding polynomials, it may be necessary to insert a placeholder to help align the terms. For example, if one of the polynomials does not have an x^2 term, add $0x^2$ to keep the terms aligned.

Go Online You can complete an Extra Example online.

Check

Find each sum. Write your answer in standard form.

a. $(12y + 20y^2 - 2) + (-13y^2 + y - 10)$

b. $(-4b - b^2 + 2) + 2(b^2 + 2b - 1)$

c. $(-f + 5f^2 + 5) + (3f^3 - f + f^2)$

Learn Subtracting Polynomials

You can subtract a polynomial by adding its additive inverse. To find the additive inverse of a polynomial, write the opposite of each term.

Select a method to find $(11x - 13 - 7x^3 - 8x^2) - (2x + 8x^2 + 20)$.

Method 1 Horizontal method

Subtract $2x + 8x^2 + 20$ by adding its additive inverse.

$(11x - 13 - 7x^3 - 8x^2) - (2x + 8x^2 + 20)$

$= (11x - 13 - 7x^3 - 8x^2) + (-2x - 8x^2 - 20)$ The additive inverse of $2x + 8x^2 + 20$ is $-2x - 8x^2 - 20$.

$= -7x^3 + [-8x^2 + (-8x^2)] + [11x + (-2x)] + [-13 + (-20)]$

$= -7x^3 - 16x^2 + 9x - 33$

Method 2 Vertical method

Align like terms in columns and subtract by adding the additive inverse.

$$
\begin{array}{r} -7x^3 - 8x^2 + 11x - 13 \\ (-)\ 0x^3 + 8x^2 + 2x + 20 \\ \hline \end{array}
\quad \rightarrow \quad
\begin{array}{r} -7x^3 - 8x^2 + 11x - 13 \\ (+)\ -0x^3 - 8x^2 - 2x - 20 \\ \hline -7x^3 - 16x^2 + 9x - 33 \end{array}
$$

Adding or subtracting integers results in an integer, so the set of integers is closed under addition and subtraction. Similarly, when you add or subtract polynomials, you are combining like terms. This results in a polynomial with the same variables and exponents as the original polynomials, but possibly different coefficients. Thus, the sum or difference of two polynomials is always a polynomial, and the set of polynomials is closed under addition and subtraction.

Example 4 Subtract Polynomials Horizontally

Find $(6x - 11) - (2x - 19)$.

Subtract $(2x - 19)$ by adding its additive inverse.

$(6x - 11) - (2x - 19) = (6x - 11) + (-2x + 19)$

$\qquad\qquad\qquad\qquad = [6x + (-2x)] + [-11 + 19]$ Group like terms.

$\qquad\qquad\qquad\qquad = 4x + 8$ Combine like terms.

Go Online You can complete an Extra Example online.

> **Think About It!**
> In the example, why is the term $0x^3$ introduced when subtracting the polynomials?

Example 5 Subtract Polynomials Vertically

Find $(x + 2) - (7x - 3x^2 + 14)$.

Align like terms in columns and subtract by adding the additive inverse.

$$
\begin{array}{r} 0x^2 + x + 2 \\ (-)\ -3x^2 + 7x + 14 \\ \hline \end{array} \rightarrow \begin{array}{r} 0x^2 + x + 2 \\ (+)\ 3x^2 - 7x - 14 \\ \hline 3x^2 - 6x - 12 \end{array}
$$

Think About It!
What is the first step for finding the difference of polynomials?

Check

a. Find $(z^2 + 2z - 5) - (9z - 3z^2)$. Write your answer in standard form.

b. Find $(8r - 14 + 7r^2) - (-16r^2 - 7r - 3)$. Write your answer in standard form.

c. Find $(h - 2h - h^2) - (5h^2 - 2 + 8h)$. Write your answer in standard form.

Study Tip
Units Pay attention to the language in the question. The equations represent the number of hard copy and digital albums sold in the thousands, so your answer should represent album sales in the thousands.

Example 6 Add and Subtract Polynomials

ALBUM SALES Today's recording artists can sell hard copies H and digital copies D of their albums. The equations $H = 9w + 53$ and $D = 13w + 126$ represent the number of albums (in thousands) one artist sold in w weeks. Write an equation that shows how many more digital albums were sold than hard copies S. Then predict how many more digital albums are sold than hard copies in 52 weeks.

To write an equation that represents how many more digital albums were sold than hard copies S, subtract the equation for the number of hard copies H sold from the equation for the number of digital albums D sold.

$$S = (13w + 126) - (9w + 53)$$

$$= 4w + 73$$

Substitute 52 for w to predict how many more digital albums are sold than hard copies in 52 weeks.

There will be 281,000 more digital albums sold than hard copies in 52 weeks.

Think About It!
What assumption did you make about the trend of sales over the 52 weeks? Can you determine the sales of hard copy and digital albums for a specific week? Explain.

Go Online You can complete an Extra Example online.

Lesson 10-1 • Adding and Subtracting Polynomials **549**

Check

COLLEGE LIVING The total number of students T who attend a college consists of two groups: students who live in dorm rooms on campus D and students who live in apartments off campus A. The number (in hundreds) of students who live in dorm rooms and the total number of students enrolled in the college can be modeled by the following equations, where n is the number of years since 2001.

$$T = 17n + 23$$
$$D = 11n + 8$$

Part A Write an equation that models the number of students who live in apartments.

Part B Predict the number of students who will live in apartments in 2020.

Part C What do you need to assume in order to predict the number of students who will live on campus in 2020?

A. The total number of students does not include students who commute.
B. Students do not share dorm rooms.
C. The number of students enrolled in the college remains the same.
D. Many students live at home during the summer.

Practice

Go Online You can complete your homework online.

Example 1

Determine whether each expression is a polynomial. If it is a polynomial, find the degree and determine whether it is a *monomial*, *binomial*, or *trinomial*.

1. $\frac{5y^3}{x^2} + 4x$

2. 21

3. $c^4 - 2c^2 + 1$

4. $d + 3d^c$

5. $a - a^2$

6. $5n^3 + nq^3$

Example 2

Write each polynomial in standard form. Identify the leading coefficient.

7. $5x^2 - 2 + 3x$

8. $8y + 7y^3$

9. $4 - 3c - 5c^2$

10. $-y^3 + 3y - 3y^2 + 2$

11. $11t + 2t^2 - 3 + t^5$

12. $2 + r - r^3$

13. $\frac{1}{2}x - 3x^4 + 7$

14. $-9b^2 + 10b - b^6$

Examples 3–5

Find each sum or difference.

15. $(2x + 3y) + (4x + 9y)$

16. $(6s + 5t) + (4t + 8s)$

17. $(5a + 9b) - (2a + 4b)$

18. $(11m - 7n) - (2m + 6n)$

19. $(m^2 - m) + (2m + m^2)$

20. $(x^2 - 3x) - (2x^2 + 5x)$

21. $(d^2 - d + 5) - (2d + 5)$

22. $(2h^2 - 5h) + (7h - 3h^2)$

23. $(5f + g - 2) + (-2f + 3)$

24. $(6k^2 + 2k + 9) + (4k^2 - 5k)$

25. $(2c^2 + 6c + 4) + (5c^2 - 7)$

26. $(2x + 3x^2) - (7 - 8x^2)$

Lesson 10-1 • Adding and Subtracting Polynomials **551**

Find each sum or difference.

27. $(3c^3 - c + 11) - (c^2 + 2c + 8)$

28. $(z^2 + z) + (z^2 - 11)$

29. $(2x - 2y + 1) - (3y + 4x)$

30. $(4a - 5b^2 + 3) + (6 - 2a + 3b^2)$

31. $(x^2y - 3x^2 + y) + (3y - 2x^2y)$

32. $(-8xy + 3x^2 - 5y) + (4x^2 - 2y + 6xy)$

33. $(5n - 2p^2 + 2np) - (4p^2 + 4n)$

34. $(4rxt - 8r^2x + x^2) - (6rx^2 + 5rxt - 2x^2)$

Example 6

35. **PROFIT** Company A and Company B both started their businesses in the same year. The profit P, in millions, of Company A is given by the equation $P = 3.2x + 12$, where x is the number of years in business. The profit P, in millions, of Company B is given by the equation $P = 2.7x + 10$, where x is the number of years in business.

    a. Write a polynomial equation to give the difference in profit D after x years.

    b. Predict the difference in profit after 10 years.

36. **ENVELOPES** An office supply company produces yellow document envelopes. The envelopes come in a variety of sizes, but the length is always 4 centimeters more than double the width, x.

    a. Write a polynomial equation to give the perimeter P of any of the envelopes.

    b. Predict the perimeter of an envelope with a width of 6 centimeters.

Mixed Exercises

Classify each polynomial according to its degree and number of terms.

37. $4x - 3x^2 + 5$

38. $11z^3$

39. $9 + y^4$

40. $3x^3 - 7x$

41. $-2x^5 - x^2 + 5x - 8$

42. $10t - 4t^2 + 6t^3$

Find each sum or difference.

43. $(4x + 2y - 6z) + (5y - 2z + 7x) + (-9z - 2x - 3y)$

44. $(5a^2 - 4) + (a^2 - 2a + 12) + (4a^2 - 6a + 8)$

45. $(3c^2 - 7) + (4c + 7) - (c^2 + 5c - 8)$

46. ROCKETS Two toy rockets are launched straight up into the air. The height, in feet, of each rocket at t seconds after launch is given by the polynomial equations shown. Write an equation to find the difference in height of Rocket A and Rocket B. Predict the difference in height after 5 seconds.

Rocket A: $D_1 = -16t^2 + 122t$

Rocket B: $D_2 = -16t^2 + 84t$

47. INDUSTRY Two identical right cylindrical steel drums containing oil need to be covered with a fire-resistant sealant. In order to determine how much sealant to purchase, George must find the surface area of the two drums. The surface area, including the top and bottom bases, is given by the formula $S = 2\pi rh + 2\pi r^2$.

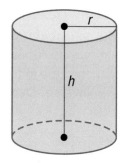

a. Write a polynomial to represent the total surface area of the two drums.

b. Find the total surface area, to the nearest tenth of a square meter, if the height of each drum is 2 meters and the radius of each is 0.5 meter.

48. MANUFACTURING A company delivers their product in cubic boxes that have volume x^3. When the company begins to manufacture a second product, manufacturing designs a new shipping box that is 3 inches longer in one dimension and 1 inch shorter in another dimension. The volume of the new box is $x^3 + 2x^2 - 3x$.

a. Write an expression to represent the total volume of 4 of each kind of box.

b. Find an expression that shows the difference in volume between the two boxes.

49. FIND THE ERROR Claudio says that polynomials are not closed under addition and gives this counterexample, which he says is not a polynomial: $(x^2 - 2x) + (-x^2 + 2x) = 0$. Is Claudio correct? Explain your reasoning.

50. VOLUME The volume of a sphere with radius x is $\frac{4}{3}\pi x^3$ units^3, and the volume of a cube with side length x is x^3 units^3.

a. Find the total combined volume of the sphere and cube.

b. How much greater is the volume of the sphere?

51. STRUCTURE Compute the following differences.

a. $(5x^2 - 3x + 7) - (18x^2 - 2x - 3)$

b. $(18x^2 - 2x - 3) - (5x^2 - 3x + 7)$

c. What do you notice about **part a** compared to **part b**? How does this relate to the structure of the integers?

52. REGULARITY In the set of integers, every integer has an additive inverse. In other words, for every integer n, there is another integer $-n$ such that $n + (-n) = 0$. Is this true in the set of polynomials? Does every polynomial have an additive inverse? Demonstrate with an example.

53. FIND THE ERROR Cheyenne and Nicolas are finding $(2x^2 - x) - (3x + 3x^2 - 2)$. Is either correct? Explain your reasoning.

Cheyenne	Nicolas
$(2x^2 - x) - (3x + 3x^2 - 2)$	$(2x^2 - x) - (3x + 3x^2 - 2)$
$= (2x^2 - x) + (-3x + 3x^2 - 2)$	$= (2x^2 - x) + (-3x - 3x^2 - 2)$
$= 5x^2 - 4x - 2$	$= -x^2 - 4x - 2$

54. ANALYZE Determine whether each of the following statements is *true* or *false*. Justify your argument.

a. A binomial can have a degree of zero.

b. The order in which polynomials are subtracted does not matter.

55. PERSEVERE Write a polynomial that represents the sum of $2n + 1$ and the next two consecutive odd integers.

56. WRITE Why would you add or subtract equations that represent real-world situations? Explain.

57. WRITE Describe how to add and subtract polynomials using both the vertical and horizontal methods.

58. CREATE Write two polynomials that can be added to have a sum of $-11x^3 - x^2 + 5x + 6$.

Lesson 10-2
Multiplying Polynomials by Monomials

Explore Using Algebra Tiles to Find Products of Polynomials and Monomials

Online Activity Use algebra tiles to complete the Explore.

> **INQUIRY** How can you use the Distributive Property to find the product of a polynomial and a monomial?

Today's Goal
- Multiply polynomials by monomials.

Learn Multiplying a Polynomial by a Monomial

To find the product of a monomial and a binomial, you can use the Distributive Property. You can also use the Distributive Property to find the product of a monomial and a longer polynomial. When polynomials are multiplied, the product is also a polynomial. Therefore, the set of polynomials is closed under multiplication. This is similar to the system of integers, which is also closed under multiplication.

Example 1 Multiply a Polynomial by a Monomial

Simplify $-2x(4x^2 + 3x - 5)$.

$-2x(4x^2 + 3x - 5)$	Original expression
$= -2x(4x^2) + (-2x)(3x) - (-2x)(5)$	Distributive Property
$= -8x^3 + (-6x^2) - (-10x)$	Multiply.
$= -8x^3 - 6x^2 + 10x$	Simplify.

Go Online An alternate method is available for this example.

Example 2 Simplify Expressions

Simplify $3n(6n^3 - 4n) - 2(9n^4 - 11)$.

$3n(6n^3 - 4n) - 2(9n^4 - 11)$	
$= 3n(6n^3) - 3n(4n) + (-2)(9n^4) + (-2)(-11)$	Distributive Property
$= 18n^4 - 12n^2 - 18n^4 + 22$	Multiply.
$= (18n^4 - 18n^4) - 12n^2 + 22$	Commutative and Associative Properties
$= -12n^2 + 22$	Combine like terms.

Watch Out!

Negatives If the monomial has a negative coefficient, remember to distribute the negative sign to each term in the polynomial.

Go Online You can complete an Extra Example online.

Example 3 Write and Evaluate a Polynomial Expression

Use a Source
Find the height of the basket building and use it to determine the area of one side.

ARCHITECTURE The world's largest basket is a building. Each face of the building is in the shape of a trapezoid, with the largest face having a height of h and two base lengths, $h + 90$ and $2h + 84$. Write and simplify an expression to represent the area of one side of the building.

Let h = the height of a trapezoid, $a = h + 90$ and $b = 2h + 84$.

$$A = \tfrac{1}{2}h(a + b) \qquad \text{Area of a trapezoid}$$
$$= \tfrac{1}{2}h[(h + 90) + (2h + 84)] \qquad a = h + 90 \text{ and } b = 2h + 84$$
$$= \tfrac{1}{2}h(3h + 174) \qquad \text{Add and simplify.}$$
$$= \tfrac{3}{2}h^2 + 87h \qquad \text{Distributive Property}$$

The area of one side of the building is $\tfrac{3}{2}h^2 + 87h$.

Example 4 Solve Equations with Polynomial Expressions

Solve each equation.

a. $-16p = -2(p + 3) + 3(6p - 30)$

$-16p = -2(p + 3) + 3(6p - 30)$	Original equation
$-16p = -2p - 6 + 18p - 90$	Distributive Property
$-16p = 16p - 96$	Combine like terms.
$-32p = -96$	Subtract $16p$ from each side.
$p = 3$	Divide each side by -32.

b. $-2q(4q - 9) = q(-8q + 15) - 3(-4q - 6)$

$-2q(4q - 9) = q(-8q + 15) - 3(-4q - 6)$	Original equation
$-8q^2 + 18q = -8q^2 + 15q + 12q + 18$	Distributive Property
$-8q^2 + 18q = -8q^2 + 27q + 18$	Combine like terms.
$18q = 27q + 18$	Add $8q^2$ to each side.
$-9q = 18$	Subt. $27q$ from each side.
$q = -2$	Divide each side by -9.

Study Tip
Check To ensure that your answer is correct, substitute it into the original equation and verify that the simplified expressions are equal.

Check

Solve each equation.

a. $2p = 3(4p - 10)$
$p = \underline{\quad ? \quad}$

b. $-3q(2q + 5) = 2(-3q^2 + 15) - 5(10q + 6)$
$q = \underline{\quad ? \quad}$

Go Online You can complete an Extra Example online.

Practice

Go Online You can complete your homework online.

Example 1

Simplify each expression.

1. $b(b^2 - 12b + 1)$

2. $f(f^2 + 2f + 25)$

3. $-3m^3(2m^3 - 12m^2 + 2m + 25)$

4. $2j^2(5j^3 - 15j^2 + 2j + 2)$

5. $2pr^2(2pr + 5p^2r - 15p)$

6. $4t^3u(2t^2u^2 - 10tu^4 + 2)$

Example 2

Simplify each expression.

7. $-3(5x^2 + 2x + 9) + x(2x - 3)$

8. $a(-8a^2 + 2a + 4) + 3(6a^2 - 4)$

9. $-4d(5d^2 - 12) + 7(d + 5)$

10. $-9g(-2g + g^2) + 3(g^3 + 4)$

11. $2j(7j^2k^2 + jk^2 + 5k) - 9k(-2j^2k^2 + 2k^2 + 3j)$

12. $4n(2n^3p^2 - 3np^2 + 5n) + 4p(6n^2p - 2np^2 + 3p)$

Example 3

13. **NUMBER THEORY** The sum of the first n whole numbers is given by the expression $\frac{1}{2}(n^2 + n)$. Expand the equation by multiplying, then find the sum of the first 12 whole numbers.

14. **COLLEGE** Troy's grandfather gave him $700 to start his college savings account. Troy's grandfather also gives him $40 each month to add to the account. Troy's mother gives him $50 each month, but has been doing so for 4 fewer months than Troy's grandfather. Write a simplified expression for the amount of money Troy has received m months after his mother started giving him money.

15. **MARKET** Sophia went to the farmers' market to purchase some vegetables. She bought peppers and potatoes. The peppers were $0.39 each and the potatoes were $0.29 each. She spent $3.88 on vegetables and bought 4 more potatoes than peppers. If $x =$ the number of peppers, write and solve an equation to find out how many of each vegetable Sophia bought.

16. **GEOMETRY** The volume of a pyramid can be found by multiplying the area of its base B by one-third of its height. The area of the rectangular base of a pyramid is given by the polynomial equation $B = x^2 - 4x - 12$.

    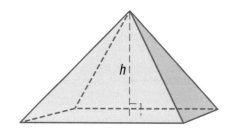

    a. Write a polynomial equation to represent the volume of the pyramid V if its height is 10 meters.

    b. Find the volume of the pyramid if $x = 12$ m.

Lesson 10-2 • Multiplying Polynomials by Monomials 557

Example 4
Solve each equation.

17. $7(t^2 + 5t - 9) + t = t(7t - 2) + 13$

18. $w(4w + 6) + 2w = 2(2w^2 + 7w - 3)$

19. $5(4z + 6) - 2(z - 4) = 7z(z + 4) - z(7z - 2) - 48$

20. $9c(c - 11) + 10(5c - 3) = 3c(c + 5) + c(6c - 3) - 30$

21. $2f(5f - 2) - 10(f^2 - 3f + 6) = -8f(f + 4) + 4(2f^2 - 7f)$

22. $2k(-3k + 4) + 6(k^2 + 10) = k(4k + 8) - 2k(2k + 5)$

Mixed Exercises
Simplify each expression.

23. $a(4a + 3)$

24. $-c(11c + 4)$

25. $x(2x - 5)$

26. $2y(y - 4)$

27. $-3n(n^2 + 2n)$

28. $4h(3h - 5)$

29. $3x(5x^2 - x + 4)$

30. $7c(5 - 2c^2 + c^3)$

31. $-4b(1 - 9b - 2b^2)$

32. $6y(-5 - y + 4y^2)$

33. $2m^2(2m^2 + 3m - 5)$

34. $-3n^2(-2n^2 + 3n + 4)$

Simplify each expression.

35. $w(3w + 2) + 5w$

36. $f(5f - 3) - 2f$

37. $-p(2p - 8) - 5p$

38. $y^2(-4y + 5) - 6y^2$

39. $2x(3x^2 + 4) - 3x^3$

40. $4a(5a^2 - 4) + 9a$

41. $4b(-5b - 3) - 2(b^2 - 7b - 4)$

42. $3m(3m + 6) - 3(m^2 + 4m + 1)$

43. $-5q^2w^3(4q + 7w) + 4qw^2(7q^2w + 2q) - 3qw(3q^2w^2 + 9)$

44. $-x^2z(2z^2 + 4xz^3) + xz^2(xz + 5x^3z) + x^2z^3(3x^2z + 4xz)$

Solve each equation.

45. $3(a + 2) + 5 = 2a + 4$

46. $2(4x + 2) - 8 = 4(x + 3)$

47. $5(y + 1) + 2 = 4(y + 2) - 6$

48. $4(b + 6) = 2(b + 5) + 2$

49. $6(m - 2) + 14 = 3(m + 2) - 10$

50. $3(c + 5) - 2 = 2(c + 6) + 2$

51. LANDSCAPING The courtyard on a college campus has a sculpture surrounded by a circle of 50 flags. The university plans to install a new sidewalk 12 feet wide around the perimeter of the outside of the circle of flags. If the outside circumference of the sidewalk is 1.10 times the circumference of the circle of flags, write an equation for the outside circumference of the sidewalk. Solve the equation for the radius of the circle of flags. Recall that the circumference of a circle is $2\pi r$.

52. STRUCTURE The base lengths of the trapezoid shown are given by polynomial expressions in terms of the trapezoid's height h.

   a. Write and simplify an expression for the area of the trapezoid.

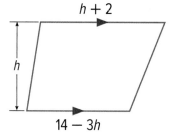

   b. If the height of the trapezoid is 4 units, what is the area of the trapezoid?

53. FIND THE ERROR Andres simplified the expression $12y^2(3y - 2y^2) - 3y^3(4 - 2y)$. Is he correct? Explain your reasoning.

$12y^2(3y - 2y^2) - 3y^3(4 - 2y)$
$36y^3 - 24y^2 - 12y^3 - 6y^4$
$24y^3 - 24y^2 - 6y^4$

54. STRUCTURE The diagram shows the dimensions of a right rectangular prism.

   a. Write and simplify an expression for the volume of the prism.

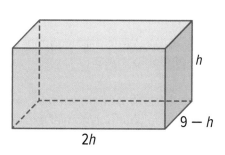

   b. If the height of the rectangular prism is 6 units, what is the volume of the rectangular prism?

Lesson 10-2 • Multiplying Polynomials by Monomials **559**

55. **USE TOOLS** Through market research, a company finds that it can expect to sell $45 - 5x$ products if each is priced at $1.25x$ dollars.

   a. Write and simplify an expression for the expected revenue.

   b. Determine the price of each product, to the nearest cent, when $x = 4.5$.

   c. Determine the revenue when the expected number of products are sold, to the nearest cent, when $x = 4.5$.

56. **STRUCTURE** An area is enclosed by the fence shown at the right, represented by the solid line segments. Write and simplify an expression for the area of the enclosure.

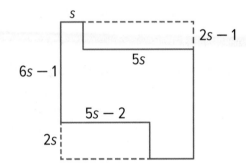

57. **FIND THE ERROR** Pearl and Ted both simplified $2x^2(3x^2 + 4x + 2)$. Is either of them correct? Explain your reasoning.

Pearl	Ted
$2x^2(3x^2 + 4x + 2)$	$2x^2(3x^2 + 4x + 2)$
$6x^4 + 8x^2 + 4x^2$	$6x^4 + 8x^3 + 4x^2$
$6x^4 + 12x^2$	

58. **PERSEVERE** Find p such that $3x^p(4x^{2p+3} + 2x^{3p-2}) = 12x^{12} + 6x^{10}$.

59. **PERSEVERE** Simplify $4x^{-3}y^2(2x^5y^{-4} + 6x^{-7}y^6 - 4x^0y^{-2})$.

60. **ANALYZE** Is there a value of x that makes the statement $(x + 2)^2 = x^2 + 2^2$ true? If so, find a value for x. Justify your argument.

61. **CREATE** Write a monomial and a polynomial using n as the variable. Find their product.

62. **WRITE** Describe the steps to multiply a polynomial by a monomial.

63. **CREATE** Write a polynomial equation, with variables on both sides, that has a solution of $t = 9$.

64. **CREATE** Write a polynomial expression that can be simplified to $-3c^3 + 74c^2 - 4c$. Be sure your polynomial expression requires the use of the Distributive Property at least two times in order to simplify.

Lesson 10-3

Multiplying Polynomials

Today's Goal
- Multiply binomials by using the Distributive Property and the FOIL Method.

Today's Vocabulary
quadratic expression

Explore Using Algebra Tiles to Find Products of Two Binomials

Online Activity Use algebra tiles to complete the Explore.

> **INQUIRY** How can you use the Distributive Property to find the product of two binomials?

Learn Multiplying Binomials

Binomials can be multiplied horizontally or vertically. Multipy $(x - 2)(x + 6)$.

Method 1 Vertical Method

$$
\begin{array}{r}
x - 2 \\
(\times) \quad x + 6 \\
\hline
6x - 12 \\
(+) \quad x^2 - 2x \\
\hline
x^2 + 4x - 12
\end{array}
$$

Multiply by 6.
Multiply by x.
Combine like terms.

Method 2 Horizontal Method

$(x - 2)(x + 6) = x(x + 6) - 2(x + 6)$ Rewrite as the sum of two products.
$ = x^2 + 6x - 2x - 12$ Distributive Property
$ = x^2 + 4x - 12$ Combine like terms.

You can also use a shortcut version of the Distributive Property, called the FOIL method, to multiply binomials.

Key Concept • FOIL Method

To multiply two binomials, find the sum of the products of **F** the *First terms*, **O** the *Outer terms*, **I** the *Inner terms*, and **L** the *Last terms*.

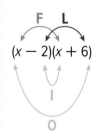

Product of First Terms	Product of Outer Terms	Product of Inner Terms	Product of Last Terms
↓	↓	↓	↓
$= x(x)$	$+ \quad x(6)$	$+ \quad -2(x)$	$+ \quad -2(6)$

$= x^2 + 6x - 2x - 12$
$= x^2 + 4x - 12$

An expression in one variable with a degree of 2 is called a **quadratic expression**.

Go Online You can complete an Extra Example online.

💭 **Think About It!**
How does the horizontal method using the Distributive Property differ from the FOIL method?

Study Tip

FOIL Method
The FOIL method can be used only when multiplying two binomials. The FOIL method cannot be used when multiplying, for example, two trinomials since six products are needed.

Example 1 Multiply Binomials by Using the Vertical Method

Find $(x - 1)(x + 7)$ by using the vertical method.

$$
\begin{array}{r}
x - 1 \\
(\times) \quad x + 7 \\
\hline
7x - 7 \\
x^2 - x \\
\hline
x^2 + 6x - 7
\end{array}
$$

Multiply by 7.
Multiply by x.
Combine like terms.

Example 2 Multiply Binomials by Using the Horizontal Method

Find $(3x - 4)(4x - 10)$ by using the horizontal method.

$(3x - 4)(4x - 10)$ $= 3x(4x - 10) + -4(4x - 10)$ Rewrite as sum of two products.

$= 12x^2 - 30x - 16x + 40$ Distributive Property

$= 12x^2 - 46x + 40$ Combine like terms.

Study Tip
Signs Notice that in the first step of the solution, you added the product of -4 and $4x$, and in the second step, that turned into a subtraction of $16x$. Remember that $-4(4x) = -16x$.

Example 3 Multiply Binomials by Using the FOIL Method

Find $(2a - 12)(5a + 3)$ by using the FOIL method.

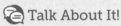

$= 2a(5a) + 2a(3) + (-12)(5a) + (-12)(3)$

$= 10a^2 + 6a - 60a - 36$

$= 10a^2 - 54a - 36$

Study Tip
FOIL Method The FOIL method is a memory device that can help you remember to find all four products when multiplying two binomials. The order in which the terms are multiplied is not important.

Talk About It!
Is your answer complete after you use the FOIL method to multiply? Explain.

Check
Find the product of $(3p - 9)(2p + 6)$ by using the FOIL Method.

Part A Find the product of each pair of terms.

First Terms: $(3p)(\underline{\quad ? \quad})$

Outer Terms: $(\underline{\quad ? \quad})(6)$

Inner Terms: $(\underline{\quad ? \quad})(2p)$

Last Terms: $(-9)(\underline{\quad ? \quad})$

Part B What is the product of $(3p - 9)(2p + 6)$ written in standard form?

Go Online You can complete an Extra Example online.

Apply Example 4 Use the FOIL Method

SNOW REMOVAL A town worker is clearing the snow from a parking lot and the surrounding sidewalk. The sidewalk extends x feet on every side of the parking lot. Write an expression for the total area of the parking lot and sidewalk.

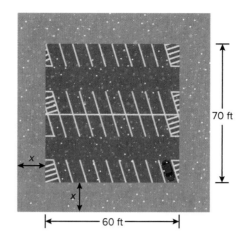

1 What is the task?

Describe the task in your own words. Then list any questions that you may have. How can you find answers to your questions?

The worker needs to shovel the snow from the parking lot and the sidewalk. We need to find the total area that needs to be shoveled.

2 How will you approach the task? What have you learned that you can use to help you complete the task?

First I'll find the total length and the total width. Then I'll multiply the length and the width to find the area. I have learned how to write and multiply binomials.

3 What is your solution?

Use your strategy to solve the problem.

Length $= 2x + 60$ Width $= 2x + 70$

The total area of the parking lot and sidewalk is $4x^2 + 260x + 4200$.

4 How can you know that your solution is reasonable?

Write About It! Write an argument that can be used to defend your solution.

Let $x = 4$, substituting 4 for x in the equation results in $(2(4) + 60)(2(4) + 70) = 4(4^2) + 260(4) + 4200$. This simplifies to $5304 = 5304$.

Check

FRAME Jacinta is framing her newest painting. The dimensions of the frame with the painting are represented by a width of $5y - 5$ and a length of $2y + 6$. Write an expression in standard form that represents the area of the frame with the painting.

Go Online You can complete an Extra Example online.

Example 5 Multiply Polynomials by Using the Distributive Property

The Distributive Property can also be used to multiply any two polynomials.

Find each product.

a. $(x + 3)(x^2 - x - 9)$

$\quad (x + 3)(x^2 - x - 9)$

$\quad = x(x^2 - x - 9) + 3(x^2 - x - 9)$ Distributive Property

$\quad = x^3 - x^2 - 9x + 3x^2 - 3x - 27$ Multiply.

$\quad = x^3 + 2x^2 - 12x - 27$ Combine like terms.

b. $(3t^2 - 5t + 9)(4t^2 - 4t + 14)$

$\quad (3t^2 - 5t + 9)(4t^2 - 4t + 14)$

$\quad = 3t^2(4t^2 - 4t + 14) - 5t(4t^2 - 4t + 14) + 9(4t^2 - 4t + 14)$

$\quad = 12t^4 - 12t^3 + 42t^2 - 20t^3 + 20t^2 - 70t + 36t^2 - 36t + 126$

$\quad = 12t^4 - 32t^3 + 98t^2 - 106t + 126$

Check

Find each product.

a. $(m - 3)(m^2 + m - 5)$

    A. $m^3 - 2m^2 - 8m + 15$

    B. $m^3 - 4m^2 - 8m + 15$

    C. $m^3 - 2m^2 + 8m + 15$

    D. $m^3 - 2m^2 - 8m - 15$

b. $(6v^2 + 4v - 3)(v^2 - v + 6)$

    A. $6v^4 + 10v^3 + 29v^2 + 27v - 18$

    B. $6v^4 - 2v^3 + 32v^2 + 27v - 18$

    C. $6v^4 - 2v^3 + 29v^2 + 27v - 18$

    D. $6v^4 - 2v^3 + 29v^2 + 24v - 18$

> **Think About It!**
> When multiplying two trinomials, do you have to apply the Distributive Property in any certain order?

Go Online You can complete an Extra Example online.

Practice

Examples 1–3
Find each product.

1. $(3c - 5)(c + 3)$
2. $(g + 10)(2g - 5)$
3. $(6a + 5)(5a + 3)$
4. $(4x + 1)(6x + 3)$
5. $(5y - 4)(3y - 1)$
6. $(6d - 5)(4d - 7)$
7. $(3m + 5)(2m + 3)$
8. $(7n - 6)(7n - 6)$
9. $(12t - 5)(12t + 5)$
10. $(5r + 7)(5r - 7)$
11. $(8w + 4x)(5w - 6x)$
12. $(11z - 5y)(3z + 2y)$

Example 4

13. **PLAYGROUND** The dimensions of a playground are represented by a width of $9x + 1$ feet and a length of $5x - 2$ feet. Write an expression that represents the area of the playground.

14. **THEATER** The Loft Theater has a center seating section with $3c + 8$ rows and $4c - 1$ seats in each row. Write an expression for the total number of seats in the center section.

15. **CRAFTS** Suppose a rectangular quilt made up of squares has a length-to-width ratio of 5 to 4. The length of the quilt is $5x$ inches. The quilt can be made slightly larger by adding a border of 1-inch squares all the way around the perimeter of the quilt. Write a polynomial expression for the area of the larger quilt.

16. **FLAG CASE** A United States flag is sometimes folded into a triangle shape and displayed in a triangular display case. If a display case has dimensions shown in inches, write a polynomial expression that represents the area of wall space covered by the display case.

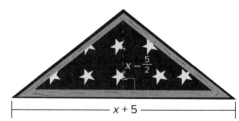

17. **NUMBER THEORY** Think of a whole number. Subtract 2. Write down this number. Take the original number and add 2. Write down this number. Find the product of the numbers you wrote down. Subtract the square of the original number. The result is always -4. Use polynomials to show how this number trick works.

Lesson 10-3 • Multiplying Polynomials 565

Example 5

Find each product.

18. $(2y - 11)(y^2 - 3y + 2)$

19. $(4a + 7)(9a^2 + 2a - 7)$

20. $(m^2 - 5m + 4)(m^2 + 7m - 3)$

21. $(x^2 + 5x - 1)(5x^2 - 6x + 1)$

22. $(3b^3 - 4b - 7)(2b^2 - b - 9)$

23. $(6z^2 - 5z - 2)(3z^3 - 2z - 4)$

Mixed Exercises

Find each product.

24. $(m + 4)(m + 1)$

25. $(x + 2)(x + 2)$

26. $(b + 3)(b + 4)$

27. $(t + 4)(t - 3)$

28. $(r + 1)(r - 2)$

29. $(n - 5)(n + 1)$

30. $(3c + 1)(c - 2)$

31. $(2x - 6)(x + 3)$

32. $(d - 1)(5d - 4)$

33. $(2\ell + 5)(\ell - 4)$

34. $(3n - 7)(n + 3)$

35. $(q + 5)(5q - 1)$

36. $(3b + 3)(3b - 2)$

37. $(2m + 2)(3m - 3)$

38. $(4c + 1)(2c + 1)$

39. $(5a - 2)(2a - 3)$

40. $(4h - 2)(4h - 1)$

41. $(x - y)(2x - y)$

42. $(w + 4)(w^2 + 3w - 6)$

43. $(t + 1)(t^2 + 2t + 4)$

44. $(k - 4)(k^2 + 5k - 2)$

45. $(m + 3)(m^2 + 3m + 5)$

46. $(2x + 1)(x^2 - 3x - 4)$

47. $(3b + 4)(2b^2 - b + 4)$

Simplify.

48. $(m + 2)[(m^2 + 3m - 6) + (m^2 - 2m + 4)]$

49. $[(t^2 + 3t - 8) - (t^2 - 2t + 6)](t - 4)$

Find each product.

50. $(a - 2b)^2$

51. $(3c + 4d)^2$

52. $(x - 5y)^2$

53. $(2r - 3t)^3$

54. $(5g + 2h)^3$

55. $(4y + 3z)(4y - 3z)^2$

56. PRECISION Write each expression as a simplified polynomial.

a. $(3c - 2)(4c^2 - c^3 + 3)$

b. $(5x - y)(3x^2 - 2xy) + (2x + y)(y^2 - 4x^2)$

c. $-2x(3 - x^2)(2x + 4)$

d. $(z - 1)(2 - z)(z + 1)$

57. ART The museum where Julia works plans to have a large wall mural painted in its lobby. First, Julia wants to paint a large frame around where the mural will be. She only has enough paint for the frame to cover 100 square feet of wall surface. The mural's length will be 5 feet longer than its width, and the frame will be 2 feet wide on all sides.

a. Write an expression for the area of the mural.

b. Write an expression for the area of the frame.

c. Write and solve an equation to find how large the mural can be.

58. STRUCTURE The dimensions of the composite figure shown are given in terms of the triangle's height, h.

a. Write and simplify a quadratic expression for the area of the figure.

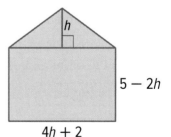

b. If $h = 1.42$ units, what is the area of the figure? Round to the nearest hundredth, if necessary.

59. STRUCTURE Consider the expression $x^{4p + 1}(x^{1 - 2p})^{2p + 3}$.

a. Use the laws of exponents to simplify the expression.

b. Find any integer values of p that make this expression equal to 1 for all values of x.

60. USE A MODEL The relationship between monthly profit P, monthly sales n, and unit price p is $P = n(p - U) - F$, where U is the unit cost per sale and F is a fixed cost that does not depend on the number of sales. For an online business advice service, the unit cost is $30 per hour-long session and the monthly fixed cost is $3000.

  a. Given a model for monthly sales of $n = 5000 - 40p$ for a given price p per session, write and simplify a quadratic expression for P in terms of p.

  b. If the unit price, p, is $77.50, what is the monthly profit, P?

61. STRUCTURE Find and simplify an expression for the volume of the rectangular prism shown.

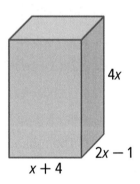

62. ANALYZE Determine if the following statement is *sometimes, always,* or *never* true. Justify your argument.

The FOIL method can be used to multiply a binomial and a trinomial.

63. PERSEVERE Find $(x^m + x^p)(x^{m-1} - x^{1-p} + x^p)$.

64. CREATE Write a binomial and a trinomial involving a single variable. Then find their product.

65. WRITE Compare and contrast the procedure used to multiply a trinomial by a binomial using the vertical method with the procedure used to multiply a three-digit number by a two-digit number.

66. WRITE Summarize the methods that can be used to multiply polynomials.

67. WHICH ONE DOESN'T BELONG? Which polynomial expression does not belong with the other expressions? Explain your reasoning.

$(2x + 11)(3x - 7)$	$(6y + 1)(y - 4)$
$(3y - 4)(2y + 5)$	$(2x - 1)(x^2 + x - 1)$

68. FIND THE ERROR Jariah and Malia are multiplying the expression $(2x + 1)(x - 4)$. Is either of them correct? Explain your reasoning.

Jariah	Malia
$(2x + 1)(x - 4) = 2x^2 + 8x + 1x - 4$	$(2x + 1)(x - 4) = 2x^2 - 8x + 1x - 4$
$= 2x^2 + 9x - 4$	$= 2x^2 - 7x - 4$

Lesson 10-4

Special Products

Today's Goals
- Multiply binomials by applying the pattern formed by squares of sums.
- Multiply binomials by applying the pattern formed by squares of differences.

Explore Using Algebra Tiles to Find the Squares of Sums

 Online Activity Use algebra tiles to complete the Explore.

 INQUIRY How can you write the square of a sum?

Learn Square of a Sum

Key Concept • Square of a Sum

Words The square of $a + b$ is the square of a plus twice the product of a and b plus the square of b.

Symbols $(a + b)^2 = (a + b)(a + b)$
$\qquad\qquad = a^2 + 2ab + b^2$

Example $(x + 3)^2 = (x + 3)(x + 3)$
$\qquad\qquad\quad = x^2 + 6x + 9$

Example 1 Square of a Sum

Find each product.

a. $(x + 6)^2$

$(a + b)^2 = a^2 + 2ab + b^2$ Square of a sum

$(x + 6)^2 = (x)^2 + 2(x)(6) + 6^2$ $a = x, b = 6$

$\qquad\qquad = x^2 + 12x + 36$ Simplify.

b. $(3g + 10h)^2$

$(a + b)^2 = a^2 + 2ab + b^2$ Square of a sum

$(3g + 10h)^2 = (3g)^2 + 2(3g)(10h) + (10h)^2$ $a = 3g, b = 10h$

$\qquad\qquad\quad = 9g^2 + 60gh + 100h^2$ Simplify.

 Think About It!

How can you check that using the square of a sum pattern gives the correct product?

Check

Find $(2x + 9)^2$. Select the correct product.

A. $2x^2 + 18x + 18$

B. $2x^2 + 36x + 81$

C. $4x^2 + 18x + 81$

D. $4x^2 + 36x + 81$

Find $(6m + 11n)^2$. Select the correct product.

A. $6m^2 + 132mn + 11n^2$

B. $12m^2 + 66mn + 121n^2$

C. $36m^2 + 66mn + 121n^2$

D. $36m^2 + 132mn + 121n^2$

Example 2 Use Squares of Sums

GENETICS A Punnett square is used to predict the probability of offspring inheriting certain genetic characteristics. In Doberman Pinschers, black fur *B* is dominant over the recessive gene for brown fur *b*. If two parents have both a dominant and a recessive gene, use the square of a sum to determine the possible combinations of their offspring.

	B	b
B	BB	Bb
b	bB	bb

Both parents can be represented by $(B + b)$, and the combinations of the offspring are the product of $(B + b)^2$.

$(a + b)^2 = a^2 + 2ab + b^2$ Square of a sum

$(B + b)^2 = B^2 + 2Bb + b^2$ $a = B, b = b$

Check

SURFACE AREA The surface area of a cube is given by $A = 6s^2$, where *s* is the length of one side. Select the expression that represents the surface area of a cube with side length $3n + 6$.

A. $6(3n^2 + 18n + 36)$

B. $6(9n^2 + 36)$

C. $6(9n^2 + 36n + 36)$

D. $6(9n^2 + 81n + 36)$

Go Online You can complete an Extra Example online.

Explore Using Algebra Tiles to Find the Squares of Differences

Online Activity Use algebra tiles to complete the Explore.

@ INQUIRY How can you write the square of a difference?

Learn Square of a Difference

Key Concept • Square of a Difference

Words The square of $a - b$ is the square of a minus twice the product of a and b plus the square of b.

Symbols $(a - b)^2 = (a - b)(a - b)$
$= a^2 - 2ab + b^2$

Example $(x - 8)^2 = (x - 8)(x - 8)$
$= x^2 - 2 \cdot x \cdot 8 + 8^2$
$= x^2 - 16x + 64$

Example 3 Square of a Difference

Find $(7d - 2f)^2$.

$(a - b)^2 = a^2 - 2ab + b^2$ Square of a difference

$(7d - 2f)^2 = (7d)^2 - 2(7d)(2f) + (2f)^2$ $a = 7d, b = 2f$

$= 49d^2 - 28df + 4f^2$ Simplify.

Check

Find $(4k - 1)^2$.

A. $4k^2 - 8k - 1$

B. $16k^2 - 1$

C. $16k^2 - 8k + 1$

D. $16k^2 - 16k + 1$

Study Tip

Watching Signs Since the square of a sum and the square of a difference vary only by the sign of the middle term, pay close attention to the sign being used within the square of the trinomial.

 Think About It!

Demarco says that the Square of a Difference pattern is a special case of the Square of a Sum pattern. Is he correct? Explain.

Go Online You can complete an Extra Example online.

Lesson 10-4 • Special Products **571**

Explore Using Algebra Tiles to Find Products of Sums and Differences

Online Activity Use algebra tiles to complete the Explore.

 INQUIRY How can you write the product of a sum and a difference?

Learn Product of a Sum and a Difference

Key Concept • Product of a Sum and a Difference	
Words	The product of $a + b$ and $a - b$ is the square of a minus the square of b.
Symbols	$(a + b)(a - b) = a^2 - ab + ab - b^2$
	$= a^2 - b^2$
Example	$(x + 6)(x - 6) = x^2 - 6x + 6x - 6^2$
	$= x^2 - 36$

Study Tip
Identical Values Note that the values of a and b must be identical within each quantity to use the pattern for the product of a sum and a difference. The only difference between the quantities is the operation between a and b.

Example 4 Product of a Sum and a Difference

Find each product.

a. $(z + 5)(z - 5)$

$(a + b)(a - b) = a^2 - b^2$ Product of a sum and a difference

$(z + 5)(z - 5) = z^2 - 5^2$ $a = z, b = 5$

$= z^2 - 25$ Simplify.

b. $(3y^3 + 4)(3y^3 - 4)$

$(a + b)(a - b) = a^2 - b^2$ Product of a sum and a difference

$(3y^3 + 4)(3y^3 - 4) = (3y^3)^2 - (4)^2$ $a = 3y^3, b = 4$

$= 9y^6 - 16$ Simplify.

Watch Out!
Power of a Power Remember to square all parts of a and b, including powers. When a power is raised to another power, multiply the exponents.

Check

Find each product.

a. $(x^2 + 3y)(x^2 - 3y)$
b. $(2x + 9y^2)(2x - 9y^2)$
c. $(x + 9y)(x - 9y)$
d. $(x + 9)(x - 9)$

 **Talk About It!**
Explain how you would tell someone to find the product of a sum and a difference.

Go Online You can complete an Extra Example online.

572 Module 10 • Polynomials

Practice

Go Online You can complete your homework online.

Examples 1 and 3
Find each product.

1. $(a + 10)(a + 10)$
2. $(b - 6)(b - 6)$
3. $(h + 7)^2$
4. $(x + 6)^2$
5. $(8 - m)^2$
6. $(9 - 2y)^2$
7. $(2b + 3)^2$
8. $(5t - 2)^2$
9. $(8h - 4n)^2$
10. $(4m - 5n)^2$

Example 2

11. **ROUNDABOUTS** A city planner is proposing a roundabout to improve traffic flow at a busy intersection. Write a polynomial equation for the area A of the traffic circle if the radius of the outer circle is r and the width of the road is 18 feet.

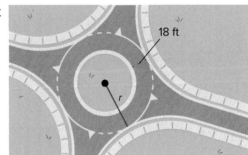

12. **NUMBER CUBES** Kivon has two number cubes. Each edge of number cube A is 3 millimeters less than each edge of number cube B. Each edge of number cube B is x millimeters. Write an equation that models the surface area of number cube A.

Number Cube A — $(x - 3)$ mm, $(x - 3)$ mm, $(x - 3)$ mm

Number Cube B — x mm, x mm, x mm

13. **PROBABILITY** The spinner has two equal sections, blue (B) and red (R). Use the square of a sum to determine the possible combinations of spinning the spinner two times.

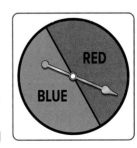

14. **BUSINESS** The Combo Lock Company finds that its profit data from 2015 to the present can be modeled by the function $y = (2n + 11)^2$, where y is the profit n years since 2015. Which special product does this polynomial demonstrate? Simplify the polynomial.

Lesson 10-4 • Special Products

Example 4

Find each product.

15. $(u + 3)(u - 3)$

16. $(b + 7)(b - 7)$

17. $(2 + x)(2 - x)$

18. $(4 - x)(4 + x)$

19. $(2q + 5r)(2q - 5r)$

20. $(3a^2 + 7b)(3a^2 - 7b)$

Mixed Exercises

Find each product.

21. $(n + 3)^2$

22. $(x + 4)(x + 4)$

23. $(y - 7)^2$

24. $(t - 3)(t - 3)$

25. $(b + 1)(b - 1)$

26. $(a - 5)(a + 5)$

27. $(p - 4)^2$

28. $(z + 3)(z - 3)$

29. $(\ell + 2)(\ell + 2)$

30. $(r - 1)(r - 1)$

31. $(3g + 2)(3g - 2)$

32. $(2m - 3)(2m + 3)$

33. $(6 + u)^2$

34. $(r + t)^2$

35. $(3q + 1)(3q - 1)$

36. $(c - d)^2$

37. $(2k - 2)^2$

38. $(w + 3h)^2$

39. $(3p - 4)(3p + 4)$

40. $(t + 2u)^2$

41. $(x - 4y)^2$

42. $(3b + 7)(3b - 7)$

43. $(3y - 3g)(3y + 3g)$

44. $(n^2 + r^2)^2$

45. $(2k + m^2)^2$

46. $(3t^2 - n)^2$

47. **GEOMETRY** The length of a rectangle is the sum of two whole numbers. The width of the rectangle is the difference of the same two whole numbers. Write a verbal expression for the area of the rectangle.

48. Find the product of $(10 - 4t)$ and $(10 + 4t)$. What type of special product does this represent?

49. STORAGE A cylindrical tank is placed along a wall. A cylindrical PVC pipe will be hidden in the corner behind the tank. See the side-view diagram shown. The radius of the tank is r inches, and the radius of the PVC pipe is s inches.

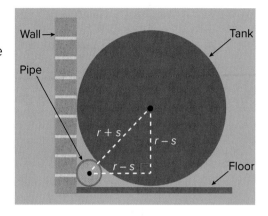

a. Use the Pythagorean Theorem to write an equation for the relationship between the two radii. Simplify your equation so that there is a zero on one side of the equal sign.

b. Write a polynomial equation you could solve to find the radius s of the PVC pipe if the radius of the tank is 20 inches.

Find each product.

50. $(2a - 3b)^2$

51. $(5y + 7)^2$

52. $(8 - 10a)^2$

53. $(10x - 2)(10x + 2)$

54. $(3t + 12)(3t - 12)$

55. $(a + 4b)^2$

56. $(3q - 5r)^2$

57. $(2c - 9d)^2$

58. $(g + 5h)^2$

59. $(6y - 13)(6y + 13)$

60. $(3a^4 - b)(3a^4 + b)$

61. $(5x^2 - y^2)^2$

62. $(8a^2 - 9b^3)(8a^2 + 9b^3)$

63. $\left(\frac{3}{4}k + 8\right)^2$

64. $\left(\frac{2}{5}y - 4\right)^2$

65. $(7z^2 + 5y^2)(7z^2 - 5y^2)$

66. $(2m + 3)(2m - 3)(m + 4)$

67. $(r + 2)(r - 5)(r - 2)(r + 5)$

Find each product.

68. $(c + d)(c + d)(c + d)$

69. $(2a - b)^3$

70. $(f + g)(f - g)(f + g)$

71. $(k - m)(k + m)(k - m)$

72. $(n - p)^2(n + p)$

73. $(q - r)^2(q - r)$

74. Consider the product $(a + b)(a - b)(a + b)(a - b)$.
   a. Show the steps required to determine the product.
   b. Evaluate the original expression for $a = 5$ and $b = 2$.
   c. Evaluate the simplified expression you wrote in **part a** for $a = 5$ and $b = 2$. Compare this to the result from **part b**.

75. STRUCTURE Tanisha is investigating growth patterns for the area of a square. She begins with a square of side length s and looks at the effects of enlarging the side length by one unit at a time.

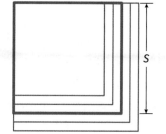

   a. How much more area does the square with side length $s + 1$ have compared to the square with side length s?
   b. How much more area does the square with side length $s + 2$ have compared to the square with side length $s + 1$?
   c. How much more area does the square with side length $s + 3$ have compared to the square with side length $s + 2$?

76. WHICH ONE DOESN'T BELONG? Which expression does not belong? Explain your reasoning.

| $(2c - d)(2c - d)$ | $(2c + d)(2c - d)$ | $(2c + d)(2c + d)$ | $(c + d)(c + d)$ |

77. PERSEVERE Does a pattern exist for the cube of a sum $(a + b)^3$?
   a. Investigate this question by finding the product $(a + b)(a + b)(a + b)$.
   b. Use the pattern you discovered in **part a** to find $(x + 2)^3$.
   c. Draw a diagram of a geometric model for $(a + b)^3$.
   d. What is the pattern for the cube of a difference $(a - b)^3$?

78. ANALYZE The square of a sum is called a *perfect square trinomial*. Find c that makes $25x^2 - 90x + c$ a perfect square trinomial.

79. CREATE Write two binomials with a product that is a binomial. Then write two binomials with a product that is not a binomial.

80. WRITE Describe how to square the sum of two quantities, how to square the difference of two quantities, and how to find the product of a sum of two quantities and a difference of two quantities.

Module 10 • Polynomials

Lesson 10-5

Using the Distributive Property

Explore Using Algebra Tiles to Factor Polynomials

Online Activity Use algebra tiles to complete the Explore.

> **INQUIRY** How is factoring polynomials related to multiplying polynomials?

Today's Goals
- Factor polynomials by using the Distributive Property
- Factor polynomials by using the Distributive Property and grouping

Today's Vocabulary
factoring

factoring by grouping

Learn Factoring by Using the Distributive Property

You can use the Distributive Property to multiply a polynomial by a monomial.

$$3y(2y + 5) = 3y(2y) + 3y(5)$$
$$= 6y^2 + 15y$$

You can also use the Distributive Property to factor a polynomial. **Factoring** is the process of expressing a polynomial as the product of monomials and polynomials.

$$6y^2 + 15y = 3y(2y) + 3y(5)$$
$$= 3y(2y + 5)$$

So, $3y(2y + 5)$ is the factored form of $6y^2 + 15y$. When factoring a polynomial, it must be factored completely. If you are using the Distributive Property to factor, one factor will be the greatest common factor (GCF) for all terms of the polynomial.

Example 1 Use the Distributive Property

Use the Distributive Property to factor each polynomial.

a. $12a^2 + 16a$

Step 1 Factor each term.

$12a^2 = 2 \cdot 2 \cdot 3 \cdot a \cdot a$

$16a = 2 \cdot 2 \cdot 2 \cdot 2 \cdot a$

Step 2 Underline the common terms.

$12a^2 = \underline{2} \cdot \underline{2} \cdot 3 \cdot \underline{a} \cdot a$

$16a = \underline{2} \cdot \underline{2} \cdot 2 \cdot 2 \cdot \underline{a}$

Step 3 Find the GCF.

GCF $= 2 \cdot 2 \cdot a$ or $4a$

> **Study Tip**
> **Greatest Common Factor** To find the GCF of a polynomial, write all factors of each term as prime numbers or variables to the first degree.

(continued on the next page)

Lesson 10-5 • Using the Distributive Property **577**

Step 4 Write each term as the product of the GCF and the remaining factors. Use the Distributive Property.

$12a^2 + 16a = 4a(3a) + 4a(4)$ Rewrite each term using the GCF.

$ = 4a(3a + 4)$ Distributive Property.

b. $20x^2y^2 - 45x^2y - 35x^2$

$20x^2y^2 = 2 \cdot 2 \cdot 5 \cdot x \cdot x \cdot y \cdot y$

$-45x^2y = -1 \cdot 3 \cdot 3 \cdot 5 \cdot x \cdot x \cdot y$

$-35x^2 = -1 \cdot 5 \cdot 7 \cdot x \cdot x$

GCF = $5x^2$

Write each term as the product of the GCF and the remaining factors. Use the Distributive Property.

$20x^2y^2 - 45x^2y - 35x^2 = 5x^2(4y^2) + 5x^2(-9y) + 5x^2(-7)$

$ = 5x^2(4y^2 - 9y - 7)$

Watch Out!
Factoring Completely Once you have factored the polynomial, check the remaining polynomial for other common factors you may have missed. *Factoring* means to factor completely.

Check
Factor each polynomial.

a. $33n^3 - 121n^2$

b. $14a^2b^2c - 6ac^2 + 10ac$

🌐 Example 2 Use Factoring

VOLCANOS The 1980 eruption of Washington's Mt. St. Helens had an initial lateral blast with a velocity of 440 feet per second. The expression $440t - 16t^2$ models the height of a rock erupted from the volcano after t seconds. Factor the expression.

$440t = 2 \cdot 2 \cdot 2 \cdot 5 \cdot 11 \cdot t$

$-16t^2 = -1 \cdot 2 \cdot 2 \cdot 2 \cdot 2 \cdot t \cdot t$

GCF = $8t$

$440t - 16t^2 = 8t(55) + 8t(-2t)$

$ = 8t(55 - 2t)$

Check

GOLF A golfer hits a golf ball with a velocity of 112 feet per second. The expression $112t - 16t^2$ represents the height of the golf ball after t seconds. Factor the expression.

🌐 **Go Online** You can complete an Extra Example online.

Learn Factor by Grouping

When a polynomial has four or more terms, you can sometimes use a method called **factoring by grouping**. Similar terms are grouped, and the Distributive Property is applied to a common binomial.

Key Concept • Factor by Grouping

Words
A polynomial can be factored by grouping only if all of the following conditions exist.
- There are four or more terms.
- Terms have common factors that can be grouped together.
- There are at least two common factors that are identical or additive inverses of each other.

Symbols
$ax + bx + ay + by = (ax + bx) + (ay + by)$
$= x(a + b) + y(a + b)$
$= (x + y)(a + b)$

Talk About It!
Explain why you cannot use factoring by grouping on a polynomial with three terms.

Example 3 Factor by Grouping

Factor $2uv + 6u + 5v + 15$.

$2uv + 6u + 5v + 15 = (2uv + 6u) + (5v + 15)$ Group terms with common factors.

$= 2u(v + 3) + 5(v + 3)$ Factor the GCF from each group.

Notice that $(v + 3)$ is common to both groups, so it becomes the GCF.

$= (2u + 5)(v + 3)$ Distributive Property

Study Tip

Creating Groups If you are unable to get identical or additive inverse binomials after factoring out the GCF, try grouping the terms in a different way.

Check

Factor $tw + 10t - 2w - 20$.

Go Online You can complete an Extra Example online.

> **Think About It!**
> Which property is used to simplify $3m[(-1)(p - 7)] + 4(p - 7)$ to $-3m(p - 7) + 4(p - 7)$?

Example 4 Factor by Grouping with Additive Inverses

It is helpful to be able to recognize when binomials are additive inverses of each other. For example, $5 - x = -1(x - 5)$.

Factor $21m - 3mp + 4p - 28$.

$21m - 3mp + 4p - 28$	Original expression
$= (21m - 3mp) + (4p - 28)$	Group terms with common factors.
$= 3m(7 - p) + 4(p - 7)$	Factor the GCF from each group.
$= 3m[(-1)(p - 7)] + 4(p - 7)$	$7 - p = -1(p - 7)$
$= -3m(p - 7) + 4(p - 7)$	Associative Property
$= (-3m + 4)(p - 7)$	Distributive Property

Alternate Method

$21m - 3mp + 4p - 28$	Original expression
$= 4p - 3mp + 21m - 28$	Rearrange the expression.
$= (4p - 3mp) + (21m - 28)$	Group terms with common factors.
$= p(4 - 3m) + 7(3m - 4)$	Factor the GCF from each group.
$= p(-3m + 4) + 7(3m - 4)$	$4 - 3m = -3m + 4$
$= p(-3m + 4) + 7[-1(-3m + 4)]$	$3m - 4 = -1(-3m + 4)$
$= p(-3m + 4) - 7(-3m + 4)$	Associative Property
$= (p - 7)(-3m + 4)$	Distributive Property
$= (-3m + 4)(p - 7)$	Commutative Property

Check

Factor $-3x^3 + 33x^2 + 4x - 44$.

> **Go Online** to learn how to prove the Elimination Method in Expand 10-5.

Go Online You can complete an Extra Example online.

Practice

Go Online You can complete your homework online.

Example 1

Use the Distributive Property to factor each polynomial.

1. $16t - 40y$
2. $30v + 50x$
3. $2k^2 + 4k$
4. $5z^2 + 10z$
5. $4a^2b^2 + 2a^2b - 10ab^2$
6. $5c^2v - 15c^2v^2 + 5c^2v^3$

Example 2

7. **PHYSICS** The distance d an object falls after t seconds is given by $d = 16t^2$ (ignoring air resistance). To find the height of an object launched upward from ground level at a rate of 32 feet per second, use the expression $32t - 16t^2$, where t is the time in seconds. Factor the expression.

8. **SWIMMING POOL** The area of a rectangular swimming pool is given by the expression $12w - w^2$, where w is the width of one side. Factor the expression.

9. **VERTICAL JUMP** Your vertical jump height is measured by subtracting your standing reach height from the height of the highest point you can reach by jumping without taking a running start. Typically, NBA players have vertical jump heights of up to 34 inches. If an NBA player jumps this high, his height in inches above his standing reach height after t seconds can be modeled by the expression $162t - 192t^2$. Factor the expression.

10. **PETS** Conner is playing with his dog. He tosses a treat upward with an initial velocity of 13.7 meters per second. His hand starts at the same height as the dog's mouth, so the height of the treat above the dog's mouth in meters after t seconds is given by the expression $13.7t - 4.9t^2$. Factor the expression.

Examples 3 and 4

Factor each polynomial.

11. $fg - 5g + 4f - 20$
12. $a^2 - 4a - 24 + 6a$
13. $hj - 2h + 5j - 10$
14. $xy - 2x - 2 + y$
15. $45pq - 27q - 50p + 30$
16. $24ty - 18t + 4y - 3$
17. $3dt - 21d + 35 - 5t$
18. $8r^2 + 12r$
19. $21th - 3t - 35h + 5$
20. $vp + 12v + 8p + 96$
21. $5br - 25b + 2r - 10$
22. $2nu - 8u + 3n - 12$
23. $b^2 - 2b + 3b - 6$
24. $2j^2 + 2j + 3j + 3$
25. $2a^2 - 4a + a - 2$

Lesson 10-5 • Using the Distributive Property **581**

Mixed Exercises

Factor each polynomial.

26. $7x + 49$

27. $8m - 6$

28. $5a^2 - 15$

29. $10q - 25q^2$

30. $a^2b^2 + a$

31. $x + x^2y + x^3y^2$

32. $3p^2r^2 + 6pr + p$

33. $4a^2b^2 + 16ab + 12a$

34. $10h^3n^3 - 2hn^2 + 14hn$

35. $48a^2b^2 - 12ab$

36. $6x^2y - 21y^2w + 24xw$

37. $x^2 + 3x + x + 3$

38. $2x^2 - 5x + 6x - 15$

39. $3n^2 + 6np - np - 2p^2$

40. $4x^2 - 1.2x + 0.5x - 0.15$

41. $9x^2 - 3xy + 6x - 2y$

42. $3x^2 + 24x - 1.5x - 12$

43. $2x^2 - 0.6x + 3x - 0.9$

44. **ARCHERY** The height, in feet, of an arrow can be modeled by the expression $80t - 16t^2$, where t is the time in seconds. Factor the expression.

45. **REGULARITY** You have factored some polynomials using the Distributive Property, and others were factored by grouping. What are the similarities between the two methods?

46. **CREATE** Write a four-term polynomial that can be factored by grouping. Then factor the polynomial.

47. **FIND THE ERROR** Given the polynomial expression $3x^2 + 7x - 18x - 42$, Theresa says you need to factor by grouping using the binomials $(3x^2 - 18x)$ and $(7x - 42)$. Akash says you need to use the binomials $(3x^2 + 7x)$ and $(-18x - 42)$. Is either of them correct? Justify your answer.

48. **PRECISION** Choose a value, q, so that $2x^2 + qx + 7x - 21$ can be factored by grouping. Then factor the expression.

49. **WRITE** Explain how to factor the polynomial $12a^2b^2 - 16a^2b^3$.

Lesson 10-6

Factoring Quadratic Trinomials

Explore Using Algebra Tiles to Factor Trinomials

Online Activity Use algebra tiles to complete the Explore.

> **INQUIRY** How can you use the constant and coefficients in a polynomial to find its factors?

Learn Factoring Trinomials with a Leading Coefficient of 1

Key Concept • Factoring Trinomials with a Leading Coefficient of 1

Words	To factor trinomials in the form $x^2 + bx + c$, find two integers, m and p, with a sum of b and a product of c. Then write $x^2 + bx + c$ as $(x + m)(x + p)$.
Symbols	$x^2 + bx + c = (x + m)(x + p)$ when $m + p = b$ and $mp = c$.
Example	$x^2 + 8x + 15 = (x + 3)(x + 5)$ because $3 + 5 = 8$ and $3 \cdot 5 = 15$.

A polynomial that cannot be written as a product of two polynomials with integer coefficients is called a **prime polynomial**.

Example 1 c Is Positive

Factor $x^2 - 9x + 18$.

In this trinomial, $b = -9$ and $c = 18$. Because c is positive and b is negative, you need to find two negative factors with a sum of -9 and a product of 18.

Complete the table to make an organized list of the factors of 18 and look for the pair of factors with a sum of -9.

Factors of 18	Sum of Factors
−1, −18	−19
−2, −9	−11
−3, −6	−9

The correct factors are −3 and −6.

$x^2 - 9x + 18 = (x + m)(x + p)$ Write the pattern.

$\qquad\qquad\quad\; = [x + (-3)][x + (-6)]$ $m = -3$ and $p = -6$

$\qquad\qquad\quad\; = (x - 3)(x - 6)$ Simplify.

Go Online You can complete an Extra Example online.

Today's Goals
- Determine the factors of trinomials with a leading coefficient of 1.
- Determine the factors of trinomials with a leading coefficient not equal to 1.

Today's Vocabulary
prime polynomial

Go Online
You may want to complete the Concept Check to check your understanding.

Think About It!
If c is negative, then what is the relationship between the signs of the factors? Explain your reasoning.

Problem-Solving Tip
Guess and Check
When factoring a trinomial, make an educated guess, check for reasonableness, and then adjust the guess until you find the correct answer.

Lesson 10-6 • Factoring Quadratic Trinomials   **583**

Example 2 c Is Negative and b Is Positive

Factor $x^2 + 5x - 14$.

In this trinomial, $b = 5$ and $c = -14$. Since c is negative, the factors m and p have opposite signs. So, either m or p is negative, but not both. Since b is positive, the factor with the greater absolute value is also positive.

Complete the table to make a list of the factors of -14, where one factor of each pair is negative and the factor with the greater absolute value is positive. Look for the pair of factors with a sum of 5.

Factors of −14	Sum of Factors
−1, 14	13
−2, 7	5

The correct factors are -2 and 7.

$$x^2 + 5x - 14 = (x + m)(x + p)$$
$$= (x - 2)(x + 7)$$

Example 3 c Is Negative and b Is Negative

Factor $x^2 - 3x - 4$.

In this trinomial, $b = -3$ and $c = -4$. Either m or p is negative, but not both. Since b is negative, the factor with the greater absolute value is also negative.

Complete the table to make a list of the factors of -4, where one factor of each pair is negative and the factor with the greater absolute value is negative. Look for the pair of factors with a sum of -3.

Factors of −4	Sum of Factors
1, −4	−3
2, −2	0

The correct factors are 1 and -4.

$$x^2 - 3x - 4 = (x + m)(x + p)$$
$$= (x + 1)(x - 4)$$

Example 4 Factor a Polynomial

Factor $x^2 - 4x + 8$, if possible. If the polynomial cannot be factored using integers, write *prime*.

In this trinomial, $b = -4$ and $c = 8$. Since b is negative, $m + p$ is negative. Since c is positive, mp is positive. So, m and p are both negative.

Next, list the factors of 8. Look for the pair with a sum of -4.

Factors of 8	Sum of Factors
−1, −8	−9
−2, −4	−6

There are no factors with a sum of -4. So, the trinomial cannot be factored using integers. Therefore, $x^2 - 4x + 8$ is prime.

Check

Write the factored form of each polynomial. If the polynomial cannot be factored using integers, write *prime*.

a. $x^2 + 7x + 6$ b. $x^2 - 8x + 12$ c. $x^2 + 3x - 40$

💭 **Think About It!**

Is $(x + 1)(x - 4)$ equal to $(x - 4)(x + 1)$? to $(x - 1)(x + 4)$? Explain your reasoning.

💬 **Talk About It**

How does the process of factoring $ax^2 + bx + c$ compare to the process of factoring $x^2 + bx + c$?

Example 5 Solve a Problem by Factoring

FLAG DESIGN Switzerland's flag has a very unique shape; it is a square. However, the flag used by the country's naval vessels is rectangular, as shown. If the area of the square flag is $x^2 - 6x + 9$ square feet, and the length is increased by 4 feet, then what is the area of the naval flag in terms of x?

Step 1 Factor $x^2 - 6x + 9$. In this trinomial, $b = -6$ and $c = 9$. Because c is positive and b is negative, you need to find two negative factors with a sum of -6 and a product of 9.

Factors of 9	Sum of Factors
−1, −9	−10
−3, −3	−6

The correct factors are -3 and -3.

$x^2 - 6x + 9 = (x + m)(x + p)$ Write the pattern.

$= (x - 3)(x - 3)$ $m = -3$ and $p = -3$

Step 2 Increase length and multiply. The length is increased by 4 feet, so the factor representing the length must be increased by 4.

$(x - 3 + 4)(x - 3) = (x + 1)(x - 3)$ Add 4 to the length.

$= x^2 - 3x + x - 3$ FOIL

$= x^2 - 2x - 3$ Simplify.

The new area is $x^2 - 2x - 3$ square feet.

> **Study Tip**
>
> **Assumptions** Because the polynomial represents the area of a flag, you can assume that it can be factored. The area of a rectangle can always be written as the product of two sides.

Learn Factoring Trinomials

Key Concept • Factoring Trinomials with a Leading Coefficient ≠ 1

To factor trinomials in the form $ax^2 + bx + c$, find two integers, m and p, with a sum of b and a product of ac. Then write $ax^2 + bx + c$ as $ax^2 + mx + px + c$, and factor by grouping.

Example 6 c Is Negative

Factor $4x^2 + 18x - 10$.

In this trinomial, a GCF of 2 can be factored out.

$4x^2 + 18x - 10 = 2(2x^2 + 9x - 5)$

Then in the trinomial $2x^2 + 9x - 5$, $a = 2$, $b = 9$, and $c = -5$. You need to find two numbers with a sum of 9 and a product of $2(-5)$ or -10.

> **Think About It!**
>
> Which property is applied to go from $2[2x(x + 5) - (x + 5)]$ to $2(2x - 1)(x + 5)$?

(continued on the next page)

 Go Online You can complete an Extra Example online.

Math History Minute

Abraham bar Hiyya (1070–1136) was a Spanish mathematician, astronomer, and philosopher. His most famous work is *Treatise on Measurement and Calculation*, which contains the first complete solution of the quadratic equation $x^2 - ax + b = 0$ known in Europe. It is said that his work influenced the later work of Leonardo Fibonacci.

 Go Online An alternate method is available for this example.

Complete the table to make a list of the factors of −10.

Factors of −10	Sum of Factors
1, −10	−9
2, −5	−3
5, −2	3
10, −1	9

Look for a pair of factors with a sum of 9. The correct factors are 10 and −1.

$4x^2 + 18x - 10 = 2(2x^2 + mx + px - 5)$ Write the pattern.

$\qquad = 2(2x^2 + 10x + (-1)x - 5)$ $m = 10$ and $p = -1$

$\qquad = 2[(2x^2 + 10x) + (-x - 5)]$ Group terms with common factors.

$\qquad = 2[2x(x + 5) - (x + 5)]$ Factor the GCFs.

$\qquad = 2(2x - 1)(x + 5)$ $x + 5$ is the common factor.

Example 7 *c* Is Positive

Factor $2x^2 - 17x + 21$.

In this trinomial, $a = 2$, $b = -17$, and $c = 21$. Since b is negative, $m + p$ will be negative. Since c is positive, mp will be positive. To determine m and p, list the negative factors of ac. The sum of m and p should be equal to b.

Complete the table to make a list of the negative factors of 42, and look for a pair of factors with a sum of −17.

Factors of 42	Sum of Factors
−1, −42	−43
−2, −21	−23
−3, −14	−17
−6, −7	−13

The correct factors are −3 and −14.

$2x^2 - 17x + 21 = 2x^2 + mx + px + 21$ Write the pattern.

$\qquad = 2x^2 + (-3)x + (-14)x + 21$ $m = -3$ and $p = -14$

$\qquad = (2x^2 - 14x) + (-3x + 21)$ Group terms with common factors.

$\qquad = 2x(x - 7) + (-3)(x - 7)$ Factor the GCFs.

$\qquad = (2x - 3)(x - 7)$ $x - 7$ is the common factor.

 Go Online You can complete an Extra Example online.

Practice

Go Online You can complete your homework online.

Examples 1–4

Factor each polynomial, if possible. If the polynomial cannot be factored using integers, write *prime*.

1. $x^2 + 17x + 42$

2. $y^2 - 17y + 72$

3. $a^2 + 8a - 48$

4. $n^2 - 2n - 35$

5. $44 + 15h + h^2$

6. $40 - 22x + x^2$

7. $-24 - 5x + x^2$

8. $-42 - m + m^2$

9. $t^2 + 8t + 12$

10. $d^2 + 5d - 13$

11. $y^2 - 6y + 17$

12. $n^2 + 7n + 12$

13. $b^2 - 12b - 101$

14. $p^2 + 9p + 20$

15. $h^2 + 9h + 18$

16. $c^2 + c + 21$

Example 5

17. **COSMETICS CASE** The top of a cosmetics case is a rectangle in which the width is 2 centimeters greater than the length. The expression $x^2 + 26x + 168$ represents the area of the top of the case. Factor the expression.

18. **CARPENTRY** Miko wants to build a crate to hold record albums. The expression $2x^2 - 6x - 80$ represents the volume of the crate. Factor the expression.

19. **BRIDGE ENGINEERING** A suspension bridge is a bridge in which the deck is supported by cables, with towers spaced throughout the span of the bridge. The height of a cable n inches above the deck measured at distance d in yards from the first tower is given by $d^2 - 36d + 324$. Factor the expression.

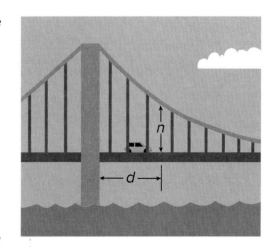

20. **FINANCE** The break-even point for a business occurs when the revenues equal the cost. A local children's museum studied their costs and revenues from paid admission. They found that their break-even point is given by the expression $2h^2 - 2h - 24$, where h is the number of hours the museum is open per day. Factor the expression.

Lesson 10-6 • Factoring Quadratic Trinomials

Examples 6 and 7

Factor each polynomial, if possible. If the polynomial cannot be factored using integers, write *prime*.

21. $5x^2 + 34x + 24$

22. $2x^2 + 19x + 24$

23. $4x^2 + 22x + 10$

24. $4x^2 + 38x + 70$

25. $2x^2 - 3x - 9$

26. $4x^2 - 13x + 10$

27. $2x^2 + 3x + 6$

28. $5x^2 + 3x + 4$

29. $12x^2 + 69x + 45$

30. $4x^2 - 5x + 7$

31. $3x^2 - 8x + 15$

32. $5x^2 + 23x + 24$

33. $2x^2 + 3x - 6$

34. $2t^2 + 9t - 5$

35. $2y^2 + y - 1$

36. $4h^2 + 8h - 5$

Mixed Exercises

Factor each polynomial, if possible. If the polynomial cannot be factored using integers, write *prime*.

37. $n^2 + 3n - 18$

38. $x^2 + 2x - 8$

39. $r^2 + 4r - 12$

40. $x^2 - x - 12$

41. $w^2 - w - 6$

42. $y^2 - 6y + 8$

43. $t^2 - 15t + 56$

44. $-4 - 3m + m^2$

45. $2x^2 + 5x + 2$

46. $3n^2 + 5n + 2$

47. $3g^2 - 7g + 2$

48. $2t^2 - 11t + 15$

49. $4x^2 - 3x - 3$

50. $4b^2 + 15b - 4$

Factor each polynomial, if possible. If the polynomial cannot be factored using integers, write *prime*.

51. $9p^2 + 6p - 8$

52. $6q^2 - 13q + 6$

53. $a^2 - 10a + 21$

54. $x^2 + 2x - 15$

55. $2x^2 + 7x + 3$

56. $6x^2 + x + 2$

57. $x^2 + x - 20$

58. $x^2 - 6x - 7$

59. $p^2 - 10p + 21$

60. $5x^2 - 6x + 1$

61. $q^2 + 11qr + 18r^2$

62. $x^2 - 14xy - 51y^2$

63. $x^2 - 6xy + 5y^2$

64. $a^2 + 10ab - 39b^2$

65. $-6x^2 - 23x - 20$

66. $-4x^2 - 15x - 14$

67. $-5x^2 + 18x + 8$

68. $-6x^2 + 31x - 35$

69. $-4x^2 + 5x - 12$

70. $-12x^2 + x + 20$

71. MONUMENTS Susan is designing a pyramidal stone monument for a local park. The design specifications tell her that width of the base must be 5 feet less than the length and there must be 2 feet of space on each side of the pyramid. The expression $x^2 + 3x - 4$ represents that area that the pyramidal stone monument will require.

  a. Factor the expression that represents the area that the pyramidal stone monument will require.

  b. What does x represent?

72. PROJECTILES The height of a projectile in feet is given by $-16t^2 + vt + h_0$, where t is the time in seconds, v is the initial upward velocity in feet per second, and h_0 is the initial height in feet. A T-shirt is propelled from 32 feet above ground level into the air at an initial velocity of 16 feet per second.

  a. Write an expression to represent how much time the T-shirt is in the air.

  b. Factor the expression that represents the amount of time the T-shirt is in the air.

73. WHICH ONE DOESN'T BELONG? Which expression does not belong with the others? Explain your reasoning.

| $x^2 + 2x - 24$ | $x^2 + 11x + 24$ | $x^2 - 10x - 24$ | $x^2 + 12x + 24$ |

74. FIND THE ERROR Jamaall and Charles have factored $x^2 + 6x - 16$. Is either of them correct? Explain your reasoning.

Jamaall	Charles
$x^2 + 6x - 16 = (x + 2)(x - 8)$	$x^2 + 6x - 16 = (x - 2)(x + 8)$

ANALYZE Find all values of k so that each polynomial can be factored using integers.

75. $x^2 + kx - 19$

76. $x^2 + kx + 14$

77. $x^2 - 8x + k, k > 0$

78. $x^2 - 5x + k, k > 0$

79. ANALYZE For any factorable trinomial, $x^2 + bx + c$, will the absolute value of b *sometimes*, *always*, or *never* be less than the absolute value of c? Justify your argument.

80. CREATE Give an example of a trinomial that can be factored using the factoring techniques presented in this lesson. Then factor the trinomial.

81. PERSEVERE Factor $(4y - 5)^2 + 3(4y - 5) - 70$.

82. WRITE Explain how to factor trinomials of the form $x^2 + bx + c$ and how to determine the signs of the factors of c.

83. ANALYZE A square has an area of $9x^2 + 30xy + 25y^2$ square inches. The dimensions are binomials with positive integer coefficients. What is the perimeter of the square? Explain.

84. PERSEVERE Find all values of k so that $2x^2 + kx + 12$ can be factored as two binomials using integers.

85. WRITE Explain how to determine which values should be chosen for m and p when factoring a polynomial of the form $ax^2 + bx + c$.

Lesson 10-7

Factoring Special Products

Explore Using Algebra Tiles to Factor Differences of Squares

Online Activity Use algebra tiles to complete the Explore.

INQUIRY How is factoring a difference of squares related to the product of a sum and a difference?

Learn Factoring Differences of Squares

The product of the sum and difference of two quantities results in a **difference of two squares.** So, the factored form of a difference of squares is the product of the sum and difference of two quantities.

Key Concept • Factoring Differences of Squares

Symbols	$a^2 - b^2 = (a + b)(a - b)$ or $(a - b)(a + b)$
Examples	$x^2 - 9 = (x + 3)(x - 3)$ or $(x - 3)(x + 3)$
	$4u^2 - 1 = (2u + 1)(2u - 1)$ or $(2u - 1)(2u + 1)$

Example 1 Factor Differences of Squares

Factor each polynomial.

a. $81v^2 - 64w^2$

$\begin{aligned} 81v^2 - 64w^2 &= (9v)^2 - (8w)^2 &&\text{Write in the form } a^2 - b^2. \\ &= (9v + 8w)(9v - 8w) &&\text{Factor the difference of squares.} \end{aligned}$

b. $1 - 144q^2$

$\begin{aligned} 1 - 144q^2 &= (1)^2 - (12q)^2 &&\text{Write in the form } a^2 - b^2. \\ &= (1 + 12q)(1 - 12q) &&\text{Factor the difference of squares.} \end{aligned}$

Check

Factor each polynomial.

a. $25x^2 - 64y^2$

b. $196 - g^4$

Today's Goals
- Factor binomials that are differences of squares.
- Factor trinomials that are perfect squares.

Today's Vocabulary
difference of two squares
perfect square trinomials

Talk About It!
How can you check that the factors of a polynomial are correct?

Go Online You can complete an Extra Example online.

Lesson 10-7 • Factoring Special Products **591**

Watch Out!

Sum of Squares The sum of squares cannot be factored; that is, $a^2 + b^2 \neq (a + b)(a + b)$. The sum of squares is a prime polynomial.

Example 2 Factor More Than Once

Factor $x^4 - 256$.

$$\begin{aligned} x^4 - 256 &= (x^2)^2 - (16)^2 & \text{Write in the form } a^2 - b^2. \\ &= (x^2 + 16)(x^2 - 16) & \text{Factor the difference of squares.} \\ &= (x^2 + 16)(x^2 - 4^2) & x^2 - 16 \text{ is also a difference of squares.} \\ &= (x^2 + 16)(x + 4)(x - 4) & \text{Factor the difference of squares.} \end{aligned}$$

Check

Factor each polynomial.

a. $81n^4 - 1$

b. $16x^8 - y^4$

Think About It!

What would the factors be if Rosita's family knew that they needed to tile a 1-meter square area near the door, but did not know the dimensions of their square living room?

🌐 Example 3 Use Factors to Find Area

AREA Rosita's family is buying carpet for their living room, which is 6 meters wide and 6 meters long. They plan to carpet the whole area except for a square area near the door, which will be tiled. Find the factors representing the area of the living room that will be carpeted.

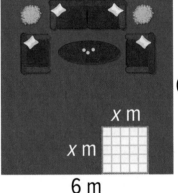

The area of the living room is 6 · 6 or 36 square meters, and the tiled area is $x \cdot x$ or x^2 square meters. So, the carpet will cover an area of $36 - x^2$.

$$\begin{aligned} 36 - x^2 &= (6)^2 - (x)^2 & \text{Write in the form } a^2 - b^2. \\ &= (6 + x)(6 - x) & \text{Factor the difference of squares.} \end{aligned}$$

Check

VOLUME A rectangular solid has a volume of $x^4 - 16$. Factor the polynomial to determine the dimensions of the rectangular solid.

A. $x^2(x + 4)(x - 4)$

B. $(x^2 + 4)(x + 2)(x - 2)$

C. $(x^2 - 4)(x + 2)(x + 2)$

D. $(x^3 + 4)(x - 2)(x - 2)$

🔵 **Go Online** You can complete an Extra Example online.

Learn Factoring Perfect Squares

Squares of binomials, such as $(a + b)^2$ and $(a - b)^2$, have special products called **perfect square trinomials.**

$(a + b)^2 = (a + b)(a + b)$
$\qquad = a^2 + ab + ab + b^2$
$\qquad = a^2 + 2ab + b^2$

$(a - b)^2 = (a - b)(a - b)$
$\qquad = a^2 - ab - ab + b^2$
$\qquad = a^2 - 2ab + b^2$

For a trinomial to be factorable as a perfect square, the following must be true:

- The first term is a perfect square.
- The last term is a perfect square.
- The middle term is two times the product of the square roots of the first and last terms.

Key Concept • Factoring Perfect Square Trinomials

Symbols $a^2 + 2ab + b^2 = (a + b)(a + b) = (a + b)^2$

$\qquad\quad a^2 - 2ab + b^2 = (a - b)(a - b) = (a - b)^2$

Examples $4x^2 - 28x + 49 = (2x - 7)(2x - 7) = (2x - 7)^2$

$\qquad\quad x^2 + 10x + 25 = (x + 5)(x + 5) = (x + 5)^2$

Example 4 Identify a Perfect Square Trinomial

Determine whether $4j^2 + 8j + 16$ is a perfect square trinomial. If so, factor it.

Is the first term a perfect square?   Yes, $4j^2 = (2j)^2$

Is the second term equal to $2(2j)(4)$?   No, $8j \neq 2(2j)(4)$

Is the last term a perfect square?   Yes, $16 = 4^2$

Because this trinomial does not satisfy all conditions, it is not a perfect square trinomial.

> **Think About It!**
> To make $?j^2 + 8j + 16$ a perfect square trinomial, what would the coefficient of the first term need to be?
>
> To make $4j^2 + ?j + 16$ a perfect square trinomial, what would the coefficient of the middle term need to be?
>
> To make $4j^2 + 8j + ?$ a perfect square trinomial, what would the constant term need to be?

Go Online You can complete an Extra Example online.

> **Study Tip**
> **Negative Constants**
> Since perfect square trinomials always have a positive constant, check the constant first. If it is negative, the trinomial will not be a perfect square.

Example 5 Recognize and Factor a Perfect Square Trinomial

Determine whether $36h^2 - 12h + 1$ is a perfect square trinomial. If so, factor it.

Is $36h^2$ a perfect square? yes

Is $-12h$ equal to $-2(6h)(1)$? yes

Is 1 a perfect square? yes

This trinomial satisfies all the conditions for a perfect square trinomial.

$36h^2 - 12h + 1 = (6h)^2 - 2(6h)(1) + (1)^2$ Write as $a^2 - 2ab + b^2$.

$\qquad\qquad\qquad\quad = (6h - 1)^2$ Factor using the pattern.

Check

If the trinomial is a perfect square trinomial, factor it. If not, write *not a perfect square trinomial*.

$36x^2 - 36x + 9$

$36x^2 - 18x + 9$

$36x^2 + 36x + 9$

> Go Online to practice what you've learned about factoring quadratic expressions in the Put It All Together over Lessons 10-6 and 10-7.

Go Online You can complete an Extra Example online.

Practice

Go Online You can complete your homework online.

Examples 1 and 2
Factor each polynomial.

1. $q^2 - 121$
2. $r^4 - k^4$
3. $w^4 - 625$
4. $r^2 - 9t^2$
5. $h^4 - 256$
6. $2x^3 - x^2 - 162x + 81$
7. $x^2 - 4y^2$
8. $3c^3 + 2c^2 - 147c - 98$
9. $f^3 + 2f^2 - 64f - 128$
10. $r^3 - 5r^2 - 100r + 500$
11. $3t^3 - 7t^2 - 3t + 7$
12. $a^2 - 49$
13. $4m^3 + 9m^2 - 36m - 81$
14. $3x^3 + x^2 - 75x - 25$

Example 3

15. TICKETING A ticketing company for sporting events analyzes the ticket purchasing patterns. The expression $9a^2 - 4b^2$ is developed to help officials calculate the likely number of people who will buy tickets for a certain sporting event. Factor the expression.

16. BASKETBALL COURT A half-court basketball court is a square of pavement with an area represented by $x^2 - 25$. Factor the expression.

17. DECORATING Marvin saw a rug in a store that he would like to purchase. It has an area represented by the expression shown on the rug. He cannot remember the length and width, but he remembers that the length and the width were the same.

    a. Factor the expression that represents the area of the rug.
    b. What do the factors in the factored expression represent?

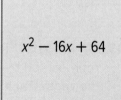

$x^2 - 16x + 64$

Examples 4 and 5
Determine whether each trinomial is a perfect square trinomial. Write *yes* or *no*. If so, factor it.

18. $4x^2 - 42x + 110$
19. $16x^2 - 56x + 49$
20. $81x^2 - 90x + 25$
21. $x^2 + 26x + 168$

Lesson 10-7 • Factoring Special Products 595

Mixed Exercises

Factor each polynomial, if possible. If the polynomial cannot be factored using integers, write *prime*.

22. $36t^2 - 24t + 4$

23. $4h^2 - 56$

24. $17a^2 - 24ab$

25. $q^2 - 14q + 36$

26. $y^2 + 24y + 144$

27. $6d^2 - 96$

28. $1 - 49d^2$

29. $-16 + p^2$

30. $k^2 + 25$

31. $36 - 100w^2$

32. $64m^2 - 9y^2$

33. $4h^2 - 25g^2$

34. $x^3 + 3x^2 - 4x - 12$

35. $8x^2 - 72p^2$

36. $20q^2 - 5r^2$

37. $32a^2 - 50b^2$

38. $16b^2 - 100$

39. $49x^2 - 64y^2$

40. $3n^4 - 42n^3 + 147n^2$

41. $8m^3 - 24m^2 + 18m$

42. **GARDEN DESIGN** Marren is planning to build a raised garden bed. The area of the rectangular plot can be represented by $x^2 - 49$. Factor the expression to determine the possible length and width of the garden bed.

43. **PARKING LOT** The area of a rectangular parking lot is represented by the expression $a^2 - 25$, where the length is longer than the width. Factor the expression to determine the possible dimensions of the length and width of the parking lot. If the length of the parking lot is 105 yards, what is the width of the parking lot?

44. **USE A SOURCE** Research the dimensions of the outside diameter and inside diameter of metal washers. Write an expression for the surface area of the top of a metal washer with outside diameter D and inside diameter d. Factor your expression. Then use your expression and the dimensions you researched to find the surface area of the top of a metal washer.

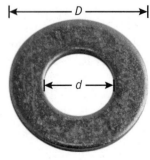

Determine whether each trinomial is a perfect square trinomial. Write *yes* or *no*. If so, factor it.

45. $m^2 - 6m + 9$

46. $r^2 + 4r + 4$

47. $g^2 - 14g + 49$

48. $2w^2 - 4w + 9$

49. $4d^2 - 4d + 1$

50. $9n^2 + 30n + 25$

51. $9z^2 - 6z + 1$

52. $36x^2 - 60x + 25$

53. $49r^2 + 14r + 4$

54. $a^2 + 14a + 49$

55. $t^2 - 18t + 81$

56. $4c^2 + 2cd + d^2$

57. **ARCHITECTURE** The drawing shows a triangular roof truss with a base measuring the same as its height. The expression $\frac{1}{2}x^2 - 98$ represents the area of the triangular roof truss, where x is the length of the base and the height. Factor the expression.

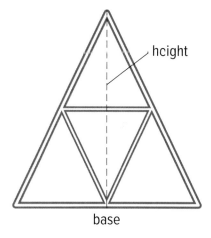

58. **MANUFACTURING** A company that manufactures cardboard boxes sells three sizes of boxes. The volume of the small box is represented by $n^3 + 4n^2 - 16n - 64$ in³. The volume of the medium box is represented by $n^3 + 6n^2 - 36n - 216$ in³. The volume of the large box is represented by $n^3 + 8n^2 - 64n - 512$ in³. The company wants to start selling an extra-large box. Predict the dimensions and the volume of the extra-large box.

   a. Find the dimensions of the small, medium, and large boxes. Explain the method you used.

   b. Look for a pattern in the areas. What pattern is represented by the areas of the small, medium, and large boxes? Use this pattern to predict the dimensions of the extra-large box.

   c. Use your prediction in **part b** to find the volume of the extra-large box.

59. **ANALYZE** Paul suggests that to factor $x^4 - 81$, you can use difference of squares twice. He says that the result will have the same factors as $x^2 - 9$. Prove or disprove his statement.

Lesson 10-7 • Factoring Special Products **597**

60. **REASONING** Debony claims the expression $9x^2 + 50x + 25$ is a perfect square trinomial. Is she correct? If she is incorrect, show how the expression can be changed so that it is a perfect square.

61. **STRUCTURE** Find a value for m that will make the expression $4x^4 - 44x^2 + m$ a perfect square. Then use the value to factor the expression completely.

62. **FIND THE ERROR** Elizabeth and Lorenzo are factoring an expression. Is either of them correct? Explain your reasoning.

Elizabeth	Lorenzo
$16x^4 - 25y^2 =$	$16x^4 - 25y^2 =$
$(4x - 5y)(4x + 5y)$	$(4x^2 - 5y)(4x^2 + 5y)$

63. **PERSEVERE** Factor and simplify $9 - (k + 3)^2$, a difference of squares.

64. **ANALYZE** Write and factor a binomial that is the difference of two perfect squares and that has a greatest common factor of $5mk$.

65. **ANALYZE** Determine whether the following statement is *true* or *false*. Give an example or counterexample to justify your answer.

    All binomials that have a perfect square in each of the two terms can be factored.

66. **CREATE** Write a binomial in which the difference of squares pattern must be repeated to factor it completely. Then factor the binomial.

67. **WRITE** Describe why the difference of squares has no middle term.

68. **WHICH ONE DOESN'T BELONG?** Identify the trinomial that does not belong. Explain.

| $4x^2 - 36x + 81$ | $25x^2 + 10x + 1$ | $4x^2 + 10x + 4$ | $9x^2 - 24x + 16$ |

69. **WRITE** Explain how to determine whether a trinomial is a perfect square trinomial.

70. **PERSEVERE** Use the difference of squares to factor and simplify the expression $121x^2y^6z^4 - 16y^2z^2$.

Module 10 • Polynomials
Review

Essential Question
How can you perform operations on polynomials and use them to represent real-world situations?

Module Summary

Lesson 10-1
Adding and Subtracting Polynomials
- The degree of a monomial is the sum of the exponents of all its variables.
- The degree of a polynomial is the greatest degree of any term in the polynomial.
- When adding polynomials, you can group like terms by using a horizontal or vertical format.
- You can subtract a polynomial by adding its additive inverse.
- Adding or subtracting polynomials results in a polynomial, so the set of polynomials is closed under addition and subtraction.

Lessons 10-2 through 10-4
Multiplying Polynomials
- To find the product of a monomial and a binomial, use the Distributive Property.
- The FOIL method helps to multiply binomials: To multiply two binomials, find the sum of the products of *F* the *First Terms*, *O* the *Outer terms*, *I* the *Inner terms*, and *L* the *Last terms*.
- Square of a sum: $(a + b)^2 = (a + b)(a + b)$
  $= a^2 + 2ab + b^2$
- Square of a difference: $(a - b)^2 =$
  $(a - b)(a - b) = a^2 - 2ab + b^2$
- The product of $a + b$ and $a - b$ is the square of a minus the square of b: $(a + b)(a - b) =$
  $a^2 - ab + ab - b^2 = a^2 - b^2$

Lessons 10-5 through 10-7
Factoring Polynomials
- By using the Distributive Property in reverse, you can factor a polynomial as the product of a monomial and a polynomial.
- To factor trinomials in the form $x^2 + bx + c$, find two integers, m and p, with a sum of b and a product of c. Then write $x^2 + bx + c$ as $(x + m)(x + p)$.
- To factor trinomials in the form $ax^2 + bx + c$, find two integers, m and p, with a sum of b and a product of ac. Then write $ax^2 + bx + c$ as $ax^2 + mx + px + c$, and factor by grouping.
- The factored form of a difference of squares is the product of the sum and difference of two quantities.
  $a^2 - b^2 = (a + b)(a - b)$ or $(a - b)(a + b)$
- Squares of binomials, such as $(a + b)^2$ and $(a - b)^2$, have special products called perfect square trinomials.
  $a^2 + 2ab + b^2 = (a + b)(a + b) = (a + b)^2$
  $a^2 - 2ab + b^2 = (a - b)(a - b) = (a - b)^2$

Study Organizer

Foldables
Use your Foldable to review this module. Working with a partner can be helpful. Ask for clarification of concepts as needed.

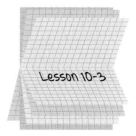
Lesson 10-3

Test Practice

1. **MULTI-SELECT** Select all of the expressions that are polynomials. (Lesson 10-1)

   A. $4xy - 3x$

   B. $4ab + \frac{5}{x^2}$

   C. $45 + 3d^2 + d - w^3$

   D. 12

   E. $\frac{x^{-3} y^2}{5}$

2. **OPEN RESPONSE** The perimeter of this triangular plot of land can be represented by $3x^2 + 10x + 20$. (Lesson 10-1)

   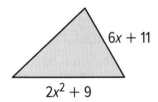

   Write a polynomial that represents the measure of the third side of the plot of land.

3. **MULTIPLE CHOICE** Find the difference. $(5d - 8) - (-2d + 3)$ (Lesson 10-1)

   A. $3d - 5$

   B. $7d - 11$

   C. $7d - 5$

   D. $7d + 11$

4. **MULTIPLE CHOICE** Find the sum. $(x^2 + 3x - 9) + (4x^2 - 6x + 1)$ (Lesson 10-1)

   A. $4x^2 - 3x - 8$

   B. $5x^2 - 3x - 8$

   C. $5x^2 + 9x - 10$

   D. $4x^2 + 9x - 10$

5. **OPEN RESPONSE** Mr. Soto is installing a new trapezoid shaped window. He plans to cover the window with a protective coating that prevents UV rays from entering the house. The height h of the window is 24 inches.

   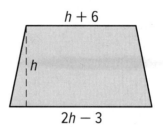

   The formula for the area of a trapezoid is $A = \frac{1}{2}h(b_1 + b_2)$ where h is the height and b_1 and b_2 are the bases of the trapezoid. How many square inches of protective coating will Mr. Soto need? (Lesson 10-2)

6. **OPEN RESPONSE** Use the Distributive Property to simplify $-3(g^2 - 4g + 1)$. (Lesson 10-2)

7. **MULTIPLE CHOICE** Find the product of $(2x + 7)$ and $(x - 5)$. (Lesson 10-3)

   A. $3x + 2$

   B. $2x^2 - 3x - 35$

   C. $2x^2 - 10x - 35$

   D. $2x^2 - 35$

8. **OPEN RESPONSE** Saurabh is designing a rectangular vegetable garden with a stone path around it. The total area of the path is 296 square feet. Use the diagram to determine the value of x. (Lesson 10-3)

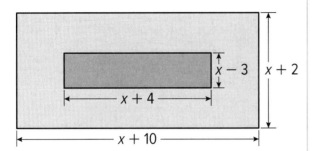

9. **OPEN RESPONSE** Find the product. $(2r - 3)(r^2 - 5r + 1)$ (Lesson 10-3)

10. **OPEN RESPONSE** *True* or *false*: $(8k^2 + 3k - 1)(k^2 - k + 1)$ is equivalent to $8k^4 - 5k^3 + 4k^2 + 4k - 1$. (Lesson 10-3)

11. **MULTIPLE CHOICE** Find $(8a + 3b)^2$. (Lesson 10-4)

   A. $73a^2b^2 + 48ab$

   B. $64a^2 + 48ab + 9b^2$

   C. $64a^2 + 9b^2$

   D. $32a^2 + 22ab + 12b^2$

12. **OPEN RESPONSE** Find the product. $(6h + 7)(6h - 7)$ (Lesson 10-4)

13. **OPEN RESPONSE** Describe a general rule for finding the square of a sum. (Lesson 10-4)

14. **MULTIPLE CHOICE** Kala is making a tile design for her kitchen floor. Each tile has sides that are 3 inches less than twice the side length of the smaller square inside the design. (Lesson 10-4)

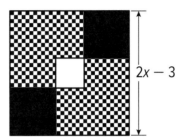

Select the polynomial that represents the area of the tile.

   A. $2x^2 - 3x$

   B. $4x^2 - 12x + 9$

   C. $4x^2 + 12x + 9$

   D. $4x^2 - 9$

15. **OPEN RESPONSE** Factor $-10xy - 15x + 12y + 18$. (Lesson 10-5)

16. **OPEN RESPONSE** Factor $36x^3 - 24x^2$. (Lesson 10-5)

17. MULTI-SELECT A golf ball is hit with a velocity of 128 feet per second. The expression $128t - 16t^2$ represents the height of the ball after t seconds. Select all that are equivalent to the expression. (Lesson 10-5)

A. $16t(8 - t)$

B. $-16t(t - 8)$

C. $t(t - 8)(8 - t)$

D. $8t(7 - 2t)$

E. $8(-2t + 16)$

18. OPEN RESPONSE Factor $x^2 + 8x + 15$. (Lesson 10-6)

19. MULTIPLE CHOICE Mrs. Torres wants to add x feet onto the length and width of an existing rectangular patio. The new patio would have an area of $x^2 + 14x + 48$ square feet. (Lesson 10-6)

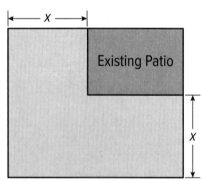

What are the dimensions of the existing patio?

A. 14 ft by 48 ft

B. 6 ft by 8 ft

C. 7 ft by 24 ft

D. 4 ft by 12 ft

20. OPEN RESPONSE Factor $9x^2 + 9x + 2$. (Lesson 10-6)

21. OPEN RESPONSE Identify if each polynomial is a difference of squares. Write *yes* or *no*. (Lesson 10-7)

a) $16y^2 - 25b^2$

b) $m^2 - 100n^2$

c) $64r^2 - 225$

d) $49z^2 - 36a^3$

e) $-4y^2 - k^2$

22. MULTI-SELECT Select all of the perfect square trinomials. (Lesson 10-7)

A. $49x^2 + 112x + 64$

B. $16x^2 - 24x + 9$

C. $49x^2 + 30x + 64$

D. $9x^2 - 6x + 16$

E. $x^2y^2 - 10xy^2 + 25y^2$

23. MULTIPLE CHOICE A square piece of cloth has an area of $4y^2 - 28y + 49$ square meters. Find the length of each side. (Lesson 10-7)

A. $(2y + 7)$ m

B. $(2y - 7)$ m

C. $(4y - 49)$ m

D. $(2y - 14)$ m

Module 11
Quadratic Functions

e Essential Question
Why is it helpful to have different methods to analyze quadratic functions and solve quadratic equations?

What Will You Learn?
How much do you already know about each topic **before** starting this module?

KEY
👎 — I don't know.   ✋ — I've heard of it.   👍 — I know it!

	Before 👎	👋	👍	After 👎	👋	👍
graph quadratic equations using a table						
graph quadratic equations using key features						
transform quadratic functions						
solve systems of linear and quadratic equations						
solve quadratic equations by factoring						
solve quadratic equations by completing the square						
find maximum and minimum values						
solve quadratic equations using the quadratic formula						
fit a quadratic function to data						
combine functions						

Foldables Make this Foldable to help you organize your notes about quadratic functions. Begin with a sheet of notebook paper.

1. **Fold** the sheet of paper along the length so that the edge of the paper aligns with the margin rule of the paper.
2. **Fold** the sheet twice widthwise to form four sections.
3. **Unfold** the sheet and cut along the folds on the front flap only.
4. **Label** each section as shown.

What Vocabulary Will You Learn?

- axis of symmetry
- coefficient of determination
- completing the square
- curve fitting
- discriminant
- double root
- maximum
- minimum
- parabola
- quadratic equation
- quadratic function
- standard form of a quadratic function
- vertex form

Are You Ready?

Complete the Quick Review to see if you are ready to start this module.
Then complete the Quick Check.

Quick Review

Example 1

Describe the translation in $g(x) = 5^x + 3$ as it relates to the graph of the parent function $f(x) = 5^x$.

A constant term is being added to the function, so it is a translation of that many units up.

The graph of $g(x) = 5^x + 3$ is the translation of the graph of the parent function $f(x) = 5^x$ 3 units up.

Determine whether $x^2 - 10x + 25$ is a perfect square trinomial. Write *yes* or *no*. If so, factor it.

Is the first term, x^2, a perfect square? yes

Is the last term, 25, a perfect square? yes

Is the middle term equal to the opposite of twice the product of the first and last terms?

yes; $2(5)(x) = 10x$ and the opposite of $10x$ is $-10x$

$x^2 - 10x + 25 = (x - 5)^2$

Quick Check

Describe the translation in $g(x)$ as it relates to the graph of the parent function $f(x) = 2^x$.

1. $g(x) = 2^{x-4}$
2. $g(x) = -2^x$
3. $g(x) = 2^x - 1$
4. $g(x) = 0.25 \cdot 2^x$

Determine whether each trinomial is a perfect square trinomial. Write *yes* or *no*. If so, factor it.

5. $a^2 + 12a + 36$
6. $w^2 + 5w + 25$
7. $m^2 - 22m + 121$
8. $5t^2 + 12t + 100$

How Did You Do?

Which exercises did you answer correctly in the Quick Check?

Lesson 11-1

Graphing Quadratic Functions

Learn Analyzing Graphs of Quadratic Functions

A second-degree polynomial, for example $x^2 + 2x - 8$, is a quadratic polynomial, which has a related quadratic function $f(x) = x^2 + 2x - 8$. A **quadratic function** has an equation of the form $y = ax^2 + bx + c$, where $a \neq 0$. The graph of a quadratic function is called a **parabola**.

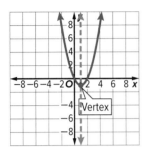

Parabolas are symmetric about a central line called the **axis of symmetry**. Every point on the parabola to the left of the axis of symmetry has a corresponding point on the right half.

The axis of symmetry intersects a parabola at the vertex. The vertex is either the lowest point or the highest point on a parabola.

When $a > 0$, the graph of $y = ax^2 + bx + c$ opens up. In this case, the graph has a **minimum** at the lowest point. When $a < 0$, the graph opens down. In this case, the graph has a **maximum** at the highest point.

The leading coefficient determines the end behavior of a quadratic function.

Example 1 Identify Characteristics: Graph with x-intercept

Identify the axis of symmetry, the vertex, and the y-intercept of the graph. Then describe the end behavior.

The equation of the axis of symmetry is $x = -2$.

The vertex is located at the maximum point, $(-2, 8)$.

The parabola crosses the y-axis at $(0, 4)$, so the y-intercept is 4.

As x increases or decreases, y decreases.

Today's Goals
- Analyze graphs of quadratic functions.
- Graph quadratic functions by using key features and tables.
- Use graphing calculators to analyze key features of quadratic functions.

Today's Vocabulary
quadratic function
parabola
axis of symmetry
minimum
maximum
end behavior
standard form of a quadratic function

💭 Think About It!
What do you notice about the equation of the axis of symmetry and the x-coordinate of the vertex?

Lesson 11-1 • Graphing Quadratic Functions **605**

Check

Identify the axis of symmetry, the vertex, and the y-intercept of the graph. Then describe the end behavior.

axis of symmetry: $x =$ ___?___

vertex: (___?___ , ___?___)

y-intercept:

end behavior:

    As x increases, y ___?___.

    As x decreases, y ___?___.

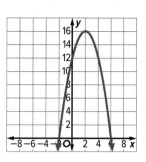

Example 2 Identify Characteristics: Graph with No x-intercept

Identify the axis of symmetry, the vertex, and the y-intercept of the graph. Then describe the end behavior of the function.

The equation of the axis of symmetry is $x = 0$.

The vertex is located at the minimum point, (0, 3).

The parabola crosses the y-axis at (0, 3), so the y-intercept is 3.

The parabola opens up.

As x increases, y increases.

As x decreases, y increases.

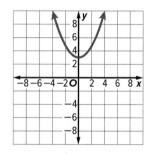

Check

Identify the axis of symmetry, the vertex, and the y-intercept of the graph. Then describe the end behavior of the function.

axis of symmetry: $x =$ ___?___

vertex: (___?___ , ___?___)

y-intercept:

end behavior:

    As x increases, y ___?___.

    As x decreases, y ___?___.

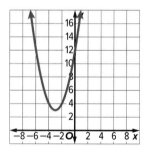

Go Online You can complete an Extra Example online.

Explore Graphing Parabolas

Online Activity Use graphing technology to complete the Explore.

INQUIRY How can you use the equation of a quadratic function to visualize its graph?

Learn Graphing Quadratic Functions

Key Concept • Standard Form of a Quadratic Function

Words
- The **standard form of a quadratic function** is $f(x) = ax^2 + bx + c$, where a, b, and c are integers and $a \neq 0$.

Examples
- In $f(x) = 3x^2 - 2x + 9$, $a = 3$, $b = -2$, and $c = 9$.
- In $f(x) = -16x^2$, $a = -16$, $b = 0$, and $c = 0$.

Key Concept • Graphing Quadratic Functions

Step 1 Find the equation of the axis of symmetry.

Step 2 Find the vertex, and determine whether it is a maximum or minimum.

Step 3 Find the y-intercept.

Step 4 Use symmetry to find additional points on the graph, if necessary.

Step 5 Connect the points with a smooth curve.

Example 3 Graph a Quadratic Function by Using Key Features

Graph $f(x) = x^2 + 2x - 6$.

Step 1 Find the axis of symmetry.

Use the formula to find the equation of the axis of symmetry.

$x = -\dfrac{b}{2a}$ Equation of the axis of symmetry

$x = -\dfrac{2}{2(1)} = -1$ $a = 1$ and $b = 2$

Step 2 Find the vertex.

Use the value for the axis of symmetry as the x-coordinate of the vertex. Find the y-coordinate using the original equation.

$f(x) = x^2 + 2x - 6$ Original equation

$= (-1)^2 + 2(-1) - 6$ $x = -1$

$= -7$ Simplify.

The vertex lies at $(-1, -7)$. Because a is positive, the graph opens up. So the vertex is a minimum.

(continued on the next page)

Study Tip

Open and Closed Intervals The interval at which a function increases or decreases is always an open interval along the x-axis and is expressed using parentheses. A closed interval is expressed using brackets.

Talk About It!

Compare and contrast the graphs of exponential and quadratic functions.

Watch Out!

Minimum and Maximum Values Don't forget to find both coordinates of the vertex (x, y). The minimum or maximum value is the y-coordinate.

Step 3 Find the y-intercept.

$$f(x) = x^2 + 2x - 6 \quad \text{Original equation}$$
$$= (0)^2 + 2(0) - 6 \quad x = 0$$
$$= -6 \quad \text{Simplify.}$$

The y-intercept is −6.

Step 4 Find additional points.

Use the axis of symmetry to find the corresponding points.

Let $x = 2$.

$$f(x) = x^2 + 2x - 6 \quad \text{Original equation}$$
$$= (2)^2 + 2(2) - 6 \quad x = 2$$
$$= 2 \quad \text{Simplify.}$$

Step 5 Connect the points.

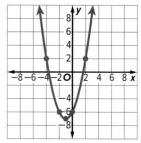

> **Study Tip**
> **Symmetry and Points**
> When locating points that are on opposite sides of the axis of symmetry, not only are the points equidistant from the axis of symmetry, they are also equidistant from the vertex.

Example 4 Graph a Quadratic Function by Using a Table

Use a table of values to graph $y = 2x^2 - 8x + 2$.

First find the x-coordinate of the vertex by using the equation for the axis of symmetry.

$$x = -\frac{b}{2a} \quad \text{Equation of the axis of symmetry}$$
$$x = \frac{-8}{-2(2)} = 2 \quad a = 2, b = -8$$

Complete the table.

x	0	1	2	3	4
y	2	−4	−6	−4	2

y-intercept (0, 2)

The vertex of this function is a minimum.

Graph the function.

The parabola extends to infinity, so the domain is all real numbers. The range is $\{y \mid y \geq -6\}$.

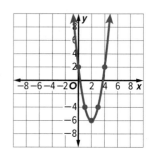

> **Go Online**
> You can watch a video to see how to graph quadratic functions by using a table.

> 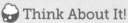 **Think About It!**
> Why do you think that the range is $\{y \mid y \geq -6\}$?

Go Online You can complete an Extra Example online.

Example 5 Use the Graph of a Quadratic Function

CATAPULT In 1304, Edward I built Warwolf, one of the largest catapults ever used in battle. It had the capacity to launch a 300-pound boulder over 900 feet. The height of a projectile launched from Warwolf can be modeled by the function $h(x) = -16x^2 + 96x + 40$, where $h(x)$ represents the height in feet of the projectile after x seconds. Graph the function. Interpret the key features of the graph in terms of the quantities.

Step 1 Find the axis of symmetry and vertex.

$x = -\dfrac{b}{2a}$ Equation of the axis of symmetry

$x = -\dfrac{96}{2(-16)} = 3$ $a = -16$ and $b = 96$

$h(x) = -16x^2 + 96x + 40$ Original equation

$ = -16(3)^2 + 96(3) + 40$ $x = 3$

$ = 184$ Simplify.

The vertex is at (3, 184).

Step 2 Complete the table by substituting each x-value.

x	0	1	2	3	4	5
h(x)	40	120	168	184	168	120

Step 3 Graph the function.

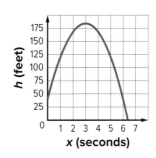

Step 4 Interpret the key features.

intercepts: $x = 6.4$, so the projectile landed about 6.4 seconds after it was launched.

vertex: (3, 184), so the projectile reached its maximum height of 184 feet 3 seconds after launch.

increasing: The function is increasing for $x < 3$, so it is gaining height up to 3 seconds after launch.

decreasing: The function is decreasing for $x > 3$, so it is falling 3 seconds after launch.

positive: intervals where the function represents the flight

end behavior: represents the projectile starting and returning to the ground

domain: all nonnegative numbers less than 6.4

range: $\{h(x) \mid 0 \leq h(x) \leq 184\}$

Watch Out!

Common Error A common error when determining the axis of symmetry of a parabola is to forget the negative sign in the equation $x = -\dfrac{b}{2a}$. Remember that if your parabola opens down, a is negative and, therefore, $2a$ will also be negative.

Study Tip

Labeling Axes When modeling a real-world quadratic function on the coordinate plane, it is important to appropriately label the axes using the given information. In this case, the independent variable is height in feet h; therefore, the y-axis is labeled h (feet).

> **Go Online** You can watch a video to see how to graph and analyze a quadratic function on a graphing calculator.

> **Go Online** to see how to use a graphing calculator with this example.

Math History Minute

During the 19th century, several designers of the Brooklyn Bridge were unable to complete it. But **Emily Warren Roebling** (1843–1903) supervised the bridge's construction until its completion. The bridge was the first major suspension bridge constructed in the U.S. Its cables form a catenary, which approximates a parabola.

Learn Analyzing Key Features of Quadratic Functions

You can use a graphing calculator to analyze key features of quadratic functions by graphing or by using the table feature.

Example 6 Interpret the Graph of a Quadratic Function

HORSES In 1949, Huaso, ridden by Alberto Larraguibel Morales, set the world record for the highest jump by a horse when he cleared a 2.47-meter-high obstacle. His path can be approximately modeled by $f(x) = -0.418x^2 + 2.033x$, where $f(x)$ is the height above the ground in meters and x is the distance from the point of take-off in meters. Find and interpret the key features of the path of Huaso's jump.

Step 1 Graph the function.

The graph only exists in Quadrant I in the context of the situation.

Step 2 Find the vertex.

Use the maximum feature from the CALC menu to find the vertex.

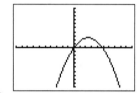

The vertex is located at about (2.43, 2.47). This represents Huaso's maximum height of 2.43 meters when he was 2.47 meters from the point of take-off.

Step 3 Find the axis of symmetry.

The axis of symmetry is located at $x = 2.43$. This means that Huaso's path before he is 2.43 meters from the point of take-off is the same as after he reaches 2.43 meters.

Step 4 Find the y-intercept.

Use the value feature from the CALC menu to find the y-intercept. Enter 0 for x. The y-intercept is located at (0, 0), so the y-intercept is 0. This means that when Huaso took off, he was 0 meters high and 0 meters from the point of take-off.

Step 5 Find the zeros.

Use the zero feature from the CALC menu to find the zeros, or x-intercepts.

The zeros are located at x equals 0 and x equals 4.86. These represent the points of take-off and landing, when Huaso's height above the ground was 0 meters.

Step 6 Examine the end behavior.

As x increases or decreases from the maximum, the value of y decreases until it reaches 0. This represents Huaso taking off and returning to the ground.

Practice

Go Online You can complete your homework online.

Examples 1 and 2

Identify the axis of symmetry, the vertex, and the y-intercept of each graph. Then describe the end behavior.

1.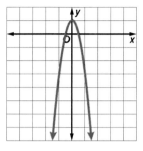

2.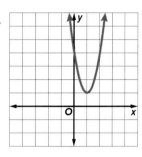

3.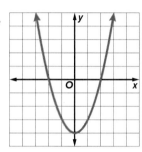

4.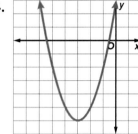

Example 3

Graph each function.

5. $y = -3x^2 + 6x - 4$

6. $y = -2x^2 - 4x - 3$

7. $y = -2x^2 - 8x + 2$

8. $y = x^2 + 6x - 6$

9. $y = x^2 - 2x + 2$

10. $y = 3x^2 - 12x + 5$

Example 4

Use a table of values to graph each function. State the domain and range.

11. $y = x^2 + 4x + 6$

12. $y = 2x^2 + 4x + 7$

13. $y = 2x^2 - 8x - 5$

14. $y = 3x^2 + 12x + 5$

15. $y = 3x^2 - 6x - 2$

16. $y = x^2 - 2x - 1$

Examples 5 and 6

17. **OLYMPICS** Olympics were held in 1896 and have been held every four years except 1916, 1940, and 1944. The winning height y in men's pole vault at any number Olympiad x can be approximated by the equation $y = 0.37x^2 + 4.3x + 126$. Copy and complete the table to estimate the winning pole vault heights in each of the Olympic Games. Round your answers to the nearest tenth. Graph the function. Interpret the key features of the graph in terms of the quantities.

Year	Olympiad (x)	Height (y inches)
1896	1	
1900	2	
1924	7	
1936	10	
1964	15	
2008	26	
2012	27	
2016	28	

18. **PHYSICS** Mrs. Capwell's physics class investigates what happens when a ball is given an initial push, rolls up, and then rolls back down an inclined plane. The class finds that $y = -x^2 + 6x$ accurately predicts the ball's position y after rolling x seconds. Graph the function. Interpret the key features of the graph in terms of the quantities.

19. **ARCHITECTURE** A hotel's main entrance is in the shape of a parabolic arch. The equation $y = -x^2 + 10x$ models the arch height y, in feet, for any distance x, in feet, from one side of the arch. Graph the function. Interpret the key features of the graph in terms of the quantities.

20. **SOFTBALL** Olympic softball gold medalist Michele Smith pitches a curveball with a speed of 64 feet per second. If she throws the ball straight upward at this speed, the ball's height h in feet after t seconds is given by $h = -16t^2 + 64t$. Graph the function. Interpret the key features of the graph in terms of the quantities.

Mixed Exercises

21. **CONSTRUCTION** Teddy is building the rectangular deck shown.

    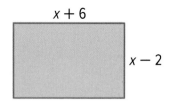

    a. Write an equation representing the area of the deck y.

    b. What is the equation of the axis of symmetry?

    c. Graph the equation and label its vertex.

22. **USE TOOLS** Write a quadratic function whose graph opens up. Write an exponential function. Graph both functions on the same coordinate plane. Compare and contrast the domains and ranges of the two functions and any symmetry of the two graphs.

23. **STRUCTURE** Consider the quadratic function $y = -x^2 - 2x + 2$.

    a. Find the equation for the axis of symmetry.

    b. Find the coordinates of the vertex and determine if it is a maximum or minimum.

    c. Graph the function.

Identify the axis of symmetry, the vertex, and the y-intercept of each graph. Then describe the end behavior.

24. $y = 2x^2 - 8x + 6$
25. $y = x^2 + 4x + 6$
26. $y = -3x^2 - 12x + 3$

27. **STRUCTURE** DeMarcus is sitting in a lifeguard's chair at the beach. He tosses a beanbag into the air and lets it land on the beach below him. The function $h(t) = -16t^2 + 8t + 12$ models the beanbag's height in feet, t seconds after DeMarcus tosses it into the air.

    a. Graph the function on a coordinate plane. Be sure to label the x- and y-axes and provide a scale for each axis.

    b. Find the intercepts of the graph. Describe what they represent in the context of the situation.

    c. What is the maximum value of the function, and what does it represent? At what time is the maximum reached?

    d. What is the value of $h(0.5)$? Describe its meaning in the context of the situation.

    e. State a reasonable domain for this function. What does it represent in the context of the situation?

28. **REGULARITY** Write the equation for a quadratic function that has a y-intercept of 0 and a minimum value at $x = 2$. Explain the steps you used to write the equation, and graph the function on a coordinate plane.

29. **CREATE** Write a quadratic function for which the graph has an axis of symmetry of $x = -\frac{3}{8}$. Summarize your steps.

30. **FIND THE ERROR** Noelia thinks that the parabolas represented by the graph and the description have the same axis of symmetry. Chase disagrees. Is either correct? Explain your reasoning.

    > a parabola that opens downward, passing through (0, 6) and having a vertex at (2, 2)

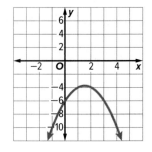

31. **PERSEVERE** Using the axis of symmetry, the y-intercept, and one x-intercept, write an equation for the graph shown.

32. **ANALYZE** The graph of a quadratic function has a vertex (2, 0). One point on the graph is (5, 9). Find another point on the graph. Explain how you found it.

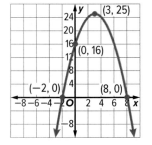

33. **CREATE** Describe a real-world situation that involves a quadratic equation. Explain what the vertex represents.

34. **ANALYZE** Provide a counterexample that shows that the following statement is false. Justify your argument.

    The vertex of a parabola is always the minimum of the graph.

35. **WRITE** Use tables and graphs to compare and contrast an exponential function $f(x) = ab^x + c$, where $a \neq 0$, $b > 0$, and $b \neq 1$, a quadratic function $g(x) = ax^2 + c$, and a linear function $h(x) = ax + c$. Include intercepts, portions of the graph where the functions are increasing, decreasing, positive, negative, relative maxima, relative minima, symmetries, and end behavior. Which function eventually exceeds the others?

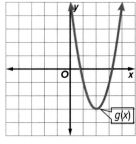

36. **ANALYZE** Consider $g(x)$ shown in the graph and $f(x) = -x^2 + 3x - 2$. Determine which function has the lesser minimum.

Lesson 11-2
Transformations of Quadratic Functions

Explore Transforming Quadratic Functions

Online Activity Use graphing technology to complete the Explore.

INQUIRY How does performing an operation on a quadratic function change its graph?

Today's Goals
- Apply translations to quadratic functions.
- Apply dilations to quadratic functions.
- Use transformations to identify quadratic functions from graphs and write equations of quadratic functions.

Today's Vocabulary
vertex form

Learn Translations of Quadratic Functions

Key Concept • Vertical Translations of Quadratic Functions

The graph of $g(x) = x^2 + k$ is the graph of $f(x) = x^2$ translated vertically.

If $k > 0$, the graph of $f(x)$ is translated k units up.

If $k < 0$, the graph of $f(x)$ is translated $|k|$ units down.

Key Concept • Horizontal Translations of Quadratic Functions

The graph of $g(x) = (x - h)^2 + k$ is the graph of $f(x) = x^2$ translated horizontally.

If $h > 0$, the graph of $f(x)$ is translated h units right.

If $h < 0$, the graph of $f(x)$ is translated $|h|$ units left.

Go Online You can watch a video to see how to translate functions.

Think About It!
Is the following statement *sometimes*, *always*, or *never* true? Explain.
The graph of $g(x) = x^2 + k$ has its vertex at the origin.

Example 1 Vertical Translations of Quadratic Functions

Describe the translation in $g(x) = x^2 - 4$ as it relates to the graph of the parent function.

Graph the parent function, $f(x) = x^2$, for quadratic functions.

Since $f(x) = x^2$, $g(x) = f(x) + k$ where $k = -4$.

The constant k is added to the function after it has been evaluated, so k affects the output, or y-values. The value of k is less than 0, so the graph of $f(x) = x^2$ is translated down 4 units.

$g(x) = x^2 - 4$ is the translation of the graph of the parent function 4 units down.

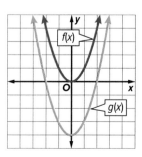

Check

Describe the transformation of $g(x) = x^2 - 8$ as it relates to the graph of the parent function.

The graph of $g(x) = x^2 - 8$ is the translation of the graph of the parent function __?__ units __?__.

Example 2 Horizontal Translations of Quadratic Functions

Describe the translation in $g(x) = (x + 2)^2$ as it relates to the graph of the parent function.

Graph the parent function, $f(x) = x^2$, for quadratic functions.

Since $f(x) = x^2$, $g(x) = f(x - h)$ where $h = -2$.

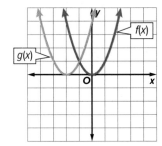

The constant h is subtracted from x before the function is performed, so h affects the input, or x-values. The value of h is less than 0, so the graph of $f(x) = x^2$ is translated $|-2|$ units left.

$g(x) = (x + 2)^2$ is the translation of the graph of the parent function 2 units left.

Check

Describe the transformation of $g(x) = (x - 6)^2$ as it relates to the graph of the parent function.

The graph of $g(x) = (x - 6)^2$ is the translation of the graph of the parent function __?__ units __?__.

Example 3 Multiple Translations of Quadratic Functions

Describe the translation in $g(x) = (x - 3)^2 + 1$ as it relates to the graph of the parent function.

Graph the parent function, $f(x) = x^2$, for quadratic functions.

Since $f(x) = x^2$, $g(x) = f(x - h) + k$ where $h = 3$ and $k = 1$.

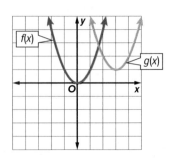

The constant h is subtracted from x before the function is performed and is greater than 0, so the graph of $f(x) = x^2$ is translated 3 units right.

The constant k is added to the function after it has been evaluated and is greater than 0, so the graph of $f(x) = x^2$ is also translated 1 unit up.

Go Online You can complete an Extra Example online.

> **Think About It!**
> What do you notice about the vertex of horizontally translated quadratic functions compared to the vertex of the parent function?

Check

Describe the translation in $g(x) = (x + 3)^2 - 5$ as it relates to the graph of the parent function.

The graph of $g(x) = (x + 3)^2 - 5$ is the translation of the graph of the parent function 3 units __?__ and 5 units __?__.

Graph $g(x) = (x + 3)^2 - 5$.

Learn Dilations of Quadratic Functions

Key Concept • Vertical Dilations of Quadratic Functions

The graph of $g(x) = ax^2$ is the graph of $f(x) = x^2$ stretched or compressed vertically by a factor of $|a|$.

If $|a| > 1$, the graph of $f(x)$ is stretched vertically away from the x-axis.

If $0 < |a| < 1$, the graph of $f(x)$ is compressed vertically toward the x-axis.

Key Concept • Horizontal Dilations of Quadratic Functions

The graph of $g(x) = (ax)^2$ is the graph of $f(x) = x^2$ stretched or compressed vertically by a factor of $\frac{1}{|a|}$.

If $|a| > 1$, the graph of $f(x)$ is compressed horizontally toward the y-axis.

If $0 < |a| < 1$, the graph of $f(x)$ is stretched horizontally away from the y-axis.

Go Online You can watch a video to see how to describe dilations of functions.

Example 4 Vertical Dilations of Quadratic Functions

Describe the dilation in $g(x) = 3x^2$ as it relates to the graph of the parent function.

Graph the parent function, $f(x) = x^2$, for quadratic functions.

Since $f(x) = x^2$, $g(x) = a \cdot f(x)$ where $a = 3$.

The function is multiplied by the positive constant a after it has been evaluated and $|a|$ is greater than 1, so the graph of $f(x) = x^2$ is stretched vertically by a factor of $|a|$, or 3.

$g(x) = 3x^2$ is a vertical stretch of the graph of the parent function.

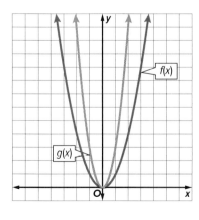

Study Tip

Vertical Dilations For a vertical dilation, if you multiply each y-coordinate of the function $f(x)$ by a, you'll get the corresponding y-coordinate of the function $g(x)$. For example, for the function above, the point (2, 4) on $f(x)$ corresponds to the point (2, 12) on $g(x)$. The y-coordinate of $f(x)$, 4, is multiplied by a, which is 3.

Check

Graph $g(x) = (2x)^2$.

Example 5 Horizontal Dilations of Quadratic Functions

Describe the dilation in $g(x) = \left(\frac{1}{2}x\right)^2$ as it relates to the graph of the parent function.

Graph the parent function, $f(x) = x^2$, for quadratic functions.

Since $f(x) = x^2$, $g(x) = f(a \cdot x)$ where $a = \frac{1}{2}$.

x is multiplied by the positive constant a before the function is performed and $|a|$ is between 0 and 1, so the graph of $f(x) = x^2$ is stretched horizontally by a factor of $\frac{1}{|a|}$, or 2.

$g(x) = \left(\frac{1}{2}x\right)^2$ is a horizontal stretch of the graph of the parent function.

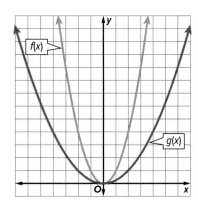

Check

Graph $g(x) = \frac{1}{4}x^2$.

> **Talk About It!**
> Could the graph of $g(x) = \left(\frac{1}{2}x\right)^2$ also be described by a vertical dilation? Justify your argument.

> **Study Tip**
> **Horizontal Dilations** For a horizontal dilation, if you multiply each x-coordinate of the function $f(x)$ by $\frac{1}{|a|}$, you'll get the corresponding x-coordinate of the function $g(x)$. For example, for the function above, the point $(-1, 1)$ on $f(x)$ corresponds to the point $(-2, 1)$ on $g(x)$. The x-coordinate of $f(x)$, -1, is multiplied by $\frac{1}{|a|}$.

Go Online You can complete an Extra Example online.

Learn Reflections of Quadratic Functions

Key Concept • Reflections of Quadratic Functions Across the x-axis

The graph of $-f(x)$ is the reflection of the graph of $f(x) = x^2$ across the x-axis.

Key Concept • Reflections of Quadratic Functions Across the y-axis

The graph of $f(-x)$ is the reflection of the graph of $f(x) = x^2$ across the y-axis.

Example 6 Vertical Reflections of Quadratic Functions

Describe how the graph of $g(x) = -\frac{2}{3}x^2$ is related to the graph of the parent function.

Graph the parent function, $f(x) = x^2$, for quadratic functions.

Because $f(x) = x^2$, $g(x) = -1 \cdot a \cdot f(x)$ where $a = \frac{2}{3}$.

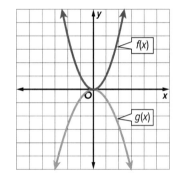

The function is multiplied by -1 and the constant a after it has been evaluated and $|a|$ is between 0 and 1, so the graph of $f(x)$ is compressed vertically and reflected across the x-axis.

$g(x) = -\frac{2}{3}x^2$ is the graph of the parent function compressed vertically and reflected across the x-axis.

Check

Graph $g(x) = -\frac{3}{5}x^2$.

Think About It!

How does a reflection of the parent function $f(x) = x^2$ across the x-axis affect the end behavior of the graph?

Go Online You can complete an Extra Example online.

Example 7 Horizontal Reflections of Quadratic Functions

Describe how the graph of $g(x) = (-3x)^2$ is related to the graph of the parent function.

Graph the parent function, $f(x) = x^2$, for quadratic functions.

Because $f(x) = x^2$, $g(x) = f(-1 \cdot a \cdot x)^2$ where $a = 3$.

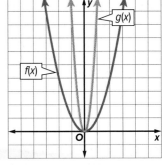

x is multiplied by -1 and the constant a before the function is performed and $|a|$ is greater than 1, so the graph of $f(x)$ is compressed horizontally and reflected across the y-axis.

$g(x) = (-3x)^2$ is the graph of the parent function compressed horizontally and reflected across the y-axis.

Check

Graph $g(x) = \left(-\frac{7}{2}x\right)^2$.

Learn Transformations of Quadratic Functions

A quadratic function in the form $f(x) = a(x - h)^2 + k$ is written in **vertex form**. Each constant in the equation affects the parent graph.

- The value of $|a|$ stretches or compresses (dilates) the parent graph.
- When the value of a is negative, the graph is reflected across the x-axis.
- The value of k shifts (translates) the parent graph up or down.
- The value of h shifts (translates) the parent graph left or right.

> **Go Online**
> You can watch a video to see how to use a graphing calculator to graph transformations of quadratic functions.

Example 8 Multiple Transformations of Quadratic Functions

Describe how the graph of $g(x) = -3(x - 1)^2 - 2$ is related to the graph of the parent function.

Graph the parent function, $f(x) = x^2$.

Since $f(x) = x^2$, $g(x) = a \cdot f(x - h) + k$ where $a = -3$.

$a < 0$ and $|a| > 1$, so the graph of $f(x) = x^2$ is compressed vertically and reflected across the x-axis and stretched vertically by a factor of $|a|$, or 3.

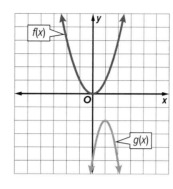

$h > 0$, so the graph is then translated h units right, or 1 unit right.

$k < 0$, so the graph is then translated $|k|$ units down, or 2 units down.

$g(x) = -3(x - 1)^2 - 2$ is the graph of the parent function stretched vertically, reflected across the x-axis, and translated 1 unit right and 2 units down.

Check
Graph $g(x) = -\frac{1}{3}(x + 4)^2 - 1$.

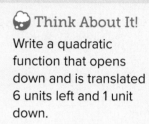

Think About It!
Write a quadratic function that opens down and is translated 6 units left and 1 unit down.

Example 9 Apply Transformations of Quadratic Functions

FOOTBALL Although they may appear flat, properly designed football fields arc to allow water to drain. Fields rise from each sideline to the center of the field, known as the crown, which should be between 1 and 1.5 feet in height. A cross section of a football field that is 160 feet wide and has a 1.5 foot crown can be modeled by $g(x) = -0.000234(x - 80)^2 + 1.5$, where $g(x)$ is the height of the field and x is the distance from the sideline in feet. Describe how $g(x)$ is related to the graph of $f(x) = x^2$.

$a < 0$, so the graph of $f(x) = x^2$ is a reflection across the x-axis.

$0 < |-0.000234| < 1$, so the graph of $f(x) = x^2$ is compressed vertically.

$80 > 0$, so the graph of $f(x) = x^2$ is translated 80 units right.

$1.5 > 0$, so the graph of $f(x) = x^2$ is translated 1.5 units up.

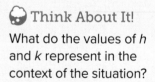

Think About It!
What do the values of h and k represent in the context of the situation?

Go Online You can complete an Extra Example online.

Check

BRIDGES The lower arch of the Sydney Harbor Bridge can be modeled by $g(x) = -0.0018(x - 251.5)^2 + 118$. Select all of the transformations that occur in $g(x)$ as it relates to the graph of $f(x) = x^2$.

A. vertical compression

B. translation down 251.5 units

C. translation up 118 units

D. reflection across the x-axis

E. vertical stretch

F. translation right 251.5 units

G. reflection across the y-axis

Example 10 Identify a Quadratic Equation from a Graph

Use the graph of the function to write its equation.

Step 1 Analyze the graph.

The graph appears to be narrower than the parent function, implying a vertical stretch, and is reflected across the x-axis. So, $a < 0$ and $|a| > 1$. The graph has also been shifted left and up from the parent graph. So, $h < 0$ and $k > 0$.

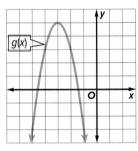

Step 2 Identify the translations.

The vertex is shifted 3 units left, so $h = -3$.

It is also shifted 5 units up, so $k = 5$.

Step 3 Identify the dilation and/or reflection.

The point $(-2, 3)$ lies on the graph. Substitute the coordinates in for x and y to solve for a.

$y = a(x + 3)^2 + 5$	Vertex form of the graph
$3 = a(-2 + 3)^2 + 5$	$(-2, 3) = (x, y)$
$3 = a(1)^2 + 5$	Add.
$3 = a + 5$	Evaluate 1^2.
$-2 = a$	Subtract 5 from each side.

So, the equation is $g(x) = -2(x + 3)^2 + 5$.

 Watch Out!

Choosing a Point When substituting for x and y in the equation, use a point other than the vertex.

Think About It! How does the equation you found compare to the prediction you made in Step 1?

 Go Online You can complete an Extra Example online.

Practice

Go Online You can complete your homework online.

Examples 1–3

Describe the translation in each function as it relates to the graph of the parent function.

1. $g(x) = -10 + x^2$
2. $g(x) = x^2 + 2$
3. $g(x) = (x - 1)^2$
4. $g(x) = x^2 - 8$
5. $g(x) = (x + 3)^2$
6. $g(x) = x^2 + 7$
7. $g(x) = (x + 2.5)^2$
8. $g(x) = (x + 5)^2 - 2$
9. $g(x) = 6 + x^2$
10. $g(x) = -4 + (x - 3)^2$
11. $g(x) = (x - 9.5)^2$
12. $g(x) = (x - 1.5)^2 + 3.5$

Examples 4 and 5

Describe the dilation in each function as it relates to the graph of the parent function.

13. $g(x) = 7x^2$
14. $g(x) = \frac{1}{5}x^2$
15. $g(x) = (4x)^2$
16. $g(x) = \left(\frac{1}{2}x\right)^2$
17. $g(x) = \left(\frac{5}{3}x\right)^2$
18. $g(x) = 5x^2$
19. $g(x) = \frac{3}{4}x^2$
20. $g(x) = \left(\frac{7}{8}x\right)^2$

Examples 6 and 7

Describe how the graph of each function is related to the graph of the parent function.

21. $g(x) = -6x^2$
22. $g(x) = (-9x)^2$
23. $g(x) = -\frac{1}{3}x^2$
24. $g(x) = \left(-\frac{2}{3}x\right)^2$
25. $g(x) = -2x^2$
26. $g(x) = \left(-\frac{6}{5}x\right)^2$

Lesson 11-2 • Transformations of Quadratic Functions 623

Example 8
Describe how the graph of each function is related to the graph of the parent function.

27. $h(x) = -7 - x^2$

28. $g(x) = 2(x - 3)^2 + 8$

29. $h(x) = 6 + \frac{2}{3}x^2$

30. $g(x) = -5 - \frac{4}{3}x^2$

31. $h(x) = 3 + \frac{5}{2}x^2$

32. $g(x) = -x^2 + 3$

Example 9

33. **SPRINGS** The potential energy stored in a spring is given by $U_s = \frac{1}{2}kx^2$, where k is a constant known as the spring constant, and x is the distance the spring is stretched or compressed from its initial position. How is the graph of the function for a spring where $k = 10$ newtons/meter related to the graph of the function for a spring where $k = 2$ newtons/meter?

34. **PHYSICS** A ball is dropped from a height of 20 feet. The function $h = -16t^2 + 20$ models the height of the ball in feet after t seconds. Compare this graph to the graph of its parent function.

35. **ACCELERATION** The distance d in feet a car accelerating at 6 ft/s² travels from the start of a race after t seconds is modeled by the function $d = 3t^2$. Suppose a second car begins the race at the same time 100 feet ahead of the first car and accelerating at 4 ft/s². The distance the second car is from the starting line after t seconds is modeled by the function $d = 2t^2 + 100$.

  a. Explain how each graph is related to the graph of $d = t^2$.

  b. After how many seconds will the first car pass the second car?

Example 10
Use the graph of each function to write its equation.

36.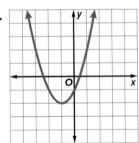

37.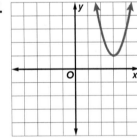

38.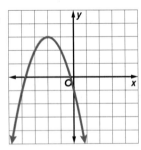

Mixed Exercises
Describe how the graph of each function is related to the graph of the parent function.

39. $g(x) = 5 - \frac{1}{5}x^2$

40. $g(x) = 4(x - 1)^2$

41. $g(x) = 0.25x^2 - 1.1$

42. $h(x) = 1.35(x + 1)^2 + 2.6$

43. $g(x) = \frac{3}{4}x^2 + \frac{5}{6}$

44. $h(x) = 1.01x^2 - 6.5$

STRUCTURE Use transformations to graph each quadratic function. Then describe the transformation.

45. $h(x) = -2(x + 2)^2 + 2$ **46.** $g(x) = \left(\frac{1}{2}x\right)^2 - 3$ **47.** $h(x) = -\frac{1}{4}(x - 1)^2 + 4$

Match each function to its graph.

A.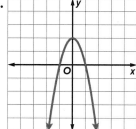

B.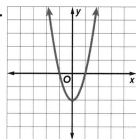

C.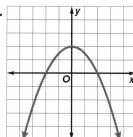

D.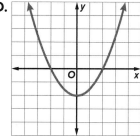

48. $f(x) = 2x^2 - 2$

49. $g(x) = \frac{1}{2}x^2 - 2$

50. $h(x) = -\frac{1}{2}x^2 + 2$

51. $j(x) = -2x^2 + 2$

52. PRECISION Write a set of instructions that a classmate could use to transform the graph of $f(x) = x^2$ and end up with the graph of $g(x) = -10(x - 17)^2$.

53. CONSTRUCT ARGUMENTS A function is an even function if $f(-x) = f(x)$ for all x in the domain of the function. A function is an odd function if $f(-x) = -f(x)$ for all x in the domain of the function. Is the parent function of a quadratic function an even function or an odd function? Justify your answer.

54. USE A MODEL An animator is using a coordinate plane to design a scene in a movie. In the scene, a comet enters the screen in Quadrant II, moves around the screen in a parabolic path, and leaves the screen in Quadrant I.

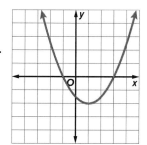

  a. The figure shows the path of the comet. What equation represents the path?

  b. The animator decides to translate the path of the comet 8 units up and 7 units left. What equation represents the new path?

Use the graph of each function to write its equation.

55.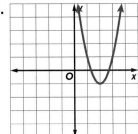

56.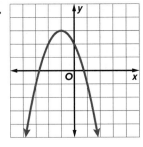

57.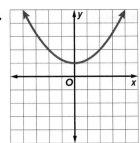

58.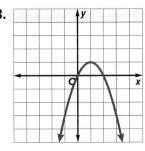

59.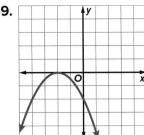

60.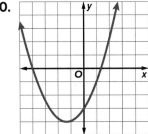

🧠 **Higher-Order Thinking Skills**

61. **ANALYZE** Are the following statements *sometimes*, *always*, or *never* true? Justify your argument.

    a. The graph of $y = x^2 + k$ has its vertex at the origin.

    b. The graphs of $y = ax^2$ and its reflection across the x-axis are the same width.

    c. The graph of $f(x) = x^2 + k$, where $k \geq 0$, and the graph of a quadratic function $g(x)$ with vertex at $(0, -3)$ have the same maximum or minimum.

62. **PERSEVERE** Write a function of the form $y = ax^2 + k$ with a graph that passes through the points $(-2, 3)$ and $(4, 15)$.

63. **ANALYZE** Determine whether all quadratic functions that are reflected across the y-axis produce the same graph. Justify your argument.

64. **CREATE** Write a quadratic function with a graph that opens down and is wider than the parent graph.

65. **WRITE** Describe how the values of a and k affect the graphical and tabular representations of the functions $y = ax^2$, $y = x^2 + k$, and $y = ax^2 + k$.

626 Module 11 • Quadratic Functions

Lesson 11-3
Solving Quadratic Equations by Graphing

Today's Goal
- Solve quadratic equations by graphing.

Today's Vocabulary
quadratic equation
double root

Explore Roots and Zeros of Quadratics

Online Activity Use graphing technology to complete the Explore.

> **INQUIRY** How can you use the x-intercepts of a quadratic function to identify the solutions of its related equation?

Learn Solving Quadratic Equations by Graphing

A **quadratic equation** can be written in the standard form $ax^2 + bx + c = 0$, where $a \neq 0$. Because the graphs of these quadratic functions have zero, one, or two zeros, the corresponding quadratic equations also have zero, one, or two solutions.

Key Concept • Solutions of Quadratic Equations

two unique real solutions	*one* unique real solution	*no* real solutions
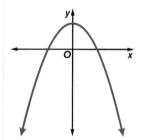	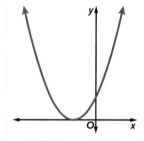	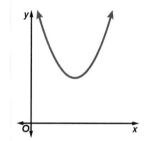

When the vertex of the related quadratic function lies on the x-axis, the solution is a **double root**, which means that there are two roots of a quadratic equation that are the same number.

💭 Think About It!
How can the solutions of a quadratic equation help you graph the related function?

Example 1 Solve a Quadratic Equation with Two Roots

Solve $-x^2 + 4x + 5 = 0$ by graphing.

Graph the related function $f(x) = -x^2 + 4x + 5$. The x-intercepts of the graph appear to be at −1 and 5, so the solutions are −1 and 5.

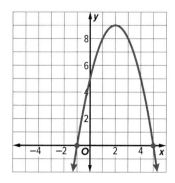

💭 Think About It!
How do you know that the solutions are correct?

Check

Use the graph of the related function to solve $x^2 + 5x + 6 = 0$ by graphing. The solutions are __?__ and __?__.

Lesson 11-3 • Solving Quadratic Equations by Graphing **627**

Example 2 Solve a Quadratic Equation with a Double Root

Solve $x^2 - 8x = -16$ by graphing.

Rewrite the equation in standard form and then graph the related function $f(x) = x^2 - 8x + 16$.

Notice that the vertex of the parabola is the only x-intercept. Therefore, there is only one solution, 4.

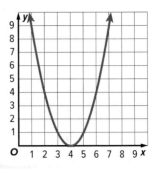

Talk About It!
The function $f(x) = x^2 - 8x + 16$ is a perfect square trinomial. Explain why there is a double root for this type of function.

Example 3 Solve a Quadratic Equation with No Real Roots

Solve $-2x^2 - x - 2 = 0$ by graphing.

Graph the related function $f(x) = -2x^2 - x - 2$.

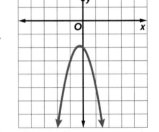

This graph has no x-intercepts. Therefore, this equation has no real number solutions.

Example 4 Approximate Roots of Quadratic Functions

BASEBALL In 1941, the Boston Red Sox's Ted Williams hit a baseball 172 meters. The hit would later be regarded as the longest home run ever hit at the team's home field, Fenway Park. The function $h = -16t^2 + 105t + 1.5$ models the height of the baseball h in meters after t seconds. If it falls to the ground, approximately how long would the ball be in the air?

You need to find the roots of the equation $-16t^2 + 105t + 1.5 = 0$. Graph the related function $h = -16t^2 + 105t + 1.5$, and use estimation to approximate the zeros.

The t-intercept that shows when the ball would hit the ground is located between 6 and 7 seconds. To find a more exact time when the ball would hit the ground, complete the table using an increment of 0.1 for the t-values between 6 and 7.

Go Online An alternate method is available for this example.

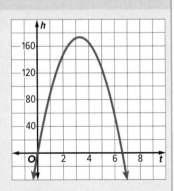

t	6.3	6.4	6.5	6.6	6.7	6.8	6.9
h	27.96	18.14	8	−2.46	−13.24	−24.34	−35.76

The function value that is closest to zero when the sign changes is −2.46. Thus, the ball would be in the air for approximately 6.6 seconds before hitting the ground.

Go Online You can complete an Extra Example online.

Practice

Go Online You can complete your homework online.

Examples 1–3

Solve each equation by graphing.

1. $x^2 + 7x + 14 = 0$
2. $x^2 + 2x - 24 = 0$
3. $x^2 + 16x + 64 = 0$
4. $x^2 - 5x + 12 = 0$
5. $x^2 + 14x = -49$
6. $x^2 = 2x - 1$
7. $x^2 - 10x = -16$
8. $-2x^2 - 8x = 13$
9. $2x^2 - 16x = -30$
10. $2x^2 = -24x - 72$
11. $-3x^2 + 2x = 15$
12. $x^2 = -2x + 80$

Example 4

13. **SOCCER** Claudia kicked a soccer ball off of a platform. The equation $y = -x^2 + 3x + 12$ models the height of the ball y in feet after x seconds. Approximately how long is the ball in the air?

14. **TRAMPOLINE** A gymnast jumped on a trampoline. The equation $y = -16x^2 + 58x$ models the height of the gymnast y in feet after x seconds for one of the jumps. Approximately how long is the gymnast in the air?

Mixed Exercises

Estimate the solution(s) to each equation by graphing the related function. Round to the nearest tenth.

15. $p^2 + 4p + 2 = 0$
16. $x^2 + x - 3 = 0$
17. $d^2 + 6d = -3$
18. $h^2 + 1 = 4h$
19. $3x^2 - 5x = -1$
20. $x^2 + 1 = 5x$

21. **FARMING** In order for Mr. Moore to decide how much fertilizer to apply to his corn crop this year, he reviews records from previous years. His crop yield y depends on the amount of fertilizer he applies to his fields x according to the equation $y = -x^2 + 4x + 12$. Graph the function, and find the point at which Mr. Moore gets the highest yield possible. Then find and interpret the zero(s) of the function.

22. **FRAMING** A rectangular photograph is 7 inches long and 6 inches wide. The photograph is framed using a material that is x inches wide. If the area of the frame and photograph combined is 156 square inches, what is the width of the framing material?

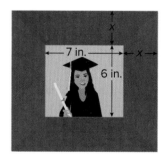

23. **WRAPPING PAPER** Can a rectangular piece of wrapping paper with an area of 81 square inches have a perimeter of 60 inches? (*Hint*: Let length = 30 − width.) Explain.

24. **ENGINEERING** The shape of a satellite dish is often parabolic because of the reflective qualities of parabolas. Suppose a particular satellite dish is modeled by the function $0.5x^2 = 2 + y$.
    a. Approximate the zeros of this function by graphing.
    b. On the coordinate plane, translate the parabola so that there is only one zero. Label this curve A.
    c. Translate the parabola so that there are no real zeros. Label this curve B.

Lesson 11-3 • Solving Quadratic Equations by Graphing **629**

25. **REASONING** The three equations below are shown on the graph.

$y = x^2 + 12x + m$; $y = 2x^2 - nx + 72$; $y = -x^2 + 3x - 10$

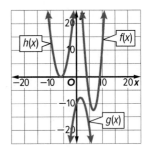

a. Describe the solutions for each function. Then, match each graph to its related equation. Explain your reasoning.
b. How could you choose an appropriate value for *m*?
c. How could you choose an appropriate value for *n*?

26. **ROCKETS** The height *h* of a model rocket launched from ground level into the air after *t* seconds can be modeled by the equation $h = -16t^2 + 160t$. The equation is graphed on the coordinate grid.

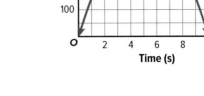

a. Where does the graph intersect the *x*-axis? What do these points represent?
b. How long does it take the rocket to reach its maximum height? Explain your reasoning.

27. **USE TOOLS** Through market research, a company finds that its profit in dollars can be modeled by the function $P(x) = -50{,}000x^2 + 300{,}000x - 250{,}000$, where *x* is the price in dollars at which they sell their product.

a. Describe how to use graphing technology to graph the function, and then provide a sketch of the graph.
b. Find the *x*-intercepts. What do these represent in the context of the situation?
c. Write an equation for the price at which the company should sell their product to make a profit of $150,000. Sketch a graph of the function relating to this equation and use it to solve the equation graphically.
d. Is there a price at which the company can sell their product to make a profit of $300,000? Explain your reasoning, and use a graph to support your answer.

28. **FIND THE ERROR** Iku and Zachary are finding the number of real zeros of the function graphed at the right. Iku says that the function has no real zeros because there are no *x*-intercepts. Zachary says that the function has one real zero because the graph has a *y*-intercept. Is either of them correct? Explain your reasoning.

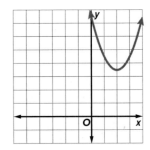

29. **CREATE** Describe a real-world situation in which a thrown object travels in the air. Write an equation that models the height of the object with respect to time, and determine how long the object travels in the air.

30. **ANALYZE** The graph shown is a *quadratic inequality*. Analyze the graph, and find 3 solutions with *y*-values greater than 2.

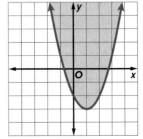

31. **PERSEVERE** Write a quadratic equation that has the roots described.
    a. one double root
    b. no real solutions
    c. two unique real solutions

32. **WRITE** Explain how to approximate the roots of a quadratic equation when the roots are not integers.

Lesson 11-4
Solving Quadratic Equations by Factoring

Learn Solving Quadratic Equations by Using the Square Root Property

Key Concept • Square Root Property

Words: To solve a quadratic equation in the form $x^2 = n$, take the square root of each side.

Symbols: For any number $n \geq 0$, if $x^2 = n$, then $x = \pm\sqrt{n}$.

Example:
$x^2 = 49$
$x = \pm\sqrt{49}$ or $x = \pm 7$

In the equation $x^2 = n$, if n is a perfect square, you will have an exact answer. If n is not a perfect square, you will need to approximate the square root. You can use a calculator or estimation to find an approximation.

Example 1 Use the Square Root Property

Solve $(y + 5)^2 = 21$.

$(y + 5)^2 = 21$ Original equation

$y + 5 = \pm\sqrt{21}$ Square Root Property

$y = -5 \pm \sqrt{21}$ Subtract 5 from each side.

$y = -5 + \sqrt{21}$ or $y = -5 - \sqrt{21}$ Separate into two equations.

The solutions are $-5 + \sqrt{21}$ or $-5 - \sqrt{21}$.
Using a calculator $-5 + \sqrt{21} \approx -0.42$ and $-5 - \sqrt{21} \approx -9.58$.

Check

Select all the solutions of $(x - 2)^2 = 16$.

A. −6
B. −4
C. −2
D. 2
E. 4
F. 6

Select all the solutions of $(y + 18)^2 = 77$.

A. ≈ −26.77
B. ≈ −9.23
C. ≈ −7.68
D. ≈ 7.68
E. ≈ 9.23
F. ≈ 26.77

 Go Online You can complete an Extra Example online.

Today's Goals
- Solve quadratic equations by using the Square Root Property.
- Solve quadratic equations by factoring and sketch graphs of quadratic functions by using the zeros.
- Write equations of quadratic functions given their graphs or roots.

💭 Think About It!
Without using a calculator, describe how can you approximate $\sqrt{30}$.

Study Tip
Reading Math $\pm\sqrt{30}$ is read as "plus or minus the square root of 30."

Watch Out!
Square Root Be careful not to confuse the *roots* of an equation with taking the *square root* of an expression. The *roots* of an equation are any values of the equation that make it true. The *square root* of an expression is one of two equal factors.

Think About It!

Why is $t \approx -1.37$ not a solution?

Example 2 Solve an Equation by Using the Square Root Property

EGG DROP During the Egg Drop Competition at Anika's school, students must create a container for an egg that prevents the egg from breaking when dropped from a height of 30 feet. The formula $h = -16t^2 + h_0$ can be used to approximate the number of seconds t it takes for the container to reach height h from an initial height of h_0 in feet. Find the time it takes the container to reach the ground.

At ground level, $h = 0$ and the initial height is 30, so $h_0 = 30$.

$h = -16t^2 + h_0$	Original equation
$0 = -16t^2 + 30$	$h = 0$ and $h_0 = 30$
$-30 = -16t^2$	Subtract 30 from each side.
$1.875 = t^2$	Divide each side by -16.
$1.37 \approx t$	Use the Square Root Property.

It takes ≈ 1.37 seconds for the container to reach the ground.

Study Tip

Modeling The function $h = -16t^2 + h_0$ does not model the path of the container after it is dropped. This function is used to model the height of the container as a function of time.

Explore Using Factors to Solve Quadratic Equations

Online Activity Use graphing technology to complete the Explore.

INQUIRY How can you use factoring to find the solutions of a quadratic equation?

Learn Solving Quadratic Equations by Factoring

Key Concept • Zero Product Property

Words: If the product of two factors is 0, then at least one of the factors must be 0.

Symbols: For any real numbers a and b, if $ab = 0$, then $a = 0$, $b = 0$, or both a and b equal zero.

Factor Using the Distributive Property

$ax + bx + zy + by$
$= x(a + b) + y(a + b)$
$= (a + b)(x + y)$

Factor Quadratic Trinomials

$ax^2 + bx + c = ax^2 + mx + px + c$ when $m + p = b$ and $mp = ac$

Factor Differences of Squares

$a^2 - b^2 = (a + b)(a - b)$

Factor Perfect Squares

$a^2 + 2ab + b^2 = (a + b)^2$

Think About It!

Why do you think that there are so many different methods for factoring an equation?

Go Online You can complete an Extra Example online.

Example 3 Solve a Quadratic Equation by Using the Distributive Property

Solve $4x^2 - 12x = 0$. Check your solution.

$4x^2 - 12x = 0$ Original equation

$4x(x - 3) = 0$ Factor by using the Distributive Property.

$4x = 0$ and $x - 3 = 0$ Zero Product Property

$x = 0$ $x = 3$ Simplify.

Check the roots by substituting them into the original equation.

$4(0)^2 - 12(0) = 0$ ✓ $4(3)^2 - 12(3) = 0$ ✓

Example 4 Solve a Quadratic Equation by Factoring a Trinomial

Part A Solve $x^2 - 2x - 8 = 0$. Check your solution.

$x^2 - 2x - 8 = 0$ Original equation
$(x + 2)(x - 4) = 0$ Factor the trinomial.
$x + 2 = 0$ and $x - 4 = 0$ Zero Product Property
$x = -2$ $x = 4$ Simplify.

Check the roots by substituting them into the original equation.

Part B Use the roots of $x^2 - 2x - 8 = 0$ to sketch the graph of the related function.

Graph $(-2, 0)$ and $(4, 0)$. The vertex is $(1, -9)$. Because a is positive, the graph opens up.

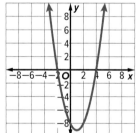

Check
Solve $x^2 - 10x + 24 = 0$.

> **Problem-Solving Tip**
>
> **Make a Chart** Making a chart to organize the factors of c can help you identify the pair of factors that have a sum of b.

Example 5 Solve a Quadratic Equation by Factoring a Difference of Squares

Solve $49 - x^2 = 0$. Check your solution.

$49 - x^2 = 0$ Original equation
$7^2 - x^2 = 0$ Write in the form $a^2 - b^2$.
$(7 + x)(7 - x) = 0$ Difference of squares
$7 + x = 0$ and $7 - x = 0$ Zero Product Property
$x = -7$ and $x = 7$ Simplify.

Check the roots by substituting them into the original equation.

> **Think About It!**
> Why does $25y^2 - 60y + 81 = 45$ have one solution instead of two?

Example 6 Solve a Quadratic Equation by Factoring a Perfect Square Trinomial

Solve $25y^2 - 60y + 81 = 45$. Check your solution.

$25y^2 - 60y + 81 = 45$	Original equation
$25y^2 - 60y + 36 = 0$	Subtract 45 from each side.
$(5y)^2 - 2(5y)(6) + 6^2 = 0$	Factor the perfect square trinomial.
$(5y - 6)^2 = 0$	Factor the trinomial.
$\sqrt{(5y - 6)^2} = \sqrt{0}$	Square Root Property
$y = \frac{6}{5}$	Simplify.

Check the root by substituting it into the original equation.

Apply Example 7 Factor a Trinomial to Solve a Problem

The area of an isosceles triangle is 108 square feet. Find the perimeter of the triangle if the length of each leg is 15 feet, the length of the base is $4x + 2$ feet, and the height is $3x$ feet.

1. What is the task?

Describe the task in your own words. Then list any questions that you may have. How can you find answers to your questions?

I need to use the given information about the area of the triangle to write and solve an equation for x. Then I need to use this information to find the perimeter of the triangle. How can I write an equation for the area of the triangle? I can review the formula for the area of a triangle and how to solve equations.

2. How will you approach the task? What have you learned that you can use to help you complete the task?

I will approach this task by first drawing a diagram. I will use what I have learned about factoring to help me solve the equation.

> **Think About It!**
> Can $x = -\frac{9}{2}$ ever be a solution of $108 = \frac{1}{2}(4x + 2)(3x)$?

3. What is your solution?

Use your strategy to solve the problem. Label the drawing and write an equation for the area of the triangle.

$108 = \frac{1}{2}(4x + 2)(3x)$

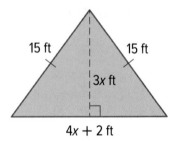

Solve your equation by factoring.

$x = -\frac{9}{2}$ or $x = 4$

The perimeter of the triangle is 48 ft.

4. How can you know that your solution is reasonable?

Write About It! Write an argument that can be used to defend your solution.

I can substitute $x = 4$ into the formula for the area to check my answer.

$108 = \frac{1}{2}[8(4) - 14][3(4)]$

> **Go Online**
> You can complete an Extra Example online.

Check

The surface area of a prism is 264 square centimeters. Find the volume of the prism if its height is 15 centimeters, its length is $x + 4$ centimeters, and its width is x centimeters. (Hint: $S = Ph + 2B$ and $V = Bh$)

A. 2 cm³

B. 26 cm³

C. 180 cm³

D. 264 cm³

Learn Writing Quadratic Functions Given the Zeroes

You can write the equation of a quadratic function if you know its zeroes.

Key Concept • Writing Equations of Quadratic Functions

Step 1 Find the factors of a related expression.

Step 2 Determine whether the value of a is positive or negative.

Step 3 Use another point on the graph to determine the value of a.

💬 **Talk About It!**

How many points on the graph of a quadratic function must be identified in order to write its equation? Explain.

Example 8 Write a Quadratic Function Given a Graph

Write a quadratic function for the given graph.

Step 1 Find the factors of a related expression.

The two zeros on the graph are −2 and 6, so $(x + 2)$ and $(x - 6)$ are factors of the related expression.
The function $f(x) = a(x + 2)(x - 6)$ represents the graph.

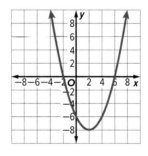

Step 2 Determine whether the value of a is positive or negative.
The graph opens upward, so a must be positive.

Step 3 Use another point on the graph to determine the value of a.
The point $(2, -8)$ is on the graph.

$f(x) = a(x + 2)(x - 6)$ Quadratic function with zeros of −2 and 6

$-8 = a(2 + 2)(2 - 6)$ $[x, f(x)] = (2, -8)$

$-8 = -16a$ Simplify.

$\frac{1}{2} = a$ Divide each side by −16.

One equation for the function is $f(x) = \frac{1}{2}(x + 2)(x - 6)$, or $f(x) = \frac{1}{2}x^2 - 2x - 6$.

🧠 **Think About It!**

How would this process have been different if the graph opened downward?

Go Online You can complete an Extra Example online.

Check

Which quadratic function represents the given graph?

A. $f(x) = -\frac{1}{4}x^2 - \frac{3}{4}x + 1$

B. $f(x) = \frac{1}{4}x^2 - \frac{3}{4}x + 1$

C. $f(x) = -\frac{1}{4}x^2 + \frac{3}{4}x + 1$

D. $f(x) = \frac{1}{4}x^2 + \frac{3}{4}x + 1$

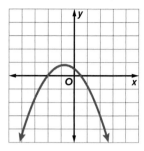

Example 9 Write a Quadratic Function Given Points

Write a quadratic function for the graph that contains (−7, 0), (−3, −32), and (1, 0).

We know that the zeros of a function occur at the x-intercepts, or when $y = 0$. Therefore, −7 and 1 are the zeros of this function, and $x + 7$ and $x - 1$ are factors of the related function. To find the value of a in the related function $f(x) = a(x + 7)(x - 1)$, use the third point $(-3, -32)$.

$f(x) = a(x + 7)(x - 1)$	Quadratic function with zeros of −7 and 1
$-32 = a(-3 + 7)(-3 - 1)$	$[x, f(x)] = (-3, -32)$
$-32 = -16a$	Simplify.
$2 = a$	Divide each side by −16.

One function that passes through the given points is
$f(x) = 2(x + 7)(x - 1)$ or $f(x) = 2x^2 + 12x - 14$.

Check

Which quadratic function has a graph that contains $\left(-\frac{5}{2}, 0\right)$, $(1, 35)$, and $\left(\frac{7}{2}, 0\right)$?

A. $f(x) = x^2 + x - \frac{35}{4}$

B. $f(x) = -4x^2 + 4x + 35$

C. $f(x) = 4x^2 + 4x - 35$

D. No quadratic function exists that passes through these points.

Go Online You can complete an Extra Example online.

Practice

Go Online You can complete your homework online.

Examples 1, 3, 5, 6

Solve each equation. Check your solutions.

1. $x^2 = 36$
2. $x^2 = 81$
3. $(k + 1)^2 = 9$
4. $81 - 4b^2 = 0$
5. $3b(9b - 27) = 0$
6. $(7x + 3)(2x - 6) = 0$
7. $b^2 = -3b$
8. $a^2 = 4a$
9. $x^2 - 6x = 27$
10. $a^2 + 11a = -18$
11. $n^2 - 120 = 7n$
12. $d^2 + 56 = -18d$
13. $y^2 - 90 = 13y$
14. $h^2 + 48 = 16h$
15. $x^2 + 9x + 18 = 0$
16. $4x^2 + 28x + 49 = 0$
17. $-9x^2 = 30x + 25$
18. $-x^2 - 12 = -8x$
19. $36w^2 = 121$
20. $100 = 25w^2$
21. $64x^2 - 1 = 0$
22. $4y^2 - \frac{9}{16} = 0$
23. $\frac{1}{4}b^2 = 16$
24. $81 - \frac{1}{25}x^2 = 0$

Example 4

25. Consider the equation $c^2 + 10c + 9 = 0$.
    a. Solve the equation by factoring.
    b. Use the roots to sketch the related function.

26. Consider the equation $x^2 - 18x = -32$.
    a. Solve the equation by factoring.
    b. Use the roots to sketch the related function.

Examples 2 and 7

27. **NUMBER THEORY** The product of the two consecutive positive integers is 11 more than their sum. What are the numbers?

28. **LADDERS** A ladder is resting against a wall. The top of the ladder touches the wall at a height of 15 feet, and the length of the ladder is one foot more than twice its distance from the wall. Find the distance from the wall to the bottom of the ladder. (*Hint*: Use the Pythagorean Theorem to solve the problem.)

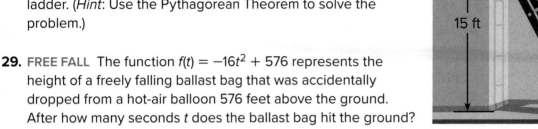

29. **FREE FALL** The function $f(t) = -16t^2 + 576$ represents the height of a freely falling ballast bag that was accidentally dropped from a hot-air balloon 576 feet above the ground. After how many seconds t does the ballast bag hit the ground?

30. **VOLUME** Catalina can make an open-topped box out of a square piece of cardboard by cutting 3-inch squares from the corners and folding up the sides to meet. The volume of the resulting box is $V = 3x^2 - 36x + 108$, where x is the original length and width of the cardboard.

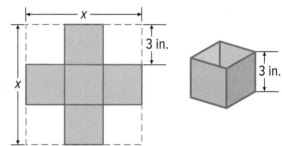

   a. Factor the polynomial expression from the volume equation.

   b. What is the volume of the box if the original length of each side of the cardboard was 9 inches?

Example 8

Write a quadratic function for the given graph.

31.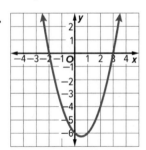

32.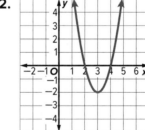

33.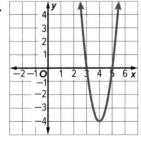

34.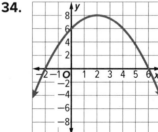

Example 9

Write a quadratic function that contains the given points.

35. (−9, 0), (−3, −30), (2, 0)

36. (−5, 0), (−4, 0), (3, 56)

37. (−4, 0), (2, −9), (8, 0)

38. (1, 0), (3, −18), (6, 0)

Mixed Exercises

39. CONSTRUCT ARGUMENTS Kayla says the zeros of the equation $2m^2 - 12m = 0$ are 0 and −12. Is she correct? Explain your reasoning.

40. STRUCTURE Write an expression for the perimeter of a rectangle that has an area $A = x^2 + 20x + 96$. Explain how you solved the problem.

Solve each equation. Check the solutions.

41. $25p^2 - 16 = 0$

42. $4p^2 + 4p + 1 = 0$

43. $n^2 - \frac{9}{25} = 0$

44. $9y^2 + 18y - 12 = 6y$

45. $6h^2 + 8h + 2 = 0$

46. $16d^2 = 4$

47. $\frac{1}{16} y^2 = 81$

48. $8p^2 - 16p = 10$

49. $x^2 + 30x + 150 = -75$

50. $10b^2 - 15b = 8b - 12$

51. $y^2 - 16y + 64 = 81$

52. $(m - 4)^2 = 9$

53. $4x^2 = 80x - 400$

54. $9 - 54x = -81x^2$

55. $4c^2 + 4c + 1 = 9$

56. $x^2 - 16x + 64 = 4$

57. STRUCTURE A triangle's height is 10 feet more than its base. If the area of the triangle is 100 square feet, use factoring to find its dimensions.

58. REGULARITY Explain how to use the Zero Product Property to solve a quadratic equation.

59. REGULARITY Consider a quadratic equation written in standard form, $ax^2 + bx + c = 0$, where $a \neq 1$. Explain how solving an equation by factoring a trinomial where $a \neq 1$ differs from solving a quadratic equation by factoring a trinomial where $a = 1$.

60. **STRUCTURE** A square has an area of $4x^2 + 16xy + 16y^2$ square inches. The dimensions are binomials with positive integer coefficients. Find the perimeter of the square. Is the perimeter a multiple of $(x + 2y)$? Explain.

61. **STRUCTURE** The equation $x^2 + qx - 12 = 6x$ can be solved by factoring. How can you use number relationships to predict values for q?

62. **USE A MODEL** The rectangle at the right has a perimeter of $14x + 30$ centimeters and an area of 225 square centimeters. Find the dimensions of the rectangle. Explain how you solved the problem.

63. **AREA** A triangle has an area of 64 square feet. If the height of the triangle is 8 feet more than its base, x, what are its height and base?

$6x + 15$ cm

64. **CONSTRUCT ARGUMENTS** Jarrod claims that the quadratic equation $x^2 - 7x + 5 = 0$ has no solution because the left side does not factor. Do you agree or disagree? Use a graph to justify your argument.

Higher-Order Thinking Skills

65. **PERSEVERE** Given the equation $(ax + b)(ax - b) = 0$, solve for x. What do you know about the values of a and b?

66. **WRITE** Explain how to solve the quadratic equation $x^2 + 16x = -64$.

67. **FIND THE ERROR** Ignatio and Samantha are solving $6x^2 - x = 12$. Is either of them correct? Explain your reasoning.

Ignatio	Samantha
$6x^2 - x = 12$	$6x^2 - x = 12$
$x(6x - 1) = 12$	$6x^2 - x - 12 = 0$
$x = 12$ or $6x - 1 = 12$	$(2x - 3)(3x + 4) = 0$
$6x = 13$	$2x - 3 = 0$ or $3x + 4 = 0$
$x = \frac{13}{6}$	$x = \frac{3}{2}$ $x = -\frac{4}{3}$

68. **ANALYZE** What should you consider when solving a quadratic equation that models a real-world situation?

69. **CREATE** Write a perfect square trinomial equation in which the coefficient of the middle term is negative and the last term is a fraction. Solve the equation.

70. **PERSEVERE** Factor the polynomial $x^3 + x^2 - 6x$. Make a table. Then sketch the graph of the related function.

Lesson 11-5
Solving Quadratic Equations by Completing the Square

Explore Using Algebra Tiles to Complete the Square

Online Activity Use algebra tiles to complete the Explore.

> **INQUIRY** How does forming a square to create a perfect square trinomial help you solve quadratic equations?

Today's Goals
- Solve quadratic equations by completing the square.
- Identify key features of quadratic functions by writing quadratic equations in vertex form.

Today's Vocabulary
completing the square

Learn Solving Quadratic Equations by Completing the Square

Key Concept • Completing the Square

To **complete the square** for any quadratic expression of the form $x^2 + bx$, follow the steps below.

Step 1 Find one-half of b, the coefficient of x.

Step 2 Square the result from Step 1.

Step 3 Add the result of Step 2 to $x^2 + bx$.

The pattern to complete the square is represented by $x^2 + bx + \left(\frac{b}{2}\right)^2 = \left(x + \frac{b}{2}\right)^2$.

Example 1 Complete the Square

Find the value of c that makes $x^2 + 8x + c$ a perfect square trinomial. Use the completing-the-square algorithm.

Step 1 Find half of 8. $\frac{8}{2} = 4$

Step 2 Square the result of Step 1. $4^2 = 16$

Step 3 Add the result of Step 2 to $x^2 + 8x$. $x^2 + 8x + 16$

Thus, $c = 16$. Notice that $x^2 + 8x + 16 = (x + 4)^2$.

Check

Find the value that makes the expression a perfect square trinomial.

$x^2 + 40x +$ ___?___

Go Online You can complete an Extra Example online.

> **Think About It!**
> To complete the square, what constant would you add to $x^2 + 5x$?

> **Think About It!**
> How is the process of completing the square different when b is even and when b is odd?

> **Think About It!**
> What do the zeros of a quadratic function tell you about its graph?

Example 2 Solve an Equation by Completing the Square

Solve $x^2 - 10x + 14 = 5$ by completing the square.

In order to solve a quadratic equation by completing the square, first isolate $x^2 - bx$ on one side.

$x^2 - 10x + 14 = 5$	Original equation
$x^2 - 10x = -9$	Subtract 14 from each side.
$x^2 - 10x + 25 = -9 + 25$	Because $\left(\frac{-10}{2}\right)^2 = 25$, add 25 to each side.
$(x - 5)^2 = 16$	Factor $x^2 - 10x + 25$.
$x - 5 = \pm 4$	Take the square root of each side.
$x = 5 \pm 4$	Add 5 to each side.
$x = 9$ or 1	Separate the solutions.

Example 3 Solve an Equation with *a* Not Equal to 1

Solve $-4x^2 + 32x - 72 = 0$ by completing the square.

To solve a quadratic equation when the leading coefficient is not 1, divide or multiply each term to eliminate the coefficient. Then, isolate $x^2 - bx$ and complete the square.

$-4x^2 + 32x - 72 = 0$	Original equation
$x^2 - 8x + 18 = 0$	Divide each side by -4.
$x^2 - 8x = -18$	Subtract 18 from each side.
$x^2 - 8x + 16 = -18 + 16$	Add $\left(\frac{8}{2}\right)^2$ to each side.
$(x - 4)^2 = -2$	Factor $x^2 - 8x + 16$.

No real number has a negative square. So, this equation has no real solutions.

> **Study Tip**
> **Multiplicative Inverse** When the leading coefficient is a fraction, remember that you can either divide all of the terms by the fraction or multiply each term by the multiplicative inverse.

Learn Finding the Maximum or Minimum Value

Key Concept • Use Vertex Form to Graph

Step 1 Complete the square to write the function in vertex form.

Step 2 Identify the axis of symmetry and extrema based on the function in vertex form. When the leading coefficient is positive, the parabola will open up and the vertex will be a minimum. When the leading coefficient is negative, the parabola will open down and the vertex will be a maximum.

Step 3 Solve for *x* to find the zeros. The zeros are the *x*-intercepts of the graph.

Step 4 Use the key features to graph the function.

> **Go Online** You can complete an Extra Example online.

Example 4 Find a Minimum

Write $y = x^2 + 2x - 5$ in vertex form. Identify the axis of symmetry, extrema, and zeros. Then, use the key features to graph the function.

Step 1 Complete the square to write the function in vertex form.

$$y = x^2 + 2x - 5 \quad \text{Original function}$$
$$y + 5 = x^2 + 2x \quad \text{Add 5 to each side.}$$
$$y + 5 + 1 = x^2 + 2x + 1 \quad \text{Add } \left(\frac{b}{2}\right)^2 \text{ to each side.}$$
$$y + 6 = (x + 1)^2 \quad \text{Factor.}$$
$$y = (x + 1)^2 - 6 \quad \text{Subtract 6 from each side to write in vertex form.}$$

Step 2 Identify the axis of symmetry and extrema.

In vertex form the vertex of the parabola (h, k) or $(-1, -6)$. Because the x^2-term is positive, the vertex is a minimum. The axis of symmetry is $x = h$ or $x = -1$.

Step 3 Solve for x to find the zeros.

$$(x + 1)^2 - 6 = 0$$
$$(x + 1)^2 = 6$$
$$x + 1 = \pm\sqrt{6}$$
$$x \approx -3.45 \text{ or } 1.45$$

Step 4 Use the key features to graph the function.

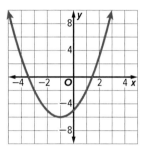

Talk About It
Explain how you can determine whether a parabola will open up or open down.

Example 5 Find a Maximum

Write $y = -2x^2 + 20x - 42$ in vertex form. Identify the axis of symmetry, extrema, and zeros. Then, use the key features to graph the function.

Step 1 Complete the square to write the function in vertex form.

$$y = -2x^2 + 20x - 42 \quad \text{Original function}$$
$$y + 42 = -2x^2 + 20x \quad \text{Add 42 to each side.}$$
$$y + 42 = -2(x^2 - 10x) \quad \text{Factor out } -2.$$
$$y + 42 - 50 = -2(x^2 - 10x + 25) \quad \text{Because } -2\left[\left(\frac{10}{2}\right)^2\right] = -50, \text{ add } -50 \text{ to each side.}$$
$$y - 8 = -2(x - 5)^2 \quad \text{Factor } x^2 - 10x + 25.$$
$$y = -2(x - 5)^2 + 8 \quad \text{Add 8 to each side to write in vertex form.}$$

(continued on the next page)

Watch Out!
Leading Coefficients When completing the square of a quadratic equation where $a \neq 1$, be sure to multiply the constant added by the coefficient before adding it to both sides.

Step 2 Identify the axis of symmetry and extrema. In vertex form the vertex of the parabola is at (h, k) or $(5, 8)$. Since the x^2-term is negative, the vertex is a maximum. The axis of symmetry is $x = h$ or $x = 5$.

Step 3 Solve for x to find the zeros.

$$0 = -2(x - 5)^2 + 8$$
$$2(x - 5)^2 = 8$$
$$(x - 5)^2 = 4$$
$$x - 5 = \pm 2$$
$$x = 3 \text{ or } 7$$

Step 4 Use the key features to graph the function.

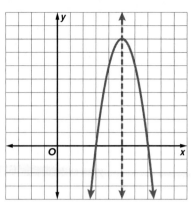

🌎 Example 6 Use Extrema and Key Features

LUNAR LANDING Suppose the path of a golf ball can be represented by $y = -0.001x^2 + 0.248x$, where y is the height of the ball in meters and x is its horizontal distance in meters. Determine the maximum height of the golf ball and the horizontal distance it traveled.

Complete the square to write the equation of the function in vertex form.

$y = -0.001x^2 + 0.248x$	Original equation
$y = -0.001(x^2 - 248x)$	Factor out -0.001.
$y - 15.376 = -0.001(x^2 - 248x + 15{,}376)$	Add $-15{,}376$ to each side.
$y - 15.376 = -0.001(x - 124)^2$	Factor $x^2 - 248x + 15{,}376$.
$y = -0.001(x - 124)^2 + 15.376$	Add 15.376 to each side.

The vertex of the parabola is at $(124, 15.376)$. Since the vertex represents the maximum, the ball reached a height of 15.376 meters.

To find the horizontal distance the ball traveled, find the zeros.

$$0 = -0.001(x - 124)^2 + 15.376$$
$$0.001(x - 124)^2 = 15.376$$
$$(x - 124)^2 = 15{,}376$$
$$x - 124 = \pm 124$$
$$x = 0 \text{ or } 248$$

Since $x = 0$ represents the golf ball's initial point, the golf ball traveled 248 meters before hitting the surface of the Moon.

🌐 **Go Online** You can complete an Extra Example online.

🍿 **Think About It!**

Suppose Alan Shepard had hit a golf ball standing on a platform 3 meters above the lunar surface. How would this affect the vertex form of the equation?

Practice

Go Online You can complete your homework online.

Example 1

Find the value of c that makes each trinomial a perfect square.

1. $x^2 + 26x + c$
2. $x^2 - 24x + c$
3. $x^2 - 19x + c$
4. $x^2 + 17x + c$
5. $x^2 + 5x + c$
6. $x^2 - 13x + c$
7. $x^2 - 22x + c$
8. $x^2 - 15x + c$
9. $x^2 + 24x + c$

Examples 2 and 3

Solve each equation by completing the square. Round to the nearest tenth, if necessary.

10. $x^2 + 6x - 16 = 0$
11. $x^2 - 2x - 14 = 0$
12. $x^2 - 8x - 1 = 8$
13. $x^2 + 3x + 21 = 22$
14. $x^2 - 11x + 3 = 5$
15. $5x^2 - 10x = 23$
16. $2x^2 - 2x + 7 = 5$
17. $3x^2 + 12x + 81 = 15$
18. $4x^2 + 6x = 12$
19. $4x^2 + 6 = 10x$
20. $-2x^2 + 10x = -14$
21. $-3x^2 - 12 = 14x$

Examples 4 and 5

Write each equation in vertex form. Identify the axis of symmetry, extrema, and zeros. Then, use the key features to graph the function.

22. $y = x^2 + 8x + 7$
23. $y = x^2 - 12x + 16$
24. $y = -x^2 - 4x + 5$
25. $y = x^2 - 8x + 10$

Example 6

26. **MARS** On Mars, the gravity acting on an object is less than that on Earth. On Earth, a golf ball hit with an initial upward velocity of 26 meters per second will hit the ground in about 5.4 seconds. The height h of an object on Mars that leaves the ground with an initial velocity of 26 meters per second is given by the equation $h = -1.9t^2 + 26t$. How much longer will it take for the golf ball hit on Mars to reach the ground? What is the maximum height of the golf ball on Mars? Round your answer to the nearest tenth.

27. **FROGS** A frog sitting on a stump 3 feet high hops off and lands on the ground. During its leap, its height h in feet is given by $h = -0.5d^2 + 2d + 3$, where d is the distance from the base of the stump. How far is the frog from the base of the stump when it lands on the ground? What is the maximum height of the frog during its leap?

28. **FALLING OBJECTS** Keisha throws a rock down an abandoned well. The distance d in feet the rock falls after t seconds can be represented by $d = 16t^2 + 64t$. If the water in the well is 80 feet below ground, how many seconds will it take for the rock to hit the water?

Mixed Exercises

29. **GARDENING** Peggy is planning a rectangular vegetable garden using 200 feet of fencing material. She only needs to fence three sides of the garden since one side borders an existing fence. Let x = the width of the rectangle. For what widths would the area of Peggy's garden equal 4800 square feet if she uses all the fencing material?

30. **REASONING** Find the value of q that makes $0.5x^2 + 0.5qx + 72$ a perfect square trinomial. Show that the same value of q makes $4x^2 + 24x + 1.5q$ a perfect square trinomial.

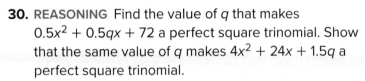

4x in.

(x + 3) in.

31. **AREA** The area of the rectangle shown is 352 square inches. What are the dimensions of the rectangle?

32. **CONSTRUCT ARGUMENTS** Use completing the square to show that no two consecutive positive even integers can have a product of 27.

33. **STRUCTURE** Lawrence writes a trinomial that can be solved by completing the square and cannot be solved by factoring. One of the solutions of his equation is between 4 and 5.

   a. Write a possible equation.

   b. What are the solutions to the equation?

   c. Explain why your equation meets Lawrence's criteria.

34. **STRUCTURE** The quadratic function $f(t) = -5000t^2 + 70{,}000t + 5000$ is used to model the number of products a company sells in years t since the product was released. Complete the square to find the zeros of $f(t)$ and interpret each in the context of the situation. Then find the maximum sales for the product.

35. **PERSEVERE** Given $y = ax^2 + bx + c$ with $a \neq 0$, derive the equation for the axis of symmetry by completing the square, and rewrite the equation in the form $y = a(x - h)^2 + k$.

36. **ANALYZE** Determine the number of solutions $x^2 + bx = c$ has if $c < -\left(\frac{b}{2}\right)^2$. Justify your argument.

37. **WHICH ONE DOESN'T BELONG?** Identify the expression that does not belong with the other three. Explain your reasoning.

$n^2 - n + \frac{1}{4}$ $\quad$ $n^2 + n + \frac{1}{4}$ $\quad$ $n^2 - \frac{2}{3}n + \frac{1}{9}$ $\quad$ $n^2 + \frac{1}{3}n + \frac{1}{9}$

38. **CREATE** Write a quadratic equation for which the only solution is 4.

39. **WRITE** Compare and contrast the following strategies for solving $x^2 - 5x - 7 = 0$: completing the square, graphing, and factoring.

Lesson 11-6
Solving Quadratic Equations by Using the Quadratic Formula

Learn Solving Quadratic Equations by Using the Quadratic Formula

Completing the square of the quadratic equation produces a formula that allows you to find the solutions of *any* quadratic equation. This formula is called the Quadratic Formula, $x = \dfrac{-b \pm \sqrt{b^2 - 4ac}}{2a}$.

Example 1 Use the Quadratic Formula

Solve $x^2 - 12x = -11$ by using the Quadratic Formula.

Step 1 Rewrite the equation in standard form.

$x^2 - 12x = -11$	Original equation
$x^2 - 12x + 11 = 0$	Add 11 to each side.

Step 2 Apply the Quadratic Formula.

$x = \dfrac{-b \pm \sqrt{b^2 - 4ac}}{2a}$	Quadratic Formula
$= \dfrac{-(-12) \pm \sqrt{(-12)^2 - 4(1)(11)}}{2(1)}$	$a = 1, b = -12, c = 11$
$= \dfrac{12 \pm \sqrt{144 - 44}}{2}$	Multiply.
$= \dfrac{12 \pm \sqrt{100}}{2}$	Subtract.
$= \dfrac{12 \pm 10}{2}$	Take the square root.
$x = \dfrac{12 + 10}{2}$ or $x = \dfrac{12 - 10}{2}$	Separate the solutions.
$= 11 \qquad\qquad = 1$	Simplify.

The solutions are 1 and 11.

Check

Solve the equation by using the Quadratic Formula.

$x^2 - 19x = -70$

$x = \underline{\ ?\ }, \underline{\ ?\ }$

Today's Goals
- Solve quadratic equations by using the Quadratic Formula.
- Use the discriminant to determine the number of solutions of a quadratic equation.

Today's Vocabulary
discriminant

💬 Talk About It!
What does the ± symbol indicate? How does that affect the number of factors determined by using the Quadratic Formula? Explain.

💀 Think About It!
If the expression under the square root simplified to 0, then what would be the relation between the two factors?

Watch Out!

Squaring a Negative For the expression b^2, the result will be positive regardless of whether b is negative or positive.

Example 2 Use the Quadratic Formula When a Is Not Equal to 1

Solve $2x^2 + 5x = 12$ by using the Quadratic Formula.

Step 1 Rewrite the equation in standard form.

$$2x^2 + 5x - 12 = 0$$

Step 2 Apply the Quadratic Formula.

$$x = \frac{-b \pm \sqrt{b^2 - 4ac}}{2a}$$ Quadratic Formula

$$= \frac{-(5) \pm \sqrt{(5)^2 - 4(2)(-12)}}{2(2)}$$ $a = 2, b = 5, c = -12$

$$= \frac{-5 \pm \sqrt{25 + 96}}{4}$$ Multiply.

$$= \frac{-5 \pm \sqrt{121}}{4}$$ Subtract.

$$= \frac{-5 \pm 11}{4}$$ Take the square root.

$$x = \frac{-5 - 11}{4} \text{ or } x = \frac{-5 + 11}{4}$$ Separate the solutions.

$$= -4 \qquad = \frac{3}{2}$$ Simplify.

The solutions are -4 and $\frac{3}{2}$.

🌐 **Example 3** Solve a Quadratic Equation with Irrational Roots

HEALTH The normal blood pressure of an adult woman P in millimeters of mercury given her age t in years can be modeled by the equation $P = 0.01t^2 + 0.5t + 107$. If Seiko's blood pressure is 120, how old might she be?

Step 1 Rewrite the equation in standard form.

$$0.01t^2 + 0.5t + 107 = 120$$ Original equation

$$0.01t^2 + 0.5t - 13 = 0$$ Subtract 120 from each side.

Step 2 Apply the Quadratic Formula.

$$x = \frac{-b \pm \sqrt{b^2 - 4ac}}{2a}$$ Quadratic Formula

$$= \frac{-0.5 \pm \sqrt{(0.5)^2 - 4(0.01)(-13)}}{2(0.01)}$$ $a = 0.01, b = 0.5, c = -13$

$$= \frac{-0.5 \pm \sqrt{0.25 + 0.52}}{0.02}$$ Multiply.

$$= \frac{-0.5 \pm \sqrt{0.77}}{0.02}$$ Simplify.

$$x = \frac{-0.5 + \sqrt{0.77}}{0.02} \text{ or } x = \frac{-0.5 - \sqrt{0.77}}{0.02}$$ Separate the solutions.

$$\approx 18.9 \qquad\qquad \approx -68.9$$ Simplify.

The solutions are 18.9 and -68.9.

▶ **Go Online** You can complete an Extra Example online.

Step 3 Eliminate unreasonable solutions.

Because *t* measures age in years, it is reasonable to assume that Seiko is 18.9 years old.

Because *t* measures age in years, which cannot be negative, −68.9 is an unreasonable solution.

> **Think About It!**
> If $P = 0$, then there are no real solutions to the equation. Does this make sense in the context of the situation?

Check

FOOTBALL The quarterbacks on a high school football team want to see who can get the most hang-time when throwing the ball. Connor knows that using his initial velocity and height, he can model the trajectory of his throw with the equation $h = -16t^2 + 95t + 5.2$, where *h* is the height of the ball and *t* is the time after throwing. Solve the equation by using the Quadratic Formula, eliminating any extraneous solutions. Round your answer(s) to the nearest tenth.

$t =$?

Explore Deriving the Quadratic Formula Algebraically

Online Activity Use an interactive tool to complete the Explore.

> **@ INQUIRY** How does having a formula make it possible to solve quadratic equations where the other methods are not easy to apply?

Explore Deriving the Quadratic Formula Visually

Online Activity Use an interactive tool to complete the Explore.

> **@ INQUIRY** Why can you use the Quadratic Formula to solve any quadratic equation?

Go Online You can complete an Extra Example online.

Learn The Discriminant

In the Quadratic Formula, $x = \dfrac{-b \pm \sqrt{b^2 - 4ac}}{2a}$, the expression under the radical sign, $b^2 - 4ac$, is called the **discriminant**. You can use the discriminant to determine the number of real solutions of a quadratic equation.

Key Concept • Using the Discriminant

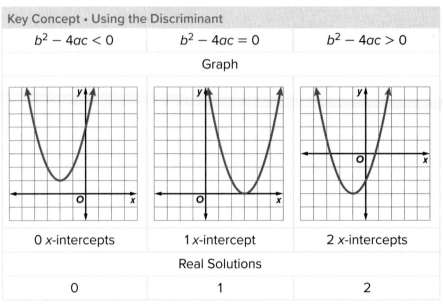

$b^2 - 4ac < 0$	$b^2 - 4ac = 0$	$b^2 - 4ac > 0$
0 x-intercepts	1 x-intercept	2 x-intercepts
Real Solutions		
0	1	2

Study Tip

Real Solutions The real solutions are the cases where $y = 0$, which means that they are values of x where the graph of the function crosses the x-axis.

Example 4 Use the Discriminant

State the value of the discriminant of $4x^2 - 3x = -1$. Then determine the number of real solutions of the equation.

Step 1 Rewrite the equation in standard form.

$4x^2 - 3x = -1$ Original equation

$4x^2 - 3x + 1 = 0$ Add 1 to each side.

Step 2 Find the discriminant.

$b^2 - 4ac = (-3)^2 - 4(4)(1)$ $a = 4, b = -3, c = 1$

$= -7$ Simplify.

Because the discriminant is negative, the equation has no real solution.

Check

State the value of the discriminant of $8x^2 - 15x = -9$.

The discriminant is __?__.

Determine the number of real solutions of the equation.

The equation has __?__ solution(s).

Go Online to practice what you've learned about solving quadratic equations in the Put It All Together over Lessons 11-3 through 11-6.

Go Online You can complete an Extra Example online.

Practice

Go Online You can complete your homework online.

Examples 1 and 2

Solve each equation by using the Quadratic Formula. Round to the nearest tenth, if necessary.

1. $x^2 - 49 = 0$
2. $x^2 - x - 20 = 0$
3. $x^2 - 5x - 36 = 0$

4. $4x^2 + 5x - 6 = 0$
5. $x^2 + 16 = 0$
6. $6x^2 - 12x + 1 = 0$

7. $5x^2 - 8x = 6$
8. $2x^2 - 5x = -7$
9. $5x^2 + 21x = -18$

10. $81x^2 = 9$
11. $8x^2 + 12x = 8$
12. $4x^2 = -16x - 16$

13. $10x^2 = -7x + 6$
14. $-3x^2 = 8x - 12$
15. $2x^2 = 12x - 18$

Example 3

16. **BUSINESS** Tanya runs a catering business. Based on her records, her weekly profit can be approximated by the function $f(x) = x^2 + 2x - 37$, where x is the number of meals she caters. If $f(x)$ is negative, it means that the business has lost money. What is the least number of meals that Tanya needs to cater in order to make a profit?

17. **FREE FALL** Josh drops his phone from a height of 50 feet. The situation is best modeled by $h = -16t^2 + 50$, where h is the height in feet and t is the time in seconds. How long will it take for the phone to hit the ground? Round to the nearest tenth.

18. **ARCHITECTURE** The Golden Ratio appears in the design of the Greek Parthenon because the width and height of the façade are related by the equation $\frac{W + H}{W} = \frac{W}{H}$. If the height of a model of the Parthenon is 16 inches, what is its width? Round your answer to the nearest tenth.

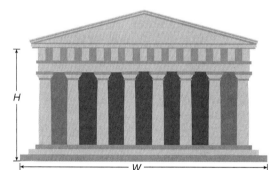

19. **CRAFTS** Ariadna cut a 60-inch chenille stem into two unequal pieces, and then she used each piece to make a square. The sum of the areas of the squares was 117 square inches. Let x be the length of one piece. Write and solve an equation to represent the situation and find the lengths of the two original pieces.

Example 4
State the value of the discriminant for each equation. Then determine the number of real solutions of the equation.

20. $0.2x^2 - 1.5x + 2.9 = 0$
21. $2x^2 - 5x + 20 = 0$
22. $x^2 - \frac{4}{5}x = 3$

23. $0.5x^2 - 2x = -2$
24. $2.25x^2 - 3x = -1$
25. $2x^2 = \frac{5}{2}x + \frac{3}{2}$

26. $x^2 + 2x + 1 = 0$
27. $x^2 - 4x + 10 = 0$
28. $x^2 - 6x + 7 = 0$

29. $x^2 - 2x - 7 = 0$
30. $x^2 - 10x + 25 = 0$
31. $2x^2 + 5x - 8 = 0$

32. $2x^2 + 6x + 12 = 0$
33. $2x^2 - 4x + 10 = 0$
34. $3x^2 + 7x + 3 = 0$

Mixed Exercises

Solve each equation by using the Quadratic Formula. Round to the nearest tenth, if necessary.

35. $x^2 + 4x = -1$
36. $x^2 - 9x + 22 = 0$
37. $x^2 + 6x + 3 = 0$

38. $2x^2 + 5x - 7 = 0$
39. $2x^2 - 3x = -1$
40. $2x^2 + 5x + 4 = 0$

41. $2x^2 + 7x = 9$
42. $3x^2 + 2x - 3 = 0$
43. $3x^2 - 7x - 6 = 0$

Without graphing, determine the number of x-intercepts of the graph of the related function for each equation.

44. $x^2 + 4x + 3 = 0$
45. $4.25x + 3 = -3x^2$

46. $x^2 + \frac{2}{25} = \frac{3}{5}x$
47. $0.25x^2 + x = -1$

48. **RECTANGLES** The base of a rectangle is 4 inches greater than the height. The area of the rectangle is 15 square inches. What are the dimensions of the rectangle to the nearest tenth of an inch?

49. **REASONING** For the equation $3x^2 + 2x + q = 0$, find all values of q so that there are two real solutions to the equation. Then find all values of q so that there are two complex solutions for the equation. Explain.

50. **SITE DESIGN** The town of Smallport plans to build a new water treatment plant on a rectangular piece of land 75 yards wide and 200 yards long. The buildings and facilities need to cover an area of 10,000 square yards. The town's zoning board wants the site designer to allow as much room as possible between each edge of the site and the buildings and facilities. Let x represent the width of the border.

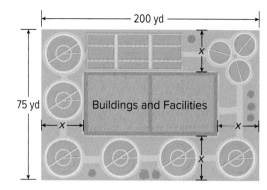

a. Write an equation to represent the area covered by the building and facilities.

b. Write the standard form of the quadratic equation.

c. Find the width of the border. Round your answer to the nearest tenth.

51. **USE A MODEL** New City's public works is putting together a fireworks presentation to celebrate the Fourth of July. They will be launching fireworks at different heights. The path of one of the fireworks is defined by the equation $h_1 = -16t^2 + 90t + 120$, with t in seconds and h_1 in feet.

a. The equation provides the height the firework is above the ground at any time between when it is launched and when it hits the ground. What does 120 represent in the context of the situation?

b. Write the equation that models when the firework will hit the ground. Solve the equation. Round to the nearest hundredth.

c. How high is the firework when it is at its highest point? How long did it take to get to this point? Round to the nearest hundredth. Explain.

d. Without graphing the equation, use the information from parts **a, b,** and **c** to describe what the graph would look like. Explain.

e. Write and solve an equation for the times at which the firework reaches a height of 200 feet. Round to the nearest tenth.

52. **STRUCTURE** A cylinder is filled completely with water. Its base has a radius of x cm, and its height is 8 cm. After Ella removes $40x$ cm³ of water, the volume of the remaining water is 250 cm³. What is the radius of the cylinder? Round to the nearest whole number, if necessary.

53. **REGULARITY** Solve the equation $x^2 + bx + c = 0$ for x in terms of b and c. Then determine whether each statement is *sometimes, always,* or *never* true.

a. If both b and c are negative, there will be at least one real solution.

b. If both b and c are positive, there will be complex nonreal solutions.

54. **REASONING** Braden wrote three quadratic equations: $-2x^2 - 3x - 8 = 0$, $-x^2 + 6x - 9 = 0$, and $-2x^2 + 4x + 3 = 0$. He graphs one of the equations as shown. He shows Ava the graph and the three equations and challenges her to match the correct equation with the graph. Does Ava need to solve all the equations to match one with the graph? Explain your reasoning.

Lesson 11-6 • Solving Quadratic Equations by Using the Quadratic Formula **653**

55. **STRUCTURE** A rectangular area is to be enclosed with its length 20 feet less than its width, as shown.

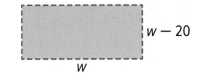

   a. What should the width be in order for the enclosure to have an area of 100 ft²?

   b. Find the width for which the area enclosed is R ft² for $R > 0$.

56. **STRUCTURE** A frame is made with width w inches and length $15 - w$ inches, as shown in the figure.

   a. Is there any width for which the frame will have an area of 150 in²? Explain your reasoning.

   b. Find the maximum area of the frame and the dimensions that produce the maximum area.

57. **USE A SOURCE** The St. Louis Arch can be approximated by a parabola. Research its dimensions and use the dimensions to write a quadratic equation representing the St. Louis Arch. Then set the equation equal to 0 and use the Quadratic Formula to solve for x. Interpret the solutions.

Determine whether there are *two*, *one*, or *no* real solutions of each equation.

58. The graph of the related quadratic function does not have an x-intercept.

59. The graph of the related quadratic function touches, but does not intersect the x-axis.

60. The graph of the related quadratic function intersects the x-axis twice.

61. Both a and b are greater than 0 and c is less than 0 in a quadratic equation.

62. **WRITE** Why can the discriminant be used to confirm the number of real solutions of a quadratic equation?

63. **ANALYZE** Use factoring techniques to determine the number of real zeros of $f(x) = x^2 - 8x + 16$. Compare this method to using the discriminant.

64. **PERSEVERE** Find all values of k such that $2x^2 - 3x + 5k = 0$ has two solutions.

65. **WRITE** Describe the advantages and disadvantages of each method of solving quadratic equations. Why are the methods equivalent? Which method do you prefer, and why?

66. **CREATE** Write a quadratic equation that has no real roots, and find its discriminant. Explain how the discriminant shows that a quadratic equation has no real roots.

67. **CREATE** Write a quadratic equation that has one real root, and find its discriminant. Explain why the Quadratic Formula yields only one solution when the discriminant of a quadratic equation is equal to zero.

68. **CREATE** Write a quadratic equation that has two real roots. Determine whether completing the square, graphing, or factoring would be the best method to use to solve your quadratic equation.

Lesson 11-7
Solving Systems of Linear and Quadratic Equations

Explore Using Algebra Tiles to Solve Systems of Linear and Quadratic Equations

Online Activity Use algebra tiles to complete the Explore.

> **INQUIRY** How can you use algebra tiles to solve systems of linear and quadratic equations?

Today's Goals
- Solve systems of linear and quadratic equations by graphing.
- Solve systems of linear and quadratic equations by using algebraic methods.

Think About It!
Can you put the steps in any other order? Explain your reasoning.

Learn Solving Systems of Linear and Quadratic Equations by Graphing

To solve a system of linear and quadratic equations by graphing, you can follow a set of steps.

Step 1 Graph the quadratic function.
Step 2 Graph the linear function.
Step 3 Find the point(s) of intersection.
Check Substitute the values in the original equation.

Talk About It!
Cordell says that the graphs of a linear function and a quadratic function always have 2 points of intersection. Do you agree or disagree? Justify your reasoning or provide a counterexample.

Example 1 Solve a System of Linear and Quadratic Equations Graphically

Solve the system of equations by graphing.

$y = x^2 + 4x - 1$

$y = 2x + 2$

Step 1 Graph $y = x^2 + 4x - 1$.

Step 2 Graph $y = 2x + 2$.

Step 3 Find the points of intersection.

The graphs appear to intersect at $(-3, -4)$ and $(1, 4)$.

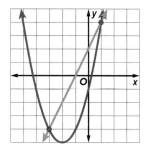

Go Online You can complete an Extra Example online.

Check

The graphs of $f(x) = -2x + 3$ and $g(x) = x^2 - 4x - 5$ are shown.

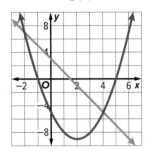

Complete the ordered pair(s) to represent each situation.

$f(x) = 0$ (__?__, __?__)

$g(x) = 0$ (__?__, __?__) (__?__, __?__)

$f(x) = g(x)$ (__?__, __?__) (__?__, __?__)

🧠 **Think About It!**
What does it mean if there is no solution when you substitute one expression for *y* in the other equation and solve?

Learn Solving Systems of Linear and Quadratic Equations Algebraically

To solve a system of linear and quadratic equations algebraically, you can follow a set of steps.

Step 1 Solve the equations for *y*.
Step 2 Substitute one expression for *y* in the other equation.
Step 3 Solve for *x*.
Step 4 Substitute the *x*-value(s) in either equation.
Step 5 Solve for *y*.
Check Graph the equations.

Example 2 Solve a System of Linear and Quadratic Equations Algebraically

Solve the system of equations algebraically.

$y = x^2 - 2x - 3$

$x + y = 3$

Step 1 Solve the equations for *y*.

The first equation is already solved for *y*.

$x + y = 3$	Original second equation
$y = -x + 3$	Subtract *x* from each side.

Step 2 Substitute an expression for *y*.

$y = -x + 3$	Second equation solved for *y*
$x^2 - 2x - 3 = -x + 3$	Substitution

 Go Online You can complete an Extra Example online.

656 Module 11 · Quadratic Functions

Step 3 Solve for x.

$$x^2 - 2x - 3 = -x + 3$$ Original equation
$$x^2 - x - 3 = +3$$ Add x to each side.
$$x^2 - x - 6 = 0$$ Subtract 3 from each side.
$$(x - 3)(x + 2) = 0$$ Factor.
$$x - 3 = 0 \quad x + 2 = 0$$ Zero Product Property
$$x = 3 \quad\quad x = -2$$ Simplify.

Step 4 Substitute the x-value(s) in either equation.

$$y = -x + 3 \quad\quad\quad y = -x + 3$$
$$y = -3 + 3 \quad\quad\quad y = -(-2) + 3$$

Step 5 Solve for y.

$$y = -3 + 3 \quad\quad\quad y = -(-2) + 3$$
$$y = 0 \quad\quad\quad\quad\quad y = 5$$

Check Graph the equations.

The graphs of the functions intersect at $(3, 0)$ and $(-2, 5)$, so $(3, 0)$ and $(-2, 5)$ are solutions of this system.

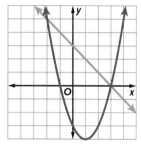

Check

Solve the system of equations algebraically.

$$y = x^2 - 8x + 19 \quad\quad x - 0.5y = 3$$

(_?_ , _?_)

> 💭 **Think About It!**
> Are there any other methods that can be used to solve the system? Explain your reasoning.

🌐 Example 3 Use a System of Linear and Quadratic Equations

SALES Bathing suit sales at a clothing store can be modeled by the function $y = -x^2 + 12x + 25$, and gift card sales can be modeled by the function $y = 5x + 7$, where x represents the number of months past January and y represents the total revenue in thousands of dollars. Solve a system of equations algebraically to find the month when the revenue from bathing suit sales is equal to the revenue from gift card sales.

(continued on the next page)

🌐 **Go Online** You can complete an Extra Example online.

> **Think About It!**
> Find the revenue from bathing suit sales in the month when the revenue from bathing suit sales is equal to the revenue from gift card sales.

Set the expressions equal to each other and solve for x.

$-x^2 + 12x + 25 = 5x + 7$	Substitute.
$12x + 25 = x^2 + 5x + 7$	Add $-x^2$ to each side.
$25 = x^2 - 7x + 7$	Subtract $12x$ from each side.
$0 = x^2 - 7x - 18$	Subtract 25 from each side.
$0 = (x - 9)(x + 2)$	Factor.
$x - 9 = 0 \quad x + 2 = 0$	Zero Product Property
$x = 9 \quad\quad x = -2$	Simplify.

Because the expressions are equivalent when $x = 9$, the revenue of bathing suit sales and gift card sales is equal 9 months past January, or in October.

Graph the equations to check your solution.

> **Watch Out!**
> **Labeling Axes** The x-axis represents the months past January, so January is represented by $x = 0$, not $x = 1$.

Check

SALES Revenue from single-day ticket sales to a local amusement park can be modeled by the function $y = -x^2 + 11x + 42$, and revenue from season pass sales can be modeled by the function $y = -5x + 81$, where x represents the number of months past January and y represents the total revenue in tens of thousands of dollars. Solve a system of equations algebraically to find the first month when the revenue from single-day tickets is equal to the revenue from season pass sales.

A. March

B. April

C. May

D. June

Go Online You can complete an Extra Example online.

Practice

Go Online You can complete your homework online.

Example 1

Solve each system of equations by graphing.

1. $y = x^2 - 4$
   $y = -3$

2. $y = x^2 + x - 2$
   $y = -x + 1$

3. $y = 2x^2 + 1$
   $y = 1$

4. $y = x^2 + 3x + 1$
   $y = x + 1$

Example 2

Solve each system of equations algebraically.

5. $y = x^2 - 2x - 5$
   $y = 3$

6. $y = x^2 + 4x - 1$
   $y = 3x + 1$

7. $y = x^2 - 6x + 5$
   $x + y = -1$

8. $y = x^2 + x + 1$
   $y - 1 = x$

9. $y + 3x = x^2 - 3$
   $y = -2x + 3$

10. $y - 1 = 2x^2 - x$
    $-2x + y = 3$

Example 3

11. **GYMNASTICS** A gymnast throws a baton up into the air during her floor routine. The height of the baton can be modeled by $y = -16x^2 + 24x + 4$, where x is the time in seconds and y is the height of the baton in feet since it was released. The ceiling of the gym can be represented by the function $y = 42$.

    a. Solve the system of equations algebraically.

    b. Will the baton reach the ceiling of the gym? Explain your reasoning.

12. **DISC GOLF** The height y in feet of a disc x seconds after it was thrown can be modeled by $y = -\frac{1}{30}x^2 + \frac{1}{2}x + 5\frac{2}{15}$. The height of Rodrigo's hands as he runs to catch the disc can be represented by the function $y = \frac{1}{2}x + 3$.

    a. Solve the system of equations algebraically.

    b. At what height will Rodrigo catch the disc?

13. **ZOO** Revenue from single-day ticket sales at a local zoo can be modeled by the function $y = -x^2 + 25x + 80$, and revenue from season pass sales can be modeled by the function $y = -4x + 200$. In both functions, x represents the number of months past January, and y represents the total revenue in thousands of dollars. Solve the system of equations algebraically to find the first month when the revenue from single-day ticket sales is equal to the revenue from season pass sales.

Lesson 11-7 • Solving Systems of Linear and Quadratic Equations 659

Mixed Exercises

Solve each system of equations.

14. $y = x^2$
    $y = 2x$

15. $y = -2x^2 + 7x - 2$
    $y = 3 - 4x$

16. $y = -x^2 + 4$
    $y = \frac{1}{5}x + 5$

17. $y = -x^2 + 4x - 4$
    $y = 2x - 3$

18. $y = -x^2 - 5x - 6$
    $y = -3x - 1$

19. $y = 2x^2 - 4$
    $y = 2x$

20. $y = x^2 + 7x + 12$
    $y = 2x + 8$

21. $y = x^2 - x - 20$
    $y = 3x + 12$

22. $y = 3x^2 - x - 2$
    $y = -2x + 2$

23. $y = x^2 - x - 18$
    $y = x - 3$

24. $y = x^2 - 3x + 1$
    $y = x + 1$

25. $y = x^2 - 4x + 6$
    $y = 2x - 3$

26. $y = -x^2 + 2x + 3$
    $y = x + 2$

27. $y = x^2 + 4x - 1$
    $y = 3x$

Solve each equation by using a system of equations.

28. $x^2 - 4x + 3 = x - 1$

29. $x^2 + 2x - 1 = x - 2$

30. $x^2 + 4x + 4 = 4$

31. $x^2 = -2x - 1$

32. $\frac{1}{2}x^2 - 4 = 3x + 4$

33. $x^2 + 5x + 5 = -x - 8$

34. **WINGSUIT** A wingsuit flyer jumps off a tall cliff. He falls freely for a few seconds before deploying the wingsuit and slowing his descent. His height during the freefall can be modeled by the function $y = -4.9x^2 + 420$, where y is the height above the ground in meters and x is the time in seconds. After deploying the wingsuit, the flyer's height is given by the function $y = -3x + 200$.

   a. To the nearest whole number, how long after jumping does the flyer deploy the wingsuit?

   b. To the nearest whole number, what was the flyer's height when he deployed the wingsuit?

35. REASONING The graph shows a quadratic function and a linear function $y = k$.

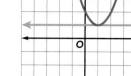

a. How many solutions are there to the system?

b. If the linear function were changed to $y = k + 5$, how many solutions would the system have?

c. If the linear function were changed to $y = k - 2$, how many solutions would the system have?

36. USE A MODEL The function $R(x) = -x^2 + 5x + 14$ represents the revenue earned by a manufacturing company, where R is the revenue in millions of dollars and x is the number of items produced in millions. It cost the company $4 million to produce 3 million items.

a. Write a function to represent the cost C in millions of dollars of producing x million items.

b. Solve the system involving the revenue function and cost function. What does the solution represent?

37. USE TOOLS In a video game, players fling virtual rubber bands at a small target that moves around the screen. The rubber bands are all launched from the point $(-4, 0)$ and follow a path along the line $y = \frac{1}{2}x + 2$. The target moves around the screen following a parabolic path represented by the equation $y = -x^2 - 2x + 4$.

a. Graph the path of the rubber bands and the path of the target.

b. If a player hits the target, at approximately what point or points on the plane will this take place? Explain.

c. Use your calculator to find the coordinates of the point or points at which the rubber bands may hit the target. Round to the nearest tenth. How do your results compare to your answer in part **b**?

38. ROCKETS Jamie is using binoculars to watch a friend set off a model rocket from a distance. Jamie's line of sight through the binoculars is a straight line from 50 meters above the launch point, decreasing $\frac{1}{2}$ a meter in height for every 1 meter of lateral distance to the point where Jamie is standing. The rocket's path follows a projectile motion formula $-4.9t^2 + vt + h$, where v is the initial velocity in meters per second and h is the launch height in meters. The rocket is launched at a velocity of 40 meters per second from the ground. At what height(s), in meters, will the rocket be in Jamie's line of sight?

a. Write a system of equations to represent the situation.

b. Find the solution to the system of equations graphically.

c. Find the solution to the system of equations algebraically.

d. Explain what the solution(s) mean in the context of the situation.

39. REGULARITY Explain a method for determining the number of solutions that exist for a system of equations from a graph of the related functions.

40. STATE YOUR ASSUMPTION Kasia found two solutions to a system of equations she used to determine whether her slinky would travel far enough to get to the next stair. One solution has a negative x-value, and one solution has a positive x-value. Which solution should Kasia use? What assumption did you make to decide?

41. **CONSTRUCT ARGUMENTS** Rafaela kicked a rock off the top of a cliff to a deserted river below. The rock's path is modeled by $h(x) = -0.1x^2 + x + 600$, where $h(x)$ is the height, in meters, of the rock after x seconds. The cliff face is modeled approximately by $h(x) = 600 - 9x$. Will the rock land on the cliff or in the river? Justify your argument.

PERSEVERE Use a graphing calculator to solve other types of systems.

42. $y = x^2 + 3x - 5$
    $y = -x^2$

43. $y = \frac{3}{4}x$
    $x^2 + y^2 = 1$ (*Hint:* Enter as two functions, $y = \sqrt{1 - x^2}$ and $y = -\sqrt{1 - x^2}$.)

44. **ANALYZE** For what value of k does the system $y = x^2 + x$ and $y = -2x + k$ have exactly one solution? Explain your reasoning.

45. **ANALYZE** For what value of k does the system $y = x^2 + 2$ and $y = 3x + k$ have exactly one solution? Explain your reasoning.

46. **CREATE** Below are a graph and a table of values that describe two equations, y_1 and y_2, one linear and one quadratic. Write a problem that could be solved using the graph and table of values, then solve your problem.

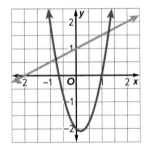

x	y_1	y_2
-2	0	12
0	1	-2
1	1.5	0

47. **WRITE** What is the maximum number of solutions that a system of one linear and one quadratic equation can have? What is the maximum number of solutions to a system of two linear functions? Write a paragraph to explain your reasoning.

48. **FIND THE ERROR** Xavier found the solution to a system of equations using the steps shown. In which step did Xavier make an error? Correct his error.

$y = x^2 - 4x - 6$
$y = 9x + 8$

Step 1: $x^2 - 4x - 6 = 9x + 8$
Step 2: $x^2 - 4x - 9x - 6 - 8 = 0$
Step 3: $x^2 - 13x - 14 = 0$
Step 4: $(x - 14)(x + 1) = 0$

Step 5: $y = 14, y = -1$
Step 6: $14 = 9x + 8; x = \frac{2}{3}$
Step 7: $-1 = 9x + 8; x = -1$

49. **WHICH ONE DOESN'T BELONG?** Analyze the three systems of equations given, and identify which system does not belong. Explain your reasoning.

$y = x^2 - 3x - 4$
$y = 5x + 1$

$y = -x^2 - x + 11$
$y = 2x - 10$

$y = 4x^2 + 4x - 2$
$y = x - 7$

Lesson 11-8

Modeling and Curve Fitting

Explore Using Differences and Ratios to Model Data

 Online Activity Use an interactive tool to complete the Explore.

> **@ INQUIRY** How can differences and ratios of successive *y*-values be used to write a model?

Learn Modeling Real-World Situations

Different types of functions can be used to model data. You can use the equation or graph of a data set to determine the type of function that represents the data.

Concept Summary • Linear and Nonlinear Functions

Linear	Quadratic	Exponential
Function		
$y = mx + b$	$y = ax^2 + bx + c$	$y = ab^x$, where $b > 0$
Graph		
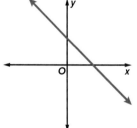	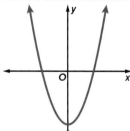	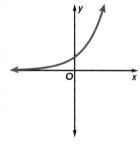
Description		
y increases or decreases at a constant rate.	*y* increases or decreases by a square of *x*.	*y* increases or decreases by a power of *x*.
Differences/Ratios		
First differences are equal.	Second differences are equal.	Successive ratios are equal.
Uses of Difference/Ratio		
The first difference is the slope *m*.	Half of the second difference is *a*.	The common ratio is the base *b* of the function.

 Go Online You can complete an Extra Example online.

Today's Goals
- Distinguish between situations that can be modeled with linear, exponential, and quadratic functions.
- Make and evaluate predictions by fitting nonlinear functions to sets of data.

Today's Vocabulary
coefficient of determination

curve fitting

🔊 Talk About It!
When determining the best model for a function, why is it important to consider the data instead of just looking at the graph?

Think About It!

How would the process of finding the function that represents the data below differ from this example?

x	9.75	10	10.25	10.5
y	−9	−5	−1	3

Example 1 Determine a Model by Using First Differences

Look for a pattern in the data table to determine the kind of model that best describes the data. Then, write the function.

x	5	6	7	8	9
y	−1	−4	−7	−10	−13

Part A Determine the model.

Notice that the x-values are increasing by 1, so the model can be determined by examining successive differences or ratios.

$$-1 \quad -4 \quad -7 \quad -10 \quad -13$$
$$\;-3\;\;\;-3\;\;\;-3\;\;\;-3$$

The first differences are equal. This means that the data can be represented by a linear function.

Part B Write the function.

Substitute for m.

$y = mx + b$ Slope-intercept form of a linear function

$y = -3x + b$ Since the first difference is −3, the slope is −3.

Find the y-intercept.

$-1 = -3(5) + b$ $(x_1, y_1) = (5, -1)$

$-1 = -15 + b$ Simplify.

$14 = b$ Add 15 to each side.

Write the equation in slope-intercept form.

$y = mx + b$ Slope-intercept form

$y = -3x + 14$ Replace m with −3 and b with 14.

The data are modeled by $y = -3x + 14$.

Go Online You can complete an Extra Example online.

Example 2 Determine a Model by Using Second Differences

Look for a pattern in the data table to determine the kind of model that best describes the data. Then, write the function.

x	−2	−1	0	1	2
y	−1	0	4	11	21

Part A Determine the model.

The x-values are increasing by 1, so the model can be determined by examining successive differences or ratios.

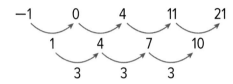

The second differences are equal. This means that the data can be modeled by a quadratic function.

Part B Write the function.

$y = ax^2 + bx + c$ Standard form of a quadratic function

$y = 1.5x^2 + bx + 4$ a is one-half of the second difference, and c is the y-intercept from the given data points.

Find b.

$-1 = 1.5(-2)^2 + b(-2) + 4$ Use data point (−2, −1) for (x, y).

$-1 = 6 - 2b + 4$ Simplify.

$-1 = 10 - 2b$ Simplify.

$-11 = -2b$ Subtract 10 from each side.

$5.5 = b$ Divide each side by −2.

Write the function in standard form.

$y = ax^2 + bx + c$ Standard form of a quadratic function

$y = 1.5x^2 + 5.5x + 4$ $a = 1.5, b = 5.5, c = 4$

The data are modeled by $y = 1.5x^2 + 5.5x + 4$.

Go Online You can complete an Extra Example online.

Example 3 Determine a Model by Using Ratios

Look for a pattern in the data table to determine the kind of model that best describes the data. Then, write the function.

x	3	4	5	6	7
y	400	100	25	6.25	1.5625

Part A Determine the model.

Since the x-values are increasing by 1, the model can be determined by examining successive differences or ratios.

400   100   25   6.25   1.5625

−300   −75   −18.75   −4.6875

225   56.25   14.0625

The first differences are not equal. This means that the data cannot be modeled by a linear function. The second differences are not equal. This means that the data cannot be modeled by a quadratic function.

400   100   25   6.25   1.5625

$\frac{100}{400} = \frac{1}{4}$ $\frac{25}{100} = \frac{1}{4}$ $\frac{6.25}{25} = \frac{1}{4}$ $\frac{1.5625}{6.25} = \frac{1}{4}$

The ratios of successive y-values are equal. Therefore, the data can be modeled by an exponential function.

Part B Write the function.

$y = a(b)^x$ — Exponential function

$y = a\left(\frac{1}{4}\right)^x$ — Since the ratio is $\frac{1}{4}$, $b = \frac{1}{4}$.

Solve for a.

$400 = a\left(\frac{1}{4}\right)^3$ — $(x_1, y_1) = (3, 400)$

$400 = \left(\frac{1}{64}\right)a$ — Simplify.

$25{,}600 = a$ — Multiply each side by 64.

Write the equation.

$y = a(b)^x$ — Exponential function

$y = 25{,}600\left(\frac{1}{4}\right)^x$ — Replace a with 25,600 and b with $\frac{1}{4}$.

The data are modeled by $y = 25{,}600\left(\frac{1}{4}\right)^x$.

Check

Determine the kind of model that best describes the data set. Then write the function.

x	0	1	2	3
y	−3	−6	−12	−24

> **Watch Out!**
>
> **Constant Difference**
> Before checking differences and ratios, check that the x-values are increasing or decreasing by a constant value. In order to use the differences or ratios in the model function, x must be increasing by 1.

Learn Curve Fitting

You can use a graphing calculator to find a regression equation for a set of data that is approximated by a function. This process is called **curve fitting**.

The **coefficient of determination**, R^2, indicates how well the function fits the data. The closer R^2 is to 1, the better the model.

The table shows a scatter plot and graphs of linear, exponential, and quadratic functions. Which type of function is the best fit for the data?

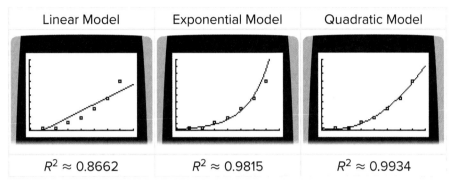

Linear Model	Exponential Model	Quadratic Model
$R^2 \approx 0.8662$	$R^2 \approx 0.9815$	$R^2 \approx 0.9934$

Based on the coefficients of determination, the data are best modeled by a quadratic function.

> **Go Online** You can watch a video to see how to fit a curve to a set of data.

Example 4 Find the Best Model

VIDEO STREAMING The table shows the number of households worldwide that subscribe to a video streaming service. Write a model that fits the data.

Year	2007	2008	2009	2010	2011	2012	2013	2014	2015
Households (millions)	8	9	12	19	23	33	44	57	75

Step 1 Enter the data.

Enter the data by pressing STAT and selecting the Edit option.

Let the year 2007 be represented by 0.

Enter the years since 2007 into List 1 (L1).

Enter the number of households in List 2 (L2).

Step 2 Make a scatter plot.

Graph the scatter plot. Turn on Plot 1 under the STAT PLOT menu and choose the scatter plot feature.

Change the viewing window so that all data are visible by pressing ZOOM and then selecting ZoomStat.

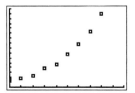

(*continued on the next page*)

> **Go Online** to see how to use a graphing calculator with this example.

> **Use a Source**
> Use an outside source to find the number of subscribers or members of other services, such as ride-sharing companies, Internet service, or social media networks, over time. Then determine whether the data can be modeled by a linear, quadratic, exponential function, or none of these.

Lesson 11-8 • Modeling and Curve Fitting **667**

Step 3 Find the regression equation.

Exponential Regression

Select ExpReg and press ENTER.

The equation is about $y = 7.306(1.342)^x$, with a coefficient of determination of approximately 0.994.

Quadratic Regression

Select Quadreg and press ENTER.

The equation is about $y = 1.076x^2 - 0.439x + 8.485$, with a coefficient of determination of approximately 0.998.

Though both models are a good fit for the data, the quadratic regression is closer to 1 and, therefore, a better fit.

Step 4 Graph the quadratic regression equation.

To copy the quadratic regression equation to the Y= list, press **Y=**, **VARS**, and choose **Statistics**.

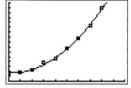

From the **EQ** menu, choose **RegEQ**.

Press GRAPH.

Check

RACING The table shows the speed in kilometers per hour of a racecar after x seconds.

Time x	0.5	1	1.5	2	2.5
Speed y	12	32	58	88	122

Part A

Select the model that best fits the data.

A. $y = -5.3x^2 + 0.2$

B. $y = 9.1x^2 + 27.8x - 4.4$

C. $y = 8.7(3.1)^x$

D. $y = 0.6(1.0)^x$

Part B

If the racecar continues to accelerate following the same trend, predict its speed at 3.5 seconds. Round to the nearest kilometer.

____?____ kph

Go Online You can complete an Extra Example online.

Go Online to learn about exponential growth patterns in Expand 11-8.

Practice

Go Online You can complete your homework online.

Examples 1–3

Look for a pattern in each table of values to determine which kind of model best describes the data. Then write an equation for the function that models the data.

1.
x	−3	−2	−1	0
y	−8.8	−8.6	−8.4	−8.2

2.
x	−2	−1	0	1	2
y	10	2.5	0	2.5	10

3.
x	−1	0	1	2	3
y	0.75	3	12	48	192

4.
x	−2	−1	0	1	2
y	0.008	0.04	0.2	1	5

5.
x	0	1	2	3	4
y	0	4.2	16.8	37.8	67.2

6.
x	−3	−2	−1	0	1
y	14.75	9.75	4.75	−0.25	−5.25

7.
x	−3	−2	−1	0	1	2
y	32	16	8	4	2	1

8.
x	−1	0	1	2	3
y	7	3	−1	−5	−9

9.
x	−3	−2	−1	0	1
y	−27	−12	−3	0	−3

10.
x	−2	−1	0	1	2
y	−8	−4	0	4	8

11.
x	0	1	2	3	4
y	0.5	1.5	4.5	13.5	40.5

12.
x	−1	0	1	2	3
y	27	9	3	1	$\frac{1}{3}$

13.
x	2	3	4	5	6
y	12	27	48	75	108

14.
x	0	1	2	3	4
y	80	73	66	59	52

Example 4

15. **DRONES** The table shows the time after a drone was launched (in seconds) x and the height of the drone, in feet, above the ground y.

   a. Make a scatter plot of the data.

   b. Which regression equation has an R^2 value closest to 1?

   c. Find an appropriate regression equation, and state the coefficient of determination. Based on the regression equation, what are the relevant domain and range?

   d. Predict the height of the drone 7 seconds after it was launched. Round to the nearest foot.

x	y
1	30
2	40
3	50
4	55
5	50
6	40

Lesson 11-8 • Modeling and Curve Fitting 669

16. BAKING Alyssa baked a cake and is waiting for it to cool so she can ice it. The table shows the temperature of the cake every 5 minutes after Alyssa took it out of the oven.

Time (min)	Temperature (°F)
0	350
5	244
10	178
15	137
20	112
25	96
30	89

a. Make a scatter plot of the data.

b. Which regression equation has an R^2 value closest to 1? Is this the equation that best fits the context of the problem? Explain your reasoning.

c. Find an appropriate regression equation, and state the coefficient of determination. Based on the regression equation, what are the relevant domain and range?

d. Alyssa will ice the cake when it reaches room temperature (70°F). Use the regression equation to predict when she can ice her cake.

Mixed Exercises

USE TOOLS Use a graphing calculator to determine whether to use a *linear*, *quadratic*, or *exponential* regression equation. State the coefficient of determination.

17.

x	y
0	1.1
2	3.3
4	2.9
6	5.6
8	11.9
10	19.8

18.

x	y
1	1.67
5	2.59
9	4.37
13	6.12
17	5.48
21	3.12

19.

x	y
−2	0.2
−1	0.5
0	1
1	2
2	4
3	7.5

20. WEATHER The San Mateo weather station records the amount of rainfall since the beginning of a thunderstorm. Data for a storm is recorded as a series of ordered pairs (2, 0.3), (4, 0.6), (6, 0.9), (8, 1.2), (10, 1.5), where the x-value is the time in minutes since the start of the storm, and the y-value is the amount of rain in inches that has fallen since the start of the storm. Determine which kind of model best describes the data.

21. INVESTING The value of a certain parcel of land has been increasing in value ever since it was purchased. The table shows the value of the land parcel over time.

Year Since Purchasing	0	1	2	3	4
Land Value (thousands $)	$1.05	$2.10	$4.20	$8.40	$16.80

Look for a pattern in the table of values to determine which model best describes the data. Then write an equation for the function that models the data.

22. **BOATS** The value of a boat typically depreciates over time. The table shows the value of a boat over a period of time.

Years	0	1	2	3	4
Boat Value ($)	8250	6930	5821.20	4889.81	4107.44

Write an equation for the function that models the data. Then use the equation to determine how much the boat is worth after 9 years.

23. **NUCLEAR WASTE** Radioactive material slowly decays over time. The amount of time needed for an amount of radioactive material to decay to half its initial quantity is known as its half-life. Consider a 20-gram sample of a radioactive isotope.

Half-Lives Elapsed	0	1	2	3	4
Amount of Isotope Remaining (grams)	20	10	5	2.5	1.25

a. Is radioactive decay a *linear* decay, a *quadratic* decay, or an *exponential* decay?

b. Write an equation to determine how many grams y of the radioactive isotope will be remaining after x half-lives.

c. How many grams of the isotope will remain after 11 half-lives?

d. Plutonium-238 is one of the most dangerous waste products of nuclear power plants. If the half-life of plutonium-238 is 87.7 years, how long would it take for a 20-gram sample of plutonium-238 to decay to 0.078 gram?

24. **USE A MODEL** The table shows the populations of two towns from 2015 to 2018.

    a. For each town, write a function that models the town's population x years after 2015.

    b. Based on the function types you used in **part a**, what can you conclude about the populations of the towns as time goes on? Explain.

Year	Population Dixon	Population Midville
2015	80,000	96,000
2016	84,000	96,070
2017	88,200	96,280
2018	92,610	96,630

25. **STRUCTURE** The table shows the height of an elevator above ground level at various times.

Time (s), x	0	1	2	3	4
Height (ft), y	142	124	106	88	70

    a. What type of function best models the data in the table? Why?

    b. The average rate of change is the change in the value of the dependent variable divided by the change in the value of the independent variable. What is the average rate of change of the function over the interval $x = 0$ to $x = 4$? What does this tell you about the motion of the elevator?

    c. Write a function that models the height of the elevator as a function of time. How is the coefficient of x related to your answers to **parts a** and **b**?

26. **SOCCER** The table shows the total number of games a soccer team played and the number of total goals scored in a season.

Games Played	Goals Scored
1	1
2	3
3	6
4	10
5	14

    a. Make a scatter plot of the data.

    b. Which regression equation has an R^2 value closest to 1? What is the value?

    c. Find the appropriate regression equation. What are the domain and range?

    d. The team's goal is to score 25 goals for the season. Use the regression equation to predict in which game they will score their 25th goal.

27. **USE A SOURCE** Look up hourly temperature data for a starting time of 6:00 A.M. and every hour after that for 8 hours. Plot the points, with x representing the time in hours after 6:00 A.M. and y representing the temperature. Then, identify what type of model best describes the data.

28. **PRECISION** Jase used a graphing calculator to find the predicted dollar value of a used vehicle after 7 years. The calculator reported the value as 5127.45928113. How should Jase decide how many digits to report?

29. **CONSTRUCT ARGUMENTS** Diego claims that no function can have equal non-zero first differences and equal non-zero second differences. Is Diego correct? Use examples or counterexamples to justify your argument.

30. **PERSEVERE** Write a function that has constant second differences, first differences that are not constant, a y-intercept of −5, and contains the point (2, 3).

31. **ANALYZE** What type of function will have constant third differences, but not constant second differences? Justify your argument.

32. **CREATE** Write a linear function that has a constant first difference of 4.

33. **PERSEVERE** Explain why linear functions grow by equal differences over equal intervals, and exponential functions grow by equal factors over equal intervals. (*Hint*: Let $y = ax$ represent a linear function, and let $y = a^x$ represent an exponential function.)

34. **WRITE** How can you determine whether a given set of data should be modeled by a *linear* function, a *quadratic* function, or an *exponential* function?

Lesson 11-9

Combining Functions

Today's Goals
- Combine standard function types by using addition and subtraction.
- Combine standard function types by using multiplication.

Explore Using Graphs to Combine Functions

Online Activity Use graphing technology to complete the Explore.

> **INQUIRY** How can you use the graphs of functions to determine their sum, difference, or product?

Learn Adding and Subtracting Functions

Some situations are best modeled by the sum or difference of functions.

Consider $f(x) = x^2 - 2x + 4$ and $g(x) = -5x + 1$. To find the value of the sum of the functions when $x = -3$ or $(f + g)(-3)$, you can find $f(-3)$ and $g(-3)$ and add them.

$f(x) = x^2 - 2x + 4$	Original function
$f(-3) = (-3)^2 - 2(-3) + 4$	Substitute -3 for x.
$f(-3) = 19$	Simplify.

$g(x) = -5x + 1$	
$g(-3) = -5(-3) + 1$	
$g(-3) = 16$	

The sum is given by $f(-3) + g(-3) = 19 + 16$ or 35.

To find a function that gives the sum for all values of the domain, add the two functions and combine the like terms.

$(f + g)(x) = f(x) + g(x)$ Addition of functions
$ = (x^2 - 2x + 4) + (-5x + 1)$ Substitution
$ = x^2 - 2x - 5x + 4 + 1$ Combine like terms.
$ = x^2 - 7x + 5$ Simplify.

Solve for $(f + g)(-3)$ to check that this method results in the same solution as $f(-3) + g(-3)$.

$(f + g)(-3) = (-3)^2 - 7(-3) + 5$ Substitute -3 for x.
$ = 9 + 21 + 5$ Simplify.
$ = 35$ Simplify.

So, $f(x) + g(x) = (f + g)(x)$.

Example 1 Add Functions

Given $f(x) = 3^x + 1$ and $g(x) = 6x^2 + x - 2$, find $(f + g)(x)$.

$(f + g)(x) = f(x) + g(x)$ Addition of functions
$ = (3^x + 1) + (6x^2 + x - 2)$ Substitution
$ = 3^x + 6x^2 + x - 1$ Simplify.

> **Think About It!**
> What similarities do you notice about $(f - g)(x)$ and $(g - f)(x)$? Will those similarities exist between any two functions? Justify your answer.

Example 2 Subtract Functions

Given $f(x) = -x^2 + 4x - 5$ and $g(x) = x - 7$, find each function.

a. $(f - g)(x)$

$$(f - g) = f(x) - g(x)$$
$$= (-x^2 + 4x - 5) - (x - 7)$$
$$= -x^2 + 4x - 5 - x + 7$$
$$= -x^2 + 3x + 2$$

b. $(g - f)(x)$

$$(g - f) = g(x) - f(x)$$
$$= (x - 7) - (-x^2 + 4x - 5)$$
$$= x - 7 + x^2 - 4x + 5$$
$$= x^2 - 3x - 2$$

Check

Given $f(x) = 5 \cdot 2^x - 1$, $g(x) = -3x^2 + 2x - 8$, and $h(x) = 4x - 5$, find each function. Write each answer in standard form.

a. $(f + h)(x)$

b. $(h - g)(x)$

Learn Multiplying Functions

Consider $f(x) = x^2 + x + 3$ and $g(x) = 4x$. To find $(f \cdot g)(2)$ you can find the product of $f(2)$ and $g(2)$.

$f(x) = x^2 + x + 3$	Original function	$g(x) = 4x$
$f(2) = (2)^2 + (2) + 3$	Substitute 2 for x.	$g(2) = 4(2)$
$f(2) = 9$	Simplify.	$g(2) = 8$

$f(2) \cdot g(2) = 9 \cdot 8$ or 72

Multiply the two functions and combine like terms.

$(f \cdot g)(x) = f(x) \cdot g(x)$ Multiplication of functions
$\qquad = (x^2 + x + 3) \cdot (4x)$ Substitution
$\qquad = x^2(4x) + x(4x) + 3(4x)$ Distributive Property
$\qquad = 4x^3 + 4x^2 + 12x$ Simplify.

Solve for $(f \cdot g)(2)$.

$(f \cdot g)(2) = 4(2)^3 + 4(2)^2 + 12(2)$ Substitute 2 for x.
$\qquad = 32 + 16 + 24$ or 72 Simplify.

So, $f(x) \cdot g(x) = (f \cdot g)(x)$.

> **Think About It!**
> How could you use the degrees of polynomials to determine the degree of their combined function?

Go Online You can complete an Extra Example online.

Example 3 Multiply Linear and Quadratic Functions

Given $f(x) = x^2 + 2x - 7$ and $g(x) = 3x - 10$, find $(f \cdot g)(x)$.

$$(f \cdot g)(x) = f(x) \cdot g(x)$$
$$= (x^2 + 2x - 7) \cdot (3x - 10)$$
$$= x^2(3x) + x^2(-10) + 2x(3x) + 2x(-10) - 7(3x) - 7(-10)$$
$$= 3x^3 - 10x^2 + 6x^2 - 20x - 21x + 70$$
$$= 3x^3 - 4x^2 - 41x + 70$$

Think About It!
Consider the degree of the polynomials $f(x)$ and $g(x)$. Does the degree of the product meet your expectations? Explain.

Example 4 Multiply Linear and Exponential Functions

Given $g(x) = 3x - 10$ and $h(x) = \left(\frac{1}{3}\right)^x + x$, find $(h \cdot g)(x)$.

$$(h \cdot g)(x) = h(x) \cdot h(x)$$
$$= \left[\left(\frac{1}{3}\right)^x + x\right] \cdot (3x - 10)$$
$$= \left(\frac{1}{3}\right)^x(3x) + \left(\frac{1}{3}\right)^x(-10) + x(3x) + x(-10)$$
$$= 3x\left(\frac{1}{3}\right)^x - 10\left(\frac{1}{3}\right)^x + 3x^2 - 10x$$

🌐 Example 5 Combine Functions

FINANCE According to the Wall Street Journal, the average student loan debt was more than $35,000 in 2015. Hugo graduated from college with $28,500 in student loan debt. He decides to defer his payments while he is in graduate school. However, he still accrues interest on his student loans at an annual rate of 5.65%.

Watch Out!
Percentages 5.65% = 0.0565

Part A Write an exponential function $r(t)$ to express the amount of money Hugo owes on his student loans after time t, where t is the number of years after interest began accruing.

$r(t) = a(1 + r)^t$ Equation for exponential growth
$ = 28{,}500(1 + 0.0565)^t$ $a = 28{,}500$ and $r = 5.65\%$ or 0.0565
$ = 28{,}500(1.0565)^t$ Simplify.

Part B While he is in graduate school, Hugo's parents also lend him $400 a month for rent. His parents decide not to charge him interest on this loan. Write a function $p(t)$ to represent this loan, where t is the time in years that Hugo borrows money from his parents.

Since t is the time in years, first find the amount of money Hugo borrows from his parents each year.

$12(400) = 4800$

So, $p(t) = 4800t$ represents the loan from his parents as a function of time.

Go Online You can complete an Extra Example online.

(continued on the next page)

Lesson 11-9 · Combining Functions **675**

> **Talk About It!**
> What assumptions did you make in **Part A**?

Part C Find $C(t) = r(t) + p(t)$. What does this new function represent?

$C(t) = 28{,}500(1 + 0.0565)^t + 4800t$

$C(t)$ represents the total amount of money Hugo has to repay after t years.

Part D If Hugo spends 3 years in graduate school, find the total amount of money he will have to repay.

Because Hugo spends 3 years in graduate school, $t = 3$.

$$C(t) = 28{,}500(1 + 0.0565)^t + 4800t$$
$$C(3) = 28{,}500(1 + 0.0565)^3 + 4800(3)$$
$$= 33{,}608.83 + 14{,}400$$
$$= \$48{,}008.83$$

Example 6 Combine Two Functions

SANDWICHES A sandwich shop charges $7 for a large sub and sells an average of 360 large subs per day. The shop predicts that they will sell 20 fewer subs for every $0.25 increase in the price.

Part A Let x represent each $0.25 price increase. Write a function $P(x)$ to represent the price of a large sub.

The price of a large sub is $7 plus $0.25 times each price increase.

$P(x) = 7 + 0.25x$

Part B Write a function $T(x)$ to represent the number of subs sold.

The number of subs sold is 360 minus 20 times the number of price increases.

$T(x) = 360 - 20x$

Part C Write a function $R(x)$ that can be used to maximize the revenue from sales of large subs.

The revenue from sales of large subs will be equal to the price times the number of subs sold.

$$R(x) = P(x) \cdot T(x)$$
$$= (7 + 0.25x)(360 - 20x)$$
$$= -5x^2 - 50x + 2520$$

> **Think About It**
> How could you determine the amount of money the sub shop should charge for a large sub?

Part D If the sandwich shop charges $8.50 for a large sub, find the revenue from sales of large subs. x represents each $0.25 price increase. So, $x = 6$.

$$R(x) = -5x^2 - 50x + 2520$$
$$R(6) = -5(6)^2 - 50(6) + 2520$$
$$= -180 - 300 + 2520 \text{ or } \$2040$$

Go Online You can complete an Extra Example online.

Practice

Go Online You can complete your homework online.

Examples 1 and 2

Given that $f(x) = x - 9$, $g(x) = 3x^2 - 2x + 5$, and $h(x) = -6x$, find each function.

1. $(f + g)(x)$
2. $(f - g)(x)$
3. $(f + h)(x)$
4. $(g - f)(x)$
5. $(g - h)(x)$
6. $(g + h)(x)$
7. $(h - g)(x)$
8. $(f - h)(x)$

Examples 3 and 4

Given that $f(x) = 11x$, $g(x) = x^2 - 6x + 3$, $h(x) = -x + 4$, and $j(x) = 2^x + 3$, find each function.

9. $(f \cdot g)(x)$
10. $(f \cdot h)(x)$
11. $(g \cdot h)(x)$
12. $(f \cdot f)(x)$
13. $(f \cdot j)(x)$
14. $(h \cdot j)(x)$

Examples 5 and 6

15. **TRANSPORTATION** Jan's walking speed can be represented by the function $f(t) = 4t + 3$, where t is the number of seconds. The moving walkway at an airport has a speed that can be represented by the function $g(t) = 3t + 1$.

    a. Find $f(t) + g(t)$ and explain what it represents.

    b. Find $f(t) - g(t)$ and explain what it represents.

16. **CLUBS** An improvisational acting club has 32 members. The manager of the club expects its membership to increase by 4 members per year. A photography club has 60 members and is expected to grow by 10% per year.

    a. Write a function $f(t)$ to represent the number of members in the acting club after t years.

    b. Write a function $g(t)$ to represent the number of members in the photography club after t years.

    c. Find $(f + g)(t)$ and explain what this function represents.

17. MAGIC SHOW A magician currently sells tickets to his shows for $15 and averages 180 spectators per show. He estimates that he can sell 10 more tickets for each $0.75 decrease in price.

  a. Let x represent the number of $0.75 price decreases. Write a function $P(x)$ to represent the price of a ticket and a function $T(x)$ to represent the number of tickets sold.

  b. Write a function $R(x)$ that can be used to find the revenue from ticket sales.

  c. If the magician decides to sell the tickets for $12, find his revenue.

18. GOLF A golf course offers a membership plan where players pay an annual fee of $250 plus a fee for each round of golf played. The function $C(x) = 15x + 250$ represents the total cost for playing x rounds of golf. Explain how you can think of $C(x)$ as the sum of two functions. What does each of these functions represent?

Mixed Exercises

Given that $p(x) = -2x$, $q(x) = x^2 + 5$, $r(x) = x^2 - x + 2$, and $t(x) = \left(\frac{1}{4}\right)^x - 5$ find each function.

19. $(q + t)(x)$

20. $(t - r)(x)$

21. $(p \cdot q)(x)$

22. $(p \cdot r)(x)$

23. $(q \cdot r)(x)$

24. $(q \cdot q)(x)$

25. $(p \cdot t)(x)$

26. $(p + q + r)(x)$

27. STORAGE Raul is making boxes out of sheets of cardboard to hold his video game collection. He begins with a rectangular sheet that is 20 inches long and 12 inches wide, as shown. Then he cuts identical squares from each corner and bends the sides to form a box.

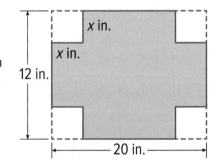

  a. If x represents the length of each side of the squares, write functions $\ell(x)$ and $w(x)$ to represent the length and width, respectively, of the resulting box.

  b. Find $(\ell \cdot w)(x)$ and explain what it represents.

  c. What is the domain of $(\ell \cdot w)(x)$? Explain.

  d. Find $(\ell \cdot w)(1.5)$ and interpret its meaning.

28. USE A MODEL Alba sells homemade granola at a farmers' market. When Alba sets the price at $4 per bag, she can sell 185 bags. She finds that for every increase of $0.25 in the price of a bag of granola, she sells 10 fewer bags.

  a. Let x represent the number of $0.25 increases in the price. Write a function $P(x)$ that gives the price of the granola as a function of x. Write a function $G(x)$ that gives the number of bags of granola Alba sells as a function of x.

  b. Explain how to combine functions to write a function $R(x)$ that gives Alba's revenue from selling granola as a function of x.

  c. Alba is considering raising the price of the granola to $4.75. Use your model to explain whether or not this is a good idea.

29. STRUCTURE Tyree makes a pattern, as shown, using shaded tiles and white tiles.

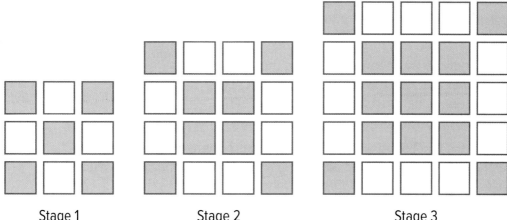

Stage 1   Stage 2   Stage 3

a. Write a function f(n) that gives the number of shaded tiles needed to make stage n of the pattern and a function g(n) that gives the number of white tiles needed to make stage n of the pattern.

b. Explain how to write a function h(n) that gives the total number of tiles needed to make stage n of the pattern. Use the function to find the total number of tiles needed to make stage 10.

c. Explain how you know your answer to **part b** is correct.

30. USE TOOLS A chemist works in a lab that maintains a constant temperature of 22°C. She heats a saline solution and then lets the solution cool. As the solution cools, she records its temperature every 3 minutes. Her data is shown in the table.

Time (minutes)	0	3	6	9	12	15
Temperature of Solution (°C)	82.0	79.1	76.3	73.6	70.9	68.4
Temperature Above the Lab Temperature	60.0	57.1	54.3	51.6	48.9	46.4

a. Write a function P(x) that models the temperature of the solution above the lab temperature in degrees Celsius x minutes after the chemist starts recording the data. Write a second function S(x) that models the temperature of the solution.

b. Explain how the function you wrote for S(x) in **part a** is a combination of two functions.

c. The chemist wants to know approximately how long it will take until the solution cools to a temperature of 45°C. Explain how to use your calculator to estimate this time to the nearest minute.

31. **CONSTRUCT ARGUMENTS** A rectangular flower bed in a garden is 12 feet long and 8 feet wide. Kazuo plans to add a gravel border around the flower bed so that the border is twice as wide along the 8-foot sides of the flower bed as it is along the 12-foot sides.

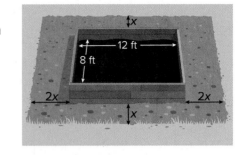

   a. Let x be the width of the gravel border along the 12-foot sides of the flower bed, as shown. Write a function $A(x)$ that gives the area of the gravel border in square feet. Explain how you can write this function as a combination of simpler functions.

   b. Kazuo said that the function $A(x)$ is a quadratic function. Therefore, as x increases, the area of the gravel border will increase up to a point, reach a maximum value, and then start to decrease. Do you agree? Justify your argument.

32. **REGULARITY** Describe the difference between the method for adding a linear and a quadratic function and the method for multiplying a linear and a quadratic function.

33. **STRUCTURE** Adding the linear function $j(x) = 5$ to the exponential function $k(x) = 3^x - 7$, gives a new exponential function $m(x)$, which is a translation of $k(x)$. What is the equation of the translated function, $m(x)$?

34. **FRAME** An 8 × 10 picture frame has a glass center that is 8 inches wide and 10 inches long. The frame that surrounds the glass can be a variety of different sizes. How much larger is the area of a framed picture than its perimeter if the frame extends x inches outside the glass in each direction?

35. **CREATE** Define $f(x)$ and $g(x)$ if $(f - g)(x) = x^2 + 14x - 12$, where $f(x)$ is a quadratic function and $g(x)$ is a linear function.

36. **CREATE** Define $f(x)$ and $g(x)$ if $(f \cdot g)(x) = x^3 - x^2 - 4x + 4$, where $f(x)$ is a quadratic function and $g(x)$ is a linear function.

37. **ANALYZE** Determine whether the following statement is *sometimes*, *always*, or *never* true. Justify your argument.

    Multiplying a linear equation with two terms by a quadratic equation with three terms will result in an equation that has three terms.

38. **WRITE** You learned that when a linear function and a quadratic function are multiplied, the product is a third-degree polynomial. What do you think would be the degree of the product of two quadratic functions? What about the degree of the sum of two quadratic functions? Explain your reasoning.

39. **FIND THE ERROR** Delfina multiplied the linear function $f(x) = 6x - 8$ by the quadratic function $g(x) = 4x^2 + 7x - 19$. She found that $(f \cdot g)(x) = 24x^3 + 42x^2 + 152$. Is Delfina's solution correct? Explain your reasoning.

40. **WHICH ONE DOESN'T BELONG?** Consider the functions $a(x)$, $b(x)$, and $c(x)$, below, and their products, $(a \cdot b)(x)$, $(a \cdot c)(x)$, and $(b \cdot c)(x)$. Determine which of the products doesn't belong. Justify your conclusion.

    $a(x) = 4x + 8$ $b(x) = -x^2 + 3x - 5$ $c(x) = 6x - 7$

Module 11 • Quadratic Functions

Review

Essential Question
Why is it helpful to have different methods to analyze quadratic functions and solve quadratic equations?

Module Summary

Lessons 11-1 and 11-2
Graphing Quadratic Functions
- The graph of a quadratic function is a parabola. The axis of symmetry intersects a parabola at the vertex. The vertex is either the lowest point or the highest point on a parabola.
- The standard form of a quadratic function is $f(x) = ax^2 + bx + c$, where a, b, and c are integers and $a \neq 0$.
- Quadratic functions can be translated like other functions.
- A quadratic function in the form $f(x) = a(x - h)^2 + k$ is in vertex form.

Lessons 11-3 through 11-6
Solving Quadratic Equations
- The solutions or roots of an equation can be identified by finding the x-intercepts of the graph of the related function.
- If the product of two factors is 0, then at least one of the factors must be 0.
- To complete the square for any quadratic expression of the form $x^2 + bx$, find one-half of b, the coefficient of x. Square the result. Then add the result to $x^2 + bx$.
- Quadratic Formula: $x = \dfrac{-b \pm \sqrt{b^2 - 4ac}}{2a}$.

Lesson 11-7
Solving Systems of Linear and Quadratic Equations
- The solution to a system of linear and quadratic equations is at the point of intersections of the graphs of each equation.
- You can use the Substitution Method or the Elimination Method to solve a system of linear and quadratic equations.

Lesson 11-8
Modeling and Curve Fitting
- The coefficient of determination, R^2, indicates how well the function fits the data.

Lesson 11-9
Combining Functions
- You can add two functions and combine the like terms.
- You can multiply two functions.

Study Organizer
Foldables
Use your Foldable to review this module. Working with a partner can be helpful. Ask for clarification of concepts as needed.

Test Practice

1. **GRAPH** Graph $f(x) = x^2 + 4x - 2$. (Lesson 11-1)

2. **MULTIPLE CHOICE** The height of a baseball is modeled by the function $h(x) = -16x^2 + 50x + 5$, where $h(x)$ represents the height, in feet, of the ball x seconds after being hit by a bat. Which of the following describes the most appropriate domain for this function? (Lesson 11-1)

   A. all integers
   B. positive integers
   C. all real numbers
   D. positive real numbers

3. **MULTIPLE CHOICE** Find the vertex of the graph of $f(x) = 4x^2 - 24x - 2$. (Lesson 11-1)

   A. $(6, -2)$
   B. $(3, -38)$
   C. $(3, -62)$
   D. $(-3, 106)$

4. **MULTIPLE CHOICE** Find the axis of symmetry of the graph of $f(x) = 4x^2 - 24x - 2$. (Lesson 11-1)

   A. $x = 6$
   B. $x = 3$
   C. $x = \frac{1}{3}$
   D. $x = -3$

5. **OPEN RESPONSE** Find the y-intercept of $f(x) = 4x^2 - 24x - 2$. (Lesson 11-1)

6. **MULTIPLE CHOICE** If $f(x) = -4x^2$, write a function $k(x)$ representing a reflection of $f(x)$ across the x-axis. (Lesson 11-2)

   A. $k(x) = -\frac{1}{4}x^2$
   B. $k(x) = \frac{1}{4}x^2$
   C. $k(x) = (-4x)^2$
   D. $k(x) = 4x^2$

7. **GRAPH** Given the parent function, $f(x) = x^2$, graph the translation 3 units to the right, $g(x)$. (Lesson 11-2)

8. **MULTI-SELECT** Use the graph of the related function to solve $x^2 + x = 20$. Select all the solutions that apply. (Lesson 11-3)

   A. -20
   B. -5
   C. 1
   D. 4
   E. 9

9. **MULTIPLE CHOICE** Which quadratic function models the graph? (Lesson 11-3)

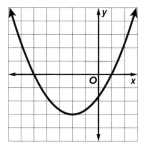

A. $y = \frac{1}{3}x^2 - \frac{4}{3}x + \frac{5}{3}$

B. $y = \frac{1}{3}x^2 - \frac{4}{3}x - \frac{5}{3}$

C. $y = \frac{1}{3}x^2 + \frac{4}{3}x - \frac{5}{3}$

D. $y = \frac{1}{3}x^2 + \frac{4}{3}x + \frac{5}{3}$

10. **OPEN RESPONSE** During a thunderstorm, a branch fell from a tree. Chantel estimates the branch fell from 25 feet above the ground.

    The formula $h = -16t^2 + h_0$ can be used to approximate the number of seconds t it takes for the branch to reach height h from an initial height of h_0 in feet. Find the time it takes the branch to reach the ground. Round to the nearest hundredth, if necessary. (Lesson 11-4)

11. **MULTIPLE CHOICE** The area of the rectangle shown is 120 square inches.

    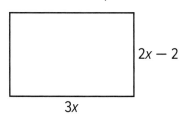

    What is the perimeter of the rectangle? (Lesson 11-4)

    A. 5 inches
    B. 23 inches
    C. 46 inches
    D. 120 inches

12. **MULTIPLE CHOICE** Solve $x^2 + 6x = 7$ by completing the square. (Lesson 11-5)

    A. 3, 9
    B. 6, 7
    C. 1, −7
    D. 3, 4

13. **OPEN RESPONSE** Write $y = 2x^2 + 4x + 6$ in vertex form. Identify the extrema, and explain whether it is a minimum or maximum. (Lesson 11-5)

14. **OPEN RESPONSE** Solve $x^2 - 20x + 27 = 8$ by completing the square. (Lesson 11-5)

15. **OPEN RESPONSE** A cat that is sitting on a boulder 5 feet high jumps off and lands on the ground. During its jump, its height h in feet is given by $h = -0.5d^2 + 2d + 5$, where d is the distance from the base of the boulder. (Lesson 11-5)

    How far is the cat from the base of the boulder when it lands on the ground? Round to the nearest tenth.

    What is the maximum height of the cat during its jump?

16. MULTIPLE CHOICE A local company manufactures brake drums. Based on their records, their daily profit can be approximated by the function $f(x) = x^2 + 3x - 19$, where x is the number of brake drums they manufacture. If $f(x)$ is negative, it means the company has lost money. What is the least number of brake drums that the company needs to manufacture each day in order to make a profit? (Lesson 11-6)

A. 3
B. 4
C. 7
D. 19

17. OPEN RESPONSE How many real solutions does the equation $2x^2 + 5x + 7 = 0$ have? (Lesson 11-6)

18. MULTIPLE CHOICE Find the solution(s) of the system shown.

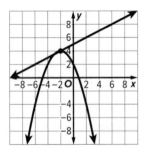

(Lesson 11-7)

A. $(-2, 4)$
B. $(-2, 4)$ and $(-4, 2)$
C. $(0, 2)$
D. $(0, 5)$

19. OPEN RESPONSE Let x represent the time in seconds and y represent the height in feet. The path of a drone can be modeled by $y = -x^2 + 19x + 1$ and the path of a projectile can be modeled by $y = 2x + 1$.

Solve a system of equations algebraically to calculate how many seconds it will take the projectile to meet the drone. (Lesson 11-7)

20. MULTIPLE CHOICE The table shows the profit earned y as it relates to the number of magazines sold x.

x	0	1	2	3	4
y	0.2	1	5	25	125

Which model best describes the data in the table? (Lesson 11-8)

A. Linear
B. Exponential
C. Quadratic
D. Square

21. MULTIPLE CHOICE Which coefficient of determination shows the best fit of the data? (Lesson 11-8)

A. $R^2 = 0.589$
B. $R^2 = 0.880$
C. $R^2 = 0.989$
D. $R^2 = 1.54$

22. OPEN RESPONSE Given $f(x) = 7x^2 + 22x - 6$ and $g(x) = 3x - 9$, find $(g - f)(x)$. (Lesson 11-9)

Module 12
Statistics

Essential Question
How do you summarize and interpret data?

What will you learn?
How much do you already know about each topic **before** starting this module?

KEY

👎 — I don't know.   👍 — I've heard of it.   👍 — I know it!

	Before			After		
find measures of center in a data set						
calculate percentiles						
represent data in dot plots, bar graphs, and histograms						
collect data and analyze bias						
represent data in box plots						
calculate standard deviation						
analyze data distributions						
transform linear data						
compare two data sets						
represent data in two-way frequency tables						
find frequencies, including marginal and conditional relative frequencies						

Foldables Make this Foldable to help you organize your notes about statistics. Begin with 8 sheets of $8\frac{1}{2}''$ by 11'' paper.

1. **Fold** each sheet of paper in half. Cut 1 inch from the end to the fold. Then cut 1 inch along the fold.

2. **Write** the lesson number and title on each page.

3. **Label** the inside of each sheet with *Definitions* and *Examples*.

4. **Stack** the sheets. Staple along the left side. Write *Statistics* on the first page.

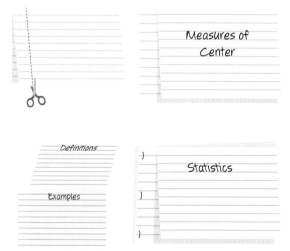

What Vocabulary Will You Learn?

- bar graph
- bias
- box plot
- categorical data
- conditional relative frequency
- distribution
- dot plot
- extreme values
- five-number summary
- histogram
- interquartile range
- joint frequencies
- linear transformation
- lower quartile
- marginal frequencies
- measurement data
- measures of center
- measures of spread
- median
- negatively skewed distribution
- outlier
- percentile
- population
- positively skewed distribution
- quartile
- range
- relative frequency
- sample
- standard deviation
- statistic
- symmetric distribution
- two-way frequency table
- two-way relative frequency table
- univariate data
- upper quartile
- variable
- variance

Are You Ready?

Complete the Quick Review to see if you are ready to start this module.
Then complete the Quick Check.

Quick Review

Example 1

Add the set of values.

12.5, 3.4, 1.75, 9

```
  12.5
   3.4
   1.75
+  9
──────
  26.65
```
Align the numbers at the decimal.

The sum is 26.65.

Example 2

Write the fraction $\frac{33}{80}$ as a percent. Round to the nearest tenth.

$\frac{33}{80} \approx 0.413$ Simplify and round.

$0.413 \cdot 100 = 41.3$ Multiply the decimal by 100.

$\frac{33}{80} \approx 41.3\%$ Write as a percent.

Quick Check

Add each set of values.

1. 13.2, 15, 17.68
2. 4.5, 1.95, 2.36, 8.1
3. $\frac{2}{3}, \frac{3}{4}, \frac{5}{6}, \frac{9}{10}$
4. −8, −4, 1, 5

Write each fraction as a percent. Round to the nearest tenth.

5. $\frac{14}{17}$
6. $\frac{7}{8}$
7. $\frac{107}{125}$
8. $\frac{625}{1024}$

How did you do?

Which exercises did you answer correctly in the Quick Check?

Lesson 12-1

Measures of Center

Learn Mean, Median, and Mode

- A **variable** is any characteristic, number, or quantity that can be counted or measured. A variable is an item of data.
- Data that have units and can be measured are called **measurement data** or quantitative data.
- Data that can be organized into different categories are called **categorical data** or qualitative data.
- Measurement data in one variable, called **univariate data**, are often summarized using a single number to represent what is average, or typical.
- Measures of what is average are called **measures of center** or central tendency. The most common measures of center are mean, median, and mode.

  Mode: the value of the elements that appear most often in a set of data

  Mean: the sum of the elements of a data set divided by the total number of elements in the set

  Median: the middle element, or the mean of the two middle elements, in a set of data when the data are arranged in numerical order

Today's Goals
- Represent sets of data by using measures of center.
- Represent sets of data by using percentiles.

Today's Vocabulary
variable
measurement data
categorical data
univariate data
measures of center
percentile

Talk About It
A set of data can have only one value for the mean and median. How many values can a set of data have for the mode? Explain your reasoning.

Example 1 Measures of Center

BASKETBALL The table shows the total number of points scored in several NCAA Championship Basketball Games. Find the mean, median, and mode of the data.

Year	Score	Year	Score
2016	150	2008	143
2015	131	2007	159
2014	114	2006	130
2013	158	2005	145
2012	126	2004	155
2011	94	2003	159
2010	120	2002	116
2009	161		

(*continued on the next page*)

Study Tip
Mean When calculating the mean, your answer will always be between the least and greatest values of the data set. It can never be less than the least value or greater than the greatest value.

Watch Out!

Median If there is an even number of values in the set of data, you will have to find the average of the two middle values to find the median. To do this, divide the sum of the two middle values by 2.

Think About It!

Carlos says that a set of data cannot have the same mean and mode. Do you agree or disagree? Explain your reasoning or provide a counterexample.

Study Tip

Tools To quickly calculate the mean $\bar{x}$ and the median **Med** of a data set, enter the data as **L1** in a graphing calculator, and then use the **1-VAR Stats** feature from the **CALC** menu.

Mean

To find the mean, find the sum of all the points and divide by the number of years in the data set.

$$= \frac{150 + 131 + 114 + 158 + 126 + 94 + 120 + 161 + 143 + 159 + 130 + 145 + 155 + 159 + 116}{15}$$

$$= \frac{2061}{15} \text{ or about } 137.4.$$

The mean is about 137 points.

Median and Mode

To find the median, order the points from least to greatest and find the middle value.

94, 114, 116, 120, 126, 130, 131, 143, 145, 150, 155, 158, 159, 159, 161

median   mode

The median is 143.

From the arrangement of data values, we can see that 159 is the only value that appears more than once. So, the mode is 159.

The mean and median are close together, so they both represent the average of the scores well. Notice that the median is greater than the mean. This indicates that the scores less than the median are more spread out than the scores greater than the median. The mode is greater than most of the scores.

Check

FOOTBALL The data show the number of interceptions thrown during one regular season for each team in the NFC. Find the mean, median, and mode. Round to the nearest whole number, if necessary.

13	17	10	12	22	14	8	11
9	12	14	18	12	8	15	11

Mean: ___?___

Median: ___?___

Mode: ___?___

Go Online You can complete an Extra Example online.

Explore Finding Percentiles

Online Activity Use a real-world situation to complete the Explore.

INQUIRY How can you describe a data value based on its position in the data set?

Go Online You can watch a video to see how to find percentile rank.

Learn Percentiles

A **percentile** is a measure that is often used to report test data, such as standardized test scores. It tells us what percent of the total scores were below a given score.

- Percentiles measure rank from the bottom.
- There is no 0 percentile rank. The lowest score is at the 1st percentile.
- There is no 100th percentile rank. The highest score is at the 99th percentile.

Key Concept • Finding Percentiles

To find the percentile rank of an element of a data set, use these steps.

Step 1 Order the data values from greatest to least.

Step 2 Find the number of data values less than the chosen element. Divide that number by the total number of values in the data set.

Step 3 Multiply the value from Step 2 by 100.

Example 2 Find Percentiles

FIGURE SKATING The table shows the total points scored by each country in the team figure skating event in the 2014 Olympic Winter Games. Find the United States' percentile rank.

Country	Score
Canada	65
China	20
France	22
Germany	17
Great Britain	8
Italy	52
Japan	51
Russia	75
Ukraine	10
United States	60

Study Tip

Percent vs Percentile *Percent* and *percentile* mean two different things. For example, a score at the 40th percentile means that 40% of the scores are either the same as the score at the 40th percentile or less than the score at that rank. It does not mean that the person scored 40% of the possible points.

(continued on the next page)

Go Online You can complete an Extra Example online.

Lesson 12-1 • Measures of Center **689**

Step 1 Order the data.

Order the data values from greatest to least.

Country	Score
Russia	75
Canada	65
United States	60
Italy	52
Japan	51
France	22
China	20
Germany	17
Ukraine	10
Great Britain	8

Step 2 Divide.

Divide the number of teams with scores lower than the United States by the total number of teams.

$$\frac{\text{number of teams below the United States}}{\text{Total number of teams}} = \frac{7}{10}$$

Step 3 Multiply by 100.

$\frac{7}{10} \cdot 100$ or 70

The United States figure skating team scored at the 70th percentile in the 2014 Olympics.

Check

DRUM CORPS The table shows the scores of the corps that competed in the Drum Corps International World Championship World Class Finals in 2015.

Corps	Score
Bluecoats	96.925
Blue Devils	97.650
Blue Knights	91.850
Blue Stars	85.150
Boston Crusaders	86.800
Carolina Crown	97.075
Crossmen	85.025
Madison Scouts	88.750
Phantom Regiment	90.325
Santa Clara Vanguard	93.850
The Cadets	95.900
The Cavaliers	88.325

The Cavaliers scored at the __?__ th percentile. The __?__ scored at the 75th percentile.

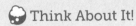 **Go Online** You can complete an Extra Example online.

Study Tip

Percentiles The team with the highest score is at the 99th percentile rank, and the team with the lowest score is at the 1st percentile rank.

Think About It!

Which team scored at the 40th percentile?

Practice

Go Online You can complete your homework online.

Example 1

Find the mean, median, and mode for each data set.

1. {17, 11, 8, 15, 28, 20, 10, 16}

2. {2.5, 6.4, 7.0, 5.3, 1.1, 6.4, 3.5, 6.2, 3.9, 4.0}

3.

2	1	1	5	7
3	2	4	6	2

4.

50	30	40	10
20	80	60	90
10	30	110	70

5. number of students helping at a booth each hour: 3, 5, 8, 1, 4, 11, 3

6. weight in pounds of boxes loaded onto a semi-truck: 201, 201, 200, 199, 199

7. car speeds in miles per hour observed by a highway patrol officer: 60, 53, 53, 52, 53, 55, 55, 57

8. number of songs downloaded by students last week in Ms. Turner's class: 3, 7, 21, 23, 63, 27, 29, 95, 23

9. ratings of an online video: 2, 5, 3.5, 4, 4.5, 1, 1, 4, 2, 1.5, 2.5, 2, 3, 3.5

Example 2

MARCHING BAND A competition was recently held for 12 high school marching bands. Each band received a score from 0 through 100, with 100 being the highest.

Band	Score	Band	Score
Freeport	78	Madison	69
Ross	85	Monmouth	67
Hamilton	88	Carlisle	65
Groveport	94	Dupont	48
Lakehurst	56	Cave City	90
Benton	77	Monroe	80

10. Find Hamilton High School's percentile rank.

11. Find Monmouth High School's percentile rank.

12. Find Freeport High School's percentile rank.

Mixed Exercises

13. **REASONING** The mean number of people at the movies on Saturday nights throughout the year is 425, and the median is 412. Explain why the mean could be slightly higher.

14. **REASONING** The mode length of time it takes to fly from New York City to Chicago is 2 hours 35 minutes, and the mean is 3 hours 15 minutes. Explain why the mode could be slightly lower.

Lesson 12-1 • Measures of Center 691

15. **FOOTBALL** Find the mean, median, and mode for the data set. The weights in pounds of 5 offensive linemen of a football team: 217, 212, 285, 245, 301.

16. **WEB SITES** The ratings for a new recipe Web site varied from very low, 1 point, to very high, 10 points, with half of the scores receiving a rating of 7. If a new rating of 7 were added to the data set, how would the mode be affected? Explain.

17. Find a mean of {16, 19, 22, 27, 33, 19, 25}.

18. **SINGING** In a singing competition that involved 50 contestants, Reina's score ranked higher than 40 of the contestants. In what percentile did Reina score?

19. **SPORTS** The table shows the number of points scored by a basketball team during their first several games. Find the mean, median, and mode of the number of points scored.

Game	1	2	3	4	5	6	7	8
Points	43	50	52	47	55	61	48	56

20. **PERFORMANCE** At a bodybuilding competition, Shawnte earned a score of 42 points. There were 19 competitors who received a lower score than Shawnte and 5 competitors who earned a higher score. What was Shawnte's percentile rank in the bodybuilding competition?

21. **GRADES** On her first four quizzes, Rachael has earned scores of 21, 24, 23, and 17 points. What score must Rachael earn on her fifth and final quiz so that both the mean and median of her quiz scores is 21?

22. **QUIZZES** Sequon scored 95, 86, 81, 83, and 95 on his math quizzes this quarter. Find the mode of his quiz scores.

23. **BAND** Out of the 30 bands at the competition, Coastal High School's band scored higher than 27 others. Find the percentile rank for Coastal High School's band.

24. **SHOPPING** The table shows the prices of comparable laptop computers at different retailers.

    a. Find the mean, median, and mode of the prices.

    b. Why are the mean and median much lower than the mode?

    c. After deliberation, Nikki is interested in buying a laptop from either retailer C, F, or J. What are the percentile ranks for the laptop at each of these retailers?

Retailer	Price ($)
A	389
B	425
C	350
D	499
E	475
F	360
G	319
H	425
I	299
J	379

25. **NOVELS** The table shows the lengths (in words) of the seven novels on the required reading list of Miguel's language arts class.

    a. Find the mean, median, and mode for the data set.

    b. Predict which novels will be lower than the 50th percentile in length. Verify your prediction.

    c. If *Lord of the Flies* comes with an additional 10,020-word online reading assignment, then how will the median length be affected? How will the mean be affected?

Novel	Number of Words
Old Yeller	35,968
Lord of the Flies	59,900
Moby Dick	206,052
Jane Eyre	183,858
Great Expectations	183,349
Call of the Wild	31,750
The Color Purple	66,556

26. **VOLUNTEERING** The table shows the number of hours different students spent volunteering as part of a community outreach program. Find the mean, median, and mode of the data set.

Volunteer Hours				
25	30	35	40	35
25	50	45	25	90

27. **USE A SOURCE** Research the total medal counts for Canada, France, Japan, Russia, Brazil, and Great Britain at the 2016 Rio de Janeiro Olympics. Make a table of the data you collect. Then find the percentile rank of each country.

28. **CONSTRUCT ARGUMENTS** The table shows the number of pet adoptions each week for a shelter over a two-month period. If there are 65 pets adopted next week, how are the mean, median, and mode affected?

Pet Adoptions			
7	19	26	20
23	21	24	20

Lesson 12-1 • Measures of Center 693

29. **STATE YOUR ASSUMPTION** A data set has a mean of 37, a median of 36.5, and a mode of 37. What assumption(s) can you make about the dataset?

30. **BOWLING** The table shows Lucinda's score for each of her last ten bowling games.
    a. Find the mean, median, and mode of the scores. Round to the nearest whole number.
    b. Why is the mean slightly higher than the median?

Game	Score
1	220
2	235
3	255
4	210
5	240
6	220
7	225
8	220
9	250
10	210

31. **DANCE COMPETITION** At a dance competition, Pascal earned a score of 73 points. There were 12 competitors who received a lower score than Pascal and 3 competitors who earned a higher score. What was Pascal's percentile rank in the dance competition?

32. **CREATE** Create a data set that has a mean of 11, a median of 10, and a mode of 8.

33. **WRITE** Describe how an outlier value that is greater than the numbers in the data set affects each measure of center.

34. **ANALYZE** Determine whether the statement is *true* or *false*. If it is false, explain how to make the statement true.

    To find percentile rank, divide the selected value by the total of all the values.

35. **PERSEVERE** Describe the effect on the mean, median, and mode of a set when all the items in the set are multiplied by the same number.

36. **WHICH ONE DOESN'T BELONG?** Analyze each situation. Which situation is NOT best described by the median of the data? Explain.

An art gallery has many items for sale that are reasonably priced, but it also carries luxury priced paintings.	Most of the students volunteered 2 hours each week, but James volunteered 8 hours per week.	The amusement park had about the same number of attendees each day. On the annual bring-a-friend-for-free day, the number of attendees tripled.

37. **FIND THE ERROR** Julio is studying botany and has been tracking the growth of 10 tomato plants each week. The first week, the plants measured the following growth: 1 in., 1.5 in., 2.2 in., 0.5 in., 1 in., 1.25 in., 1.4 in., 2 in., 2.1 in., 1.9 in. In his research paper, Julio includes the median growth value for the week. Has Julio chosen the best measure of center to describe the plant growth? Explain.

38. **STRUCTURE** Explain how you determine that a data set is best described by the mean.

39. **WRITE** Explain in your own words the process for finding a percentile rank.

694 Module 12 · Statistics

Lesson 12-2

Representing Data

Learn Dot Plots

One way to represent data is by using a **dot plot**, which is a diagram that shows the frequency of data on a number line.

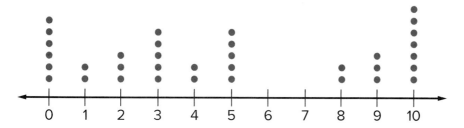

Key Concept • Making Dot Plots

Step 1 Write the data points in order from least to greatest.

Step 2 Make a number line that starts at the least data point and ends at the greatest data point. Choose an appropriate scale.

Step 3 Plot the dots on the number line. Stack the points when there is more than one data point with the same number.

Step 4 If appropriate, include a label for the number line and title for the dot plot.

Example 1 Make a Dot Plot

Represent the data as a dot plot.

11, 12, 14, 15, 12, 13, 15, 13, 9, 15, 12, 13, 15, 15, 11

Step 1 Write the data points in order from least to greatest.

9, 11, 11, 12, 12, 12, 13, 13, 13, 14, 15, 15, 15, 15, 15

Step 2 Make a number line.

The data are whole numbers ranging from 9 to 15. So, make a number line starting at 9 with intervals of 1.

Step 3 Plot the dots on the number line.

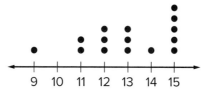

Step 4 If appropriate, include a label for the number line and title for the dot plot.

Because no information is given regarding what these data represent, no title is needed for this dot plot.

 Go Online You can complete an Extra Example online.

Today's Goals
- Represent sets of data by using dot plots.
- Determine whether discrete or continuous graphical representations are appropriate, and represent sets of data by using bar graphs or histograms.

Today's Vocabulary
dot plot
bar graph
histogram

💭 Think About It!
What would be the benefit of representing data in a dot plot?

Study Tip
Accuracy Count the listed data and check that the number of data points matches the sum of the frequency table. Missing just one or two pieces of data can change the values of statistics.

Check

Represent the data as a dot plot.

8, 6, 0, 2, 7, 1, 8, 1, 4, 8, 0, 1, 2, 8, 4, 7, 1, 5, 9, 1

🌐 Example 2 Make a Dot Plot by Using a Scaled Number Line

INTERNET USAGE The data show Internet users from Middle Eastern countries as a percentage of their total population. Represent the data as a dot plot.

96.4	57.2	33.0	74.7	86.1	78.7	80.4
78.6	64.6	91.9	65.9	28.1	93.2	22.6

Step 1 Write the data points in order from least to greatest.

22.6, 28.1, 33.0, 57.2, 64.6, 65.9, 74.7, 78.6, 78.7, 80.4, 86.1, 91.9, 93.2, 96.4

Step 2 Make a number line.

The data range from 22.6 to 96.4. Since these data represent a broad range with specific values, it is unlikely that any data point is represented more than once. To represent the data in a meaningful way, scale the number line.

Step 3 Plot the dots on the number line.

Internet Usage from Middle Eastern Countries

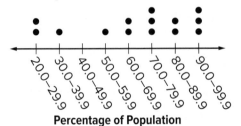

Percentage of Population

Step 4 If appropriate, include a label for the number line and title for the dot plot.

▶ **Go Online** You can complete an Extra Example online.

💬 **Talk About It!**
How would these data appear if the number line were scaled by 1 instead of 10? Do you think a scale of 1 would be a good way to represent the data? Explain.

Check

MOUNTAINS The data give the elevation of the highest mountain peaks in the United States. Create a dot plot that best represents the data.

| 20,308 | 18,009 | 17,402 | 16,421 | 16,391 |
| 16,237 | 15,325 | 14,951 | 14,829 | 14,573 |

Learn Bar Graphs and Histograms

A **bar graph** is a graphical display that compares categories of data using bars of different heights. Bar graphs are used when the data are discrete. To indicate this, there is a space between each of the bars.

A **histogram** is a graphical display that uses bars to display numerical data that have been organized in equal intervals. A histogram represents continuous data, so the bins have no spaces between them.

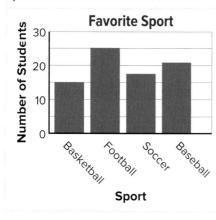

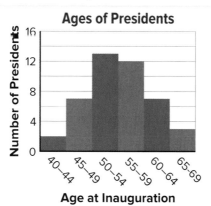

Key Concept • Making a Bar Graph or Histogram

Step 1 Determine whether the data should be represented as a bar graph or histogram.

Step 2 Determine appropriate categories or bins, and tally the data, if necessary.

Step 3 Draw bars to represent each category or bin.

Step 4 Label the axes. If appropriate, include a title for the graph.

🌐 Apply Example 3 Determine an Appropriate Graph for Discrete Data

OLYMPICS The table shows the total number of Olympic medals won by U.S. athletes competing in selected events from the first Summer Olympics in 1896 through 2012. Make a graph of the data to show the total medals won for each sport.

1 What is the task?

Describe the task in your own words. Then list any questions that you may have. How can you find answers to your questions?

Determine whether the data should be represented as a bar graph or histogram. These data represent discrete, categorical data, so use a bar graph.

2 How will you approach the task? What have you learned that you can use to help you complete the task?

I will tally the number of medals won for each sport and then I'll create the bar graph. Each event will be represented by one bar. I have learned how to make a bar graph.

3 What is your solution?

Use your strategy to solve the problem.

Event	Gold	Silver	Bronze	Total
Boxing	49	23	39	111
Diving	48	41	43	132
Swimming	230	164	126	520
Track & Field	319	247	193	759
Wrestling	52	43	34	129

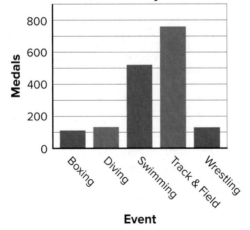

4 How can you know that your solution is reasonable?

✏️ Write About It! Write an argument that can be used to defend your solution.

Because the graph is supposed to represent the total number of medals in each event, it makes sense for the data to be organized in categories. Categorical data should be represented by a bar graph.

Check

VIDEO GAMES The table shows the number of active video game players in each country. Make a graph that best displays the data.

Country	Australia	Brazil	France	Germany	Italy	Poland	Spain	Turkey	UK	US
Players (millions)	9.5	40.2	25.3	38.5	18.6	11.8	17	21.8	33.6	157

🌐 Example 4 Determine an Appropriate Graph for Continuous Data

MARATHON The results of the top finishers of the 2015 New York City Marathon, wheelchair division, are given below. Determine whether the data are *discrete* or *continuous*. Then make a graph.

1:30:54   1:30:55   1:34:05   1:35:19   1:35:21   1:35:37   1:35:38   1:36:45
1:36:59   1:38:39   1:39:22   1:39:22   1:39:27   1:39:27   1:40:36   1:43:04

Step 1 Because racers can finish with any time, the data are continuous and you can use a histogram.

(continued on the next page)

> **Think About It!**
> Describe the histogram. What does it show you about the racers' times? What do the gaps in the graph represent?

Step 2 Because the data are spread over several minutes, group the data by the minute. Then, tally each interval.

Time (h:m:s)	Frequency
1:30:00–1:30:59	2
1:31:00–1:31:59	0
1:32:00–1:32:59	0
1:33:00–1:33:59	0
1:34:00–1:34:59	1
1:35:00–1:35:59	4
1:36:00–1:36:59	2

Time (h:m:s)	Frequency
1:37:00-1:37:59	0
1:38:00-1:38:59	1
1:39:00-1:39:59	3
1:40:00-1:40:59	1
1:41:00-1:41:59	0
1:42:00-1:42:59	0
1:43:00-1:43:59	1

Steps 3 and 4 Draw a bar to represent each bin. Label the axes. Include a title for the graph.

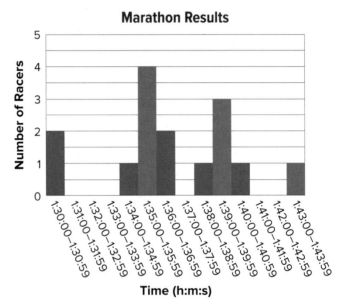

Check

PHOTO SHARING The table shows the users of a photo sharing app by age group. Make a graph that best displays the data.

Age	18-24	25-34	35-44	45-54	55-64	65+
Users (%)	45	26	13	10	6	1

Go Online You can complete an Extra Example online.

Practice

Go Online You can complete your homework online.

Examples 1 and 2

1. **READING** The table shows the number of books read by students in a summer reading program. Make a dot plot of the data.

Number of Books Read				
3	8	5	6	5
4	5	5	4	5
5	6	8	8	4

2. **QUIZ SCORES** Represent the quiz scores as a dot plot. Scale the number line as needed.

   50, 45, 24, 28, 27, 38, 21, 22, 23, 42, 41, 35, 37, 25, 43

Examples 3 and 4

3. **SURVEY** A survey was conducted among students in Mr. Dalton's science class to determine a field trip destination. The results are shown in the table at the right. Make a graph to display the data.

Destination	Number of Votes
zoo	6
museum	4
observatory	11
state park	7

4. **MOVIES** In a survey, students were asked to name their favorite type of movie. Of those surveyed, 8 chose action movies, 6 chose comedies, 5 chose horror movies, 3 chose dramas, and 7 chose science fiction movies (sci-fi). Determine whether the data are discrete or continuous. Then make a graph.

5. **CONCERT** The table shows the number of attendees by age at a concert. Determine whether the data should be shown in a *bar graph* or *histogram*. Then make an appropriate graph for the data.

Concert Attendees	
0–9	400
10–19	1440
20–29	2400
30–39	2000
40–49	960
50–59	560
60–69	240

Mixed Exercises

6. **PRIZES** The table shows the number of prizes won by customers at a carnival game each of the past several days. Determine whether the data are discrete or continuous. Then make an appropriate graph for the data.

Prizes Won				
37	29	53	32	42
21	41	45	17	27
44	34	24	34	31
19	51	48	35	54
46	38	39	49	25

7. **JOGGING** The number of miles Lisa jogged each of the last 10 days are 3, 4, 6, 2, 5, 8, 7, 6, 4, and 5.

   a. Choose the most appropriate type of data display and graph the data.

   b. How many days did Lisa jog at least 4 miles?

   c. What was the greatest number of miles she jogged in a day?

8. **MOVIES** The number of movies that are released theatrically each year are shown in the table.

   a. Select an appropriate display for the data. Explain your reasoning.

Year	2010	2011	2012	2013	2014	2015	2016
Number of Movies Released	563	609	678	661	709	708	718

   Source: comScore, MPAA

   b. Make a graph of the data.

Lesson 12-2 • Representing Data

9. **RUNNING** The ages of the participants in a 10K race at Masonville are 65, 47, 23, 70, 41, 55, 32, 29, 56, 39, 12, 57, 25, 33, 15, 18, 35, 22, 63, 49, 23, 30, 37, 40, and 50.

   a. Construct an appropriate data display for the data.
   b. How many participants are less than 30 years old?
   c. In what interval is the most frequent age?

10. **ORCHESTRA** The ages of the members of an orchestra are 39, 43, 31, 53, 41, 25, 35, 46, 27, 34, 37, 26, 51, 29, 36, 40, 33, 28, 48, 26, 42, and 38 years. Make a graph of the data.

11. **PRECISION** A scientific research study tracks the growth of an insect in millimeters. The growth data for each insect in the study during week 1 are 1.1, 1.25, 1.3, 1.67, 1.9, 2.35, 2.1, 2.3, 1.5, 1.7, 2.25, 2.1, 2.45, 1.37, 1.83. The scientist is preparing a histogram to show the distribution of growth across the population. How should the scientist break down his data into categories?

12. **PETS** The pets owned by Liza's classmates are rabbit: 2, dog: 6, cat: 3, horse: 2, bird: 5, mouse: 1, fish: 3, and other: 1.

    a. Make a dot plot of the data.
    b. How many types of pets are represented by the dot plot?
    c. Which pet is the most popular?

13. **ANALYZE** Make two conclusions about a product that received the ratings shown in the dot plot. Justify your conclusions.

14. **REGULARITY** Explain when a histogram is the best model for data, and describe the process of creating a histogram.

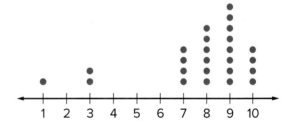

15. **WRITE** Explain why it may be necessary to scale the number line of a dot plot.

16. **PERSEVERE** Using the data provided in the double bar graph about peanut butter, what are two conclusions the grocery store could infer?

17. **STRUCTURE** How is a bar graph similar to a histogram? How is it different?

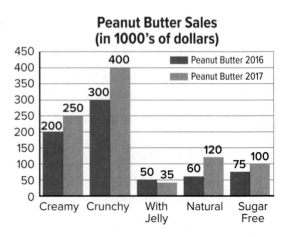

Lesson 12-3

Using Data

Explore Phrasing Questions

Online Activity Use a real-world situation to complete the Explore.

> **INQUIRY** How can the way you collect data affect the results?

Today's Goals
- Identify potential bias in sampling methods and questions.
- Identify potential bias in statistics and representations of data.

Today's Vocabulary
population
sample
bias
statistic

Learn Collecting Data

A **population** consists of all the members of a group of interest about which data will be collected. Since it may be impractical to examine every member of a population, a subset of the group, called a **sample**, is sometimes selected to represent the population. The sample can then be analyzed to draw conclusions about the entire population.

Sample data are often used to estimate a characteristic of a population. Therefore, a sample should be selected so that it closely represents the entire population. Also, the larger the sample size, or the more samples taken, the better it represents the population.

A **bias** is an error that results in a misrepresentation of a population. If a sample favors one conclusion over another, the sample is biased and the data are invalid.

Talk About It!
Some polls use both landlines and cell phones. How might this alleviate the issue of bias with the landline-only sampling method?

Example 1 Sample Bias

POLLS Before the 2010 elections for members of the U.S. House of Representatives, pollsters called American households on their landline phones to see how they planned to vote. What kind of sample bias might have affected the poll?

Step 1 Identify the intended population.

The population is all likely voters.

Step 2 Identify the sample method.

The data for this poll were collected over landline phones, so the sample consists of likely voters who have a landline.

Step 3 Determine potential bias.

Because not all likely voters have landline phones, the results could be skewed because not all likely voters are available for this sample.

Go Online You can complete an Extra Example online.

Check

SOCIAL MEDIA Shia wants to determine the age of the average internet user. He posts a poll to his friends on a social media site, asking their age. Is this a good sample? If not, what kind of sample bias might have affected the poll?

A. This is a good sample.

B. This is not a good sample. He asked only people on the Internet.

C. This is not a good sample. He asked only about users' ages.

D. This is not a good sample. He asked only his friends on a specific Web site.

Example 2 Question Bias

SOFT DRINKS A survey organization wants to see what percent of New York City citizens support a ban on soft drinks. The question posed is, "Do you support a ban on soft drinks, which contribute to heart disease and tooth decay?"

Part A Identify bias.

Step 1 Identify the purpose of the question: To find the percent of New York citizens who support a ban on soft drinks.

Step 2 Identify potential bias in the question: The question lists some of the health risks of soft drinks. This might make respondents more likely to respond that they do support a ban.

Part B Identify interests.

The bias in the question might make respondents more likely to support a ban. This bias could serve the interests of health groups who want to ban soft drinks or companies who sell competing drinks, like juices.

Check

FILM One of your friends wants to determine whether people in your class prefer to watch movies or television. She asks, "Do you prefer to watch movies or television?"

Part A Does this question potentially bias the results?

A. No; the question is as neutral as possible.

B. Yes; your friend asked only people in your class.

C. Yes; your friend provided only two options.

D. Yes; the framing of the question influences the respondent to choose *television*.

Part B Whose interests might be served by asking the question in this way?

Go Online You can complete an Extra Example online.

Think About It!
Rewrite the question to try to remove the bias.

Study Tip
Assumptions When you identify and try to remove bias, you will make assumptions about what information is important for the question and what might create undue bias. For example, you need to consider whether details of the plan are relevant and whether the question includes persuasive language.

Learn Using Statistics and Representations

A **statistic** is a measure that describes a characteristic of a sample. Like data, statistics and representations of data are nonneutral. When the average of a set of data is discussed, it uses a measure of center: mean, median, or mode. However, depending on the data set and what information is being conveyed, one measure of center might not give the whole picture of the data. Even if the data is being discussed in whole, it can also be misrepresented. For example, a person might manipulate the scales of the axes of a graph or how the data are represented graphically to misrepresent the data.

Example 3 Data Summaries

TEACHING A teacher wants to tell his students how the average student did on an exam, so he looks at the scores in his gradebook. Two students scored a 0 because they stopped showing up for class in the last month and did not take the exam. He uses the mean, 71, as the measure of center. Does the mean accurately represent these data?

0, 0, 82, 83, 85, 87, 88, 88, 91, 91, 91

Step 1 Identify the other measures of center. Round your answer to the nearest unit.

median = 87

mode = 91

Step 2 Analyze the measures of center and how they align with the information the teacher wants to convey.

Mean: The mean, 71, is affected by the two 0 scores. However, no one who showed up for the exam scored below an 82, so the mean does not do a good job of indicating the performance of the students who took the exam.

Median: The median, 87, is not affected by the extreme values. It provides a more accurate average for how students performed on the test because it includes the scores of the two students who did not take the exam at all.

Mode: The mode, 91, is both the score most students received and the highest score received on the exam, but it does not accurately portray how students performed on average.

Because the teacher wants to discuss the performance of students who took the exam, the mean is not the best measure of center. It indicates that all students who took the exam performed worse than their actual scores.

Math History Minute

With M. A. Girschick, **David Blackwell** (1919–2010) authored the classic book *Theory of Games and Statistical Decisions*. In 1965, he became the first African American president of the American Statistical Society.

Go Online You can complete an Extra Example online.

Check

READING Karen writes down the number of books she has read for each of her classes so far: 5, 6, 4, 5, 5, 6, 5, 4, 5. Using the mean, 5, she says that she reads around 5 books on average for an English course. Does the mean accurately represent the data for the situation? Explain.

A. No; the mean is overly influenced by the low numbers.

B. No; the mean is overly influenced by the high numbers.

C. No; the mean doesn't tell how many pages are in the average book.

D. Yes; the mean accurately represents the data for the situation.

Example 4 Data Representation

SOCCER A group compares two soccer players who play the same position for different teams. They make a graph of the number of goals scored throughout the season for each player. Do the graphs misrepresent the data? Whose interests might be served by the representation?

Part A Identify misleading representations.

Step 1 Identify the purpose of the graphs. The purpose of the graphs is to compare the number of goals each soccer player scored.

Step 2 Identify differences in the graphs. The graphs appear to be the same in terms of the data being represented, but the *y*-axis for the second player goes up to 90 in increments of 10, whereas the first goes up to 45 in increments of 5.

Step 3 Identify how this affects the representation of the data. Although the numbers are the same, the scale for Player 1 makes it look like he is scoring more goals than Player 2.

Part B Identify interests.

The misleading representation of the data makes it appear that Player 1 scores more goals than Player 2. This bias could serve the interests of the team, sponsors of Player 1's team, or Player 1's agent.

Go Online You can complete an Extra Example online.

Think About It!

If the two graphs had the same scales but were comparing a goalkeeper and a forward, how might the data be misleading when comparing the skill of the two players?

Use a Source

Find a graph or set of statistics online. Ask yourself, are the data accurately represented? Whose interests are served by the graph or statistics?

Practice

Example 1

1. **SPORTS** Awan wants to know what the favorite sport is among students. To find out, he asks everyone he sees leaving school after basketball practice. Identify the intended population and determine the potential sample bias.

2. **STORES** Raya wants to conduct a survey at a nearby mall to determine which are the mall's most popular stores. How could she choose a sample that is unbiased?

Example 2

3. **MUSIC** Shea is shopping online, and a survey question pops up that says, "Music education enriches student learning. Do you support music education in schools?"

   a. Identify potential bias in the question.

   b. Identify whose interests may be served by the question.

4. **CANDIDATES** There are three candidates for mayor. To investigate how the townspeople feel about the candidates, a newspaper posts a poll that lists the three candidates and asks which candidate people support. The poll appears on the same page as an opinion piece in support of one of the candidates.

   a. Identify potential bias in the question.

   b. Identify whose interests may be served by the question.

Example 3

5. **BUTTERFLIES** Tania recorded the number of butterflies she saw on her daily runs each day for a week. The numbers are: 1, 8, 2, 2, 5, 6, and 4. Find the mean, median, and mode of the data. Which measure(s) are appropriate to accurately summarize the data?

6. **OUTLIERS** In a data set with an outlier, which measure of center, mean or median, is the better measure to use to describe the center of the data? Explain your reasoning.

Example 4

7. **SALES** The graphs show the number of T-shirts sold at a baseball tournament for two years by two different vendors. The tournament director wants to compare the vendors. Do the graphs misrepresent the data? How does that difference affect the interpretation?

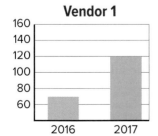
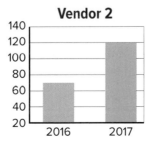

8. **SCALE** If the same set of data is graphed with a scale of 0 to 10 on the y-axis and then with a scale of 0–100 on the y-axis, what effect does that have on the representation of the data?

Mixed Exercises

9. **SCIENCE** A school wants to know which area of science; physics, biology, or chemistry, is most interesting to its students. Would it be better to survey students in a class that is an elective or required to get a sample with the least bias? Explain your reasoning.

10. **TAX** Before surveying people about whether they favor or oppose a proposed tax, the surveyors want to present information about the tax. Suppose the surveyors give facts about the tax without giving opinions. How could the facts given by the surveyors introduce bias?

11. **CONSTRUCT ARGUMENTS** The weights, in pounds, of several dolphins at a sea animal care facility are 185, 222, 755, 801, 835, 990, and 1104. Which measure of center best represents the data? Justify your conclusion.

12. **FOOD DRIVE** The chart shows the number of canned goods collected by Valley High School in 2012 and 2017. Is the graph misleading? Explain.

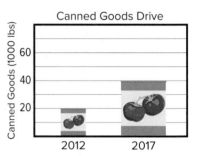

13. **REASONING** The number of participants at reading club for six weeks are 11, 12, 10, 13, 10, and 10. Without calculating the measures of center, how would adding an outlier of 24 participants affect which measure of center most appropriately represents the data?

14. **PRECISION** A community garden has 8 tomato plants with heights ranging from 0.4 to 0.9 meters. Regina found the median to be 0.7 meters, which she rounded and reported as 1 meter. Is Regina's report of the median accurate? Explain your reasoning.

15. **REGULARITY** Describe a general method for assessing a sample for bias.

16. **STRUCTURE** How are the median and mean scores affected if all data values in a set are increased by a specific value, such as 10?

17. **CREATE** Create two sets of data and display them in a graph or chart that shows bias toward one of the sets of data.

18. **WRITE** Write two scenarios that have different examples of sample bias. Have a classmate rewrite your statements without bias.

19. **CREATE** Think of a topic about which you can survey the teachers at your school. Conduct the survey. Explain whether your survey question(s) introduce bias.

20. **ANALYZE** Is a biased sample *sometimes*, *always*, or *never* valid? Justify your argument.

21. **PERSEVERE** If the mean, median, and mode of a data set are equal, the data set is symmetric. If a data set has a mean that is less than its median, what does that tell you about the data set?

22. **FIND THE ERROR** Two students collected data on the sizes of box turtle shells. Olivia measured 8 turtles from a pond near her school. Caleb measured 2 turtles from each of 4 ponds around town. Which is more likely to be free of sample bias? Explain your reasoning.

Lesson 12-4

Measures of Spread

Explore Using Measures of Spread to Describe Data

 Online Activity Use a real-world situation to complete the Explore.

> **INQUIRY** Why might you describe a data set with more than the mean?

Learn Range and Interquartile Range

Statisticians use **measures of spread** or variation to describe how widely data values vary. One such measure is the **range**, which is the difference between the greatest and least values in a set of data.

Quartiles divide a data set arranged in ascending order into four groups, each containing about one-fourth, or 25%, of the data. A **five-number summary** contains the minimum, quartiles, and maximum of a data set.

The **median** marks the second quartile, Q_2, and separates the data into upper and lower halves.

The **lower quartile**, Q_1, is the median of the lower half.

The **upper quartile**, Q_3, is the median of the upper half.

A **box plot**, or box-and-whisker plot, is a graphical representation of the five-number summary of a data set. A box is drawn from Q_1 to Q_3 with a vertical line at the median. This box represents the **interquartile range**, or IQR, which is the difference between the upper and lower quartiles. The whiskers of the box plot are drawn from Q_1 to the minimum and from Q_3 to the maximum.

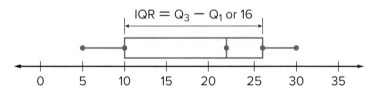

Example 1 Range

GRADES What is the range of the scores?

79, 83, 88, 62, 91, 99, 70

Step 1 Arrange the data in ascending order. 62, 70, 79, 83, 88, 91, 99

Step 2 Determine the range.

range = greatest value − least value

= 99 − 62 or 37

Go Online You can complete an Extra Example online.

Today's Goals
- Determine measures of spread, including the range and interquartile range, of a set of data.
- Determine the standard deviation of a data set.

Today's Vocabulary
measures of spread
range
quartiles
five-number summary
median
lower quartile
upper quartile
box plot
interquartile range
standard deviation

Study Tip

Ordering Because the range involves only the greatest and least values in a data set, it can be determined without ordering the numbers. However, it is often useful to order the data to avoid missing a number.

Think About It!

If Q_1 were located between two numbers, how would you determine its value?

Example 2 Make a Box Plot

BOX OFFICE A financial analyst for a movie studio wants to determine how much most of the top-earning movies have grossed to compare his studio's recent grosses. The worldwide grosses, in millions of dollars, for the top 10 highest-grossing films of all time are given. Determine the five-number summary and draw a box plot of the data to see the spread of the data.

| 2788 | 2187 | 2060 | 1670 | 1520 |
| 1516 | 1405 | 1342 | 1277 | 1215 |

Part A Determine the five-number summary.

Step 1 Arrange the data in ascending order.

1215, 1277, 1342, 1405, 1516, 1520, 1670, 2060, 2187, 2788

Step 2 Determine the five-number summary of the data.

1215, 1277, 1342, 1405, 1516, 1520, 1670, 2060, 2187, 2788

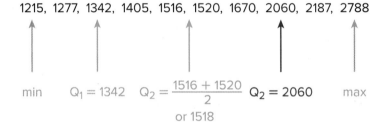

min $Q_1 = 1342$ $Q_2 = \dfrac{1516 + 1520}{2}$ $Q_2 = 2060$ max
or 1518

Part B Construct a box plot.

Step 1 Construct a number line.

Because the minimum is 1215 and the maximum is 2788, your number line must include those values.

Step 2 Draw the box.

Draw and label a box from Q_1 to Q_3, with a vertical line at the median.

Step 3 Draw the whiskers.

Draw a line from the minimum to Q_1. Draw a line from Q_3 to the maximum.

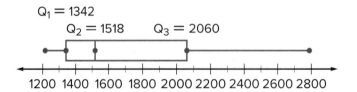

Go Online You can complete an Extra Example online.

Check

MARRIAGE The average age at which women first get married differs by country. The ages for eight countries are shown. Select the box plot for the data.

25, 26, 21, 27, 31, 31, 29, 20

A. Average Age of Women at First Marriage

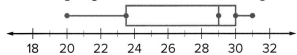

B. Average Age of Women at First Marriage

C. Average Age of Women at First Marriage

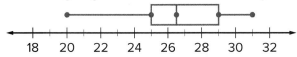

D. Average Age of Women at First Marriage

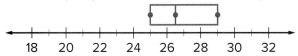

Example 3 Interquartile Range

AUDIO Sarah wants to upload a song that she recorded and share it with her friends. She wants to know whether her song, which is currently −12 decibels on average, is the right volume compared to other songs online so listeners do not have to adjust their volume. She writes down the average volume of seven songs. Find the interquartile range of these average volumes.

−18, −20, −8, −13, −14, −15, −18

Step 1 Order the data: −20, −18, −18, −15, −14, −13, −8

Step 2 Determine Q_1 and Q_3.

Step 3 Determine the *IQR*.

$IQR = Q_3 - Q_1 = -13 - (-18)$ or 5

−20, −18, −18 | −15 | −14, −13, −8

Q_1 Q_2 Q_3

Check

WRITING Cora is writing a novel and tracks the number of pages she writes each day for a week. The number of pages she wrote each day for a week is shown. Find the interquartile range of the data set.

6, 5, 0, 3, 8, 1, 4

Go Online You can complete an Extra Example online.

Think About It!
If all the data in a set were the same value, then what would the standard deviation be?

Learn Standard Deviation

In a data set, the **standard deviation** shows how the data deviate from the mean. A majority of the data in a set, approximately two-thirds, is contained within 1 standard deviation below and above the mean. So, if two data sets have the same mean, but one has a greater standard deviation, then the data in that set is more spread out from the mean.

Key Concept • Standard Deviation

Step 1 Find the mean μ.

Step 2 Find the square of the difference between each data value x_n and the mean, $(\mu - x_n)^2$.

Step 3 Find the sum of all the values in Step 2.

Step 4 Divide the sum by the number of values in the set of data n. This value is the variance.

Step 5 Take the square root of the variance.

Formula $\sigma = \sqrt{\dfrac{(\mu - x_1)^2 + (\mu - x_2)^2 + \cdots + (\mu - x_n)^2}{n}}$

Example 4 Calculate Standard Deviation

UNIVERSITY The number of students accepted at the main campus of each university in the Big Ten East Division are shown. Find and interpret the standard deviation of the data set.

27,300	12,333	15,570	21,610
17,413	25,772	18,230	

Step 1 Find the mean μ.

$$\mu = \frac{27{,}300 + 12{,}333 + 15{,}570 + 21{,}610 + 17{,}413 + 25{,}772 + 18{,}230}{7} \text{ or } 19{,}747$$

Step 2 Find the square of the differences, $(\mu - x_n)^2$.

$(19{,}747 - 27{,}300)^2 = 57{,}047{,}809$

$(19{,}747 - 12{,}333)^2 = 54{,}967{,}396$

$(19{,}747 - 15{,}570)^2 = 17{,}447{,}329$

$(19{,}747 - 21{,}610)^2 = 3{,}470{,}769$

$(19{,}747 - 17{,}413)^2 = 5{,}447{,}556$

$(19{,}747 - 25{,}772)^2 = 36{,}300{,}625$

$(19{,}747 - 18{,}230)^2 = 2{,}301{,}289$

Step 3 Find the sum.

$57{,}047{,}809 + 54{,}967{,}396 + \ldots + 2{,}301{,}289 = 176{,}982{,}773$

Step 4 Divide by the number of values. This value is the variance.

$\dfrac{176{,}982{,}773}{7} \approx 25{,}283{,}253$

Step 5 Find the standard deviation.

$\sqrt{25{,}283{,}253} \approx 5028$

Think About It!
If another data point less than 14,719 or greater than 24,775 is added to the data set, how would that change the standard deviation?

Go Online
You can complete an Extra Example online.

Practice

Go Online You can complete your homework online.

Example 1

Find the range of each data set.

1. 12, 27, 43, 52, 43, 18, 45, 53, 26
2. 132, 127, 129, 130, 141, 125, 138, 129
3. 56, 101, 78, 49, 55, 108, 111, 64
4. 5.9, 6.2, 3.9, 3.7, 8.5, 6.2, 9.0, 8.7, 4.5, 9.3

5. **EXERCISE** Kent tracked his daily number of minutes of exercise. Find the range of the data set.

Number of Minutes of Exercise					
30	35	25	28	40	38
36	29	34	45	42	39

Example 2

Determine the five-number summary and draw a box plot of the data.

6. prices in dollars of smartphones: 311, 309, 312, 314, 399, 312
7. attendance at an event for the last nine years: 68, 99, 73, 65, 67, 62, 80, 81, 83
8. books a student checks out of the library: 17, 9, 10, 17, 18, 5, 2
9. ounces of soda dispensed into 36-ounce cups: 36.1, 35.8, 35.2, 36.5, 36.0, 36.2, 35.7, 35.8, 35.9, 36.4, 35.6
10. ages of riders on a roller coaster: 45, 17, 16, 22, 25, 19, 20, 21, 32, 37, 19, 21, 24, 20, 18, 22, 23, 19

Example 3

Find the interquartile range of each data set.

11. 43, 36, 51, 68, 50, 27, 38, 81, 33
12. 201, 225, 217, 240, 232, 252, 228, 231
13. 94, 87, 105, 99, 118, 97, 102, 85
14. 8.4, 7.1, 6.3, 6.8, 9.2, 7.3, 8.8, 7.9, 5.3, 8.2

15. **HEART RATE** A nurse tracked the heart rates of several patients. Find the interquartile range (IQR) of the data set: 108, 88, 119, 75, 96, 88, 100, 99, 125, 81.

Example 4

Find the standard deviation.

16. {10, 9, 11, 6, 9}
17. {6, 8, 2, 3, 2, 9}
18. {23, 18, 28, 36, 15}
19. {44, 35, 40, 37, 43, 38, 40}

20. **PARKING** A city councilor wants to know how much revenue the city would earn by installing parking meters on Main Street. He counts the number of cars parked on Main Street each weekday: {64, 79, 81, 53, 63}. Find the standard deviation.

Mixed Exercises

21. **REASONING** A hockey team keeps track of how many goals it scores each game: {2, 4, 0, 3, 7, 2}. Find and interpret the standard deviation of the data.

Lesson 12-4 • Measures of Spread   713

22. FOOTBALL The table shows information about the number of carries a running back had over a number of years. Find and interpret the standard deviation of the number of carries.

Year	Number of Carries
2006	31
2007	90
2008	105
2009	115
2010	162

23. MOVIES The manager at a movie theater kept track of the age of each person in a matinee movie: 67, 62, 65, 38, 69, 67, 59, 41, 43, 36, 45, 22, 69, 68, 18, 15, 9, 60, 64.

  a. Determine the five-number summary for the data set.

  b. Draw a box plot of the data.

24. GAS PRICES Renee is planning a road trip to her aunt's house. To estimate how much the trip will cost, she goes online and finds the price of a gallon of gasoline for 5 randomly selected gas stations along the route: $2.09, $2.19, $3.99, $2.39, $2.29.

  a. Determine the five-number summary for the data set.

  b. Draw a box plot of the data.

Find the range, five-number summary, interquartile range, and standard deviation for each data set. Then draw a box plot of the data.

25. SEASHELLS Jorja collected the following number of seashells for the last nine trips to the beach: 5, 11, 7, 12, 13, 17, 3, 15, 14.

26. SHOE SIZE The following shoe sizes of students at a high school were randomly recorded for one hour: 6, 8, 8.5, 10, 12, 6.5, 7, 8, 8.5, 7.5, 9, 11.5, 10, 13, 5.5, 6.5, 5, 9.5.

27. FIND THE ERROR Jennifer and Megan are determining one way to decrease the size of the standard deviation of a set of data. Is either correct? Explain your reasoning.

> **Jennifer**
>
> Remove the outliers from the data set.

> **Megan**
>
> Add data values to the data set that are equal to the mean.

28. ANALYZE Determine whether the statement *Two random samples taken from the same population will have the same mean and standard deviation* is *sometimes, always,* or *never* true. Justify your argument.

29. CREATE Write your own survey question and collect data about your question from 8 classmates. Use that data to find the range, five-number summary, interquartile range, and standard deviation for the data set. Then draw a box plot of the data.

30. WRITE What does the interquartile range tell you about how data clusters around the median of the data?

Lesson 12-5
Distributions of Data

Learn Shapes of Distributions

Analyzing the shape of a **distribution** can help you learn a lot about the data it represents. When data are graphed, the shape of the distribution can be seen.

Key Concept • Symmetric and Skewed Distributions

Histograms	Box Plots	Dot Plots

In a **symmetric distribution**, the mean and median are approximately equal.

The data are evenly distributed.

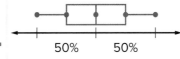

50%   50%

The whiskers are the same length. The median is in the center of the data.

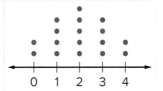

0 1 2 3 4

The data are evenly distributed.

A **negatively skewed distribution** typically has a median greater than the mean.

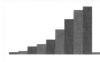

Fewer data on the left.

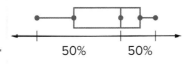

50%   50%

The left whisker is longer than the right. The median is closer to the shorter whisker.

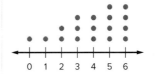

0 1 2 3 4 5 6

Fewer data on the left.

A **positively skewed distribution** typically has a mean greater than the median.

Fewer data on the right.

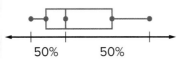

50%   50%

The right whisker is longer than the left. The median is closer to the shorter whisker.

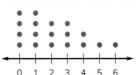

0 1 2 3 4 5 6

Fewer data on the right.

Analyzing Distribution

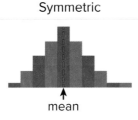

Negatively Skewed	Symmetric	Positively Skewed
mean median	mean	median mean
Use the five-number summary.	Use mean and standard deviation.	Use the five-number summary.

Today's Goals
- Interpret differences in the shapes of distributions.
- Account for the possible effects of extreme data points.

Today's Vocabulary
distribution

symmetric distribution

negatively skewed distribution

positively skewed distribution

outlier

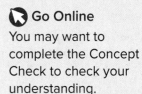

Go Online
You may want to complete the Concept Check to check your understanding.

Study Tip

Distribution Shape
For a histogram, drawing a curve over the data bars may help you see the distribution.

Talk About It!

How could different bin widths of a histogram affect the shape of the distribution? Explain your reasoning.

Example 1 Analyze Distribution by Using Technology

Use a graphing calculator to construct a histogram and box plot for the data. Then describe the shape of the distribution.

78, 53, 24, 75, 76, 83, 78, 60, 64, 53, 36, 47, 32, 75, 54, 68, 68, 74, 85, 42

Method 1 Histogram

Steps 1 and 2 Enter the data and then graph the histogram.

Step 3 Analyze the histogram.

The histogram is higher on the right and has a tail on the left. Therefore, the distribution is negatively skewed.

Method 2 Box Plot

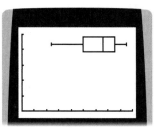

[20, 90] scl: 10 by [0, 8] scl: 1

Steps 1 and 2 Enter the data and graph the box plot.

Step 3 Analyze the box plot.

The left whisker is longer than the right and the median is closer to the shorter whisker. Therefore, the distribution is negatively skewed.

[0, 90] scl: 10 by [0, 5] scl: 1

Check

Use a histogram or box plot to determine the shape of the data.

61, 135, 217, 388, 354, 459, 512, 243, 440, 307

Study Tip:

Window Settings On a TI-84, use **ZoomStat** from the **Zoom** menu to get a basic fitting view window. Then, adjust the window parameters and bin width.

Go Online

to see how to use a graphing calculator with these examples.

Example 2 Choose Appropriate Statistics by Using a Histogram

Describe the center and spread of the data using either the mean and standard deviation or the five-number summary. Justify your choice by constructing a histogram for the data.

18, 3, 28, 17, 13, 18, 11, 22, 21, 14, 12, 7, 9, 24, 17, 28

Step 1 Graph the histogram.

Use a graphing calculator to create a histogram. Adjust the parameters of the graph to appropriately display the data.

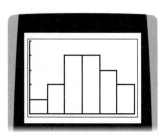

[0, 30] scl: 5 by [0, 5] scl: 1

Go Online You can complete an Extra Example online.

This graph shows that the frequency of the data in the middle is high while frequency of data to the left and right are low. Therefore, the distribution is symmetric.

Step 2 Calculate statistics.

The distribution is symmetric, so the mean and standard deviation are good statistics to represent the data.

To display the statistics, press **STAT**, access the **CALC** menu, select **1-VAR Stats**, and press **ENTER**.

The mean $\bar{x}$ is about 16.4 with a standard deviation σ of about 7.0.

> 💭 **Think About It!**
> Why is mean an appropriate statistic to represent the center for these data?

Example 3 Choose Appropriate Statistics by Using a Box Plot

Describe the center and spread of the data using either the mean and standard deviation or the five-number summary. Justify your choice by constructing a box plot for the data.

202, 148, 21, 60, 74, 140, 462, 157, 225, 23, 88, 241, 59, 139, 351

Step 1 Graph the box plot.

Use a graphing calculator to create a box plot. Adjust the parameters of the graph to appropriately display the data.

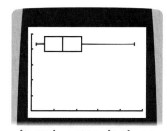

[0, 500] scl: 100 by [0, 5] scl: 1

The right whisker is longer than the left and the median is slightly closer to the right whisker. So, this distribution is positively skewed.

Step 2 Calculate statistics.

The distribution is positively skewed, so use the five-number summary.

To display the statistics, press stat, access the **CALC** menu, select **1-VAR Stats**, and press enter. Use the down arrow key to display more statistics.

Maximum: 462

Minimum: 21

Median: 140

Lower Quartile: 60

Upper Quartile: 225

 Go Online You can complete an Extra Example online.

Learn Extreme Data Points

The least and greatest values in a set of data are called **extreme values**. An **outlier** is a value that is more than 1.5 times the interquartile range above the third quartile or below the first quartile. Outliers can significantly skew the mean and standard deviation.

Example 4 Choose Appropriate Statistics with Extreme Data Points

SHARKS **The lengths, in feet, of adult sharks of various species are shown. Describe the center and spread of the data using appropriate statistics, and identify the effect of extreme data points.**

33	5.2	12.5	11.5	12	0.6	11	20	23	10
6.5	18	13	5	12	4	1	20	12	46

Step 1 Make a box plot.

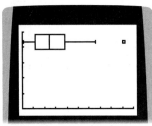

[0, 50] scl: 5 by [0, 5] scl: 1

Step 2 Analyze the graph.

Notice that the right whisker is longer than the left. So, this distribution is positively skewed. The plot also shows that there is an outlier.

Step 3 Calculate statistics.

Include the mean with the five-number summary to see the effect of the outlier.

Mean: 13.82   Median: 12

Max: 46   Min: 0.6

Lower Quartile: 5.85

Upper Quartile: 19

The interquartile range is 13.15. Since 46 is more than 19 + 1.5(13.15), 46 is an outlier.

Step 4 Describe the effect of the outlier.

Since 46 is an outlier, it has affected the mean. To see how much, remove 46 from the set of data and display the statistics again.

Mean: 12.12   Median: 12

Notice that the median did not change when the extreme data point was removed, but the mean did. Without the outlier, the mean and median are closer to the same value.

Think About It!

Suppose a new species of shark is discovered that has an average length of 50 feet. How would two extreme data points affect the measures of center?

Go Online

to see how to use a graphing calculator with this example.

Check

PRECIPITATION The table shows the annual rainfall in Death Valley, CA.

Year	2006	2007	2008	2009	2010	2011	2012	2013
Rainfall (in.)	0.85	0.18	1.04	0.26	2.41	0.98	0.40	0.51

Part A What year(s) represent outlier(s)?

Part B Because of the outlier, the mean is ___?___.

Go Online You can complete an Extra Example online.

Practice

Go Online You can complete your homework online.

Example 1

Use a graphing calculator to construct a histogram and a box plot for the data. Then describe the shape of the distribution.

1. 55, 65, 70, 73, 25, 36, 33, 47, 52, 54, 55, 60, 45, 39, 48, 55, 46, 38, 50, 54, 63, 31, 49, 54, 68, 35, 27, 45, 53, 62, 47, 41, 50, 76, 67, 49

2. 42, 48, 51, 39, 47, 50, 48, 51, 54, 46, 49, 36, 50, 55, 51, 43, 46, 37, 50, 52, 43, 40, 33, 51, 45, 53, 44, 40, 52, 54, 48, 51, 47, 43, 50, 46

Example 2

Describe the center and spread of the data using either the mean and standard deviation or the five-number summary. Justify your choice by constructing a histogram for the data.

3. 32, 44, 50, 49, 21, 12, 27, 41, 48, 30, 50, 23, 37, 16, 49, 53, 33, 25, 35, 40, 48, 39, 50, 24, 15, 29, 37, 50, 36, 43, 49, 44, 46, 27, 42, 47

4. 82, 86, 74, 90, 70, 81, 89, 88, 75, 72, 69, 91, 96, 82, 80, 78, 74, 94, 85, 77, 80, 67, 76, 84, 80, 83, 88, 92, 87, 79, 84, 96, 85, 73, 82, 83

Example 3

Describe the center and spread of the data using either the mean and standard deviation or the five-number summary. Justify your choice by constructing a box plot for the data.

5. 47, 16, 70, 80, 28, 33, 91, 55, 60, 45, 86, 54, 30, 98, 34, 87, 44, 35, 64, 58, 27, 67, 72, 68, 31, 95, 37, 41, 97, 56, 49, 71, 84, 66, 45, 93

6. 64, 36, 32, 65, 41, 38, 50, 44, 39, 34, 47, 35, 46, 36, 53, 35, 68, 40, 36, 62, 34, 38, 59, 46, 63, 38, 67, 39, 59, 43, 39, 36, 60, 47, 52, 45

Example 4

7. **FLYING** The various prices of a flight from Los Angeles to New York are shown. $182, $234, $264, $271, $277, $314, $317, $455
   a. Make a box plot of the data.
   b. Calculate the statistics that best represent the data.
   c. Describe the effect of the outlier.

8. **EXERCISE** Yoshiko tracked her minutes of exercise each day for 10 days as shown. 57, 60, 53, 59, 57, 61, 61, 54, 62, 10
   a. Make a box plot of the data.
   b. Calculate the statistics that best represent the data.
   c. Describe the effect of the outlier.

Mixed Exercises

USE TOOLS Use a graphing calculator to construct a histogram and a box plot for the data. Then describe the shape of the distribution.

9. 14, 71, 63, 42, 24, 76, 34, 77, 37, 69, 54, 64, 47, 74, 59, 43, 76, 56, 78, 52, 18, 54, 39, 28, 56, 74, 68, 36, 20, 49, 67, 47, 69, 68, 72, 69

10. 53, 34, 36, 38, 43, 49, 52, 36, 39, 37, 58, 45, 37, 38, 46, 52, 45, 39, 55, 39, 40, 55, 38, 40, 42, 38, 45, 36, 46, 39, 35, 41, 49, 43, 52, 34

11. 51, 19, 46, 64, 29, 51, 58, 30, 55, 31, 34, 31, 50, 37, 40, 39, 40, 41, 42, 32, 24, 48, 43, 45, 38, 43, 58, 47, 34, 36, 50, 54, 46, 28, 60, 22

12. **TRACK** Daryn recorded the number of laps he walked around the track each week. Use a graphing calculator to construct a histogram for the data, and describe the shape of the distribution.

    17, 21, 23, 26, 27, 28, 28, 27, 33, 34, 33, 27, 29, 22, 19, 28, 35

13. **GOLF** Mr. Swatsky's geometry class's miniature golf scores are shown below. Use a graphing calculator to construct a box plot for the data, and describe the shape of the distribution.

Scores
36, 38, 38, 39, 40, 42, 44, 46, 46, 47, 48, 48, 50, 52, 52, 53, 54, 55, 56, 56, 56, 60, 57, 58, 63

14. **HAIR LENGTH** Ruth recorded the lengths, in centimeters, of hair of students in her school. Describe the center and spread of the data using either the mean and standard deviation or the five-number summary. Justify your choice by creating a box plot for the data.

    40, 39, 37, 26, 25, 40, 35, 34, 26, 39, 42, 33, 26, 25, 34, 38, 41, 34, 37, 39, 32, 30, 22, 38, 36, 28, 27, 39, 34, 26, 36, 38, 25, 39, 23, 8

15. **PRESIDENTS** The ages of the presidents of the United States at the time of their inaugurations are shown. Describe the center and spread of the data using either the mean and standard deviation or the five-number summary. Justify your choice by creating a box plot for the data.

Ages of Presidents
57, 61, 57, 57, 58, 57, 61, 54, 68, 51, 49, 64, 50, 48, 65, 52, 56, 46, 54, 49, 51, 47, 55, 55, 54, 42, 51, 56, 55, 51, 54, 51, 60, 62, 43, 55, 56, 61, 52, 69, 64, 46, 54, 47

16. **AUTOMOTIVE** A service station tracks the number of cars they service per day.

Cars Serviced
40, 47, 37, 42, 46, 31, 50, 41, 17, 43, 36, 45, 21, 43, 45, 23, 49, 50, 48, 26, 42, 46, 35, 52, 27, 51, 31, 44, 35, 27, 46, 39, 33, 50, 45, 50

    a. Use a graphing calculator to construct a histogram for the data, and describe the shape of the distribution.
    b. Describe the center and spread of the data using either the mean and standard deviation or the five-number summary. Justify your choice.

17. **COMMUTE** The number of miles that Armando drove each week during a 15-week period is shown.

Distance (miles)
62, 110, 92, 430, 73, 84, 525, 123, 86, 290, 114, 98, 103, 312, 71

    a. Use a graphing calculator to construct a box plot. Describe the center and spread of the data.
    b. Armando visited four colleges during this period, and these visits account for the four highest weekly totals. Remove these four values from the data set. Use a graphing calculator to construct a box plot that reflects this change. Then describe the center and spread of the new data set.
    c. Calculate and compare the mean and median for the original data set to the mean and median for the data set from **part b**.

18. **ELEVATION** The table contains data about 10 elevations in the United States.

Elevations in the US	
Mt McKinley, AK	20,237
Mt Whitney, CA	14,494
Mt Elbert, CO	14,433
Mt Rainier, WA	14,410
Gannett Peak, WY	13,804
Mauna Kea, HI	13,796
Kings Peak, UT	13,528
Wheeler Peak NM	13,161
Boundary Peak, NV	13,140
Granite Peak, MT	12,799

    a. Use a graphing calculator to construct a box plot for the data, and describe the shape of the distribution.
    b. Describe the center and spread of the data using either the mean and standard deviation or the five-number summary. Justify your choice.
    c. If there is an outlier, describe its effect on the statistics.

19. **USE A MODEL** The histograms show the weight of sample boxes of two brands of pasta.

    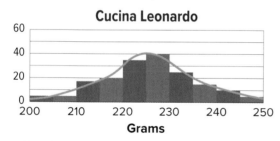
    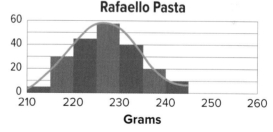

    a. Do the two packages of pasta likely have the same advertised weight? Which manufacturer's quantity control appears better? Explain your answers based on the distributions.
    b. Infer the two population distribution shapes by analyzing the smooth curves across the tops of the histograms. Describe the shapes you observe.

20. **STRUCTURE** The United States has been sending astronauts up in the Space Shuttle since 1981. The table provides data regarding the duration of Space Shuttle flights from 1981 to 1985, and then from 2005 to 2011.

Length of Flights from 1981–1985 (days)	Length of Flights from 2005–2011 (days)
Days: 2, 2, 8, 7, 5, 5, 6, 6, 10, 8, 7, 6, 8, 8, 3, 7, 7, 7, 8, 7, 4, 7, 7	Days: 14, 13, 12, 13, 14, 13, 15, 13, 16, 14, 15, 13, 13, 16, 14, 11, 14, 15, 12, 13, 16, 13

    Choose and calculate the statistics appropriate for the distribution of the data sets. Use the statistics to compare the two sets.

21. **REASONING** Gerardo live streams with 15 of his friends. Most of his streams have lasted 10-15 days so far, however he has two streams that have lasted 93 days. Describe what Gerardo's data distribution would look like currently and how it would be affected if he lost his longest streams.

22. **CONSTRUCT ARGUMENTS** Examine the two box plots shown. Without knowing the data points but assuming the same scale, what conclusion can be made? Justify your argument.

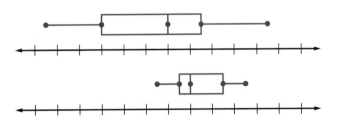

23. **SUPREME COURT** The table gives the ages of the Supreme Court Justices in 2017.
    a. Use a graphing calculator to construct a histogram for the data, and describe the shape of the distribution.
    b. Describe the center and spread of the data using appropriate statistics. Justify your choice.
    c. If there is an outlier, describe its effect on the statistics.

Supreme Court Justices	
Neil Gorsuch	49
Elena Kagan	57
Sonia Sotomayor	62
Samuel Anthony Alito	67
Stephen G. Breyer	78
Ruth Bader Ginsburg	84
Clarence Thomas	68
Anthony M Kennedy	80
John G. Roberts Jr.	62

24. **PERSEVERE** Identify the box plot that corresponds to each of the following histograms.

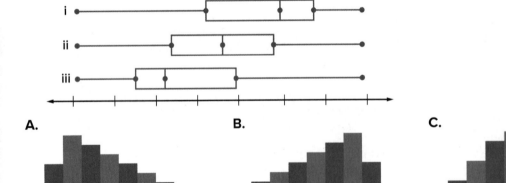

25. **ANALYZE** Research and write a definition for a *bimodal distribution*. How can the measures of center and spread of a bimodal distribution be described?

26. **CREATE** Give an example of a set of real-world data with a distribution that is symmetric and one with a distribution that is not symmetric.

27. **WRITE** Explain why the mean and standard deviation are used to describe the center and spread of a symmetrical distribution and the five-number summary is used to describe the center and spread of a skewed distribution.

Lesson 12-6
Comparing Sets of Data

Explore Transforming Sets of Data by Using Addition

 Online Activity Use graphing technology to complete the Explore.

> **INQUIRY** How can you find the measures of center and spread of a set of data that has been transformed using addition?

Explore Transforming Sets of Data by Using Multiplication

 Online Activity Use graphing technology to complete the Explore.

> **INQUIRY** How can you find the measures of center and spread of a set of data that has been transformed using multiplication?

Learn Linear Transformations of Data

A **linear transformation** is one or more operations performed on a set of data that can be written as a linear function. Common linear transformations are adding a constant to or multiplying a constant by every value in the set of data.

Key Concept • Linear Transformations of Data

Transformations Using Addition	Transformations Using Multiplication
A real number k is added to every value in a set of data, $k \neq 0$.	Every value in a set of data is multiplied by a constant k, $k > 0$.
Measures of Center	
The mean, median, and mode of the new set of data can be found by adding k to the mean, median, and mode of the original set of data.	The mean, median, and mode of the new set of data can be found by multiplying each original statistic by k.
Measures of Spread	
The range and standard deviation of the new set of data will be unchanged.	The range and standard deviation of the new set of data can be found by multiplying each original statistic by k.

Today's Goal
- Describe the effects that linear transformations have on measures of center and spread.

Today's Vocabulary
linear transformation

Study Tip

1-Var Stats To quickly calculate the mean $\bar{x}$, median **Med**, standard deviation σ, and range of a data set, enter the data as **L1** in a graphing calculator, and then use the **1-VAR Stats** feature from the **CALC** menu. Subtract **minX** from **maxX** to find the range.

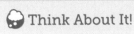 **Think About It!**

Compare and contrast the two methods.

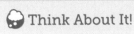 **Think About It!**

Use a calculator and the data set to examine the effect of multiplying by a constant k, $k<0$. How can the mean, median, mode, range, and standard deviation of the new data set be found?

Example 1 Transformations Using Addition

Find the mean, median, mode, range, and standard deviation of the data set obtained after adding 6 to each value.

8, 11, 3, 6, 15, 3, 5, 7, 14, 3, 5, 4

Method 1 Add k to the measures of center and spread of the original set of data.

Find the mean, median, mode, range, and standard deviation of the original data set.

| Mean | 7 | Mode | 3 | Standard Deviation | 4 |
| Median | 5.5 | Range | 12 | | |

Add 6 to the mean, median, and mode. The range and standard deviation are unchanged.

| Mean | 13 | Mode | 9 | Standard Deviation | 4 |
| Median | 11.5 | Range | 12 | | |

Method 2 Add k to each data value of the original set of data. Add 6 to each data value.

14, 17, 9, 12, 21, 9, 11, 13, 20, 9, 11, 10

| Mean | 13 | Mode | 9 | Standard Deviation | 4 |
| Median | 11.5 | Range | 12 | | |

Example 2 Transformations Using Multiplication

Find the mean, median, mode, range, and standard deviation of the data set obtained after multiplying each value by 4.

12, 18, 20, 12, 14, 18, 11, 21, 13, 18, 11, 24

Find the measures of center and spread for the original data set.

Mean	Median	Mode	Range	Standard Deviation
16	16	18	13	4.2

Multiply the measures of center and spread by 4.

Mean	Median	Mode	Range	Standard Deviation
64	64	72	52	16.8

Check

Find the mean, median, mode, range, and standard deviation of the data set obtained after multiplying each value by 0.6. Round to the nearest tenth, if necessary.

45, 33, 43, 51, 39, 48, 34, 39, 30, 39, 47, 44

Go Online You can complete an Extra Example online.

Example 3 Compare Symmetric Distributions of Data

RESTAURANTS The numbers of customers eating at a restaurant during breakfast, lunch, and dinner each day are shown below.

Breakfast: 71, 58, 65, 48, 44, 56, 68, 64, 51, 67, 74, 62, 59, 53, 62, 73, 54, 49, 63, 55

Lunch: 115, 105, 87, 108, 117, 110, 92, 101, 114, 91, 109, 96, 100, 98, 103, 111, 95, 94, 102, 106

Dinner: 76, 62, 91, 76, 79, 68, 65, 89, 81, 76, 90, 82, 79, 74, 71, 73, 84, 87, 81, 64

Part A Construct a histogram or box plot for each set of data. Then describe the shape of each distribution.

Method 1 Histogram

Enter the data in **L1**, **L2**, and **L3**. From the **STAT PLOT** menu, enter **L1** as the **Xlist** for Plot 1, **L2** for Plot 2, and **L3** for Plot 3. Select 📊 as the plot type for each Plot. View each histogram by turning on Plot 1, Plot 2, and then Plot 3. Use the same window dimensions and bin width for each graph.

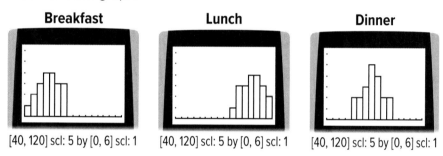

[40, 120] scl: 5 by [0, 6] scl: 1   [40, 120] scl: 5 by [0, 6] scl: 1   [40, 120] scl: 5 by [0, 6] scl: 1

For each time of day, the distribution is high in the middle and low on the left and right. Therefore, all of the distributions are symmetric.

Method 2 Box Plot

Enter the data using the same process. Select 📦 as the plot type for each set of data. To view all of the box plots at once, turn on Plot 1, Plot 2, and Plot 3 and graph.

For each time of day, the lengths of the whiskers are approximately equal, and the median is in the middle of the data. The left and right sides are approximately mirror images of one another. Therefore, all of the distributions are symmetric.

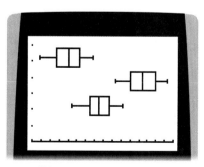

[40, 120] scl: 5 by [0, 6] scl: 1

(continued on the next page)

> **Study Tip**
> **Window and Bin Settings** When setting the window dimensions for multiple sets of data, try setting the minimum and maximum as the least and greatest values of all the sets. When selecting a bin width, consider the context of the situation. For example, if the data does not include fractional numbers, as would be the case with number of people, use a whole number as the bin width.

Go Online You can complete an Extra Example online.

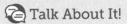

Talk About It!

Describe how this data could be used to make decisions about the restaurant.

Go Online

to see how to use a graphing calculator with this example.

Part B Compare the data sets using the means and standard deviations.

All of the distributions are symmetric, so use the means and standard deviation to describe the centers and spreads.

Breakfast

Lunch

Dinner

The means vary, with breakfast having the lowest average number of customers and lunch having the highest average number of customers. However, the standard deviations are approximately equal. This means that, while the average number of customers for each time of day is very different, the number of customers for each time of day generally varies by the same amount from day to day.

Check

DOGS The weights, in pounds, for a sample of the three most popular breeds of dogs are shown below.

Labrador Retriever: 75, 59, 63, 68, 67, 59, 69, 63, 60, 76, 70, 74, 67, 68, 71, 65, 62, 74, 66, 78

German Shepherd: 53, 61, 58, 74, 85, 80, 72, 57, 64, 69, 81, 75, 73, 64, 76, 68, 66, 51, 67, 73

Golden Retriever: 62, 59, 67, 72, 64, 67, 69, 76, 63, 64, 73, 69, 71, 75, 59, 64, 69, 59, 74, 68

Part A Use a graphing calculator to construct a histogram or box plot for each set of data. Then complete the statement about the shape of each distribution.

All of the distributions are _____?_____.

Part B Compare the data sets using the means and standard deviations. What conclusion(s) can you make about the sets of data? Select all that apply.

A. The average weight of each breed is about the same.

B. The weights of all three breeds are very close to their means.

C. The weights of the German shepherds vary more than the other breeds.

D. On average, the golden retrievers weigh much more than the other breeds.

E. The means of the weights differ by less than 1.5 pounds.

F. The weights of the Labrador retrievers and golden retrievers are generally closer to their means than the German shepherds' weights are to their mean.

Example 4 Compare Skewed Distributions of Data

SPORTS The numbers of high school boys and girls, in hundred thousands, participating in tennis from 2001–2015 are shown below.

Boys (hundred thousands)	Girls (hundred thousands)
144, 139, 145, 153, 149, 153, 157, 156, 157, 163, 161, 160, 157, 161, 157	164, 160, 163, 168, 169, 174, 177, 172, 178, 182, 182, 181, 181, 184, 183

Part A Construct a histogram or box plot for each set of data. Then describe the shape of each distribution.

Method 1 Histogram

Enter the data in **L1** and **L2**. From the **STATPLOT** menu, enter **L1** as the **Xlist** for Plot 1 and **L2** for Plot 2. Select [icon] as the plot type for each Plot. View each histogram by turning on Plot 1, and then Plot 2. Use the same window dimensions and bin width for each graph.

Boys

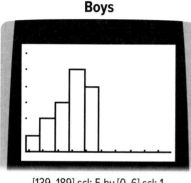

[139, 189] scl: 5 by [0, 6] scl: 1

Girls

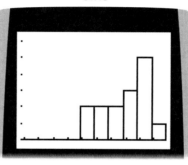

[139, 189] scl: 5 by [0, 6] scl: 1

Both distributions are high on the right and have tails on the left. Therefore, both distributions are negatively skewed.

Method 2 Box Plot

Enter the data using the same process. Select [icon] as the plot type for each set of data. To view both box plots at once, turn on Plot 1 and Plot 2 and graph.

For each distribution, the left whisker is longer than the right, and the median is closer to the right whisker. Therefore, both distributions are negatively skewed.

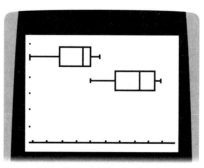

[139, 189] scl: 5 by [0, 6] scl: 1

Part B Compare the data sets using the five-number summaries.

Both distributions are skewed, so use the five-number summary to compare the data.

(continued on the next page)

 Go Online You can complete an Extra Example online.

Go Online to see how to use a graphing calculator with this example.

The upper quartile for the number of boys that participated in tennis is 160, while the minimum number of girls that participated is 160. This means there were only 160,000 or more boys participating in tennis for 25% of the years, while at least 160,000 girls participated every year.

We can conclude that many more girls participated in tennis from 2001 to 2015 than boys.

Boys

Girls

Think About It!
Compare two other statistics from the five-number summaries. What does this tell you about the number of girls and boys that participated in tennis?

Check

FUNDRAISING The number of raffle tickets sold by Darius and Makya each day are shown below.

Darius: 5, 1, 15, 4, 10, 23, 9, 3, 17, 2, 6, 21, 5, 13, 28, 10, 14, 7, 5, 19, 9, 22, 10, 8, 15, 9, 13, 19, 22, 30

Makya: 18, 1, 17, 10, 19, 3, 7, 20, 9, 22, 12, 13, 16, 18, 16, 5, 17, 15, 6, 11, 18, 14, 16, 18, 1, 16, 18, 23, 15, 10

Part A Use a graphing calculator to construct a histogram or box plot for each set of data. Then complete the statement about the shape of each distribution.

The distribution of Darius' raffle ticket sales is ———?———.

The distribution of Makya's raffle ticket sales is ———?———.

Part B Compare the data sets using the five-number summaries. What conclusion(s) can you make about the sets of data? Select all that apply.

A. The median number of tickets Darius sold is much higher than the median number of tickets Makya sold.

B. The median number of tickets Darius sold is the same as the lower quartile of Makya's sales.

C. The data from Darius' sales is spread over a wider range than the data from Makya's sales.

D. The median number of tickets each student sold was the same.

E. The fewest number of tickets each student sold in a day was 1.

F. The upper 50% of Darius' data spans from 10 to 30, while the upper 75% of Makya's data spans from 10 to 23.

Go Online You can complete an Extra Example online.

Practice

Go Online You can complete your homework online.

Example 1

Find the mean, median, mode, range, and standard deviation of each data set that is obtained after adding the given constant to each value.

1. 52, 53, 49, 61, 57, 52, 48, 60, 50, 47; +8

2. 101, 99, 97, 88, 92, 100, 97, 89, 94, 90; +(−13)

3. 27, 21, 34, 42, 20, 19, 18, 26, 25, 33; +(−4)

4. 72, 56, 71, 63, 68, 59, 77, 74, 76, 66; +16

Example 2

Find the mean, median, mode, range, and standard deviation of each data set that is obtained after multiplying each value by the given constant.

5. 11, 7, 3, 13, 16, 8, 3, 11, 17, 3; ×4

6. 64, 42, 58, 40, 61, 67, 58, 52, 51, 49; ×0.2

7. 33, 37, 38, 29, 35, 37, 27, 40, 28, 31; ×0.8

8. 1, 5, 4, 2, 1, 3, 6, 2, 5, 1; ×6.5

Examples 3 and 4

9. **BASEBALL** The total wins per season for the first 17 seasons of the Marlins are shown. The total wins over the same time period for the Cubs are also shown.

Marlins
64, 51, 67, 80, 92, 54, 64, 79, 76, 79, 91, 83, 83, 78, 71, 84, 87

Cubs
84, 49, 73, 76, 68, 90, 67, 65, 88, 67, 88, 89, 79, 66, 85, 97, 83

  a. Use a graphing calculator to construct a box plot for each set of data. Then describe the shape of each distribution.

  b. Compare the data sets using either the means and standard deviations or the five-number summaries. Justify your choice.

10. **HEALTH CLUBS** To plan their future equipment purchases, the Northville Health Club randomly chooses 8 patrons and tracks how many minutes they spend on the treadmill.

    a. Use a graphing calculator to construct a histogram for each set of data. Then describe the shape of each distribution.

    b. Compare the data sets using either the means and standard deviations or the five-number summaries. Justify your choice.

Minutes on Treadmill Last Week	Minutes on Treadmill This Week
30	20
30	30
45	45
20	45
60	30
30	60
30	50
45	45

Lesson 12-6 • Comparing Sets of Data 729

Mixed Exercises

Find the mean, median, mode, range, and standard deviation of each data set that is obtained after adding or multiplying each value by the given constant(s).

11. 98, 95, 97, 89, 88, 95, 90, 81, 87, 95; +2

12. 32, 30, 27, 29, 25, 33, 38, 26, 23, 31; ×1.6

13. 14, 17, 13, 9, 15, 7, 12, 16, 8, 9; ×5

14. 5, 12, 7, 3, 8, 5, 7, 1, 4, 7, 3, 9; +22

15. 12, 15, 16, 12, 12, 15, 17, 19, 22, 27, 42, 42; +5

16. 49, 43, 26, 39, 40, 30, 33, 64, 26, 45, 23, 26; ×3, +(−8)

17. 71, 72, 68, 70, 72, 67, 68, 72, 65, 70; ×0.2

18. 112, 91, 108, 129, 80, 99, 78, 80; +(−15)

19. 57, 38, 42, 51, 39, 44, 33, 55; +(−7), ×2

20. 55, 50, 58, 52, 56, 57, 50, 55, 50; ×2, +5

21. BOWLING The scores of 15 bowlers are shown in the table.

Score
211, 123, 183, 176, 224, 115, 109, 136, 152, 177, 127, 196, 143, 166, 170

  a. Find the mean, median, mode, range, and standard deviation of the scores.

  b. The handicap of the bowling team will add 56 points to each score. Find the statistics of the scores while including the handicap.

22. COMPETITION The distances that 18 participants threw a football are shown in the table.

Distance (feet)
96, 94, 114, 85, 96, 109, 90, 109, 67, 82, 98, 79, 69, 70, 106, 96, 112, 84

  a. Find the mean, median, mode, range, and standard deviation of the participants' distances.

  b. Find the statistics of the participants' distances in yards.

23. TEMPERATURE The monthly average high temperatures for Lexington, Kentucky, are shown in the table.

Temperature (°F)
40, 45, 55, 65, 74, 82, 86, 85, 78, 67, 55, 44

  a. Find the mean, median, mode, range, and standard deviation of the temperatures.

  b. Find the statistics of the temperatures in degrees Celsius. Recall that $C = \frac{5}{9}(F - 32)$.

24. **FANTASY SPORTS** The weekly total points of Scott's and Azumi's fantasy baseball teams are shown in the tables.

Scott's Team
109, 99, 121, 137, 131, 141, 77, 83, 139, 92, 42, 133, 98, 153, 124, 102, 113, 117, 112, 128, 107, 147

Azumi's Team
113, 121, 98, 104, 106, 123, 175, 141, 109, 129, 49, 110, 112, 144, 106, 119, 127, 88, 132, 93, 137, 123

a. Use a graphing calculator to construct a box plot for each set of data. Then describe the shape of each distribution.

b. Compare the data sets using either the means and standard deviations or the five-number summaries. Justify your choice.

c. How does eliminating the outliers of each data set affect the statistics and comparison from **part b**?

25. **BUSINESS** Saeed owns an electronics store. He is revising his pricing for phone accessories. His current prices for an assortment of accessories are listed at the right. He has also determined that the mean price for the same assortment of accessories at a rival store is $10.99.

Saeed's Price Data ($)		
14.99	4.49	9.99
18.49	12.99	6.99
8.49	21.99	13.49
13.99	9.99	10.99
12.49	4.49	12.99

a. Saeed wants to match his rival's prices. Make a table to list the new prices. Explain.

b. Compare the mean and standard deviation of the current prices to the new prices.

26. **REASONING** Two different samples on the shell diameter of a species of snail are shown.

Sample A (mm)		
45	35	37
40	42	40
28	38	31

Sample B (mm)		
26	44	40
27	35	28
26	39	31

a. Use the median and interquartile range to compare the samples.

b. Based on your findings and on the data points in each sample, which sample appears to be more representative? Explain your reasoning.

27. **STRUCTURE** Height data samples of 17-year-old male and female students are shown. Use the mean and standard deviation to compare the samples.

Heights of Male Students (inches)		
71	69	67
68	69	70
72	74	68
71	69	72

Heights of Female Students (inches)		
67	62	69
65	71	66
63	65	68
66	63	70

28. CONSTRUCT ARGUMENTS Francisca is planning a two-week vacation to one of two cities and wants to base her decision on the weather history for the same dates as her vacation. She has collected the number of days that it has rained during this two-week period for each city over the past 10 years. The results are shown.

City A	
5	0
7	6
5	6
6	6
3	2

City B	
4	4
6	5
3	7
4	3
5	7

a. Determine the shape of each distribution, and use the appropriate statistics to find the center and spread for each set of data.

b. Which city do you think Francisca should visit on her vacation? Justify your argument.

29. WRITE Compare and contrast the benefits of displaying data using histograms and box plots.

30. ANALYZE If every value in a set of data is multiplied by a constant k, $k < 0$, then how can the mean, median, mode, range, and standard deviation of the new data set be found?

31. PERSEVERE A salesperson has 15 SUVs priced between $33,000 and $37,000 and 5 luxury cars priced between $44,000 and $48,000. The average price for all of the vehicles is $39,250. The salesperson decides to reduce the prices of the SUVs by $2000 per vehicle. What is the new average price for all of the vehicles?

32. ANALYZE If k is added to every value in a set of data, and then each resulting value is multiplied by a constant m, $m > 0$, how can the mean, median, mode, range, and standard deviation of the new data set be found? Justify your argument.

33. WRITE Explain why the mean and standard deviation are used to compare the center and spread of two symmetrical distributions, and the five-number summary is used to compare the center and spread of two skewed distributions or a symmetric distribution and a skewed distribution.

Lesson 12-7

Summarizing Categorical Data

Explore Categorical Data

Online Activity Use a real-world situation to complete the Explore.

> **INQUIRY** What is the advantage of organizing data in a two-way table?

Learn Two-Way Frequency Tables

A **two-way frequency table** or *contingency table* is used to show the frequencies of data from a survey or experiment classified according to two categories, with the rows indicating one category and the columns indicating the other.

Suppose you are constructing a two-way frequency table based on two categories, grade level and employment. The table is constructed below for sample values.

Grade	Employed	Unemployed	Totals
Junior	8	12	20
Senior	15	10	25
Totals	23	22	45

Subcategories: The subcategories are the column and row headers that represent the two different types of categories. In this case, Employed, Unemployed, Junior, and Senior are the subcategories.

Joint frequencies: **Joint frequencies** are the values for every combination of subcategories. So, 8 is a joint frequency that represents the number of students who are employed and juniors.

Marginal frequencies: **Marginal frequencies** are the totals of each subcategory. So, 20 is a marginal frequency that represents the total number of juniors.

Today's Goals
- Organize categorical data in a two-way frequency table.
- Determine and interpret the values in a two-way relative frequency table.

Today's Vocabulary
two-way frequency table

joint frequencies

marginal frequencies

relative frequency

two-way relative frequency table

conditional relative frequency

Study Tip

Check For each cell, you can see if your calculations are correct by calculating the value for that cell using different data from the table. For example, you could calculate the total number of female participants in the study either by adding the number of women named Casey or Riley, or by subtracting the number of men from the total number of people named Casey or Riley. Either way, you should get the same number.

Use a Source

Create your own two-way frequency table. Find data online that divides a group of subjects into two categories, with each subject fitting into one subcategory of each. For example, in the data shown the categories are whether each person is male or female and whether each person's name is Casey or Riley. Determine the subcategories, enter the given data, and fill in any cells for which values are not provided.

Example 1 Use a Two-Way Frequency Table

NAMES Unisex names are names often used for both males and females. At one point the most common unisex names in the U.S. were Casey and Riley, with 176,544 Caseys and 154,861 Rileys. During that time, there were 104,161 males with the name Casey and 75,882 females with the name Riley. Organize the data in a two-way frequency table.

Steps 1 and 2 Enter the given data in a table. Then use the information given to fill in the rest of the cells.

Top Unisex Names in the U.S.			
	Casey	Riley	Totals
Male	104,161	78,979	183,140
Female	72,383	75,882	148,265
Totals	176,544	154,861	331,405

Male Rileys: $154{,}861 - 75{,}882 = 78{,}979$

Total Males: $104{,}161 + 78{,}979 = 183{,}140$

Female Caseys: $176{,}544 - 104{,}161 = 72{,}383$

Total Females: $72{,}383 + 75{,}882 = 148{,}265$

Totals: $176{,}544 + 154{,}861 = 331{,}405$

Check

TECHNOLOGY Pew Research Center released a survey that asked whether participants thought technological advancements in the future will make people's lives better or worse. Of the people interviewed, 423 earned less than $50,000 per year and 328 earned $50,000 or more. Of those earning less than $50,000 per year, 262 thought that people's lives would get better, and 240 of those who earned $50,000 or more thought the same. Copy and complete the two-way frequency table.

Will technological advancements in the future make people's lives better or worse?			
	Better	Worse	Totals
<			
≥			
Totals			

Go Online You can complete an Extra Example online.

734 Module 12 • Statistics

Learn Two-Way Relative Frequency Tables

A **relative frequency** is the ratio of the number of observations in a category to the total number of observations. A **two-way relative frequency table** can help you see patterns of association in the data. To create a two-way relative frequency table, divide each of the values by the total number of observations and replace them with their corresponding decimals or percents.

A **conditional relative frequency** is the ratio of the joint frequency to the marginal frequency. Because each two-way frequency table has two categories, each two-way relative frequency table can provide two different conditional relative frequency tables.

🌐 Example 2 Use a Two-Way Relative Frequency Table

PARENTING Many parents monitor their teenagers' Internet usage. The Pew Research Center conducted a survey of whether parents do or do not check what sites their teens had visited and whether they are the parent of a teen between the ages of 13 and 14 or between the ages of 15 and 17. The results of the survey are shown. Organize the data in a relative frequency table by age group, and interpret the data.

How Parents Monitor Teenagers' Internet Usage			
	Does Check	Does Not Check	Totals
13 to 14	299	140	439
15 to 17	348	273	621
Totals	647	413	1060

Part A Organize the data in a relative frequency table.

How Parents Monitor Teenagers' Internet Usage			
	Does Check	Does Not Check	Totals
13 to 14	$\frac{299}{1060} \approx 28.2\%$	$\frac{140}{1060} \approx 13.2\%$	$\frac{439}{1060} \approx 41.4\%$
15 to 17	$\frac{348}{1060} \approx 32.8\%$	$\frac{273}{1060} \approx 25.8\%$	$\frac{621}{1060} \approx 58.6\%$
Totals	$\frac{647}{1060} \approx 61.0\%$	$\frac{413}{1060} \approx 39.0\%$	$\frac{1060}{1060} = 100\%$

Part B Interpret the data.

Do more parents check what sites their teens have visited, or do more parents not check?

61% of parents do check the sites their teens have visited compared to 39% who do not.

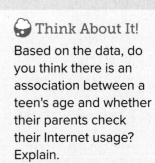

Think About It!
Based on the data, do you think there is an association between a teen's age and whether their parents check their Internet usage? Explain.

Lesson 12-7 • Summarizing Categorical Data **735**

> **Think About It!**
>
> If the condition for the relative frequency table were whether a person had voted or not rather than age group, the joint frequency for people between the ages of 18 and 24 who voted would be 52.5%. If someone claims that this indicates 52.5% of all people who voted in the 2012 election were between the ages of 18 and 24, would they be correct? Justify your argument.

Example 3 Use a Two-Way Conditional Relative Frequency Table

VOTING According to the U.S. Census Bureau, voter turnout describes how many eligible voters show up to vote in an election. The table shows the number of eligible voters who did and did not vote in 2012 for the oldest and youngest eligible age groups. Organize the data in a conditional relative frequency table by age group, and interpret the data.

Voter Turnout			
	Voted	Did Not Vote	Totals
18 to 24	12,515	13,275	25,790
75 and over	11,344	5380	16,724
Totals	23,859	18,655	42,514

Part A Organize the data in a conditional relative frequency table by age group.

Step 1 Determine which marginal frequencies to use.

The conditional relative frequency relates the number of voters or nonvoters to the age group, so the relevant marginal frequencies are the total numbers of voters for each age group.

Step 2 Determine the ratios of the joint frequencies to the marginal frequencies.

Voter Turnout			
	Voted	Did Not Vote	Totals
18 to 24	$\frac{12,515}{25,790} \approx 48.5\%$	$\frac{13,275}{25,790} \approx 51.5\%$	$\frac{25,790}{25,790} = 100\%$
75 and over	$\frac{11,344}{16,724} \approx 67.8\%$	$\frac{5380}{16,724} \approx 32.2\%$	$\frac{16,724}{16,724} = 100\%$

Part B Interpret the data.

Which age group has the higher voter turnout?

The percent of eligible voters aged 18 to 24 that voted is 48.5%, and the percent for those aged 75 and over is 67.8%. Based on the data, there is an association between age and whether a person voted. People aged 18 to 24 were more likely to not have voted than people aged 75 and over.

Go Online You can complete an Extra Example online.

Practice

Go Online You can complete your homework online.

Example 1

TREATS The owner of a snow cone stand keeps track of the sizes and flavors sold one afternoon. He sold 125 snow cones in all. Of these, 40% were large snow cones, 32% were grape, and 12% were small watermelon snow cones. The stand sold 15 more cherry snow cones than grape. The most popular snow cone of the day was small cherry, with a total of 35 sales.

1. Construct a two-way frequency table to organize the data.
2. How many large grape snow cones were sold?
3. How many watermelon snow cones were sold in all?
4. How many more small snow cones were sold than large snow cones?

Example 2

FOREIGN LANGUAGE Christy surveyed several students at her school and asked each person what foreign language he or she is studying. The results are shown in the table.

	Male	Female	Total
Spanish	18	20	38
French	16	12	28
German	6	8	14
Total	40	40	80

5. Construct a relative frequency table by converting the data in the table to percentages. Round to the nearest tenth, if necessary.
6. Find the joint relative frequency of a female student who is studying French.
7. Interpret the data.

Example 3

CLASS PRESIDENT In a poll for senior class president, 68 of the 145 male students said they planned to vote for Santiago. Out of 139 female students, 89 planned to vote for his opponent, Measha.

8. Construct a conditional relative frequency table based on voter preference. Show your calculations.
9. What does each conditional relative frequency represent?
10. What is the probability that a vote for Measha will come from a female student? How is this different from the probability that a female student intends to vote for Measha?

Mixed Exercises

VETERINARIAN The two-way frequency table shows the number of dogs and cats that were seen at a veterinarian's office and the primary purpose of their visit.

	Dog	Cat	Total
Exam	12	5	17
Shots	6	3	9
Grooming	7	2	9
Total	25	10	35

11. How many dogs were seen for an exam today?

12. How many more dogs than cats were seen at the veterinarian's office?

BIRD WATCHING A group of bird-watchers has been tracking the number of tree swallows, cardinals, and goldfinches in a region. Over the weekend, a total of 40 birds were observed. Of those, 45% were male, 37.5% were cardinals, and 12.5% were male tree swallows. Twice as many female cardinals were observed as male cardinals. There were 5 female goldfinches spotted.

13. Construct a two-way frequency table to organize the data.

14. How many more female tree swallows were seen than male cardinals?

15. How many male goldfinches and female cardinals were seen?

16. How many more female birds were seen than male birds?

SCHOOL ACTIVITIES The two-way frequency table shows the number of students who participate in school sports or clubs at Monroe High School.

	Sports or Clubs	No Sports or Clubs	Total
Freshmen	48	60	108
Sophomores	60	72	132
Juniors	51	69	120
Seniors	57	63	120
Total	216	264	480

17. Construct a relative frequency table by converting the data in the table to percentages. Round to the nearest tenth, if necessary.

18. Find the joint relative frequency of a sophomore who participates in school sports or clubs.

19. What percentage of freshmen do not participate in school sports or clubs? Round to the nearest tenth percent, if necessary.

20. What percentage of seniors participate in school sports or clubs? Round to the nearest tenth percent, if necessary.

SCHOOL MASCOT The freshmen and sophomores at Lakeview High School are tasked with adopting a new school mascot next school year. The district asked a representative group of students to vote for one of the three mascot finalists and to indicate to which grade they belong. The results are shown in the table.

School Mascot Vote Results			
	Freshmen	Sophomores	Total
Panthers	30	36	66
Hornets	17	33	50
Lions	28	31	59
Total	75	100	175

21. How many students voted for Panthers?

22. How many students voted for Lions?

23. How many sophomores were in the representative group?

24. Of the students who voted for Hornets, how many of them are freshmen?

25. Of the students who voted for Lions, how many of them are sophomores?

26. To the nearest whole, what percent of all the students voted for Lions?

27. To the nearest whole, what percent of all the students voted for Panthers?

28. To the nearest whole, what percent of all the students who voted were freshmen?

THANKSGIVING PIE An online poll collected a sample of Thanksgiving pie preferences for different U.S. regions.

	Apple	Sweet Potato	Pumpkin	Totals
West	77	4	13	
Midwest	32		54	
South		63	24	
Northeast	92	2		
Total	213	75	117	

29. PRECISION Copy and complete the table. Then find each relative frequency to the nearest tenth of a percent.

30. USE A MODEL Assuming the poll is representative of the whole population, what is a reasonable estimate of the probability that a family will be from the northeast and will be eating pumpkin pie on Thanksgiving?

31. STRUCTURE Construct a table of conditional relative frequencies based on pie preference. Round each percent to the nearest tenth. Interpret the meaning of the probabilities in the context of the problem.

32. REGULARITY If we had found the conditional relative frequencies by dividing by the total replies from each region, what would be the meaning of the probability in each cell?

VEHICLES The table shows the relative frequencies of drive systems for different vehicle types in a school parking lot. There are 215 vehicles in the lot.

	2WD	AWD	Totals
Hatchbacks	42%	4%	
Sedans	28%	6%	
SUVs	1%	19%	
Total			215

33. **USE TOOLS** Construct a table to show the joint and marginal frequencies.

34. **REASONING** Without calculating individual frequencies, how many times greater will the conditional relative frequencies based on drive systems for AWD be than the relative frequencies for AWD, and why?

35. **PERSEVERE** Len conducted a survey among a random group of 1000 families in his home state of California. He wanted to determine whether there is an association between gasoline prices and distances traveled on family vacations. He collected the following information. According to Len's two-way frequency table, does there appear to be an association between gasoline prices and vacation distances traveled? Explain.

	$1.75–$3.24 per gallon	$3.25-$4.74 per gallon	Total
Less than 250 miles	109	255	364
More than 250 miles	329	86	415
No vacation travel	34	187	221
Total	472	528	1000

36. **CREATE** Select your own data for a two-way frequency table, write a question related to the data in the table, and provide the solution.

37. **WRITE** Compare two-way relative frequency tables and two-way conditional relative frequency tables.

38. **FIND THE ERROR** Magdalena took a survey of students in her school to find out what snack was most popular.

Favorite Snack Vote Results			
	Freshmen	Sophomores	Total
Fruit Snack	65	61	126
Granola	27	21	48
Yogurt	21	18	39
Total	113	100	213

a. Interpret the data based on the conditional relative frequency related to age groups.

b. Magdalena claims that fruit snack is the most popular snack for freshmen and sophomores, and Ben claims that a higher percentage of sophomores prefer fruit snack than do freshmen. Is either correct? Explain your reasoning.

Module 12 • Statistics
Review

Essential Question
How do you summarize and interpret data?

Module Summary

Lessons 12-1 and 12-4

Measures of Center and Spread

- The mean of a data set is the sum of the elements of the data set divided by the total number of elements in the set.
- The median of a data set is the middle element or the mean of the two middle elements in the set of data when the data are arranged in numerical order.
- The mode of a data set is the value of the elements that appear most often in the set of data.
- The formula for standard deviation, with mean $\bar{x}$ and n terms is
$$\sigma = \sqrt{\frac{(\bar{x} - x_1)^2 + (\bar{x} - x_2)^2 + \ldots + (\bar{x} - x_n)^2}{n}}.$$

Lessons 12-2 and 12-3

Representing and Using Data

- Dot plots, bar graphs, and histograms are commonly used to represent data.
- Bar graphs are used with discrete data, and histograms are used with continuous data.
- A population is all members of a group of interest about which data will be collected. A sample is a subset of the population.
- A bias is an error that results in a misrepresentation of a population.

Lesson 12-5

Distributions of Data

- In a symmetric distribution, the mean and median are approximately equal.
- A negatively skewed distribution typically has a median greater than the mean. A positively skewed distribution typically has a mean greater than the median.
- An outlier is a value that is more than 1.5 times the interquartile range above the third quartile or below the first quartile.

Lesson 12-6

Comparing Sets of Data

- A linear transformation is one or more operations performed on a set of data that can be written as a linear function.
- Common linear transformations are adding a constant to or multiplying a constant by every value in the set of data.

Lesson 12-7

Two-Way Frequency Tables

- A two-way frequency table shows the frequencies of data classified according to two or more categories.

Study Organizer

 Foldables

Use your Foldable to review this module. Working with a partner can be helpful. Ask for clarification of concepts as needed.

Test Practice

1. GRAPH Make a dot plot of the quiz scores of Ms. Perez's third period class.

Quiz Scores			
85	88	75	100
90	90	88	72
72	79	88	85

(Lesson 12-2)

2. OPEN RESPONSE When is it a good idea to scale the number line when making a dot plot?

(Lesson 12-2)

3. MULTI-SELECT Which of the statements are true regarding dot plots, bar graphs, and histograms? Select all that apply.

Lesson 12-2

A. Dot plots use a number line and dots to represent very large amounts of data.

B. Bar graphs are used to represent data that is continuous.

C. Histograms are used to represent data that is continuous.

D. Bar graphs are used to represent data that is discrete.

E. Histograms are used to represent data that is discrete.

4. MULTIPLE CHOICE Which dot plot correctly models these data values?

36, 38, 42, 36, 36, 40, 42, 38, 38, 39, 40, 38, 38, 38, 40 (Lesson 12-2)

A.
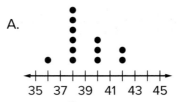

B.
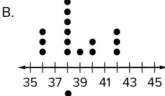

C.
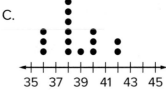

D.
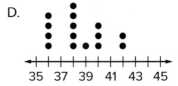

5. OPEN RESPONSE Given the set of data in the table, describe what size intervals could be used when making a histogram. (Lesson 12-2)

Ages of Guests at a Picnic
74, 26, 32, 4, 61, 56, 16, 15, 17, 28, 39, 42, 47, 72, 66, 12, 16, 38, 35, 8, 16, 11, 10, 41, 47, 5, 13, 77, 24, 30, 9, 62

6. GRAPH A survey was conducted among students in Mr. Sadiq's history class to determine their favorite major topic covered in class this semester. The results are shown in the table. Make a bar graph to display the data. (Lesson 12-2)

Topic	Number of Votes
Civil War	12
Revolutionary War	5
The Industrial Revolution	8
Westward Expansion	10

7. OPEN RESPONSE Akeem wants to determine how long it took students in his class to complete a 1-mile run.

Running Time (min)
18.5, 8.4, 10.2, 27.1, 9.5, 10.9, 17.0, 5.3, 6.1, 8.4, 8.4, 9.9, 10.0, 7.4, 8.4

State two types of displays Akeem could use to appropriately display his data. (Lesson 12-2)

8. MULTIPLE CHOICE Which box plot correctly models these data values? 95, 72, 84, 98, 87, 75, 100, 86, 90, 81, 93, 90 . (Lesson 12-4)

A.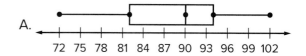

B.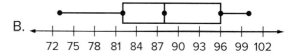

C.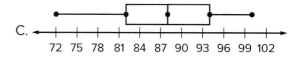

D.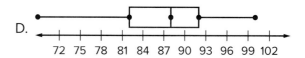

9. GRAPH The table shows the number of pages each student read in one night.

Pages Read
13, 15, 8, 22, 11, 17, 15, 9, 14, 16, 13

Create a box plot to represent the set of data. (Lesson 12-4)

10. OPEN RESPONSE Fifteen people in their fifties were surveyed about the number of apps they have on their cell phone. (There was an assumption that all 15 of them owned a cell phone). The results are listed, below.
6, 0, 11, 8, 9, 6, 7, 3, 1, 2, 10, 7, 22, 5, 13
(Lesson 12-4)

A box plot to represent this data would have to begin at ? because that is the minimum value, and would have to extend to ? because that is the maximum value.

The most appropriate scale to display the data in the box plot should be ? or 2.

11. MULTIPLE CHOICE The table shows the annual snowfall amounts for several towns.

Town	Snowfall (in.)
Westfield	246
Brattleboro	73
Cambridge	54
Danville	73
Shelburne	86
Lowell	67

Which measure(s) of center and measure(s) of spread best describe the set of data? (Lesson 12-5)

A. mean

B. median

C. standard deviation

D. five-number summary

12. OPEN RESPONSE *True* or *false*: Histogram B has more variability than Histogram A. (Lesson 12-6)

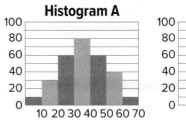
Histogram A

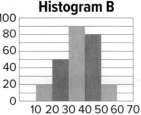
Histogram B

13. MULTIPLE CHOICE The junior varsity dance team is selecting the color of their new uniforms. The team consists of 28 freshmen and sophomores. Of the 16 freshmen, 7 want red uniforms and 9 want black uniforms. Only 4 of the sophomores want black uniforms. How many total team members want red uniforms? (Lesson 12-7)

A. 7

B. 13

C. 15

D. 21

14. OPEN RESPONSE The table shows the frequencies of positions for different offensive players on a school football team. There are 38 offensive players on the team.

	Senior	Junior	Sophomore
Quarterback	1	1	0
Running Back	2	1	1
Receiver	3	2	2
Lineman	13	6	6

Suppose a junior player was picked at random. What is the probability that player is a receiver? (Lesson 12-7)

Selected Answers

Module 1

Quick Check

1. $\frac{2}{3}$ **3.** $\frac{1}{13}$ **5.** 4 **7.** $\frac{3}{2}$ or $1\frac{1}{2}$

Lesson 1-1

1. $2 + 12 \div 4$ **3.** $3 \times 11 + 7$ **5.** $6 - 3 - 1$
7. $24 \div 6 + 7$ **9.** $116 \div 4 + 28 - 33$ **11.** $\frac{3+7}{2}$
13. $(4 + 9) \times 3$ **15.** $\frac{10}{4 \times 5}$ **17.** $\frac{1+2}{20}$ **19.** $\frac{12+16}{3+4}$
21. $\frac{36+14}{2 \times 5}$ **23.** $\frac{6+15}{13-9}$ **25.** 365×85
27. $2 \times 35 - 5$ **29.** 12×7 **31.** 2744 **33.** 11
35. 42 **37.** 7 **39.** 56 **41.** 22 **43.** 8
45. 324 **47.** 29 **49a.** $20 \times 5 + 9$
49b. 109 flies **51a.** $2 \times \frac{1}{2}(30 + 50)24$
51b. 1920 in² **53.** $8^4 + 6$ **55.** Tamara; when evaluating, first perform the multiplication and division from left to right, and then the addition and subtraction from left to right.
57. The cashier; Kelly should have entered the expression into her calculator as $3(18.95 - 2) + 2(11.50)$. **59.** $10 \times 18 + 8 \times 18$; You can also find the length of each side of the apartment, 18 and $10 + 8$, and then multiply: $18(10 + 8)$.
61. The student should have added/subtracted from left to right. The correct value is 30.

Lesson 1-2

1. four times a number q **3.** 15 plus r
5. 3 times x squared **7.** two times a plus six
9. twenty-five plus six times a number squared **11.** three times a number raised to the fifth power divided by two **13.** 5 times g to the sixth power **15.** four minus five times h
17. 1 less than 7 times x cubed **19.** 3 times n squared minus x **21.** 18 times the quantity p plus 5 **23.** $n - 35$ **25.** $\frac{1}{3}n$ **27.** $\frac{45}{r}$
29. $18 - 3d$ **31.** $\frac{20}{t^5}$ **33.** $(k + 2) - 15$
35. $2m + 6$ **37.** $\frac{6136}{y}$ **39.** $19.95 \times t - 10$ or $19.95t - 10$ **41.** $5f + 45h$ **43.** 1 **45.** 7
47. 149 **49.** $\frac{65}{4}$ **51.** 44 **53.** $\frac{1}{2}$ **55.** 18
57. 10 **59.** 16 **61.** 13 **63a.** $5t - 100$
63b. 1400 students **65a.** $1.75 + 3.45m$
65b. $29.35 **67a.** Sample answer: the quotient of $5x$ and 2 plus y cubed; $5x$ divided by 2 plus y to the third power **67b.** 18
69. 89 **71.** 52 **73.** Sample answer: the quotient of x minus 1 and 2; $\frac{x-1}{2}$
75a. $x - (36 \times 4)$ **75b.** $\frac{x - (36 \times 4)}{0.20}$
75c. 350 mi **77.** Sample answer: An algebraic expression is a math phrase that contains one or more numbers or variables. To write an algebraic expression from a real-world situation, first assign variables. Then determine arithmetic operations done on the variables. Finally, put the terms in order. **79.** Sample answer: Movie tickets cost $10 and a box of popcorn cost $5.25. You buy t movie tickets and a box of popcorn. What is the greatest number of movies tickets you can purchase with $50?

Lesson 1-3

1. Symmetric Property of Equality
3. Symmetric Property of Equality **5.** 14
7. 34 **9a.** Exit 15 to Exit 8 **9b.** Symmetric Property of Equality

11.
$= (3 \div 2)\frac{2}{3}$	Multiplicative Identify
$= \frac{3}{2} \cdot \frac{2}{3}$	Substitution
$= 1$	Multiplicative Identify

13.
$= 2(5 - 5)$	Substitution
$= 2(0)$	Additive Inverse
$= 0$	Multiplicative Property of Zero

15.
$= 2(2 - 1) \cdot \frac{1}{2}$	Substitution
$= 2(1) \cdot \frac{1}{2}$	Substitution
$= 2 \cdot \frac{1}{2}$	Multiplicative Identity
$= 1$	Multiplicative Inverse

Selected Answers SA1

17.
$= 4 + \frac{4}{9} + 7 + \frac{2}{9}$	Substitution
$= 4 + 7 + \frac{4}{9} + \frac{2}{9}$	Commutative (+)
$= 4 + 7 + \left(\frac{4}{9} + \frac{2}{9}\right)$	Associative (+)
$= 11 + \frac{6}{9}$	Substitution
$= 11\frac{6}{9}$	Substitution
$= 11\frac{2}{3}$	Substitution

19.
$= (2 \cdot 8) \cdot (10 \cdot 2)$	Associative (×)
$= 16 \cdot 20$	Substitution
$= 320$	Substitution

21.
$= \left(2\frac{3}{4} \cdot 1\frac{1}{8}\right) \cdot 32$	Associative (×)
$= \left(\frac{11}{4} \cdot \frac{9}{8}\right) \cdot 32$	Substitution
$= \frac{99}{32} \cdot 32$	Substitution
$= 99$	Substitution

23.
$= 2 \cdot 5 \cdot 4 \cdot 3$	Commutative (×)
$= (2 \cdot 5) \cdot (4 \cdot 3)$	Associative (×)
$= 10 \cdot 12$ or 120	Substitution

25.
$= \frac{4}{3} \cdot 3 \cdot 7 \cdot 10$	Commutative (×)
$= \left(\frac{4}{3} \cdot 3\right) \cdot (7 \cdot 10)$	Associative (×)
$= 4 \cdot 70$ or 280	Substitution

27. -64 **29.** -5 **31.** -9 **33.** Sample answer: Multiplicative Identity and Multiplicative Inverse **35.** 0; Additive Identity **37.** 1; Multiplicative Identity **39.** 5; Additive Identity **41.** 1; Multiplicative Inverse **43.** 3; Reflexive Property **45.** Yes; the Commutative and Associative Properties of Multiplication allow it to be rewritten. **47.** Sample answer: $126 + 28 + 52 = 126 + (28 + 52) = 126 + 80 = 206$ **49.** Sample answer: $5 = 3 + 2$ and $3 + 2 = 4 + 1$, so $5 = 4 + 1$; $5 + 7 = 8 + 4$ and $8 + 4 = 12$, so $5 + 7 = 12$. **51.** $4 \div 8 \neq 8 \div 4$ because $4 \div 8 = \frac{1}{2}$ and $8 \div 4 = 2$, so there is no Commutative Property for division. $16 \div (8 \div 4) \neq (16 \div 8) \div 4$ because $16 \div (8 \div 4) = 16 \div 2 = 8$ and $(16 \div 8) \div 4 = 2 \div 4 = \frac{1}{2}$, so there is no Associative Property for division. As long as neither number is 0, when the order of division of two numbers is switched, the results are multiplicative inverses of each other. **53a.** False; sample answer: $3 - 4 = -1$, which is not a whole number. **53b.** True **53c.** False; sample answer: $2 \div 3 = \frac{2}{3}$, which is not a whole number. **55.** $(2j)k = 2(jk)$; The other three equations illustrate the Commutative Property of Addition or Multiplication. This equation represents the Associative Property of Multiplication.

Lesson 1-4

1. $4(6) + 5(6)$; 54 **3.** $6(6) - 6(1)$; 30 **5.** $14(8) - 14(5)$; 42 **7a.** $39(23 + 2)$ **7b.** $975 **9a.** $10\left(3\frac{3}{5}\right)$ **9b.** $10\left(3\frac{3}{5}\right) = 10\left(3 + \frac{3}{5}\right) = 10(3) + 10\left(\frac{3}{5}\right) = 30 + 6 = 36$ yards of fabric **11.** $7(500 - 3)$; 3479 **13.** $36\left(3 + \frac{1}{4}\right)$; 117 **15.** $5(90 - 1)$; 445 **17.** $15(100 + 4)$; 1560 **19.** $12(100 - 2)$; 1176 **21.** $3(10 + 0.2)$; 30.6 **23.** $2(x) + 2(4)$; $2x + 8$ **25.** $4(8) + (-3m)(8)$; $32 - 24m$ **27.** $2(17) + (-4n)(17)$; $34 - 68n$ **29.** $\frac{1}{3}(27) + (-2b)(27)$; $9 - 54b$ **31.** $6(2c) + 6(-cd^2) + 6(d)$; $12c - 6cd^2 + 6d$ **33.** $3(m) + 3(n)$; $3m + 3n$ **35.** $\frac{1}{2}(14) + (6a)(14)$; $7 + 84a$ **37.** $0.3(9) + (-6x)(9)$; $2.7 - 54x$ **39.** $18r$ **41.** $2m + 7$ **43.** $13m + 5p$ **45.** $14m + 11g$ **47.** $12k^3 + 12k$ **49.** $18g$ **51.** $5a^2$ **53.** $2q^2 + q$ **55a.** $3a + 5(a - b)$

55b.
$3a + 5(a - b) = 3a + 5a - 5b$	Distributive Property
$= (3a + 5a) - 5b$	Associative (+)
$= 8a - 5b$	Substitution

57. $24x + 28$ **59.** $18d + 20$ **61.** $7y^3 + y^4$ **63.** $4b$ **65.** $20x + 37y$ **67.** $2n + 2m$ and $2(n + m)$ **69.** No; sample answer: 10 pounds 5 ounces is $10(16) + 5 = 165$ ounces, but Ariana used the Distributive Property incorrectly. She should have written $8(20 + 2) = 8(20) + 8(2) = 160 + 16 = 176$ ounces. **71.** Sample answer: Algebraic expressions are helpful because they are easier to interpret and apply than verbal expressions. They can also be written in a more simplified form.

Lesson 1-5

1. $|p - t|$ and $|t - p|$ **3.** $|r - w|$ and $|w - r|$ **5.** 15 **7.** 22 **9.** 37 **11.** 32 **13.** -62 **15.** 11 **17.** 10 **19.** 5 **21.** 5 **23.** 6 **25.** -7.4 **27.** 8.4 **29.** -15 **31.** 22 **33.** 14.5 **35a.** $|g - d|$ and $|d - g|$ **35b.** 5 meters **37.** Sample answer: A meteorologist says that the high temperature is going to be 89 degrees. If the actual high temperature that day is x, then $|x - 89|$ represents the number of degrees the meteorologist is away from the actual high temperature.

39. False; sample answer: Suppose $a = 5$ and $b = -3$, then $|a + b| = |5 + (-3)| = |5 - 3| = |2| = 2$ and $|a| + |b| = |5| + |-3| = 5 + 3 = 8$. $2 \neq 8$, so Diaz's claim is not correct.

Lesson 1-6

1. 32   **3.** 11.4 seconds   **5.** 0.512   **7.** Automatic Method: $6000; Exact Method: $5625
9. $2\frac{1}{3}$ snack bars   **11.** 6   **13.** $333.33
15. $10,500   **17.** Because the number of students enrolled at Hartgrove High School can be counted, giving an exact enrollment is accurate.   **19.** The map maker is probably accurate because the number of traffic lights in New York City is not very specific.   **21.** Sample answer: Steve Nash would be selected as a free throw shooter. Michael Jordan would be selected as a free throw shooter. Shaquille O'Neal would not be selected as a free throw shooter.   **23.** light-years; Sample answer: The distance from Earth to the star is very great so using the largest distance unit is appropriate in this situation.   **25.** Sample answer: The line represents the most accurate number of visitors at the zoo for a given temperature.
27. Sample answer: The number of visitors does not increase at the same rate for each average daily temperature.   **29.** Sample answer: An employer might consider the number of sick days an employee takes or the amount of sales an employee generates.
31. Sample answer: $28.43 because $3.299 \times 8.618 = 28.430782$. The answer could be accurate to the thousandths place, but it is only necessary to round to the nearest hundredths place because the penny is the smallest unit of money.

Module 1 Review

1. D   **3.** $5(x + 7) - 4^3$ **5.** A   **7.** B
9.

Statement	True	False
$4(6 - 2 \times 3) = 0$	X	
$11(3^2 - 9) + 2\left(\frac{1}{2}\right) = 0$		X
$4 \cdot 0 + 4^2 - 2^3 - (2 + 2 \cdot 3) = 0$	X	

11.

Expression	Yes	No
$-7(m^3 - 11)$		X
$-7(3m) - 7(-11)$	X	
$-21m - 77$		X
$-21m + 77$	X	
$-21m - 11$		X
$-7m^3 + 77$		X

13. C, E   **15.** 67
17. No; sample answer: The manager rounded down, but actually spent much more than $4000. It would have been better to report a greater amount so that it was clear her budget was not overspent.

Module 2

Quick Check

1. $6n + 2$ **3.** $4b + 9$ **5.** 8 **7.** 32 **9.** 36

Lesson 2-1

1. $3m + 2 = 18$ **3.** $\frac{24}{x} = 14 - 2x$ **5.** $2 + 3h = 6$ **7.** $(48 + 33) + n = 107$ **9.** $2a + a^3 = b$ **11.** $x + x^2 = yz$ **13.** $A = \ell^2$ **15.** $P = 2\ell + 2w$ **17.** $I = prt$ **19.** The sum of j and sixteen is thirty-five. **21.** Seven times the sum of p and twenty-three is the same as one hundred two. **23.** Two-fifths of v plus three-fourths is identical to two-thirds of x squared. **25.** g plus 10 is the same as 3 times g. **27.** 4 times the sum of a and b is 9 times a. **29.** Half of the sum of f and y is f minus 5. **31.** Sample answer: The volume equals π times the radius squared times the height. The base is a circle so the expression πr^2 represents the area of the base. **33.** Sample answer: The interest equals the product of the principal, the rate, and the time. **35.** Sample answer: Force equals mass times acceleration. The expression ma represents the force on an object with mass m that is accelerating. **37.** B **39.** A **41.** $y^2 - 12 = 5x$ **43.** $100 - 3b = 6b$ **45.** Four times n equals x times the difference of five and n. **47.** The sum of y and the product of 3 and the square of x is 5 times x. **49.** $V = \ell wh$ **51.** $m + 2m = 24$ or $3m = 24$ **53.** $c = 10w + 0.1(10w)$ or $c = 11w$
55a. It is correct. The product is squared, so parentheses are needed. **55b.** It is not correct. One-half of a number means to multiply, not divide, by one-half. It should be $\frac{1}{2}n + 3 = n - 2$. **57.** Sample answer: A teacher ordered 188 math books. The algebra books were packed in boxes of 12. The geometry books were packed in boxes of 10. He ordered one more box of algebra books than geometry books. How many books of each type book did he order? Let $a = $ number of algebra books.
59. $S = 6\ell^2$ **61.** Sample answer: First, you should identify the unknown quantity or quantities for which you are trying to solve, and assign variables. Then, you should look for key words or phrases that can help you to determine operations that are being used. You can then write the equation using the numbers that you are given and the variables and operations that you assigned.

Lesson 2-2

1. 23 **3.** -43 **5.** -12 **7.** 73 **9.** -15
11. -54 **13.** $\frac{7}{20}$ **15.** $-\frac{7}{15}$ **17.** -937
19. -147 **21.** -25 **23.** $-\frac{9}{2}$ **25.** 15
27. 10 **29.** 64 **31.** 28 **33.** 18 **35.** 24
37. 27 **39.** 39 **41.** 64 **43.** 9 **45.** -12
47. 7 **49.** 64 **51.** -252 **53.** -52
55. $x + 33 = 2005$; $x = 1972$ **57.** $x - 21 = -9$; $x = 12°C$ **59a.** Let $p = $ the number of players who signed up for the soccer league. If 13% of the players who signed up for the soccer league dropped out, then 100% -13%, or 87% of the players finished the season. So, $0.87p$ represents the number of players who finished the season. **59b.** $0.87p = 174$ **59c.** $p = 200$; 200 players signed up for the soccer league
61. $\frac{2}{3} = -8n$; $-\frac{1}{12}$ **63.** $\frac{4}{5} = \frac{10}{16}n$; $\frac{32}{25}$
65. $4\frac{4}{5}n = 1\frac{1}{5}$; $\frac{1}{4}$ **67.** -77 **69.** $\frac{16}{3}$
71. -10 **73.** $-\frac{10}{7}$ or $-1\frac{3}{7}$ **75.** 18
77. 225 **79.** -14 **81.** 4 **83.** -49
85. 40 **87.** -15 **89.** $-\frac{8}{15}$ **91a.** $12x = 780$; $x = 65$ **91b.** $20 **93.** $x = 216$; Multiplication Property of Equality **95.** $y = -224$; Subtraction Property of Equality
97. $15 = b$; Division Property of Equality
99. $n - 16 = 29$ does not belong because for the other three, $n = 13$, and for this one $n = 45$.
101. Sample answer: $x - 4 = 10$ **103.** Sample answer: To solve $5x = 35$, I would divide each side by 5 to get $x = 7$. To solve $5 + x = 35$, I would subtract 5 from each side to get $x = 30$. In both equations I used properties of equality to isolate the variable. In the first equation I used the Division Property of Equality and I used the Subtraction Property of Equality in the second equation.

Lesson 2-3

1. -5 **3.** -5 **5.** 70 **7.** 27 **9.** 16
11. -61 **13.** $\frac{1}{2}a - 5.25 = 22.50$; $55.50
15. $\frac{t-10}{15} = 4$; 70 treats
17. $71 = 2h - 1$; 36 inches
19. $\frac{18}{a}$ **21.** $\frac{-35}{a}$ **23.** $\frac{-24}{a}$ **25.** $\frac{-14}{a}$
27. 7 **29.** 10 **31.** -16 **33.** -2
35. 18 **37.** $(n-2) \div 3 = 30$; 92 **39.** Sample answer: Both are correct. Dividing by a number and multiplying by that number's reciprocal are equivalent operations. **41a.** $x = \frac{-2}{a}$
41b. $x = 13a$ **41c.** $x = \frac{10}{a}$ **43.** Never; whenever three odd integers are added together, the sum is always odd.

Lesson 2-4

1. 6 **3.** 1 **5.** -2 **7.** -2 **9.** 14 **11.** 4
13. -5 **15.** 0 **17.** $7 + F = 4F + 1$; France won 2 gold medals and the U.S won 9 gold medals. **19.** $38 + 4x = 45.5 + 2.5x$; 5 years **21.** $180 - x = 10 + 2(90 - x)$; 10°
23. $9(5 + x) = 15\frac{3}{7}x$; 7 **25.** no solution
27. identity **29.** no solution **31.** one solution
33. identity **35.** no solution **37.** no solution
39. all numbers **41.** -25 **43.** 3
45. -2 **47.** 15 **49a.** Let $n =$ the first odd integer; $2(n + 2) = 3n - 13$ **49b.** 17 and 19
51a. Let $k =$ the number; $4k - 3 = 2k + 5$
51b. $k = 4$ **51c.** Substitute 4 for k in the expression for the perimeter of Figure 2, $2k + 5$. So the perimeter of Figure 2 is $2(4) + 5 = 8 + 5 = 13$. **51d.** Substitute 4 for k in the expression for the perimeter of Figure 1, $4k - 3$. So the perimeter of Figure 1 is $4(4) - 3 = 16 - 3 = 13$. The perimeter for Figure 1 and Figure 2 is the same, so the value of k is correct.
53. Anthony is correct. When Patty added m to each side, she subtracted the terms instead of adding them. **55.** Sample answer: $2(3x + 6) = 3(2x + 5)$ **57a.** Incorrect; the 2 must be distributed over both g and 5, then 10 must be subtracted from each side; 6.
57b. Correct; the Subtraction Property was used to combine the variable terms on the left side of the equation. The Division Property was used to isolate the variable on one side.
57c. Incorrect; to eliminate $-6z$ on the left side of the equal sign, $6z$ must be added to each side of the equation; 1. **59.** Sample answer: $2x + 1 = x + 9$

Lesson 2-5

1. $\{-2, 8\}$

3. ∅

5. $\{-3.25, 2\}$

7. ∅

9. $\{-8, 16\}$

11. $\{2, 5\}$ **13.** $\{-6, 4\}$ **15.** $\{0, 4\}$
17. $\{-5, -1\}$ **19.** $|t - 400| = 15$; min = 385°F; max = 415°F **21a.** $|t - 20.9| = 5.3$ **21b.** 15.6 to 26.2; 10.3 to 31.5

23. $|x - 35| = 0.5$ Absolute value equation
Case 1: $x - 35 = 0.5$ Definition of absolute value.
$x = 35.5$ Simplify
Case 2: $x - 35 = -0.5$ Definition of absolute value.
$x = 34.5$ Simplify
The bags of rock salt weigh no less than 34.5 pounds and no more than 35.5 pounds.
25. $|x| = 6$ **27.** $|x + 2| = 4$ **29.** $|x + 3| = 2$
31. $|x| = 4$
33. $\left\{-\frac{3}{2}, \frac{9}{2}\right\}$

35. $\{5.5, -5.5\}$

37. $\{2, -2\}$

39. $|x| = 1\frac{1}{2}$ **41.** $\left|x - \frac{1}{4}\right| = \frac{1}{4}$ **43.** $\left|x + \frac{1}{3}\right| = 1$
45a. $|x - 38| = 2$ **45b.** 40°F, 36°F
47. $|x - 3| = 1$ **49.** $|a - b| + |b - c| = |a - c|$ **51.** Cami; The absolute value of a number cannot be negative. **53.** Sample answer: Let $x =$ the temperature at night. Then the temperature is 4 ± 10 degrees.

Lesson 2-6

1. 40 **3.** 29.25 **5.** 9.8 **7.** 1.32 **9.** 0.84
11. 0.57 **13.** 6 **15.** 11 **17.** 18 **19.** 0.8
21. 11 **23.** −2 **25.** 1.44 **27.** −2.29
29. −2.2 **31.** 10 **33.** 3 **35.** −8.4
37. 12.5 gal **39.** $46.27 **41a.** 60 free throws **41b.** Sample answer: I assumed that Brent continues to make free throws at the same rate. **43.** 22.5 in. **45.** $3333.33
47. 204.55 mL **49.** 6 **51.** 10 **53.** 21 **55.** 8
57. 42 **59.** 27 **61.** 3 **63.** 15 **65.** −3
67. −0.4 **69.** −6 **71a.** Sample answer: 2.2 cm **71b.** Sample answer: about 6.6 miles
71c. Sample answer: about 453.7 mi^2
73a. $4.50; because 8 potatoes cost $1.50, multiply by 3 to get a cost of $4.50 for 24 potatoes. **73b.** $4.13; Sample answer: $4.13 is slightly less than $4.50, which aligns with my estimate. **73c.** 37 **73d.** $0.19 **75.** $\frac{2}{4}$ or $\frac{1}{2}$
77. Ratios and rates each compare two numbers by using division. However, rates compare two measurements that involve different units of measure. **79.** $\frac{x}{100} = \frac{z}{y}$

Lesson 2-7

1. $y = \frac{x-1}{2}$ **3.** $f = \frac{5-g}{7}$ **5.** $t = \frac{x}{7}$ **7.** $r = \frac{q}{2}$
9. $a = -\frac{b}{8}$ **11.** $v = \frac{u-z}{w}$ **13.** $g = \frac{10j + 9h}{f}$
15. $t = \frac{3}{2}(r - v)$ **17.** $a = \frac{-33 + x}{10c}$
19a. $\ell = \frac{P - 2w}{2}$ **19b.** 14 m
21a. $g = \frac{c - p}{13.50}$ **21b.** 6 games
25. ≈ 12.96 trillion pounds **27.** 0.44 ft
29. 24 miles **31.** 90.2 gallons
33. 82 students **35.** $c = \frac{2k - 3g}{b}$
37. $c = \frac{5p - 6j}{8}$ **39.** $c = x - 2d$
41. $t = \frac{w - 11v}{31}$ **43.** $c = \frac{-13 + f}{10 - d}$ **45.** $r = \frac{A}{P} - 1 = \frac{2182.25}{2150} - 1 = 1.015 - 1 = 0.015$; The interest rate is 1.5%. **47a.** $y = \frac{mx + mt - z}{r}$
47b. Division by 0 is undefined, so in the original equation $m \neq 0$, and in the final equation $r \neq 0$. **49.** No; Sasha does not have a correct solution. When she multiplied F by $\frac{5}{9}$, she should have multiplied 32 by $\frac{5}{9}$. To avoid multiplying each term by a fraction, she could have subtracted 32 from both sides, then multiplied both sides by $\frac{5}{9}$.

Module 2 Review

1. A **3.** D **5.** D **7.** C **9.** B, D
11. C **13.** Let d = age of dogwood tree (in years); $3d - 2 = \frac{1}{2}(d + 8 + 8)$; The gingko tree is 10 years old, and the dogwood tree is 4 years old. **15.** A **17.** $b = 8$ **19.** B
21. D **23A.** $h = 3\frac{V}{B}$ **23B.** 12

Module 3

Quick Check

1, 3.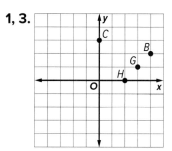

5. 20 **7.** −3

Lesson 3-1

1.

x	y
−1	−1
1	1
2	1
3	2

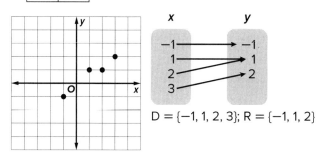

D = {−1, 1, 2, 3}; R = {−1, 1, 2}

3.

x	y
3	−2
1	0
−2	4
3	1

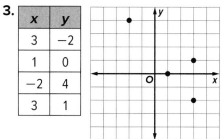

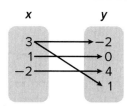

D = {−2, 1, 3}; R = {−2, 0, 1, 4}

5a. independent: price of item, dependent: number of items purchased **5b.** As the price of an item increases, the number of items purchased decreases.

7.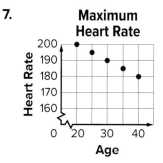

9. The x-axis represents the time in seconds. The y-axis represents the height of the elevator in feet. The x-axis has a scale of 1 mark = 1 second. The y-axis has a scale of 1 mark = 10 feet. The origin (0, 0) represents a height of 0 feet in 0 seconds. **11.** {(1, 7), (3, 45), (5, 11), (13, 15)}
13. {(2, 5), (5, 0), (7, 8), (7, 10), (10, 2)} **15.** {(2, 80), (3, 120), (6, 240), (8, 320)}; The x-axis represents the number of gallons of syrup. The y-axis represents the number of gallons of sap. The x-axis has a scale of 1 mark = 1 gallon of syrup. The y-axis has a scale of 1 mark = 40 gallons of sap. The origin (0, 0) represents 0 gallons of sap makes 0 gallons of syrup.
17a. {(0, 12), (1, 8), (2, 23), (3, 28), (4, 11), (5, 11)}
17b. {0, 1, 2, 3, 4, 5}
17c. {8, 11, 12, 23, 28}
19. sample answer:

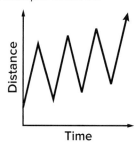

21. Tim drives away from the pizzeria, stops to make a delivery, continues to drive away from the pizzeria, stops to make another delivery, and then returns to the pizzeria. **23.** Disagree; The intersection point represents a time when Tim and Lauren were both at the same distance from the pizzeria. **25.** Disagree; sample counterexample: In the relation {(1, 2), (1, 3)}, the domain is {1}, so it has one element, while the range is {2, 3}, which has two elements. **27a.** Sample answer: {(−1, −3), (0, −3), (0, −1), (1, 4), (2, 5)}

27b. sample answer:

x	y
−1	−3
0	−1
0	−3
1	4
2	5

29. Sample answer: A dependent variable is determined by the independent variable for a given relation.

Lesson 3-2

1. Yes; for each element of the domain, there is only one element of the range. **3.** No; the element 4 in the domain is paired with both 2 and 5 in the range. **5.** No; the element 5 in the domain is paired with both −3 and 2 in the range. **7.** Yes; for each element of the domain, there is only one element of the range. **9.** Yes; for any value x, the vertical line passes through no more than one point on the graph. **11.** Yes; for any value x, the vertical line passes through no more than one point on the graph. **13.** No; for x > 0, the vertical line passes through more than one point on the graph.

15a.

Year	2014	2015	2016	2017
Value ($)	254,000	293,000	338,000	372,000

15b. Domain: {2014, 2105, 2016, 2017}; Range: {254,000; 293,000; 338,000; 372,000}
15c. For each element of the domain, there is only one element of the range. So, this relation is a function. **17.** 26 **19.** 2 **21.** 42 **23.** 4 **25.** 6 **27.** $9b^2 - 3b$ **29.** $f(3.5) = 12.25$, which means the area of a square with a side of length 3.5 units is 12.25 square units. **31.** $f(12) = \$435$, which is the cost of a gym membership for 12 months, or 1 year. **33.** −1 **35.** 14 **37.** −4 **39.** $-8y - 3$ **41.** $-2c - 8$ **43.** $-10d - 15$

45a.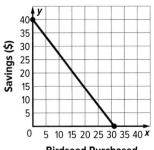

45b. Yes; for any value x, the vertical line passes through no more than one point on the graph. **45c.** $f(3) = 36.25$, which means if Aisha buys 3 pounds of birdseed, she saves $36.25; $f(18) = 17.50$, which means if Aisha buys 18 pounds of birdseed, she saves $17.50; $f(36) = -5$, which means if Aisha wants to buy 36 pounds of birdseed, she needs $5 extra. **45d.** 8 pounds **47a.** $h(20) = 46$; The height of the balloon 20 seconds after it is released is 46 feet. **47b.** 2 minutes is 2(60) = 120 seconds, so calculate $h(120)$ by substituting $t = 120$ in the equation; $h(120) = 2(120) + 6 = 246$; The height of the balloon is 246 feet. **47c.** 6 feet; $t = 0$ before the balloon is released, and $h(0) = 6$. **47d.** Sample answer: The values of t must be greater than or equal to zero because a negative value for the time does not make sense for the given situation. The graph would start at the vertical axis and go only to the right. **49.** Sample answer: You can determine whether each element of the domain is paired with exactly one element of the range. For example, if given a graph, you could use the vertical line test; if a vertical line intersects the graph more than once, then the relation that the graph represents is not a function. **51.** $f(g + 3.5) = -4.3g - 17.05$ **53.** Sample answer: $f(x) = 3x + 2$

Lesson 3-3

1. Neither; because the function has continuous sections but is not a single line or curve, it is neither continuous or discrete **3.** Discrete; because the function is made up entirely of individual points, it is discrete. **5.** Continuous; because the function is graphed with a single line, it is continuous. **7.** Discrete; because the function is made up entirely of individual points, it is discrete. **9.** discrete **11.** discrete

13. linear **15.** nonlinear **17.** nonlinear
19. nonlinear **21.** nonlinear **23.** linear
25a. linear

25b.
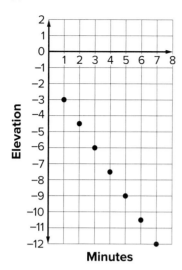

27a. nonlinear

27b.
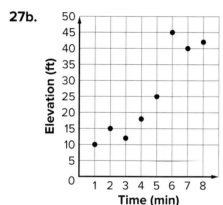

29. continuous; nonlinear **31.** neither; nonlinear **33.** discrete; nonlinear
35. Sample answer: A studio charges musicians to use the space and recording equipment by the hour, rounding a fraction of an hour up. So, for up to 1 hour, the studio charges $100, but for up to 2 hours, the studio charges $200, and so on. The function that models this situation is neither discrete nor continuous.

Lesson 3-4

1. x-intercept: $(-0.75, 0)$ y-intercept: $(0, 3)$ positive: when $x > -0.75$ negative: when $x < -0.75$ **3.** x-intercepts: $(0, 0)$ and $(2, 0)$ y-intercept: $(0, 0)$ positive: when $x < 0$ and when $x > 2$ negative: $0 < x < 2$
5. x-intercepts: $(-2, 0)$ y-intercept: $(0, 4)$ positive: when $x > -2$ negative: when $x < -2$

7. x-intercepts: $(-5, 0)$ and $(3, 0)$ y-intercept: $(0, 3)$ positive: $-5 < x < 3$ negative: $x < -5$ and when $x > 3$
9. x-intercepts: none y-intercept: $(0, -3)$ positive: never negative: always
11. The x-intercept is 0. The y-intercept is 0. This means that Ryan earns $0 for working 0 hours. The function is positive when x is greater than 0, which means that Ryan earns money for working. No portion of the graph shows that the function is negative. **13a.** The x-intercept is 6. The y-intercept is 1950.
13b. The x-intercept means that after 6 months, Javier's remaining balance will be $0, or it will take Javier 6 months to repay his parents. The y-intercept means that Javier owes his parents $1950 after 0 months, or Javier initially borrowed $1950 from his parents.

15. Sample graph; no solution

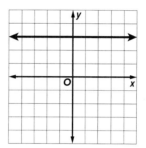

17. -2

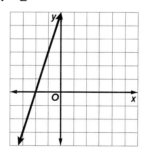

19. 1

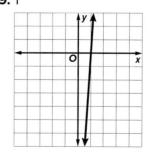

21. The zero of the function is at 32. This represents that after tying ribbon on 32 gift bags, Juanita will have no ribbon left.

23. x-intercepts: (−2, 0) and (2, 0) y-intercept: (0, 4) positive: −2 < x < 2 negative: when x < −2 and when x > 2 **25.** The x-intercepts are 3 and 7. That means that the bird will be at sea level at 3 seconds and at 7 seconds. The y-intercept is 4.5. This means that at time 0, the bird was at a height of 4.5 feet. The function is positive when x is less than 3 and when x is greater than 7, which means the bird is above sea level from 0 to 3 seconds and after 7 seconds. The function is negative when x is between 3 and 7, which means that the bird is below sea level, or under water, for 4 seconds.

27. To find the x-intercept in a graph, find the place where the function crosses the x-axis. To find the y-intercept in a graph, find the place where the function crosses the y-axis. To find the x-intercept in a table, find the x-value when the y-value is 0. To find the y-intercept in a table, find the y-value when the x-value is 0. **29.** Find the related function. Subtract 16 from each side: $0 = x + 4 + (2^4 − 6) − 16$. Evaluate the exponent: $0 = x + 4 + (16 − 6) − 16$. Evaluate the expression in parentheses: $0 = x + 4 + 10 − 16$. Add and subtract: $0 = x − 2$. Replace 0 for f(x). The related function is $f(x) = x − 2$. The graph of the related function intersects the x-axis at 2. This is the x-intercept, or zero. So the solution of the equation is 2. Check the solution by solving the equation algebraically. Evaluate the exponent: $16 = x + 4 + (16 − 6)$. Evaluate the expression in parentheses: $16 = x + 4 + 10$. Add: $16 = x + 14$. Subtract 14 from each side: $2 = x$.

Lesson 3-5

1. This function is symmetric in the line $x = −1$.
3. This function is symmetric in the line $x = 2.5$.
5. The graph is symmetric in the line $x = 5$. In the context of the situation, the symmetry of the graph tells you that the area is the same when the width is a number less than or greater than 5. **7.** always decreasing
9. decreasing: $x < 1.5$; increasing: $x > 1.5$
11. extrema: B and D; rel min: D; rel max: B
13. extrema: B and D; rel min: B; rel max: D

15. Point A is a relative maximum. Point A represents the greatest height of the golf ball given the distance from the tee. **17.** As x decreases, y increases. As x increases, y increases. **19.** As x decreases, y decreases. As x increases, y increases. **21.** no line symmetry; always decreasing; extrema: none; As x decreases, y increases. As x increases, y decreases. **23.** The approximate point (2.5, 114) is a relative maximum. This represents the greatest height of the rock given the time.
25. The graph has one relative minimum at about (−2.25, −16); This statement is not true because there are two relative minimums: one at about (−2.25, −16) and one at about (2.25, −16).

Lesson 3-6

1.

3.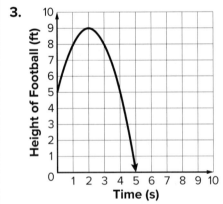

SA10 Selected Answers

5.

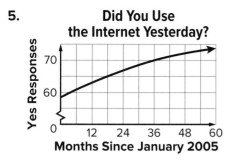

Did You Use the Internet Yesterday?

7. Sample answer: Internet use at home initially has a higher number of users than Internet use away from home. Both Internet use at home and Internet use away from home increase after 36 months since March 2004. Neither Internet use at home nor Internet use away from home reaches 0 users.

9.

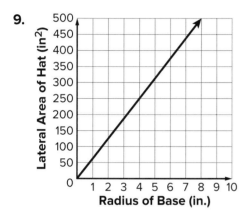

11. Sample answer: The graph on the calculator and the graph I sketched are both linear, increasing, and have an x- and y-intercept at 0. **13.** $P(x) = 28x - 840$; $P(x)$ is Aidan's profit from fixing and selling x bicycles. **15.** The x-intercept; To find the x-intercept, locate the point on the graph when $P(x) = 0$, which is 30. So, when 30 bicycles are bought and sold, Aidan makes a profit of $0.

Module 3 Review

1. B, C, F
3. C
5. Sample answer: The element −4 in the domain is paired with both 8 and 13 in the range. This relation is not a function. **7.** B
9. (−1, 0) and (0, −1) **11.** 40; 60 **13.** Sample answer: In 1900 the population of Ohio was nearly 4 million more than the population of Florida. Both populations grew between 1900 and 1950. At this point, the population of Ohio exceeded that of Florida by approximately 5 million, indicating a greater growth rate for Ohio than Florida during those decades. Then from 1950 to 2000, the population of Ohio grew by about 3.4 million, whereas the population of Florida grew by about 13 million, indicating a significantly greater growth rate for Florida during those decades. In fact, by 2000, the population of Florida surpassed Ohio by more than 4 million people.

Module 4

Quick Check

1, 3, 5.

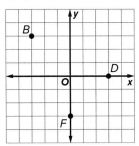

7. $y = -3x + 1$ **9.** $y = \frac{5}{2}x - 6$ **11.** $y = -10x + 6$

Lesson 4-1

1.

x	y
−2	0
−2	1
−2	2

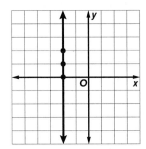

3.

x	y
−1	8
0	0
1	−8

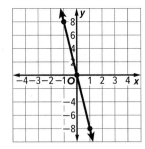

5.

x	y
0	8
1	7
2	6

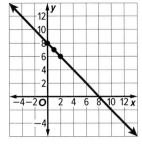

7.

x	y
0	1
2	2
4	3

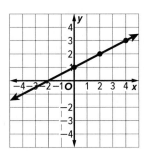

9.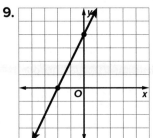

11.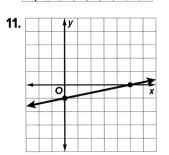

13.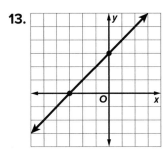

15a. The x-intercept is 6. This means that after 6 weeks, Amanda will have $0 in her school lunch account. The y-intercept is 210. This means that there was initially $210 in Amanda's school lunch account.

15b.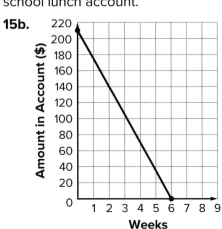

17.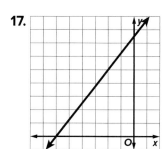

19.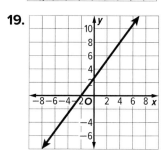

21. *x*-int: 7; *y*-int: −2

23. *x*-int: $1\frac{1}{3}$; *y*-int: 4

25. *x*-int: $-1\frac{1}{2}$; *y*-int: 1

27. $y = 1.7x + 40$; The *y*-intercept is 40. This means that it would cost $40 to hook up the car.

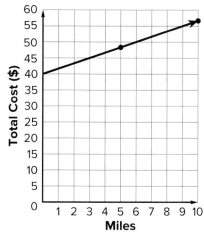

29. Sample answer: The *x*-intercept is −4. The *x*-intercept is not reasonable because the football team cannot lose −4 games. The *y*-intercept is 4. The *y*-intercept is reasonable because the *y*-intercept means that if the football team won 4 games, they lost 0 games. **31.** No; sample answer: A horizontal line only has a *y*-intercept and a vertical line only has an *x*-intercept. **33.** In the equation, let $y = 0$ to find the *x*-intercept: $2x + (0) = 4$. So the *x*-intercept is 2. In the equation, let $x = 0$ to find the *y*-intercept: $2(0) + y = 4$. So the *y*-intercept is 4. Robert graphed points at (2, 0) and (0, 4) and connected the points with a line. **35.** Sample answer: $y = 8$; horizontal line **37.** Sample answer: $x - y = 0$; line through (0, 0)

Lesson 4-2

1. $\frac{1}{5}$ **3.** increased about 1.9 people per square mile **5a.** −5; This means the temperature decreased 5°F per hour from 6 A.M. to 7 A.M. **5b.** −5; This means the temperature decreased 5°F per hour from 1 P.M. to 2 P.M. **7.** linear; $-\frac{1}{1}$ or −1 **9.** not linear **11.** $-\frac{3}{5}$ **13.** 1 **15.** 0 **17.** $\frac{1}{6}$ **19.** $\frac{4}{3}$ **21.** undefined **23.** 1 **25.** undefined **27.** 2 **29.** undefined **31.** −1 **33.** undefined **35.** $-\frac{7}{2}$ **37.** $\frac{5}{2}$ **39.** $\frac{3}{4}$ **41.** 6 **43.** 8 **45.** 11 **47.** $\frac{1}{20}$ **49.** $-\frac{1}{2}$ **51.** $\frac{1}{3}$ **53.** $\frac{1}{2}$ **55.** −1 **57.** $\frac{7}{4}$ **59.** 3 **61.** After drawing a graph, use the two points on the graph to determine the slope. This can be done by counting squares for the rise and run of the line or by using the coordinates of the points in the Slope Formula. **63.** The rate of change is $2\frac{1}{4}$ inches of growth per week. **65.** Step 1; she reversed the order of the *x*-coordinates in the formula. **67.** The difference in the *x*-values is always 0, and division by 0 is undefined.

Lesson 4-3

1. $y = 5x - 3$ **3.** $y = -6x - 2$ **5.** $y = 3x + 2$ **7.** $y = x - 12$ **9.** $y = 5x + 6$ **11.** $y = \frac{1}{3}x - 2$ **13.** $y = -0.25x - 3$ **15.** $y = 25x + 100$ **17.** $y = 0.12x + 9$

19.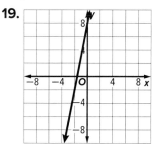

21.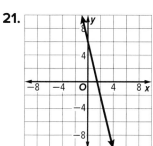

23.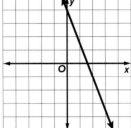

25.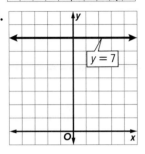

27.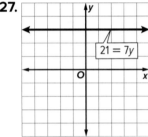

29a. $c = 13 + 8p$

29b.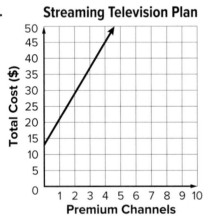

29c. $37 **31.** $y = \frac{1}{2}x - 3$

33.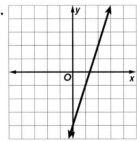

35.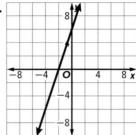

37. $y = 2x - 3$ **39.** $y = -x - 1$

41a. $T = 10x + 80$

41b.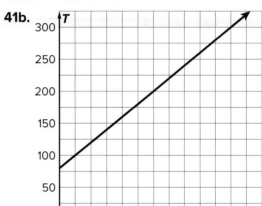

41c. 300°F   **41d.** $T = -5x + 300$

43a. $y = -4.25x + 25.5$

43b.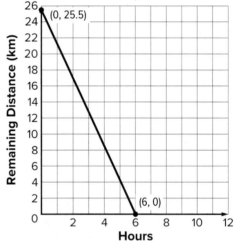

43c. The x-intercept (6) represents the number of hours it will take Jazmin to complete the walk. The y-intercept (25.5) represents the length of the walk.   **43d.** Using the graph, I can determine the value of x when y equals $-17 + 25.5$ or 8.5 km, and use the value of the x-intercept. The value of x is 4 when $y = 8.5$ and the x-intercept is 6. Therefore, Jazmin has $6 - 4$ or 2 hours more to walk.   **45.** Yes; you can find the value of x on the graph when $y = 0$; $x = \frac{1}{2}$.

47. Sample answer: $y = 25x + 200$; I have $200 in savings and will save $25 per week until I have enough money to buy a new phone. I can predict how much money I'll have after x number of weeks.

Lesson 4-4

1. $g(x)$ is a translation of the parent function 11 units up   **3.** $g(x)$ is a translation of the parent function 7 units right   **5.** $g(x)$ is a translation of the parent function 10 units left and 1 unit down   **7.** $g(x) = 4x$; $g(x)$ is the translation of $f(x)$ 3.5 units down.   **9.** $g(h) = 8h + 15$; $g(h)$ is the translation of $f(h)$ 5 units up.   **11.** $g(x)$ is a vertical compression of the parent function by a factor of $\frac{1}{3}$ **13.** $g(x)$ is a horizontal compression of the parent function by a factor of $\frac{1}{3}$ **15.** $g(x)$ is a horizontal stretch of the parent function by a factor of 2.5   **17.** $g(x)$ is a vertical stretch of the parent function by a factor of 8 and a reflection across the x-axis   **19.** $g(x)$ is a horizontal stretch of the parent function by a factor of $\frac{5}{4}$ and a reflection across the y-axis   **21.** $g(x)$ is a horizontal compression of the parent function by a factor of $\frac{2}{3}$ and a reflection across the y-axis   **23.** $g(x)$ is a translation of the parent function 2 units right and 8 units down   **25.** $g(x)$ is a vertical compression of the parent function by a factor of $\frac{1}{5}$ **27.** $g(x)$ is a horizontal compression of the parent function by a factor of 0.4   **29.** $g(x) = x - 7$ **31.** $g(x) = 1.5x$; The graph of $g(x) = 1.5x$ is the graph of $f(x) = 0.50x$ stretched vertically by a factor of 3.   **33a.** $g(x) = 1.29x$ **33b.** The graph of $g(x) = 1.29x$ is the graph of $f(x) = x$ stretched vertically by a factor of 1.29.   **35.** $y = \frac{1}{a}x$; The function is horizontally stretched by a factor of a.

Lesson 4-5

1. This sequence has a common difference of 4 between its terms. This is an arithmetic sequence.   **3.** This sequence does not have a common difference between its terms. This is not an arithmetic sequence.   **5.** This sequence does not have a common difference between its terms. This is not an arithmetic sequence.   **7.** This sequence has a common difference of 3 between its terms. This is an arithmetic sequence.   **9.** 1.06; 4.26, 5.32, 6.38
11. -2; 13, 11, 9   **13.** $\frac{1}{3}$; $3\frac{2}{3}$, 4, $4\frac{1}{3}$
15. 4; 19, 23, 27   **17.** 2; -5, -3, -1
19. $a_n = -5n + 2$; -33
21. $a_n = -4n - 7$; -35
23a. $f(n) = -4n + 128$

23b.
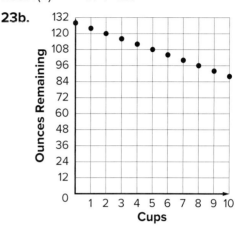

23c. 72 ounces   **25a.** $f(n) = 0.17n + 0.71$

25b.
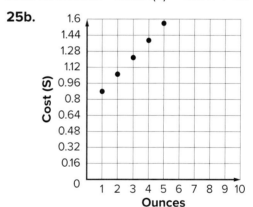

25c. 8 ounces   **27a.** $f(n) = 2n + 28$

27b.
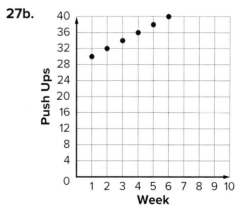

27c. 11th week   **29.** This sequence does not have a common difference between its terms. This is not an arithmetic sequence.
31. This sequence has a common difference of 2 between its terms. This is an arithmetic sequence.

33. $a_n = -4n + 34$

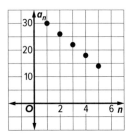

35a. $a_n = 3000 + 500n$ **35b.** $15,000
37a. $a_n = 3n - 1$ **37b.** 59   **39.** Sample answer: 5, 3, 8, 6, 11, 9, 14, …; The pattern is to subtract 2 from the first term to find the second term, then add 5 to the second term to find the third term. **41.** Sample answer: 2, −8, −18, −28, …   **43a.** Sample answer: $a_n = -2 - 3n$ **43b.** $a_n = -19 + 7n$ **43c.** $a_n = 12 - 2n$ **45.** On day 9, Andre has read 270 pages while Sam has 270 pages left to read. The table shows that both functions have a value of 270 when $x = 9$.

Lesson 4-6

1.

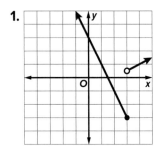

D = all real numbers,
R = $f(x) \geq -3$

3.

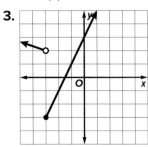

D = all real numbers,
R = $f(x) \geq -3$

5.

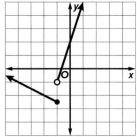

D = all real numbers,
R = $f(x) \geq -2.5$

7.

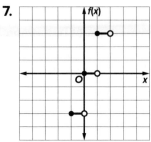

D = all real numbers,
R = all integer multiples of 3

9.

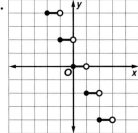

D = all real numbers,
R = all even integers

11.

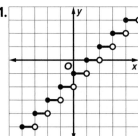

D = all real number,
R = all integers

13.

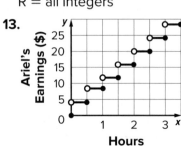

15.
$$f(x) = \begin{cases} 16.20 & \text{if } 0 < x \leq 1 \\ 19.30 & \text{if } 1 < x \leq 2 \\ 22.40 & \text{if } 2 < x \leq 3 \\ 25.50 & \text{if } 3 < x \leq 4 \\ 28.60 & \text{if } 4 < x \leq 5 \end{cases}$$

$D = \{x \mid 0 < x \leq 5\}$;
$R = \{16.20, 19.30, 22.40, 25.50, 28.60\}$

17. $g(x) = \begin{cases} 2x + 1 & \text{if } x \leq 2 \\ x - 2 & \text{if } x > 2 \end{cases}$

19. $133.00

21a.

x	0	2	4	6	8
f(x)	0	75	175	275	375

21b. $f(x) = 25 + 50[\![x]\!]$

21c.

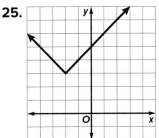

21d. $2 < x \leq 3$

23. Sample answer:
$$y = \begin{cases} -x & \text{if } x < -4 \\ 2x & \text{if } -4 \leq x \leq 2 \\ x - 2 & \text{if } x > 2 \end{cases}$$

25. A step function has different constants over different intervals of its domain. A piecewise-defined function can have different algebraic rules over different intervals of its domain.

27. $f(x) = \begin{cases} \frac{1}{2}x - 3 & x > 6 \\ -\frac{1}{2}x + 3 & x \leq 6 \end{cases}$

29. $R = f(x) \geq 0$

31. 2.4

7. $f(x) = |x + 2|$ **9.** $f(x) = |x| - 3$
11. $f(x) = |x| + 1$ **13.** The graph of $g(x)$ is a horizontal compression of the parent function.
15. The graph of $g(x)$ is a vertical stretch of the parent function.   **17.** The graph of $g(x)$ is a horizontal stretch of a parent function.
19. The graph of $g(x)$ is a reflection of the parent function across the x-axis and a vertical stretch.   **21.** The graph of $g(x)$ is a reflection of the parent function across the y-axis and a horizontal stretch.   **23.** The graph of $g(x)$ is a reflection of the parent function across the y-axis and a horizontal compression.

25.

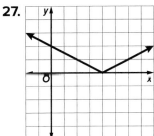

D = all real numbers,
R = $g(x) \geq 3$

27.

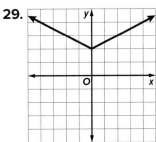

D = all real numbers,
R = $f(x) \geq 0$

29.

D = all real numbers,
R = $f(x) \geq 2$

Lesson 4-7

1. The graph of $g(x)$ is the parent function translated 5 units down.   **3.** The graph of $g(x)$ is the parent function translated 2 units right and 7 units up.   **5.** The graph of $g(x)$ is the parent function translated 1 unit up.

31.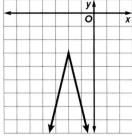
D = all real numbers,
R = f(x) ≤ −3

33.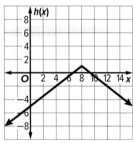
D = all real numbers,
R = h(x) ≤ 1

35. $y = 65|10 - x|$

37.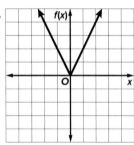
D = all real numbers,
R = f(x) ≥ 0
The graph of f(x) is the parent function horizontally compressed by a factor of $\frac{1}{2}$.

39. $f(x) = |-3x - 5|$ **41.** $f(x) = \left|\frac{1}{3}x + 2\right|$

43. $x = |s - 16|$

45. $x = |t - 21.7|$; The range of times is twice the value of x, 3.2(2) = 6.4 s; The solution to the equation is 24.9 and 18.5, which has a range of 24.9 − 18.5 = 6.4 s.

47. $x = |b - 12|$

49. To get the graph of h(x), the parent absolute value function is reflected in the x-axis, then translated 2 units left and 3 units down.

51. $f(x) = \begin{cases} -x + 5 & \text{if } x < 3 \\ x - 1 & \text{if } x \geq 3 \end{cases}$

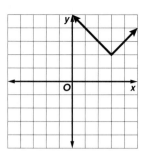

Module 4 Review

1.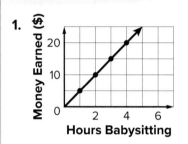

3. −500 gallons/hr **5.** B **7.** A

9. dilation **11.** C

13. $f(n) = 9n - 8$

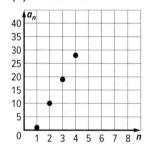

15.

Hours Worked, x	Money Earned, f(x)
30	270
35	315
40	360
45	427.5
50	495

17. B

19. Sample answer: It is translated 5 units up.

21. $f(x) = -|x - 4| + 3$

Module 5

Quick Check

1. $y = 5 - x$ **3.** $y = x + 5$ **5.** $(4, 2)$ **7.** $(2, -4)$
9. $(-3, -3)$

Lesson 5-1

1. $y = \frac{1}{2}x$ **3.** $-\frac{3}{4}x + \frac{17}{2}$ **5.** $y = \frac{1}{2}x + 1$
7. $d = 3t + 12$ **9.** $C = 2.54y + 62.38$
11. $y = -4$ **13.** $y = \frac{4}{3}x - \frac{1}{3}$ **15.** $y = -\frac{3}{2}x - \frac{9}{2}$
17. $y = -\frac{4}{11}x + \frac{58}{11}$ **19.** $y = -\frac{1}{2}x - \frac{9}{2}$
21. $y = \frac{1}{6}x + \frac{19}{24}$ **23.** $C = 10d + 12$
25. $T = -4.5x + 103$ **27.** $y = 3x - 1$
29. $y = -x - 4$ **31.** $y = -x + 3$ **33.** No; substituting 3 and -1 for x and y results in an equation that is not true. **35.** Yes; substituting 15 and -13 for x and y results in an equation that is true. **37.** Sample answer: $(3, -3)$
39. Sample answer: $(0, -5)$ **41.** Sample answer: $(0, 4)$ **43.** C; x represents the number of plane tickets per order and y represents the total cost of an order. **45.** A; x represents the number of hours and y represents the oil level in the tank, in inches. **47a.** $y = x + 2.5$
47b. 10 **47c.** Sample answer: $y = x + 1.5$
49a. $y = 7.5x + 1$ **49b.** 1; Koby's puppy weighed 1 pound at birth (0 months) **49c.** 7.5; Koby's puppy gained 7.5 pounds a month for the first 6 months. **51.** Jacinta; Tess switched the x- and y-coordinates on the point that she entered in Step 3. **53.** Sample answer: Let y represent the number of quarts of water in a pitcher, and let x represent the time in seconds that water is pouring from the pitcher. As time increase by 1 second, the amount of water in the pitcher decrease by $\frac{1}{2}$ qt. An equation is $y = -\frac{1}{2}x + 4$. The slope is the rate at which the water is leaving the pitcher, $\frac{1}{2}$ quart per second. The y-intercept represents the amount of water in the pitcher when it is full, 4 qt.

Lesson 5-2

1. $y + 3 = -1(x + 6)$

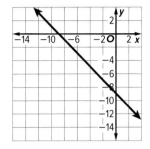

3. $y - 11 = \frac{4}{3}(x + 2)$

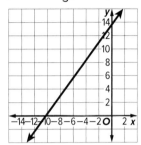

5. Sample answer: $y + 3 = -4(x - 1)$
7. Sample answer: $y - 3 = \frac{4}{3}(x - 3)$
9. $y = -6x - 47$ **11.** $y = \frac{1}{6}x - \frac{8}{3}$
13. $y - 18 = 3.5(x - 5)$ **15.** $2x - y = 6$
17. $x + 6y = -7$ **19.** $x - y = -1$
21. $2x + 3y = -13$
23. $3x + y = -3$
25. Sample answer: $y = x - 5$; $y = -x + 1$
27. Sample answer: $y = -5x + 2$; $y = \frac{1}{5}x + 2$
29. Sample answer:
$y = -\frac{3}{4}x + \frac{3}{2}$; $y = \frac{4}{3}x + \frac{17}{3}$
31. neither
33. perpendicular **35.** neither
37. $5x + 4y = 20$
39. $y = 9x + 5$; $9x - y = -5$
41. $y = -6x - 45$; $6x + y = -45$
43. $y = \frac{9}{10}x - 4\frac{3}{10}$; $9x - 10y = 43$
45. Yes; sample answer: The line that represents one of the ceiling walls has a slope of $-\frac{1}{4}$ and the line that represents the other ceiling wall has a slope of 4.

47a. Sample answer: $y - 0 = 0.5(x - 0)$
47b. $y = 0.5x$ **47c.** $x - 2y = 0$
49. Sample answer: You need to know the slope of the line and the y-intercept of the line, the slope and the coordinates of another point on the line, or the coordinates of two points on the line.

Selected Answers **SA19**

51. No; the line through (7, −10) and (3, −2) has a slope of −2 and $2x − y = −5$ has a slope of 2.
53. Sample answer: $y − g = \frac{j − g}{h − f}(x − f)$
55. Sample answer: Jocari spent $18 to go to a carnival and play 5 games. The price she paid included admission. The games cost $2 each; $y − 18 = 2(x − 5)$, $y = 2x + 8$.

Lesson 5-3

1. Positive; as time spent exercising increases, the more Calories are burned.   **3.** Negative; as weight increases, the number of repetitions decreases.   **5a.** $y = −328.275x + 3142.15$
5b. about 187.675 million
7a. There is a positive correlation between the child's age and annual cost.

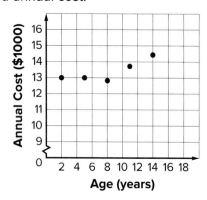

7b. $y = 270x + 10,640$ **7c.** about $15,230
9. no correlation
11a.

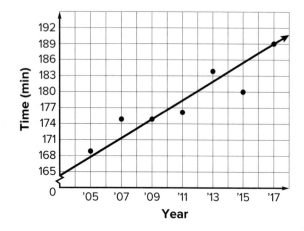

11b. Sample answer: x represents the number of years since 2005, so year 2005 is represented by $x = 0$ and year 2020 is represented by $x = 15$. Two points on the line of fit are (4, 175) and (12, 189). Use these two points to find the slope to be 1.75 and the equation of the line of fit to be $y = 1.75x + 168$.
11c. Sample answer: about 196 minutes
11d. Sample answer: Not all of the data points are close to the line of fit, so there is not a consistent trend regarding the length of games. Therefore, the predicted game length may or may not be accurate.   **13a.** positive correlation; As the number of years since 2007 increases, the price of a ticket increases.
13b. $y = 2.87x + 64.24$ **13c.** about $133.12
15a.

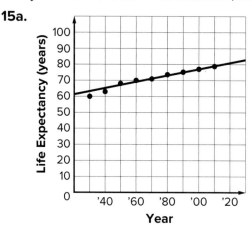

15b. Sample answer: About 81.8; The data show a positive correlation, so as the years increase the life expectancy also increases. Therefore, the life expectancy should be higher than that of a baby born in 2010.   **15c.** Sample answer: I assumed that the trend continues, so as the year increases, the life expectancy also increases.
17. Sample answer: The salary of an individual and the years of experience that he or she has could be modeled using a scatter plot. This would be a positive correlation because the more experience an individual has, the higher the salary would likely be.
19. Neither; line g has the same number of points above the line and below the line. Line f is close to 2 of the points; but for the rest of the data, there are 3 points above and 3 points below the line.
21. Sample answer: You can visualize a line to determine whether the data has a positive or negative correlation. The graph shows

the ages and heights of people. To predict a person's age given his or her height, write a linear equation for the line of fit. Then substitute the person's height and solve for the corresponding age. You can use the pattern in the scatter plot to make decisions.

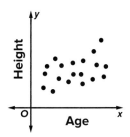

Lesson 5-4

1a.

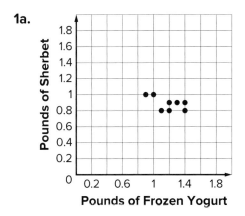

1b. Negative; as the number of pounds of frozen yogurt consumed increases, the number of pounds of sherbet consumed decreases.
1c. The relationship may be a causation. Since both are frozen desserts, eating more frozen yogurt may cause people to decrease the amount of sherbet they eat. Other things that might influence the data are an increase in frozen yogurt stores and a decrease in popularity or availability of sherbet.
3. Correlation, sample answer: Having a wider palm does not cause someone to watch less television. **5.** Causation; sample answer: An increase in the price of cereal likely causes customers to buy less cereal.

7a.

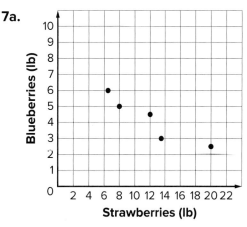

7b. Negative; as the number of pounds of strawberries produced increases, the number of pounds of blueberries produced decreases.
7c. The relationship is a correlation, but not a causation. A better yield of strawberries does not cause the blueberries to grow poorly. Other factors, such as temperature and rain, could be affecting the plants that week.
9. positive correlation and causation; Sample answer: Because pizzas are topped with cheese, an increase in the number of pizzas made cause more cheese to be used.
11. Sample answer: Two elements can have a strong correlation, but it does not mean that one causes the other. There could be an unknown factor affecting the elements.
13. Sample answer: Correlation does not mean causation. Even though there is a strong correlation that does not mean buying swimsuits causes the use of air conditioners. Another factor, like the temperature, could be affecting both swimsuit sales and use of air conditioners.

Lesson 5-5

1a. $y = -1.31x + 50.95$ **1b.** $r \approx -0.714$;
The equation models the data fairly well. Its negative value means that as the years since 2010 increase, the total number of goals the soccer team scores each season decreases.
3a. $y = 8.52x + 3.18$
3b. $r \approx 0.999$; The equation models the data very well. Its positive value means that as the years since 2010 increase, sales, in millions of dollars, increase.

5a. $y = 103.77x + 108.06$ **5b.** about $3221.16
7a. $y = 0.59x + 1.51$ **7b.** The residuals are randomly scattered and are centered about the line $y = 0$. So, the best-fit line models that data well. **9a.** $y = 0.26x + 21.21$
9b. $r \approx 0.359$; The equation does not model the data well. Its value means that as the years since the 2011–2012 school year increase, the percentage of students in public school who met all six of California's physical fitness standards each year varies. **9c.** Because the data on the students who meet all six standards is reported as a percentage, it cannot exceed 100. **11a.** $y = 140.4x + 13.8$
11b. $r \approx 0.999$; The equation models the data very well. Its positive value means that as the number of games increases, the cumulative number of yards increases.
11c. Sample answer: Because the data have a positive correlation, the total number of yards will increase as the number of games increases. So, the running back will have run for 950 yards between games 6 and 9.
11d. during game 7
13a. $y = 9619x + 443{,}918.8$ **13b.** $r \approx 0.999$; The equation models the data very well. Its positive value means that as the number of years since the 2010-2011 school year increases, the number of student athletes participating in college athletics each year increases. **13c.** The residuals are randomly scattered and are centered about the line $y = 0$. So, the best-fit line models that data well.
13d. about 684,394
15. Apply a linear regression model to the data. Use the number of each test as the independent variable. If there is no correlation, the r-value will not be close enough to 1 or -1. If this is the case, the line of fit could not be used to predict the scores of the other students.
17a. $y = 84{,}345.0x + 5{,}003{,}868.3$
17b. about 7,365,528

Lesson 5-6

1. $\{(-1, -9), (-4, -7), (-7, -5), (-10, -3), (-13, -1)\}$

3. $\{(-2, -4), (-1, -2), (1, 0), (0, 2), (2, 4)\}$
5. $\{(-3, 5), (-9, 2), (-15, -1), (-21, -4)\}$
7. $\{(16, -1), (12, -2), (8, -3), (4, -4)\}$
9. $\{(-49, -4), (35, 8), (-28, -1), (7, 4)\}$
11.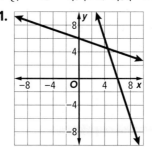

13.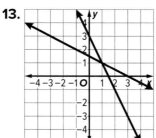

15.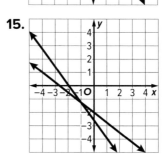

17. $f^{-1}(x) = \frac{x}{6} - 7$
19. $f^{-1}(x) = \frac{5}{2}(x + 16)$ **21.** $f^{-1}(x) = \frac{1 - 5x}{4}$
23a. $P^{-1}(x) = \frac{(x + 36)}{7.6}$ **23b.** x represents Alisha's profit and represents the number of dozens of brownies sold. **23c.** 5
25a. $C^{-1}(x) = \frac{x - 125}{16}$ **25b.** 108 feet
27. $f^{-1}(x) = \frac{1}{4}x + 6$ **29.** $f^{-1}(x) = 6x - 42$
31. $f^{-1}(x) = \frac{7}{2}x - 14$ **33.** $f^{-1}(x) = \frac{1}{7}x - \frac{6}{7}$
35. $f^{-1}(x) = 2x - 22$ **37.** B **39.** A
41. $\{(-k, b), (p, -g), (-m, -w), (q, r)\}$ **43.** The slopes are reciprocals. For example, if the slope of one line is $\frac{2}{3}$, then the slope of the inverse function is $\frac{3}{2}$. **45.** Sample answer: This claim is incorrect. The -1 in the inverse function notation is not an exponent. As an example, the inverse function for $y = x + 1$ is found by switching x and y and solving for y, which gives $y = x - 1$. $y = x - 1$ is not the same as $y = \frac{1}{(x + 1)}$, which is not a line. This method does not work.

47. $a = 2; b = 14$

49. sometimes; Sample answer: $f(x)$ and $g(x)$ do not need to be inverse functions for $f(a) = b$ and $g(b) = a$. For example, if $f(x) = 2x + 10$, then $f(2) = 14$ and if $g(x) = x - 12$, then $g(14) = 2$, but $f(x)$ and $g(x)$ are not inverse functions. However, if $f(x)$ and $g(x)$ are inverse functions, then $f(a) = b$ and $g(b) = a$.

51. Sample answer: A situation may require substituting values for the dependent variable into a function. By finding the inverse of the function, the dependent variable becomes the independent variable. This makes the substitution an easier process.

Module 5 Review

1. A **3.** $y = 1.5x + 11$ **5.** A
7. $y - 4 = 2.5(x - 2)$ **9.** A
11. a positive correlation **13.** C
15. $f^{-1}(x) = -2x + 1$;

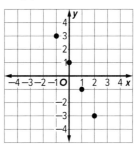

17. A

Module 6

Quick Check

1. 4 **3.** −2 **5.** {−29, 7} **7.** {−1, 15}

Lesson 6-1

1.

3.

5.

7. $t < -1$ **9.** $w < 5$ **11.** $b \geq -5$
13. $\{m \mid m < 7\}$ **15.** $\{r \mid r \leq 15\}$ **17.** $\{b \mid b \geq 2\}$
19. $\{c \mid c \leq -4\}$ **21.** $\{m \mid m \geq 4\}$
23. $\{r \mid r \geq 22\}$ **25.** $\{a \mid a \leq -4\}$
27. $\{w \mid w \geq -5\}$ **29.** $\{x \mid x \leq 5\}$
31. $\frac{3}{10}x \leq 4.50$, $x \leq \$15$ **33.** no more than 2.1 pounds per day **35.** at least 500 pieces
37. $\{m \mid m \leq -68\}$

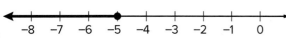

39. $\{c \mid c > 121\}$

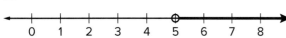

41. $\{x \mid x \leq 20\}$

43. $\{h \mid h > 21\}$

45. $\{n \mid n \geq 108\}$

47. $\{r \mid r < 16\}$

49. $\{t \mid t > -1\}$

51. $\{z \mid z \geq 11\}$

53. $\{d \mid d > -2\frac{1}{2}\}$

55. d **57.** a **59.** b
61. Sample answer: Let n = the number. $n + 7 \leq -18$; $\{n \mid n \leq -25\}$
63. Sample answer: Let n = the number. $n + 2 \leq 1$; $\{n \mid n \leq -1\}$
65. Sample answer: Let n = the number. $-12n \leq 84$; $\{n \mid n \geq -7\}$
67. $\{g \mid g > 4\}$

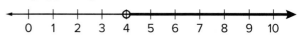

69. $\{x \mid x < 36\}$

71. $\{m \mid m < 5.4\}$

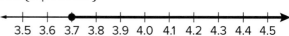

73. $\{c \mid c \geq 3.7\}$

75. $22.23.
77. Sample answer: Let x represent the decibel level of the calls of a blue whale; $x - 83 \leq 105$; $x \leq 188$. The calls of a blue whale are less than or equal to 188 decibels.
79. $-\frac{x}{2} < 1$ **81a.** $x < \frac{7}{a}$ **81b.** $x \geq \frac{12}{a}$
81c. $x > 3$ **81d.** $x \geq \frac{1}{4}$

Lesson 6-2

1a. $15 + 2h \leq 35$ **1b.** $h \leq 10$; 10 hours
3a. $1.50 + 0.25(5x - 1) \leq 3.75$ **3b.** $x \leq 2$; 2 mi
3c. Because the service charges per $\frac{1}{a}$ mile, multiply a by the number of miles x to find the number of $\frac{1}{a}$ miles. Subtract 1 from the total number of $\frac{1}{a}$ miles, ax, to find the number of additional $\frac{1}{a}$ miles. Multiply the difference by the cost per additional $\frac{1}{a}$ mile, $0.25, and add

the cost for the first $\frac{1}{a}$ mile, $1.50. This sum is less than or equal to $3.75, so
$1.50 + 0.25(ax - 1) \leq 3.75$.
5a. $100 + 40x \leq 250$
5b. $x \leq 3.75$; 3 people
7. $21 > 15 + 2x$; $x < 3$

9. $\frac{x}{8} - 13 > -6$; $x > 56$

11. $37 < 7 - 10x$; $x < -3$

13. $-\frac{5}{4}x + 6 < 12$; $x > -\frac{24}{5}$

15. $15x + 30 < 10x - 45$; $x < -15$

17. $\{a \mid a \leq 11\}$

19. $\{b \mid b$ is a real number.$\}$

21. $\{a \mid a \geq -9\}$

23. $\{x \mid x \geq \frac{1}{2}\}$ **25.** $\{m \mid m \geq 18\}$
27. $\{w \mid w > -2\}$ **29.** $\{x \mid x \leq 8\}$
31. $\{x \mid x > -6\}$ **33.** $\{x \mid x \geq 1.5\}$
35. $\{p \mid p \leq 1\frac{1}{9}\}$

37a. $2x + 4 \leq 13$; $x \leq 4.5$
37b. 4.5 ft
37c. 5 ft
39. Eric does not have any pencils. Based on his statement, the inequality is $6p + 15 < 20$, where p is the number of pencils. The solution of the inequality is $p < \frac{5}{6}$. However, the number of pencils must be a whole number, so $p = 0$.

41. $10n - 7(n + 2) > 5n - 12$
(Original inequality)
$10n - 7n - 14 > 5n - 12$
(Distributive Property)
$3n - 14 > 5n - 12$
(Combine like terms.)
$3n - 14 - 5n > 5n - 12 - 5n$
(Subtract $5n$ from each side.)
$-2n - 14 > -12$
(Simplify.)
$-2n - 14 + 14 > -12 + 14$
(Add 14 to each side.)
$-2n > 2$
(Simplify.)
$\frac{-2n}{-2} < \frac{2}{-2}$
Divide each side by -2. Change $>$ to $<$.
$n < -1$
(Simplify.)
The solution set is $\{n \mid n < -1\}$.
43. $\frac{76 + 80 + 78 + x}{4} \geq 82$; $x \geq 94$; Mei needs a score of at least 94 on the next exam.
45. Sample answer: $2(2x - 1) < 10$
47. Let c = the number of baseball cards Ted has; $4c > 5c - 15$; $15 > c$; Ted has fewer than 15 cards. **49.** ∅; If the inequality is always true, the opposite inequality will always be false.
51. Sample answer: The solution set for the inequality that results in a false statement is the empty set, as in $12 \geq 15$. The solution set for an inequality in which any value of x results in a true statement is all real numbers, as in $12 \leq 12$.

Lesson 6-3

1. $\{f \mid 6 < f < 11\}$

3. $\{y \mid y \geq 8$ or $y < -4\}$

5. $\{p \mid -4 < p \leq 5\}$

7. $\{h \mid 2 \leq h < 3\}$

9. $\{y \mid y < -3\}$

11. $\{b \mid 4 < b \leq 5\}$

13. $\{m \mid m < -6 \text{ or } m > -1\}$

15. $\{m \mid 2 \leq m < 4\}$

17a. $x + 8 < 20$ or $x + 8 > 35$ **17b.** $0 < x < 12$ or $x > 27$; Because the combined height of the sign and pole cannot be negative, the value of x must be greater than 0.

17c.
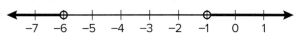

19. $-3 < x \leq 3$ **21.** $x < -2$ or $x \geq 1$
23. $b > 3$ or $b \leq 0$ **25.** $y < -1$ or $y \geq 1$
27. $f \mid -2 < f < -1\}$

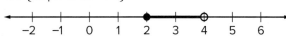

29. $\{b \mid -2 < b < 6\}$

31. $\{a \mid -2 \leq a < 5\}$

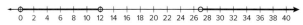

33. Sample answer: Let $n =$ the number. $n - 2 \leq 4$ or $n - 2 \geq 9$; $\{n \mid n \leq 6 \text{ or } n \geq 11\}$
35. $54° \leq x \leq 68°$ **37.** The minimum is 67, since the solution of the inequality $2000 \leq 1000 + 15x$ is $66\frac{2}{3} \leq x$, and the number of students must be a whole number. The maximum is 100, since the solution of the inequality $1000 + 15x \leq 3000$ is $x \leq 133\frac{1}{3}$, but the bus can only hold 100 students. **39a.** The side lengths must be 5, x, and $9 - x$. Using the Triangle Inequality results in the compound inequality $x + 5 > 9 - x$ and $14 - x > x$.

39b. The solution of the compound inequality is $2 < x < 7$, so each of the lengths must be greater than 2 m but less than 7 m. The sum of the two lengths must be 9 m.

41. $\{x \mid -2 < x < 5\}$

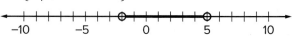

43a. $\$400 \leq x \leq \$800$ **43b.** $\$428 \leq x \leq \$856$
45. B
47. $x > -1$ or $x \geq 4$; This can be written as $x > -1$ because this is the union of two graphs.
49. The union of the two graphs is the graph on the left, so the graph on the left is the graph of the solution set for Exercise 47. The intersection of the two graphs is the graph on the right, so the graph on the right is the graph of the solution set for Exercise 48.
51a. $x > -\frac{4}{a}$ and $x \leq \frac{4}{a}$
51b. $x < -\frac{6}{a}$ or $x > 5a$
53. Sometimes; The graph of $x > 2$ or $x < 5$ includes the entire number line.

Lesson 6-4

1. $\{x \mid -24 < x < 8\}$

3. $\{c \mid -3 \leq c \leq 4\}$

5. $\{\emptyset\}$

7. $\{r \mid r < -8 \text{ or } r > 4\}$

9. $\{h \mid h \leq -3 \text{ or } h \geq 6\}$

11. $\{v \mid v \text{ is a real number.}\}$

13. $\left\{n \mid n \leq -5\frac{1}{4} \text{ or } n \geq 3\frac{3}{4}\right\}$

15. $\{h \mid -5\frac{2}{3} < h < 5\}$

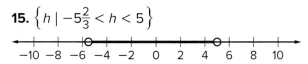

17. $\{\emptyset\}$

19. $\{r \mid -2 < r < \frac{2}{3}\}$

21. $\{x \mid 58.5 \leq x \leq 61.5\}$ **23a.** $|p - 130| \leq 3.05$
23b.

25a. $|x - 515| \leq 114$ **25b.** 287 to 743
27. $|n + 2| \geq 1$ **29.** $|w - 2| < 2$ **31.** $|x| > 1$
33. d   **35.** c
37. $|x - 92| \leq 8$; $\{x \mid 84 \leq x \leq 100\}$

39. $\{x \mid -1 \leq x \leq 3\}$

41. $\{x \mid -7 \leq x \leq 3\}$

43. $\{x \mid x > 18 \text{ or } x < -17\}$

45. By definition, the absolute value is always greater than a negative number. Therefore, no matter what number is chosen, it will always be greater than −1 when evaluated in the absolute value inequality given.   **47a.** Set the absolute value of an unknown variable, x, minus the recommended weight, 516, to be less than or equal to the variance of 4. So, the inequality $|x - 516| \leq 4$ represents the situation.
47b. Write two inequalities, one for each case: $x - 516 \leq 4$ and $-(x - 516) \leq 4$. For the first case, add 516 to both sides: $x \leq 520$. For the second case, distribute the negative on the left side, subtract 516 from both sides and divide by a negative 1 remembering to switch the inequality sign: $x \geq 512$. This means a box of cereal should have a minimum weight of 512 g and a maximum weight of 520 g.

49. The solution set for $|x - 2| > 4$ is $\{x \mid x < -2 \text{ or } x > 6\}$. The solution set for $-2x < 4$ or $x > 6$ is $\{x \mid x > -2\}$. One includes numbers greater than −2, and the other includes numbers less than −2 or greater than 6. These solution sets are not the same.
51. Jordan is correct. Chloe did not distribute the negative to both x and 3.
53. $(-8 \leq n < -3)$ or $(1 < n \leq 6)$. To solve this compound inequality, split it into two inequalities. The first one to solve is $|n + 1| > 2$ and the second one is $|n + 1| \leq 7$. The solution set of the entire problem is the overlap of the individual solutions.

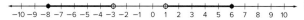

55. No; Sample answer: Lucita forgot to change the direction of the inequality sign for the negative case of the absolute value.
57. Sample answer: If $t = 0$, then the absolute value is equal to 0, not greater than 0.
59. Sample answer: When an absolute value is on the left and the inequality symbol < or ≤, the compound sentence uses *and*, and if the inequality symbol is > or ≥, the compound sentence uses *or*. To solve, if $|x| < n$, then set up and solve the inequalities $x < n$ and $x > -n$, and if $|x| > n$, then set up and solve the inequalities $x > n$ or $x < -n$.

Lesson 6-5

1.

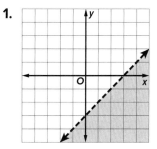

3.

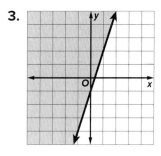

5.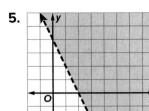

7.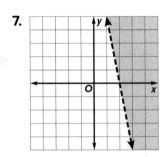

9.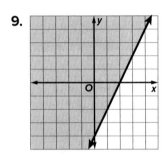

11.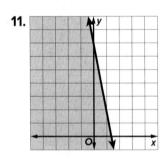

13a. $y < 1240x + 48{,}200$

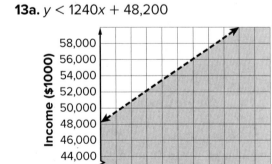

13b. no, no, yes, no

15.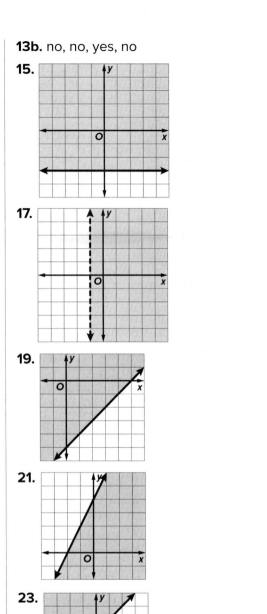

17.

19.

21.

23.

25a. $2.25p + 2b > 90$; $p \geq 0$ and $b \geq 0$

25b. **Profits from Smoothies**

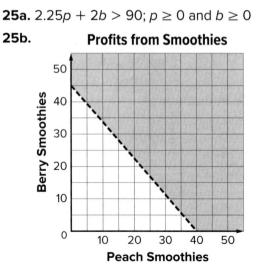

25c. The café sold more than 22 berry smoothies.   **25d.** 41; Sample explanation: 40 peach smoothies results in a profit of exactly $90, so to make a profit of more than $90, the café must have sold 41 smoothies.   **27.** The value of c must be positive. Since $(0, 0)$ is a solution of the inequality, $a(0) + b(0) < c$ must be a true statement, so $0 < c$.   **29.** Sample answer: $y < -x + 1$ **31.** Sample answer: The inequality $y > 10x + 45$ represents the cost of a monthly smartphone data plan with a one-time fee of $45, plus $10 per GB of data used. Both the domain and range are nonnegative real numbers because the GB used, and the total cost cannot be negative.

Module 6 Review

1. A
3. 8 rows
5. C
7. $\{t \mid t < -3\}$
9. A, B
11. $\{g \mid -5 \leq g\}$
13A. $\{h \mid -8 < h < 2\}$
13B.

Number line from −10 to 10 with open circles at −8 and 2 and segment between them shaded.

15. B
17. C
    D
    B
    A

Module 7

Quick Check

1. (4, 0)   **3.** (0, 0)   **5.** $x = 6 - 2y$ **7.** $m = 2n + 6$

Lesson 7-1

1. 1; consistent; independent

3. 0; inconsistent

5. 1; consistent; independent

7. 1; consistent; independent

9. 1; consistent; independent

11. 1 solution; (0, −3)

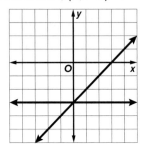

13. no solution

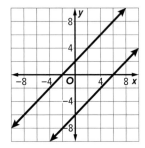

15. infinitely many solutions

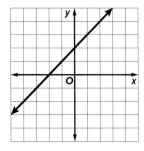

17a. $y = 400x + 1000$; $y = 5900 - 300x$

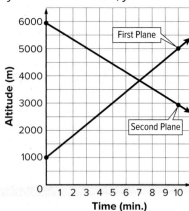

17b. After 7 minutes the planes will be at the same altitude.

19. $y = 3x + 6$ and $y = 6$; (0, 6)

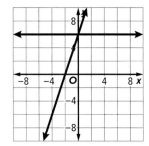

21. $y = -12x + 90$ and $y = 30$; (5, 30)

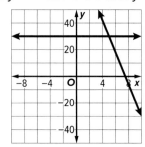

23. $y = 2x + 5$ and $y = 2x + 5$; infinitely many solutions

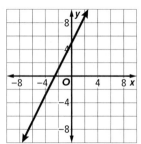

25. approximately (2.68, 1.01)

27. approximately (2.67, −0.88)

29. Sample answer: $x + y = 260$; $2.5x + 0.75y = 450$; approximately (145.71, 114.29); The bookstore will make a weekly profit of $450 with total weekly sales of 260 items when about 146 books and about 114 magazines are sold.

31. no solution; inconsistent

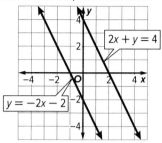

33. 1 solution; (1, −3); consistent; independent

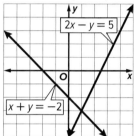

35. infinitely many solutions; consistent; dependent

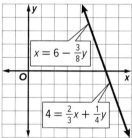

37. infinitely many solutions; consistent; dependent

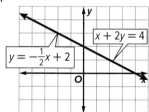

39. 1 solution; (2, 1); consistent; independent

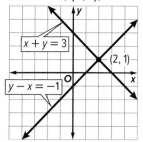

41a. $x =$ time, in seconds, $y =$ distance from where Olivia started to the finish line, in feet; $y = 20x$; $y = 15x + 150$

41b.

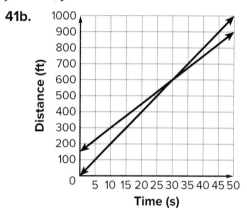

41c. 600 ft

43a. $x =$ time walking in minutes, $y =$ time on bike in minutes; $3x + 2y = 70$, $x = y + 15$

43b.

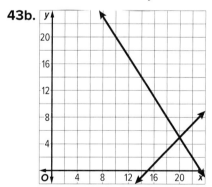

43c. 20 minutes

45. (−2, 3)

47. Sample answer: $4x + 2y = 14$, $12x + 6y = 18$; This system is inconsistent, while the others are consistent and independent.

49. Graphing clearly shows whether a system of equations has one solution, no solution, or infinitely many solutions. However, finding the exact value of x and y from a graph can be difficult.

51. Francisca; If the item is less than $100, then $10 off is better. Of the item is more than $100, then the 10% is better.

Lesson 7-2

1. (1, 6) **3.** (29, 53) **5.** (1, 1) **7.** infinitely many
9. no solution **11.** (0, 1) **13.** (2, 5)
15. infinitely many
17a. Sample answer: $a + b = 5$; $0.7a + 0.2b = 0.65(5)$
17b. 4.5 mL from Beaker A and 0.5 mL from Beaker B
19. $\left(\frac{1}{2}, -\frac{3}{8}\right)$
21. Sample answer: In 2011, the population of Ecuador was about 15,180,000 and the population of Chile was about 17,150,000. The population of Ecuador increased by 1,210,000 and the population of Chile increased by 760,000 from 2011 to 2016. Let x = the number of 5-year periods and y = population. The system is $y = 15{,}180{,}000 + 1{,}210{,}000x$ and $y = 17{,}150{,}000 + 760{,}000x$. Solve by substitution to find that $x \approx 4.4$, or $4.4 \times 5 = 22$ years. So, the population of Ecuador and Chile will be equal in about $2011 + 22 = 2033$. (Source: World Bank)
23. Let x = tens digit and y = units digit of the original number; $10y + x = 10x + y - 45$; $x = 3y + 1$; (7, 2); The original number is 72.
25. Neither; Guillermo substituted incorrectly for b. Cara solved correctly for b, but misinterpreted the pounds of apples bought.
27. Sample answer: The solutions found by each of these methods should be the same. However, it may be necessary to estimate when using a graph. So, when a precise solution is needed, you should use substitution.
29. An equation containing a variable with a coefficient of 1 can easily be solved for the variable. That expression can then be substituted into the second equation for the variable.

Lesson 7-3

1. (−3, 4) **3.** (−3, 1) **5.** (4, −2) **7.** (8, −7)
9. (4, 7) **11.** (4, 1.5) **13.** (2, 1) **15.** (11, 0)
17. (−3, 7) **19.** (2, −1) **21.** (−3, −5)
23. (10, 4) **25.** (7, 5) **27.** (2, −3)
29. −2 and −4
31a. $r + s = 181$ and $r - s = 119$
31b. 31 state senators and 150 state representatives
33. (4, −1) **35.** $\left(-1, 3\frac{1}{3}\right)$
37. (−36, −4) **39.** 34 games
41a. Sample answer: $4p + 2n = 18.50$, $7p + 2n = 26.75$, where p is the price of a bag of popcorn and n is the price of a plate of nachos
41b. (2.75, 3.75); A bag of popcorn costs $2.75 and a plate of nachos costs $3.75.
43a. Add the equations because this will eliminate the variable y, and then you can solve for x.
43b. (5, −2)
43c.

Sample answer: The point of intersection on the graph will match the solution (5, −2).
43d. The equations would be equivalent. There would be infinitely many solutions, all real numbers x and y satisfying the equation $x - 5y = 15$.
43e. There would be no solution because the lines would be parallel and would never intersect.
45. Sample answer: $x + y = 1$ and $-x - y = 1$; This system of equations has no solutions.

47. Sample answer: $-x + y = 5$, I used the solution to create another equation with the coefficient of the x-term being opposite of its corresponding coefficient.

49. Sample answer: It would be most beneficial to use elimination to solve a system of equations when one variable has either the same coefficient in both equations or one variable has coefficients that are additive inverses in the equations.

Lesson 7-4

1. $(-1, 3)$ **3.** $(-3, 4)$ **5.** $(-2, 3)$ **7.** $(3, 5)$
9. $(1, -5)$ **11.** $(0, 1)$

13a. $2x + y = 592.30$ and $x + 2y = 691.31$, where x is the number of MLB games and y is the number of NBA games

13b. MLB: $164.43, NBA: $263.44

15. 8 and -1

17. wash: $6, vacuum: $2

19a.

	Tropical Breeze	Kona Cooler	Total
Amount of Juice (qt)	t	k	10
Amount of Pineapple Juice (qt)	$0.2t$	$0.5k$	4

19b. $\left(3\frac{1}{3}, 6\frac{2}{3}\right)$; The owner should mix $3\frac{1}{3}$ qt of Tropical Breeze and $6\frac{2}{3}$ qt of Kona Cooler.

19c. $3\frac{1}{3}$ qt $+ 6\frac{2}{3}$ qt $= 10$ qt, so the total amount is correct, and $0.2\left(3\frac{1}{3} \text{ qt}\right) + 0.5\left(6\frac{2}{3} \text{ qt}\right) = 4$ qt, so the amount of pineapple juice in the new drink is correct.

21. Jason; In order to eliminate the t-terms, you can multiply the second equation by 2 and then subtract, or multiply the equation by -2 and then add. Daniela did not subtract the equations correctly.

23. Sample answer: $2x + 3y = 6$ and $4x + 9y = 5$

25. Sample answer: It is more helpful to use substitution when one of the variables has a coefficient of 1 or if a coefficient can be reduced to 1 without turning other coefficients into fractions. Otherwise, elimination is more helpful because it will avoid the use of fractions when solving the system.

Lesson 7-5

1.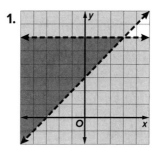

3.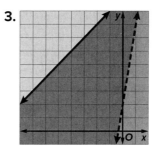

5. no solutions
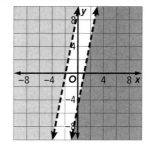

7.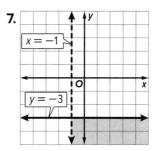

9.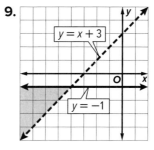

11. no solution

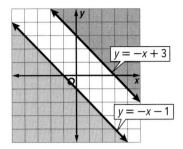

13a. Sample answer: Let $x =$ miles of walking and $y =$ hours at gym; $x \geq 9$, $x \leq 12$, $y \geq 4.5$, $y \leq 6$

13b.

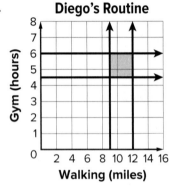

13c. Sample answers: gym 5 h, walk 9 mi; gym 6 h, walk 10 mi, gym 5.5 h, walk 11 mi

15. $y \leq x + 2, y \geq x - 3$

17. $y \geq x + 1, y < 1$

19. The solution set is the region where the graphs of the inequalities overlap. The point (2.5, 1) is not in the overlapping region, so it is not a solution. A solution must make all of the inequalities in the system true statements: $4x - 5y \geq 2 \rightarrow 4(2.5) - 5(1) \geq 2 \rightarrow 10 - 5 \geq 2 \rightarrow 5 \geq 2$; $2x + 3y > 8 \rightarrow 2(2.5) + 3(1) > 8 \rightarrow 5 + 3 > 8 \rightarrow 8 > 8$; The first inequality is true, but the second inequality is false. So, (2.5, 1) is not a solution.

21. Let $x =$ tins of popcorn and $y =$ tins of peanuts; $x + y \leq 200$; $x \geq y$; $3x + 4y \leq 900$; $x \geq 0$ and $y \geq 0$.

23. Sample answer: (3, 3)

25. Sometimes; sample answer: $y > 3, y < -3$ will have no solution, but $y < -3, y < 3$ will have solutions.

27. 9 units^2

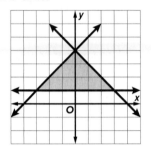

Module 7 Review

1. B, C, D   **3.** B

5. 4 wooden frames and 3 plastic frames

7. one solution; $(-2, 7)$

9. (10, 4)

11. A, B   **13.** $r = 6, t = 5$ **15.** D   **17.** C

Module 8

Quick Check

1. 4^5 **3.** m^3p^5 **5.** 32 **7.** $\frac{1}{16}$

Lesson 8-1

1. $2q^6$ **3.** $9w^8x^{12}$ **5.** $7b^{14}c^8d^6$ **7.** $j^{20}k^{28}$
9. 2^8 or 256 **11.** $4096r^{12}t^6$ **13.** y^3z^3
15. $-15m^{11}$ **17.** $9p^2r^4$ **19.** 1.8×10^5 flowers
21. 242 pounds **23.** 2.9465×10^{12} ml
25. $25x^6y^{10}$ **27.** $64m^{24}n^{12}$ **29.** $9c^{16}$
31. $32{,}000k^{10}m^{19}$ **33.** $800x^8y^{12}z^4$
35. $30x^5y^{11}z^6$ **37.** $0.064h^{15}$ **39.** $\frac{16}{25}a^4$
41. $8m^3p^6$ **43.** $288a^{31}b^{26}c^{30}$ **45.** no **47.** no
49. yes **51a.** $20x^7$ **51b.** The power is one-fourth the previous power: $\frac{1}{4}(20x^7)$, or $5x^7$.
53. $12x$ ft^2 **55a.** If the process of raising a number to an exponent were commutative, then $a^b = b^a$ for all numbers a and b. This is not true, because $2^3 = 8$, whereas $3^2 = 9$.
55b. If the process of raising a number to an exponent were associative, then $(a^b)^c = a^{(b^c)}$ for all numbers a, b, and c. This is not true, because $(4^3)^2 = 64^2 = 4096$, whereas $4^{(3^2)} = 4^9 = 262{,}144$. **55c.** If the process of raising a number to an exponent were to distribute over addition, then $(a + b)^c = a^c + b^c$ for all numbers a, b, and c. This is not true, because $(1 + 2)^3 = 3^3 = 27$, but $1^3 + 2^3 = 1 + 8 = 9$. **55d.** If the process of raising a number to an exponent were to distribute over multiplication, then $(ab)^c = a^cb^c$ for all numbers a, b, and c. This is true. It is a Power of a Product Property.
57. Sample answer: Use the Power of a Power Property to simplify the expression to $\frac{a^{2tm}}{b^{2t^2}}$.
59. Sample answer: $x^4 \cdot x^2$; $x^5 \cdot x$; $(x^3)^2$
61. Jade is correct; Sample answer: Only the h is raised to the power of 6. The exponent of the second g is 1.

Lesson 8-2

1. m^2p **3.** c^2 **5.** $\frac{p^6t^{21}}{1000}$ **7.** $\frac{9n^2p^6}{49q^4}$ **9.** $\frac{81m^{20}r^{12}}{256p^{32}}$
11. k^2mp^2 **13.** $-4x^2yz^3$ **15.** $3d$ **17.** x^2
19. $25^3 = 15{,}625$ **21.** $26^2 = 676$ **23.** $-\frac{w^9}{3}$
25. $\frac{4a^8c^9}{25b^6d^6}$
27. $\frac{64a^{15}b^{30}}{27}$ **29.** $\frac{8x^6}{3y^3}$ **31.** $\frac{8x^6y^3z^{12}}{27}$
33. $\frac{c^{24}}{64d^{12}}$ **35.** 10^7 **37.** $3x^2y$
39a. $n(1.04^t)$
39b. 26.5%; $\frac{n(1.04^8)}{n(1.04^2)} = 1.04^{(8-2)} = 1.04^6 \approx 1.2653$; $126.5\% - 100\% = 26.5\%$
41. 8
43. No; sample answer: The denominator of both expressions is y^6, but the numerators are different. The first numerator is x^3 and the second is x^2.
45. C
47. a^x
49a. Sample answer: $h = 3$ and $k = 1$
49b. Yes; sample answer: As long as $h - k = 2$, then any numbers would work.
51. Sometimes; sample answer: The equation is true when $x = 0$ or 1, or when $y = 2$ and $z = 2$, but it is false in all other cases.
53. Sample answer: The Quotient of Powers Property is used when dividing two powers with the same base. The exponents are subtracted. The Power of a Quotient Property is used to find the power of a quotient. You find the power of the numerator and the power of the denominator.
55. $\frac{(-3)^8}{(-2)^4}$; The numerator and denominator do not have the same base.

Lesson 8-3

1. $\frac{r^2}{n^9}$ **3.** $\frac{1}{f^{11}}$ **5.** $\frac{g^4h^2}{f^5}$ **7.** $-\frac{3}{u^4}$ **9.** $\frac{5m^6y^3r^5}{7}$
11. $\frac{r^4p^2}{4m^3t^4}$ **13.** $\frac{y^9z^2}{x^4}$ **15.** $\frac{p^4r^2}{t^3}$ **17.** $\frac{-f}{4}$
19. 5 **21.** 6 **23.** 1
25. $1600k^{13}$ **27.** $\frac{5q}{r^6t^3}$ **29.** $\frac{4g^{12}}{h^4}$ **31.** $\frac{4x^8y^4}{z^6}$
33. $\frac{16z^2}{y^8}$ **35.** 3 **37a.** 12,500
37b. 6.25 petabytes
37c. $\frac{1}{10^9}$
39. No; sample answer: The exponent does not affect the coefficient in $4n^{-5}$, so the two expressions simplify to $4n^5$ and $\frac{4}{n^5}$.

41. $\frac{1}{p^{-3}} = \frac{1}{(p^3)^{-1}} = \frac{1}{\left(\frac{1}{p^3}\right)} = 1 \cdot \frac{p^3}{1} = p^3$

43. 750

45a. $\left(\frac{4a^3}{2a^{-2}}\right)^4 = (2a^5)^4 = 16a^{20}$

45b. $\left(\frac{4a^3}{2a^{-2}}\right)^4 = \frac{(4a^3)^4}{(2a^{-2})^4} = \frac{256a^{12}}{16a^{-8}} = 16a^{20}$

45c. Sample answer: When simplifying monomials, the order of applying the Quotient of Powers Property and Power of a Quotient Property does not matter. **47a.** Both; sample answer: Colleen applied the Power of a Power Property first, while Tyler applied the Quotient of a Power Property first. Their answers are equivalent. **47b.** Sample answer: Colleen's answer is in simplest form. Tyler's answer uses negative exponents, so it is not considered simplest form. **49.** Sample answer: Lola is correct. If we are asked to simplify $\frac{a^7}{a^2}$, instead of using the Division Property of Exponents ($a^{7-2} = a^5$), we could rewrite the fraction as $a^7 a^{-2}$, and then use the Multiplication Property of Exponents to get $a^{7+(-2)} = a^5$. **51.** Sample answer: Students measure the width of their classroom and the width of a pencil, then write and solve a question such as: How many times wider is the classroom than the pencil?
53. Each power is 1 less than the previous power. Each value is half the previous value. So, each time 1 is subtracted from the power, divide the value by 2. Since $2^1 = 2$, then 2^{1-1}, or $2^0 = 2 \div 2$, or 1.

Lesson 8-4

1. $\sqrt{15}$ **3.** $4\sqrt{k}$ **5.** $26^{\frac{1}{2}}$ **7.** $2(ab)^{\frac{1}{2}}$ **9.** $\frac{1}{2}$ **11.** 9
13. 4 **15.** $\frac{2}{5}$ **17.** 7 **19.** $\frac{1}{3}$ **21.** 243 **23.** 625
25. $\frac{27}{1000}$ **27.** 96 ft/s
29. 9 astronomical units **31.** $51,000 **33.** $\sqrt[3]{17}$
35. $7\sqrt[3]{b}$ **37.** $29^{\frac{1}{3}}$ **39.** $2a^{\frac{1}{3}}$ **41.** 0.3 **43.** a
45. 16 **47.** $\frac{1}{3}$ **49.** $\frac{1}{27}$ **51.** $\frac{1}{\sqrt{k}}$ **53a.** 107
53b. 236 **55.** 4.1 m **57.** Sample answer: The expressions in a, c, and d are equivalent; The expression in b is not equivalent to the others. Let $p = 4$, $f = 3$, and $k = 2$, then $p^{\frac{f}{k}} = 8$, $p^{\frac{f}{5}} \approx 2.30$, $\sqrt[k]{p^f} = 8$, and $(\sqrt[k]{p})^f = 8$.
59. 112π cm² **61.** Sample answer: $2^{\frac{1}{2}}$ and $4^{\frac{1}{4}}$ **63.** $-1, 0, 1$ **65.** Sample answer: Both Sachi and Makayla are correct. Sachi addressed the numerator of the rational exponent first and then the denominator. Makayla addressed the denominator of the rational exponent first and then the numerator. Both methods will lead to the correct answer, 9.

Lesson 8-5

1. $2\sqrt{13}$ **3.** $6\sqrt{2}$ **5.** $9\sqrt{3}$ **7.** $5\sqrt{2}$ **9.** $2\sqrt{5}$
11. $4\sqrt{6}$ **13.** $\frac{5\sqrt{3}}{7}$ **15.** $\frac{2\sqrt{2}}{3}$ **17.** $\frac{2\sqrt{21}}{11}$ **19.** $4b^2$
21. $9a^6d^2$ **23.** $2a^2b\sqrt{7b}$ **25.** $40\sqrt{3}$ ft/s
27. 0.9 second **29.** 2 **31.** 6 **33.** 3 **35.** $3\sqrt[3]{9}$
37. $3\sqrt[3]{6}$ **39.** $2\sqrt[3]{10}$ **41.** $2\sqrt[3]{12}$ **43.** $2\sqrt[3]{9}$ **45.** 3
47. $\frac{3\sqrt[3]{2}}{5}$ **49.** $\frac{\sqrt[3]{9}}{4}$ **51.** 4 **53.** $3b\sqrt[3]{2b^2}$ **55.** $\frac{4}{7x}$
57. $3x^2y\sqrt[4]{10x}$ **59.** $7m^4n^2\sqrt{5mn}$ **61.** $\frac{4df^4\sqrt{2d}}{5e^2}$
63. $4x^4y^5z^6\sqrt[3]{5x^2y^2z^2}$ **65.** $\frac{2y^3\sqrt{11y}}{x}$ **67.** $18\sqrt{3}$ m²
69. $8\sqrt{7}$ **71.** 27.6 in.
73a. The Product Property of Square Roots states that the square root of a product is equal to the product of the square roots of the factors. For this property, $a \geq 0$ and $b \geq 0$.
73b. The Quotient Property of Square Roots states that the square root of a quotient is equal to the quotient of the square roots of the numerator and denominator. For this property, $a \geq 0$ and $b > 0$. **73c.** Both properties state that finding a square root can be done before or after certain other operations.
75. Sample answer: $\sqrt{5b} \cdot \sqrt{15b^4} = 5b^2\sqrt{3b}$
77. Sample answer: 0.25. Any number between 0 and 1 will have a square root larger than itself.
79. Sample answer: Because $4\sqrt{12} \times 6\sqrt{15} = 144\sqrt{5}$, $4\sqrt{12}$ in. is a possible length and $6\sqrt{15}$ in. is a possible width.

Lesson 8-6

1. $5\sqrt{7}$ **3.** $11\sqrt{15}$ **5.** $3\sqrt{r}$ **7.** $-\sqrt{5}$
9. $-3\sqrt{13} + 5\sqrt{2}$ **11.** $12\sqrt{3} + \sqrt{2}$
13. $115\sqrt{149}$ m **15.** $26\sqrt{37} + 10\sqrt{41}$ in.
17. $\frac{20 - 10\sqrt{2}}{\sqrt{\pi}}$ cm
19. $18\sqrt{5}$ **21.** $24\sqrt{14}$ **23.** $198\sqrt{2}$
25. $4 + 2\sqrt{3}$ **27.** $5\sqrt{10}$ **29.** $60 + 32\sqrt{10}$
31. $\frac{1 - 5\sqrt{5}}{5}$ **33.** 2 **35.** $2\sqrt{6} - 13\sqrt{3} + 4\sqrt{2}$

37. $15\sqrt{15}$ **39.** $\frac{3\sqrt{2}}{2} + 2\sqrt{15}$
41. $2ac\sqrt{6} + ad\sqrt{15} + 2bc\sqrt{10} + 5bd$
43. $63\sqrt{15}$ ft² **45a.** $2\sqrt{2}$ s **45b.** about 2.8 s
47. $11.25\sqrt{2}$ ft **49.** $\frac{5}{6}\sqrt{5}$ in.
51. True; $(x + y)^2 > \left(\sqrt{x^2 + y^2}\right)^2 \to x^2 + 2xy + y^2 > x^2 + y^2 \to 2xy > 0$; Because $x > 0$ and $y > 0$, $2xy > 0$ is always true. So, $(x + y)^2 > \left(\sqrt{x^2 + y^2}\right)^2$ is always true for all $x > 0$ and $y > 0$.
53. Sample answer: $\sqrt{12} + \sqrt{27} = 5\sqrt{3}$; Simplify $\sqrt{12}$ to get $2\sqrt{3}$. Simplify $\sqrt{27}$ to get $3\sqrt{3}$.
Because $2\sqrt{3}$ and $3\sqrt{3}$ have the same radicand, they can be combined.
55. $6\sqrt{12}$; Sample answer: $\sqrt{12}$ is the only radical that does not simplify to $k\sqrt{5}$.

Lesson 8-7

1. 9 **3.** 6 **5.** 8 **7.** 8 **9.** 5 **11.** 2 **13.** 16 ohms
15. 20 months **17.** 3 **19.** $-\frac{1}{4}$ **21.** 11 **23.** 2
25. 2 **27.** 37.52 ft **29.** Use 4 as the common base. Replace 8 with $4^{\frac{3}{2}}$ and 16 with 4^2 because $4^{\frac{3}{2}} = (\sqrt{4})^3$, or 8 and $4^2 = 16$. So, $4^{\frac{3}{2}(2 + x)} = 4^{2(2.5 - 0.5x)}$. Distribute and get $4^{(3 + 1.5x)} = 4^{(5 - x)}$. Set the exponents equal to each other: $3 + 1.5x = 5 - x$. Solve for x: $x = 0.8$, or $\frac{4}{5}$.
31. Sample answer: There is no way to make a common base between 2 and 5. **33.** $x = 0$ and $x = -4$ **35.** $3^{2x + 1} = 243$; This equation has a solution of $x = 2$. The other three equations have a solution of $x = 3$.

Module 8 Review

1. A, B **3.** B **5.** 9 **7.** C **9.** 0.000001
11. Sample answer: Because $b^{\frac{1}{n}} = \sqrt[n]{b}$, rewrite $729^{\frac{1}{6}}$ as $\sqrt[6]{729}$. Because $729 = 3 \cdot 3 \cdot 3 \cdot 3 \cdot 3 \cdot 3$, you can write $\sqrt[6]{729}$ as $\sqrt[6]{3 \cdot 3 \cdot 3 \cdot 3 \cdot 3 \cdot 3}$, which equals 3. **13.** B **15.** 27
17. A **19.** $2\sqrt{5} + 6\sqrt{7}$ **21.** B **23.** 6

Module 9

Quick Check

1. −196
3. 0.25
5. 32
7. −3

Lesson 9-1

1. No; the domain values are at regular intervals and the range values have a common difference 3.

3. Yes; the domain values are at regular intervals and the range values have a common factor 2.

5. No; there is no common factor between the picture areas.

7a. y-intercept = 50; D = {all real numbers}, R = $\{y \mid y > 0\}$

7b.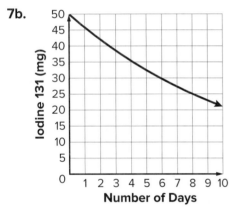

7c. Because time cannot be negative, the relevant domain is $\{x \mid x \geq 0\}$. Because the amount of Iodine 131 cannot be negative, and the amount when $x = 0$ is 50 mg, the relevant range is $\{y \mid 0 < y \leq 50\}$.

9.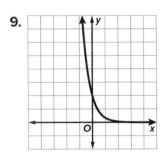

2; D = {all real numbers}, R = $\{y \mid y > 0\}$; $y = 0$

11.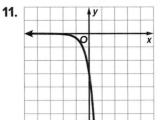

−3; D = {all real numbers}, R = $\{y \mid y < 0\}$; $y = 0$

13.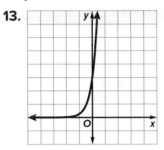

3; D = {all real numbers}, R = $\{y \mid y > 0\}$; $y = 0$

15a. 1038 millibars

15b. about 794 millibars

15c. It decreases.

17. $f(x) = 3(2^x)$

19. Sample answer: The number of teams competing in a basketball tournament can be represented by $y = 2^x$, where the number of teams competing is y and the number of rounds is x. The y-intercept of the graph is 1. The graph increases rapidly for $x > 0$. With an exponential model, each team that joins the tournament will play all of the other teams. If the scenario were modeled with a linear function, each team that joined would play a fixed number of teams.

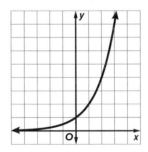

Lesson 9-2

1. translated up 8 units **3.** compressed horizontally **5.** reflected across the *x*-axis; translated 1 unit right **7.** reflected across the *y*-axis; translated 4 units up **9.** stretched vertically; shifted 2 units up **11.** translated right 3 units **13.** $y = -2^x$ **15.** $y = 2^{-x} + 5$
17. stretched vertically by a factor of 2000
19. stretched vertically by a factor of 20
21. translated up 6 units **23.** reflected across the *x*-axis; compressed vertically
25. reflected across the *y*-axis
27. $g(x) = 2^x + 3$
29. $g(x) = 5^{x-2}$ **31.** $g(x) = 6^x + 5$
33. $g(x) = \frac{1}{2}(4^x)$ **35.** $g(x) = 2^{3x}$
37. $g(x) = 5^x - 2$ **39.** $g(x) = 5^{x-4}$
41a. translated up 500 units **41b.** $500

43. The graph has been reflected over the *x*-axis and reflected over the *y*-axis. It has been stretched vertically by a factor of 3 and shifted up 1 unit.

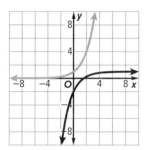

45. The graphs of these two exponential functions are the same. $f(x) = 4^{x+2} = 4^x \cdot 4^2 = 16 \cdot 4^x = g(x)$.

47. Jennifer is correct. Sample answer: As it is written, the function is multiplied by 2, which causes the graph to rise more rapidly than the parent graph, so Jennifer is correct. However, $g(x) = 2(2^x)$ is equivalent to $g(x) = 2^{x+1}$. This graph is the parent graph of $f(x) = 2^x$ shifted to the left one unit, but it still rises at the same rate.

49. The first pair; $g(x)$ is shifted right 3 units instead of left 3 units.

Lesson 9-3

1. $y = 4 \cdot 2^x$ **3.** $y = 10 \cdot 3^x$ **5.** $y = 3 \cdot 4^x$
7. $y = 3 \cdot 2^x$ **9.** $y = \left(\frac{1}{4}\right)^x$ **11.** $f(x) = 50 \cdot 2^x$, where *x* is the number of 30-minute time periods **13.** $f(x) = 43 \cdot (1.23)^x$, where *x* is the number of years since 2010.
15a. $P = 8{,}192{,}426(1.009)^t$ **15b.** about 9,370,872
17a. $Z = 60{,}000(0.90)^t$ **17b.** about $31,886
19. $2200 **21.** 360 million
23. $y = 2.6 \cdot 4^x$ **25.** about 77,529

27. Sample answer: The equation can be rewritten in the form $y = a(1 + r)^x$ to find the amount of original investment, *a*, and the rate of increase or decrease. Because $a = 2400$, he invested $2400. Because $1 + r = 0.95$ and is less than 1, his investment is decreasing in value. A graphing calculator can be used to find that the investment will be worth $1200 in about 13.5 years.

29a. $P(t) = 128(1.25)^t$

29b. an increase of approximately 41 deer per year **29c.** No; the amount of increase is exponential, not linear. **29d.** There is no common difference over equal intervals (differences are 32, 40 and 50). There is a common factor (factor is 1.25 in each case.)

31. about 9.2 years **33.** Sample answer: Exponential models can grow without bound, which is usually not the case for the situation that is being modeled. For instance, a population cannot grow without bound due to space and food constraints. Therefore, the situation that is being modeled should be carefully considered when used to make decisions.

35a. Sample answer: 5%; about 14.2 years

35b. Sample answer: 10%; about 6.6 years

35c. Sample answer: about 10.4 years; about $8320

Lesson 9-4

1a. $A(t) = (1.021)^t$; $A(t) = (1.0052)^{4t}$

1b. Bank B has the better plan because the effective quarterly interest rate is 0.8%, which is greater than the quarterly interest rate of about 0.52% for Bank A.

1c. About 3.2%; sample answer: This confirms the result of part **b** because 3.2% is greater than the annual interest rate at Bank A, so Bank B has the better plan.

3. Bank A; Bank A has a quarterly interest rate of 0.95%. Bank B has a quarterly interest rate of about 0.92%. Bank A's quarterly interest rate is higher.

5. Species B; the population of Species A is decreasing at a rate of about 0.25% per quarter. The population of Species B is decreasing at a rate of about 0.34% per quarter. The population of Species B is decreasing at a faster rate.

7. Plan A

9. Account A; Account A has a semi-annual interest rate of 2.3%. Account B has a semi-annual interest rate of about 2.1%. Account A's semi-annual interest rate is greater.

11. Account A; Account A has a monthly interest rate of 0.5%. Account B has a monthly interest rate of about 0.21%. Account A's monthly interest rate is greater.

13. $T(t) = 72 + 140(0.67)^t$

15. Sample answer: Bank A offers a savings account with a 0.6% interest rate compounded quarterly. Bank B offers a savings account with a 2% interest rate compounded annually. Bank A offers the better interest rate because it has a higher effective annual interest rate of about 2.4%.

Lesson 9-5

1. The ratios are not the same, so the sequence is not geometric.

3. Since the ratio is the same for all of the terms, 5, the sequence is geometric.

5. The ratios are not the same, so the sequence is not geometric.

7. Because the ratio is the same for all of the terms, $\frac{1}{2}$, the sequence is geometric.

9. The ratios are not the same, so the sequence is not geometric.

11. The ratios are not the same, so the sequence is not geometric.

13. −250, 1250, −6250

15. 108, 324, 972

17. −2058; −14,406; −100,842

19. 54, 162, 486

21. $\frac{1}{10}, \frac{1}{20}, \frac{1}{40}$

23. $\frac{1}{3}, \frac{1}{18}, \frac{1}{108}$

25. 387,420,489

27. 177,147

29. $a_n = 4 \cdot \left(\frac{3}{2}\right)^{n-1}$

31. $1310.72

33a. $a_n = P \cdot 1.005^n$

33b. $538.84

35. $a_n = \frac{9}{16}\left(\frac{2}{3}\right)^{n-1}; \frac{4}{81}$

37. $a_n = -8\left(\frac{1}{4}\right)^{n-1}; -\frac{1}{2048}$

39. Sample answer: The average annual salary is about $39,416, and the average annual rate of increase is about 3%. $a_n = 39{,}416(1.03)^{n-1}$; $\approx 69{,}116.19$; This means that after 20 years of employment the average annual salary will be about $69,116.19.

41. −3, −12, −48

43a. The first method provides a starting salary of $100 and an $8 per month raise. The second method provides a starting salary of $0.01 and doubles it each month.

43b. The first situation is linear because there is a common difference of $8. The equation is $y = 8x + 92$. The second situation is exponential because it is a geometric sequence with a common ratio of 2. The equation is $y = 0.01(2)^{x-1}$.

43c. Sample answer: As long as I do not need money immediately, I would use the second method. In the last month, I would make $y = 0.01(2)^{23} = \$83{,}886.08$ due to the fact that the payment is growing exponentially. In the last month, in the first method I would make $y = 8(24) + 92 = \$284$.

45. If the values fit a geometric sequence, then $r = \sqrt{\frac{540}{180}} = \sqrt{3}$. This would mean that the interior angles of a square would have a sum of $180\sqrt{3} \approx 312°$. Since the sum of the angles in a square is 360°, this is not a geometric sequence.

47. Neither; Haro calculated the exponent incorrectly. Matthew did not calculate $(-2)^8$ correctly.

49. Sample answer: When graphed, the terms of a geometric sequence lie on a curve that can be represented by an exponential function. They are different in that the domain of a geometric sequence is the set of natural numbers, while the domain of an exponential function is all real numbers. Thus, geometric sequences are discrete, while exponential functions are continuous.

51. Sample answer: In the geometric sequence 6, 3, 1.5, …, the value of r is 0.5 and the absolute value of a_{n+1} will be closer to zero than the value of a_n.

Lesson 9-6

1. 23, 30, 37, 44, 51

3. 8, 20, 50, 125, 312.5

5. 13, −29, 55, −113, 223

7. $a_1 = 12, a_n = a_{n-1} - 13, n \geq 2$

9. $a_1 = 2, a_n = a_{n-1} + 9, n \geq 2$

11. $a_1 = 40, a_n = -1.5a_{n-1}, n \geq 2$

13. $a_1 = 3, a_n = a_{n-1} - 1, n \geq 2$

15. $a_1 = 2, a_n = a_{n-1} + 1, n \geq 2$

17. $a_1 = \frac{5}{2}, a_n = a_{n-1} - 1, n \geq 2$

19a. 875, 1050, 1225, 1400, 1575

19b. $a_1 = 175, a_n = a_{n-1} + 175, n \geq 2$

19c. $a_n = 175n$

21a. $a_1 = 6, a_n = 0.9a_{n-1}, n \geq 2$

21b. $a_n = 6(0.9)^{n-1}$

23. $a_n = -12n + 10$

25. $a_1 = 45, a_n = a_{n-1} - 7, n \geq 2$

27. $a_1 = -11, a_n = a_{n-1} + 5, n \geq 2$

29. $a_n = 16(4)^{n-1}$

31. $a_n = 500(1.05)^{n-1}$

33. Ramon has 2 parents, 4 grandparents, 8 great-grandparents, and so on. We can write a geometric sequence to count the number of ancestors in a given generation. The recursive formula is $a_1 = 2, a_n = 2a_{n-1}, n >= 2$. The explicit formula is $a_n = 2^n$. Ramon's claim is about the 8th generation back: $a_8 = 2^8 = 256$. Ramon is correct.

35a. Sample answer: B3 = B2 + B1 and C2 = B2 ÷ B1

35b. The ratio approaches a constant value of 1.618034… . For larger values of n, the Fibonacci numbers behave like a geometric sequence with a common ratio of 1.618034… .

37. Both; Sample answer: The sequence can be written as the recursive formula $a_1 = 2$, $a_n = (-1)a_{n-1}, n \geq 2$. The sequence can also be written as the explicit formula $a_n = 2(-1)^{n-1}$.

39. False; sample answer: A recursive formula for the sequence 1, 2, 3, … can be written as $a_1 = 1, a_n = a_{n-1} + 1, n \geq 2$ or as $a_1 = 1, a_2 = 2$, $a_n = a_{n-2} + 2, n \geq 3$.

41. Sample answer: In an explicit formula, the nth term a_n is given as a function of n. In a recursive formula, the nth term a_n is found by performing operations to one or more of the terms that precede it.

Module 9 Review

1.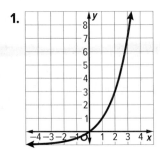

3. As x increases, y increases; and, as x decreases, y approaches 0.

5. A

7. A

9. D

11. Local Credit Union; sample answer: The monthly interest rate is 0.12% higher than at First & Loan, and the annual interest rate is 1.6% higher than at First & Loan.

13. 146 people

15. $a_1 = 20, a_n = a_{n-1} + 15, n \geq 2$

17. row 2: 12; row 3: 12, 48; row 4: 48, 192

Module 10

Quick Check

1. $a^2 + 5a$ **3.** $n^2 - 3n^3 + 2n$ **5.** $13x$
7. simplified

Lesson 10-1

1. no **3.** yes; 4; trinomial **5.** yes; 2; binomial
7. $5x^2 + 3x - 2$; 5 **9.** $-5c^2 - 3c + 4$; -5
11. $t^5 + 2t^2 + 11t - 3$; 1 **13.** $-3x^4 + \frac{1}{2}x + 7$; -3
15. $6x + 12y$ **17.** $3a + 5b$ **19.** $2m^2 + m$
21. $d^2 - 3d$ **23.** $3f + g + 1$ **25.** $7c^2 + 6c - 3$
27. $3c^3 - c^2 - 3c + 3$ **29.** $-2x - 5y + 1$
31. $-x^2y - 3x^2 + 4y$ **33.** $-6p^2 + 2np + n$
35a. $D = 0.5x + 2$ **35b.** $7,000,000
37. quadratic trinomial **39.** quartic binomial
41. quintic polynomial **43.** $9x + 4y - 17z$
45. $2c^2 - c + 8$ **47a.** $4\pi rh + 4\pi r^2$
47b. 15.7 m² **49.** Both $0x^2 + 0x$ and 0 are polynomials. **51a.** $-13x^2 - x + 10$;
$(5x^2 - 3x + 7) - (18x^2 - 2x - 3) = (5x^2 - 18x^2) + (-3x + 2x) + (7 + 3)$ **51b.** $13x^2 + x - 10$;
$(18x^2 - 2x - 3) - (5x^2 - 3x + 7) = (18x^2 - 5x^2) + (2x + 3x) + (-3 - 7)$ **51c.** The second result is the negative of the first. Reversing the order of subtraction with integers yields the negative of the original difference. For a and b integers, $(a - b) = -(b - a)$. **53.** No; neither of them found the additive inverse correctly. All terms should have been multiplied by -1. **55.** $6n + 9$ **57.** Sample answer: To add polynomials in a horizontal method, combine like terms. For the vertical method, write the polynomials in standard form, align like terms in columns, and combine like terms. To subtract polynomials in a horizontal method, find the additive inverse of the polynomial that is being subtracted, and then combine like terms. For the vertical method, write the polynomials in standard form, align like terms in columns, and subtract by adding the additive inverse.

Lesson 10-2

1. $b^3 - 12b^2 + b$
3. $-6m^6 + 36m^5 - 6m^4 - 75m^3$
5. $4p^2r^3 + 10p^3r^3 - 30p^2r^2$ **7.** $-13x^2 - 9x - 27$
9. $-20d^3 + 55d + 35$ **11.** $14j^3k^2 + 2j^2k^2 - 17jk + 18j^2k^3 - 18k^3$ **13.** $\frac{n^2}{2} + \frac{n}{2}$; 78
15. $3.88 = x(0.39) + (x + 4)(0.29)$; 4 peppers and 8 potatoes **17.** 2 **19.** $\frac{43}{6}$ **21.** $\frac{30}{43}$
23. $4a^2 + 3a$ **25.** $2x^2 - 5x$ **27.** $-3n^3 - 6n^2$
29. $15x^3 - 3x^2 + 12x$ **31.** $-4b + 36b^2 + 8b^3$
33. $4m^4 + 6m^3 - 10m^2$ **35.** $3w^2 + 7w$
37. $-2p^2 + 3p$ **39.** $3x^3 + 8x$
41. $-22b^2 + 2b + 8$
43. $-q^3w^3 - 35q^2w^4 + 8q^2w^2 - 27qw$ **45.** -7
47. -5 **49.** -2 **51.** $1.10(2\pi r) = 2\pi(r + 12)$;
$r = 120$ ft **53.** No; sample answer: The Distributive Property has been misapplied. The second term should be $12y^2(-2y^2) = -24y^4$, and the fourth term should be $-3y^3(-2y) = 6y^4$. So, it simplifies to $24y^3 - 18y^4$. **55a.** $1.25x(45 - 5x) = 56.25x - 6.25x^2$ **55b.** $5.63
55c. $126.56 **57.** Ted; Pearl used the Distributive Property incorrectly. **59.** $8x^2y^{-2} + 24x^{-10}y^8 - 16x^{-3}$ **61.** Sample answer: $3n$, $4n + 1$; $12n^2 + 3n$ **63.** Sample answer: $4t(t - 5) - 2t(3t + 2) + 108 = 3t(2t - 5) - t(8t - 3)$

Lesson 10-3

1. $3c^2 + 4c - 15$ **3.** $30a^2 + 43a + 15$
5. $15y^2 - 17y + 4$ **7.** $6m^2 + 19m + 15$
9. $144t^2 - 25$ **11.** $40w^2 - 28wx - 24x^2$
13. $45x^2 - 13x - 2$ **15.** $20x^2 + 18x + 4$ in²
17. $(x - 2)(x + 2) - x^2 = -4$; $x^2 - 2x + 2x - 4 - x^2 = -4$; $-4 = -4$
19. $36a^3 + 71a^2 - 14a - 49$
21. $5x^4 + 19x^3 - 34x^2 + 11x - 1$
23. $18z^5 - 15z^4 - 18z^3 - 14z^2 + 24z + 8$
25. $x^2 + 4x + 4$ **27.** $t^2 + t - 12$
29. $n^2 - 4n - 5$ **31.** $2x^2 - 18$
33. $2\ell^2 - 3\ell - 20$ **35.** $5q^2 + 24q - 5$
37. $6m^2 - 6$ **39.** $10a^2 - 19a + 6$
41. $2x^2 - 3xy + y^2$ **43.** $t^3 + 3t^2 + 6t + 4$
45. $m^3 + 6m^2 + 14m + 15$
47. $6b^3 + 5b^2 + 8b + 16$ **49.** $5t^2 - 34t + 56$
51. $9c^2 + 24cd + 16d^2$
53. $8r^3 - 36r^2t + 54rt^2 - 27t^3$
55. $64y^3 - 48y^2z - 36yz^2 + 27z^3$

57a. Let w represent the width of the mural; $w(w + 5)$ ft² **57b.** $(w + 9)(w + 4) - (w + 5)w$ ft²
57c. $(w + 9)(w + 4) - (w + 5)w = 100$; 8 ft by 13 ft

59a. $x^{4p+1}(x^{1-2p})^{2p+3} = x^{4p+1+(1-2p)(2p+3)} = x^{4p+1+(2p+3)-2p(2p+3)} = x^{6p+4-2p(2p)-2p(3)} = x^{6p+4-4p^2-6p} = x^{-4p^2+4}$ **59b.** $x^0 = 1$ for all values of x, so need to find p such that $-4p^2 + 4 = 0$. So, $p = \pm 1$. **61.** $8x^3 + 28x^2 - 16x$ units^3; Volume $= 4x(2x - 1)(x + 4) = 4x[2x(x + 4) - 1(x + 4)] = 4x[2x^2 + 8x - x - 4] = 4x(2x^2 + 7x - 4) = 4x(2x^2) + 4x(7x) + 4x(-4) = 8x^3 + 28x^2 - 16x$ **63.** $x^{2m-1} - x^{m-p+1} + x^{m+p} + x^{m+p-1} - x + x^{2p}$

65. The three monomials that make up a trinomial are similar to the three digits that make up the 3-digit number. The single monomial is similar to a 1-digit number. With each procedure, you perform 3 multiplications. The difference is that polynomial multiplication involves variables and the resulting product is often the sum of two or more monomials, while numerical multiplication results in a single number. **67.** The expression, $(2x - 1)(x^2 + x - 1)$, does not belong with the other three polynomial expressions because the product of this expression has 4 terms, and the products of the other three polynomial expressions have 3 terms.

Lesson 10-4

1. $a^2 + 10a + 100$ **3.** $h^2 + 14h + 49$
5. $64 - 16m + m^2$ **7.** $4b^2 + 12b + 9$
9. $64h^2 - 64hn + 16n^2$ **11.** $A = \pi r^2 - \pi(r - 18)^2$ or $36\pi(r - 9)$ **13.** $B^2 + 2BR + R^2$ **15.** $u^2 - 9$
17. $4 - x^2$ **19.** $4q^2 - 25r^2$ **21.** $n^2 + 6n + 9$
23. $y^2 - 14y + 49$ **25.** $b^2 - 1$ **27.** $p^2 - 8p + 16$
29. $\ell^2 + 4\ell + 4$ **31.** $9g^2 - 4$ **33.** $36 + 12u + u^2$
35. $9q^2 - 1$ **37.** $4k^2 - 8k + 4$ **39.** $9p^2 - 16$
41. $x^2 - 8xy + 16y^2$ **43.** $9y^2 - 9g^2$
45. $4k^2 + 4km^2 + m^4$ **47.** Sample answer: The area of the rectangle is the square of the larger number minus the square of the smaller number. **49a.** $0 = r^2 - 6rs + s^2$
49b. $0 = s^2 - 120s + 400$ **51.** $25y^2 + 70y + 49$
53. $100x^2 - 4$ **55.** $a^2 + 8ab + 16b^2$
57. $4c^2 - 36cd + 81d^2$ **59.** $36y^2 - 169$
61. $25x^4 - 10x^2y^2 + y^4$ **63.** $\frac{9}{16}k^2 + 12k + 64$
65. $49z^4 - 25y^4$ **67.** $r^4 - 29r^2 + 100$
69. $8a^3 - 12a^2b + 6ab^2 - b^3$
71. $k^3 - k^2m - km^2 + m^3$

73. $q^3 - 3q^2r + 3qr^2 - r^3$
75a. $2s + 1$; $(s + 1)^2 - s^2 = (s^2 + 2(s)(1) + 1^2) - s^2 = 2s + 1$ **75b.** $2s + 3$; $(s + 2)^2 - (s + 1)^2 = (s^2 + 2(s)(2) + 2^2) - (s^2 + 2s + 1) = s^2 + 4s + 4 - s^2 - 2s - 1 = 2s + 3$ **75c.** $2s + 5$; $(s + 3)^2 - (s + 2)^2 = (s^2 + 2(s)(3) + 3^2) - (s^2 + 4s + 4) = s^2 + 6s + 9 - s^2 - 4s - 4 = 2s + 5$
77a. $a^3 + 3a^2b + 3ab^2 + b^3$
77b. $x^3 + 6x^2 + 12x + 8$
77c. Sample answer:

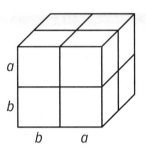

77d. $a^3 - 3a^2b + 3ab^2 - b^3$ **79.** Sample answer: $(x - 2)(x + 2) = x^2 - 4$ and $(x - 2)(x - 2) = x^2 - 4x + 4$

Lesson 10-5

1. $8(2t - 5y)$ **3.** $2k(k + 2)$ **5.** $2ab(2ab + a - 5b)$
7. $16t(2 - t)$ **9.** $6t(27 - 32t)$ **11.** $(g + 4)(f - 5)$
13. $(h + 5)(j - 2)$ **15.** $(9q - 10)(5p - 3)$
17. $(3d - 5)(t - 7)$ **19.** $(3t - 5)(7h - 1)$
21. $(r - 5)(5b + 2)$ **23.** $(b + 3)(b - 2)$
25. $(2a + 1)(a - 2)$ **27.** $2(4m - 3)$
29. $5q(2 - 5q)$ **31.** $x(1 + xy + x^2y^2)$
33. $4a(ab^2 + 4b + 3)$ **35.** $12ab(4ab - 1)$
37. $(x + 1)(x + 3)$ **39.** $(3n - p)(n + 2p)$
41. $(3x + 2)(3x - y)$ **43.** $(2x + 3)(x - 0.3)$
45. In both cases you are trying to build an expression equivalent to the original expression that has two or more factors.

47. Both are correct. Akash's grouping: $3x^2 + 7x - 18x - 42 = (3x^2 + 7x) - (18x + 42) = x(3x + 7) - 6(3x + 7) = (x - 6)(3x + 7)$. Theresa's grouping: $3x^2 - 18x + 7x - 42 = (3x^2 - 18x) + (7x - 42) = 3x(x - 6) + 7(x - 6) = (3x + 7)(x - 6)$. **49.** Find the GCF of $12a^2b^2$ and $-16a^2b^3$, which is $4a^2b^2$. Write each term as the product of the GCF and the remaining factors using the Distributive Property: $4a^2b^2(3) + 4a^2b^2(-4b) = 4a^2b^2(3 - 4b)$.

Lesson 10-6

1. $(x + 3)(x + 14)$ **3.** $(a - 4)(a + 12)$
5. $(h + 4)(h + 11)$ **7.** $(x + 3)(x - 8)$
9. $(t + 2)(t + 6)$ **11.** prime   **13.** prime
15. $(h + 6)(h + 3)$ **17.** $(x + 14)(x + 12)$
19. $(d - 18)(d - 18)$ **21.** $(5x + 4)(x + 6)$
23. $2(2x + 1)(x + 5)$ **25.** $(2x + 3)(x - 3)$
27. prime   **29.** $3(4x + 3)(x + 5)$
31. prime   **33.** prime   **35.** $(2y - 1)(y + 1)$
37. $(n + 6)(n - 3)$ **39.** $(r + 6)(r - 2)$
41. $(w - 3)(w + 2)$ **43.** $(t - 8)(t - 7)$
45. $(2x + 1)(x + 2)$ **47.** $(3g - 1)(g - 2)$
49. prime   **51.** $(3p - 2)(3p + 4)$
53. $(a - 3)(a - 7)$ **55.** $(2x + 1)(x + 3)$
57. $(x - 4)(x + 5)$ **59.** $(p - 3)(p - 7)$
61. $(q + 2r)(q + 9r)$ **63.** $(x - y)(x - 5y)$
65. $-(2x + 5)(3x + 4)$ **67.** $-(x - 4)(5x + 2)$
69. prime   **71a.** $(x + 4)(x - 1)$ **71b.** possible lengths of the pyramidal stone monument
73. $x^2 + 12x + 24$; The other 3 expressions can be factored, but this expression cannot be factored.   **75.** $-18, 18$ **77.** 7, 12, 15, 16
79. Sometimes; sample answer: The trinomial $x^2 + 10x + 9 = (x + 1)(x + 9)$ and $10 > 9$. The trinomial $x^2 + 7x + 10 = (x + 2)(x + 5)$ and $7 < 10$.   **81.** $(4y - 5)^2 + 3(4y - 5) - 70 = [(4y - 5) + 10][(4y - 5) - 7] = (4y + 5)(4y - 12) = 4(4y + 5)(y - 3)$
83. $(12x + 20y)$ in.; The area of the square equals $(3x + 5y)(3x + 5y)$ in², so the length of one side is $(3x + 5y)$ in. The perimeter is $4(3x + 5y)$ or $(12x + 20y)$ in.
85. Sample answer: Find two numbers, m and p, with a product of ac and a sum of b.

Lesson 10-7

1. $(q + 11)(q - 11)$ **3.** $(w^2 + 25)(w + 5)(w - 5)$
5. $(h^2 + 16)(h + 4)(h - 4)$ **7.** $(x + 2y)(x - 2y)$
9. $(f + 8)(f - 8)(f + 2)$ **11.** $(t + 1)(t - 1)(3t - 7)$
13. $(m + 3)(m - 3)(4m + 9)$
15. $(3a + 2b)(3a - 2b)$ **17a.** $(x - 8)(x - 8)$
17b. the length and width of the rug

19. yes; $(4x - 7)^2$ **21.** no   **23.** $4(h^2 - 14)$
25. prime   **27.** $6(d + 4)(d - 4)$
29. $(-4 + p)(4 + p)$ **31.** $(6 + 10w)(6 - 10w)$
33. $(2h + 5g)(2h - 5g)$ **35.** $8(x + 3p)(x - 3p)$
37. $2(4a + 5b)(4a - 5b)$ **39.** $(7x + 8y)(7x - 8y)$
41. $2m(2m - 3)^2$ **43.** $(a - 5)(a + 5)$; 95 yards
45. yes; $(m - 3)^2$ **47.** yes; $(g - 7)^2$
49. yes; $(2d - 1)^2$ **51.** yes; $(3z - 1)^2$ **53.** no
55. yes; $(t - 9)^2$ **57.** $\frac{1}{2}(x + 14)(x - 14)$
59. $x^4 - 81$ can be factored by using the difference of squares twice. First $(x^2 - 9)(x^2 + 9)$, then $(x + 3)(x - 3)(x^2 + 9)$. $x^2 - 9$ can be factored using the difference of squares: $(x^2 - 9) = (x + 3)(x - 3)$. The factors of $x^4 - 81$ include the factors of $x^2 - 9$, but there is an additional factor of $x^2 + 9$.
61. $m = 121$; Let $n^2 = m$. Then $2(2x^2)(n) = -44x^2$, so $n = -11$. Because $n^2 = m$, $(-11)^2 = 121$, so $m = 121$. $4x^4 - 44x^2 + 121 = (2x^2 - 11)(2x^2 - 11)$
63. $[3 + (k + 3)][3 - (k + 3)] = (k + 6)(-k) = -k^2 - 6k$ **65.** false; $a^2 + b^2$

67. Sample answer: When the difference of squares is multiplied together using the FOIL method, the outer and inner terms are opposites of each other. When these terms are added together, the sum is zero.

69. Determine if the first and last terms are perfect squares. Then determine if the middle term is equal to ± 2 times the product of the principal square roots of the first and last terms. If these three criteria are met, the trinomial is a perfect square trinomial.

Module 10 Review

1. A, C, D   **3.** B   **5.** 900 square inches   **7.** B
9. $2r^3 - 13r^2 + 17r - 3$ **11.** B   **13.** Sample answer: The square of a sum is the square of the first term plus two times the product of the first and second terms plus the square of the second term.
15. $(-5x + 6)(2y + 3)$ **17.** A, B   **19.** B
21. yes, yes, yes, no, no   **23.** B

Module 11

Quick Check

1. translation right 4 units

3. translation down 1 unit

5. yes; $(a + 6)^2$

7. yes; $(m - 11)^2$

Lesson 11-1

1. axis of symmetry: $x = 0$; vertex: (0, 1); y-intercept: 1; end behavior: As x increases or decreases, y decreases.

3. axis of symmetry: $x = 0$; vertex: (0, −4); y-intercept: −4; end behavior: As x increases or decreases, y increases.

5.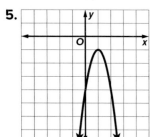

7.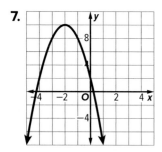

9.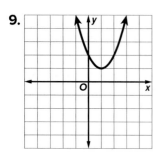

11. D = {all real numbers}; R = $\{y \mid y \geq 2\}$

x	−4	−3	−2	−1	0
y	6	3	2	3	6

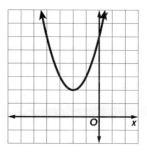

13. D = {all real numbers}; R = $\{y \mid y \geq -13\}$

x	0	1	2	3	4
y	−5	−11	−13	−11	−5

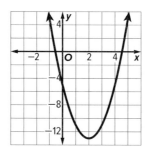

15. D = {all real numbers}; R = $\{y \mid y \geq -5\}$

x	−1	0	1	2	3
y	7	−2	−5	−2	7

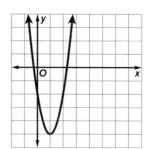

17.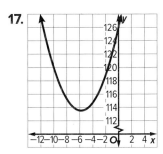

The vertex is located at about $(-5.8, 113.5)$. The vertex does not make sense in this case because there cannot be an Olympiad before the first Olympiad. The axis of symmetry is located at $x = -5.8$. The axis of symmetry does not make sense in this case because there cannot be an Olympiad before the first Olympiad. The y-intercept is located at $(0, 126)$, so the y-intercept is 126. This does not make sense in this case because there cannot be an Olympiad before the first Olympiad. As x increases from the minimum, the value of y increases. This represents that the winning pole vault height increases for each Olympiad. The graph is always positive, which means the winning pole vault height is always greater than 0 inches. The relevant domain for the situation is all nonnegative whole numbers greater than 0. This represents each Olympiad since the Olympic Games started. The relevant range is all numbers greater than or equal to 130.7. This means that the winning pole vault height is greater than or equal to 130.7 inches.

19.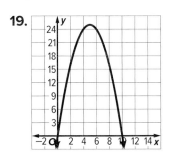

The vertex is located at $(5, 25)$. The vertex represents the arch's height of 25 feet when the distance from one side of the arch is 5 feet. The axis of symmetry is located at $x = 5$. This means that the arch's height from a distance 0 feet to 5 feet away from one side of the arch is the same as the arch's height from a distance 5 feet to 10 feet away from one side of the arch. The y-intercept is located at $(0, 0)$, so the y-intercept is 0. This means that the initial height of the arch is 0 feet when the distance from one side of the arch is 0 feet. As x increases and decreases from the maximum, the value of y decreases. This represents the height of the arch based on the distance from one side of the arch. The zeros of the graph are 0 and 10. This means the bases of the arch are a distance of 0 feet and 10 feet from one side of the arch. The relevant domain for the situation is all nonnegative numbers greater than or equal to 0 and less than or equal to 10. This represents the distance from one side of the arch to the other side of the arch. The relevant range is all numbers greater than or equal to 0 and less than or equal to 25. This means that the height of the arch is between 0 feet and 25 feet.

21a. $y = (x - 2)(x + 6)$ or $y = x^2 + 4x - 12$

21b. $x = -2$

21c.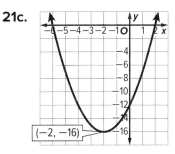

23a. $x = -1$

23b. $(-1, 3)$; maximum

23c.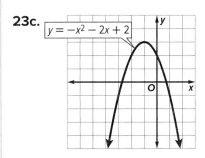

25. axis of symmetry: $x = -2$; vertex: $(-2, 2)$; y-intercept: 6; end behavior: As x increases or decreases, y increases.

27a.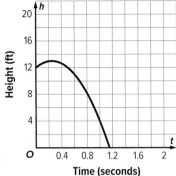

27b. The y-intercept is 12. This represents the height, in feet, at which the beanbag is tossed into the air. The x-intercept is approximately 1.15. This represents the time, in seconds, at which the beanbag lands on the beach.

27c. The maximum value is 13. The maximum height of the beanbag is 13 feet. It reaches this height 0.25 second after it is tossed into the air.

27d. $h(0.5) = 12$; This means it takes 0.5 second for the beanbag to come back down to the height at which it was tossed. **27e.** The domain is $\{t \mid 0 \le t \le 1.15\}$. The domain represents the length of time the beanbag is in the air.

29. Sample answer: $y = 4x^2 + 3x + 5$; Write the equation of the axis of symmetry, $x = -\frac{b}{2a}$. From the equation, $b = 3$ and $2a = 8$, so $a = 4$. Substitute these values for a and b into the equation $y = ax^2 + bx + c$.

31. $y = -x^2 + 6x + 16$

33. Sample answer: A quadratic equation can be used to model the path of a football that is kicked during a game. The vertex represents the maximum height of the ball.

35. Sample answer: Suppose $a = 1$, $b = 2$, and $c = 1$.

x	$f(x) = 2^x + 1$	$g(x) = x^2 + 1$	$h(x) = x + 1$
−10	1.00098	101	−9
−8	1.00391	65	−7
−6	1.01563	37	−5
−4	1.0625	17	−3
−2	1.25	5	−1
0	2	1	1
2	5	5	3
4	17	17	5
6	65	37	7
8	257	65	9
10	1025	101	11

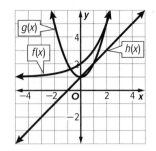

Intercepts: $f(x)$ and $g(x)$ have no x-intercepts, $h(x)$ has one at −1 because $c = 1$. $g(x)$ and $h(x)$ have one y-intercept at 1 and $f(x)$ has one y-intercept at 2. The graphs are all translated up 1 unit from the graphs of the parent functions because $c = 1$.

Increasing/Decreasing: $f(x)$ and $h(x)$ are increasing on the entire domain. $g(x)$ is increasing to the right of the vertex and decreasing to the left.

Positive/Negative: The function values for $f(x)$ and $g(x)$ are all positive. The function values of $h(x)$ are negative for $x < -1$ and positive for $x > 1$.

Maxima/Minima: $f(x)$ and $h(x)$ have no maxima or minima. $g(x)$ has a minimum at (0, 1).

Symmetry: $f(x)$ and $h(x)$ have no symmetry. $g(x)$ is symmetric about the y-axis.

End behavior: For $f(x)$ and $h(x)$, as x increases, y increases and as x decreases, y decreases. For $g(x)$ as x increases, y decreases and as x decreases, y increases.

The exponential function $f(x)$ eventually exceeds the others.

Lesson 11-2

1. translated down 10 units **3.** translated right 1 unit **5.** translated left 3 units **7.** translated left 2.5 units **9.** translated up 6 units **11.** translated right 9.5 units **13.** stretched vertically **15.** compressed horizontally **17.** compressed horizontally **19.** compressed vertically **21.** stretched vertically and reflected across the x-axis **23.** compressed vertically and reflected across the x-axis **25.** stretched vertically and reflected across the x-axis **27.** reflected across the x-axis and translated down 7 units **29.** compressed vertically and translated up 6 units

31. stretched vertically and translated up 3 units   **33.** The graph of $U_s = 5x^2$ is a vertical stretch of the graph of $U_s = x^2$.   **35a.** $d = 3t^2$ is a vertical stretch of $d = t^2$; $d = 2t^2 + 100$ is a vertical stretch of $d = t^2$ translated up 100 units (feet).   **35b.** after 10 seconds   **37.** $f(x) = 2(x - 3)^2 + 1$
39. compressed vertically, reflected across the x-axis, and translated up 5 units
41. compressed vertically and translated down 1.1 units   **43.** compressed vertically and translated up $\frac{5}{6}$ unit

45. Apply a vertical stretch of 2 to the parent function. Reflect across the x-axis. Translate left 2 units and up 2 units.

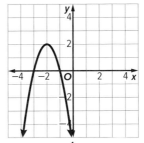

47. Apply a vertical compression of $\frac{1}{4}$ to the parent function. Reflect over the x-axis. Translate right 1 and up 4 units.

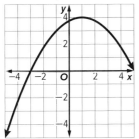

49. D   **51.** A
53. The parent quadratic function is an even function since $f(-x) = (-x)^2 = x^2$, so $f(-x) = f(x)$.
55. $y = 2(x - 2)^2 - 1$ **57.** $y = 0.25x^2 + 1$
59. $y = -0.5(x + 2)^2$ **61a.** Sometimes; sample answer: This occurs only if $k = 0$. For any other value, the graph will be translated up or down.
61b. Always; sample answer: The reflection does not affect the width. Both graphs are dilated by a factor of a.
61c. Never; sample answer: Because $k >= 0$, the graph of $f(x)$ has a minimum point with a y-coordinate greater than or equal to 0. The graph of $g(x)$ has a vertex with a y-coordinate of -3. Even if the graph of $g(x)$ opens up, the graphs can never have the same minimum.   **63.** Sample answer: Not all reflections over the y-axis produce the same graph. If the vertex of the original graph is not on the y-axis, the graph will not have the y-axis as its axis symmetry, and its reflection across the y-axis will be a different parabola.
65. Sample answer: For $y = ax^2$, the parent graph is stretched vertically if $a > 1$ or compressed vertically if $0 < a < 1$. The y-values in the table will all be multiplied by a factor of a. For $y = x^2 + k$, the parent graph is translated up if k is positive and moved down if k is negative. The y-values in the table will all have the constant k added to them or subtracted from them. For $y = ax^2 + k$, the graph will either be stretched vertically or compressed vertically based upon the value of a and then will be translated up or down depending on the value of k. The y-values in the table will be multiplied by a factor of a and the constant k will be added to them.

Lesson 11-3

1. no solutions

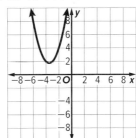

3. -8

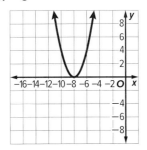

5. -7

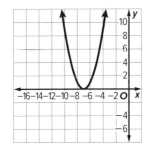

7. 2, 8

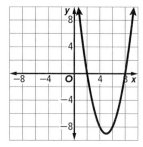

9. 3, 5

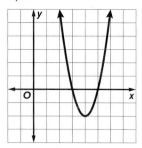

11. no solutions

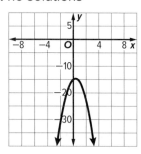

13. 5.3 seconds

15. −3.4, −0.6

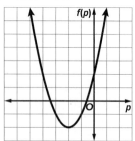

17. −5.4, −0.6

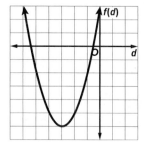

19. 0.2, 1.4

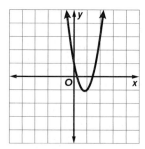

21. point of highest yield: (2, 16); There is a zero at (6, 0), which means that 6 units of fertilizer will yield 0 crops.

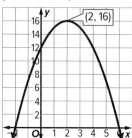

23. Yes; solving the equation $(30 - w)w = 81$ gives $w = 3$, or 27. A 3 in. by 27 in. sheet of paper has an area of 81 in² and a perimeter of 60 in. **25a.** Sample answer: $f(x)$ has two real solutions at $x = 4$ and $x = 9$. $g(x)$ has no real solutions. $h(x)$ has one real solution at $x = -6$. Because $h(x)$ has one solution, it could be a perfect square trinomial; $y = x^2 + 12x + m$ is the only equation that could be a perfect square. $g(x)$ has no real solutions, and $y = -x^2 + 3x - 10$ is not factorable, so it is likely $g(x)$. I can double check that by substituting x-values into the equation and seeing if the resulting ordered pair is on the graph. $f(x)$ has two real solutions, and since $y = 2x^2 - nx + 72$ could be made into a factorable equation that is not a perfect square, it is likely $f(x)$. **25b.** For $h(x)$, because the solution is $x = -6$, the first term of the equation has a coefficient of 1, and the middle term is positive, the factored equation could be $y = (x + 6)(x + 6)$, which would mean that $m = 36$. **25c.** For $f(x)$, the solutions are $x = 9$ and $x = 4$, and the first term of the equation has a coefficient of 2. So, the first term of one of the factors must be x and the other must be $2x$. I can multiply 2 by 72 and find suitable factor pairs, and then apply trial and error until a pair results in the solutions of 4 and 9, $n = 26$.

27a. Sample answer: Press Y= and enter the function $-50{,}000x^2 + 300{,}000x - 250{,}000$.

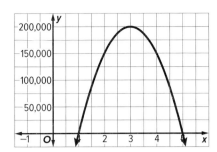

27b. The x-intercepts are $x = 1$ and $x = 5$. Both represent selling prices of the product that result in zero profit.

27c. $-50{,}000x^2 + 300{,}000x - 250{,}000 = 150{,}000$. The function relating to this equation is $f(x) = -50{,}000x^2 + 300{,}000x - 400{,}000$. The x-values where the graph intersects the x-axis are solutions to the equation, $x = 2$ and $x = 4$.

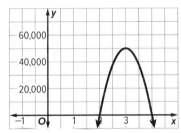

27d. No; sample answer: To make a profit of $300,000, the product should be sold at a price that is a solution to $-50{,}000x^2 + 300{,}000x - 550{,}000 = 0$, but the function $f(x) = -50{,}000x^2 + 300{,}000x - 550{,}000$ has no real solutions.

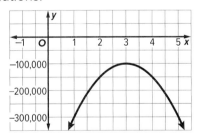

29. Sample answer: An athlete hits a tennis ball into the air. The height of the tennis ball can be modeled by the equation $h = -16t^2 + 25t + 2$. The ball is in the air for about 1.64 seconds.

31a. Sample answer: $x^2 + 8x + 16 = 0$
31b. Sample answer: $x^2 + 25 = 0$
31c. Sample answer: $x^2 - 7x + 12 = 0$

Lesson 11-4

1. ± 6 **3.** $-4, 2$ **5.** $0, 3$ **7.** $0, -3$ **9.** $-3, 9$
11. $-8, 15$ **13.** $-5, 18$ **15.** $-3, -6$ **17.** $-\frac{5}{3}$
19. $\frac{11}{6}, -\frac{11}{6}$ **21.** $\frac{1}{8}, -\frac{1}{8}$ **23.** $-8, 8$ **25a.** $-1, -9$

25b.

27. 4 and 5 **29.** after 6 seconds
31. $f(x) = x^2 - x - 6$ **33.** $f(x) = 4x^2 - 32x + 60$
35. $f(x) = x^2 + 7x - 18$
37. $f(x) = \frac{1}{4}x^2 - x - 8$ **39.** No; Kayla factored incorrectly. The correct solutions are 0 and 6. **41.** $\pm\frac{4}{15}$
43. $\pm\frac{3}{5}$ **45.** $-1, -\frac{1}{3}$ **47.** ± 36 **49.** -15
51. $-1, 17$ **53.** 10 **55.** $-2, 1$
57. Sample answer: $\frac{1}{2}(x)(x + 10) = 100$; $\frac{1}{2}(x^2 + 10x) = 100$; $x^2 + 10x = 200$; $x^2 + 10x - 200 = 0$; $(x - 10)(x + 20) = 0$. $x = 10$ or $x = -20$. Use $x = 10$. So, the base is 10 feet and the height is 20 feet. **59.** Sample answer: When $a \neq 1$, the coefficients of x in each factor may not be 1 and must have a product equal to a. The product of m and n must be equal to ac rather than c.
61. Write the equation as $x^2 - 6x + qx - 12 = 0$. Find factor pairs of -12 that have -6 as a factor. The other factor is q since $-12 = (-6)(2)$. So, $q = 2$. **63.** 8 ft by 16 ft; $\frac{1}{2}(x)(x + 8) = 64$, $\frac{1}{2}(x^2 + 8x) = 64$, $x^2 + 8x = 128$, $x^2 + 8x - 128 = 0$; The factor pair with a sum of 8 is -8 and 16, so $(x - 8)(x + 16) = 0$ and $x = 8$ or $x = -16$. Because the base cannot be negative, use $x = 8$. So, the base is 8 feet and the height is 16 feet. **65.** Because the solutions are $-\frac{b}{a}$ and $\frac{b}{a}$, $a \neq 0$ and b is any real number.
67. Samantha; sample answer: She rewrote the equation to have zero on one side. Then she factored and used the Zero Product Property. Ignatio should have subtracted 12 from each side before factoring. Then he could have used the Zero Product Property. **69.** Sample answer: $x^2 - 3x + \frac{9}{4} = 0$; $\frac{3}{2}$

Lesson 11-5

1. 169 **3.** $\frac{361}{4}$ **5.** $\frac{25}{4}$ **7.** 121 **9.** 144
11. −2.9, 4.9 **13.** −3.3, 0.3 **15.** −1.4, 3.4
17. ∅ **19.** 1, 1.5 **21.** −3.5, −1.1
23. vertex form: $y = (x − 6)^2 − 20$; axis of symmetry: $x = 6$; extrema: minimum at (6, −20); zeros: 1.5, 10.5

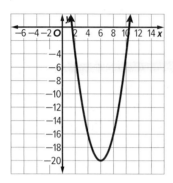

25. vertex form: $y = (x − 4)^2 − 6$; axis of symmetry: $x = 4$; extrema: minimum at (4, −6); zeros: 1.6, 6.4

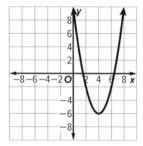

27. about 5.2 ft; 5 ft **29.** The area of the garden will be 4800 ft² for a width of 40 ft or a width of 60 ft. $(x)(200 − 2x) = 4800$, $−2x^2 + 200x − 4800 = 0$, $x = 40$ or $x = 60$.
31. $4x(x + 3) = 352$; $4x^2 + 12x = 352$; $x^2 + 3x = 88$; $\left(x + \frac{3}{2}\right)^2 = 88 + \frac{9}{4}$; $x + \frac{3}{2} = \pm\frac{19}{2}$; $x = −11, 8$; Because the dimensions must be positive, use $x = 8$. So, the dimensions are 32 inches by 11 inches. **33a.** Sample answer: $4x^2 + 28x = 226$ or $4x^2 + 28x − 226 = 0$
33b. Sample answer: $4x^2 + 28x + c = 226 + c$. $\left(\frac{28}{2}\right)^2 = 4c$, so $c = 49$. $4x^2 + 28x + 49 = 226 + 49$. $(2x + 7)^2 = 275$. Take the square root and solve the two resulting equations: $2x + 7 \approx 16.58$; $x \approx 4.79$. $2x + 7 \approx −16.58$; $x \approx −11.79$.
33c. Sample answer: $(4)(−226) = −904$, there are no factor pairs of −904 that sum to 28. So, the equation cannot be factored. Because one solution is approximately 4.79, then the equation has one solution that is between 4 and 5.
35. $y = ax^2 + bx + c$
$y = a\left(x^2 + \frac{b}{a}x\right) + c$
$y = \left[ax^2 + \frac{b}{a}x + \left(\frac{b}{2a}\right)^2\right] + c − a\left(\frac{b}{2a}\right)^2$
$y = a\left[x − \left(−\frac{b}{2a}\right)\right]^2 + \frac{4ac − b^2}{4a}$
If $h = \frac{b}{2a}$ and $k = \frac{4ac − b^2}{4a}$, then $y = a(x − h)^2 + k$. The axis of symmetry is $x = −\frac{b}{2a}$ or $x = h$.
37. $n^2 + \frac{1}{3}n + \frac{1}{9}$; It is the trinomial that is not a perfect square. **39.** Sample answer: Because the leading coefficient is 1, completing the square is simpler. Graphing the related function with a graphing calculator allows you to estimate the solution. Factoring is not possible.

Lesson 11-6

1. −7, 7 **3.** −4, 9 **5.** ∅ **7.** 2.2, −0.6
9. −3, −$\frac{6}{5}$ **11.** 0.5, −2 **13.** 0.5, −1.2 **15.** 3
17. 1.8 s **19.** $\left(\frac{60 − x}{4}\right)^2 + \left(\frac{x}{4}\right)^2 = 117$; 24 in. and 36 in. **21.** −135; no real solution
23. 0; one real solution
25. 18.25; two real solutions **27.** −24; no real solutions **29.** 32; two real solutions
31. 89; two real solutions **33.** −64; no real solutions **35.** −3.7, −0.3 **37.** −5.4, −0.6
39. $\frac{1}{2}$, 1 **41.** −4$\frac{1}{2}$, 1 **43.** −$\frac{2}{3}$, 3 **45.** 0
47. 1 **49.** Sample answer: The value under the radical is $2^2 − 4 \cdot 3q$. For the solutions to be real, that value must be positive. So $2^2 − 4 \cdot 3q \geq 0$. Solve for q; $q < \frac{1}{3}$. The solutions are real when $q < \frac{1}{3}$. For the solutions to be complex and nonreal, $2^2 − 4 \cdot 3q < 0$. Solve for q; $q > \frac{1}{3}$. The solutions are complex and nonreal when $q > \frac{1}{3}$. **51a.** the starting height **51b.** Solve the equation $0 = −16t^2 + 90t + 120$. Find t when the firework is 0 ft off the ground. Use the Quadratic Formula, where $a = −16$, $b = 90$, and $c = 120$. $t = −1.11$ and 6.74. Given that time cannot be negative, the firework hits the ground 6.74 seconds after launch.
51c. 246.56 feet; 2.81 seconds; Sample answer: Determine the vertex of the parabola. It is on the line of symmetry, which is approximately $t = 2.81$ (midway between the zeros found in part b). Substitute 2.81 for t in the original equation, $−16(2.81)^2 + 90(2.81) + 120 = 246.56$.

51d. Because of the −16, the parabola opens down. It has x-intercepts of −1.11 and 6.74. It has a y-intercept of 120. It has a vertex of (2.81, 246.56)

51e. $-16t^2 + 90t + 120 = 200$ or $-16t^2 + 90t - 80 = 0$. $a = -16$, $b = 90$, and $c = -80$, so $t = \dfrac{-90 \pm \sqrt{90^2 - 4(-16)(-80)}}{2(-16)} = \dfrac{-90 \pm \sqrt{90^2 - 5120}}{-32} = \dfrac{45 \pm \sqrt{745}}{16}$, or approximately 1.1 seconds and 4.5 seconds.

53a. always **53b.** sometimes

55a. 24.1 feet; The area of the enclosure is $w(w - 20) = w^2 - 20w$. Solve the equation $w^2 - 20w = 100$, or $w^2 - 20w - 100 = 0$. Using the Quadratic Formula, $a = 1$, $b = -20$, and $c = -100$, so $w = \dfrac{-(-20) \pm \sqrt{(-20)^2 - 4(1)(-100)}}{2(1)} = \dfrac{20 \pm \sqrt{800}}{2} = 10 \pm 10\sqrt{2}$. The solution $10 - 10\sqrt{2}$ is not valid since it is negative. So, the width should be approximately 24.1 feet.

55b. $10 + \sqrt{100 + R}$; Solve the equation $w^2 - 20w = R$, or $w^2 - 20w - R = 0$. Using the Quadratic Formula, $a = 1$, $b = -20$, and $c = -R$, so $w = \dfrac{-(-20 \pm \sqrt{(-20)^2 - 4(1)(-R)}}{2(1)} = \dfrac{20 \pm \sqrt{400 + 4R}}{2} = \dfrac{20 \pm 2\sqrt{100 + R}}{2} = 10 \pm \sqrt{100 + R}$. The solution $10 - \sqrt{100 + R}$ is not valid since it is negative. So, the width must be $10 + \sqrt{100 + R}$. **57.** The St. Louis Arch is 630 feet wide and 630 feet high. Using the points (0, 0), (315, 630), and (630, 0), the equation $y = -\dfrac{2}{315}x^2 + 4x$ represents the height x feet from one base. The solutions of the equation are $x = 0$ and $x = 630$. This means that the two bases of the St. Louis Arch are 630 feet apart. **59.** one real solution **61.** two real solutions **63.** Sample answer: The polynomial can be factored to get $f(x) = (x - 4)^2$, so the only real zero is 4. The discriminant is 0, so there is 1 real zero. The discriminant can be used to determine how many real zeros there are. Factoring can be used to determine what they are. **65.** Sample answer: Factoring is easy if the polynomial is factorable and complicated if it is not. Not all equations are factorable. Graphing gives only approximate answers, but it is easy to see the number of solutions. Using square roots is easy when there is no x-term. Completing the square can be used for any quadratic equation and exact solutions can be found, but the leading coefficient has to be 1 and the x^2- and x-terms must be isolated. It is also easier if the coefficient of the x-term is even; if not, the calculation becomes harder when dealing with fractions. The Quadratic Formula will work for any quadratic equation, and exact solutions can be found. This method can be time consuming, especially if an equation is easily factored.

67. Sample answer: $x^2 + 2x + 1 = 0$; 0; The zero has only one square root. Thus, when the determinant is zero, the Quadratic Formula yields only one value when you simplify the numerator of the formula.

Lesson 11-7

1. $(-1, -3)$ and $(1, -3)$

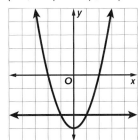

3. $(0, 1)$

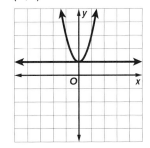

5. $(4, 3)$ and $(-2, 3)$ **7.** $(2, -3)$ and $(3, -4)$
9. $(3, -3)$ and $(-2, 7)$ **11a.** no solution
11b. There is no solution, so the baton will not reach the height of the ceiling. **13.** June
15. $\left(\dfrac{1}{2}, 1\right)$ and $(5, -17)$
17. $(1, -1)$
19. $(-1, -2)$ and $(2, 4)$
21. $(-4, 0)$ and $(8, 36)$ **23.** $(-3, 6)$ and $(5, 2)$
25. $(3, 3)$ **27.** $\left(\dfrac{-1 + \sqrt{5}}{2}, \dfrac{-3 + 3\sqrt{5}}{2}\right)$ and $\left(\dfrac{-1 - \sqrt{5}}{2}, \dfrac{-3 - 3\sqrt{5}}{2}\right)$ **29.** no solution **31.** -1
33. no solution **35a.** one **35b.** two

35c. none

37a.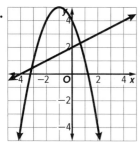

37b. Sample answer: The player will hit the target where the graphs intersect. Based on the graphs, this appears to happen at approximately (−3.1, 0.4) and (0.6, 2.3).
37c. (−3.1, 0.4) and (0.6, 2.3); these match the approximations in part b. **39.** A graph of a system shows the solutions to the system where the two lines or curves intersect. If the lines or curves intersect twice, there are two solution; if they intersect once, there is one solutions if they do not intersect, there are no solutions.

41. The rock will land in the river. A graph of the functions shows that the path of the rock does not intersect with the cliff face before the river (the x-axis).

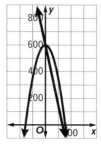

43. (0.8, 0.6) and (−0.8, −0.6) **45.** $k = -\frac{1}{4}$; Sample answer: If the system has one solution, then $x^2 + 2 = 3x + k$ and $x^2 - 3x + (2 - k) = 0$. This equation has exactly one solution if $b^2 - 4ac = 0$, or $9 - 4(2 - k) = 0$. Solving this equation for k shows that $k = \frac{1}{4}$. **47.** Sample answer: The maximum number of solutions of a quadratic and linear equation is 2. The linear function cannot change direction, and the quadratic function only changes direction once, so the two can intersect when the quadratic function is increasing and again when it is decreasing. Two linear functions have a maximum of one solution if their graphs only intersect once. They have infinitely many solutions if the two functions have the same graph. **49.** The third system does not belong because it has no solution. The other systems have two solutions.

Lesson 11-8

1. linear; $y = 0.2x - 8.2$ **3.** exponential; $y = 3 \cdot 4^x$ **5.** quadratic; $y = 4.2x^2$
7. exponential; $y = 4 \cdot 0.5^x$
9. quadratic; $y = -3x^2$
11. exponential; $y = 0.5 \cdot 3^x$
13. quadratic; $y = 3x^2$

15a.

[0, 10] scl: 1 by [0, 60] scl: 5

15b. quadratic **15c.** $y = -2.857x^2 + 22.429x + 9$; $R^2 = 0.969$; $D = \{x \mid 0 \leq x \leq 8.233\}$; $R = \{y \mid 0 \leq y \leq 53.016\}$
15d. 26 feet **17.** quadratic; $R^2 \approx 0.980$
19. exponential; $R^2 \approx 0.997$
21. exponential; $y = 1.05 \cdot 2^x$ **23a.** exponential
23b. $y = 20 \cdot 0.5^x$ **23c.** about 0.0098 g
23d. $\approx$ 701.6 years **25a.** Linear; the first differences are constant **25b.** −18 ft/s; the elevator is descending at 18 ft/s.
25c. $y = -18x + 142$; the coefficient of x is −18, which is the value of the first differences in part **a** and the constant rate of change for the function in part **b**.
27. Look up data on the temperature at any place from 6:00 A.M. to 2:00 P.M. and put that data in a scatter plot. Then find the coefficient of determination to determine which model is the best fit. Sample: temperature data from Aug 10 from Iowa City, IA (plot below) shows a quadratic model is best ($R^2 = 0.955$).

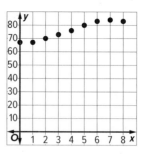

29. Sample answer: Diego is correct. Any function with equal non-zero first differences would have to have second differences of 0. For example if the first differences are all 4, then the second differences are $4 - 4 = 0$ for all terms. **31.** Linear functions have a constant

first difference and quadratic functions have a constant second difference, so cubic equations would have a constant third difference.
33. Sample answer: If one linear term is ax, the next term is $a(x + 1)$, and the difference between the terms is $a(x + 1) - ax = ax + a - ax$, or a. If one exponential term is a^x, the next term is a^{x+1}, and the ratio of terms is $\frac{a^{x+1}}{a^x}$, or a.

Lesson 11-9

1. $(f + g)(x) = 3x^2 - x - 4$
3. $(f + h)(x) = -5x - 9$
5. $(g - h)(x) = 3x^2 + 4x + 5$
7. $(h - g)(x) = -3x^2 - 4x - 5$
9. $(f \cdot g)(x) = 11x^3 - 66x^2 + 33x$
11. $(g \cdot h)(x) = -x^3 + 10x^2 - 27x + 12$
13. $(f \cdot j)(x) = 11x(2^x) + 33x$ **15a.** $f(t) + g(t) = 7t + 4$ is Jan's speed while walking with the moving walkway.   **15b.** $f(t) - g(t) = t + 2$ is Jan's speed while walking against the moving walkway.
17a. $P(x) = 15 - 0.75x$; $T(x) = 180 + 10x$
17b. $R(x) = -7.5x^2 + 15x + 2700$
17c. $2640   **19.** $\left(\frac{1}{4}\right)^x + x^2$
21. $(p \cdot q)(x) = -2x^3 - 10x$
23. $(q \cdot r)(x) = x^4 - x^3 + 7x^2 - 5x + 10$
25. $(p \cdot t)(x) = -2x\left(\frac{1}{4}\right)^x + 10x$
27a. $\ell(x) = -2x + 20$, $w(x) = -2x + 12$
27b. $(\ell \cdot w)(x) = 4x^2 - 64x + 240$ represents the area of the base of the completed box.
27c. $D = \{x \mid 0 < x < 6\}$; The side lengths of the squares must be positive numbers that are less than half the width of the rectangle to form a box.   **27d.** $(\ell \cdot w)(1.5) = 153$; If the squares are 1.5 inches per side, the area of the base of the box is 153 square inches.
29a. $f(n) = n^2 + 4$; $g(n) = 4n$ **29b.** $h(n)$ is the sum of $f(n)$ and $g(n)$, so $h(n) = n^2 + 4 + 4n$; $h(10) = 144$.   **29c.** The pattern forms a square with sides of length $n + 2$, so the total number of tiles is $(n + 2)^2$. When $n = 10$, this equals 12^2 or 144, so the answer to part **b** is correct.
31a. $A(x) = (12 + 4x)(8 + 2x) - 96$; The area of the gravel border is the product of the function that gives its length, $L(x) = 12 + 4x$, and the function that gives its width, $W(x) = 8 + 2x$, minus the constant function that gives the area of the flower bed, $F(x) = 96$.

31b. Disagree; sample answer: the function may be written as $A(x) = 8x^2 + 56x$, which shows that this is a quadratic function with $a > 0$ and $c = 0$, so the graph is an upward-opening parabola through the origin. As x increases, the area increases.
33. $m(x) = 3^x - 2$ **35.** Sample answer: $f(x) = x^2 + 12x - 7$ and $g(x) = -2x + 5$
37. Sometimes; sample answer: A linear equation with two terms has an x-term and a constant. A quadratic equation with three terms has an x^2-term, an x-term, and a constant. Multiplying the equations gives an x^3-term, an x^2-term, an x-term, and a constant. However, sometimes one or more of the terms will cancel. So, the product may have only 3 terms sometimes.   **39.** Delfina's solution is incorrect. Sample answer: She only multiplied the x-terms by other x-terms and constants by other constants. The correct answer is $(f \cdot g)(x) = 24x^3 + 10x^2 - 170x + 152$.

Module 11 Review

1.

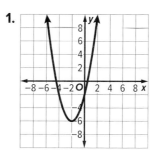

3. B   **5.** -2

7.

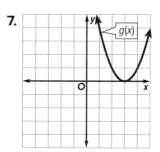

9. C   **11.** C
13. Vertex form: $y = 2(x + 1)^2 + 4$; extrema: $(-1, 4)$; because the x^2-term is positive, the parabola opens upward and the extrema is a minimum.   **15.** 5.7 ft   **17.** No real solutions   **19.** 17   **21.** C

Module 12

Quick Check

1. 45.88   **3.** $3\frac{3}{20}$ **5.** 82.4%   **7.** 85.6%

Lesson 12-1

1. mean: 15.625, median: 15.5, mode: none
3. mean: 3.3, median: 2.5, mode: 2
5. mean: 5 students; median: 4 students; mode: 3 students   **7.** mean: 54.75 mph; median: 54 mph; mode: 53 mph   **9.** mean: about 2.8; median: 2.75; mode: 2   **11.** 25th percentile
13. Sample answer: The mean could be slightly higher because on a few of Saturday nights throughout the year, there were a very large number of people at the movies, which caused the mean to increase but did not affect the median.   **15.** mean: 252; median: 245; mode: none   **17.** 23   **19.** mean: 51.5, median: 51, mode: none   **21.** 20 points   **23.** 90th percentile   **25a.** mean = 109,633; median = 66,556, no mode   **25b.** Sample answer: The novels lower than the 50th percentile would be those consisting of words in the thirty-thousands and in the upper fifty-thousands. My prediction is correct because those three books are in the 47th percentile, which is just under the 50th percentile.   **25c.** The median will change from 66,556 to 69,920, a difference of 3364 words. The mean will change from 109,633 to 111,065, a difference of 1432 words.   **27.** Canada: 20th percentile; France: 50th percentile; Japan: 40th percentile; Russia: 60th percentile; Brazil: 10th percentile; Great Britain: 70th percentile

Olympic Medal Counts	
Country	Total Medals
Australia	29
Brazil	19
Canada	22
China	70
France	42
Great Britain	67
Japan	41
New Zealand	18
Russia	56
United States	121

29. Sample answer: I can assume that the data is tightly clustered around 37 because all three measures of center are close.   **31.** 75th percentile
33. Because the mean is an average of all the numbers in the data set, it is most affected by outliers. An outlier on the high end will cause the mean to increase. The median is the middle value in the dataset, adding one high number should not have much effect on the median unless the dataset has values, which are widely spread. The mode is the most frequent number so the outlier will have no effect on the mode unless the outlier is the same as the mode.
35. The mean, median, and mode will all be multiplied by the number.   **37.** Julio should have chosen the mean because all the growth values are close together.   **39.** Sample answer: To find a percentile rank, order the data set in decreasing order. Count the number of items below the item you are ranking, and divide that by the total number of items. Multiply this answer by 100 to arrive at the percentile rank.

Lesson 12-2

1. Summer Reading Program

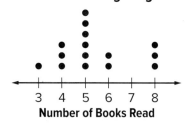

3.

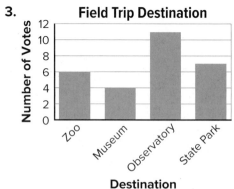

5. histogram

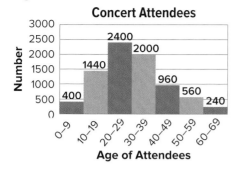

7a.

7b. 8   **7c.** 8

9a.

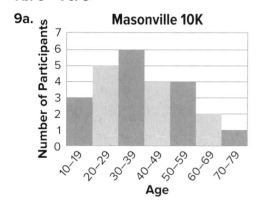

9b. 8   **9c.** 30–39

11. Sample answer: The scientist should break down the data into increments of two-tenths starting at 1 and going through 2.6.

13. Sample answer: 1) Because the data is clustered around ratings 7-10, it can be concluded that the product is well-liked by most customers and may have minor inconsistencies that certain people did not like. 2) Because there are only three low ratings, it can be concluded that dissatisfaction with the product could be a result of personal preference or manufacturer defect in a specific item.

15. Sample answer: If the range of the data is broad with specific, unrepeating values, then it makes the dot plot more meaningful if the range is divided up into equal intervals.

17. Sample answer: Bar graphs and histograms are similar because each displays data with bars. They are different because a bar graph is best used with data that are discrete and a histogram represents data that are continuous. For this reason, the bars in a bar graph do not touch and represent single values while the bars in a histogram touch and represent a range of values.

Lesson 12-3

1. Sample answer: The intended population is all students. By asking only students leaving basketball practice, Awan is not getting a representative example of the entire student body.   **3.** Sample answer: The first sentence states a positive outcome of music education, which may bias the respondent toward support. This bias may serve people trying to keep music education in schools.   **5.** Mean: 4, median: 4, mode: 2; The mean and median are appropriate measures to use to accurately summarize the data.   **7.** Sample answer: The scale for Vendor 1 starts at 40, and because of the size of the bars, it looks like their sales doubled in one year, when they increased about 50%. Vendor 2 had the same sales figures as Vendor 1 but it appears that they had more sales than Vendor 1.

9. Sample answer: The required class would be better because it is more likely to contain a representative sample of students. The elective class might not be representative of the whole student body because these courses are chosen for reasons such as personal preference or future career aspirations.

11. Median; sample answer: The two lowest weights are much lower than the others, so the mean will be affected by those outliers.

13. Sample answer: The original data are very close together, so it is likely that the measures of center will all be the same or very close. Adding an outlier of 24 to the data set will cause the mean to go up, but the median and mode would likely stay unchanged or very close to the original number. So, in this case the median or mode would best represent the center of data.

15. Sample answer: To assess a sample for bias, identify the intended population and sample method; then, based on this information assess whether there is potential sample bias.   **21.** There are more extreme values in the lower end, which cause the mean to be lower than the median.

Lesson 12-4

1. 41   **3.** 62   **5.** 20
7. 62, 66, 73, 82, 99

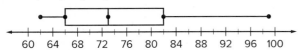

9. 35.2, 35.7, 35.9, 36.2, 36.5

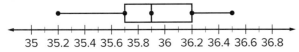

11. 25   **13.** 13   **15.** 20   **17.** 2.83   **19.** 2.97
21. 2.16; Because the standard deviation is large compared to the mean of 3, the number of goals scored each game is not relatively close to the mean.   **23a.** 9, 36, 59, 67, 69   **23b.**

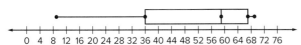

25. range: 14; minimum: 3; lower quartile: 6; median: 12; upper quartile: 14.5; maximum: 17; interquartile range: 8.5; standard deviation: 4.5

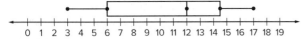

27. Both; sample answer; When an outlier is removed from a set of data, the spread and standard deviation of the data will decrease. When more values that are equal to the mean of a data set are added to the data set, the mean will be stronger and outliers will have less influence.

Lesson 12-5

1. symmetric

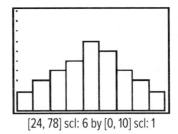

[24, 78] scl: 6 by [0, 10] scl: 1

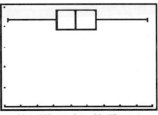

[24, 78] scl: 6 by [0, 5] scl: 1

3. Sample answer: The distribution is skewed, so use the five-number summary. The range is 53 − 12, or 41. The median is 39.5, and half of the data are between 28 and 48.

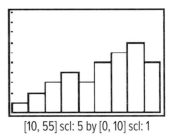

[10, 55] scl: 5 by [0, 10] scl: 1

5. Sample answer: The distribution is symmetric, so use the mean and standard deviation. The mean is about 58.7 with standard deviation of about 22.8.

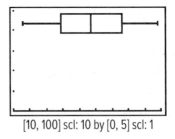

[10, 100] scl: 10 by [0, 5] scl: 1

7a.

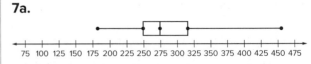

7b. min: 182, Q_1: 249, median: 274, Q_3: 315.5, max: 455   **7c.** The outlier mainly affects the mean. When the outlier is removed, the median decreases, but only $3 to $271. However, the mean changes from $289 to $266, which is more representative of the data as a whole.

9. negatively skewed

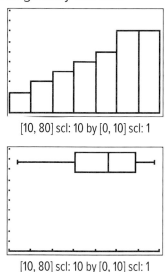

11. symmetric

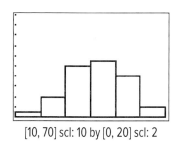

13. symmetric

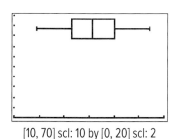

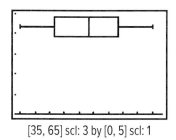

15. Sample answer: The distribution is approximately symmetric, so use the mean and standard deviation. The mean is about 54.7 years with standard deviation of about 6.2 years.

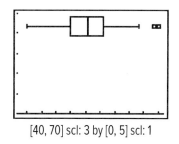

17a. Sample answer: The distribution is skewed, so use the five-number summary. min: 62, max: 525, med: 103, Q1: 84, Q3: 290
17b. Sample answer: The distribution is symmetric, so use the mean and standard deviation. The mean is about 92.4 with standard deviation of about 18.4.

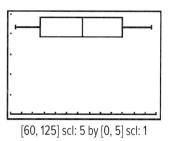

17c. Original: mean 171.5, median 103; altered: mean about 92.4, median 92. The means differ by about 79.1, while the medians differ by 11.
19a. Sample answer: 225–230 g would be a reasonable advertised weight for either brand, so it is quite likely that they have the same advertised weight. Rafaello appears to have better control over the exact quantity in each package because its distribution is grouped more closely about the mean.   **19b.** Sample answer: Both distributions have an inverted, symmetric U-shape with "tails" on either side. Leonardo's distribution is lower and wider.
21. Currently, Gerardo's distribution would be positively skewed. If he lost his longest streaks, the data would represent a symmetric distribution.

Selected Answers   SA59

23a. negatively skewed

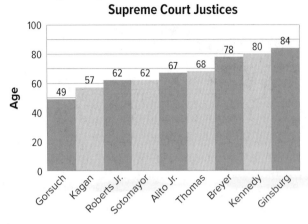

23b. The data is skewed, so use the five-number summary; min: 49, Q_1: 59.5, median: 67, Q_3: 79, max: 84. **23c.** There are no outliers in the data. **25.** Sample answer: A bimodal distribution is a distribution of data that is characterized by having data divided into two clusters, thus producing two modes and having two peaks. The distribution can be described by summarizing the center and spread of each cluster of data. **27.** Sample answer: In a symmetrical distribution, the majority of the data are located near the center of the distribution. The mean of the distribution is also located near the center of the distribution. Therefore, the mean and standard deviation should be used to describe the data. In a skewed distribution, the majority of the data lie either on the right or left side of the distribution. Because the distribution has a tail or may have outliers, the mean is pulled away from the majority of the data. The median is less affected. Therefore, the five-number summary should be used to describe the data.

Lesson 12-6

1. 60.9; 60; 60; 14; 4.7 **3.** 22.5; 21.5; no mode; 24; 7.4 **5.** 36.8; 38; 12; 56; 20.0 **7.** 26.8; 27.2; 29.6; 10.4; 3.5

9a. both negatively skewed

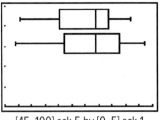

[45, 100] scl: 5 by [0, 5] scl: 1

9b. Sample answer: The distributions are skewed, so use the five-number summaries. The medians for both teams are 79. The upper quartile and maximum for the Marlins are 83.5 and 92. The upper quartile and maximum for the Cubs are 88 and 97. This means that the upper 50% of data for the Cubs is slightly higher than the upper 50% of data for the Marlins. Overall, we can conclude that the Cubs were slightly more successful than the Marlins during this time period. **11.** 93.5; 94.5; 97; 17; 5.1 **13.** 60; 62.5; 45; 50; 16.9 **15.** 25.9; 21.5; 17; 30; 10.3 **17.** 13.9; 14; 14.4; 1.4; 0.5 **19.** 75.8; 72; no mode; 48, 16.1 **21a.** 160.5; 166; no mode; 115; 33.9 **21b.** 216.5; 222; no mode; 115; 33.9 **23a.** 64.7, 66, 55, 46, 15.9 **23b.** 18.1, 18.9, 12.8, 25.6, 8.9

25a. Sample answer: The mean of Saeed's prices is $11.79, which is $0.80 more than his rival's mean price. The new prices come from subtracting $0.80 from each price, which will reduce the mean price to be the same as his rival's.

New Prices				
14.19	3.69	9.19	17.69	12.19
6.19	7.69	21.19	12.69	13.19
9.19	10.19	11.69	3.69	12.19

25b. Current prices: $\mu = 11.79$, $\sigma = 4.60$ New prices: $\mu = 10.99$, $\sigma = 4.60$ The mean has dropped by 0.8, but the standard deviation has remained constant. **27.** Sample answer: Male students, $\bar{x} = 70.0$ in., $\sigma = 2.0$ in. Female students, $\bar{x} = 66.3$ in., $\sigma = 2.7$ in. Sample answer: For male students, the mean is 70.0 in., and the standard deviation is 2.0 in. For female students, the mean is 66.3 in., and the standard deviation is 2.7 in. On average, males are taller. However, because the standard deviation of males is smaller than that of females, the heights of females are more spread out.

29. Sample answer: Histograms show the frequency of values occurring within set intervals. This makes the shape of the distribution easy to recognize. However, no specific values of the data set can be identified from looking at the histogram, and the overall spread of the data can be difficult to determine. The box plot shows the data divided into four sections. This aids when comparing the spread of one set of data to another. However, the box plots are limited because they cannot display the data any more specifically than showing it divided into four sections.   **31.** $37,750
33. Sample answer: When two distributions are symmetric, determine how close the averages are and how spread out each set of data is. The mean and standard deviation are the best values to use for this comparison. When distributions are skewed, determine which direction the data is skewed and the degree to which the data is skewed. The mean and standard deviation cannot provide information in this regard, but get this information by comparing the range, quartiles, and medians found in the five-number summaries. So if one or both sets of data are skewed, it is best to compare their five-number summaries.

Lesson 12-7

1.

	Small	Large	Total
Cherry	35	20	55
Grape	25	15	40
Watermelon	15	15	30
Total	75	50	125

3. 30

5.

	Male	Female	Total
Spanish	22.5%	25%	47.5%
French	20%	15%	35%
German	7.5%	10%	17.5%
Total	50%	50%	100%

7. Sample answer: Most of the students are studying Spanish.   **9.** Sample answer: Each conditional relative frequency represents the proportion of each candidate's support from each gender.   **11.** 12

13.

	Male	Female	Total
Tree Swallow	5	7	12
Cardinal	5	10	15
Goldfinch	8	5	13
Total	18	22	40

15. 18

17.

	Sports or Clubs	No Sports or Clubs	Total
Freshmen	10%	12.5%	22.5%
Sophomores	12.5%	15%	27.5%
Juniors	10.6%	14.4%	25%
Seniors	11.9%	13.1%	25%
Total	45%	55%	100%

19. 55.6%   **21.** 66   **23.** 100   **25.** 31   **27.** 38%

29.

Region	Apple	Sweet Potato	Pumpkin	Total
West	77 ≈ 19.0%	4 ≈ 1.0%	13 ≈ 3.2%	94 ≈ 23.2%
Midwest	32 ≈ 7.9%	6 ≈ 1.5%	54 ≈ 13.3%	92 ≈ 22.7%
South	12 ≈ 3.0%	63 ≈ 15.6%	24 ≈ 5.9%	99 ≈ 24.4%
Northeast	92 ≈ 22.7%	2 ≈ 0.5%	26 ≈ 6.4%	120 ≈ 29.6%
Total	213 ≈ 52.6%	75 ≈ 18.5%	117 ≈ 28.9%	405 = 100%

31. Sample answer: The conditional relative frequencies based on pie preference give the probability of a person preferring a particular pie choice being from one of the U.S. regions. For example, there is an 84% probability that a person who prefers sweet potato pie is from the south.

Region	Apple	Sweet Potato	Pumpkin
West	36.2%	5.3%	11.1%
Midwest	15.0%	8.0%	46.2%
South	5.6%	84%	20.5%
Northeast	43.2%	2.7%	22.2%
Total	100%	100%	100%

33.

	2WD	AWD	Total
Hatchbacks	90	9	99
Sedans	60	13	73
SUVs	2	41	43
Total	152	63	215

35. Sample answer: Yes, there does appear to be an association. When the gasoline prices are higher, the distances traveled appear to be lower; when the gasoline prices are lower, the distances traveled appear to be higher.

37. Sample answer: A relative frequency is the ratio of the number in a category to the overall total of both categories. A conditional relative frequency is the ratio of the joint frequency to the marginal frequency. Therefore, it is important to understand what relationship is being analyzed because each two-way relative frequency table can provide two different conditional relative frequency tables.

Module 12 Review

1.

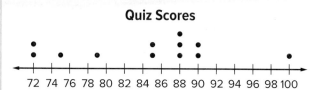

Quiz Scores

3. C, D   **5.** The data could be separated into intervals of 10, from 0–9, 10–19, 20–29, and so on through 70–79.   **7.** scaled dot plot; histogram

9.

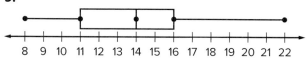

11. D   **13.** C

Glossary

English	Español

A

30°-60°-90° triangle A right triangle with two acute angles that measure 30° and 60°.

45°-45°-90° triangle A right triangle with two acute angles that measure 45°.

absolute value The distance a number is from zero on the number line.

absolute value function A function written as $f(x) = |x|$, in which $f(x) \geq 0$ for all values of x.

accuracy The nearness of a measurement to the true value of the measure.

additive identity Because the sum of any number a and 0 is equal to a, 0 is the additive identity.

additive inverses Two numbers with a sum of 0.

adjacent angles Two angles that lie in the same plane and have a common vertex and a common side but have no common interior points.

adjacent arcs Arcs in a circle that have exactly one point in common.

algebraic expression A mathematical expression that contains at least one variable.

algebraic notation Mathematical notation that describes a set by using algebraic expressions.

alternate exterior angles When two lines are cut by a transversal, nonadjacent exterior angles that lie on opposite sides of the transversal.

alternate interior angles When two lines are cut by a transversal, nonadjacent interior angles that lie on opposite sides of the transversal.

altitude of a parallelogram A perpendicular segment between any two parallel bases.

triángulo 30°-60°-90° Un triángulo rectángulo con dos ángulos agudos que miden 30° y 60°.

triángulo 45°-45°-90° Un triángulo rectángulo con dos ángulos agudos que miden 45°.

valor absoluto La distancia que un número es de cero en la línea numérica.

función del valor absoluto Una función que se escribe $f(x) = |x|$, donde $f(x) \geq 0$, para todos los valores de x.

exactitud La proximidad de una medida al valor verdadero de la medida.

identidad aditiva Debido a que la suma de cualquier número a y 0 es igual a, 0 es la identidad aditiva.

inverso aditivos Dos números con una suma de 0.

ángulos adyacentes Dos ángulos que se encuentran en el mismo plano y tienen un vértice común y un lado común, pero no tienen puntos comunes en el interior.

arcos adyacentes Arcos en un circulo que tienen un solo punto en común.

expresión algebraica Una expresión matemática que contiene al menos una variable.

notación algebraica Notación matemática que describe un conjunto usando expresiones algebraicas.

ángulos alternos externos Cuando dos líneas son cortadas por un ángulo transversal, no adyacente exterior que se encuentran en lados opuestos de la transversal.

ángulos alternos internos Cuando dos líneas son cortadas por un ángulo transversal, no adyacente interior que se encuentran en lados opuestos de la transversal.

altitud de un paralelogramo Un segmento perpendicular entre dos bases paralelas.

altitude of a prism or cylinder A segment perpendicular to the bases that joins the planes of the bases.	**altitud de un prisma o cilindro** Un segmento perpendicular a las bases que une los planos de las bases.
altitude of a pyramid or cone A segment perpendicular to the base that has the vertex as one endpoint and a point in the plane of the base as the other endpoint.	**altitud de una pirámide o cono** Un segmento perpendicular a la base que tiene el vértice como un punto final y un punto en el plano de la base como el otro punto final.
altitude of a triangle A segment from a vertex of the triangle to the line containing the opposite side and perpendicular to that side.	**altitud de triángulo** Un segmento de un vértice del triángulo a la línea que contiene el lado opuesto y perpendicular a ese lado.
ambiguous case When two different triangles could be created or described using the given information.	**caso ambiguo** Cuando dos triángulos diferentes pueden ser creados o descritos usando la información dada.
amplitude For functions of the form $y = a \sin b\theta$ or $y = a \cos b\theta$, the amplitude is $|a|$.	**amplitud** Para funciones de la forma $y = a \operatorname{sen} b\theta$ o $y = a \cos b\theta$, la amplitud es $|a|$.
analytic geometry The study of geometry that uses the coordinate system.	**geometría analítica** El estudio de la geometría que utiliza el sistema de coordenadas.
angle The intersection of two noncollinear rays at a common endpoint.	**ángulo** La intersección de dos rayos no colineales en un extremo común.
angle bisector A ray or segment that divides an angle into two congruent angles.	**bisectriz de un ángulo** Un rayo o segmento que divide un ángulo en dos ángulos congruentes.
angle of depression The angle formed by a horizontal line and an observer's line of sight to an object below the horizontal line.	**ángulo de depresión** El ángulo formado por una línea horizontal y la línea de visión de un observador a un objeto por debajo de la línea horizontal.
angle of elevation The angle formed by a horizontal line and an observer's line of sight to an object above the horizontal line.	**ángulo de elevación** El ángulo formado por una línea horizontal y la línea de visión de un observador a un objeto por encima de la línea horizontal.
angle of rotation The angle through which a figure rotates.	**ángulo de rotación** El ángulo a través del cual gira una figura.
apothem A perpendicular segment between the center of a regular polygon and a side of the polygon or the length of that line segment.	**apotema** Un segmento perpendicular entre el centro de un polígono regular y un lado del polígono o la longitud de ese segmento de línea.
approximate error The positive difference between an actual measurement and an approximate or estimated measurement.	**error aproximado** La diferencia positiva entre una medida real y una medida aproximada o estimada.
arc Part of a circle that is defined by two endpoints.	**arco** Parte de un círculo que se define por dos puntos finales.
arc length The distance between the endpoints of an arc measured along the arc in linear units.	**longitude de arco** La distancia entre los extremos de un arco medido a lo largo del arco en unidades lineales.

area The number of square units needed to cover a surface.

arithmetic sequence A pattern in which each term after the first is found by adding a constant, the common difference d, to the previous term.

asymptote A line that a graph approaches.

auxiliary line An extra line or segment drawn in a figure to help analyze geometric relationships.

average rate of change The change in the value of the dependent variable divided by the change in the value of the independent variable.

axiom A statement that is accepted as true without proof.

axiomatic system A set of axioms from which theorems can be derived.

axis of symmetry A line about which a graph is symmetric.

axis symmetry If a figure can be mapped onto itself by a rotation between 0° and 360° in a line.

área El número de unidades cuadradas para cubrir una superficie.

secuencia aritmética Un patrón en el cual cada término después del primero se encuentra añadiendo una constante, la diferencia común d, al término anterior.

asíntota Una línea que se aproxima a un gráfico.

línea auxiliar Una línea o segmento extra dibujado en una figura para ayudar a analizar las relaciones geométricas.

tasa media de cambio El cambio en el valor de la variable dependiente dividido por el cambio en el valor de la variable independiente.

axioma Una declaración que se acepta como verdadera sin prueba.

sistema axiomático Un conjunto de axiomas de los cuales se pueden derivar teoremas.

eje de simetría Una línea sobre la cual un gráfica es simétrico.

eje simetría Si una figura puede ser asignada sobre sí misma por una rotación entre 0° y 360° en una línea.

B

bar graph A graphical display that compares categories of data using bars of different heights.

base In a power, the number being multiplied by itself.

base angles of a trapezoid The two angles formed by the bases and legs of a trapezoid.

base angles of an isosceles triangle The two angles formed by the base and the congruent sides of an isosceles triangle.

base edge The intersection of a lateral face and a base in a solid figure.

base of a parallelogram Any side of a parallelogram.

base of a pyramid or cone The face of the solid opposite the vertex of the solid.

gráfico de barra Una pantalla gráfica que compara las categorías de datos usando barras de diferentes alturas.

base En un poder, el número se multiplica por sí mismo.

ángulos de base de un trapecio Los dos ángulos formados por las bases y patas de un trapecio.

ángulo de la base de un triángulo isosceles Los dos ángulos formados por la base y los lados congruentes de un triángulo isosceles.

arista de la base La intersección de una cara lateral y una base en una figura sólida.

base de un paralelogramo Cualquier lado de un paralelogramo.

base de una pirámide o cono La cara del sólido opuesta al vértice del sólido.

bases of a prism or cylinder The two parallel congruent faces of the solid.

bases of a trapezoid The parallel sides in a trapezoid.

best-fit line The line that most closely approximates the data in a scatter plot.

betweenness of points Point C is between A and B if and only if A, B, and C are collinear and $AC + CB = AB$.

bias An error that results in a misrepresentation of a population.

biconditional statement The conjunction of a conditional and its converse.

binomial The sum of two monomials.

bisect To separate a line segment into two congruent segments.

bivariate data Data that consists of pairs of values.

boundary The edge of the graph of an inequality that separates the coordinate plane into regions.

bounded When the graph of a system of constraints is a polygonal region.

box plot A graphical representation of the five-number summary of a data set.

bases de un prisma o cilindro Las dos caras congruentes paralelas de la figura sólida.

bases de un trapecio Los lados paralelos en un trapecio.

línea de ajuste óptimo La línea que más se aproxima a los datos en un diagrama de dispersión.

intermediación de puntos El punto C está entre A y B si y sólo si A, B, y C son colineales y $AC + CB = AB$.

sesgo Un error que resulta en una tergiversación de una población.

declaración bicondicional La conjunción de un condicional y su inverso.

binomio La suma de dos monomios.

bisecar Separe un segmento de línea en dos segmentos congruentes.

datos bivariate Datos que constan de pares de valores.

frontera El borde de la gráfica de una desigualdad que separa el plano de coordenadas en regiones.

acotada Cuando la gráfica de un sistema de restricciones es una región poligonal.

diagrama de caja Una representación gráfica del resumen de cinco números de un conjunto de datos.

C

categorical data Data that can be organized into different categories.

causation When a change in one variable produces a change in another variable.

center of a circle The point from which all points on a circle are the same distance.

center of a regular polygon The center of the circle circumscribed about a regular polygon.

center of dilation The center point from which dilations are performed.

datos categóricos Datos que pueden organizarse en diferentes categorías.

causalidad Cuando un cambio en una variable produce un cambio en otra variable.

centro de un círculo El punto desde el cual todos los puntos de un círculo están a la misma distancia.

centro de un polígono regular El centro del círculo circunscrito alrededor de un polígono regular.

centro de dilatación Punto fijo en torno al cual se realizan las homotecias.

center of rotation The fixed point about which a figure rotates.

center of symmetry A point in which a figure can be rotated onto itself.

central angle of a circle An angle with a vertex at the center of a circle and sides that are radii.

central angle of a regular polygon An angle with its vertex at the center of a regular polygon and sides that pass through consecutive vertices of the polygon.

centroid The point of concurrency of the medians of a triangle.

chord of a circle or sphere A segment with endpoints on the circle or sphere.

circle The set of all points in a plane that are the same distance from a given point called the center.

circular function A function that describes a point on a circle as the function of an angle defined in radians.

circumcenter The point of concurrency of the perpendicular bisectors of the sides of a triangle.

circumference The distance around a circle.

circumscribed angle An angle with sides that are tangent to a circle.

circumscribed polygon A polygon with vertices outside the circle and sides that are tangent to the circle.

closed If for any members in a set, the result of an operation is also in the set.

closed half-plane The solution of a linear inequality that includes the boundary line.

codomain The set of all the *y*-values that could possibly result from the evaluation of the function.

coefficient The numerical factor of a term.

coefficient of determination An indicator of how well a function fits a set of data.

centro de rotación El punto fijo sobre el que gira una figura.

centro de la simetría Un punto en el que una figura se puede girar sobre sí misma.

ángulo central de un círculo Un ángulo con un vértice en el centro de un círculo y los lados que son radios.

ángulo central de un polígono regular Un ángulo con su vértice en el centro de un polígono regular y lados que pasan a través de vértices consecutivos del polígono.

baricentro El punto de intersección de las medianas de un triángulo.

cuerda de un círculo o esfera Un segmento con extremos en el círculo o esfera.

círculo El conjunto de todos los puntos en un plano que están a la misma distancia de un punto dado llamado centro.

función circular Función que describe un punto en un círculo como la función de un ángulo definido en radianes.

circuncentro El punto de concurrencia de las bisectrices perpendiculares de los lados de un triángulo.

circunferencia La distancia alrededor de un círculo.

ángulo circunscrito Un ángulo con lados que son tangentes a un círculo.

polígono circunscrito Un polígono con vértices fuera del círculo y lados que son tangentes al círculo.

cerrado Si para cualquier número en el conjunto, el resultado de la operación es también en el conjunto.

semi-plano cerrado La solución de una desigualdad linear que incluye la línea de limite.

codominar El conjunto de todos los valores y que podrían resultar de la evaluación de la función.

coeficiente El factor numérico de un término.

coeficiente de determinación Un indicador de lo bien que una función se ajusta a un conjunto de datos.

cofunction identities Identities that show the relationships between sine and cosine, tangent and cotangent, and secant and cosecant.

collinear Lying on the same line.

combination A selection of objects in which order is not important.

combined variation When one quantity varies directly and/or inversely as two or more other quantities.

common difference The difference between consecutive terms in an arithmetic sequence.

common logarithms Logarithms of base 10.

common ratio The ratio of consecutive terms of a geometric sequence.

common tangent A line or segment that is tangent to two circles in the same plane.

complement of A All of the outcomes in the sample space that are not included as outcomes of event A.

complementary angles Two angles with measures that have a sum of 90°.

completing the square A process used to make a quadratic expression into a perfect square trinomial.

complex conjugates Two complex numbers of the form $a + bi$ and $a - bi$.

complex fraction A rational expression with a numerator and/or denominator that is also a rational expression.

complex number Any number that can be written in the form $a + bi$, where a and b are real numbers and i is the imaginary unit.

component form A vector written as $\langle x, y \rangle$, which describes the vector in terms of its horizontal component x and vertical component y.

identidades de cofunción Identidades que muestran las relaciones entre seno y coseno, tangente y cotangente, y secante y cosecante.

colineal Acostado en la misma línea.

combinación Una selección de objetos en los que el orden no es importante.

variación combinada Cuando una cantidad varía directamente y / o inversamente como dos o más cantidades.

diferencia común La diferencia entre términos consecutivos de una secuencia aritmética.

logaritmos comunes Logaritmos de base 10.

razón común El razón de términos consecutivos de una secuencia geométrica.

tangente común Una línea o segmento que es tangente a dos círculos en el mismo plano.

complemento de A Todos los resultados en el espacio muestral que no se incluyen como resultados del evento A.

ángulo complementarios Dos ángulos con medidas que tienen una suma de 90°.

completar el cuadrado Un proceso usado para hacer una expresión cuadrática en un trinomio cuadrado perfecto.

conjugados complejos Dos números complejos de la forma $a + bi$ y $a - bi$.

fracción compleja Una expresión racional con un numerador y / o denominador que también es una expresión racional.

número complejo Cualquier número que se puede escribir en la forma $a + bi$, donde a y b son números reales e i es la unidad imaginaria.

forma de componente Un vector escrito como $\langle x, y \rangle$, que describe el vector en términos de su componente horizontal x y componente vertical y.

composite figure A figure that can be separated into regions that are basic figures, such as triangles, rectangles, trapezoids, and circles.

composite solid A three-dimensional figure that is composed of simpler solids.

composition of functions An operation that uses the results of one function to evaluate a second function.

composition of transformations When a transformation is applied to a figure and then another transformation is applied to its image.

compound event Two or more simple events.

compound inequality Two or more inequalities that are connected by the words *and* or *or*.

compound interest Interest calculated on the principal and on the accumulated interest from previous periods.

compound statement Two or more statements joined by the word *and* or *or*.

concave polygon A polygon with one or more interior angles with measures greater than 180°.

concentric circles Coplanar circles that have the same center.

conclusion The statement that immediately follows the word *then* in a conditional.

concurrent lines Three or more lines that intersect at a common point.

conditional probability The probability that an event will occur given that another event has already occurred.

conditional relative frequency The ratio of the joint frequency to the marginal frequency.

conditional statement A compound statement that consists of a premise, or hypothesis, and a conclusion, which is false only when its premise is true and its conclusion is false.

figura compuesta Una figura que se puede separar en regiones que son figuras básicas, tales como triángulos, rectángulos, trapezoides, y círculos.

solido compuesta Una figura tridimensional que se compone de figuras más simples.

composición de funciones Operación que utiliza los resultados de una función para evaluar una segunda función.

composición de transformaciones Cuando una transformación se aplica a una figura y luego se aplica otra transformación a su imagen.

evento compuesto Dos o más eventos simples.

desigualdad compuesta Dos o más desigualdades que están unidas por las palabras *y* u *o*.

interés compuesto Intereses calculados sobre el principal y sobre el interés acumulado de períodos anteriores.

enunciado compuesto Dos o más declaraciones unidas por la palabra *y* o *o*.

polígono cóncavo Un polígono con uno o más ángulos interiores con medidas superiores a 180°.

círculos concéntricos Círculos coplanarios que tienen el mismo centro.

conclusión La declaración que inmediatamente sigue la palabra *entonces* en un condicional.

líneas concurrentes Tres o más líneas que se intersecan en un punto común.

probabilidad condicional La probabilidad de que un evento ocurra dado que otro evento ya ha ocurrido.

frecuencia relativa condicional La relación entre la frecuencia de la articulación y la frecuencia marginal.

enunciado condicional Una declaración compuesta que consiste en una premisa, o hipótesis, y una conclusión, que es falsa solo cuando su premisa es verdadera y su conclusión es falsa.

cone A solid figure with a circular base connected by a curved surface to a single vertex.

confidence interval An estimate of the population parameter stated as a range with a specific degree of certainty.

congruent Having the same size and shape.

congruent angles Two angles that have the same measure.

congruent arcs Arcs in the same or congruent circles that have the same measure.

congruent polygons All of the parts of one polygon are congruent to the corresponding parts or matching parts of another polygon.

congruent segments Line segments that are the same length.

congruent solids Solid figures that have exactly the same shape, size, and a scale factor of 1:1.

conic sections Cross sections of a right circular cone.

conjecture An educated guess based on known information and specific examples.

conjugates Two expressions, each with two terms, in which the second terms are opposites.

conjunction A compound statement using the word *and*.

consecutive interior angles When two lines are cut by a transversal, interior angles that lie on the same side of the transversal.

consistent A system of equations with at least one ordered pair that satisfies both equations.

constant function A linear function of the form $y = b$; The function $f(x) = a$, where a is any number.

constant of variation The constant in a variation function.

cono Una figura sólida con una base circular conectada por una superficie curvada a un solo vértice.

intervalo de confianza Una estimación del parámetro de población se indica como un rango con un grado específico de certeza.

congruente Tener el mismo tamaño y forma.

ángulo congruentes Dos ángulos que tienen la misma medida.

arcos congruentes Arcos en los mismos círculos o congruentes que tienen la misma medida.

poligonos congruentes Todas las partes de un polígono son congruentes con las partes correspondientes o partes coincidentes de otro polígono.

segmentos congruentes Línea segmentos que son la misma longitud.

sólidos congruentes Figuras sólidas que tienen exactamente la misma forma, tamaño y un factor de escala de 1:1.

secciones cónicas Secciones transversales de un cono circular derecho.

conjetura Una suposición educada basada en información conocida y ejemplos específicos.

conjugados Dos expresiones, cada una con dos términos, en la que los segundos términos son opuestos.

conjunción Una declaración compuesta usando la palabra *y*.

ángulos internos consecutivos Cuando dos líneas se cortan por un ángulo transversal, interior que se encuentran en el mismo lado de la transversal.

consistente Una sistema de ecuaciones para el cual existe al menos un par ordenado que satisfice ambas ecuaciones.

función constante Una función lineal de la forma $y = b$; La función $f(x) = a$, donde a es cualquier número.

constante de variación La constante en una función de variación.

constant term A term that does not contain a variable.

constraint A condition that a solution must satisfy.

constructions Methods of creating figures without the use of measuring tools.

continuous function A function that can be graphed with a line or an unbroken curve.

continuous random variable The numerical outcome of a random event that can take on any value.

contrapositive A statement formed by negating both the hypothesis and the conclusion of the converse of a conditional.

convenience sample Members that are readily available or easy to reach are selected.

converse A statement formed by exchanging the hypothesis and conclusion of a conditional statement.

convex polygon A polygon with all interior angles measuring less than 180°.

coordinate proofs Proofs that use figures in the coordinate plane and algebra to prove geometric concepts.

coplanar Lying in the same plane.

corollary A theorem with a proof that follows as a direct result of another theorem.

correlation coefficient A measure that shows how well data are modeled by a regression function.

corresponding angles When two lines are cut by a transversal, angles that lie on the same side of a transversal and on the same side of the two lines.

corresponding parts Corresponding angles and corresponding sides of two polygons.

cosecant The ratio of the length of a hypotenuse to the length of the leg opposite the angle.

término constante Un término que no contiene una variable.

restricción Una condición que una solución debe satisfacer.

construcciones Métodos de creación de figuras sin el uso de herramientas de medición.

función continua Una función que se puede representar gráficamente con una línea o una curva ininterrumpida.

variable aleatoria continua El resultado numérico de un evento aleatorio que puede tomar cualquier valor.

antítesis Una afirmación formada negando tanto la hipótesis como la conclusión del inverso del condicional.

muestra conveniente Se seleccionan los miembros que están fácilmente disponibles o de fácil acceso.

recíproco Una declaración formada por el intercambio de la hipótesis y la conclusión de la declaración condicional.

polígono convexo Un polígono con todos los ángulos interiores que miden menos de 180°.

pruebas de coordenadas Pruebas que utilizan figuras en el plano de coordenadas y álgebra para probar conceptos geométricos.

coplanar Acostado en el mismo plano.

corolario Un teorema con una prueba que sigue como un resultado directo de otro teorema.

coeficiente de correlación Una medida que muestra cómo los datos son modelados por una función de regresión.

ángulos correspondientes Cuando dos líneas se cortan transversalmente, los ángulos que se encuentran en el mismo lado de una transversal y en el mismo lado de las dos líneas.

partes correspondientes Ángulos correspondientes y lados correspondientes.

cosecante Relación entre la longitud de la hipotenusa y la longitud de la pierna opuesta al ángulo.

cosine The ratio of the length of the leg adjacent to an angle to the length of the hypotenuse.

cotangent The ratio of the length of the leg adjacent to an angle to the length of the leg opposite the angle.

coterminal angles Angles in standard position that have the same terminal side.

counterexample An example that contradicts the conjecture showing that the conjecture is not always true.

critical values The z-values corresponding to the most common degrees of certainty.

cross section The intersection of a solid and a plane.

cube root One of three equal factors of a number.

cube root function A radical function that contains the cube root of a variable expression.

curve fitting Finding a regression equation for a set of data that is approximated by a function.

cycle One complete pattern of a periodic function.

cylinder A solid figure with two congruent and parallel circular bases connected by a curved surface.

coseno Relación entre la longitud de la pierna adyacente a un ángulo y la longitud de la hipotenusa.

cotangente La relación entre la longitud de la pata adyacente a un ángulo y la longitud de la pata opuesta al ángulo.

ángulos coterminales Ángulos en posición estándar que tienen el mismo lado terminal.

contraejemplo Un ejemplo que contradice la conjetura que muestra que la conjetura no siempre es cierta.

valores críticos Los valores z correspondientes a los grados de certeza más comunes.

sección transversal Intersección de un sólido con un plano.

raíz cúbica Uno de los tres factores iguales de un número.

función de la raíz del cubo Función radical que contiene la raíz cúbica de una expresión variable.

ajuste de curvas Encontrar una ecuación de regresión para un conjunto de datos que es aproximado por una función.

ciclo Un patron completo de una función periódica.

cilindro Una figura sólida con dos bases circulares congruentes y paralelas conectadas por una superficie curvada.

D

decay factor The base of an exponential expression, or $1 - r$.

decomposition Separating a figure into two or more nonoverlapping parts.

decreasing Where the graph of a function goes down when viewed from left to right.

deductive argument An argument that guarantees the truth of the conclusion provided that its premises are true.

factor de decaimiento La base de una expresión exponencial, o $1 - r$.

descomposición Separar una figura en dos o más partes que no se solapan.

decreciente Donde la gráfica de una función disminuye cuando se ve de izquierda a derecha.

argumento deductivo Un argumento que garantiza la verdad de la conclusión siempre que sus premisas sean verdaderas.

deductive reasoning The process of reaching a specific valid conclusion based on general facts, rules, definitions, or properties.

define a variable To choose a variable to represent an unknown value.

defined term A term that has a definition and can be explained.

definitions An explanation that assigns properties to a mathematical object.

degree The value of the exponent in a power function; $\frac{1}{360}$ of the circular rotation about a point.

degree of a monomial The sum of the exponents of all its variables.

degree of a polynomial The greatest degree of any term in the polynomial.

density A measure of the quantity of some physical property per unit of length, area, or volume.

dependent A consistent system of equations with an infinite number of solutions.

dependent events Two or more events in which the outcome of one event affects the outcome of the other events.

dependent variable The variable in a relation, usually y, with values that depend on x.

depressed polynomial A polynomial resulting from division with a degree one less than the original polynomial.

descriptive modeling A way to mathematically describe real-world situations and the factors that cause them.

descriptive statistics The branch of statistics that focuses on collecting, summarizing, and displaying data.

diagonal A segment that connects any two nonconsecutive vertices within a polygon.

razonamiento deductivo El proceso de alcanzar una conclusión válida específica basada en hechos generales, reglas, definiciones, o propiedades.

definir una variable Para elegir una variable que represente un valor desconocido.

término definido Un término que tiene una definición y se puede explicar.

definiciones Una explicación que asigna propiedades a un objeto matemático.

grado Valor del exponente en una función de potencia. $\frac{1}{360}$ de la rotación circular alrededor de un punto.

grado de un monomio La suma de los exponents de todas sus variables.

grado de un polinomio El grado mayor de cualquier término del polinomio.

densidad Una medida de la cantidad de alguna propiedad física por unidad de longitud, área o volumen.

dependiente Una sistema consistente de ecuaciones con un número infinito de soluciones.

eventos dependientes Dos o más eventos en que el resultado de un evento afecta el resultado de los otros eventos.

variable dependiente La variable de una relación, generalmente y, con los valores que depende de x.

polinomio reducido Un polinomio resultante de la división con un grado uno menos que el polinomio original.

modelado descriptivo Una forma de describir matemáticamente las situaciones del mundo real y los factores que las causan.

estadística descriptiva Rama de la estadística cuyo enfoque es la recopilación, resumen y demostración de los datos.

diagonal Un segmento que conecta cualquier dos vértices no consecutivos dentro de un polígono.

diameter of a circle or sphere A chord that passes through the center of a circle or sphere.

difference of squares A binomial in which the first and last terms are perfect squares.

difference of two squares The square of one quantity minus the square of another quantity.

dilation A nonrigid motion that enlarges or reduces a geometric figure; A transformation that stretches or compresses the graph of a function.

dimensional analysis The process of performing operations with units.

direct variation When one quantity is equal to a constant times another quantity.

directed line segment A line segment with an initial endpoint and a terminal endpoint.

directrix An exterior line perpendicular to the line containing the foci of a curve.

discontinuous function A function that is not continuous.

discrete function A function in which the points on the graph are not connected.

discrete random variable The numerical outcome of a random event that is finite and can be counted.

discriminant In the Quadratic Formula, the expression under the radical sign that provides information about the roots of the quadratic equation.

disjunction A compound statement using the word *or*.

distance The length of the line segment between two points.

distribution A graph or table that shows the theoretical frequency of each possible data value.

domain The set of the first numbers of the ordered pairs in a relation; The set of *x*-values to be evaluated by a function.

diámetro de un círculo o esfera Un acorde que pasa por el centro de un círculo o esfera.

diferencia de cuadrados Un binomio en el que los términos primero y último son cuadrados perfectos.

diferencia de dos cuadrados El cuadrado de una cantidad menos el cuadrado de otra cantidad.

dilatación Un movimiento no rígido que agranda o reduce una figura geométrica; Una transformación que estira o comprime el gráfico de una función.

análisis dimensional El proceso de realizar operaciones con unidades.

variación directa Cuando una cantidad es igual a una constante multiplicada por otra cantidad.

segment de línea dirigido Un segmento de línea con un punto final inicial y un punto final terminal.

directriz Una línea exterior perpendicular a la línea que contiene los focos de una curva.

función discontinua Una función que no es continua.

función discreta Una función en la que los puntos del gráfico no están conectados.

variable aleatoria discreta El resultado numérico de un evento aleatorio que es finito y puede ser contado.

discriminante En la Fórmula cuadrática, la expresión bajo el signo radical que proporciona información sobre las raíces de la ecuación cuadrática.

disyunción Una declaración compuesta usando la palabra *o*.

distancia La longitud del segmento de línea entre dos puntos.

distribución Un gráfico o una table que muestra la frecuencia teórica de cada valor de datos posible.

dominio El conjunto de los primeros números de los pares ordenados en una relación; El conjunto de valores *x* para ser evaluados por una función.

dot plot A diagram that shows the frequency of data on a number line.

double root Two roots of a quadratic equation that are the same number.

gráfica de puntos Una diagrama que muestra la frecuencia de los datos en una línea numérica.

raíces dobles Dos raíces de una función cuadrática que son el mismo número.

E

e An irrational number that approximately equals 2.7182818....

edge of a polyhedron A line segment where the faces of the polyhedron intersect.

elimination A method that involves eliminating a variable by combining the individual equations within a system of equations.

empty set The set that contains no elements, symbolized by { } or ∅.

end behavior The behavior of a graph at the positive and negative extremes in its domain.

enlargement A dilation with a scale factor greater than 1.

equation A mathematical statement that contains two expressions and an equal sign, $=$.

equiangular polygon A polygon with all angles congruent.

equidistant A point is equidistant from other points if it is the same distance from them.

equidistant lines Two lines for which the distance between the two lines, measured along a perpendicular line or segment to the two lines, is always the same.

equilateral polygon A polygon with all sides congruent.

equivalent equations Two equations with the same solution.

equivalent expressions Expressions that represent the same value.

evaluate To find the value of an expression.

e Un número irracional que es aproximadamente igual a 2.7182818 ....

arista de un poliedro Un segmento de línea donde las caras del poliedro se cruzan.

eliminación Un método que consiste en eliminar una variable combinando las ecuaciones individuales dentro de un sistema de ecuaciones.

conjunto vacío El conjunto que no contiene elementos, simbolizado por { } o ∅.

comportamiento extremo El comportamiento de un gráfico en los extremos positivo y negativo en su dominio.

ampliación Una dilatación con un factor de escala mayor que 1.

ecuación Un enunciado matemático que contiene dos expresiones y un signo igual, $=$.

polígono equiangular Un polígono con todos los ángulos congruentes.

equidistante Un punto es equidistante de otros puntos si está a la misma distancia de ellos.

líneas equidistantes Dos líneas para las cuales la distancia entre las dos líneas, medida a lo largo de una línea o segmento perpendicular a las dos líneas, es siempre la misma.

polígono equilátero Un polígono con todos los lados congruentes.

ecuaciones equivalentes Dos ecuaciones con la misma solución.

expresiones equivalentes Expresiones que representan el mismo valor.

evaluar Calcular el valor de una expresión.

even functions Functions that are symmetric in the y-axis.

event A subset of the sample space.

excluded values Values for which a function is not defined.

experiment A sample is divided into two groups. The experimental group undergoes a change, while there is no change to the control group. The effects on the groups are then compared; A situation involving chance.

experimental probability Probability calculated by using data from an actual experiment.

exponent When n is a positive integer in the expression x^n, n indicates the number of times x is multiplied by itself.

exponential decay Change that occurs when an initial amount decreases by the same percent over a given period of time.

exponential decay function A function in which the independent variable is an exponent, where $a > 0$ and $0 < b < 1$.

explicit formula A formula that allows you to find any term a_n of a sequence by using a formula written in terms of n.

exponential equation An equation in which the independent variable is an exponent.

exponential form When an expression is in the form x^n.

exponential function A function in which the independent variable is an exponent.

exponential growth Change that occurs when an initial amount increases by the same percent over a given period of time.

exponential growth function A function in which the independent variable is an exponent, where $a > 0$ and $b > 1$.

incluso funciones Funciones que son simétricas en el eje y.

evento Un subconjunto del espacio de muestra.

valores excluidos Valores para los que no se ha definido una función.

experimento Una muestra se divide en dos grupos. El grupo experimental experimenta un cambio, mientras que no hay cambio en el grupo de control. A continuación se comparan los efectos sobre los grupos; Una situación de riesgo.

probabilidad experimental Probabilidad calculada utilizando datos de un experimento real.

exponente Cuando n es un entero positivo en la expresión x^n, n indica el número de veces que x se multiplica por sí mismo.

desintegración exponencial Cambio que ocurre cuando una cantidad inicial disminuye en el mismo porcentaje durante un período de tiempo dado.

función exponenciales de decaimiento Una ecuación en la que la variable independiente es un exponente, donde $a > 0$ y $0 < b < 1$.

fórmula explícita Una fórmula que le permite encontrar cualquier término a_n de una secuencia usando una fórmula escrita en términos de n.

ecuación exponencial Una ecuación en la que la variable independiente es un exponente.

forma exponencial Cuando una expresión está en la forma x^n.

función exponencial Una función en la que la variable independiente es el exponente.

crecimiento exponencial Cambio que ocurre cuando una cantidad inicial aumenta por el mismo porcentaje durante un período de tiempo dado.

función de crecimiento exponencial Una función en la que la variable independiente es el exponente, donde $a > 0$ y $b > 1$.

exponential inequality An inequality in which the independent variable is an exponent.

exterior angle of a triangle An angle formed by one side of the triangle and the extension of an adjacent side.

exterior angles When two lines are cut by a transversal, any of the four angles that lie outside the region between the two intersected lines.

exterior of an angle The area outside of the two rays of an angle.

extraneous solution A solution of a simplified form of an equation that does not satisfy the original equation.

extrema Points that are the locations of relatively high or low function values.

extreme values The least and greatest values in a set of data.

desigualdad exponencial Una desigualdad en la que la variable independiente es un exponente.

ángulo exterior de un triángulo Un ángulo formado por un lado del triángulo y la extensión de un lado adyacente.

ángulos externos Cuando dos líneas son cortadas por una transversal, cualquiera de los cuatro ángulos que se encuentran fuera de la región entre las dos líneas intersectadas.

exterior de un ángulo El área fuera de los dos rayos de un ángulo.

solución extraña Una solución de una forma simplificada de una ecuación que no satisface la ecuación original.

extrema Puntos que son las ubicaciones de valores de función relativamente alta o baja.

valores extremos Los valores mínimo y máximo en un conjunto de datos.

F

face of a polyhedron A flat surface of a polyhedron.

factored form A form of quadratic equation, $0 = a(x - p)(x - q)$, where $a \neq 0$, in which p and q are the x-intercepts of the graph of the related function.

factorial of n The product of the positive integers less than or equal to n.

factoring The process of expressing a polynomial as the product of monomials and polynomials.

factoring by grouping Using the Distributive Property to factor some polynomials having four or more terms.

family of graphs Graphs and equations of graphs that have at least one characteristic in common.

feasible region The intersection of the graphs in a system of constraints.

finite sample space A sample space that contains a countable number of outcomes.

cara de un poliedro Superficie plana de un poliedro.

forma factorizada Una forma de ecuación cuadrática, $0 = a(x - p)(x - q)$, donde $a \neq 0$, en la que p y q son las intercepciones x de la gráfica de la función relacionada.

factorial de n El producto de los enteros positivos inferiores o iguales a n.

factorización por agrupamiento Utilizando la Propiedad distributiva para factorizar polinomios que possen cuatro o más términos.

factorización El proceso de expresar un polinomio como el producto de monomios y polinomios.

familia de gráficas Gráficas y ecuaciones de gráficas que tienen al menos una característica común.

región factible La intersección de los gráficos en un sistema de restricciones.

espacio de muestra finito Un espacio de muestra que contiene un número contable de resultados.

finite sequence A sequence that contains a limited number of terms.

five-number summary The minimum, quartiles, and maximum of a data set.

flow proof A proof that uses boxes and arrows to show the logical progression of an argument.

focus A point inside a parabola having the property that the distances from any point on the parabola to them and to a fixed line have a constant ratio for any points on the parabola.

formula An equation that expresses a relationship between certain quantities.

fractional distance An intermediary point some fraction of the length of a line segment.

frequency The number of cycles in a given unit of time.

function A relation in which each element of the domain is paired with exactly one element of the range.

function notation A way of writing an equation so that $y = f(x)$.

secuencia finita Una secuencia que contiene un número limitado de términos.

resumen de cinco números El mínimo, cuartiles y máximo de un conjunto de datos.

demostración de flujo Una prueba que usa cajas y flechas para mostrar la progresión lógica de un argumento.

foco Un punto dentro de una parábola que tiene la propiedad de que las distancias desde cualquier punto de la parábola a ellos ya una línea fija tienen una relación constante para cualquier punto de la parábola.

fórmula Una ecuación que expresa una relación entre ciertas cantidades.

distancia fraccionaria Un punto intermediario de alguna fracción de la longitud de un segmento de línea.

frecuencia El número de ciclos en una unidad del tiempo dada.

función Una relación en que a cada elemento del dominio de corresponde un único elemento del rango.

notación functional Una forma de escribir una ecuación para que $y = f(x)$.

G

geometric means The terms between two nonconsecutive terms of a geometric sequence; The nth root, where n is the number of elements in a set of numbers, of the product of the numbers.

geometric model A geometric figure that represents a real-life object.

geometric probability Probability that involves a geometric measure such as length or area.

geometric sequence A pattern of numbers that begins with a nonzero term and each term after is found by multiplying the previous term by a nonzero constant r.

geometric series The indicated sum of the terms in a geometric sequence.

medios geométricos Los términos entre dos términos no consecutivos de una secuencia geométrica; La enésima raíz, donde n es el número de elementos de un conjunto de números, del producto de los números.

modelo geométrico Una figura geométrica que representa un objeto de la vida real.

probabilidad geométrica Probabilidad que implica una medida geométrica como longitud o área.

secuencia geométrica Un patrón de números que comienza con un término distinto de cero y cada término después se encuentra multiplicando el término anterior por una constante no nula r.

series geométricas La suma indicada de los términos en una secuencia geométrica.

glide reflection The composition of a translation followed by a reflection in a line parallel to the translation vector.

greatest integer function A step function in which $f(x)$ is the greatest integer less than or equal to x.

growth factor The base of an exponential expression, or $1 + r$.

reflexión del deslizamiento La composición de una traducción seguida de una reflexión en una línea paralela al vector de traslación.

función entera más grande Una función del paso en que $f(x)$ es el número más grande menos que o igual a x.

factor de crecimiento La base de una expresión exponencial, o $1 + r$.

H

half-plane A region of the graph of an inequality on one side of a boundary.

height of a parallelogram The length of an altitude of the parallelogram.

height of a solid The length of the altitude of a solid figure.

height of a trapezoid The perpendicular distance between the bases of a trapezoid.

histogram A graphical display that uses bars to display numerical data that have been organized in equal intervals.

horizontal asymptote A horizontal line that a graph approaches.

hyperbola The graph of a reciprocal function.

hypothesis The statement that immediately follows the word *if* in a conditional.

semi-plano Una región de la gráfica de una desigualdad en un lado de un límite.

altura de un paralelogramo La longitud de la altitud del paralelogramo.

altura de un sólido La longitud de la altitud de una figura sólida.

altura de un trapecio La distancia perpendicular entre las bases de un trapecio.

histograma Una exhibición gráfica que utiliza barras para exhibir los datos numéricos que se han organizado en intervalos iguales.

asíntota horizontal Una línea horizontal que se aproxima a un gráfico.

hipérbola La gráfica de una función recíproca.

hipótesis La declaración que sigue inmediatamente a la palabra *si* en un condicional.

I

identity An equation that is true for every value of the variable.

identity function The function $f(x) = x$.

if-then statement A compound statement of the form *if p, then q*, where *p* and *q* are statements.

image The new figure in a transformation.

imaginary unit *i* The principal square root of -1.

incenter The point of concurrency of the angle bisectors of a triangle.

identidad Una ecuación que es verdad para cada valor de la variable.

función identidad La función $f(x) = x$.

enunciado si-entonces Enunciado compuesto de la forma *si p, entonces q*, donde *p* y *q* son enunciados.

imagen La nueva figura en una transformación.

unidad imaginaria *i* La raíz cuadrada principal de -1.

incentro El punto de intersección de las bisectrices interiors de un triángulo.

included angle The interior angle formed by two adjacent sides of a triangle.

included side The side of a triangle between two angles.

inconsistent A system of equations with no ordered pair that satisfies both equations.

increasing Where the graph of a function goes up when viewed from left to right.

independent A consistent system of equations with exactly one solution.

independent events Two or more events in which the outcome of one event does not affect the outcome of the other events.

independent variable The variable in a relation, usually x, with a value that is subject to choice.

index In nth roots, the value that indicates to what root the value under the radicand is being taken.

indirect measurement Using similar figures and proportions to measure an object.

indirect proof One assumes that the statement to be proven is false and then uses logical reasoning to deduce that a statement contradicts a postulate, theorem, or one of the assumptions.

indirect reasoning Reasoning that eliminates all possible conclusions but one so that the one remaining conclusion must be true.

inductive reasoning The process of reaching a conclusion based on a pattern of examples.

inequality A mathematical sentence that contains $<$, $>$, $\leq$, $\geq$, or $\neq$.

inferential statistics When the data from a sample is used to make inferences about the corresponding population.

infinite sample space A sample space with outcomes that cannot be counted.

infinite sequence A sequence that continues without end.

ángulo incluido El ángulo interior formado por dos lados adyacentes de un triángulo.

lado incluido El lado de un triángulo entre dos ángulos.

inconsistente Una sistema de ecuaciones para el cual no existe par ordenado alguno que satisfaga ambas ecuaciones.

creccient e Donde la gráfica de una función sube cuando se ve de izquierda a derecha.

independiente Un sistema consistente de ecuaciones con exactamente una solución.

eventos independientes Dos o más eventos en los que el resultado de un evento no afecta el resultado de los otros eventos.

variable independiente La variable de una relación, generalmente x, con el valor que sujeta a elección.

índice En enésimas raíces, el valor que indica a qué raíz está el valor bajo la radicand.

medición indirecta Usando figuras y proporciones similares para medir un objeto.

demostración indirecta Se supone que la afirmación a ser probada es falsa y luego utiliza el razonamiento lógico para deducir que una afirmación contradice un postulado, teorema o uno de los supuestos.

razonamiento indirecto Razonamiento que elimina todas las posibles conclusiones, pero una de manera que la conclusión que queda una debe ser verdad.

razonamiento inductive El proceso de llegar a una conclusión basada en un patrón de ejemplos.

desigualdad Una oración matemática que contiene uno o más de $<$, $>$, $\leq$, $\geq$, o $\neq$.

estadísticas inferencial Cuando los datos de una muestra se utilizan para hacer inferencias sobre la población correspondiente.

espacio de muestra infinito Un espacio de muestra con resultados que no pueden ser contados.

secuencia infinita Una secuencia que continúa sin fin.

informal proof A paragraph that explains why the conjecture for a given situation is true.

initial side The part of an angle that is fixed on the *x*-axis.

inscribed angle An angle with its vertex on a circle and sides that contain chords of the circle.

inscribed polygon A polygon inside a circle in which all of the vertices of the polygon lie on the circle.

intercept A point at which the graph of a function intersects an axis.

intercepted arc The part of a circle that lies between the two lines intersecting it.

interior angle of a triangle An angle at the vertex of a triangle.

interior angles When two lines are cut by a transversal, any of the four angles that lie inside the region between the two intersected lines.

interior of an angle The area between the two rays of an angle.

interquartile range The difference between the upper and lower quartiles of a data set.

intersection A set of points common to two or more geometric figures; **intersection** The graph of a compound inequality containing *and*.

intersection of *A* and *B* The set of all outcomes in the sample space of event *A* that are also in the sample space of event *B*.

interval The distance between two numbers on the scale of a graph.

interval notation Mathematical notation that describes a set by using endpoints with parentheses or brackets.

inverse A statement formed by negating both the hypothesis and conclusion of a conditional statement.

prueba informal Un párrafo que explica por qué la conjetura para una situación dada es verdadera.

lado inicial La parte de un ángulo que se fija en el eje *x*.

ángulo inscrito Un ángulo con su vértice en un círculo y lados que contienen acordes del círculo.

polígono inscrito Un polígono dentro de un círculo en el que todos los vértices del polígono se encuentran en el círculo.

interceptar Un punto en el que la gráfica de una función corta un eje.

arco intersecado La parte de un círculo que se encuentra entre las dos líneas que se cruzan.

ángulo interior de un triángulo Un ángulo en el vértice de un triángulo.

ángulos interiores Cuando dos líneas son cortadas por una transversal, cualquiera de los cuatro ángulos que se encuentran dentro de la región entre las dos líneas intersectadas.

interior de un ángulo El área entre los dos rayos de un ángulo.

rango intercuartil La diferencia entre el cuartil superior *y* el cuartil inferior de un conjunto de datos.

intersección Un conjunto de puntos communes a dos o más figuras geométricas; **intersección** La gráfica de una desigualdad compuesta que contiene la palabra *y*.

intersección de *A* y *B* El conjunto de todos los resultados en el espacio muestral del evento *A* que también se encuentran en el espacio muestral del evento *B*.

intervalo La distancia entre dos números en la escala de un gráfico.

notación de intervalo Notación matemática que describe un conjunto utilizando puntos finales con paréntesis o soportes.

inverso Una declaración formada negando tanto la hipótesis como la conclusión de la declaración condicional.

inverse cosine The ratio of the length of the hypotenuse to the length of the leg adjacent to an angle.

inverse functions Two functions, one of which contains points of the form (a, b) while the other contains points of the form (b, a).

inverse relations Two relations, one of which contains points of the form (a, b) while the other contains points of the form (b, a).

inverse sine The ratio of the length of the hypotenuse to the length of the leg opposite an angle.

inverse tangent The ratio of the length of the leg adjacent to an angle to the length of the leg opposite the angle.

inverse trigonometric functions Arcsine, Arccosine, and Arctangent.

inverse variation When the product of two quantities is equal to a constant k.

isosceles trapezoid A quadrilateral in which two sides are parallel and the legs are congruent.

isosceles triangle A triangle with at least two sides congruent.

inverso del coseno Relación de la longitud de la hipotenusa con la longitud de la pierna adyacente a un ángulo.

funciones inversas Dos funciones, una de las cuales contiene puntos de la forma (a, b) mientras que la otra contiene puntos de la forma (b, a).

relaciones inversas Dos relaciones, una de las cuales contiene puntos de la forma (a, b) mientras que la otra contiene puntos de la forma (b, a).

inverso del seno Relación de la longitud de la hipotenusa con la longitud de la pierna opuesta a un ángulo.

inverso del tangente Relación de la longitud de la pierna adyacente a un ángulo con la longitud de la pierna opuesta a un ángulo.

funciones trigonométricas inversas Arcsine, Arccosine y Arctangent.

variación inversa Cuando el producto de dos cantidades es igual a una constante k.

trapecio isósceles Un cuadrilátero en el que dos lados son paralelos y las patas son congruentes.

triángulo isósceles Un triángulo con al menos dos lados congruentes.

J

joint frequencies Entries in the body of a two-way frequency table. In a two-way frequency table, the frequencies in the interior of the table.

joint variation When one quantity varies directly as the product of two or more other quantities.

frecuencias articulares Entradas en el cuerpo de una tabla de frecuencias de dos vías. En una tabla de frecuencia bidireccional, las frecuencias en el interior de la tabla.

variación conjunta Cuando una cantidad varía directamente como el producto de dos o más cantidades.

K

kite A convex quadrilateral with exactly two distinct pairs of adjacent congruent sides.

cometa Un cuadrilátero convexo con exactamente dos pares distintos de lados congruentes adyacentes.

L

lateral area The sum of the areas of the lateral faces of the figure.

área lateral La suma de las áreas de las caras laterales de la figura.

lateral edges The intersection of two lateral faces.

lateral faces The faces that join the bases of a solid.

lateral surface of a cone The curved surface that joins the base of a cone to the vertex.

lateral surface of a cylinder The curved surface that joins the bases of a cylinder.

leading coefficient The coefficient of the first term when a polynomial is in standard form.

legs of a trapezoid The nonparallel sides in a trapezoid.

legs of an isosceles triangle The two congruent sides of an isosceles triangle.

like radical expressions Radicals in which both the index and the radicand are the same.

like terms Terms with the same variables, with corresponding variables having the same exponent.

line A line is made up of points, has no thickness or width, and extends indefinitely in both directions.

line of fit A line used to describe the trend of the data in a scatter plot.

line of reflection A line midway between a preimage and an image; The line in which a reflection flips the graph of a function.

line of symmetry An imaginary line that separates a figure into two congruent parts.

line segment A measurable part of a line that consists of two points, called endpoints, and all of the points between them.

line symmetry A graph has line symmetry if it can be reflected in a vertical line so that each half of the graph maps exactly to the other half.

linear equation An equation that can be written in the form $Ax + By = C$ with a graph that is a straight line.

aristas laterales La intersección de dos caras laterales.

caras laterales Las caras que unen las bases de un sólido.

superficie lateral de un cono La superficie curvada que une la base de un cono con el vértice.

superficie lateral de un cilindro La superficie curvada que une las bases de un cilindro.

coeficiente líder El coeficiente del primer término cuando un polinomio está en forma estándar.

patas de un trapecio Los lados no paralelos en un trapezoide.

patas de un triángulo isósceles Los dos lados congruentes de un triángulo isósceles.

expresiones radicales semejantes Radicales en los que tanto el índice como el radicand son iguales.

términos semejantes Términos con las mismas variables, con las variables correspondientes que tienen el mismo exponente.

línea Una línea está formada por puntos, no tiene espesor ni anchura, y se extiende indefinidamente en ambas direcciones.

línea de ajuste Una línea usada para describir la tendencia de los datos en un diagrama de dispersión.

línea de reflexión Una línea a medio camino entre una preimagen y una imagen; La línea en la que una reflectión voltea la gráfica de una función.

línea de simetría Una línea imaginaria que separa una figura en dos partes congruentes.

segmento de línea Una parte medible de una línea que consta de dos puntos, llamados extremos, y todos los puntos entre ellos.

simetría de línea Un gráfico tiene simetría de línea si puede reflejarse en una línea vertical, de modo que cada mitad del gráfico se asigna exactamente a la otra mitad.

ecuación lineal Una ecuación que puede escribirse de la forma $Ax + By = C$ con un gráfico que es una línea recta.

linear extrapolation The use of a linear equation to predict values that are outside the range of data.

linear function A function in which no independent variable is raised to a power greater than 1; A function with a graph that is a line.

linear inequality A half-plane with a boundary that is a straight line.

linear interpolation The use of a linear equation to predict values that are inside the range of data.

linear pair A pair of adjacent angles with noncommon sides that are opposite rays.

linear programming The process of finding the maximum or minimum values of a function for a region defined by a system of linear inequalities.

linear regression An algorithm used to find a precise line of fit for a set of data.

linear transformation One or more operations performed on a set of data that can be written as a linear function.

literal equation A formula or equation with several variables.

logarithm In $x = b^y$, y is called the logarithm, base b, of x.

logarithmic equation An equation that contains one or more logarithms.

logarithmic function A function of the form $f(x) = \log$ base b of x, where $b > 0$ and $b \neq 1$.

logically equivalent Statements with the same truth value.

lower quartile The median of the lower half of a set of data.

extrapolación lineal El uso de una ecuación lineal para predecir valores que están fuera del rango de datos.

función lineal Una función en la que ninguna variable independiente se eleva a una potencia mayor que 1; Una función con un gráfico que es una línea.

desigualdad lineal Un medio plano con un límite que es una línea recta.

interpolación lineal El uso de una ecuación lineal para predecir valores que están dentro del rango de datos.

par lineal Un par de ángulos adyacentes con lados no comunes que son rayos opuestos.

programación lineal El proceso de encontrar los valores máximos o mínimos de una función para una región definida por un sistema de desigualdades lineales.

regresión lineal Un algoritmo utilizado para encontrar una línea precisa de ajuste para un conjunto de datos.

transformación lineal Una o más operaciones realizadas en un conjunto de datos que se pueden escribir como una función lineal.

ecuación literal Un formula o ecuación con varias variables.

logaritmo En $x = b^y$, y se denomina logaritmo, base b, de x.

ecuación logarítmica Una ecuación que contiene uno o más logaritmos.

función logarítmica Una función de la forma $f(x) =$ base log b de x, donde $b > 0$ y $b \neq 1$.

lógicamente equivalentes Declaraciones con el mismo valor de verdad.

cuartil inferior La mediana de la mitad inferior de un conjunto de datos.

M

magnitude The length of a vector from the initial point to the terminal point.

magnitud La longitud de un vector desde el punto inicial hasta el punto terminal.

magnitude of symmetry The smallest angle through which a figure can be rotated so that it maps onto itself.

major arc An arc with measure greater than 180°.

mapping An illustration that shows how each element of the domain is paired with an element in the range.

marginal frequencies In a two-way frequency table, the frequencies in the totals row and column; The totals of each subcategory in a two-way frequency table.

maximum The highest point on the graph of a function.

maximum error of the estimate The maximum difference between the estimate of the population mean and its actual value.

measurement data Data that have units and can be measured.

measures of center Measures of what is average.

measures of spread Measures of how spread out the data are.

median The beginning of the second quartile that separates the data into upper and lower halves.

median of a triangle A line segment with endpoints that are a vertex of the triangle and the midpoint of the side opposite the vertex.

metric A rule for assigning a number to some characteristic or attribute.

midline The line about which the graph of a function oscillates.

midpoint The point on a line segment halfway between the endpoints of the segment.

midsegment of a trapezoid The segment that connects the midpoints of the legs of a trapezoid.

midsegment of a triangle The segment that connects the midpoints of the legs of a triangle.

minimum The lowest point on the graph of a function.

magnitud de la simetria El ángulo más pequeño a través del cual una figura se puede girar para que se cargue sobre sí mismo.

arco mayor Un arco con una medida superior a 180°.

cartografía Una ilustración que muestra cómo cada elemento del dominio está emparejado con un elemento del rango.

frecuencias marginales En una tabla de frecuencias de dos vías, las frecuencias en los totales de fila y columna; Los totales de cada subcategoría en una tabla de frecuencia bidireccional.

máximo El punto más alto en la gráfica de una función.

error máximo de la estimación La diferencia máxima entre la estimación de la media de la población y su valor real.

medicion de datos Datos que tienen unidades y que pueden medirse.

medidas del centro Medidas de lo que es promedio.

medidas de propagación Medidas de cómo se extienden los datos son.

mediana El comienzo del segundo cuartil que separa los datos en mitades superior e inferior.

mediana de un triángulo Un segmento de línea con extremos que son un vértice del triángulo y el punto medio del lado opuesto al vértice.

métrico Una regla para asignar un número a alguna caracteristica o atribuye.

linea media La línea sobre la cual oscila la gráfica de una función periódica.

punto medio El punto en un segmento de línea a medio camino entre los extremos del segmento.

segment medio de un trapecio El segmento que conecta los puntos medios de las patas de un trapecio.

segment medio de un triángulo El segmento que conecta los puntos medios de las patas de un triángulo.

mínimo El punto más bajo en la gráfica de una función.

minor arc An arc with measure less than 180°.

mixture problems Problems that involve creating a mixture of two or more kinds of things and then determining some quantity of the resulting mixture.

monomial A number, a variable, or a product of a number and one or more variables.

monomial function A function of the form $f(x) = ax^n$, for which a is a nonzero real number and n is a positive integer.

multi-step equation An equation that uses more than one operation to solve it.

multiplicative identity Because the product of any number a and 1 is equal to a, 1 is the multiplicative identity.

multiplicative inverses Two numbers with a product of 1.

multiplicity The number of times a number is a zero for a given polynomial.

mutually exclusive Events that cannot occur at the same time.

N

natural base exponential function An exponential function with base e, written as $y = e^x$.

natural logarithm The inverse of the natural base exponential function, most often abbreviated as ln x.

negation A statement that has the opposite meaning, as well as the opposite truth value, of an original statement.

negative Where the graph of a function lies below the x-axis.

negative correlation Bivariate data in which y decreases as x increases.

negative exponent An exponent that is a negative number.

---

arco menor Un arco con una medida inferior a 180°.

problemas de mezcla Problemas que implican crear una mezcla de dos o más tipos de cosas y luego determinar una cierta cantidad de la mezcla resultante.

monomio Un número, una variable, o un producto de un número y una o más variables.

función monomial Una función de la forma $f(x) = ax^n$, para la cual a es un número real no nulo y n es un entero positivo.

ecuaciones de varios pasos Una ecuación que utiliza más de una operación para resolverla.

identidad multiplicativa Dado que el producto de cualquier número a y 1 es igual a, 1 es la identidad multiplicativa.

inversos multiplicativos Dos números con un producto es igual a 1.

multiplicidad El número de veces que un número es cero para un polinomio dado.

mutuamente exclusivos Eventos que no pueden ocurrir al mismo tiempo.

---

función exponencial de base natural Una función exponencial con base e, escrita como $y = e^x$.

logaritmo natural La inversa de la función exponencial de base natural, más a menudo abreviada como ln x.

negación Una declaración que tiene el significado opuesto, así como el valor de verdad opuesto, de una declaración original.

negativo Donde la gráfica de una función se encuentra debajo del eje x.

correlación negativa Datos bivariate en el cual y disminuye a x aumenta.

exponente negativo Un exponente que es un número negativo.

negatively skewed distribution A distribution that typically has a median greater than the mean and less data on the left side of the graph.

net A two-dimensional figure that forms the surfaces of a three-dimensional object when folded.

no correlation Bivariate data in which x and y are not related.

nonlinear function A function in which a set of points cannot all lie on the same line

nonrigid motion A transformation that changes the dimensions of a given figure.

normal distribution A continuous, symmetric, bell-shaped distribution of a random variable.

nth root If $a^n = b$ for a positive integer n, then a is the nth root of b.

nth term of an arithmetic sequence The nth term of an arithmetic sequence with first term a_1 and common difference d is given by $a_n = a_1 + (n-1)d$, where n is a positive integer.

numerical expression A mathematical phrase involving only numbers and mathematical operations.

distribución negativamente sesgada Una distribución que típicamente tiene una mediana mayor que la media y menos datos en el lado izquierdo del gráfico.

red Una figura bidimensional que forma las superficies de un objeto tridimensional cuando se dobla.

sin correlación Datos bivariados en los que x e y no están relacionados.

función no lineal Una función en la que un conjunto de puntos no puede estar en la misma línea

movimiento no rígida Una transformación que cambia las dimensiones de una figura dada.

distribución normal Distribución con forma de campana, simétrica y continua de una variable aleatoria.

raíz enésima Si $a^n = b$ para cualquier entero positive n, entonces a se llama una raíz enésima de b.

enésimo término de una secuencia aritmética El enésimo término de una secuencia aritmética con el primer término a_1 y la diferencia común d viene dado por $a_n = a_1 + (n-1)d$, donde n es un número entero positivo.

expresión numérica Una frase matemática que implica sólo números y operaciones matemáticas.

O

oblique asymptote An asymptote that is neither horizontal nor vertical.

observational study Members of a sample are measured or observed without being affected by the study.

octant One of the eight divisions of three-dimensional space.

odd functions Functions that are symmetric in the origin.

one-to-one function A function for which each element of the range is paired with exactly one element of the domain.

onto function A function for which the codomain is the same as the range.

asíntota oblicua Una asíntota que no es ni horizontal ni vertical.

estudio de observación Los miembros de una muestra son medidos o observados sin ser afectados por el estudio.

octante Una de las ocho divisiones del espacio tridimensional.

funciones extrañas Funciones que son simétricas en el origen.

función biunívoca Función para la cual cada elemento del rango está emparejado con exactamente un elemento del dominio.

sobre la función Función para la cual el codomain es el mismo que el rango.

open half-plane The solution of a linear inequality that does not include the boundary line.

opposite rays Two collinear rays with a common endpoint.

optimization The process of seeking the optimal value of a function subject to given constraints.

order of symmetry The number of times a figure maps onto itself.

ordered triple Three numbers given in a specific order used to locate points in space.

orthocenter The point of concurrency of the altitudes of a triangle.

orthographic drawing The two-dimensional views of the top, left, front, and right sides of an object.

oscillation How much the graph of a function varies between its extreme values as it approaches positive or negative infinity.

outcome The result of a single event; The result of a single performance or trial of an experiment.

outlier A value that is more than 1.5 times the interquartile range above the third quartile or below the first quartile.

medio plano abierto La solución de una desigualdad linear que no incluye la línea de limite.

rayos opuestos Dos rayos colineales con un punto final común.

optimización El proceso de valor óptimo de una función sujeto a restricciones dadas.

orden de la simetría El número de veces que una figura se asigna a sí misma.

triple ordenado Tres números dados en un orden específico usado para localizar puntos en el espacio.

ortocentro El punto de concurrencia de las altitudes de un triángulo.

dibujo ortográfico Las vistas bidimensionales de los lados superior, izquierdo, frontal y derecho de un objeto.

oscilación Cuánto la gráfica de una función varía entre sus valores extremos cuando se acerca al infinito positivo o negativo.

resultado El resultado de un solo evento; El resultado de un solo rendimiento o ensayo de un experimento.

parte aislada Un valor que es más de 1,5 veces el rango intercuartílico por encima del tercer cuartil o por debajo del primer cuartil.

P

parabola A curved shape that results when a cone is cut at an angle by a plane that intersects the base; The graph of a quadratic function.

paragraph proof A paragraph that explains why the conjecture for a given situation is true.

parallel lines Coplanar lines that do not intersect; Nonvertical lines in the same plane that have the same slope.

parallel planes Planes that do not intersect.

parallelogram A quadrilateral with both pairs of opposite sides parallel.

parábola Forma curvada que resulta cuando un cono es cortado en un ángulo por un plano que interseca la base; La gráfica de una función cuadrática.

prueba de párrafo Un párrafo que explica por qué la conjetura para una situación dada es verdadera.

líneas paralelas Líneas coplanares que no se intersecan; Líneas no verticales en el mismo plano que tienen pendientes iguales.

planos paralelas Planos que no se intersecan.

paralelogramo Un cuadrilátero con ambos pares de lados opuestos paralelos.

parameter A measure that describes a characteristic of a population; A value in the equation of a function that can be varied to yield a family of functions.

parent function The simplest of functions in a family.

Pascal's triangle A triangle of numbers in which a row represents the coefficients of an expanded binomial $(a + b)^n$.

percent rate of change The percent of increase per time period.

percentile A measure that tells what percent of the total scores were below a given score.

perfect cube A rational number with a cube root that is a rational number.

perfect square A rational number with a square root that is a rational number.

perfect square trinomials Squares of binomials.

perimeter The sum of the lengths of the sides of a polygon.

period The horizontal length of one cycle.

periodic function A function with y-values that repeat at regular intervals.

permutation An arrangement of objects in which order is important.

perpendicular Intersecting at right angles.

perpendicular bisector Any line, segment, or ray that passes through the midpoint of a segment and is perpendicular to that segment.

perpendicular lines Nonvertical lines in the same plane for which the product of the slopes is −1.

phase shift A horizontal translation of the graph of a trigonometric function.

pi The ratio $\frac{cricumference}{diameter}$.

parámetro Una medida que describe una característica de una población; Un valor en la ecuación de una función que se puede variar para producir una familia de funciones.

función basica La función más fundamental de un familia de funciones.

triángulo de Pascal Un triángulo de números en el que una fila representa los coeficientes de un binomio expandido $(a + b)^n$.

por ciento tasa de cambio El porcentaje de aumento por período de tiempo.

percentil Una medida que indica qué porcentaje de las puntuaciones totales estaban por debajo de una puntuación determinada.

cubo perfecto Un número racional con un raíz cúbica que es un número racional.

cuadrado perfecto Un número racional con un raíz cuadrada que es un número racional.

trinomio cuadrado perfecto Cuadrados de los binomios.

perimetro La suma de las longitudes de los lados de un polígono.

periodo La longitud horizontal de un ciclo.

función periódica Una función con y-valores aquella repetición con regularidad.

permutación Un arreglo de objetos en el que el orden es importante.

perpendicular Intersección en ángulo recto.

mediatriz Cualquier línea, segmento o rayo que pasa por el punto medio de un segmento y es perpendicular a ese segmento.

líneas perpendiculares Líneas no verticales en el mismo plano para las que el producto de las pendientes es −1.

cambio de fase Una traducción horizontal de la gráfica de una función trigonométrica.

pi Relación $\frac{circunferencia}{diámetro}$.

piecewise-defined function A function defined by at least two subfunctions, each of which is defined differently depending on the interval of the domain.

piecewise-linear function A function defined by at least two linear subfunctions, each of which is defined differently depending on the interval of the domain.

plane A flat surface made up of points that has no depth and extends indefinitely in all directions.

plane symmetry When a plane intersects a three-dimensional figure so one half is the reflected image of the other half.

Platonic solid One of five regular polyhedra.

point A location with no size, only position.

point discontinuity An area that appears to be a hole in a graph.

point of concurrency The point of intersection of concurrent lines.

point of symmetry The point about which a figure is rotated.

point of tangency For a line that intersects a circle in one point, the point at which they intersect.

point symmetry A figure or graph has this when a figure is rotated 180° about a point and maps exactly onto the other part.

polygon A closed plane figure with at least three straight sides.

polyhedron A closed three-dimensional figure made up of flat polygonal regions.

polynomial A monomial or the sum of two or more monomials.

polynomial function A continuous function that can be described by a polynomial equation in one variable.

función definida por piezas Una función definida por al menos dos subfunciones, cada una de las cuales se define de manera diferente dependiendo del intervalo del dominio.

función lineal por piezas Una función definida por al menos dos subfunciones lineal, cada una de las cuales se define de manera diferente dependiendo del intervalo del dominio.

plano Una superficie plana compuesta de puntos que no tiene profundidad y se extiende indefinidamente en todas las direcciones.

simetría plana Cuando un plano cruza una figura tridimensional, una mitad es la imagen reflejada de la otra mitad.

sólido platónico Uno de cinco poliedros regulares.

punto Una ubicación sin tamaño, solo posición.

discontinuidad de punto Un área que parece ser un agujero en un gráfico.

punto de concurrencia El punto de intersección de líneas concurrentes.

punto de simetría El punto sobre el que se gira una figura.

punto de tangencia Para una línea que cruza un círculo en un punto, el punto en el que se cruzan.

simetría de punto Una figura o gráfica tiene esto cuando una figura se gira 180° alrededor de un punto y se mapea exactamente sobre la otra parte.

polígono Una figura plana cerrada con al menos tres lados rectos.

poliedros Una figura tridimensional cerrada formada por regiones poligonales planas.

polinomio Un monomio o la suma de dos o más monomios.

función polinómica Función continua que puede describirse mediante una ecuación polinómica en una variable.

polynomial identity   A polynomial equation that is true for any values that are substituted for the variables.

population   All of the members of a group of interest about which data will be collected.

population proportion   The number of members in the population with a particular characteristic divided by the total number of members in the population.

positive   Where the graph of a function lies above the x-axis.

positive correlation   Bivariate data in which y increases as x increases.

positively skewed distribution   A distribution that typically has a mean greater than the median.

postulate   A statement that is accepted as true without proof.

power function   A function of the form $f(x) = ax^n$, where a and n are nonzero real numbers.

precision   The repeatability, or reproducibility, of a measurement.

preimage   The original figure in a transformation.

prime polynomial   A polynomial that cannot be written as a product of two polynomials with integer coefficients.

principal root   The nonnegative root of a number.

principal square root   The nonnegative square root of a number.

principal values   The values in the restricted domains of trigonometric functions.

principle of superposition   Two figures are congruent if and only if there is a rigid motion or series of rigid motions that maps one figure exactly onto the other.

prism   A polyhedron with two parallel congruent bases connected by parallelogram faces.

identidad polinomial   Una ecuación polinómica que es verdadera para cualquier valor que se sustituya por las variables.

población   Todos los miembros de un grupo de interés sobre cuáles datos serán recopilados.

proporción de la población   El número de miembros en la población con una característica particular dividida por el número total de miembros en la población.

positiva   Donde la gráfica de una función se encuentra por encima del eje x.

correlación positiva   Datos bivariate en el cual y aumenta a x disminuye.

distribución positivamente sesgada   Una distribución que típicamente tiene una media mayor que la mediana.

postulado   Una declaración que se acepta como verdadera sin prueba.

función de potencia   Una ecuación polinomial que es verdadera para una función de la forma $f(x) = ax^n$, donde a y n son números reales no nulos.

precisión   La repetibilidad, o reproducibilidad, de una medida.

preimagen   La figura original en una transformación.

polinomio primo   Un polinomio que no puede escribirse como producto de dos polinomios con coeficientes enteros.

raíz principal   La raíz no negativa de un número.

raíz cuadrada principal   La raíz cuadrada no negativa de un número.

valores principales   Valores de los dominios restringidos de las funciones trigonométricas.

principio de superposición   Dos figuras son congruentes si y sólo si hay un movimiento rígido o una serie de movimientos rígidos que traza una figura exactamente sobre la otra.

prisma   Un poliedro con dos bases congruentes paralelas conectadas por caras de paralelogramo.

probability The number of outcomes in which a specified event occurs to the total number of trials.	**probabilidad** El número de resultados en los que se produce un evento especificado al número total de ensayos.
probability distribution A function that maps the sample space to the probabilities of the outcomes in the sample space for a particular random variable.	**distribución de probabilidad** Una función que mapea el espacio de muestra a las probabilidades de los resultados en el espacio de muestra para una variable aleatoria particular.
probability model A mathematical representation of a random event that consists of the sample space and the probability of each outcome.	**modelo de probabilidad** Una representación matemática de un evento aleatorio que consiste en el espacio muestral y la probabilidad de cada resultado.
projectile motion problems Problems that involve objects being thrown or dropped.	**problemas de movimiento del proyectil** Problemas que involucran objetos que se lanzan o caen.
proof A logical argument in which each statement is supported by a statement that is accepted as true.	**prueba** Un argumento lógico en el que cada sentencia está respaldada por una sentencia aceptada como verdadera.
proof by contradiction One assumes that the statement to be proven is false and then uses logical reasoning to deduce that a statement contradicts a postulate, theorem, or one of the assumptions.	**prueba por contradicción** Se supone que la afirmación a ser probada es falsa y luego utiliza el razonamiento lógico para deducir que una afirmación contradice un postulado, teorema o uno de los supuestos.
proportion A statement that two ratios are equivalent.	**proporción** Una declaración de que dos proporciones son equivalentes.
pure imaginary number A number of the form bi, where b is a real number and i is the imaginary unit.	**número imaginario puro** Un número de la forma bi, donde b es un número real e i es la unidad imaginaria.
pyramid A polyhedron with a polygonal base and three or more triangular faces that meet at a common vertex.	**pirámide** Poliedro con una base poligonal y tres o más caras triangulares que se encuentran en un vértice común.
Pythagorean identities Identities that express the Pythagorean Theorem in terms of the trigonometric functions.	**identidades pitagóricas** Identidades que expresan el Teorema de Pitágoras en términos de las funciones trigonométricas.
Pythagorean triple A set of three nonzero whole numbers that make the Pythagorean Theorem true.	**triplete Pitágorico** Un conjunto de tres números enteros distintos de cero que hacen que el Teorema de Pitágoras sea verdadero.

Q

quadrantal angle An angle in standard position with a terminal side that coincides with one of the axes.	**ángulo de cuadrante** Un ángulo en posición estándar con un lado terminal que coincide con uno de los ejes.
quadratic equation An equation that includes a quadratic expression.	**ecuación cuadrática** Una ecuación que incluye una expresión cuadrática.

quadratic expression An expression in one variable with a degree of 2.

quadratic form A form of polynomial equation, $au^2 + bu + c$, where u is an algebraic expression in x.

quadratic function A function with an equation of the form $y = ax^2 + bx + c$, where $a \neq 0$.

quadratic inequality An inequality that includes a quadratic expression.

quadratic relations Equations of parabolas with horizontal axes of symmetry that are not functions.

quartic function A fourth-degree function.

quartiles Measures of position that divide a data set arranged in ascending order into four groups, each containing about one fourth or 25% of the data.

quintic function A fifth-degree function.

expresión cuadrática Una expresión en una variable con un grado de 2.

forma cuadrática Una forma de ecuación polinomial, $au^2 + bu + c$, donde u es una expresión algebraica en x.

función cuadrática Una función con una ecuación de la forma $y = ax^2 + bx + c$, donde $a \neq 0$.

desigualdad cuadrática Una desigualdad que incluye una expresión cuadrática.

relaciones cuadráticas Ecuaciones de parábolas con ejes horizontales de simetría que no son funciones.

función cuartica Una función de cuarto grado.

cuartiles Medidas de posición que dividen un conjunto de datos dispuestos en orden ascendente en cuatro grupos, cada uno de los cuales contiene aproximadamente un cuarto o el 25% de los datos.

función quíntica Una función de quinto grado.

R

radian A unit of angular measurement equal to $\frac{180°}{\pi}$ or about 57.296°.

radical equation An equation with a variable in a radicand.

radical expression An expression that contains a radical symbol, such as a square root.

radical form When an expression contains a radical symbol.

radical function A function that contains radicals with variables in the radicand.

radicand The expression under a radical sign.

radius of a circle or sphere A line segment from the center to a point on a circle or sphere.

radius of a regular polygon The radius of the circle circumscribed about a regular polygon.

radián Una unidad de medida angular igual o $\frac{180°}{\pi}$ alrededor de 57.296°.

ecuación radical Una ecuación con una variable en un radicand.

expresión radicales Una expresión que contiene un símbolo radical, tal como una raíz cuadrada.

forma radical Cuando una expresión contiene un símbolo radical.

función radical Función que contiene radicales con variables en el radicand.

radicando La expresión debajo del signo radical.

radio de un círculo o esfera Un segmento de línea desde el centro hasta un punto en un círculo o esfera.

radio de un polígono regular El radio del círculo circunscrito alrededor de un polígono regular.

range The difference between the greatest and least values in a set of data; The set of second numbers of the ordered pairs in a relation; The set of y-values that actually result from the evaluation of the function.

rate of change How a quantity is changing with respect to a change in another quantity.

rational equation An equation that contains at least one rational expression.

rational exponent An exponent that is expressed as a fraction.

rational expression A ratio of two polynomial expressions.

rational function An equation of the form $f(x) = \frac{a(x)}{b(x)}$, where $a(x)$ and $b(x)$ are polynomial expressions and $b(x) \neq 0$.

rational inequality An inequality that contains at least one rational expression.

rationalizing the denominator A method used to eliminate radicals from the denominator of a fraction or fractions from a radicand.

ray Part of a line that starts at a point and extends to infinity.

reciprocal function An equation of the form $f(x) = \frac{n}{b(x)}$, where n is a real number and $b(x)$ is a linear expression that cannot equal 0.

reciprocal trigonometric functions Trigonometric functions that are reciprocals of each other.

reciprocals Two numbers with a product of 1.

rectangle A parallelogram with four right angles.

recursive formula A formula that gives the value of the first term in the sequence and then defines the next term by using the preceding term.

reduction A dilation with a scale factor between 0 and 1.

reference angle The acute angle formed by the terminal side of an angle and the x-axis.

rango La diferencia entre los valores de datos más grande or menos en un sistema de datos; El conjunto de los segundos números de los pares ordenados de una relación; El conjunto de valores y que realmente resultan de la evaluación de la función.

tasa de cambio Cómo cambia una cantidad con respecto a un cambio en otra cantidad.

ecuación racional Una ecuación que contiene al menos una expresión racional.

exponente racional Un exponente que se expresa como una fracción.

expresión racional Una relación de dos expresiones polinomiales.

función racional Una ecuación de la forma $f(x) = \frac{a(x)}{b(x)}$, donde $a(x)$ y $b(x)$ son expresiones polinomiales y $b(x) \neq 0$.

desigualdad racional Una desigualdad que contiene al menos una expresión racional.

racionalizando el denominador Método utilizado para eliminar radicales del denominador de una fracción o fracciones de una radicand.

rayo Parte de una línea que comienza en un punto y se extiende hasta el infinito.

función recíproca Una ecuación de la forma $f(x) = \frac{n}{b(x)}$, donde n es un número real y $b(x)$ es una expresión lineal que no puede ser igual a 0.

funciones trigonométricas recíprocas Funciones trigonométricas que son recíprocas entre sí.

recíprocos Dos números con un producto de 1.

rectángulo Un paralelogramo con cuatro ángulos rectos.

formula recursiva Una fórmula que da el valor del primer término en la secuencia y luego define el siguiente término usando el término anterior.

reducción Una dilatación con un factor de escala entre 0 y 1.

ángulo de referencia El ángulo agudo formado por el lado terminal de un ángulo en posición estándar y el eje x.

reflection   A function in which the preimage is reflected in the line of reflection; A transformation in which a figure, line, or curve is flipped across a line.

regression function   A function generated by an algorithm to find a line or curve that fits a set of data.

regular polygon   A convex polygon that is both equilateral and equiangular.

regular polyhedron   A polyhedron in which all of its faces are regular congruent polygons and all of the edges are congruent.

regular pyramid   A pyramid with a base that is a regular polygon.

regular tessellation   A tessellation formed by only one type of regular polygon.

relation   A set of ordered pairs.

relative frequency   In a two-way frequency table, the ratios of the number of observations in a category to the total number of observations; The ratio of the number of observations in a category to the total number of observations.

relative maximum   A point on the graph of a function where no other nearby points have a greater y-coordinate.

relative minimum   A point on the graph of a function where no other nearby points have a lesser y-coordinate.

remote interior angles   Interior angles of a triangle that are not adjacent to an exterior angle.

residual   The difference between an observed y-value and its predicted y-value on a regression line.

rhombus   A parallelogram with all four sides congruent.

rigid motion   A transformation that preserves distance and angle measure.

reflexión   Función en la que la preimagen se refleja en la línea de reflexión; Una transformación en la que una figura, línea o curva se voltea a través de una línea.

función de regresión   Función generada por un algoritmo para encontrar una línea o curva que se ajuste a un conjunto de datos.

polígono regular   Un polígono convexo que es a la vez equilátero y equiangular.

poliedro regular   Un poliedro en el que todas sus caras son polígonos congruentes regulares y todos los bordes son congruentes.

pirámide regular   Una pirámide con una base que es un polígono regular.

teselado regular   Un teselado formado por un solo tipo de polígono regular.

relación   Un conjunto de pares ordenados.

frecuencia relativa   En una tabla de frecuencia bidireccional, las relaciones entre el número de observaciones en una categoría y el número total de observaciones; La relación entre el número de observaciones en una categoría y el número total de observaciones.

máximo relativo   Un punto en la gráfica de una función donde ningún otro punto cercano tiene una coordenada y mayor.

mínimo relativo   Un punto en la gráfica de una función donde ningún otro punto cercano tiene una coordenada y menor.

ángulos internos no adyacentes   Ángulos interiores de un triángulo que no están adyacentes a un ángulo exterior.

residual   La diferencia entre un valor de y observado y su valor de y predicho en una línea de regresión.

rombo   Un paralelogramo con los cuatro lados congruentes.

movimiento rígido   Una transformación que preserva la distancia y la medida del ángulo.

root A solution of an equation.

rotation A function that moves every point of a preimage through a specified angle and direction about a fixed point.

rotational symmetry A figure can be rotated less than 360° about a point so that the image and the preimage are indistinguishable.

raíz Una solución de una ecuación.

rotación Función que mueve cada punto de una preimagen a través de un ángulo y una dirección especificados alrededor de un punto fijo.

simetría rotacional Una figura puede girar menos de 360° alrededor de un punto para que la imagen y la preimagen sean indistinguibles.

S

sample A subset of a population.

sample space The set of all possible outcomes.

sampling error The variation between samples taken from the same population.

scale The distance between tick marks on the *x*- and *y*-axes.

scale factor of a dilation The ratio of a length on an image to a corresponding length on the preimage.

scatter plot A graph of bivariate data that consists of ordered pairs on a coordinate plane.

secant Any line or ray that intersects a circle in exactly two points; The ratio of the length of the hypotenuse to the length of the leg adjacent to the angle.

sector A region of a circle bounded by a central angle and its intercepted arc.

segment bisector Any segment, line, plane, or point that intersects a line segment at its midpoint.

self-selected sample Members volunteer to be included in the sample.

semicircle An arc that measures exactly 180°.

semiregular tessellation A tessellation formed by two or more regular polygons.

sequence A list of numbers in a specific order.

muestra Un subconjunto de una población.

espacio muestral El conjunto de todos los resultados posibles.

error de muestreo La variación entre muestras tomadas de la misma población.

escala La distancia entre las marcas en los ejes *x* e *y*.

factor de escala de una dilatación Relación de una longitud en una imagen con una longitud correspondiente en la preimagen.

gráfica de dispersión Una gráfica de datos bivariados que consiste en pares ordenados en un plano de coordenadas.

secante Cualquier línea o rayo que cruce un círculo en exactamente dos puntos; Relación entre la longitud de la hipotenusa y la longitud de la pierna adyacente al ángulo.

sector Una región de un círculo delimitada por un ángulo central y su arco interceptado.

bisectriz del segmento Cualquier segmento, línea, plano o punto que interseca un segmento de línea en su punto medio.

muestra auto-seleccionada Los miembros se ofrecen como voluntarios para ser incluidos en la muestra.

semicírculo Un arco que mide exactamente 180°.

teselado semiregular Un teselado formado por dos o más polígonos regulares.

secuencia Una lista de números en un orden específico.

series The indicated sum of the terms in a sequence.

set-builder notation Mathematical notation that describes a set by stating the properties that its members must satisfy.

sides of an angle The rays that form an angle.

sigma notation A notation that uses the Greek uppercase letter S to indicate that a sum should be found.

significant figures The digits of a number that are used to express a measure to an appropriate degree of accuracy.

similar polygons Two figures are similar polygons if one can be obtained from the other by a dilation or a dilation with one or more rigid motions.

similar solids Solid figures with the same shape but not necessarily the same size.

similar triangles Triangles in which all of the corresponding angles are congruent and all of the corresponding sides are proportional.

similarity ratio The scale factor between two similar polygons.

similarity transformation A transformation composed of a dilation or a dilation and one or more rigid motions.

simple random sample Each member of the population has an equal chance of being selected as part of the sample.

simplest form An expression is in simplest form when it is replaced by an equivalent expression having no like terms or parentheses.

simulation The use of a probability model to imitate a process or situation so it can be studied.

sine The ratio of the length of the leg opposite an angle to the length of the hypotenuse.

serie La suma indicada de los términos en una secuencia.

notación de construción de conjuntos Notación matemática que describe un conjunto al declarar las propiedades que sus miembros deben satisfacer.

lados de un ángulo Los rayos que forman un ángulo.

notación de sigma Una notación que utiliza la letra mayúscula griega S para indicar que debe encontrarse una suma.

dígitos significantes Los dígitos de un número que se utilizan para expresar una medida con un grado apropiado de precisión.

polígonos similares Dos figuras son polígonos similares si uno puede ser obtenido del otro por una dilatación o una dilatación con uno o más movimientos rígidos.

sólidos similares Figuras sólidas con la misma forma pero no necesariamente del mismo tamaño.

triángulos similares Triángulos en los cuales todos los ángulos correspondientes son congruentes y todos los lados correspondientes son proporcionales.

relación de similitud El factor de escala entre dos polígonos similares.

transformación de similitud Una transformación compuesto por una dilatación o una dilatación y uno o más movimientos rígidos.

muestra aleatoria simple Cada miembro de la población tiene la misma posibilidad de ser seleccionado como parte de la muestra.

forma reducida Una expresión está reducida cuando se puede sustituir por una expresión equivalente que no tiene ni términos semejantes ni paréntesis.

simulación El uso de un modelo de probabilidad para imitar un proceso o situación para que pueda ser estudiado.

seno La relación entre la longitud de la pierna opuesta a un ángulo y la longitud de la hipotenusa.

sinusoidal function A function that can be produced by translating, reflecting, or dilating the sine function.

skew lines Noncoplanar lines that do not intersect.

slant height of a pyramid or right cone The length of a segment with one endpoint on the base edge of the figure and the other at the vertex.

slope The rate of change in the y-coordinates (rise) to the corresponding change in the x-coordinates (run) for points on a line.

slope criteria Outlines a method for proving the relationship between lines based on a comparison of the slopes of the lines.

solid of revolution A solid figure obtained by rotating a shape around an axis.

solution A value that makes an equation true.

solve an equation The process of finding all values of the variable that make the equation a true statement.

solving a triangle When you are given measurements to find the unknown angle and side measures of a triangle.

space A boundless three-dimensional set of all points.

sphere A set of all points in space equidistant from a given point called the center of the sphere.

square A parallelogram with all four sides and all four angles congruent.

square root One of two equal factors of a number.

square root function A radical function that contains the square root of a variable expression.

square root inequality An inequality that contains the square root of a variable expression.

standard deviation A measure that shows how data deviate from the mean.

función sinusoidal Función que puede producirse traduciendo, reflejando o dilatando la función sinusoidal.

líneas alabeadas Líneas no coplanares que no se cruzan.

altura inclinada de una pirámide o cono derecho La longitud de un segmento con un punto final en el borde base de la figura y el otro en el vértice.

pendiente La tasa de cambio en las coordenadas y (subida) al cambio correspondiente en las coordenadas x (carrera) para puntos en una línea.

criterios de pendiente Describe un método para probar la relación entre líneas basado en una comparación de las pendientes de las líneas.

sólido de revolución Una figura sólida obtenida girando una forma alrededor de un eje.

solución Un valor que hace que una ecuación sea verdadera.

resolver una ecuación El proceso en que se hallan todos los valores de la variable que hacen verdadera la ecuación.

resolver un triángulo Cuando se le dan mediciones para encontrar el ángulo desconocido y las medidas laterales de un triángulo.

espacio Un conjunto tridimensional ilimitado de todos los puntos.

esfera Un conjunto de todos los puntos del espacio equidistantes de un punto dado llamado centro de la esfera.

cuadrado Un paralelogramo con los cuatro lados y los cuatro ángulos congruentes.

raíz cuadrada Uno de dos factores iguales de un número.

función raíz cuadrada Función radical que contiene la raíz cuadrada de una expresión variable.

square root inequality Una desigualdad que contiene la raíz cuadrada de una expresión variable.

desviación típica Una medida que muestra cómo los datos se desvían de la media.

standard error of the mean The standard deviation of the distribution of sample means taken from a population.

standard form of a linear equation Any linear equation can be written in this form, $Ax + By = C$, where $A \geq 0$, A and B are not both 0, and A, B, and C are integers with a greatest common factor of 1.

standard form of a polynomial A polynomial that is written with the terms in order from greatest degree to least degree.

standard form of a quadratic equation A quadratic equation can be written in the form $ax^2 + bx + c = 0$, where $a \neq 0$ and a, b, and c are integers.

standard normal distribution A normal distribution with a mean of 0 and a standard deviation of 1.

standard position An angle positioned so that the vertex is at the origin and the initial side is on the positive x-axis.

statement Any sentence that is either true or false, but not both.

statistic A measure that describes a characteristic of a sample.

statistics An area of mathematics that deals with collecting, analyzing, and interpreting data.

step function A type of piecewise-linear function with a graph that is a series of horizontal line segments.

straight angle An angle that measures 180°.

stratified sample The population is first divided into similar, nonoverlapping groups. Then members are randomly selected from each group.

substitution A process of solving a system of equations in which one equation is solved for one variable in terms of the other.

supplementary angles Two angles with measures that have a sum of 180°.

error estandar de la media La desviación estándar de la distribución de los medios de muestra se toma de una población.

forma estándar de una ecuación lineal Cualquier ecuación lineal se puede escribir de esta forma, $Ax + By = C$, donde $A \geq 0$, A y B no son ambos 0, y A, B y C son enteros con el mayor factor común de 1.

forma estándar de un polinomio Un polinomio que se escribe con los términos en orden del grado más grande a menos grado.

forma estándar de una ecuación cuadrática Una ecuación cuadrática puede escribirse en la forma $ax^2 + bx + c = 0$, donde $a \neq 0$ y a, b, y c son enteros.

distribución normal estándar Distribución normal con una media de 0 y una desviación estándar de 1.

posición estándar Un ángulo colocado de manera que el vértice está en el origen y el lado inicial está en el eje x positivo.

enunciado Cualquier oración que sea verdadera o falsa, pero no ambas.

estadística Una medida que describe una característica de una muestra.

estadísticas El proceso de recolección, análisis e interpretación de datos.

función escalonada Un tipo de función lineal por piezas con un gráfico que es una serie de segmentos de línea horizontal.

ángulo recto Un ángulo que mide 180°.

muestra estratificada La población se divide primero en grupos similares, sin superposición. A continuación, los miembros se seleccionan aleatoriamente de cada grupo.

sustitución Un proceso de resolución de un sistema de ecuaciones en el que una ecuación se resuelve para una variable en términos de la otra.

ángulos suplementarios Dos ángulos con medidas que tienen una suma de 180°.

surface area   The sum of the areas of all faces and side surfaces of a three-dimensional figure.

survey   Data are collected from responses given by members of a group regarding their characteristics, behaviors, or opinions.

symmetric distribution   A distribution in which the mean and median are approximately equal.

symmetry   A figure has this if there exists a rigid motion—reflection, translation, rotation, or glide reflection—that maps the figure onto itself.

synthetic division   An alternate method used to divide a polynomial by a binomial of degree 1.

synthetic geometry   The study of geometric figures without the use of coordinates.

synthetic substitution   The process of using synthetic division to find a value of a polynomial function.

system of equations   A set of two or more equations with the same variables.

system of inequalities   A set of two or more inequalities with the same variables.

systematic sample   Members are selected according to a specified interval from a random starting point.

área de superficie   La suma de las áreas de todas las caras y superficies laterales de una figura tridimensional.

encuesta   Los datos se recogen de las respuestas dadas por los miembros de un grupo con respecto a sus características, comportamientos u opiniones.

distribución simétrica   Un distribución en la que la media y la mediana son aproximadamente iguales.

simetría   Una figura tiene esto si existe una reflexión-reflexión, una traducción, una rotación o una reflexión de deslizamiento rígida-que mapea la figura sobre sí misma.

división sintética   Un método alternativo utilizado para dividir un polinomio por un binomio de grado 1.

geometría sintética   El estudio de figuras geométricas sin el uso de coordenadas.

sustitución sintética   El proceso de utilizar la división sintética para encontrar un valor de una función polynomial.

sistema de ecuaciones   Un conjunto de dos o más ecuaciones con las mismas variables.

sistema de desigualdades   Un conjunto de dos o más desigualdades con las mismas variables.

muestra sistemática   Los miembros se seleccionan de acuerdo con un intervalo especificado desde un punto de partida aleatorio.

tangent   The ratio of the length of the leg opposite an angle to the length of the leg adjacent to the angle.

tangent to a circle   A line or segment in the plane of a circle that intersects the circle in exactly one point and does not contain any points in the interior of the circle.

tangent to a sphere   A line that intersects the sphere in exactly one point.

term   A number, a variable, or a product or quotient of numbers and variables.

tangente   La relación entre la longitud de la pata opuesta a un ángulo y la longitud de la pata adyacente al ángulo.

tangente a un círculo   Una línea o segmento en el plano de un círculo que interseca el círculo en exactamente un punto y no contiene ningún punto en el interior del círculo.

tangente a una esfera   Una línea que interseca la esfera exactamente en un punto.

término   Un número, una variable, o un producto o cociente de números y variables.

term of a sequence A number in a sequence.

terminal side The part of an angle that rotates about the center.

tessellation A repeating pattern of one or more figures that covers a plane with no overlapping or empty spaces.

theorem A statement that can be proven true using undefined terms, definitions, and postulates.

theoretical probability Probability based on what is expected to happen.

transformation A function that takes points in the plane as inputs and gives other points as outputs. The movement of a graph on the coordinate plane.

translation A function in which all of the points of a figure move the same distance in the same direction; A transformation in which a figure is slid from one position to another without being turned.

translation vector A directed line segment that describes both the magnitude and direction of the slide if the magnitude is the length of the vector from its initial point to its terminal point.

transversal A line that intersects two or more lines in a plane at different points.

trapezoid A quadrilateral with exactly one pair of parallel sides.

trend A general pattern in the data.

trigonometric equation An equation that includes at least one trigonometric function.

trigonometric function A function that relates the measure of one nonright angle of a right triangle to the ratios of the lengths of any two sides of the triangle.

trigonometric identity An equation involving trigonometric functions that is true for all values for which every expression in the equation is defined.

término de una sucesión Un número en una secuencia.

lado terminal La parte de un ángulo que gira alrededor de un centro.

teselado Patrón repetitivo de una o más figuras que cubre un plano sin espacios superpuestos o vacíos.

teorema Una afirmación o conjetura que se puede probar verdad utilizando términos, definiciones y postulados indefinidos.

probabilidad teórica Probabilidad basada en lo que se espera que suceda.

transformación Función que toma puntos en el plano como entradas y da otros puntos como salidas. El movimiento de un gráfico en el plano de coordenadas.

traslación Función en la que todos los puntos de una figura se mueven en la misma dirección; El movimiento de un gráfico en el plano de coordenadas.

vector de traslación Un segmento de línea dirigido que describe tanto la magnitud como la dirección de la diapositiva si la magnitud es la longitud del vector desde su punto inicial hasta su punto terminal.

transversal Una línea que interseca dos o más líneas en un plano en diferentes puntos.

trapecio Un cuadrilátero con exactamente un par de lados paralelos.

tendencia Un patrón general en los datos.

ecuación trigonométrica Una ecuación que incluye al menos una función trigonométrica.

función trigonométrica Función que relaciona la medida de un ángulo no recto de un triángulo rectángulo con las relaciones de las longitudes de cualquiera de los dos lados del triángulo.

identidad trigonométrica Una ecuación que implica funciones trigonométricas que es verdadera para todos los valores para los cuales se define cada expresión en la ecuación.

trigonometric ratio A ratio of the lengths of two sides of a right triangle.

trigonometry The study of the relationships between the sides and angles of triangles.

trinomial The sum of three monomials.

truth value The truth or falsity of a statement.

two-column proof A proof that contains statements and reasons organized in a two-column format.

two-way frequency table A table used to show frequencies of data classified according to two categories, with the rows indicating one category and the columns indicating the other.

two-way relative frequency table A table used to show frequencies of data based on a percentage of the total number of observations.

relación trigonométrica Una relación de las longitudes de dos lados de un triángulo rectángulo.

trigonometría El estudio de las relaciones entre los lados y los ángulos de los triángulos.

trinomio La suma de tres monomios.

valor de verdad La verdad o la falsedad de una declaración.

prueba de dos columnas Una prueba que contiene declaraciones y razones organizadas en un formato de dos columnas.

tabla de frecuencia bidireccional Una tabla utilizada para mostrar las frecuencias de los datos clasificados de acuerdo con dos categorías, con las filas que indican una categoría y las columnas que indican la otra.

tabla de frecuencia relativa bidireccional Una tabla usada para mostrar las frecuencias de datos basadas en un porcentaje del número total de observaciones.

U

unbounded When the graph of a system of constraints is open.

undefined terms Words that are not formally explained by means of more basic words and concepts.

uniform motion problems Problems that use the formula $d = rt$, where d is the distance, r is the rate, and t is the time.

uniform tessellation A tessellation that contains the same arrangement of shapes and angles at each vertex.

union The graph of a compound inequality containing *or*.

union of *A* and *B* The set of all outcomes in the sample space of event *A* combined with all outcomes in the sample space of event *B*.

unit circle A circle with a radius of 1 unit centered at the origin on the coordinate plane.

univariate data Measurement data in one variable.

no acotado Cuando la gráfica de un sistema de restricciones está abierta.

términos indefinidos Palabras que no se explican formalmente mediante palabras y conceptos más básicos.

problemas de movimiento uniforme Problemas que utilizan la fórmula $d = rt$, donde d es la distancia, r es la velocidad y t es el tiempo.

teselado uniforme Un teselado que contiene la misma disposición de formas y ángulos en cada vértice.

unión La gráfica de una desigualdad compuesta que contiene la palabra *o*.

unión de *A* y *B* El conjunto de todos los resultados en el espacio muestral del evento *A* combinado con todos los resultados en el espacio muestral del evento *B*.

círculo unitario Un círculo con un radio de 1 unidad centrado en el origen en el plano de coordenadas.

datos univariate Datos de medición en una variable.

upper quartile   The median of the upper half of a set of data.

cuartil superior   La mediana de la mitad superior de un conjunto de datos.

valid argument   An argument is valid if it is impossible for all of the premises, or supporting statements, of the argument to be true and its conclusion false.

argumento válido   Un argumento es válido si es imposible que todas las premisas o argumentos de apoyo del argumento sean verdaderos y su conclusión sea falsa.

variable   A letter used to represent an unspecified number or value; Any characteristic, number, or quantity that can be counted or measured.

variable   Una letra utilizada para representar un número o valor no especificado; Cualquier característica, número, o cantidad que pueda ser contada o medida.

variable term   A term that contains a variable.

término variable   Un término que contiene una variable.

variance   The square of the standard deviation.

varianza   El cuadrado de la desviación estándar.

vertex   Either the lowest point or the highest point of a function.

vértice   El punto más bajo o el punto más alto en una función.

vertex angle of an isosceles triangle   The angle between the sides that are the legs of an isosceles triangle.

ángulo del vértice de un triángulo isósceles   El ángulo entre los lados que son las patas de un triángulo isósceles.

vertex form   A quadratic function written in the form $f(x) = a(x - h)^2 + k$.

forma de vértice   Una función cuadrática escribirse de la forma $f(x) = a(x - h)^2 + k$.

vertex of a polyhedron   The intersection of three edges of a polyhedron.

vértice de un polígono   La intersección de tres bordes de un poliedro.

vertex of an angle   The common endpoint of the two rays that form an angle.

vértice de un ángulo   El punto final común de los dos rayos que forman un ángulo.

vertical angles   Two nonadjacent angles formed by two intersecting lines.

ángulos verticales   Dos ángulos no adyacentes formados por dos líneas de intersección.

vertical asymptote   A vertical line that a graph approaches.

asíntota vertical   Una línea vertical que se aproxima a un gráfico.

vertical shift   A vertical translation of the graph of a trigonometric function.

cambio vertical   Una traducción vertical de la gráfica de una función trigonométrica.

volume   The measure of the amount of space enclosed by a three-dimensional figure.

volumen   La medida de la cantidad de espacio encerrada por una figura tridimensional.

work problems   Problems that involve two people working at different rates who are trying to complete a single job.

problemas de trabajo   Problemas que involucran a dos personas trabajando a diferentes ritmos que están tratando de completar un solo trabajo.

X

x-intercept The x-coordinate of a point where a graph crosses the x-axis.

intercepción x La coordenada x de un punto donde la gráfica corte al eje de x.

Y

y-intercept The y-coordinate of a point where a graph crosses the y-axis.

intercepción y La coordenada y de un punto donde la gráfica corte al eje de y.

Z

z-value The number of standard deviations that a given data value is from the mean.

valor z El número de variaciones estándar que separa un valor dado de la media.

zero An x-intercept of the graph of a function; a value of x for which $f(x) = 0$.

cero Una intercepción x de la gráfica de una función; un punto x para los que $f(x) = 0$.

Index

absolute value
  equations, 101
  expressions, 45
  functions, 267
  inequalities, 367
accuracy, 52
additive identity, 25
additive inverses, 25
Algebra Tiles, 35, 39, 69, 75, 85, 351, 399, 547, 555, 561, 569, 571, 572, 577, 583, 591, 641, 655
algebraic expressions, 13
Analyze See *Practice* in the lessons.
Apply Example, 36, 79, 173, 290, 513, 563, 634, 698
arithmetic sequences, 251
asymptotes, 491
axis of symmetry, 605

bar graphs, 697
bases, 5
best-fit lines, 319
bias, 703
binomials, 545
  difference of squares, 591
  factoring, 591
  multiplying, 561, 569, 571, 572
bivariate data, 307
boundaries, 375
box plots, 709

C

categorical data, 687
causation, 315
closed, 34
closure
  of polynomials, 548, 555
coefficients, 38
coefficients of determination, 667

common differences, 251
common ratios, 523
completing the square, 641
compound inequalities, 357
compound interest, 511
constant functions, 231
constant terms, 13
constraints, 67
continuous functions, 157
correlation
  coefficients, 319
  negative, 307
  no, 307
  positive, 307
Create See *Practice* in the lessons.
cube roots, 465
curve fitting, 667

define a variable, 13
degrees
  of a monomial, 545
  of a polynomial, 545
dependent variable, 137
descriptive modeling, 49
differences of two squares, 591
dilations, 243
dimensional analysis, 121
discrete functions, 157
discriminants, 650
distributions, 715
  negatively skewed, 715
  positively skewed, 715
  symmetric, 715
domains, 135
dot plots, 695
double roots, 627

E

elimination method
  proving the, See the digital lesson, Expand 10-5.

  using addition, 405
  using multiplication, 413
  using subtraction, 407
end behavior, 185
equations, 65
  absolute value, 101
  equivalent, 75
  exponential, 479
  linear, 160
  literal, 117
  multi-step, 85
  quadratic, 627
  solve, 75
  systems of, 387
  with coefficients represented by letters, 87
equivalent equations, 75
equivalent expressions, 38
Estimation, 58, 450, 628, 631
evaluate, 5
Expand, 30, 234, 474, 494, 580, 668
exponents, 5
  negative, 447-448
  rational, 455
exponential
  decay functions, 491, 494
  equations, 479
  functions, 489
  growth functions, 491
expressions
  absolute value, 45
  algebraic, 13
  equivalent, 38
  numerical, 3
  quadratic, 561
  radical, 463
extrema, 182
extreme values, 718

factoring, 577
  binomials, 591
  by grouping, 579
  differences of squares, 591
  perfect square trinomials, 593
  to solve quadratic equations, 631

Index **IN1**

trinomials with a leading coefficient of 1, 583
trinomials with a leading coefficient not equal to 1, 585
using the Distributive Property, 577

factors, 2, 3

families of graphs, 239

Find the Error, 12, 44, 48, 58, 61, 90, 108, 128, 218, 228, 266, 294, 306, 334, 350, 366, 374, 380, 398, 404, 418, 438, 446, 454, 462, 478, 482, 508, 522, 530, 538, 553, 554, 559, 560, 568, 582, 590, 598, 614, 630, 640, 662, 680, 694, 708, 714, 740

five-number summaries, 709

FOIL method, 561

formulas, 117
explicit, 531
Quadratic Formula, 647
recursive, 531
Slope Formula, 222, 223, 224

Four-Step Problem-Solving Plan, 7

frequencies
conditional relative, 735
joint, 733
marginal, 733
relative, 735

functions, 147
absolute value, 267
constant, 231
continuous, 157
decreasing, 182
discrete, 157
exponential, 489
exponential decay, 491, 494
exponential growth, 491
greatest integer, 260
identity, 239
increasing, 182
inverse, 328, 329
linear, 160
negative, 167
nonlinear, 160
notation, 150
parent, 239
piecewise-defined, 259
piecewise-linear, 259
positive, 167
step, 260

geometric sequences, 523
greatest integer functions, 260
growth patterns
exponential, See the digital lesson, Expand 11-8.
linear, See the digital lesson, Expand 4-3.

half-planes, 375
closed, 375, 376
open, 375
histograms, 697

identities, 25, 96
identity functions, 239
independent variables, 137
indexes, 455
inequalities, 341
absolute value, 367
compound, 357
solve, 342
systems of, 419
interquartile ranges, 709
intersections, 357
intervals, See the digital lesson, Expand 4-3.
inverse functions, 328-329
inverse relations, 327
irrational numbers, See the digital lesson, Expand 8-6.

leading coefficients, 545
like terms, 38
linear
equations, 160
extrapolation, 308
functions, 160
interpolation, 308
regression, 319
transformations, 723
linear equations
point-slope form, 295
slope-intercept form, 229
standard form, 160
writing, 65, 229, 287, 295

lines of fit, 308
line symmetry, 179
literal equations, 117
lower quartiles, 709

mapping, 135
Math History Minutes, 8, 94, 196, 262, 320, 352, 415, 448, 493, 533, 586, 705
maximum, 182, 605
measurement data, 687
measures of center, 687
mean, 687-688
median, 687, 688
mode, 687, 688
measures of spread, 709
interquartile range, 709
range, 709
standard deviation, 712
variance, 712
medians, 687, 688
metrics, 49
minimum, 182, 605
Modeling See *Practice* in the lessons.
monomials, 431, 545
degree of, 545
multiplying by polynomials, 555
multiplicative identity, 26
multiplicative inverses, 26
multi-step equations, 85

negative correlations, 307
negative exponents, 447-448
negatively skewed distributions, 715
no correlation, 307
nonlinear functions, 160
***n*th roots,** 455
numerical expressions, 3

Order of Operations, 5
outliers, 718

parabolas, 605

parallel lines, 300

parameters, 231

parent functions, 239

percentiles, 689

perfect cubes, 465

perfect squares, 463

perfect square trinomials, 593

perpendicular lines, 300

Persevere See *Practice* in the lessons.

piecewise-defined functions, 259

piecewise-linear functions, 259

polynomials, 545
  degree of, 545
  prime, 583
  standard form of, 545

populations, 703

positive correlations, 307

positively skewed distributions, 715

powers, 5

product, 3

properties
  Addition Property of Equality, 75
  Addition Property of Inequalities, 342
  Additive Identity Property, 25
  Additive Inverse Property, 25
  Distributive Property, 35
  Division Property of Equality, 78
  Division Property of Inequalities, 344
  Multiplication Property of Equality, 78
  Multiplication Property of Inequalities, 344
  Multiplicative Identity Property, 26
  Multiplicative Inverse Property, 26
  Multiplicative Property of Zero, 26
  Negative Exponent Property, 448
  Power of a Power Property, 433
  Power of a Product Property, 433
  Power of a Quotient Property, 441
  Power Property of Equality, 479
  Product of Powers Property, 431
  Product Property of Radicals, 465
  Product Property of Square Roots, 463
  Quotient of Powers Property, 439
  Quotient Property of Radicals, 465
  Quotient Property of Square Roots, 463
  Reflexive Property, 23
  Square Root Property, 631
  Substitution Property, 16
  Subtraction Property of Equality, 75
  Subtraction Property of Inequalities, 342
  Symmetric Property, 23
  Transitive Property, 23
  Zero Exponent Property, 447
  Zero Product Property, 632

proportions, 109

quadratic
  equations, 627
  expressions, 561
  functions, 605

quadratic equations, solving by
  completing the square, 641
  factoring, 631
  graphing, 627
  taking square roots, 632
  using the Quadratic Formula, 647

Quadratic Formula, 647
  derivation of, 647

quartiles, 709

radical expressions, 463

radicands, 455

ranges, 135

ranges, 709

rates of change, 219

rational exponents, 455

rational numbers, See the digital lesson, Expand 1-3.

Reasoning See *Practice* in the lessons.

reflections, 245

relations, 135

relative maximum, 182

relative minimum, 182

residuals, 322

roots, 171

samples, 703

scale, 139

scatter plots, 307

sequences, 251
  arithmetic, 251
  as explicit formulas, 531
  as functions, 252, 525
  as recursive formulas, 531
  geometric, 523
  terms of, 251, 523

set-builder notation, 342

simplest form, 38

slope, 222

solutions, 75

solve an equation, 75

special products
  product of a sum and difference, 572
  square of a difference, 571
  square of a sum, 569

square roots, 463

standard deviations, 712

standard form
  of a linear equation, 160
  of a polynomial, 545
  of a quadratic function, 607

statistics, 705

step functions, 260

substitution method, 399

symmetric distributions, 715

systems of equations, 387
  consistent, 387
  dependent, 387
  inconsistent, 387
  independent, 387

systems of inequalities, 419

T

terms, 13
terms of a sequence, 251, 523
transformations, 239
translations, 239
trinomials, 545
    factoring, 583
    perfect square, 593
two-way frequency tables, 733
two-way relative frequency tables, 735

U

unions, 359
univariate data, 687
upper quartiles, 709

V

variables, 13, 710
variable terms, 13
variances, 712
vertex forms, 621
Vertical Line Test, 147
vertices, 267, 605

W

Write See *Practice* in the lessons.

X

x-intercepts, 167

Y

y-intercepts, 167

Z

zeros, 171